Mastering Autodesk® Revit® Building

Mastering Autodesk® Revit® Building

PAUL F. AUBIN

Autodesk®

THOMSON
DELMAR LEARNING

Australia • Canada • Mexico • Singapore • Spain • United Kingdom • United States

Mastering Autodesk® Revit® Building

Paul F. Aubin

Autodesk®

Autodesk Press Staff:

Vice President, Technology and Trades ABU:
David Garza

Director of Learning Solutions:
Sandy Clark

Senior Acquisitions Editor:
James Devoe

Senior Product Manager:
John Fisher

Marketing Director:
Deborah S. Yamell

Channel Manager:
Dennis Williams

Marketing Coordinator:
Stacey Wiktorek

Production Manager:
Andrew Crouth

Senior Production Editor:
Stacy Masucci

Technology Project Manager:
Kevin Smith

Technology Project Specialist:
Linda Verde

Editorial Assistant:
Tom Best

Printed in the United States
1 2 3 4 5 XX 08 07 06

For more information contact
Thomson Delmar Learning
Executive Woods
5 Maxwell Drive,
PO Box 8007,
Clifton Park, NY
12065-8007
Or find us on the World Wide Web at
www.delmarlearning.com

Library of Congress Cataloging-in-Publication Data:
Card Number:

ISBN: 1-4180-2053-2

NOTICE TO THE READER

Publisher does not warrant or guarantee any of the products described herein or perform any independent analysis in connection with any of the product information contained herein. Publisher does not assume, and expressly disclaims, any obligation to obtain and include information other than that provided to it by the manufacturer.

The reader is expressly warned to consider and adopt all safety precautions that might be indicated by the activities herein and to avoid all potential hazards. By following the instructions contained herein, the reader willingly assumes all risks in connection with such instructions.

The publisher makes no representation or warranties of any kind, including but not limited to, the warranties of fitness for particular purpose or merchantability, nor are any such representations implied with respect to the material set forth herein, and the publisher takes no responsibility with respect to such material. The publisher shall not be liable for any special, consequential, or exemplary damages resulting, in whole or part, from the readers' use of, or reliance upon, this material.

CONTENTS

WELCOME

Within the pages of this book you will find a comprehensive introduction to the methods, philosophy and procedures of Autodesk Revit Building software. Revit Building is an advanced and powerful architectural design and documentation software package. In the detailed tutorials contained in this book, you will become immersed in its workings and functionality.

WHO SHOULD READ THIS BOOK?

The primary audience for this book is users new to Autodesk Revit Building. However, it is also appropriate for existing Revit Building users that wish to expand there knowledge. You needn't be an experienced computer operator to use this book. Only basic knowledge of the Windows operating system and basic use of a mouse and keyboard are assumed. No prior Computer Aided Design software knowledge is required. If part of your job requires that you design buildings and produce Architectural Construction Documentation or Design Drawings, Facilities Layouts, or Interior Design studies and documentation, then this book is intended for you. Architects, Interior Designers, Design Build Professionals, Facilities Planners and Building Industry CAD Professionals will benefit from the information contained within. Mastering Autodesk Revit Building was authored on Release 8.1 of the software, which was current at the time of publication.

FEATURES IN THIS EDITION

Mastering Autodesk Revit Building is a concise manual focused squarely on the rationale and practicality of the Revit Building process. The book emphasizes the process of creating projects in Revit Building rather than a series of independent commands and tools. The goal of each lesson is to help readers complete building design projects successfully. Tools are introduced together in a focused process with a strong emphasis on "why" as well as "how." The text and exercises seek to give the reader a clear sense of the value of the tools, and a clear indication of each tool's potential. Mastering Revit Building is a resource designed to shorten your learning curve, raise your comfort level, and most importantly give you real-life tested practical advice on the usage of the software to create architecture.

WHAT YOU WILL FIND INSIDE

Section I of this book is focused on the underlying theory and user interface of Autodesk Revit Building. The section is intended to get you acquainted with the software and put you in the proper mindset. Section II relies heavily on tutorial-based exercises to present the process of creating a building model in Revit Building, relying on the software's powerful Building Information Modeling functionality. Two projects are developed concurrently throughout the tutorial section: one residential and one commercial. Detailed explanations are included throughout the tutorials to clearly identify why each step is employed. Annotation and other features specific to construction documentation are covered in Section III. Section IV contains Appendices.

WHAT YOU WON'T FIND INSIDE

This book is not a command reference. This book approaches the subject of learning Revit Building by both exposing conceptual aspects of the software and extensive tutorial coverage. No attempt is made to give a comprehensive explanation of every command or every method available to execute commands. Instead, explanations cover broad topics of how to perform various tasks in Revit Building, with specific examples coming from architectural practice. References are made within the text wherever appropriate to the extensive on-line help and reference materials available on the Web. The focus of this book is the Design Development and Construction Documentation phases of architectural design. Conceptual Design tools are not extensively covered in this edition.

STYLE CONVENTIONS

Style Conventions used in this text are as follows:

Text	Autodesk Revit Building
Step-by-Step Tutorials	1. Perform these steps.
Menu picks	**View > Zoom > Zoom In Region**
On screen input	For the length type **10′-0″ [3000]**.
File and Directory Names	*C:\MasterRevitBuilding\Chapter01\Sample File.rvt*

UNITS

This book references both Imperial and Metric units. Symbol names, scales, references and measurements are given first in Imperial units, followed by the Metric equivalent in square brackets [ ]. For example, when there are two versions of the same file, they will appear like this within the text:

Curtain Wall Dbl Glass.rfa [*M_Curtain Wall Dbl Glass.rfa*].

When the scale varies, a note like this will appear: **1/8″=1′-0″** [**1:100**].

If a measurement must be input, the values will appear like this: **10′-0″** [**3000**]. Please note that in many cases, the closest logical corresponding Metric value has been chosen, rather than a "direct" mathematical translation. For instance, 10′-0″ in Imperial drawings translates to 3048 millimeters; however, a value of 3000 will be used in most cases as a more logical value

Note: Every attempt has been made to make these decisions in an informed manner. However, it is hoped that readers in countries where Metric units are the standard will forgive the American author for any poor choices or translations made in this regard.

All project files are included in both Imperial and Metric units on the included CD ROM. See the "Files Included on the CD ROM" topic below for information on how to install the dataset in your preferred choice of units.

HOW TO USE THIS BOOK

The order of chapters has been carefully thought out with the intention of following a logical flow and architectural process. If you are relatively new to Revit Building, it is recommended that you complete the entire book in order. However, if there are certain chapters that do not pertain to the type of work performed by you or your firm, feel free to skip those topics. However, bear in mind that not every procedure will be repeated in every chapter. For the best experience, it is recommended that you read the entire book, cover to cover. Most importantly, even after you have completed your initial pass of the tutorials in this book, keep Mastering Autodesk Revit Building handy, as it will remain a valuable resource in the weeks and months to come.

FILES INCLUDED ON THE CD ROM

Files used in the tutorials throughout this book, in various stages of completion, are located on the included CD ROM. You will need a licensed copy of Autodesk Revit Building 8.1 or later to open and save files from the included CD ROM. Revit Building software is not included on the CD. Therefore, you will be able to load the file for a given chapter and begin working immediately. When you install the files from the CD, the files for all chapters will be installed automatically. The files will install into a folder on your C drive named *MRB* by default. If you wish, you can choose a different root location such as your "My Documents" folder. Whatever location you choose, a folder named *MRB* will be created. Inside this folder will be a folder for each chapter. Please note that in some cases, a particular chapter or subfolder will not have any Revit files. This is usually indicated by a text file (TXT) within this folder. For example, the *Chapter01\Complete* folder contains no Revit files and instead contains a text document named: *There is no Complete version of Chapter 1.txt.* This text file simply explains that this folder was left empty intentionally.

INSTALLING CD FILES

Locate the Mastering Revit Building CD ROM in the back cover of your book. Read the license agreement before breaking the seal to the CD. To install the dataset files, do the following:

1. Place the CD in your CD drive.

   An installer window should appear on screen after a moment or two.

2. To install the dataset files in Imperial units, click the Imperial Dataset button. To install the dataset files in Metric units, click the Metric Dataset button.
3. In the "WinZip Self-Extractor" dialog, click the Browse button to locate your desired installation folder (such as *My Documents*).

If you do not wish to change the location for the files, you can simply accept the default C Drive location. To do so, do not click Browse and continue to the next step.

4. Click the Unzip button to commence installation.
5. When all files are extracted, a dialog will appear. Click OK and then click Close to finish.

If you do not intend to perform the tutorials in certain chapters, it is OK to delete the files for those chapters. Simply delete the entire folder for the chapter(s) that you wish to skip. If you wish to install both the Imperial and Metric datasets, return to the installer and repeat the steps above for the other units. Installation requires approximately 275 MB of disk space per unit type (550 MB if you install both). If you install both datasets, some files will be the same. Click OK if WinZip asks to overwrite any files.

ONLINE COMPANION

Additional resources related to the content in this book are available online. Log on to our Web site for complete information at:

http://www.autodeskpress.com/onlinecompanion.html

KEEP YOUR SOFTWARE CURRENT

It is important to keep your software current. Be sure to check online at **www.autodesk.com** on a regular basis for the latest updates and service packs to the Autodesk Revit Building software. Having the latest version installed will ensure that you benefit from the latest features and enhancements. If you are on the Autodesk Subscription program, you will be entitled to new releases as they become available. Visit the Autodesk website or talk to your local reseller for more information.

WE WANT TO HEAR FROM YOU

We welcome your comments and suggestions regarding *Mastering Autodesk Architectural Desktop 2006*. Please forward your comments and questions to:

The CADD Team
Delmar Learning
Executive Woods
5 Maxwell Drive
Clifton Park, NY 12065-8007
Website: www.autodeskpress.com

ABOUT THE AUTHOR

Paul F. Aubin is the author of several books on Autodesk Architectural Desktop including Mastering Autodesk Architectural Desktop and Mastering VIZ Render – a Resource for Autodesk Architectural Desktop Users (co-authored with James D. Smell). Paul has a background in the architectural profession spanning over eighteen years. He began his career in traditional architectural practice working on small and mid-sized residential and commercial projects. He eventually became a CAD Manager for a architectural and interior design firm in downtown Chicago. Paul is currently an independent Consultant offering training and implementation services to Architectural Firms using Autodesk building solutions. He is Moderator for CADalyst magazine's online CAD Questions forum and has spoken at Autodesk University for many years. Paul frequently speaks at Autodesk events and local users' group meetings. The combination of his experiences in architectural practice, as a CAD Manager and an Instructor give his writing and his classroom instruction a fresh and credible focus. Paul is an associate member of the American Institute of Architects. He received his Bachelor of Science in Architecture and his Bachelor of Architecture from The Catholic University of America. Paul lives outside Chicago with his wife Martha, their sons Marcus and Justin, and daughter Sarah Gemma.

ABOUT THE TECHNICAL EDITOR

Robert Guarcello Mencarini, AIA is a registered architect with over 20 years of experience in architecture covering a wide spectrum of project types. In 1999, Robert was recruited for his architectural and technology expertise into a small startup software company that had a goal of creating new software that would change the AEC industry. Robert was the first Revit Client Services Architect (CSA) managing, supporting, and training thousands of architects at hundreds of firms around the world. Robert was instrumental in driving the group's award winning record of client satisfaction. In addition to the position of CSA, he has created training materials and workshops, and worked as a liaison between the Revit community and the Autodesk Revit Development team, including designing key portions of Revit. Robert is an expert in Autodesk Revit Building and provides consulting and implementation services to the Revit community. In addition to providing Revit Building consulting services, Robert practices architecture and is currently working on residential and commercial projects in the Northeast United States.

DEDICATION

This book is dedicated to my son Marcus. I look forward to your future debut on the big screen!

ACKNOWLEDGEMENTS

The author would like to thank several people for their assistance and support throughout the writing of this book.

A very special thank you to Christie Landry. It has been a long journey, but we have finally arrived at the destination. I couldn't have done it without your help.

Thanks to Jim Devoe, John Fisher, and all of the Delmar team. It continues to be a pleasure to work with so dedicated a group of professionals.

Thanks to: Robert Guarcello Mencarini, Architect, AIA for technical editing, Sue Gaines for copy editing, both Joanne Sprott and Sue Gaines for indexing and Mike Boyd and the folks at Atlis, for composition.

Technical Contributors to portions of this text include Mark Schmieding and Robert Guarcello Mencarini, Architect, AIA. A special thanks to both of you for your invaluable contributions. Additional contributions and quotations have been noted within the text. Thank you to those contributors as well.

A special acknowledgement is due the following instructors who reviewed the chapters in detail:

Matt Dillon–DC CADD Company

Mel Persin, Coordinator–Chicago Autodesk Revit Users Group

Stephen K. Stafford II–Stafford Consulting Services (Thanks for the Workset/Library analogy)

For taking the time to discuss this project personally and offer suggestions and feedback, thanks to Jeff Millett, AIA—Vice President and Director of Information Technology, Eddie Barnett—LEED, Interior Designer, Sarah Vekasy—LEED, Architect, Marc Gabriel—LEED, Architect, John Jackson—LEED, Architect and Marwan Bakri, Stubbins Associates, Boston, MA and also to Mark Dietrick—CIO and Senior Associate and Michael DeOrsey—Gradaute Architect of Burt Hill Kosar Rittlemann Associates, Boston, MA.

There are far too many folks in Autodesk Building Solutions Division to mention. Thanks to all of them, but in particular, Jason Winstanley, Tatjana Dzambazova, David Conant, Matthew Jezyk, David Mills, Lillian Smith, Steve Crotty, Erik Egbertson, Greg Demchak, Chico Membreno, Trey Klein, Michael Juros, Brian Fitzpatrick, Kelcy Lemon, Susan Champney and all of the folks at Autodesk Tech Support.

I am ever grateful for blessings I have received in my many friends and family. Finally, I am most grateful for the constant love and support of my wife, Martha and three wonderful children.

SECTION I

Introduction and Methodology

This section introduces the methodology of Autodesk Revit Building. The concept of "Building Information Modeling" (BIM) is introduced and defined as are many other important topics and concepts. Within this section you will gain valuable experience using Revit Building and exploring its interface and overall conceptual underpinnings.

Section I is organized as follows:

Quick Start

General Autodesk Revit Building Overview

INTRODUCTION

This Quick Start provides a simple tutorial designed to give you a quick tour of some of the most common elements and features of Autodesk Revit Building. You should be able to complete the entire exercise in 30 minutes or fewer. At the completion of this tutorial, you will have experienced a first-hand look at what Revit Building has to offer.

OBJECTIVES

In this chapter you will:

- Experience an overview of the software.
- Create your first Revit Building model
- Receive a first-hand glimpse at many Revit Building tools and methods.

CREATE A SMALL BUILDING

Let's get started using Revit Building right away. For the next several minutes, we will take a "whirlwind" tour of the Autodesk Revit Building tool set. All of the tools covered in the following steps use simple, and often default, settings. The chapters that follow cover each of these tools and settings in detail. This book was authored using Microsoft Windows XP Professional but the exercises and tutorials should perform equally well in Microsoft Windows 2000 Professional. Please refer to the Preface for complete details on prerequisite assumptions.

INSTALL THE CD FILES AND OPEN A PROJECT

The lessons that follow require the dataset included on the Mastering Revit Building CD ROM. If you have already installed all of the files from the CD, simply skip down to step 3 below to open the project. If you need to install the CD files, start at step 1.

1. If you have not already done so, install the dataset files located on the Mastering Revit Building CD ROM.

   Refer to "Files Included on the CD ROM" in the Preface for instructions on installing the dataset files included on the CD.

2. Launch Autodesk Revit Building from the icon on your desktop or from the **Autodesk** group in **All Programs** on the Windows Start menu.

   - From the File menu, choose **Close**.

This closes the empty new project that Revit Building creates automatically upon launch.

3. On the Standard toolbar, click the Open icon.

TIP The keyboard shortcut for Open is CTRL + O. **Open** is also located on the File menu.

   - In the "Open" dialog box, click the *My Documents* icon on the left side.
   - Double-click on the *MRB* folder, and then the *Quick Start* folder.

   If you installed the dataset files to a different location than the one listed here, use the "Look in" drop down list to browse to that location instead.

4. Double-click *Pavilion.rvt.*

   You can also select it and then click the Open button.

NOTE In this Quick Start Tutorial, only one project has been provided and it uses Imperial units. The remainder of the book provides a Metric dataset as well.

The project will open in Revit Building with the last opened View visible on screen. In this case that is the *Level 1* floor plan View. This project has been started already and contains a Property Line element (dashed square) in the middle of the screen. There is also a Toposurface terrain model element in this file that represents the site for the building. Let's start by displaying this so we know where to place the Walls of our building.

BEGIN A NEW MODEL

To get started, we need to begin with the basics: Walls, Doors and Windows. These elements are the basic building blocks of any architectural model. Adding these elements in Revit Building is simple and straight forward.

CREATE AN UNDERLAY

On the left side of the screen running vertically is a panel named: Project Browser. In it are listed several representations of our project including drawings, schedules and sheets. Four floor plan Views are provided here: *Level 1*, *Level 2*, *Site* and *Roof.* The *Level 1* first floor plan View is bold indicating that it is the currently active View and open on screen. The screen will look something like Figure 2–1 in Chapter 2.

1. On the Project Browser, double-click to open the *Site* plan View.

Notice that the Site Plan includes contours and a shape in the middle of the plan representing the building footprint and its entrance patio.

- On the Project Browser, double-click *Level 1* to return to the first floor plan View.

We can display any one of the other levels (such as the *Site*) as an underlay to this View to help us coordinate elements at different levels.

2. On the Project Browser, right-click on the *Level 1* Floor Plan View and choose **Properties** (see Figure Q–1).

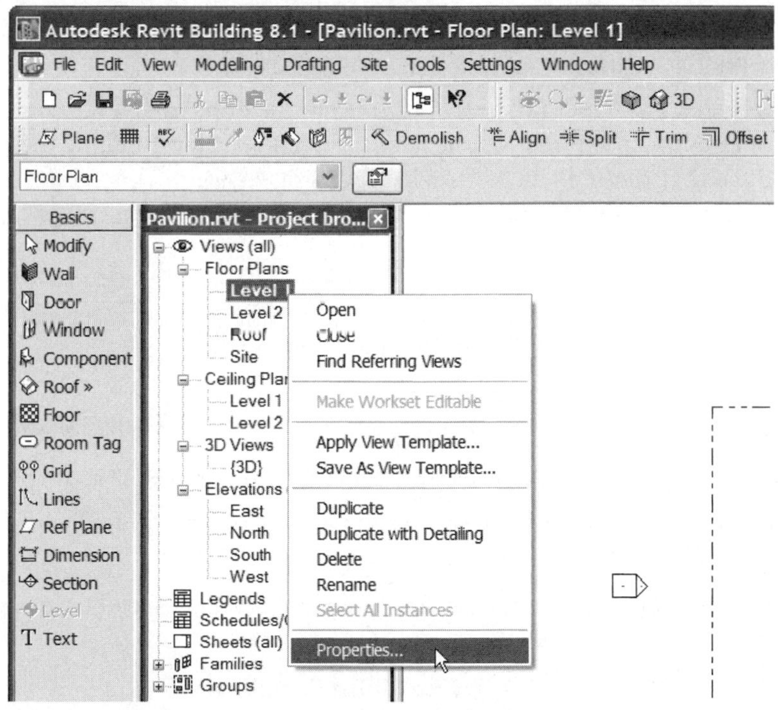

Figure Q–1 *Edit the Properties of the* ***Level 1*** *Floor Plan View*

3. In the "Element Properties" dialog, beneath the "Graphics" grouping (near the bottom) click the word None next to "Underlay."

    ◗ Open the pop-up menu that appears and then choose **Site** (see Figure Q–2).

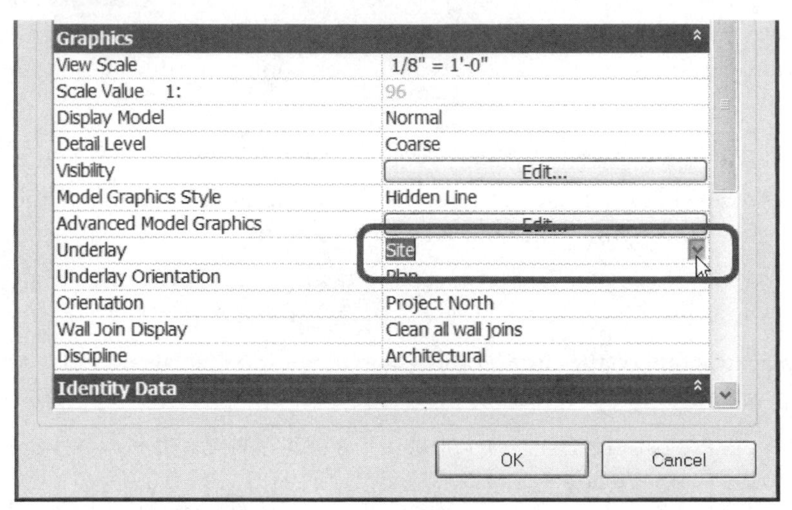

Figure Q–2 *Assign the Site plan View as an underlay*

◗ Click OK to dismiss the dialog and see the results.

Notice that only the patio and building footprint outline appeared. This is because they are the only parts of the Site that intersect the current level height (more on this later). Notice that they appear in 50% halftone gray as well. This reinforces visually that this is simply an underlay.

CREATE WALLS

We begin our building model with some simple Walls.

Locate the Design Bar on the left side of the screen. The "Basics" tab should currently be active.

4. On the Design Bar, click the Basics tab and then click the Wall tool.

Several options will appear across the bar at the top of the screen just beneath the toolbar icons. This is the Options Bar.

5. From the drop-down list on the left (know as the Type Selector) choose Basic Wall : Generic - 8'.

    ◗ From the "Loc Line" list, choose **Finish Face:Exterior**.

    ◗ On the right side of the Options Bar, click the rectangle icon (see Figure Q–3).

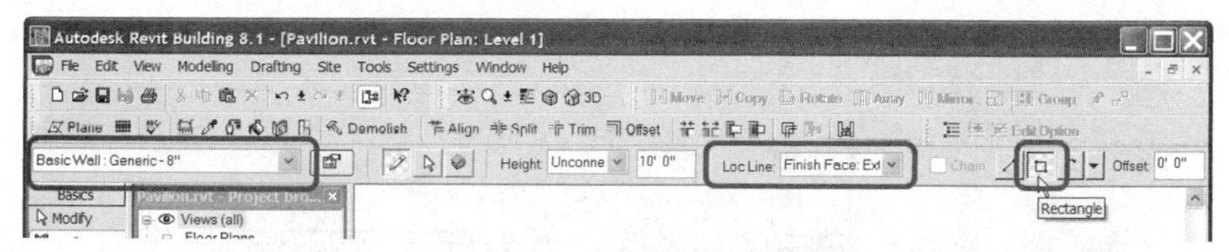

Figure Q–3 *Pick the Wall tool and set it to draw Basic 8" Walls in a rectangular shape*

6. With the mouse pointer (now shaped like a small pencil) click the lower right corner of the gray shape on screen (see the left side of Figure Q–4).

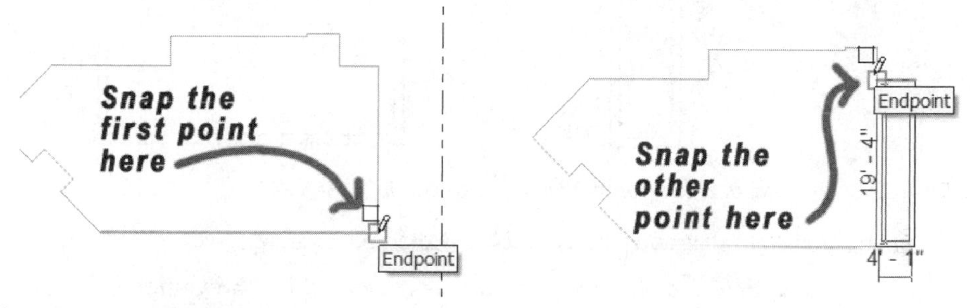

Figure Q–4 *Click the start of the Walls*

- Pick the Endpoint of the short horizontal edge indicated on the right side of Figure Q–4.

You will now have four Walls on screen. However, the "room" they define is very narrow. We can easily adjust this. Before we can manipulate the Walls however, we must cancel the current Wall creation command.

7. On the Design Bar, click the Modify tool or press the ESC key twice.

NOTE Either method can be used anytime in Revit Building to cancel the current command and return to the Modify (selection pointer) tool. In Revit Building there is always one active tool. The default tool is the "Modify" tool, which is really just the standard mouse pointer.

8. Click on the vertical Wall on the left to select it.

   In Revit Building, objects turn red on screen when they are selected.

9. Click the mouse directly on the blue text of the dimension that appears between the selected Wall and the other vertical one (see Figure Q–5).

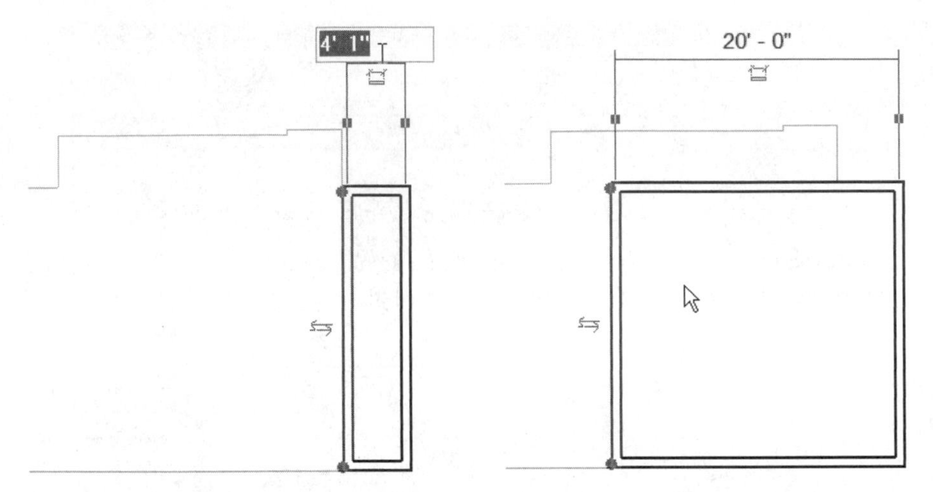

Figure Q–5 *Click the text of a temporary dimension to edit it*

10. In the text field that appears, type **20** and then press ENTER.

Notice that the Wall moved to the new location as indicated by the value we input and that the two horizontal Walls stretched with it to remain attached. Please note that when you edit this way, the selected Wall moves. The dimensions that we used for this edit are referred to as "temporary dimensions."

11. On the Design Bar, click the Wall tool again.
    - From the Type Selector choose Basic Wall : Generic - 5'.
    - From the "Loc Line" list, choose **Wall Centerline**.
    - Draw a vertical wall from top to bottom of the room, dividing it approximately into thirds (see Figure Q–6).

NOTE The exact dimensions are unimportant at this point, we will move the Wall next.

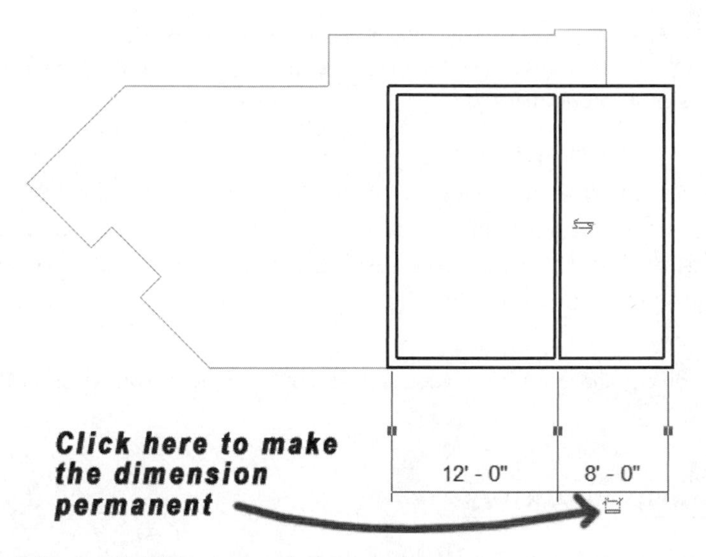

Figure Q–6 *Draw a Wall in random location in the space*

- Beneath the temporary dimension, click the small icon (indicated in Figure Q–6) to make the dimension permanent.
- On the Design Bar, click the Modify tool or press the ESC key twice.

Now that the dimension is permanent, notice that it remains on screen when the Wall is no longer selected.

12. Click to select the dimension.

- Click the small "EQ" (Equidistant) icon beneath the dimension (see Figure Q–7).

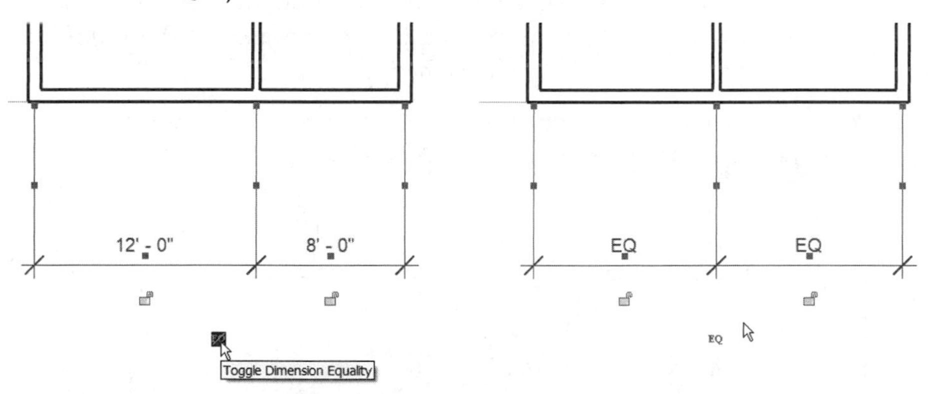

Figure Q–7 *Toggle the Dimension Equality for the indicated Walls*

NOTE This icon is part of the Revit Building constraint system. The constraint system is used to "lock-in" design intent. This notion is an integral part of the underlying concepts inherent to Revit Building.

13. From the File menu, choose **Save**.

It is important to remember to save every so often to preserve your work. Autodesk Revit Building is configured by default to remind you to save at regular intervals. You can edit the interval, but if the message asking you to save appears, you should always perform the save.

NOTE You can find more information and tutorials on working with Walls in Chapter 3.

INSERT DOORS AND WINDOWS

Next we'll add some openings in our Walls.

14. On the Design Bar, click the Basics tab and then click the Door tool.
    - Accept all of the defaults on the Options Bar.
    - Move the pointer near the top horizontal Wall to begin placing the Door.

Move the mouse around without clicking it yet. Notice how the Door follows the cursor and also stays attached to the Wall as it does. This is because elements such as Doors are "hosted" elements. This Door is "hosted" by the wall. In other words, it is not possible to place a Door on its own (freestanding) without a Wall as a host. Also notice that moving the mouse from one side of the Wall to the other will flip the Door in or out relative to the Wall. The gray underlay we added above indicates a patio shape to the left of the plan and wrapping around the top. Our first door will be out to that passageway along the top.

15. Position the mouse on the top Wall near the right side of the passageway so it swings out and then click (see Figure Q–8).

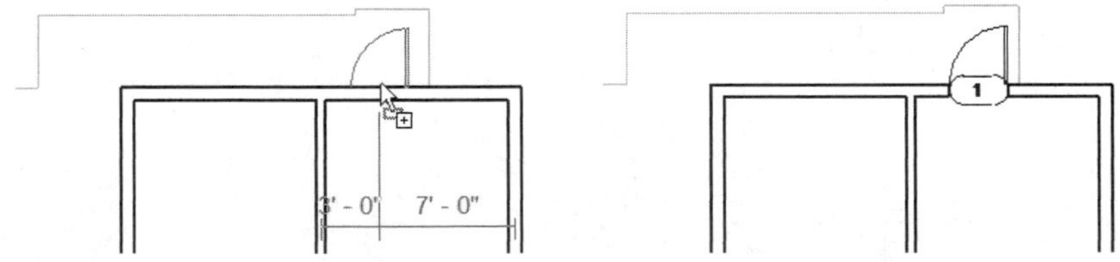

Figure Q–8 *Place a Door to the outside near the passageway*

16. Place another near the top of the interior vertical Wall.

Notice that Door tags have automatically appeared and the numbers have filled in sequentially. Sometimes after placing a Door, it is not positioned or oriented correctly. Just like the Walls above, we can select a Door, and then edit its temporary dimensions to move it to the desired location. There are also small flip control icons on the Door to control its orientation.

17. Click the Flip control to change one of the Door's orientation (see Figure Q–9).

    Repeat if desired on the other Door.

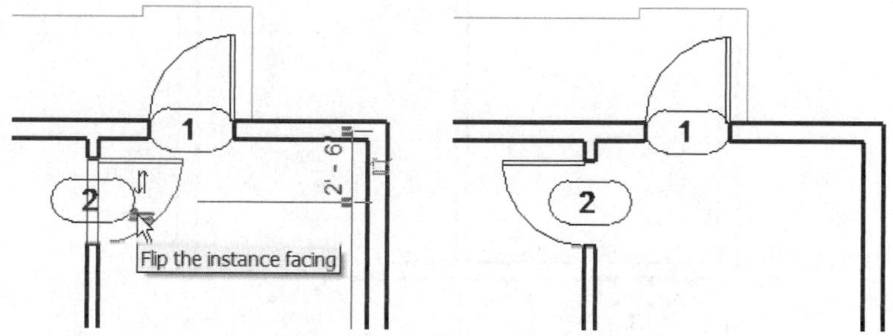

Figure Q–9 *Use the flip controls to change Door orientation*

Adding Windows works the same way as adding Doors.

18. On the Design Bar, click the Basics tab and then click the Window tool.
    - Accept all of the defaults on the Options Bar.
    - Move the pointer near the top horizontal Wall and move up then down.

Again notice how this controls the placement orientation of the Window. As with the Door, you can always flip it later if you make an error.

19. Place Windows in the two horizontal Walls only (see Figure Q–10).

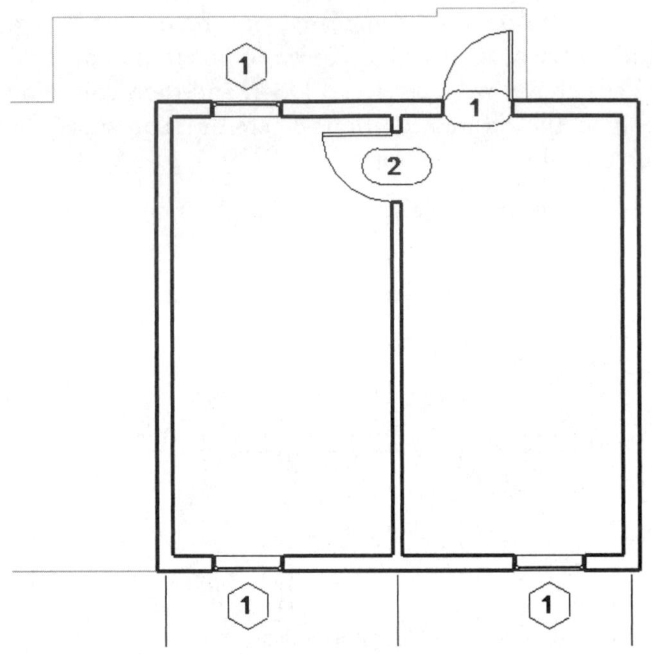

Figure Q–10 *Place Windows*

Let's place one more Door in the Wall at the left. This one we will load from an external library, however.

20. On the Design Bar, click the Basics tab and then click the Door tool.

- On the Options Bar, click the Load button.
- Navigate to the *Autodesk Web Library\Doors* sub-folder of the folder where you installed the Mastering Autodesk Revit Building CD ROM files.
- Select *Double-Glass 2.rfa* and then click Open.

This action loads the Door component into the currently active project making it available to place in the model.

- Place the Door in the center of the left Wall swinging out. Use the temporary dimensions to assist you in placement (see Figure Q–11).

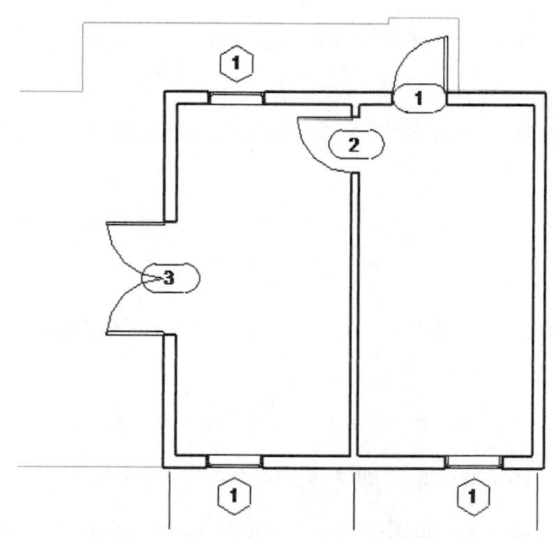

Figure Q–11 *Place a double entry Door*

- On the Design Bar, click the Modify tool or press the ESC key twice.

21. Save the project.

NOTE You can find more information and tutorials on working with Doors and Windows in Chapter 3.

WORKING IN OTHER VIEWS

We can work in many types of Views in Revit Building; not just floor plans. Our project includes elevation Views and ceiling plan Views already. We can also add section Views and 3D Views. Many other View types are also available and discussed in future chapters. One thing to keep in mind is that all of these Views representing the project are contained in one Revit Building (RVT) project file.

VIEW THE MODEL IN 3D

Opening a three-dimensional (3D) View will reveal that our Wall height needs adjustment.

1. On the Project Browser, double-click to open the *West* elevation View.

- On the Project Browser, double-click to open the *{3D}* 3D View.

This is the default three-dimensional View in Revit Building. You can modify it as you like or create others from it. We will use this one for our tutorial, but make some simple adjustments to its vantage point. Notice that the *{3D}* View is an isometric view of our building model. We can see the Walls, Doors and Windows we added from a bird's eye vantage point. We also see the Toposurface terrain model that was included in this project. You can change the vantage point of a 3D View interactively on screen.

2. On the View toolbar, click the Dynamically Modify View icon. (You can also press F8).

   A "Dynamic View" toolbox will appear in the lower left corner of the screen.

   - Click the Spin button.
   - Drag the mouse in the View window to spin the model around interactively.

   Drag side-to-side to move around the building.

   Drag up or down to change height of the vantage point.

   - Spin the model around so that the front double Door is visible (see Figure Q–12).

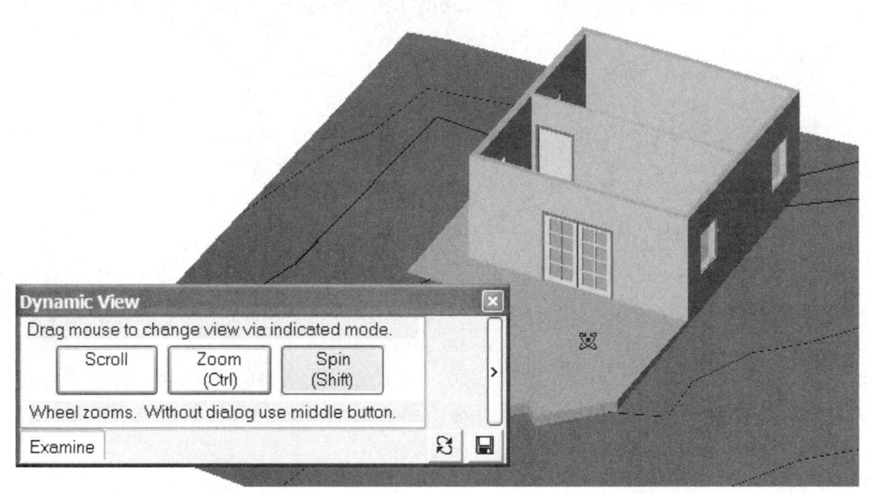

Figure Q–12 *Dynamically Modify the* **{3D}** *View window*

There are two other buttons on the "Dynamic View" toolbox, one to zoom in and out, and one to scroll, which simply slides the model View around on screen without changing the viewing angle.

- Close the "Dynamic View" toolbox.

CREATE A SECTION VIEW

Let's open one more View of the model before we proceed. This will allow us to understand the relationships built into our Revit Building model very clearly as we make some simple edits. We are going to cut a section through the model and look at the resulting View.

3. On the Project Browser, double-click to open the *Level 1* floor plan View.
4. On the Design Bar, click the Basics tab and then click the Section tool.
   - Click to the left of the double Door.

   Move the pointer through the model to the right keeping the section line horizontal.

   - Click outside the model to the right to complete the section line (see Figure Q–13).

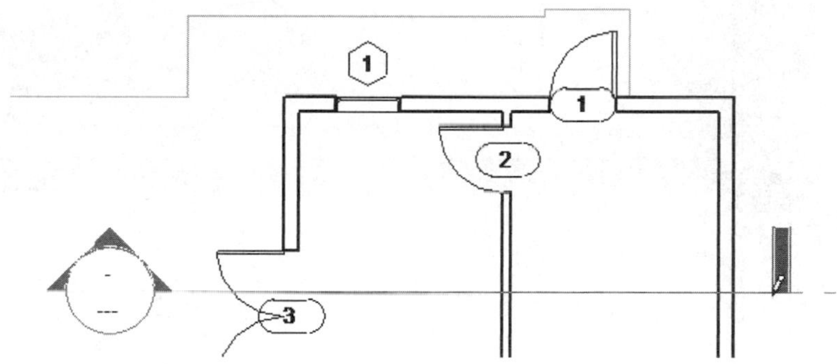

Figure Q–13 *Cut a section through the model*

A section line, with section head and tail will appear. Three green dashed lines with drag handles will also appear. The Section Head will currently be red indicating that it is selected.

5. Click next to the section line in the white area of the view background being careful not to click on any geometry.

TIP This is a quick way to deselect the selected element(s). You can also just press the ESC key.

Notice that the Section Head now turns blue. This indicates that it has a linked View associated with it and acts as a "hotlink" to the associated section view. In the Project Browser, notice that there is now a *Sections* category included in the list. Revit Building automatically creates such nodes in the Project Browser as needed.

- Double-click the blue Section Head to open the associated View.

You should now see the *Section 1* View on screen. We now have four Views open. If you click on the Window menu, you will see all open Views listed near the bottom of the menu. You can choose them off this list to bring them to the front of the pile, or you can simply double-click the View name in Project Browser again to display them. You can also tile them all on screen at once. Let's do that now.

6. From the Window menu, choose **Tile** (see Figure Q–14).

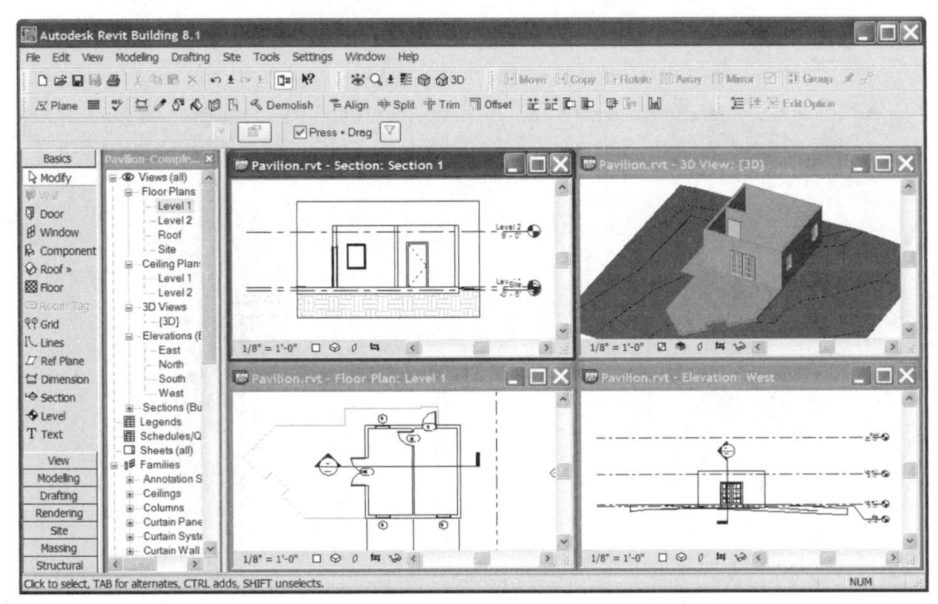

Figure Q–14 *Tile the Views on screen to view them all at once*

NOTE You can find more information and tutorials on working with Views in Chapter 4.

EDIT IN ANY VIEW

When you wish to edit a Revit Building model, you may perform the edits in *any* View. Changes will automatically be applied to *all* Views. This is the power of Autodesk Revit Building! You are describing a single virtual building model or *Building Information Model* (BIM) (see Chapter 1 for complete details). You can "view" it in an unlimited number of ways. Regardless of where you make the edit—plan, section, elevation, 3D, or even schedules, all Views are completely coordinated. These graphical and tabular Views are visual reports of the data contained within the Building Information Model.

EDITING LEVELS

With our current screen configuration, take a look at the elevation and section Views in particular. This project has been set up to have two stories plus a roof. Currently our Walls only go up one story and in fact do not even coincide with the second floor level at all. Let's fix both problems.

7. Click in the floor plan View, *Level 1* to make it active.
8. Place your mouse pointer (the Modify tool) over one of the exterior Walls.

Notice the way that it highlights under the cursor. (If you move the mouse away without clicking, the Wall will no longer highlight.) This is called "pre-highlight" and is a useful aid to proper selection.

- Pre-highlight one exterior Wall—do not click yet.
- Press the TAB key.

Notice how all of the exterior Walls now pre-highlight.

- Click the left mouse button to select the pre-highlighted elements.

Notice how all four exterior Walls are now shaded red in all open Views.

9. On the Options Bar, click the Properties icon (see Figure Q–15).

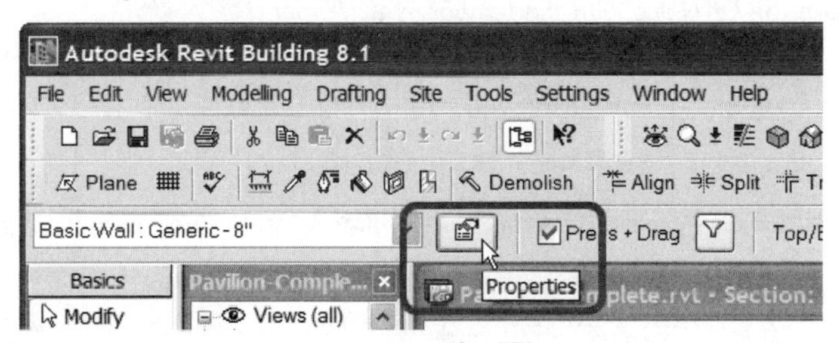

Figure Q–15 *Click the Properties icon to access the "Element Properties" dialog*

- In the "Element Properties" dialog, from the "Top Constraint" list, choose **Up to level: Roof** and then click OK.

Notice how the Walls project up to the Roof Level line in all Views. The Walls are now set relative to this Level in the project. If we were to change the height of the Roof Level, the Walls would also adjust accordingly. Let's try that now.

NOTE The section View will likely not show the top of the Walls as it is currently cropped to the first floor. You can adjust this with the round drag control at the top of the Crop Boundary. Click the rectangular box surrounding the section. Click the small blue circle at the top edge of the box and drag it upward. The Walls should now show.

10. Click anywhere in the *West* elevation View.

Zoom In Region to get a better look if you need to (right-click to access Zoom In Region).

- Click to select the second floor Level line.

Notice the temporary dimensions that appear. Like the Walls and other elements drawn so far, we can edit the blue dimension value to move the Level lines to a new location. We will move both Level 2 and the Roof level; starting with Level 2.

- Click the blue text of the temporary dimension between Level 1 and Level 2, type **10** and press ENTER (see Figure Q–16).

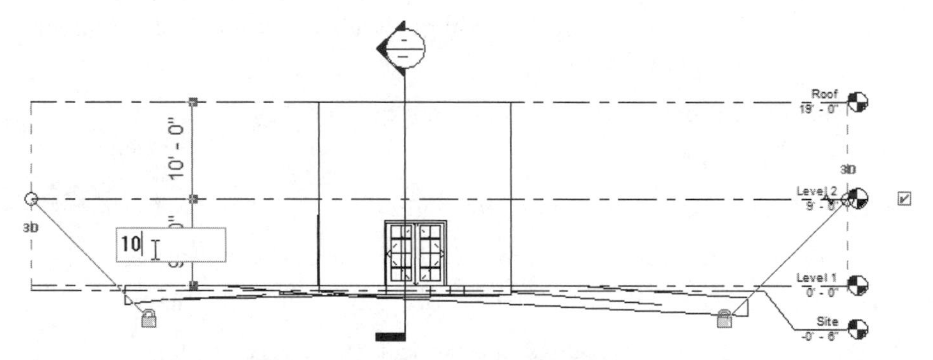

Figure Q–16 *Move the Level line with the temporary dimension*

Notice that this moves Level 2 relative to Level1 but that the distance between Level 2 and the Roof has changed as well.

- Repeat for the Roof—select the Roof Level line.
- Click the blue text of the temporary dimension between Level 2 and Roof, type **10** and press ENTER.

Notice that not only does the Roof Level line move, but since we constrained the Walls to the Roof Level, the top edge of the Walls adjusts as well!

NOTE You can find more information and tutorials on working with Levels in Chapter 4.

MODIFY A WINDOW

Let's edit a Window next. Again, we can edit in whatever View is convenient with confidence that the edit will appear in all appropriate Views automatically.

11. Spin the 3D model View to show the Windows on the north Wall.
12. Select the Window on the north Wall.

- Click on the titlebar of the plan View, then the section View.

Not only is the Window highlighted red in all Views, but in each of these orthographic Views where it is visible, the temporary dimensions appear.

- Edit the temporary dimension values to move the Window.

Notice how it moves instantly in *all* Views. You will never have to worry about chasing down a change in several different drawings to be certain that it has been coordinated. This will boost productivity and help reduce the number of costly change orders.

Move additional Windows in the same way if you wish.

At this point you may wish to line up the Window on the North Wall with the one on the South. You can do this and have Revit Building maintain the relationship with the Align tool.

13. Click the Align tool on the Tools toolbar.

You first indicate thc point of reference. We'll use the Window we just moved.

Click near the center of the Window you just moved to set the point of alignment.

Click near the center of the opposite Window to align it to the reference point (see Figure Q–17).

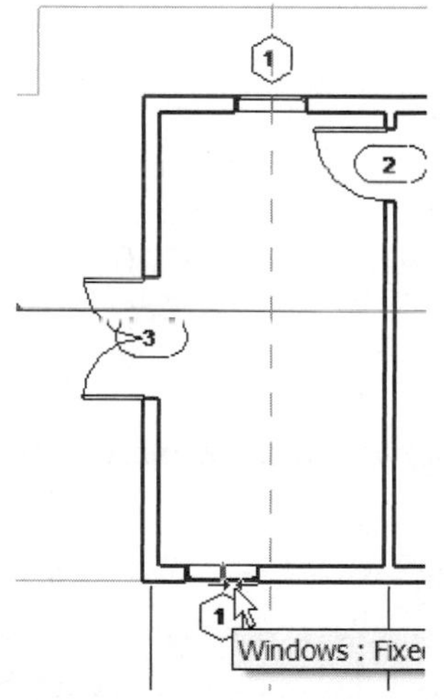

Figure Q–17 *Use the Align Tool to align Windows to one another*

Click the lock icon to constrain the alignment of the two Windows together.

On the Design Bar, click the Modify tool or press the ESC key twice.

If you now move either Window, they will move together.

ADD OPENINGS ON THE SECOND FLOOR

Now that our Walls span the height of both floors, we should add some fenestration on the second floor.

14. On the Project Browser, double-click to open the *Level 2* floor plan View.
    - Following the procedures above, add Windows and a Door to *Level 2* as shown in Figure Q–18.

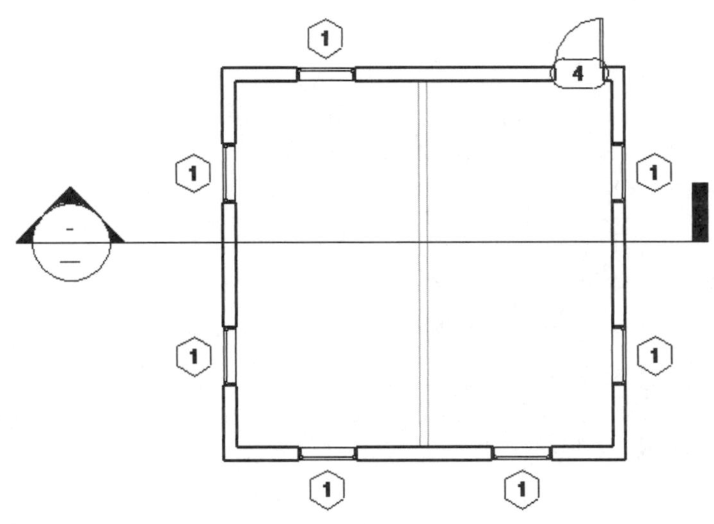

Figure Q–18 *Add Doors and Windows to Level 2*

ROUND OUT THE PROJECT

Our project needs more than just Walls, Doors and Windows. Let's enclose it with a Floor and Roof and look at how to extract data from our model with Schedules.

ADD A FLOOR

The second floor of our building will have an interior balcony on the right overlooking the space to the left.

15. On the Design Bar, click the Basics tab and then click the Floor tool.

When you click the Floor tool, the floor plan will turn gray. The Design Bar will also change to include a series of "Sketch mode" tools. Sketch mode is a special mode in Revit Building used when the element that you are creating has a shape that Revit Building cannot easily "guess." In this case, it would not be possible for Revit Building to assume the size and shape of the Floor that we want, so instead, we sketch it. This is easy to do, given that we already have several Walls.

On the Design Bar, the "Pick Walls" mode will be enabled (selected).

- Click one of the horizontal Walls, then the other.
- Click the vertical exterior Wall on the right.

Notice that with each Wall you click a magenta sketch line will appear on the Wall. Also notice that the sketch will appear on either the inside face or the outside face depending on which side of the Wall the Pick Walls tool was on when you clicked the Wall. The Floor will only cover the right half of the plan, so for the last Wall, we will use the vertical one in the center rather than the exterior one on the left.

If the sketch lines appear on the out side face of a Wall use the Modify tool and select the sketch line. Then click the double arrow flip control to reposition the sketch line to the other face. Repeat as required to locate all three sketch lines on the inside face as indicated in figure Q–19.

- Click on the vertical Wall in the center (see Figure Q–19).

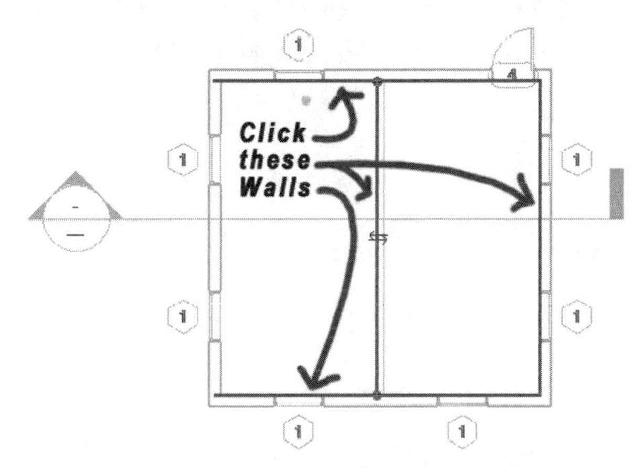

Figure Q–19 *Create Floor sketch lines from the existing Walls (Sketch lines in the figure enhanced for clarity)*

16. On the Tools toolbar click the Trim/Extend tool.

- Click the vertical sketch line in the center of the plan, and then click the right side of the horizontal line at the top.

NOTE When using the Trim/Extend tool you always select the portion of the lines that you want to keep.

- Repeat by clicking the vertical again, then the right side of the horizontal one on the bottom (see Figure Q–20).

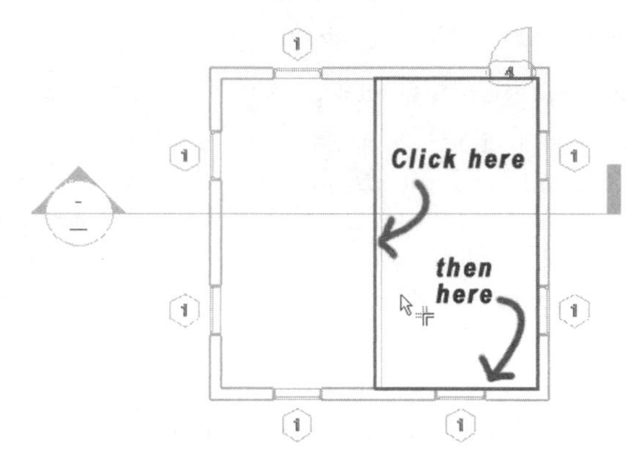

Figure Q–20 *Use the Trim/Extend tool to close the sketch*

- On the Design Bar, click the Finish Sketch button.
- In the dialog that appears, click Yes.

17. On the Project Browser, double-click to open the *{3D}* View.

NOTE If you are still working with four tiled View windows, simply click the titlebar of the *{3D}* View to make it active. Double click the View's titlebar to maximize the view if desired.

- Spin the model around to see the new Floor.

You can also study it in the section View (see Figure Q–21).

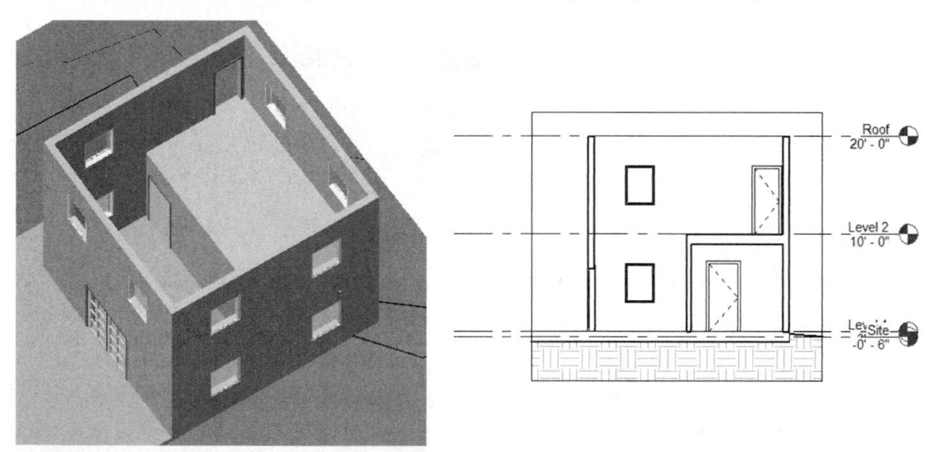

Figure Q–21 *Study the Floor in the* **{3D}** *and* **Section 1** *Views*

NOTE You can find more information and tutorials on working with Floors in Chapter 5.

ADD A ROOF

We can sketch a roof in much the way as we sketched the Floor. There are a few Roofs techniques available. For this exercise, we will make a Roof by Extrusion.

18. On the Project Browser, double-click to open the *West* elevation View.

NOTE If you are still working with four tiled View windows, simply click the titlebar of the *West* elevation View to make it active.

19. On the Design Bar, click the Roof tool. From the flyout that appears, choose **Roof by Extrusion**.

Revit Building will allow us to sketch a simple 2D shape that will become the shape of the Roof's section. The Roof will extrude this shape along its length. To do this, it needs us to establish the plane in which we wish to work.

- In the "Work Plane" dialog, accept the defaults and click OK.

If you pause your mouse for a moment, a tooltip will appear that reads: "Pick a vertical plane." The same message will appear in the Status Bar at the bottom of your screen.

- Click the Wall facing us (see Figure Q–22).

TIP To click the face of the wall, or any plane in Revit, you need to position the Pick Plane tool over the edge of the element. The perimeter of the plane will pre-highlight indicating which plane will be selected when you click.

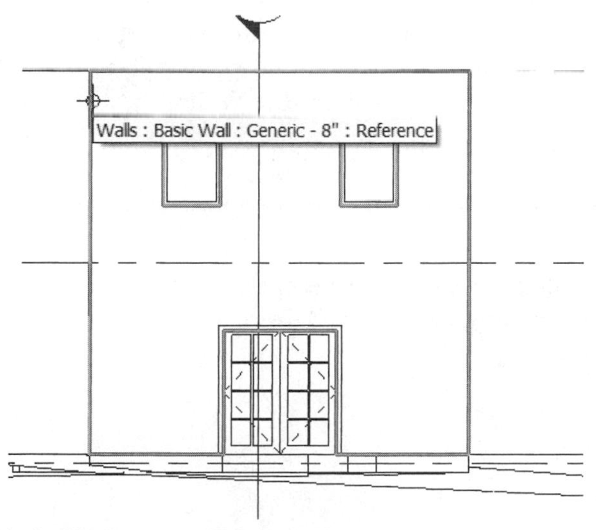

Figure Q–22 *Click the Wall to set a Reference Plane*

- In the "Roof Reference Level and Offset" dialog that appears, accept the defaults and click OK.

We are returned to sketch mode. We will create a curved shape for the Roof.

On the Design Bar, the Lines tool should be active.

- On the Options Bar, click the "Arc passing through three points" icon.
- Click the first point of the arc at the top left corner of the Wall.
- Set the next point about 6° below the right top corner (see Figure Q–23).

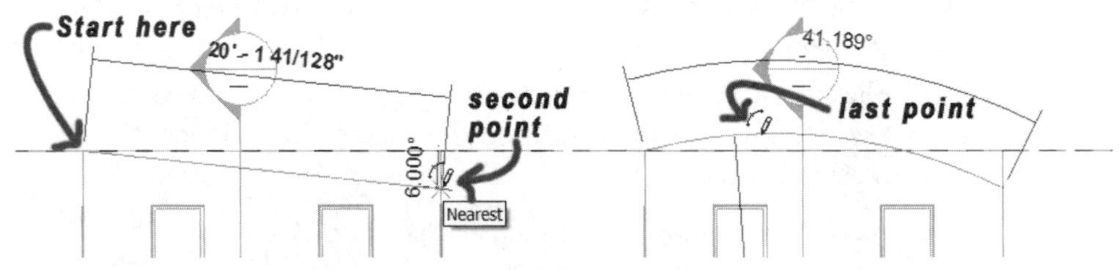

Figure Q–23 *Create a three-point arc shape for the Roof Extrusion*

- Set the final point approximately where indicated in the figure.
- On the Design Bar, click the Finish Sketch button.
- On the Project Browser, double-click to open the {*3D*} View.

NOTE If you are still working with four tiled View windows, simply click the titlebar of the {*3D*} View to make it active.

The Roof automatically spanned over the entire building model. However, its eaves are flush with the Walls and the Walls did not attach to the Roof. Let's add some overhang to the Roof, then we will fix the Walls.

20. Using the Modify tool select the Roof in the {*3D*} View.

    Notice the small arrow handles pointing away from the Roof on two ends of the extrusion.

    - Click and drag each of these handles slightly away from the Walls to create an overhang (see Figure Q–24).

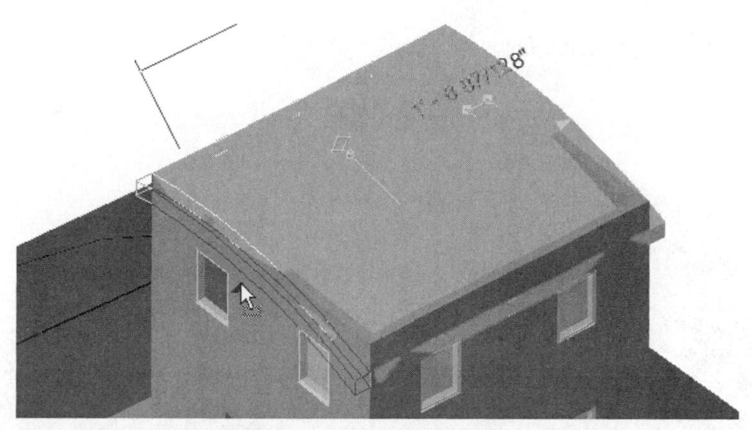

Figure Q–24 *Stretch Roof Extrusion to make overhangs*

To add overhangs in the other direction, we can edit the sketch again.

- With the Roof still selected, on the Options Bar, click the Edit button.

The sketch line will reappear and the Roof will temporarily disappear. You can switch back to the elevation View or edit the sketch directly in 3D. Remember, you can edit in any View and the change will occur in all Views.

- Drag the ends of the line away from the Walls slightly as shown in Figure Q–25.

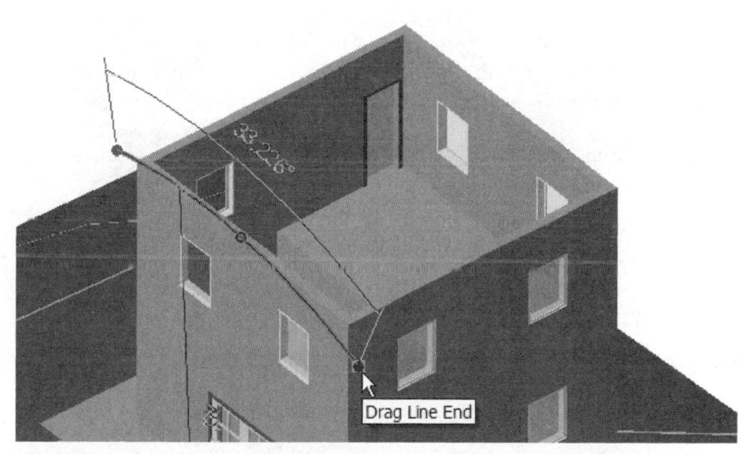

Figure Q–25 *Edit the sketch line to re-shape the Roof to include overhangs*

- On the Design Bar, click the Finish Sketch button.

21. Use the Modify tool and the TAB select method above to chain-select all the exterior Walls.

    - On the Options Bar (running horizontally across the top of the View window), click the Attach button.

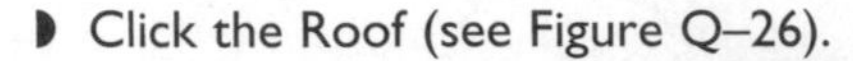

- Click the Roof (see Figure Q–26).

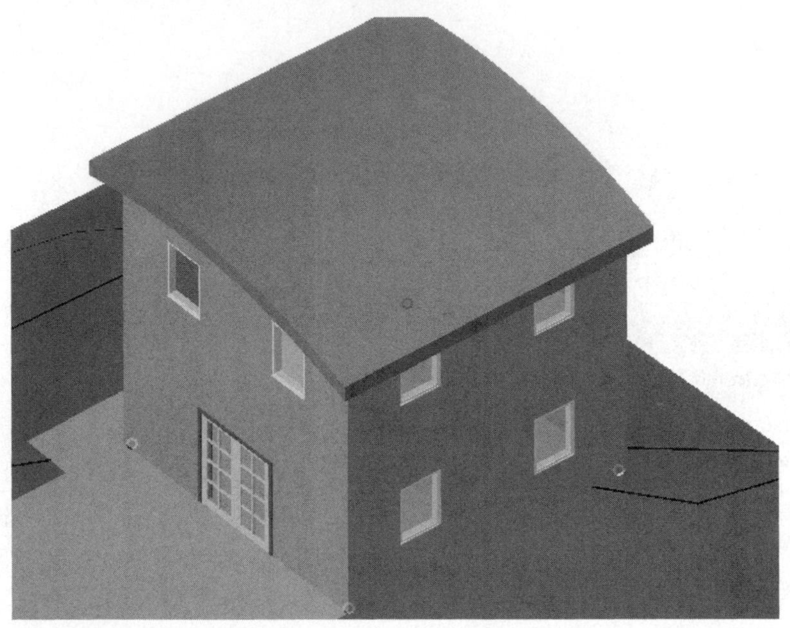

Figure Q–26 *Attach the Walls to the Roof*

NOTE If you edit the shape of the Roof, the Walls will remain attached. Try it out if you like. Select the Roof, and on the Options bar click the Edit button. Undo any change before continuing.

- Save the model.

NOTE You can find more information and tutorials on working with Roofs in Chapter 7.

ADD A STAIR

We have no way to reach our second floor balcony. Let's add an exterior Stair.

22. On the Project Browser, double-click to open the *Level 1* floor plan View.
23. On the Design Bar, click the Modeling tab and then click the Stair tool.

The Stair will go to the north (top) side of the plan on the right of the building. There is a little bump out on the patio for this purpose.

- Click near the middle of the patio bump out and drag up (see Figure Q–27).

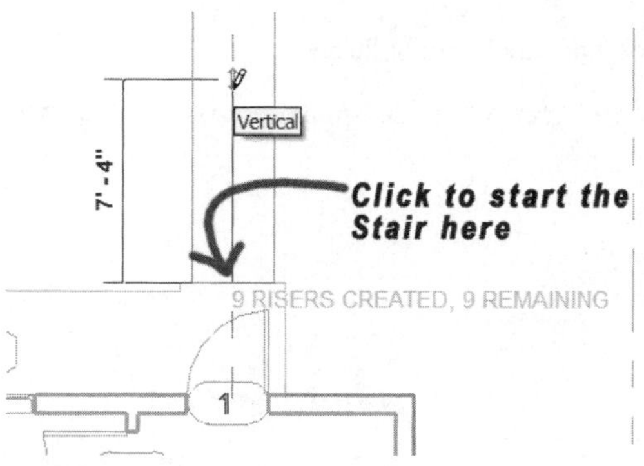

Figure Q–27 *Create half the risers*

A small label will appear on screen indicating how many risers have been created and how many remain.

- Drag straight up until the gray label reads "9 Risers created, 9 remaining" and then click.
- Click a point next to the first run of stairs at the location indicated in Figure Q–28.

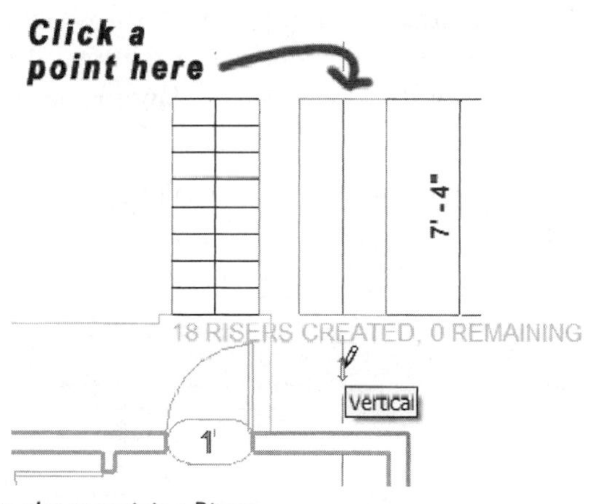

Figure Q–28 *Create the remaining Risers*

- Drag straight down until the message indicates that zero risers remain and then click.

This will give us the basic Stair but it will not "hook up" with the second floor. We need to extend the top riser to make it a landing at the top.

24. Select the riser line at the bottom right (the last riser of the Stair) with the Modify tool.

- Drag it down until it snaps to the building.

- On the Tools toolbar, click the Split tool.
- Click on each of the green lines to split them where indicated in Figure Q–29.

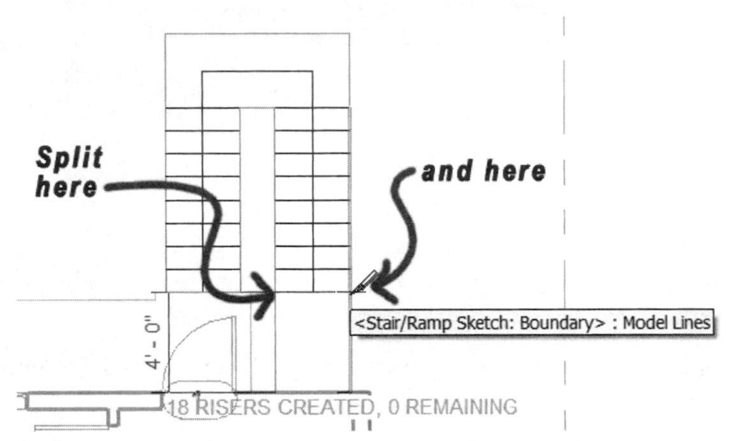

Figure Q–29 *Split the Stringer Lines to make a landing*

The green lines represent the stringers of the Stair. It is necessary to split them (break them into two segments) so that the top part in this case can slope with the Stair and the bottom part can be flat and follow the landing. We also need to change the slope of these lines so they represent a landing and not additional stringers.

- Select one of the stringer lines (the ones we just split) and on the Options Bar; choose **Flat** from the Slope list.
- Repeat for the other side (see Figure Q–30).

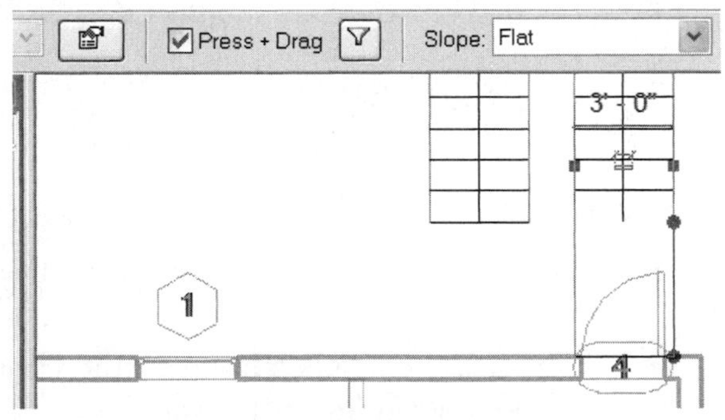

Figure Q–30 *Set the slope of the sketch lines to Flat*

- On the Design Bar, click the Finish Sketch button.

25. On the Project Browser, double-click to open the {*3D*} View.
    - Spin the model around to see the Stair (see Figure Q–31).

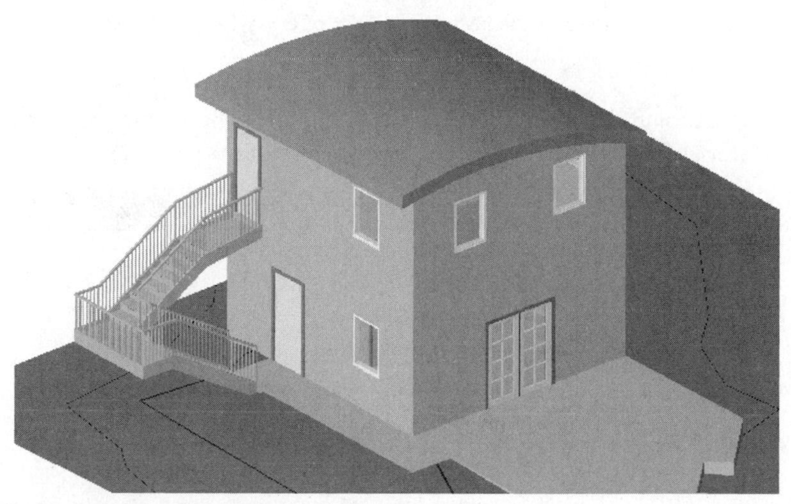

Figure Q–31 *Study the results in the {3D} View*

NOTE You can find more information and tutorials on working with Stairs and Railings in Chapter 6.

This completes the basic geometry of the model. We could add many more embellishments but let's finish this quick start by creating some construction documentation items like a Door and Window Schedule and some Sheets for printing.

CREATE A SCHEDULE

We can create automated schedules of anything in Revit Building. All we need to do is generate a Schedule View which, while not graphical like the plan, section and elevation Views, are just like the other Views in Revit Building. Plans, sections, elevations, and 3D views, etc. are graphical views. Schedules are tabular views. You can view information related to the model and even edit it directly from a schedule View.

26. On the Design Bar, click the View tab and then click the Schedules/ Quantities tool.

    The "New Schedule" dialog will appear.

    - From the "Category" list, choose Doors and then click OK.

    The "Schedule Properties" dialog will appear.

    - In the "Schedule Properties" dialog, on the "Fields" tab, click Mark in the "Available Fields" list and then click the Add × button.

NOTE "Mark" represents the instance number of every Door element.

- Repeat for the following Fields: Level, Width, Height, Frame Type, Frame Material, Family and Type, and Comments (see Figure Q–32).

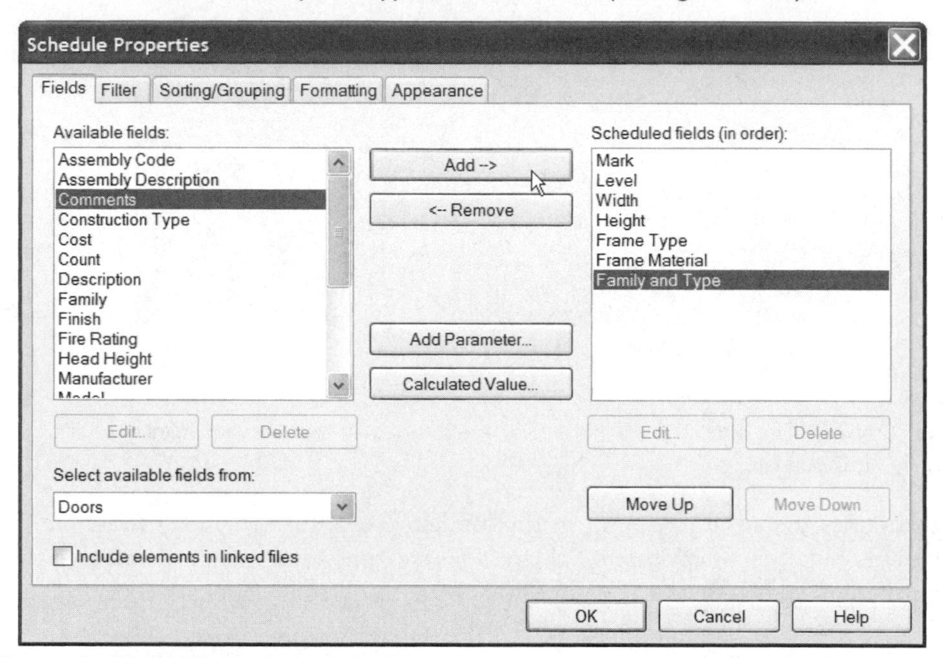

Figure Q–32 *Add Fields to the Door Schedule*

- Click OK to create the Schedule.

A Schedule View will appear on screen. The Schedule View appears much like an Excel spreadsheet. Let's look at the Schedule tiled next to one of the floor plans. However, before we tile, let's close some of the other Views.

27. From the Window menu, choose **Close Hidden Windows**.

NOTE If this command is not available, you must maximize the current window first. To do this, double-click the titlebar of the current window.

- On the Project Browser, double-click to open the *Level 1* floor plan View.
- From the Window menu, choose **Tile**.

You should now have just the *Level 1* floor plan View and the *Door Schedule* View open on screen side by side.

28. In the Door Schedule View, click on Door number 3.

The Door number will highlight in the Schedule and the Door itself will highlight in the plan (see Figure Q–33).

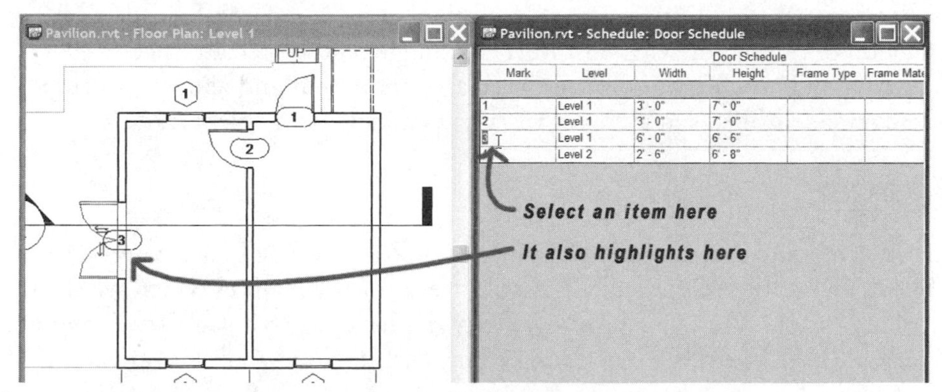

Figure Q–33 *Selected elements highlight in graphical Views and Schedules*

- Highlight the value in the Width field, type **5** and then press ENTER.
- In the message that appears, click OK.

Notice that the size of the Door changes in both the Schedule and the floor plan. The only problem with that change is that we actually edited a "Type" parameter (as indicated in the warning message). This means that all Doors of this Type would be affected. Since there is only one of these currently in the model, it was not obvious. Undo this change and choose a different Type from the "Family and Type" column of the schedule instead. This is the proper approach to changing the size of a single Door instance. Experiment with other changes if you wish.

29. Create another Schedule View (Design Bar, click the Schedule/Quantities tool).
    - In the "New Schedule" dialog, from the "Category" list, choose Windows and then click OK.
    - In the "Schedule Properties" dialog, on the "Fields" tab, add the Mark, Level, Width, Height, Count and Comments fields.
    - Click OK to create the Schedule.
30. In the Level 1 floor plan View, add Windows to the right Wall.

Notice that the new Windows appear immediately in the Schedule.

31. Save the model.

NOTE You can find more information and tutorials on working with Schedules in Chapter 11.

PREPARING OUTPUT

At some point in every architectural project, you will need to output your designs and produce some form of deliverable. Currently the most common format for this is a collection of printed drawings. In Revit Building you use special "Sheet" Views for this purpose. These Views emulate the final paper output and allow us to compose the completed Sheets in any way we wish complete with titleblocks.

ADD A SHEET

Before we can print out documents from our project, we need to create one or more Sheet Views. Sheet Views are basically pieces of paper upon which we drag and drop the various Views for printing. Both graphical and tabular Views can be combined in Sheet Views. On the Design Bar, click the View tab and then click the Sheet tool.

- In the "Select a Titleblock" dialog click OK.
- Double click the titlebar of the new window that appears to maximize it.
- From the View menu, choose **Zoom > Zoom To Fit**.

A blank Sheet View with a Titleblock appears and is ready to receive Views. There are two ways to do this. Right-click the Sheet View name on the Project Browser and choose **Add View** to receive a dialog listing all available Views, or simply drag and drop them from Project Browser. We'll use drag and drop here.

32. From the Project Browser, drag the *Level 1* floor plan View and drop it on the Sheet.

An outline of the View will appear attached to the cursor. This is called the Viewport Boundary and automatically sizes itself to the extents of the dragged View's contents. You can use this to place it on the sheet where desired.

- Position the View in the upper left corner of the Sheet and then click.
- Repeat the drag and drop process for each of the remaining floor plans (see Figure Q–34).
- Drag each of the Schedule Views to the Sheet as well.

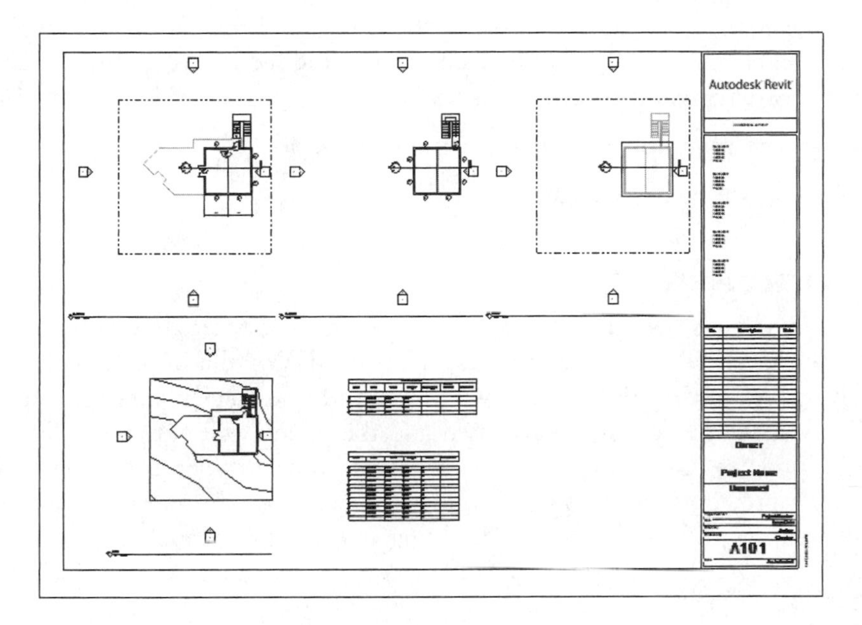

Figure Q–34 *Drag all the plans and schedules to the Sheet view and position them*

If you need to re-position a View after you drag it, you can click on it directly on the Sheet with the Modify tool, and then drag it again to move it or use the arrow keys on your keyboard to nudge it slightly. When you drag the Site plan in, notice that it is a little smaller than the others. Each View has its own scale setting that is used when it is added to a Sheet.

33. Repeat the entire process to create another Sheet and add the elevations and section Views.

Suppose that you wish to change the scale of a View after it is added. For instance after adding all of the elevations and the section View to this new Sheet, you may wish to enlarge the Section View.

34. Select the Section View on the Sheet, right-click and choose **Activate View**.

This makes the View editable as if you had opened it from the Project Browser and edit the original section View.

35. Right-click in the Section View and choose View Properties.
    - In the "Element Properties" dialog, beneath the "graphics" grouping, change the "View Scale" to **1/4″=1′-0″** and then click OK.
    - Right click in the View again and choose **Deactivate View**.
    - Reposition the View as required.

Notice that the change in scale only affected the graphics of the View. The graphics of the model elements such as walls, doors, and windows enlarged but the text and annotations remained the same size.

36. Perform any other edits and explorations you wish.
37. Save and close the project.

GOING FURTHER

Feel free to print your two Sheets out to your printer or plotter. If you have installed the Autodesk DWF Writer, you can plot to a DWF file instead. You can further edit and refine this model if you wish to add additional elements and annotations. Everything will remain coordinated and the Sheets will automatically receive all the updates. Congratulations! You have completed your first Autodesk Revit Building project! Read on in the coming chapters to continue your journey into Revit Building and the promise of Building Information Modeling.

SUMMARY

Getting started with Revit Building is easy—click a tool on the Design Bar and place the item in the View window.

Walls, Doors and other elements interact with each other as you place them in the model.

Relationships and constraints are maintained automatically as you work.

Build or edit your model from any View and changes are fully coordinated in all Views.

Views include graphical representations like plans and sections and non-graphical tabular representations like Schedules.

Drag Views to Sheets for printing.

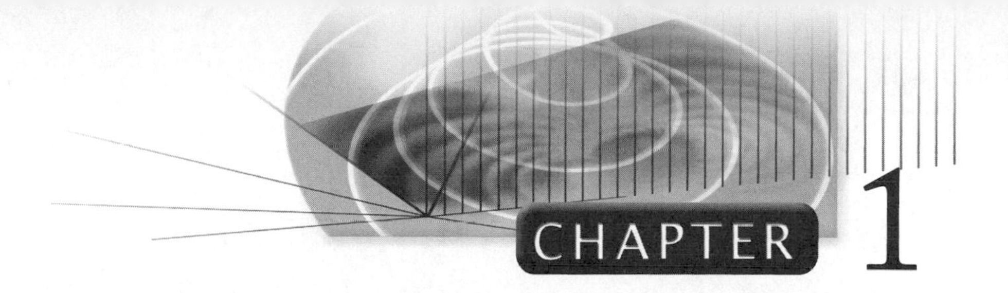

Conceptual Underpinnings of Autodesk Revit Building

INTRODUCTION

Autodesk Revit Building is an object-based software package designed to assist architects and building designers in their architectural undertakings. It seeks to facilitate the creation of "Building Information Models" (BIM) in which plans, sections, elevations, 3D models, quantities and other data are fully coordinated and can be readily manipulated and accessed. From the BIM database, one can perform design tasks, query quantities and takeoffs and generate drawing sheets for construction documentation needs. The advantages of this approach are many. From a 2D production point of view, it means less time drafting and coordinating building data, because plans, sections and elevations are automatically generated from the same source data. If the data changes, all "Views" of that data, whether plans, sections or elevations, reflect the change immediately. Schedules and data reports of quantities, component sizes, materials used, and scores of other data are also "live" Views of the model and readily available. To achieve this level of functionality, it is important to understand a bit about what it means to create and work in a Building Information Model. That is the primary goal of this chapter.

OBJECTIVES

In this chapter, we will explore the meanings of parametric design, Building Information Modeling and take a high level look at the Autodesk Revit Building software package. Working in a provided dataset, you will learn how to view a single model in many different ways which serve a variety of architectural drawing and documentation needs. Topics we will explore include:

- Building Information Modeling
- The fully coordinated nature of a Revit Building Project File
- The basics of various Revit Building Elements
- An introduction to Families and Types
- Core concepts within the context of a Project

BUILDING INFORMATION MODELING

In the Autodesk Revit Curriculum & Student Workbook, author Simon Greenwold concisely and elegantly presents the concept of Building Information Modeling. A portion of that text is reproduced here to help explain BIM and how it compares to more traditional computer aided design (or drafting) (CAD) technology. The following topic is excerpted from the above mentioned publication which is © 2005 Autodesk, Inc. It is used here with permission.

> Building Information Modeling [BIM] is a process that fundamentally changes the role of computation [and delineation] in architectural design. It means that rather than using a computer to help produce a series of drawings and schedules that together describe a building; you use the computer to produce a single, unified representation of the building so complete that it can generate all necessary documentation. The primitives from which you compose these models are not the same ones used in CAD (points, lines, curves). Instead you model with building components such as walls, doors, windows, ceilings, and roofs. The software you use to do this recognizes the form and behavior of these components, so it can ease much of the tedium of their manipulation. Walls, for instance, join and miter automatically, connecting structure layers to structure layers, and finish layers to finish layers.
>
> Many of the advantages are obvious—for instance, changes made in elevation propagate automatically to every plan, section, callout, and rendering of the project. Other advantages are subtler and take some investigation to discover. The manipulation of parametric relationships to model coarsely and then refine is a technique that has more than one career's worth of depth to plumb.
>
> BIM design marks a fundamental advance in computer-aided design. As the tools improve, ideas spread, and [practitioners] become versed in the principles, it is inevitable that just as traditional CAD has secured a deserved place in every office, so will BIM design.
>
> ### CAD VERSUS BUILDING INFORMATION MODELING
>
> ***Modeling Is Not CAD***—BIM is entirely unlike the CAD tools that emerged over the last 50 years and are still in wide use in today's architectural profession. BIM methodologies, motivations, and principles represent a shift away from the kind of assisted drafting systems that CAD offers.
>
> To arrive at a working definition of Building Information Modeling first requires an examination of the basic principles and assumptions behind this type of tool.
>
> ***Why Draw Anything Twice?***—You draw things multiple times for a variety of reasons. In the process of design refinement, you may want to use an old de-

sign as a template for a new one. You are always required to draw things multiple times to see them in different representations. Drawing a door in plan does not automatically place a door in your section. So the traditional CAD program requires that you draw the same door several times.

Why Not Draw Anything Twice?—There is more to the idea of not drawing anything than just saving time in your initial work of design representation. Suppose you have drawn a door in plan and have added that same door in two sections and one elevation. Now should you decide to move that door, you suddenly need to find every other representation of that door and change its location too. In a complicated set of drawings, the likelihood that you can find all the instances of that door the first time is slim unless you are using a good reference tracking system.

Reference and Representation—But doesn't reference tracking sound like something a computer ought to be good at? In fact, that's one of about three fundamental things a computer does at all. And it is exceedingly good at it. The basic principle of BIM design is that you are designing not by describing the forms of objects in a *specific* representation, but by placing a *reference* to an object into a holistic model. When you place a door into your plan, it automatically appears in any section or elevation or other plan in which it ought to be visible. When you move it, all these views update because there is only *one* instance of that door in the model. But there may be *many* representations of it.

You Are Not Making Drawings—That means that as you create a model in modeling software, you are not making a drawing. You are asked to specify only as much geometry as is necessary to locate and describe the building components you place into the model. For instance, once a wall exists, to place a door into it, you need to specify only the type of door it is (which automatically determines all its internal geometry) and how far along the wall it is to be placed. That same information places a door in as many drawings as there are. No further specification work is required.

That means that the act of placing a door into a model is not at all like drawing a door in plan or elevation, or even modeling it in 3D. You make no solids, draw no lines. You simply choose the type of door from a list and select the location in a wall. You don't draw it. A drawing is an artifact that can be automatically generated from the superior description you are making.

Even More Than a 3D Model—You are making a model—a *full* description of a building. This should *not* be confused with making a full three-dimensional (3D) model of a building. A 3D model is just another representation of a building model with the same incompleteness as a plan or section. A full 3D model can be cut to reveal the basic outlines for sections and plans, but there are drawing conventions in these representations that cannot be captured this way. How will a door swing be encoded into a 3D model? For a system to intelligently place a door swing into a plan but not into a 3D model, you need a high-level description of the building model separate from a 3D description of its form. This is the model in BIM design.

Encoding of Design Intent—This model encodes more than form; it encodes high-level design intent. A staircase is modeled not as a rising series of 3D solids, but as a staircase. That way if a level changes height, the stair automatically adjusts to the new criterion.

Specification of Relationships and Behavior—When a design changes, BIM software attempts to maintain design intent. The model implicitly encodes the behavior necessary to keep all relationships relative as the design evolves. Therefore the modeler is required to specify enough information that the system can apply the best changes to maintain design intent. When you move an object, it is placed at a location relative to specific data (often a floor level). When this [datum] moves, the object moves with it. This kind of relativity information is not necessary to add to CAD models, which are brittle to change.

Objects and Parameters—You may be troubled by the idea that the only doors you are allowed to place into a wall are the ones that appear on a predefined list. Doesn't this limit the range of possible doors? To allow variability in objects, they are created with a set of parameters that can take on arbitrary values. If you want to create a door that is nine feet high, it is only necessary to modify the height parameter of an existing door. Every object has parameters—doors, windows, walls, ceilings, roofs, floors, even drawings themselves. Some have fixed values, and some are modifiable. In advanced modeling you will also learn how to create custom object types with parameters of your choosing.

HOW DO BIM TOOLS DIFFER FROM CAD TOOLS?

Clearly, because modeling is different from CAD, you are obliged to learn and use different tools.

Modeling tools don't offer such low-level geometry options—As a general rule modeling deals with higher-level operations than CAD does. You are placing and modifying entire objects rather than drawing and modifying sets of lines and points. Occasionally you must do this in BIM, but not frequently. Consequently, the geometry is generated from the model and is therefore not open to direct manipulation.

Modeling tools are frustrating to people who really need CAD tools—For users who are not skilled modelers, modeling can feel like a loss of control. This is much the same argument stick-shift car drivers make about control and feel for the road. But automatic transmission lets you eat a sandwich and drive, so the choice is yours. There are also ways to layer on low-level geometric control as a post-modeling operation, so you can regain control without destroying all the benefits of a full building model.

Modeling entails a great deal of domain-specific knowledge—Many of the operations in the creation of a Building Information Model have semantic content that comes directly from the architectural domain. The list of default door types is taken from a survey of the field. Whereas CAD gets its power from being entirely syntactical and agnostic to design intent, BIM design is the opposite. When you place a component in a model, you must tell the model what it is, *not* what it looks like.

Or else requires you to build it in yourself—Adding custom features and components to a BIM design is possible but requires more effort to specify than it does in CAD. Not only must geometry be specified, but also the meanings and relationships inherent in the geometry.

IS MODELING ALWAYS BETTER THAN CAD?

As in anything, there are trade-offs.

A model requires much more information—A model comprises a great deal more information than CAD drafting. This information must come from somewhere. As you work in a modeling tool, it makes a huge number of simplifying assumptions to add all the necessary model information. For instance, as you lay down walls, they are all initially the same height. You can change these default values before object creation or later, but there are a near infinitude of parameters and possible values, so the program makes a great many assumptions as you work. The same thing happens whenever you read a sketch, in fact. That sketch does not contain enough information to fully determine a building. The viewer fills in the rest according to tacit assumptions.

Flouting of convention makes for tough modeling—This method works well when the building being modeled accords reasonably well with the assumptions the modeler is making. For instance, if the modeler makes an assumption that walls do not cant in or out but instead go straight up and down, that means that vertical angle does not need to be specified at the time of modeling. But if the designer wants tilted walls, it's going to require more work—potentially more work than it would to create these forms in a CAD program. It is even possible that the internal model that the software maintains does not have the flexibility to represent what the designer has in mind. Tilted walls may not even be representable in this piece of software. Then a workaround is required that is a compromise at best. Therefore unique designs are difficult to model.

Editor's Note: The author of this passage used the example of tilted walls simply to make the accompanying point. Tilted Walls are possible in Autodesk Revit Building, but admittedly with a little more effort than common vertical ones.

Whereas CAD doesn't care—In CAD, geometry is geometry. CAD doesn't care what is or isn't a wall. You are still bound by the geometric limitations of the software (some CAD software supports nonuniform rational b-splines [NURBS] curves and surfaces, and others do not, for instance), but for the most part there is always a way to construct arbitrary forms if you want.

Modeling can help project coordination—Having a single unified description of a building can help coordinate a project. Because drawings *cannot* ever get out of sync, there is no need for concern that updates have not reached a certain party.

A single model could drive the whole building lifecycle—Increasingly, there is interest in the architectural community for designing the entire lifecycle of a building, which lasts much longer than the design and construction phases. A full building description is invaluable for such design. Energy use can be calculated from the building model, or security practices can be prototyped, for instance. Building systems can be made aware of their context in the whole structure.

BIM potentially expands the role of the designer—Clearly this has implications for the role of architects. They may become the designers of more than the building form, but also specifiers of use patterns and building services.

Modeling may not save time while it's being learned—It is likely that while designers are learning to model rather than to draft, the technique will not save time in an office. That is to be expected. The same is true of CAD. Switching offices from hand drafting to CAD occurred only as students became trained in CAD and did not therefore have to learn the techniques on the job. The same is likely to be true of BIM design. But students are beginning to learn it, so it is bound to enter offices soon.

Potential hazards exist in BIM that do not exist in CAD—Because design *intelligence* is embedded into a model, it is equally possible to embed design *stupidity* .

Editor's Note: For example, if the computer operator inputs a floor to floor height of only 4 feet [1200 millimeters], BIM software will not automatically correct or even flag this as an error. The knowledge of the architect is still the driving force behind the design intent that creates the model. An intelligent Building Information Modeler does not replace the knowledge and experience of an Architect.

Improperly structured models that look fine can be unusable—It is possible to make a model that looks fine but is created in such a way that it is essentially unusable. For instance, it may be possible to create something that looks like a window out of a collection of extremely tiny walls. But then the program's rules for the behavior of windows would be wasted. Further, its rules for the behavior of walls would cause it to do the wrong things [with the "windows" modeled this way] when the design changed.

What Is an Engineering Technique Doing in Architecture?—BIM design comes from engineering techniques that have been refined for many years. Many forces are acting together to bring engineering methodologies like BIM into architecture. First, the computing power and the basic ability to use computers have become commonly available. Second, efficiencies of time and money are increasingly part of an architect's concern. BIM offers a possible edge in efficiency of design and construction.

DEFINING BIM

The preceeding quoted passage does a wonderful job at outlining the high level concepts involved in BIM and comparing and contrasting its tenets and techniques to traditional CAD and drafting methods. Using the points raised in this article, let's try to synthesize them into a definition of Building Information Modeling.

Unfortunately, Building Information Model(ing) is perhaps one of the most misunderstood terms in the architectural industry today. "BIM" is often assumed to be synonymous with simply generating a 3D model of a building—whether that model has any useful non-graphical information or not, and regardless of the level to which the model is detailed. Part of the problem is that Building Information Modeling *is* an evolving concept; one that will continue to change as the capabilities of technology and our own ability to manipulate technology improve. These issues make it difficult to formulate a simple definition for BIM. However, as the popularity of BIM is growing, reaching a consensus on its meaning and intent is increasingly important.

Summarizing all of the points made so far, the most important issue is that emphasis belongs on the "I—Information" in BIM. That information can be either graphical or non-graphical; either contained directly in the building model or accessible from the building model through linked data that is stored elsewhere. As has already been mentioned in the quoted passage above, when you exercise BIM, you are making a model which is a *full* description of a building—*not* just a 3D model. A data model is just as valid a model as a geometric model. (This is not a new concept in Architecture. Consider the existing requirement of both a set of drawings *and* a written specification to complete a construction documents package). Despite the importance of these distinctions, when we think of a Building Information Model, a three-dimensional geometric model of the building is often what comes to mind. So the first step to fully understanding BIM is to realize that "BIM" and "3D Model" is *not* the same thing.

In simple terms, a Building Information Model is a complete representation or depiction of a building that aids in its design, construction and potentially ongoing management. Such representation will often employ any combination of 3D graphics, 2D abstractions and/or non-graphical data as required for conveying full intent. The delivery of information and intent is the most important goal in BIM.

Some of the content of the previous topic is paraphrased from Matt Dillon's Web Log (Blog). You can find the complete article at the following URL: http://modocrmadt.blogspot.com/2005/01/bim-what-is-it-why-do-i-care-and-how.html Portions used here were used with permission.

REVIT BUILDING KEY CONCEPTS

So now that you've got a good idea of the BIM concept you may be wondering how it specifically relates to Revit Building. Even more importantly, you many also be wondering how BIM will improve the way you work. Throughout the course of this book, and even more as you begin working with Revit Building on your own projects, you will gain comfort and familiarity with key Revit Building concepts. In this topic, we will identify and describe some of the most important Revit Building concepts, including Revit Building Elements, Families & Types, and Editing Modes.

ONE PROJECT FILE – EVERYTHING RELATES

In a Revit Building project (which is typically contained within a single computer file), you will notice that no matter where in the project you work, no matter what kind of view you are working in, all elements retain their relationships with each other. This complete "bi-directional" coordination is perhaps the most significant benefit to using Autodesk Revit Building. Technically, full bi-directional coordination is not a requirement of BIM. However, to completely realize the potential of a BIM solution, careful coordination of all depictions and representations of the building model is essential to satisfy its requirement to convey intent. This is best achieved by complete bi-directional coordination of elements in the model. In Revit Building, you can make a change in any View (plan, section or schedule) with complete confidence that the change is reflected instantly throughout the entire Project File.

NOTE Changes to "Detail" Views occur only in the edited View; more on this in Chapter 10.

The most talked about examples of the bi-directional coordination usually have to do with the 'physical' aspects of your building project. If you move a door in a plan view, for example, wherever that same door shows up, in an interior elevation or perspective view, the door will also be shown as having moved.

Another example is in the 'informational' aspects of that door. If you go into a schedule view where that door is listed and change it from wood door to a glass door, not only will the calculations like quantities or costs for that door change in the schedule, but also the change will be reflected in all graphical views—for example, in shaded Views, the Door will now appear transparent. Likewise, if this data is linked to cost estimation or green building calculations, the change to a glass door Type will have other important impacts as well. This example gives a good example of how the "I" in BIM deserves more emphasis than the "M."

Another important aspect of Revit Building is the Project Browser. Every View, Family and Group is clearly listed and organized in the Project Browser. The advantage of this is that all pieces of the project are always accessible and neatly organized automatically. It is not possible for a team member to accidentally save project data in the wrong folder or location.

AUTODESK REVIT BUILDING ELEMENTS

In Revit Building you create a representation, or model of your project, using three types of elements. An element in Revit Building is simply a discrete piece of data like an object or drawing sheet. The three basic types of elements are Model Elements (Walls, Doors, Roofs), View Elements (Plans, Sections, Schedules) and Annotation Elements (Tags, Text, Dimensions). Your graphical model is derived primarily from Model Elements. View Elements are used to display, study, and edit the model in depictions that represent traditional architectural drawing types. All Views automatically appear in logical categories in the Project Browser tree. To prepare Views in Revit Building to appear on and print from Sheets, Annotation Elements such as text, dimensions and tags can be used to notate and clarify the information shown in the various views. Specific Architectural Scale, level of detail and other display characteristics are also the province of Views. Views can be easily arranged on sheets and plotted to produce presentation drawings or drawing sets.

The overall types of Revit Building Elements are illustrated with examples of each category in Figure 1–1.

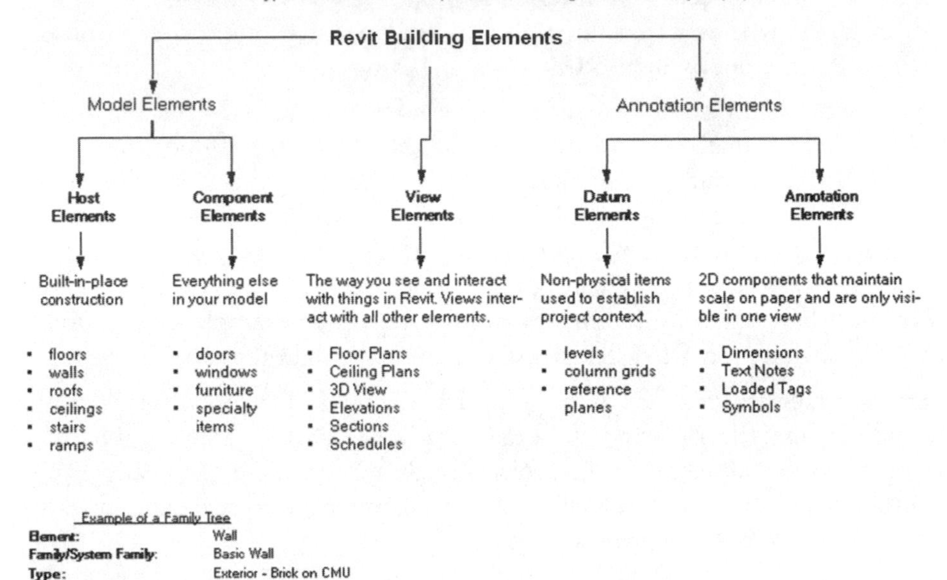

Figure 1–1 *Autodesk Revit Building Elements Flow Chart*
Image courtesy of Autodesk

Model Elements

You can think of Model Elements as the items you use to describe the physical aspects of your building project. These elements in Revit Building can be separated into two sub-categories (as shown in Figure 1–1)—Host Elements and Component Elements. Host Elements are components that are constructed in place in the actual building like Walls and Roofs. Component Elements are placed or installed in the building like Doors and Furniture. The figure shows further examples of each of these element types. For the purpose of our discussion here, let's consider the most common element of each category: a Wall (Host Element) and a Door (Component Element).

Host Elements	**Component Elements**
Here are the similarities:	
They are both considered Families.	
Wall – Basic	Door – Single-Flush
They can both have multiple Types.	
Generic – 4″ Brick	30″ x 84″
Exterior – Brick on CMU	36″ x 80″
Here are the differences:	
Their Basic Role in Projects	
Define Spaces and Enclosures	Modify/Detail Spaces and Enclosures
Their Relationship in Hosting	
Can *Host* Components (A Wall can *Host* a Door)	Can be *Hosted* (A Door can be *Hosted* by a Wall)
They are saved in different places	
Saved in Project Files (*transferred* to new projects)	Saved in Family Files (*loaded* in new projects)

The primary concept to understand about Model Elements is that they *are* the elements that you use to create a physical model which depicts your building, and these same elements are the ones that appear within *any* of the views of your project.

When adding a wall to your project, you start by deciding the type of Wall you are creating and in what location the Wall will be added. Determining how the lines and patterns that represent that wall will look on a particular drawing is handled automatically by the software. In this way, you are actually constructing a model of the required Walls rather than drafting a specific representation of them as you might in manual or CAD drafting.

NOTE The specific settings can be modified manually if required in the Object Styles dialog box.

To get a complete listing of all of the Model Elements available in Revit Building, and how they will appear graphically in your project, open the Object Styles dialog box (Settings menu), and study the list on the 'Model Objects' tab (see Figure 1–2).

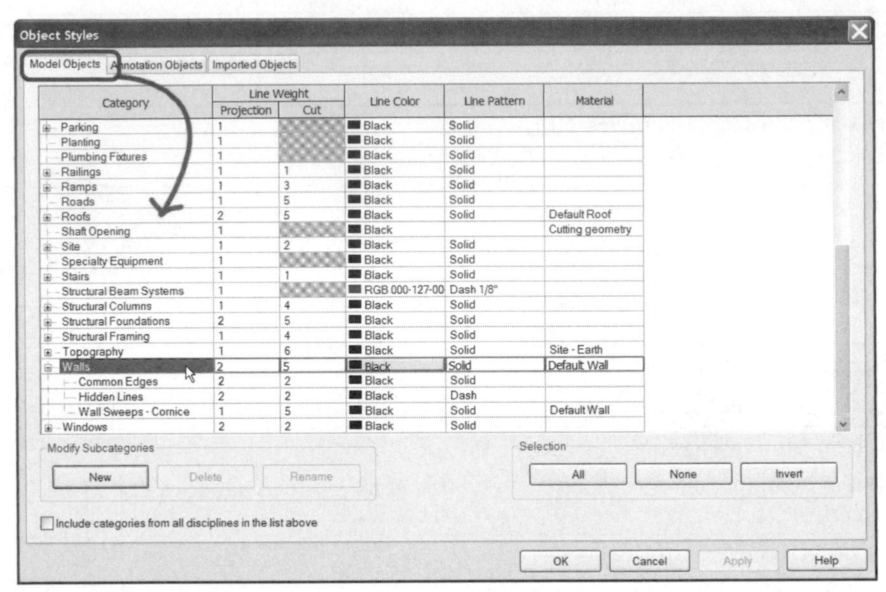

Figure 1–2 *A list of Model Elements can be seen in the Object Styles dialog*

View Elements

As you develop your project in Revit Building, you will work with and create several Views. Every Revit Building project begins with at least some Views already in the Project Browser. The specific Views that are available is a function of the template Project that was used to create the project. (More information on Project Templates can be found in Chapter 4). To open and work in a View, you simply double-click its name in the Project Browser. View elements are available for every architectural drawing type traditionally included in Architectural documentation sets. Like the drawings they represent, Views allow you to interface with your model and edit its contents and composition. Unlike traditional drawings, (as mentioned above) an edit in one View is instantly reflected in all appropriate Views throughout the project.

Multiple Views-One Model—Imagine two friends living on opposite sides of the same street. Let's assume that the street runs north-south and that one friend lives on the west side of the street while the other lives on the east side. If both friends were looking out their window at the same time, as a car was passing by on the street below from the south traveling north, which way would the friends say that the car was driving relative to their respective vantage point? The friend in the house on the west side would describe the car as traveling from his right to his left, while the friend on the east side would say the car traveled from her left to her right. Which friend was correct? What if both friends snapped a photo at the same time? The two photos would show a different "view" of the same car and its travel pattern. This illustrates how Views work in Revit Building. We frequently switch from View to View to edit and create elements; and although the specific graphics displayed on screen may vary (like showing the driver's or passenger's side of the car), they convey aspects of the same model in *all* Views. Therefore, where you make a change is irrelevant. A change to the model occurs in only one place—the model. You can study the change from any number of vantage points as represented in various Views.

To see a complete list of View Elements in Revit Building, you can look at the elements listed on the **View** > **New** Menu (see Figure 1–3).

NOTE Revit Building includes both graphical Views like plans, sections and elevations and tabular Views like Schedules. Both types provide the means to study and manipulate models. Examples occur throughout this book.

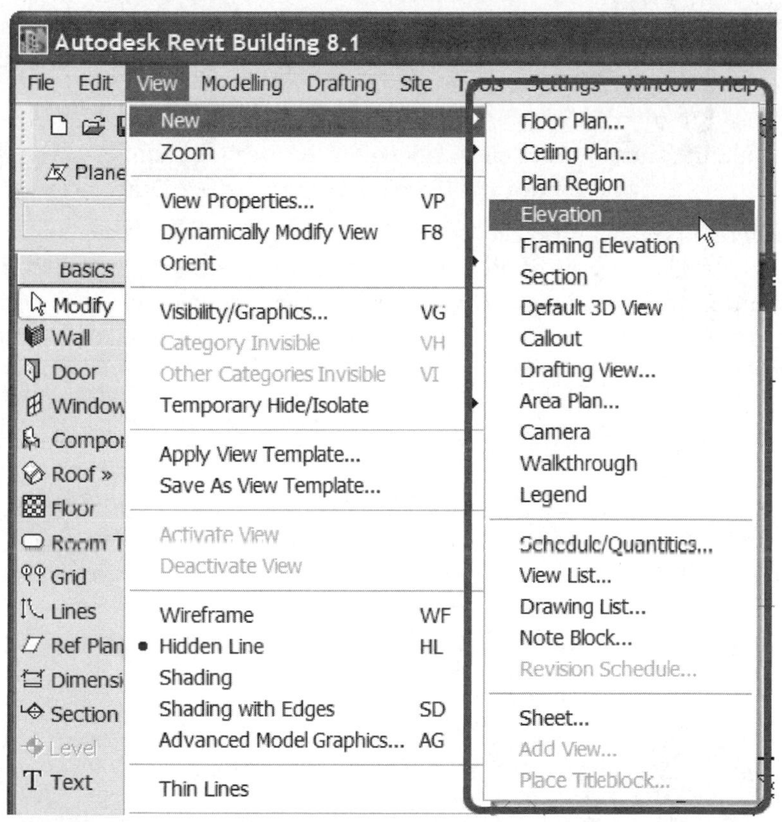

Figure 1–3 *Available View Types are seen on the* View > New *menu*

Annotation Elements

As you can see in Figure 1–1, the overall Annotation Elements category of Revit Building Elements contains two sub-categories like the Model Elements category does. The first sub-category is Datum Elements and the other is simply named Annotation Elements again. While first category of Datum elements does provide graphic annotations for views used as drawings, their behavior in terms of the project as a whole is very different than the latter category of Annotation elements. This is because changes to Datum Elements affect the project as a whole, and will potentially cause changes to both Model and View Elements.

On the other hand, Annotation Elements *only* affect the Views in which they are used. This makes their behavior very unique in relation to the other types of Revit Building elements. Annotation Elements include all of the text and other descriptive architectural symbology that is typically added to drawings to explain and clarify the intent of the graphics, such as tags, dimensions, and View-specific detail components.

Annotation Elements are the closest thing to what you might consider 'drafting' in Revit Building. This is because they rarely effect changes in other views, or to the project model as a whole. However, as you might expect, even Annotation Elements can be used to manipulate the objects from within the View where they are placed. Manipulating those model elements will in turn cause changes throughout the project. But this is typically not the primary purpose of Annotation Elements.

Annotation Elements include objects such as Dimensions, Text, Tags and many more. To get a complete listing of all of the Annotation Elements available in Revit Building, and how they will appear graphically in your project, open the Object Styles dialog box (Settings menu), and study the list on the 'Annotation Objects' tab (see Figure 1–4).

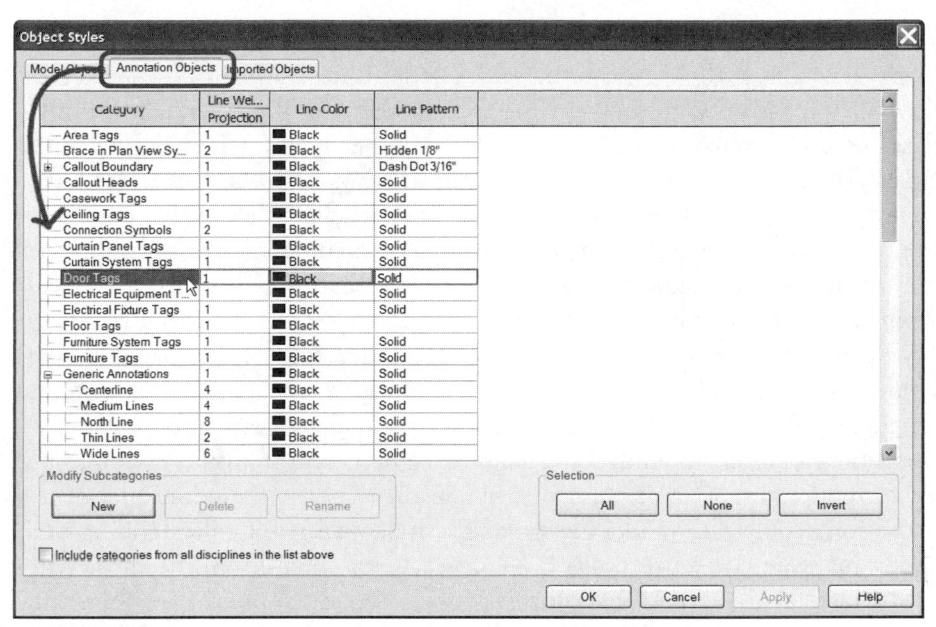

Figure 1–4 *A list of Annotation Elements can be seen in the Object Styles dialog*

Components, Model Lines and Detail Lines

Most elements in Revit Building are purpose-built elements which have obvious functions based upon their names like Wall, Door, Room Tag and Section. The function of each of these elements is easily inferred from their respective names. However, there are also more generic elements whose function is not as specific that serve important functions in Revit Building as well. These elements include items like components, text and lines. A Component is a Model Element that does not have a pre-defined function like a Door or a Window. Components are employed to create Furniture, Fixtures and other items that are placed in models. Components are based on predefined templates so that while there is no specific "Furniture" element per se, there is a Furniture Family Template. (The term Family will be discussed in more detail below).

Text and Lines on the other hand can actually be either Model or Annotation Elements. The same distinction (discussed above) between Model and Annotation applies to these elements as well—Model Text and Model Lines are actually part of the model and used to represent real items in the model. Model Text for instance could be used to model signage in a model and Model Lines might be employed to create inlaid patterns on Walls or Floors that might not be necessary to model three-dimensionally.

While the creation process and graphical appearance on screen of Model and Detail lines may appear very similar and therefore make it difficult to distinguish them from one another, they are in fact very different. For example, if you draw a floor pattern using Model Lines, it will appear in *all* appropriate Views like the floor plan View, and a 3D View of the space. Model Lines added to the surface of a Wall would appear in both elevation and section Views as if you actually painted these lines on the surfaces of the model. On the other hand, a Detail Line, like other annotation, will appear in *only* the View in which it is added. It is treated like other annotation as a simple embellishment to that particular View only. The most common use of such embellishment would be on enlarged details created from the model. Rather than meticulously model components that would only be practical to show in large scale drawings of a design, Detail Lines can be employed to represent those elements that would otherwise take too much time and effort to model throughout and would also add unnecessary overhead to the model. Such items might include building paper in a wall or roof section, nails, screws or other fasteners, reinforcing, or even moldings and trim in some cases. It would certainly be possible to model any of these elements, but in many cases, the additional overhead and effort required to model them would not be justified. Understanding this very important issue is critical to using Autodesk Revit Building in the most efficient and practical manor. More information on this concept can be found in Chapter 10.

The Revit Building interface is covered in detail in the next chapter. However in general, when you access the "Lines" tool from the "Basics" and "Modeling" tabs of the Design Bar you create a Model Line. This creates a line in your model as if someone physically at the site painted a line on a wall or floor in the building. To create a Detail Line, you must click on the "Drafting" tab of the Design Bar and choose the Detail Lines tool. Doing so will create a purely annotative line as described above. Once again, Detail Lines are only visible in the View window in which they are created. Adding a Detail Line is like drawing on a piece of paper covering the view of your project model, and will not be added *physically* to your building model. Model and

FAMILIES & TYPES

One of the most common terms in Revit Building is the term Family. A Family in Revit Building is an object that has a specific collection of parameters and behaviors. Within the limits established by the Family and its parameters, an endless number of "Types" can be spawned. Where a Family can contain variable parameters, a Type is simply a specific version of the Family. The term "Family" was selected to characterize objects in Revit Building that have an inherited-property relationship. The main idea comes from the notion of a Parent-Child relationship—which is a central idea in object-based computer programming.

In order to illustrate how this concept works in Revit Building, we will look at some common categories of Families and describe how they are created and used in a project. You will notice these follow along similar lines as the Revit Building

Elements covered above, with the addition of an additional Element class called "System Families."

Model Families

Model Families include all of the elements in the Model Elements category described above. This means that we have Host Families and Component Families in Autodesk Revit Building. The most common example of a Model Family is the Wall Family. A Wall Family is a Host Family. The progression from global to specific with regard to any Family is called a "Family Tree." Here is an example of a Wall Family Tree:

Category	Item
Element	Walls
Family/System Family	Basic Wall
Type	Exterior—Brick on CMU
Instance	Specific Instance of a Wall in the project

Here is this Wall Family Tree illustrated in a Revit Building project (see Figure 1–5):

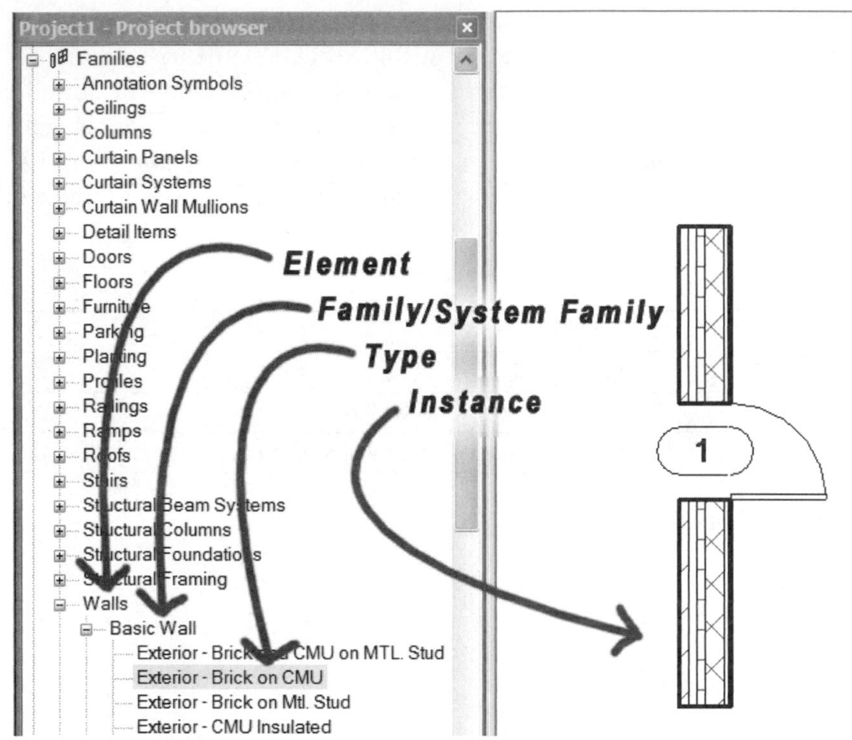

Figure 1–5 *A typical Wall Family Tree*

Recalling our discussion of Element categories above, a Wall Family is a Host Family. Host Elements as we mentioned are "built in place" construction. In Revit Building this means that Host Families must be part of a project. This means that to use a particular Type of Wall, it must already be present in the project file. If the Type that you wish to use is not present, you must either duplicate and modify a similar Type that is resident in the file or transfer one from another project. Wall Types and other Host Element Types cannot be saved into their own individual files.

Component Families include all of the Component Elements and as defined above, are "placed" or "installed" elements in a model. Perhaps the most common Component Element Family is a Door Family. Here is an example of a Door Family Tree:

Category	Item
Element	Doors
Family/System Family	Single-Flush
Type	36″ x 84″
Instance	Specific Instance of a Door in the project

Here is this Door Family Tree illustrated in a Revit Building project (see Figure 1–6):

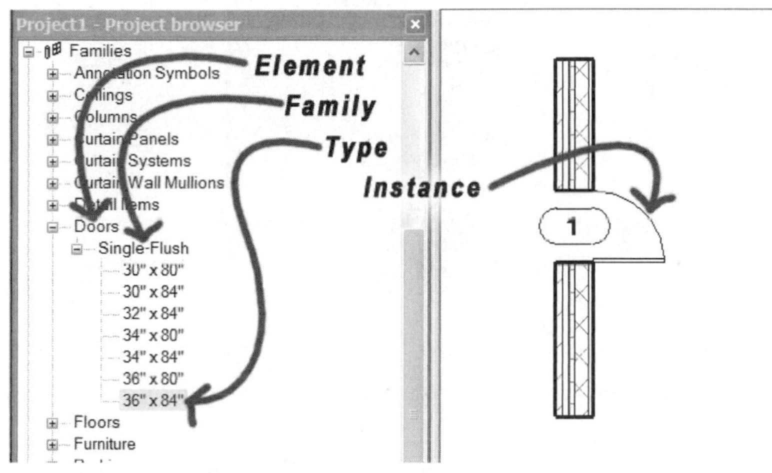

Figure 1–6 *A typical Door Family Tree*

Unlike Host Families, Component Families do not have to be resident in the project file before you can use them. You can load Component Families from external files. Doing so however will load all Types associated with that Family into the current project (unless the Family file has a "Type Catalog" associated with it like Structural shapes—see the online Help for more information). You can also create new Types within the project just as you can with Host Families. Baring these subtle differences, Host and Component Families are very similar to one another. Model Families will be covered in more detail in the coming chapters.

Annotation Families

Families do not only control the parameters of Model Elements. Every element in Revit Building belongs to a Family. Annotation Families are similar to Model Families in that they are also objects that have a specific collection of parameters and behaviors. However, in this case, those parameters and behaviors are specific to Annotation Elements rather than Model Elements. For example, there are families for Dimensions, Text, Tags and Datums. You can see most Annotation families loaded in your project by expanding the Annotation Symbols branch in the Project Browser (see Figure 1–7).

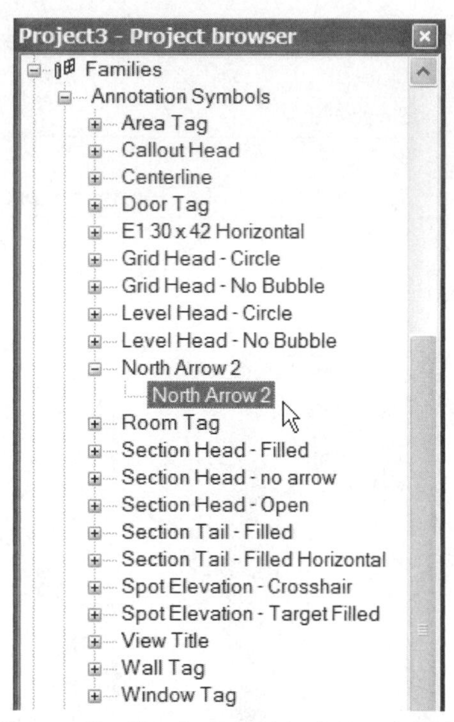

Figure 1–7 *Annotation Element Families Available in a Project appear on the Project Browser*

To see and edit the Dimension or Text System Families, select a text or dimension element and then click the Properties icon. In the "Element Properties" dialog, you will see the Family listed at the top with the Type listed beneath it. Next to the Type list you can click the Edit/New button to edit the current Type or duplicate it to create a new one (see Figure 1–8).

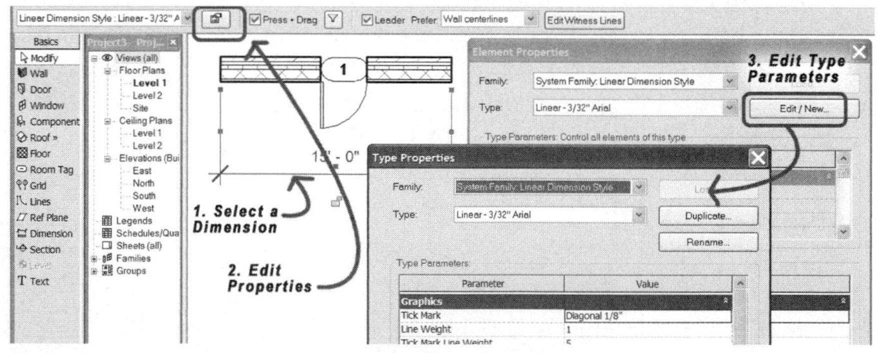

Figure 1–8 *Accessing Text and Dimension Family parameters*

Many Annotation Families serve special purposes in a project. Some are simple symbols that you can add to any View. A Title block for instance is a special kind of Annotation Family that is used when you create a Sheet. It can include text fields that automatically report project data and can also contain company logos and other graphics. Some Annotation Families, like Level Heads, Section Heads and Elevation Heads are typically associated directly to some other View in the project and provide a means to navigate from one View to another. For example, add a Section line in a plan to indicate where the section is cut. Double-click the Section head to open the associated section View.

System Families

Families even help to expand options about how to work with the software. One example of this is Browser Organization. You'll notice if you select the Views branch of the Project Browser, the Type Selector field activates with the current Browser Organization Family Type. The default type is Browser - Views: all (see Figure 1–9).

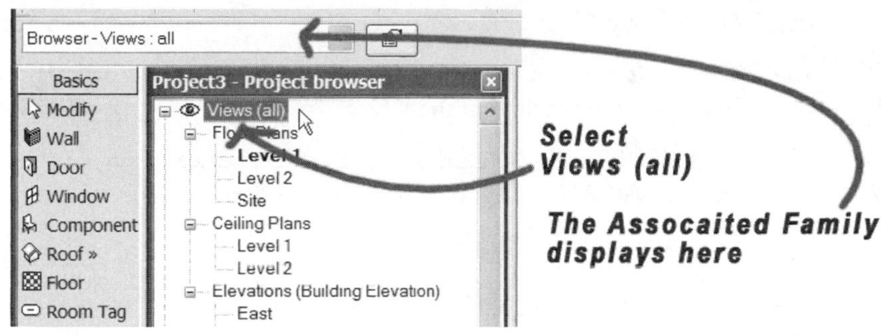

Figure 1–9 *Even the organization of the Project Browser is governed by a Family*

Like all Families, the Browser Family includes other Types. If you open the drop-down list (on the Type Selector) you will see the other Types in this Family. Many of the Types available in the list are quite useful depending on the kind of project on which you are working and/or the phase or composition of the design team (see Figure 1–10).

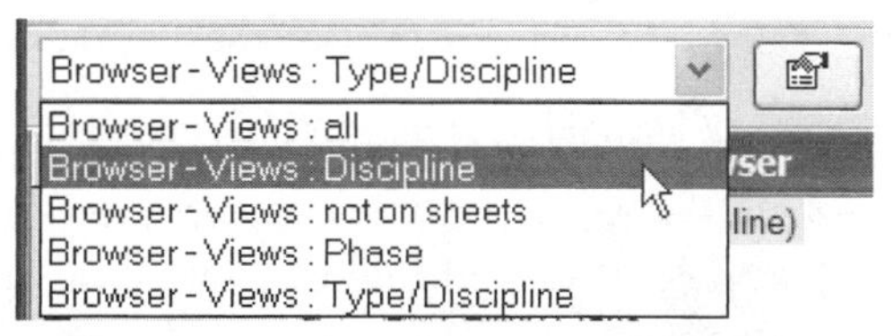

Figure 1–10 *The Browser Family contains other very useful Types*

If the Type of Browser Organization that you wish to use is not included in the Type Selector list, you can choose **Browser Organization** from the Settings menu to create new Types to help manage and display the Views created in your Revit Building project.

EXPLORE AN EXISTING PROJECT

We have covered a lot of important concepts in this chapter so far. To help reinforce and solidify the concepts presented let's open and explore a dataset provided with the files installed from the Mastering Autodesk Revit Building CD ROM.

INSTALL THE CD FILES AND OPEN A PROJECT

The lessons that follow require the dataset included on the Mastering Revit Building CD ROM. If you have already installed all of the files from the CD, simply skip down to step 3 below to open the project. If you need to install the CD files, start at step 1.

1. If you have not already done so, install the dataset files located on the Mastering Revit Building CD ROM.

   Refer to "Files Included on the CD ROM" in the Preface for instructions on installing the dataset files included on the CD.

2. Launch Autodesk Revit Building from the icon on your desktop or from the **Autodesk** group in **All Programs** on the Windows Start menu.

   - From the File menu, choose **Close**.

This closes the empty new project that Revit Building creates automatically upon launch.

3. On the Standard toolbar, click the Open icon.

TIP The keyboard shortcut for Open is CTRL + O. **Open** is also located on the File menu.

- In the "Open" dialog box, click the *My Documents* icon on the left side.
- Double-click on the *MRB* folder, and then the *Chapter01* folder.

If you installed the dataset files to a different location than the one listed here, use the "Look in" drop down list to browse to that location instead.

4. Double-click *MRB Chapter01.rvt* to open the project.

   You can also select it and then click the Open button.

NOTE For this brief tutorial, only an Imperial units dataset has been provided. For the remainder of the book, Metric datasets are provided as well.

The project will open in Revit Building with the last opened View visible on screen.

GETTING ACQUAINTED WITH THE PROJECT

For this tutorial, we will explore a series of Sheet Views included in the project. A Sheet View is a special kind of View in Revit Building that emulates a sheet of paper from which drawing sets can be printed to output devices. Sheet Views typically include a title block which includes project and drawing information.

Autodesk Revit Building remembers the last View that was open when the project was saved. In this case, it is a three-dimensional aerial view of the entire project. This is a small one-floor project for a youth center. It includes offices, exam and counseling rooms, a multipurpose room and media rooms. Let's take a look at the project (see Figure 1–11).

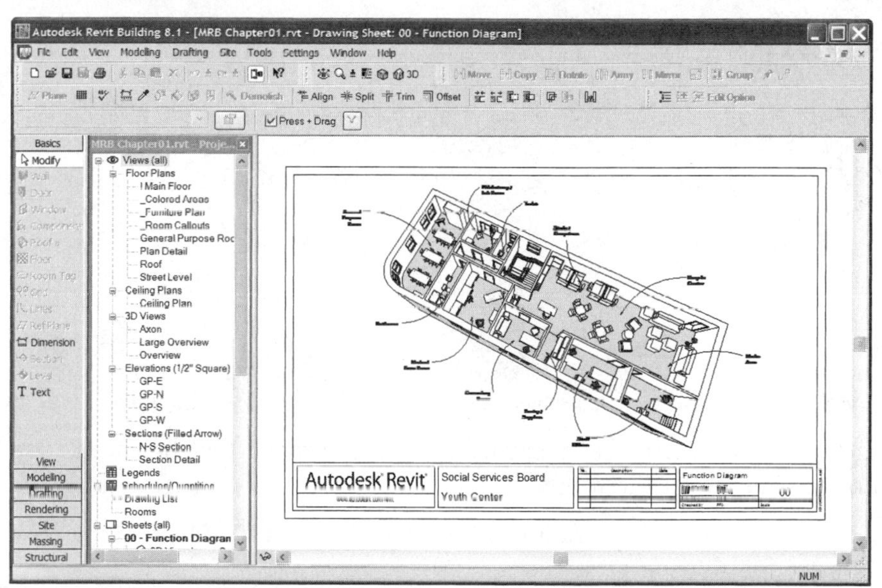

Figure 1–11 *The Youth Center dataset shown from the "Function Diagram" Sheet*

You can use the wheel on your mouse to zoom in and out in any View. You can hold the wheel in and drag to pan the screen. You can also use the scroll bars at the right and bottom for this purpose. If you do not have a wheel mouse, you should consider purchasing one. However, without a wheel mouse, you can use the commands on the **View > Zoom** menu to get a better look at the model in any View. Most of these commands will be available in all Views, like Zoom To Fit (which fits the screen to the extent of the model) and Zoom In Region (which allows you to drag a rectangular region on screen to zoom). We also have the handy **Zoom > Sheet Size** available. When you are in a Sheet, this will zoom the screen to the scale of the Sheet and give you a pretty good preview of how the Sheet will look when printed.

5. From the View menu, choose **Zoom > Sheet Size**.

   ◗ Hold in the wheel on the mouse and drag around to pan the model (see Figure 1–12).

   If you prefer, or if you don't have a wheel, use the scroll bars instead.

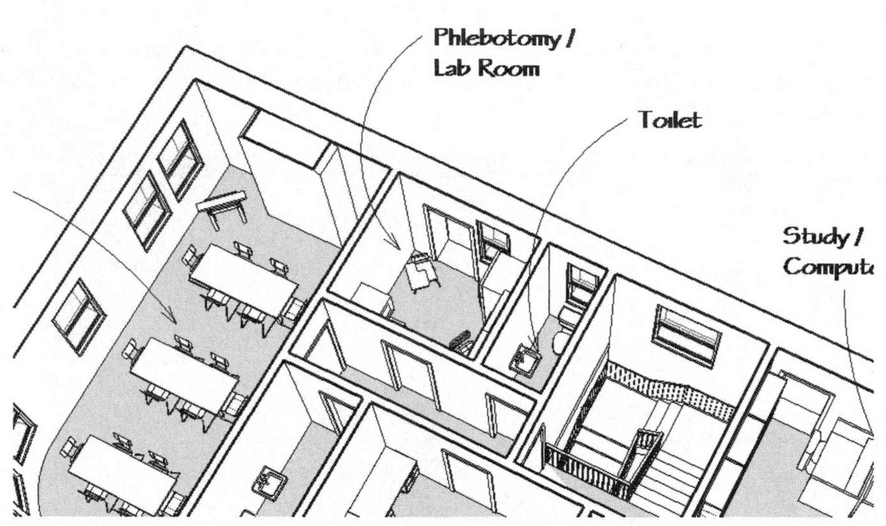

Figure 1–12 *Zoom the Sheet to Sheet Size and pan around to see it as it will print*

The image you see on screen is an actual View in the project that has been added to the current Sheet in a "Viewport."

6. Zoom back out. The easiest way is to choose **Zoom To Fit** from the View menu.

UNDERSTANDING SCREEN TOOL TIPS

Let's return the screen to showing us the full image.

7. Move your mouse pointer into the middle of the screen and pause it there—pause over the drawing, not a text note.

   Do not click the mouse.

Notice how a rectangular border highlights around the image. As you pause the mouse, an onscreen tool tip should appear as well. In this case, this tip will read: Viewports : Viewport : Viewport 1 (see Figure 1–13).

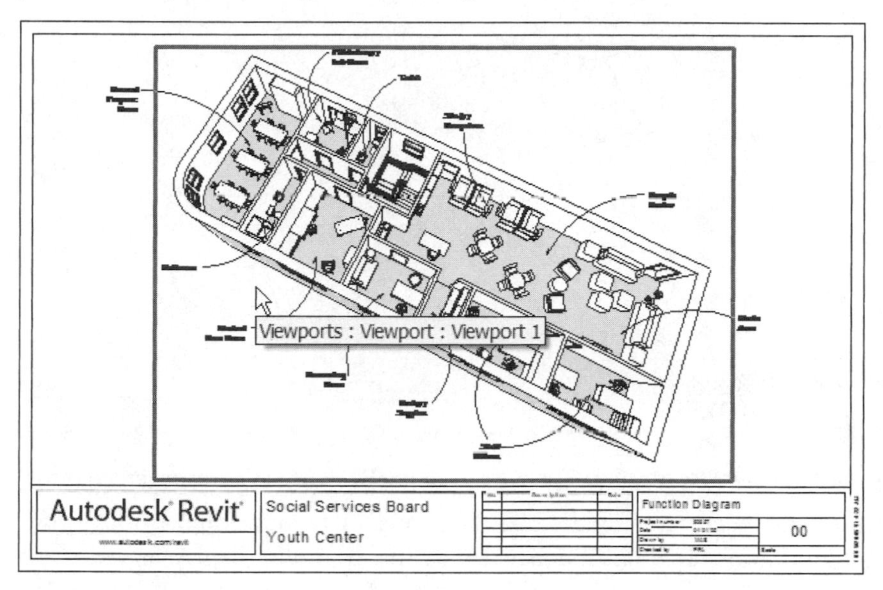

Figure 1–13 *Tool Tips will indicate the Element, Family and Type*

The tool tip conveys three bits of information about the element highlighted—Element : Family : Type. So in this case, the Element is Viewports, the Family is Viewport and the Type is Viewport 1. Now hover the Modify tool over a piece of text. This is called "Pre-highlighting." The tool tip for a piece of text will read—Text Notes : Text : 3D Notes. Here, Text Notes it the Element, Text is the Family and 3D Notes is the Type.

Since Sheets are primarily intended for printing, you do not initially see the elements within the model pre-highlighting. However, you can choose to "Activate" a Viewport which will give you access to the building model elements shown in the View. Editing them from a Viewport is no different than opening the View on Project Browser and editing them there. Let's take a look.

8. Pre-highlight the Viewport, and then click to select it this time.

- Right-click and choose **Activate Viewport**.

Notice that the Sheet title block and the text labels have grayed out. While they are still visible, this graying effect indicates that it is currently inactive.

- Move the mouse around the model.

Notice that the elements within the model now pre-highlight (see Figure 1–14).

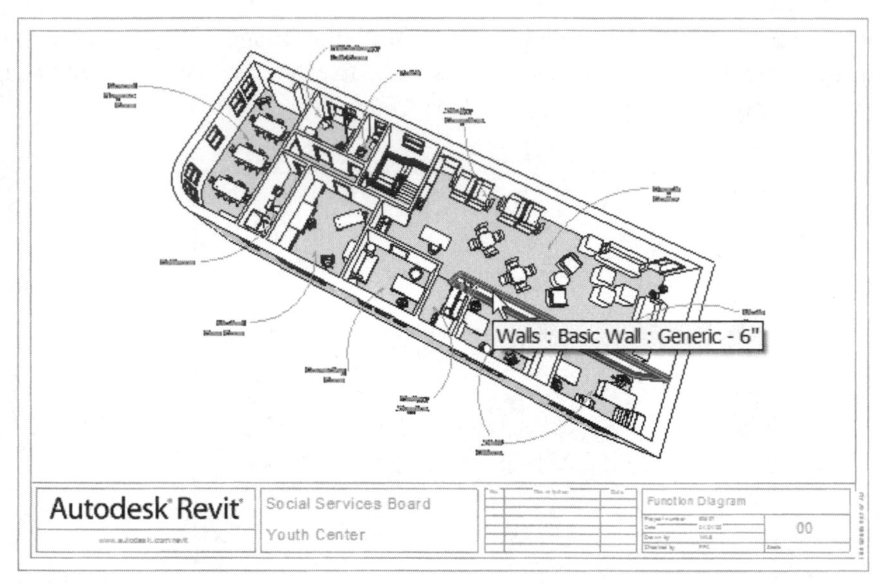

Figure 1–14 *Once the Viewport is activated, you can pre-highlight the elements in the model*

We will not actually edit any model objects in this View, but do take notice of the tool tips at this level. The interior partitions, for example, display Walls : Basic Wall : Generic - 6″.

9. When you are done exploring in the model, right-click in the Viewport again and choose **Deactivate View**.

VIEWS AND DETAILING

We have discussed above several distinctions between model and annotation in Revit Building. Using this dataset, let's explore these concepts a bit further.

10. On the Project Browser, double-click to open the _ *Main Floor* plan View.

This is the basic floor plan View for this project.

11. On the Project Browser, double-click to open the *_Room Callouts* plan View.

This plan is very similar to the _ *Main Floor* View except that it also includes callouts around the General Purpose Room on the left and some elevation and section markers. A Sheet has been provided showing this View.

12. On the Project Browser, double-click to open the *05 - Room Callout Sheet* Sheet View.

Notice how the only visual difference here is that the plan appears on a title block sheet in this View.

13. On the Project Browser, double-click to open the *02 – Floor Plan* Sheet View.

This is the Sheet presentation of the _ *Main Floor* plan View. In other words, this Sheet composes the _ *Main Floor* plan View on a title block for printing. You can easily see which Views appear on a Sheet in the Project Browser.

14. On the Project Browser, beneath the *Sheets* node, expand the tree (click the small plus (+) sign) beneath the *01 – Shaded Plan* Sheet (see Figure 1–15).

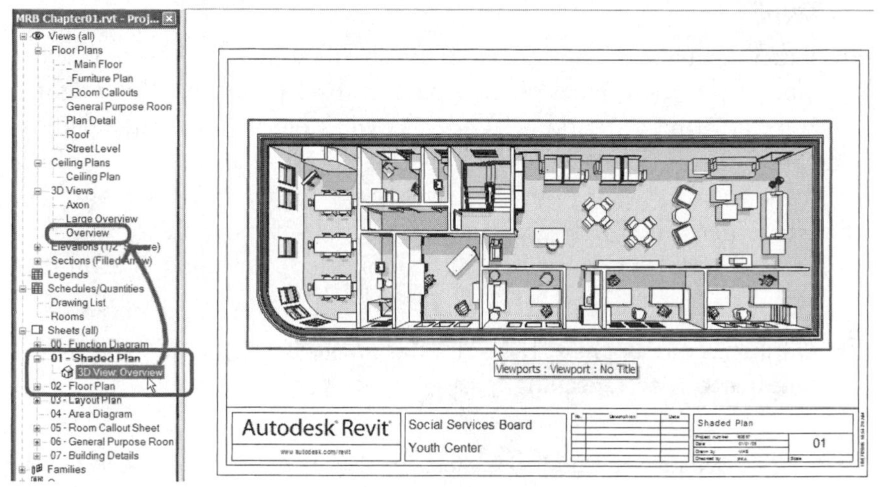

Figure 1–15 *expand the Sheet entries in the Project Browser to see the Views they contain*

This provides an easy way to see which Views are used on particular Sheets. Another useful tool noted above gives us a way to see which Views have not yet been placed on Sheets.

15. On the Project Browser, click on the top node labeled *Views (All).*

- From the Type Selector, choose **Browser – Views : not on Sheets** (see Figure 1–16).

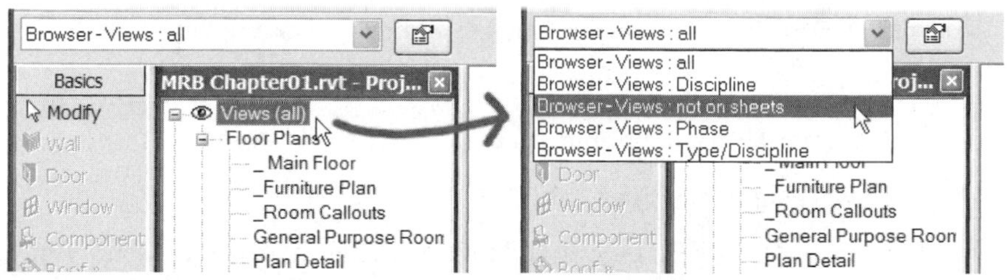

Figure 1–16 *Change the Browser Organization to "not on Sheets"*

Notice that the list of Views on the Project Browser filters to show only those Views (in any category) that are not yet assigned to a Sheet.

- From the Type Selector, return to **Browser – Views : all.**

16. On the Project Browser, double-click to return to the _ *Main Floor* plan View.

Suppose that we needed to create another floor plan that was similar to this one, but that was to convey a different type of information on the printed Sheet or that we were planning to use simply as a convenient place in which to edit the model with no intention of using it to print. To do this, we simply duplicate the View.

- On the Project Browser, right-click the _ *Main Floor* plan View and choose **Duplicate**.

A new floor plan View named *Copy of _ Main Floor* will appear and become active. Notice that none of the room labels were copied in this operation. This might be useful if you were creating a "working" View. A "working" View is intended as a View in which you manipulate the model only and do not plan to add to a Sheet for printing. Bear in mind that nothing prevents the working View from being used on a Sheet; rather it is simply not intended for that purpose. If we wanted to duplicate the View and have the labels also duplicated, we choose a different command.

- On the Project Browser, right-click the _ *Main Floor* plan View and choose **Duplicate with Detailing**.

NOTE "Duplicate with Detailing" is short for "Duplicate with View Specific Detailing elements and Annotation elements." Remember that "Detailing" is being copied, while the model is the same.

Be sure to right-click on _ *Main Floor* and not *Copy of _ Main Floor* in this step.

A new floor plan View named *Copy (2) of _ Main Floor* will appear and become active.

- Right-click *Copy (2) of _ Main Floor* and choose **Rename**.
- In the "rename View" dialog, type **Area Diagram** and then click OK.

17. On the Design Bar, click the Drafting tab and then click the Color Fill tool.

A series of small squares will appear attached to the cursor.

- Click a point above the plan to place the Color Fill Legend.
- In the warning that appears, click OK.
- Click on the Color Fill Legend and then drag the small round Control at the bottom to make the legend two columns (see Figure 1–17).

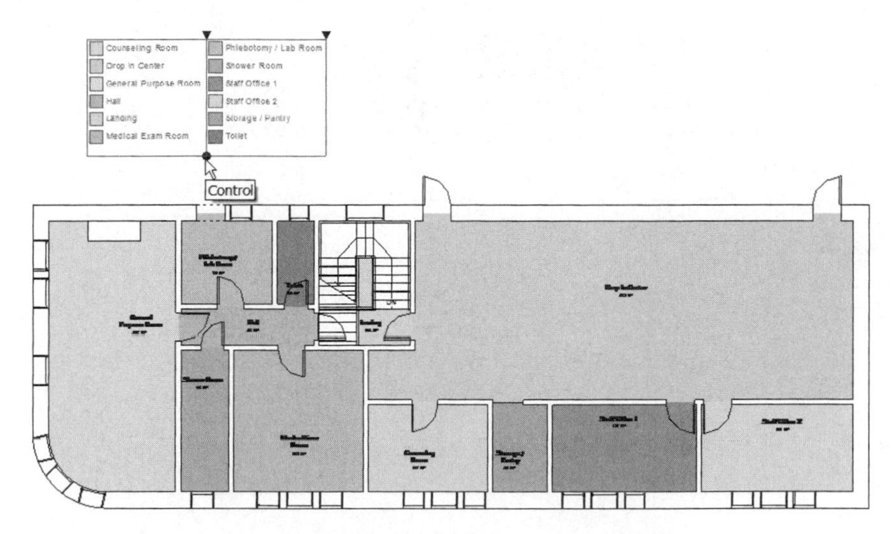

Figure 1–17 *Add a Color Fill Legend and then drag it to two columns*

18. On the Project Browser, double-click to open the *04 – Area Diagram* Sheet View.

A Sheet appears on screen which does not yet have a drawing on it. Let's add our new shaded plan to this Sheet.

- On the Project Browser, right-click the *04 – Area Diagram* Sheet and choose **Add View**.
- From the "Views" dialog, choose *Floor Plan : Area Diagram* View and then click the Add View to Sheet button.
- Click to place the View on the Sheet.

Notice that the View is a little too big for the Sheet. We can adjust the scale of the View and it will update on the Sheet.

19. On the Project Browser, reopen the *Area Diagram* View.

- At the bottom of the screen, choose **1/8″=1′-0″** from the scale pop-up menu (see Figure 1–18).

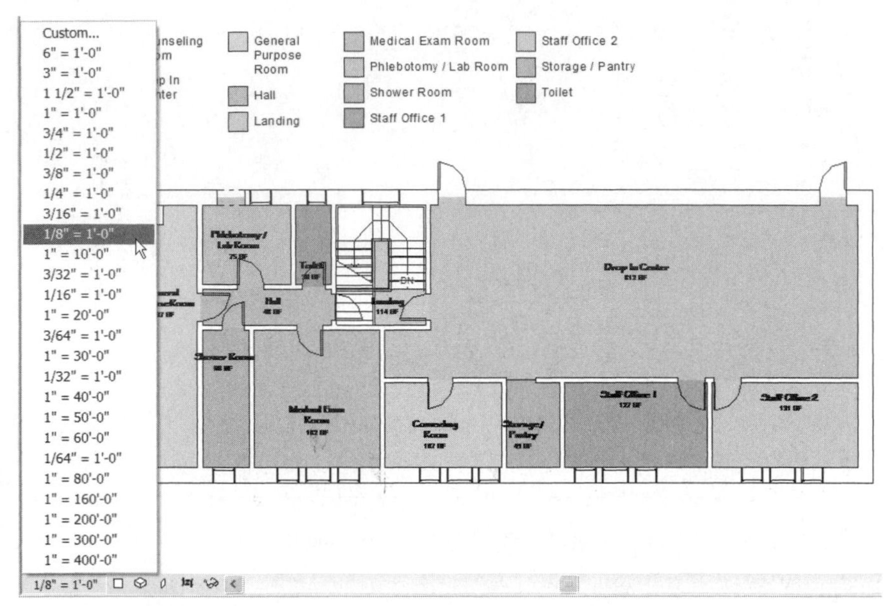

Figure 1–18 *Change the scale of the View*

20. Return to the *04 – Area Diagram* Sheet to see the change.

You should also take a look at the *_Furniture Plan* floor plan View and the *03 – Layout Plan* Sheet next. In this View and Sheet, you will notice that the plan is displayed with furniture. Therefore creating plans with and without detailing (text and other annotation) is *not* the only way to vary the specifics of what we see. We can also control the visibility of each type of element in *any* Revit Building View. The visibility settings are a parameter of the View itself. This is how we can choose to display the furniture in the *_Furniture Plan* floor plan View and not display it in the *_ Main Floor* View. While we will save the specifics of how this is achieved for later chapters, the important point for this exercise is that this sort of control *is* possible and extremely useful.

EDIT IN ANY VIEW

Perhaps the most powerful feature of Autodesk Revit Building is the ability to edit in any View and see the results instantly in all Views.

21. On the Project Browser, double-click to open the *04 – General Purpose Room* Sheet View.

This Sheet shows the Views that are associated to the callouts we saw on the *05 - Room Callout Sheet* Sheet above.

- Select the plan View on the left, right-click and choose **Activate View**.

- On the Design Bar, click the Basics tab and then click the Window tool.
- Click a point on the exterior Wall at the left to add a new Window (see Figure 1–19).

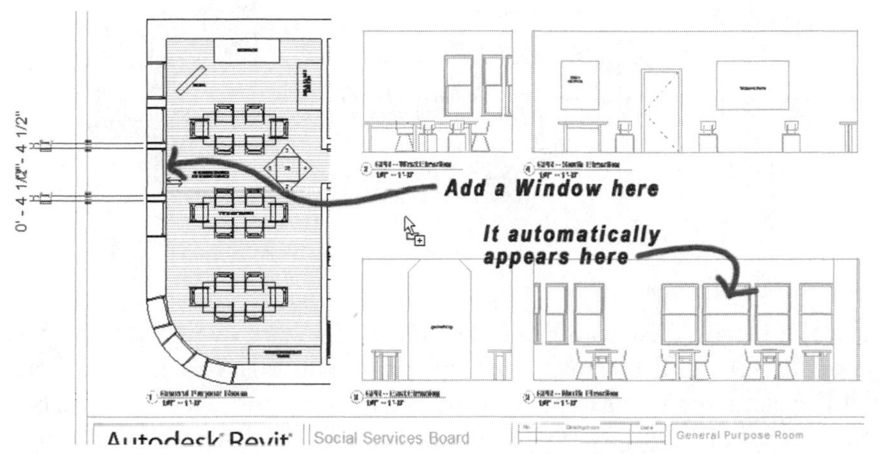

Figure 1–19 *Add a Window and it appears in all appropriate Views automatically*

- Right-click in the plan View again and choose **Deactivate View**.

Continue to explore the dataset further and get comfortable with the key Revit Building concepts that it helps to showcase. When you are finished exploring, close the project file. It is not necessary to save.

EXPLORE A DETAIL VIEW

As we have noted above, a Detail View is a little different than the other Views. Typically it will include a live View of the model—usually a callout of some part of a section or plan—and various types of annotation and other embellishments on top. One such detail View has been included in this sample dataset.

22. On the Project Browser, expand (click the plus [+] sign) the *07 – Building Details* Sheet View.

    Beneath this Sheet entry in the Project Browser will appear a listing of three Views that are already placed on the Sheet.

- Beneath the *07 – Building Details* Sheet View entry, double-click to open the *Section : Section Detail* View.
- Pre-highlight some of the elements in this View (see Figure 1–20).

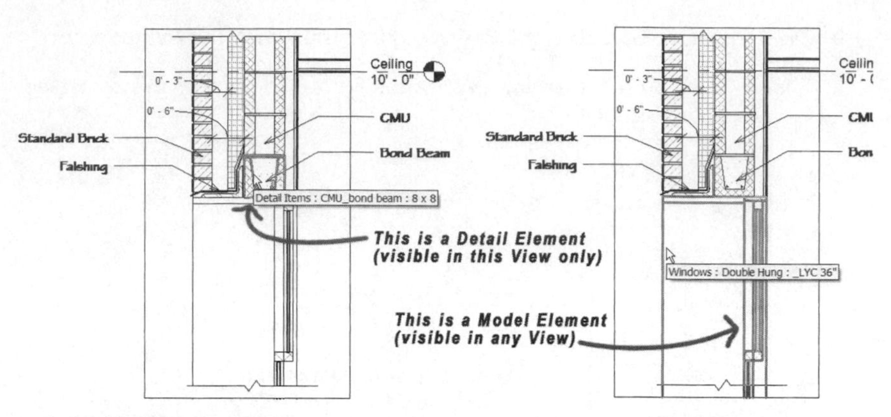

Figure 1–20 *Explore a Detail View*

Notice that the Detail View contains both Model Elements (which would appear in all Views) and Detail Elements (which appear only in this View). Even though the Detail Elements represent items like concrete blocks, brick, flashing and bond beams, these types of items are typically drawn as Detail Elements on top of the Model View geometry in this way. This saves on computer resources by not requiring us to construct the elements in our model to this level of detail. Complete details on this technique can be found in Chapter 10.

Continue to explore in this dataset as much as you wish to get a better feel of how the various elements and Views in a Revit Building project interact. Close Revit Building when you are finished exploring. You do not need to save the file.

SUMMARY

Building Information Modeling is the process of creating an accurate representation of the building from which detailed and accurate information can be extracted.

Using BIM successfully requires a firm understanding of the concepts and techniques enabling its use.

In Autodesk Revit Building, all Views relate back to a single source model. If changes are made in one View, the change is immediately available in all Views.

Model Elements are those items that represent real building components and are visible in all Views.

Annotation Elements appear only in the View to which they are added and include notes, dimensions and Detail embellishments.

All items in Revit Building belong to Families.

There are System Families, Model Families and Annotation Families.

Revit Building User Interface

INTRODUCTION

This chapter is designed to acquaint you with the user interface and work environment of Autodesk Revit Building. The Revit Building user interface is logically organized and easy to learn. In this chapter we will explore its major features. While many features follow standard Microsoft Windows™ conventions, other aspects of the user Interface are unique to Revit Building. In this overview, our goal is to make you comfortable with all aspects of interacting with and receiving feedback from Autodesk Revit Building. While some of the lessons that follow are tutorial-based, many are more descriptive in nature. Feel free to follow along in Revit Building as you read the descriptions.

OBJECTIVES

To get you quickly acquainted with the Autodesk Revit Building user interface, topics we will explore include:

- An overview of the Revit Building User Interface
- Interface terminology
- Design and Options bars
- Moving around a Revit Building model

UNIT CONVENTIONS

Throughout this book, Imperial units and files will be listed first, followed by metric in brackets, for example, **Imperial** [**Metric**]. See the Preface for complete details on style conventions used throughout this book.

Imperial dimensions throughout this text appear in the "Feet and Inch" format for clarity. However, when typing these values into Revit Building, neither the foot symbol (′), when typing whole feet, nor the hyphen separating the feet from inches, when typing both, is required. Therefore, to type values of whole feet, simply type the number. To type values of both feet and inches, type the number of feet with the foot symbol (′) followed immediately by the inches without the inch symbol. When typing just inches, the inch (″) symbol *is* required unless you preface the value with a leading 0′. For example, 4′-0″ can be typed in Revit Building as simply: **4** (or **48″**). To type four feet six inches, type: **4.5**, **4′6** or **54″**. You can also type **4 6** (that is **4** SPACE **6**). To type ten inches, type: **10″** or **0′10** or **0 10** (that is **0** SPACE **10**). Hyphens are not required. When separating inches from fractions, use a SPACE. If using metric units, all values in this text are in millimeters and can be typed in directly with no unit designation required. More information on style conventions used in this book can be found in the Preface.

Table 2–1 *Acceptable Imperial Unit Input Formats*

Value Required	Type This:	Or This
Four feet	4	48″
Six inches	.5 or 0′6 or 0 6	6″
Four feet six inches	4′6 or 4.5 or 4 6	54″
Four feet six and one half inches	4′6 1/2 or 4 6.5 or 4 6 1/2	54.5″

NOTE Typing the inch (′) mark is acceptable when typing whole feet as well; however, it is not required. Dimensions throughout this text are shown in the Feet and Inch format for clarity. However, feel free to enter dimension values in whatever acceptable formats you prefer. Eliminating the inch or foot marks where possible reduces keystrokes and is recommended despite their inclusion in this text.

UNDERSTANDING THE REVIT BUILDING USER INTERFACE

Autodesk Revit Building offers a clean and streamlined work environment designed to put the tools and features that you need to use most often within easy reach. In addition to the many onscreen tools and controls, many of the most common Revit Building tools also have keyboard shortcuts. These will be explored in a topic below. Figure 2–1 shows the Revit Building screen with each of the major interface elements labeled for your convenience.

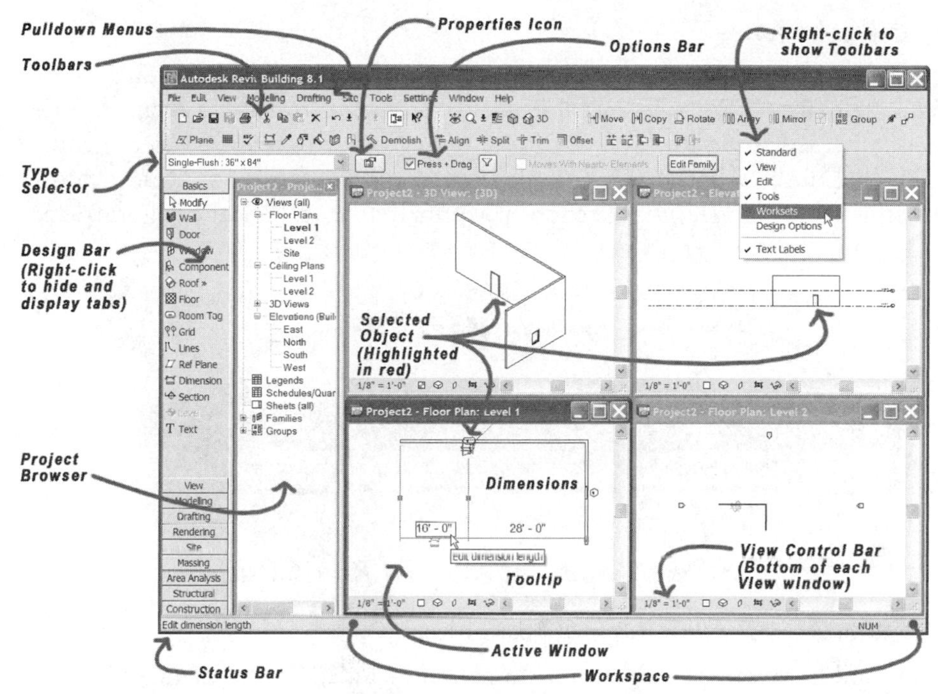

Figure 2–1 *The Autodesk Revit Building User Interface*

Consistent with most Windows software, the Revit Building screen is framed with pull-down menus along the top edge, the Windows minimize, maximize and close icons in the top right corner, a status bar along the bottom edge, and a variety of toolbars at the top edge of the screen. In addition to these Windows standards, the center of the Revit Building screen is divided into three major sections: the Design Bar, the Project Browser and the Workspace. The Design Bar is organized into several tabs. Clicking on a tab reveals a collection of tools contained within. You can right-click the Design Bar to display hidden tabs. The Project Browser can be thought of as the "table of contents" for your Revit Building project. It reveals all of the various representations of your project data—referred to in Revit Building as "Views." Views in Autodesk Revit Building can be graphical (drawings, sketches, diagrams) or non-graphical (schedules, legends, takeoffs) and offer the means to both query your project (output) and to manipulate and edit it (input). The Project Browser also lists the Families and Groups that are within the project. The Workspace is where Views are displayed and manipulated. The Workspace can present one or more Views of the project at the same time using the standard Windows minimize, maximize and tile functionality on the Window menu. Finally, stretched across the top of the Design Bar, the Project Browser and the Workspace are the Type Selector menu, Properties icon and the Options Bar. These controls are used to both create new objects and manipulate selected objects in the project (all of these items are labeled in Figure 2–1).

PULLDOWN MENUS

The most common way to issue commands in Autodesk Revit Building is by choosing them from the pull-down menus across the top of the screen. This is consistent with the interface in most of today's software. In addition to the common "File," "Edit," "Window" and "Help" menus, Revit Building includes additional menus to categorize its collection of commands into logical categories (see Figure 2–2).

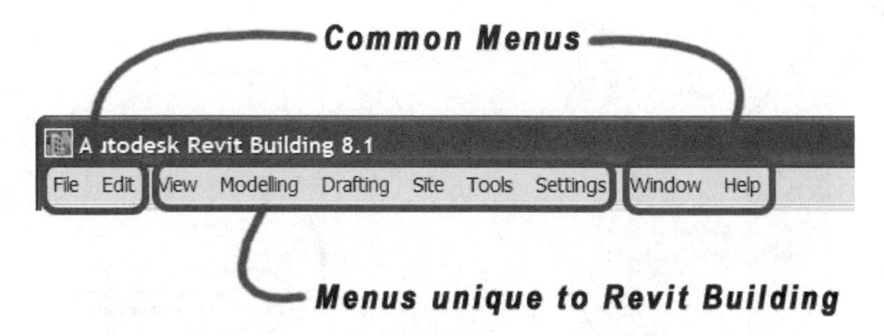

Figure 2–2 *Understanding Revit Building Pull-down Menus*

All Revit Building commands and functions are included in the menu structure. Most commands are accessible via other means as well, but the pull-down menus are the only place you will find *all* Autodesk Revit Building commands. This makes the menus a good place to look as you are learning Revit Building. In addition to the commands themselves, you can also use the pull-down menus as a way to discover if the particular command has a keyboard shortcut. Keyboard shortcuts are simple keystroke combinations that can be typed as an alternative way to issue a command (see Figure 2–3).

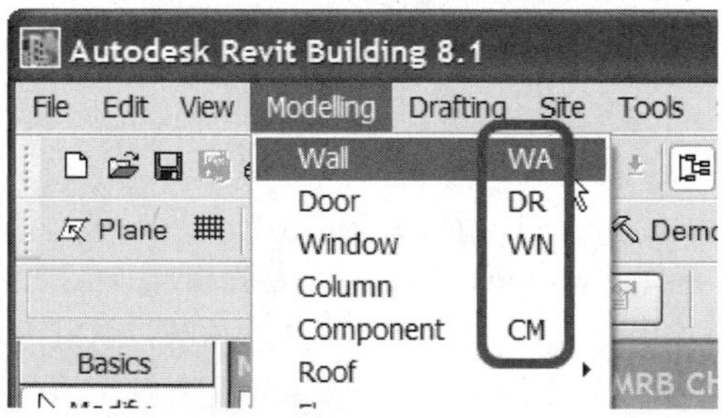

Figure 2–3 *Keyboard shortcuts shown next to commands on the pull-down menus*

Keyboard shortcuts are discussed in more detail in Appendix D. Next to some of the menu commands is a small arrow like the one next to the "Roof" command in Figure 2–3. This indicates that a cascading menu appears for this item. This is another standard Windows convention. Moving your cursor over this small arrow will cause the menu to "cascade" revealing a sub-menu (see Figure 2–4).

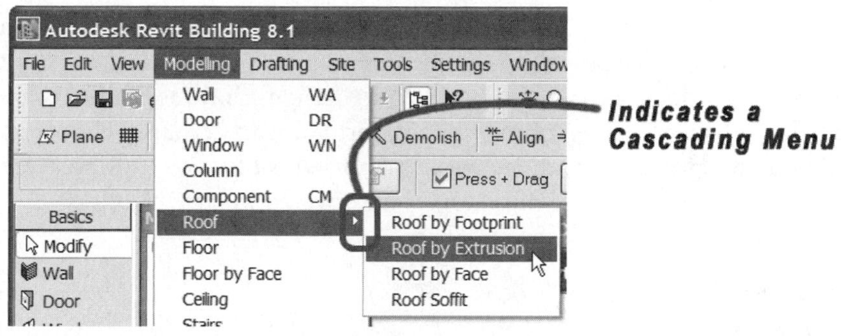

Figure 2–4 *Example of a* Cascading menu

Another Windows convention supported by Revit Building is the ability to issue menu commands with the keyboard using the ALT key and a key letter from the desired command. To try this, press the ALT key. Doing so will place a small underline beneath one letter of every menu. Next, still holding the alt key, press the underlined key for the desired menu and then finally press the underlined key for the desired command. For example, to issue the Door command from the Modeling menu, you could press ALT + M THEN D (see Figure 2–5). If a cascading menu is involved, continue pressing the appropriate keystrokes to arrive at the desired command.

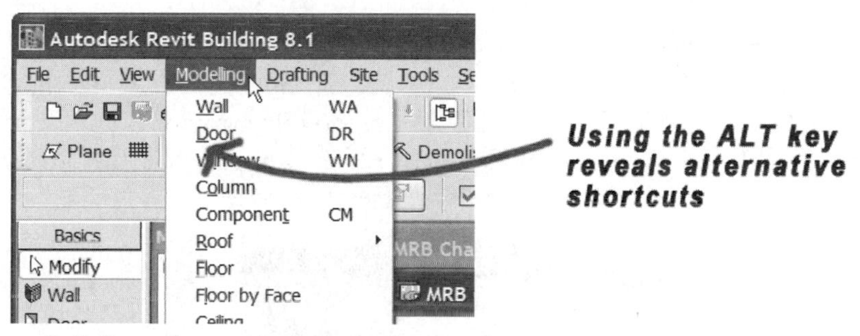

Figure 2–5 *Press the* **ALT** *key to reveal menu shortcuts*

Menu items will appear grayed out if the particular command is not available in the current context. For instance, if you have a Sheet View active on screen, most commands such as Wall, Door or Roof on the Modeling menu will not be available. Similarly, items on the Edit menu will typically not be available unless an element in the model is selected (see below for more on selection).

TOOLBARS

The toolbars in Revit Building are located just below the pull-down menus across the top of the screen. There are six toolbars available. You can choose which ones you wish to have displayed at any given time. Revit Building will remember your choices the next time you launch the program. Toolbars include commands that are typically used for editing elements in your models and modifying your work environment. To display or hide toolbars, right-click on any toolbar item for a context menu. Displayed toolbars show in the context menu with a checkmark next to their name. By default the "Worksets" and "Design Options" toolbars are not displayed. If you have a large computer monitor, you can certainly choose to display these toolbars. However, you may wish to wait to enable them until you are ready to use their functions. Both of these topics are covered in later chapters. Figure 2–6 shows all of the Revit Building toolbars.

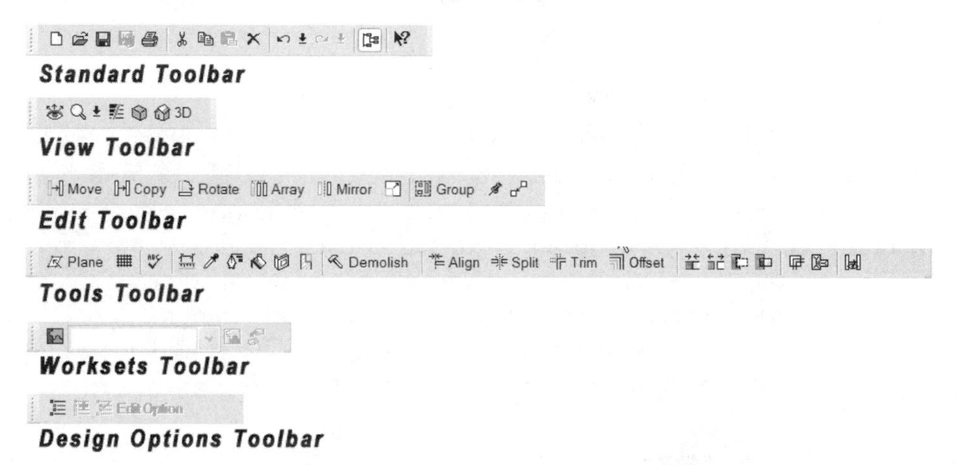

Figure 2–6 *Revit Building toolbars*

Toolbars can be "collapsed" if screen space is limited. To do this, click the small bar on the left side of the toolbar. In this state, a small double arrow will appear to indicate that some of the icons on the toolbar are hidden. Simply click this arrow to reveal a pop-up menu of the hidden commands (see Figure 2–7).

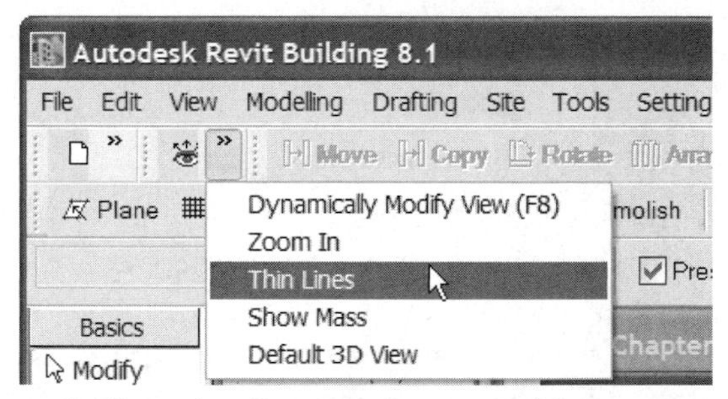

Figure 2–7 *The icons of collapsed toolbars can be accessed from a pop-up menu*

NOTE Even though you can move the toolbars into different locations, the physical arrangement of toolbars reverts to the default setting every time you restart Revit Building.

Like menu items above, toolbar icons will appear grayed out if the particular command is not available in the current context. For instance, like the Edit menu, icons on the Edit toolbar will not be available unless an element in the model is selected (see below for more on selection).

DESIGN BAR

The Design Bar is located vertically along the left side of the screen. Think of the Design Bar as an organized toolbox for the most common Revit Building functions, particularly those involving adding or creating elements and Views in your Revit Building sessions. Look to the toolbars for tools to edit *existing* elements in your workspace; look to the Design Bar for commands to create geometry and Views. As mentioned above, all functions on the Design Bar can also be accessed through the Pull-down menus, and often by keyboard shortcuts.

The Design Bar is organized into groups called "tabs." Each tab has a name that generally categorizes all commands contained on that tab (see Figure 2–8).

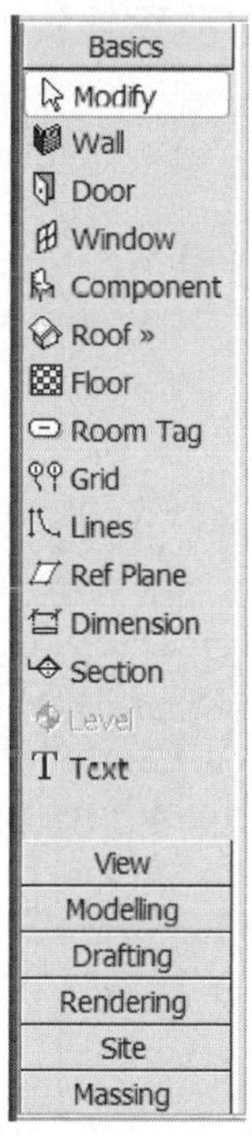

Figure 2–8 *The Revit Building Design Bar*

As with toolbars, there is a fixed quantity of Design Bar tabs included with the software and you can choose to hide or display each tab. By default, most tabs are displayed. You can display the hidden ones with a right-click. To access a command from the Design Bar, first click on the tab to reveal its tools (only one tab can be displayed at a time; the other tabs will collapse automatically). Next click on the tool you wish to execute.

NOTE At low screen resolution (e.g. 1024x768 or lower) some of the Design Bar tools may be cropped at the bottom of the screen. Unfortunately, there is no way of scrolling the Design Bar. The only workaround is to close other tabs until the tools come into view.

If you have difficulty reading the full names of some of the tools, you can widen the Design Bar. To do this, place your cursor over the border line on the right edge of the Design Bar and drag it to the right.

Also, the design bar will help guide us as to which functions are available to use based on the particular view or editing mode we are in. As you switch from general input mode to a specialized working mode (such as Family or Sketch mode to be covered later), the Design bar will change to show you what mode you are in as well as what functions are available to you in that mode.

Like menu and toolbar items above, Design Bar tools will appear grayed out if the particular command is not available in the current context. For instance, if you have a Sheet open and active on screen, most commands such as Wall, Door or Roof on the Modeling Design Bar tab will not be available. Similarly, many of the items on the View tab will typically not be available unless a viewport is activated on the Sheet.

PROJECT BROWSER

Menus, Toolbars and the Design Bar are a fixed part of the Revit Building user interface. While we can manipulate which of these items displays and how, they are defined by the software and remain consistent regardless of the project that we happen to be editing. The Project Browser on the other hand is very specific to the currently open project. When you open a Revit Building document (a file with an RVT extension and referred to as a "Project") its contents will be displayed in the Project Browser. Think of the Project Browser as the table of contents for your project. It is the primary organizational tool for a Revit Building project. It is typically docked between the Design Bar and the Workspace windows near the left side of the screen (see Figure 2–9).

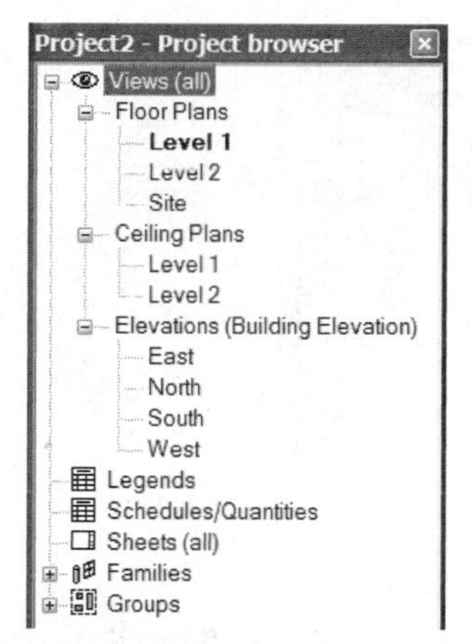

Figure 2–9 *The Revit Building Project Browser*

You can re-size and move the Project Browser depending on your screen resolution and space needs. However, please be aware that when you launch a Revit Building session, the Project Browser will reset to the default location and width. While it is possible to close the Project Browser, it is *not* recommended. Since the Project Browser is the primary means of interacting with and navigating between your project's Views, closing it makes these functions difficult and inefficient. Should you inadvertently close the Project Browser you can restore it by clicking on the Project Browser icon on the toolbar (see Figure 2–10). You can restore it with the command on the Window menu as well.

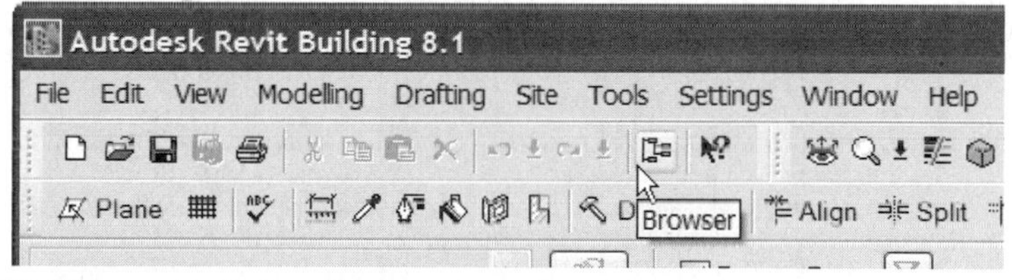

Figure 2–10 *Restore the Project Browser using the icon*

Like the Design Bar, longer names in the Project Browser can get truncated. It is therefore recommend that you widen the Project Browser window by stretching the right edge as much as your screen size will permit and to the extent of you own personal preferences. This will make it easier to read the full names of your Views. However, if you are not able to widen the Project Browser or you prefer not to, you can simply hover your mouse over a View to see a tooltip of its full name (see Figure 2–11).

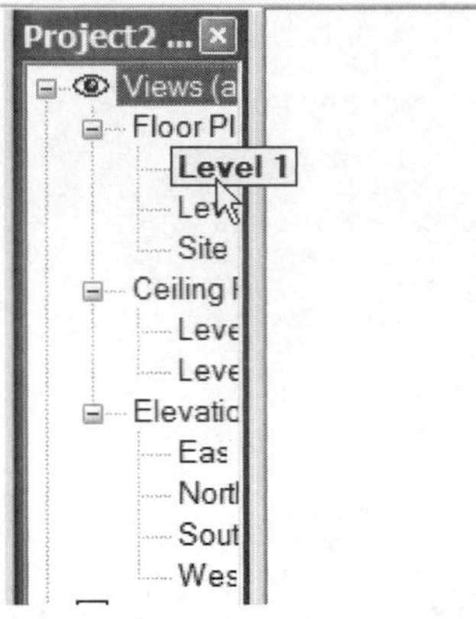

Figure 2–11 *Partially hidden names can be seen with tooltips*

The Project Browser enables you to view and work with two distinct parts of a Revit Building project: Views of your model and Elements within your model. Views are listed at the top. In the previous chapter, we explored this node of the Project Browser and its organization options. Please refer to that discussion for more details. Beneath Views are items like *Schedules/Quantities* and *Sheets*. These are specialized Views of the project. To interact directly with Elements within your model from the Project Browser, you will work within the *Families* and *Groups* nodes of the Project Browser tree. Every Family and Group in a project is listed among these items. From these nodes, you can edit existing items, or even create new ones (see Figure 2–12).

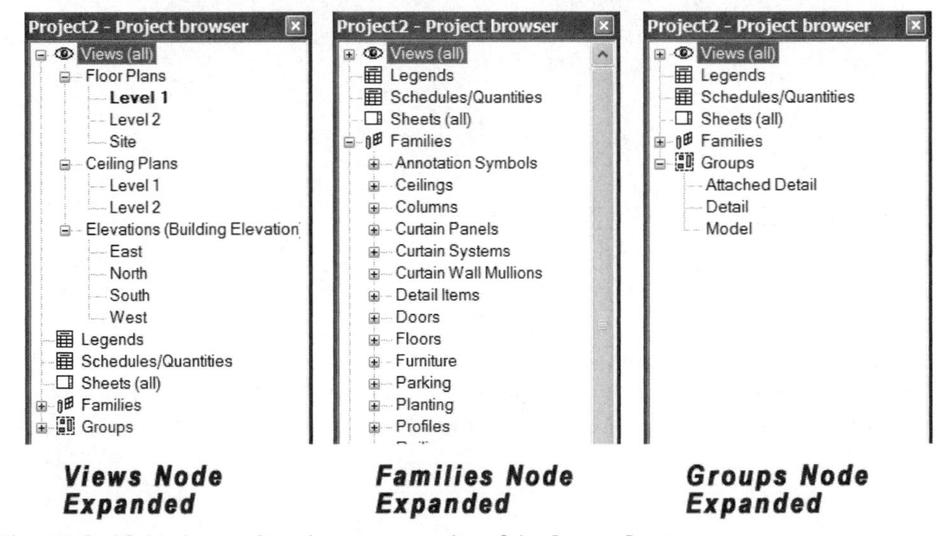

Figure 2–12 *Understanding the various nodes of the Project Browser*

Beneath the Views node of the tree, certain categories will appear automatically as various Views are added to the project. For instance, a *Floor Plans* node is automatically created to house the various floor plan Views of your project. Likewise, "Elevation" and "Section" nodes will appear as the View types are added. If you delete all Views in a particular category, the category itself will also disappear. You cannot make your own categories. These are created and maintained automatically by Revit Building. Three view categories have their own nodes on the tree: *Legends*, *Schedules/Quantities*, & *Sheets*.

The name of the currently active View will appear bold in the Project Browser. You can open any View by simply double-clicking on its name in the Project Browser. This is the most common function of the Project Browser. In addition you can right-click Views in the Browser to access commands specific to that item (see below for more on right-clicking). It is also important to note that more than one item may be selected on the Project Browser at the same time. To select multiple Views, select the first View, then hold down the CTRL key and select additional Views. Once you have more than one View selected, you may right-click on any of the highlighted View names to access a menu that will apply to the entire selection of Views. This might be useful if you wish to make a global change such as editing the scale of several Views at once (see Figure 2–13). (Scale would be accessed from the Properties item on the right-click).

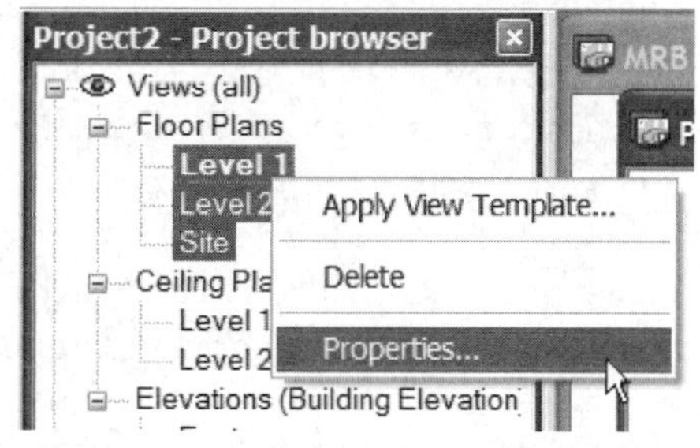

Figure 2–13 *Select multiple items in the Project Browser with the* **CTRL key**

Another important aspect of the Project Browser is its presentation of the project's Sheets. Although Sheets are technically classified as Views in Revit Building they have some unique properties not shared by other Views. They are also presented in a special node in the Project Browser. Expand the Sheets node to see the Sheet Views in your project. Since Sheets can actually have other Views placed on them, a plus (+) will appear next to the sheet name. Expanding this will reveal the names of the Views that have been placed upon the Sheet. A Sheet appearing in the Project Browser without a plus sign indicates that the Sheet does not yet contain references to any other Views. Like other Views, you can open a Sheet view by double-clicking on its name in the Project Browser. In addition, you can open referenced Views placed on Sheets by double-clicking their names beneath the expanded Sheet name (see Figure 2–14).

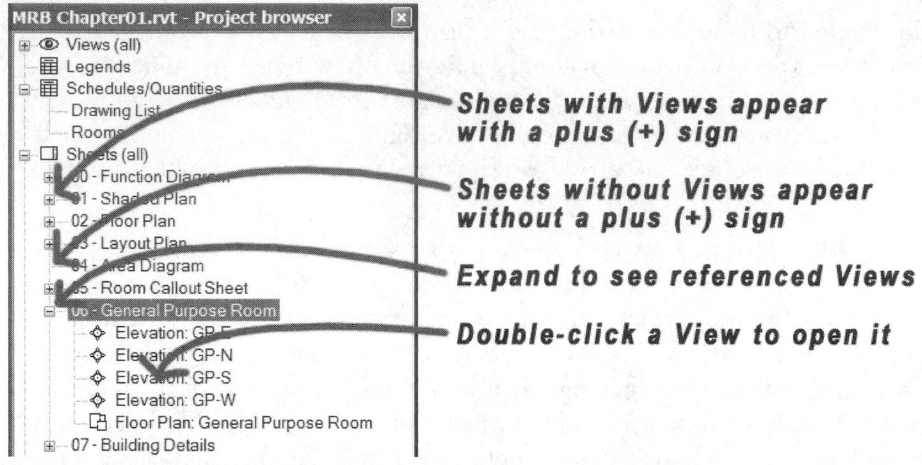

Figure 2–14 *Understanding the Sheets node on the Project Browser*

When using these techniques to navigate your project, recall from the previous chapter our discussion of changing the way that the Project Browser is organized. In that exercise, we noted how we could organize our Views by discipline, phase or even by those not yet placed on Sheets. Look back to Figure 1–10 and the accompanying discussion for more details.

When you choose to organize the Browser this way, you do not loose anything. You simple change the way things are displayed. If you wish to work on a View that is already on a Sheet, you first expand the Sheet on which it is placed, and then double-click the View name beneath it. If the View is not yet on a Sheet, it will be listed beneath the Views node on the Browser instead. Once you add a View to a Sheet, it will disappear from the Views node and appear instead beneath its host Sheet.

Views are specific to the project's geometry and configuration. They show the building design in one of many industry standard drawing types like plan and section. However your project file contains many additional items that are not necessarily specific to the project. These are like resources that are readily available to the project. These resources come in the form of Families and Groups. The Families and Groups nodes of the Project Browser show all Families and Groups, both the ones already used in the project's geometry and those that have not been used yet. Families and Groups are defined and discussed in more detail in other chapters. This section will focus only on their placement and access via the Project Browser.

Like the categories used to organize the Views in the Project Browser, the categories to which Families and Groups are assigned are created and maintained automatically by Revit Building (see Figure 2–12 above for examples). If you expand a category, you will see Families or Groups within that category. The items shown are those that are already loaded and therefore part of the current project. You can further expand the Family itself to reveal the Types that it contains (see Figure 2–15).

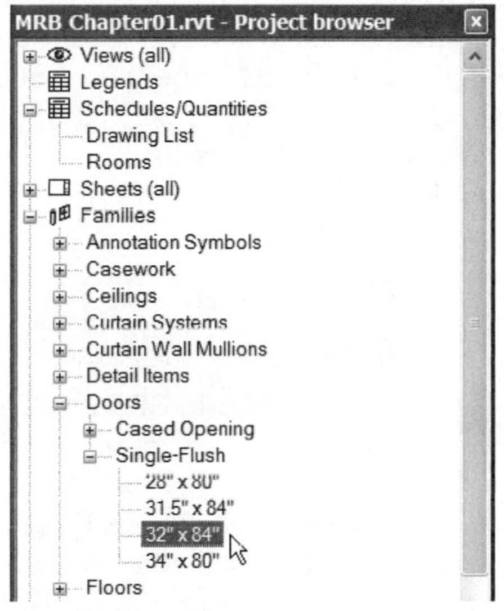

Figure 2–15 *Expanding an item in the Families node of the Project Browser*

You will get different options depending on whether you right-click on the Family itself, or its Types listed beneath it (see Figure 2–16).

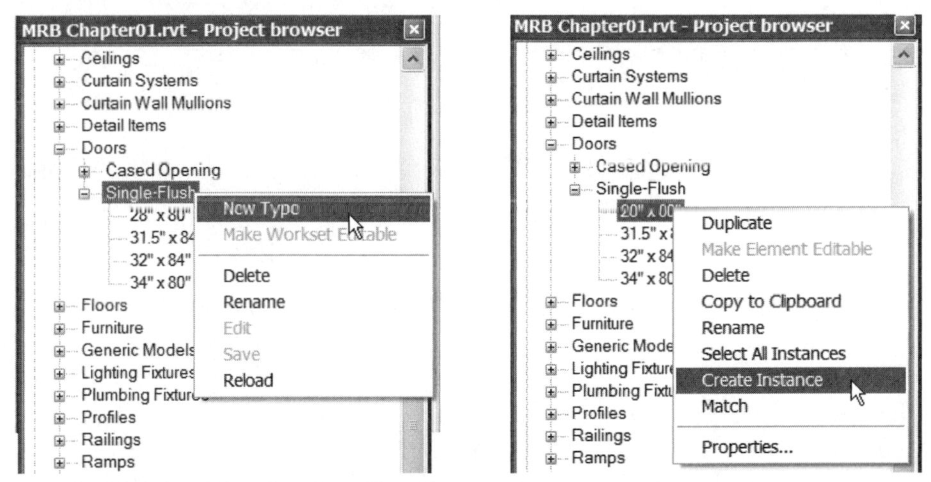

Figure 2–16 *Right-click a Family or Type for options*

To place an instance of a Family or Group in the model, you can right-click and choose Create Instance or you can simply drag it from the Project Browser to the View window. Be sure you perform this action on a Type, not the Family itself.

TYPE SELECTOR/ PROPERTIES ICON

Just above the Design Bar and beneath the toolbars on the left side of the Options Bar is the Type Selector. The Type Selector is a drop-down list of Types available for whatever item is currently selected in your project. If you have nothing selected the Type Selector will be grayed out and unavailable. As its name implies, the Type Selector is used to select a Type for an element in your project. Whenever you create elements, you will choose an appropriate Type from this list. You can also use the Type Selector to change the Type of elements already in the model. Simply select an element and then choose a different Type from the list (see Figure 2–17).

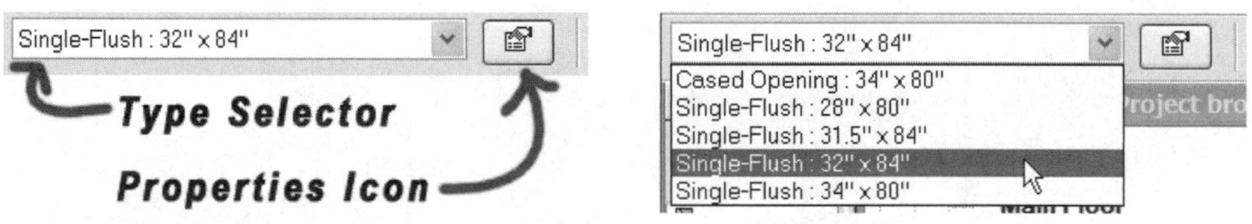

Figure 2–17 *The Type Selector and Properties Icon*

Next to the Type Selector is the Properties Icon. You can use the Properties icon to call the "Element Properties" dialog. This will allow you to explore or edit the parameters of the selected element. Every Element in Revit Building, from Model Elements like Walls and Floors, to annotation Elements like text, tags, and Levels, and even Views themselves have instance and or type properties. You will access these properties frequently from this dialog. From the "Element Properties" dialog, you can use the Edit/New button to access the "Type Properties" dialog to explore or edit the parameters of the element's Type (see Figure 2–18).

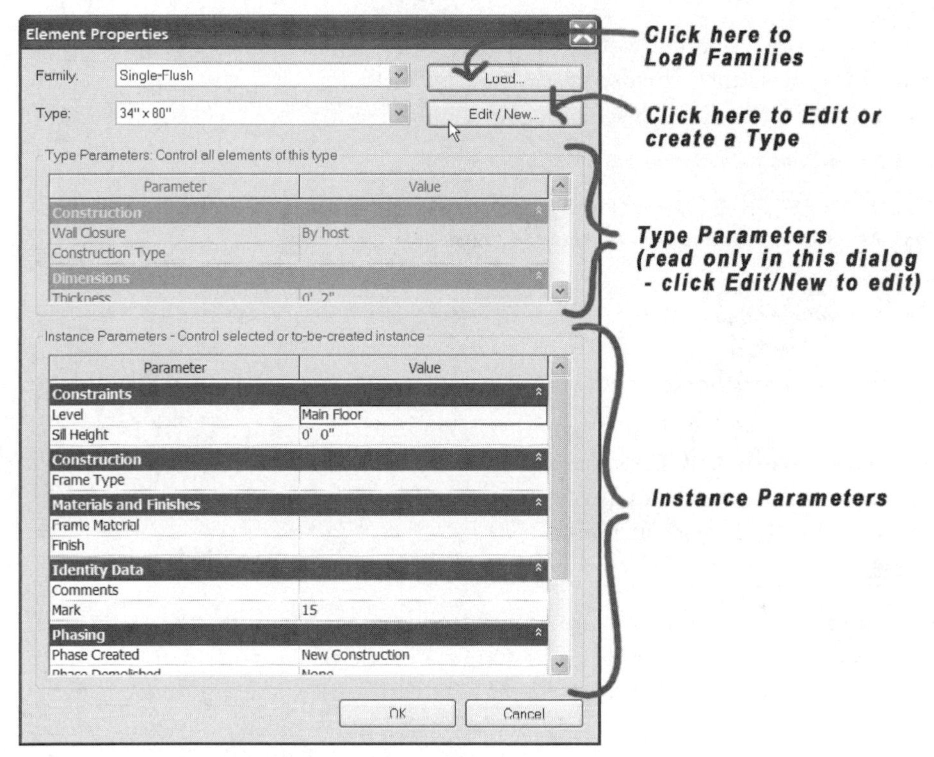

Figure 2–18 *Element Properties dialog*

You will interact with both the Type Selector and the Properties icon when creating or modifying elements. When you are creating an element, you first choose the tool for the item you wish to create on the Design Bar. Next choose a Type from the Type Selector and if necessary click the Properties icon to view or edit additional parameters. If you are modifying an existing element, simply select the element (or elements) and then choose the desired Type from the Type Selector or click the Properties icon to edit parameters in the "Element Properties" dialog.

When you select multiple elements, the Type Selector displays the common Type of the group of items you have selected. If the selection of items is the same class of element (all Doors for example) but are not currently the same Type, then the Type Selector will remain active, but will display a blank entry (all white) rather than show a Type. If the Type Selector box appears grayed out, this indicates that you have made a selection of dissimilar elements (like a Wall and a Door), and you cannot make changes to them using the Type Selector or Properties icon.

OPTIONS BAR

The Options Bar is located directly to the right of the Properties Button, above the workspace area. The Options Bar is constantly changing, depending on what element you have selected or what operation is being performed. In this way, you will only see those options relating to the current situation. This helps simplify the interface and keep tasks more focused (see Figure 2–19).

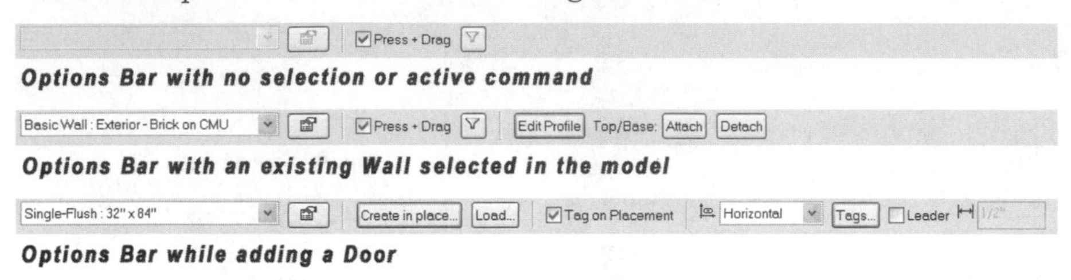

Figure 2–19 *The Options Bar constantly adjusts to show options appropriate to the current task*

You will interact with the Options Bar constantly as you work in Autodesk Revit Building. Keep this in mind and always remember to look there for context-specific options relating to the task at hand.

THE WORKSPACE

The most interactive portion of the Revit Building interface is the workspace where you work with the Views of your building model. When you open a View, it appears in a window in the workspace area. At least one View window must be open to work in a project. You can however open several Views for a project at one time. If the View windows are maximized, you can see which are opened on the Window menu. You can also choose to tile the Windows so that they appear side by side in the Workspace. This command is also on the Window menu. If you have many Views open at once, tiling can make the actual windows very small and hard to work with. To prevent this, be sure to close windows when you no longer need them. An easy way to do this is to maximize the current View window and then choose **Close Hidden Windows** from the Window menu.

STATUS BAR

The Status Bar is the gray bar along the bottom edge of the Revit Building screen. If you glance down at the Status Bar, you will notice a constant readout of feedback appears there. In some cases, the information provided prompts and clues as to what actions are required within a particular command, in other cases; the feedback may simply describe an action taking place or an element beneath the cursor (see Figure 2–20 for examples).

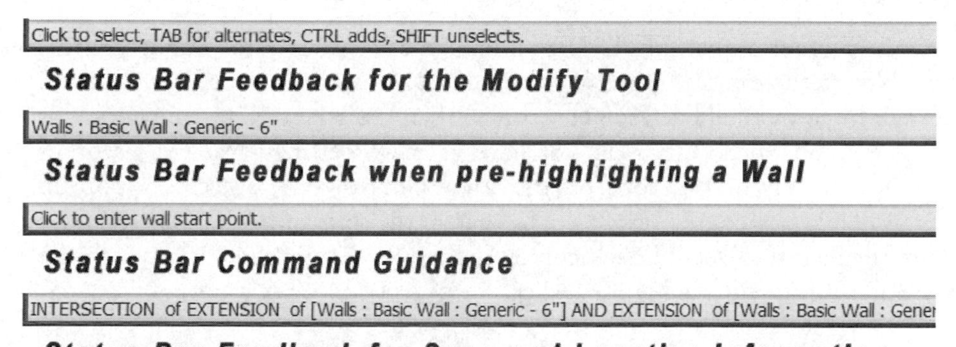

Figure 2–20 *The Status Bar provides ongoing feedback and guidance as you work*

In many cases, the same or similar feedback is available on screen in the form of tool tips. The extent to which tooltips appear is a setting that you can modify. You can opt for a high level of tooltip prompting, a moderate level or none at all. To edit the degree of Tooltip Assistance, choose **Options** from the Settings menu and edit the "Tooltip Assistance" item on the General tab. You cannot edit the Status Bar messages in any way.

VIEW CONTROL BAR

Every graphical View window has a View Control Bar located at the bottom edge of the window. When windows are maximized, this will appear directly above the Status Bar adjacent to the horizontal scroll bar. If the View windows are tiled, it will appear in the lower left corner of the window (see Figure 2–21). The View Control bar serves two purposes—it displays at a glance the most common view settings of the window and provides a simple and convenient way to change them if required.

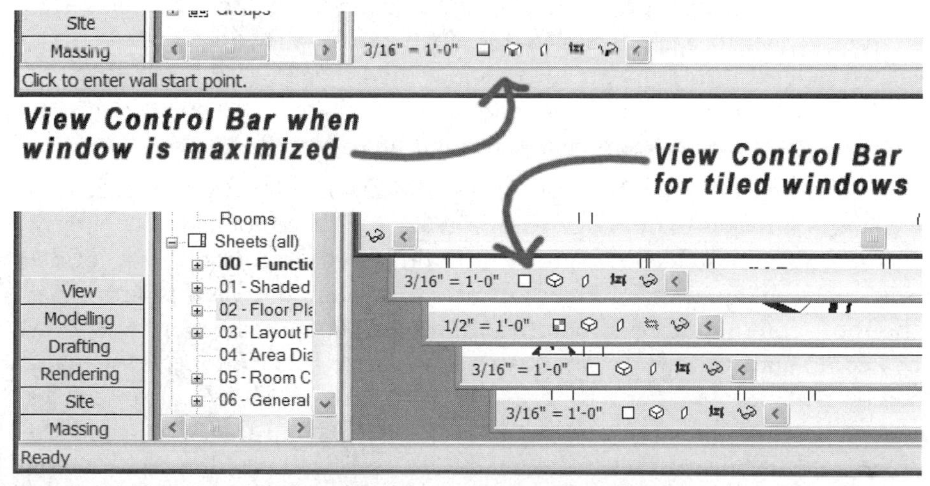

Figure 2–21 *The View Control Bar maximized (top) and tiled (bottom)*

Like so many other aspects of Revit Building, the View Control Bar is context sensitive. Notice the top tiled window in Figure 2–21 has a different View Control Bar (containing only the Hide/Isolate settings) than the others. The top window is a Sheet View and Sheet Views do not have as many settings available as other Views. Click on any of the icons on the View Control Bar to access a pop-up menu of available choices. For example, to change the scale of a View, click on the scale item, or pick the Detail Level icon to change the level of detail in which the View is being displayed (see Figure 2–22).

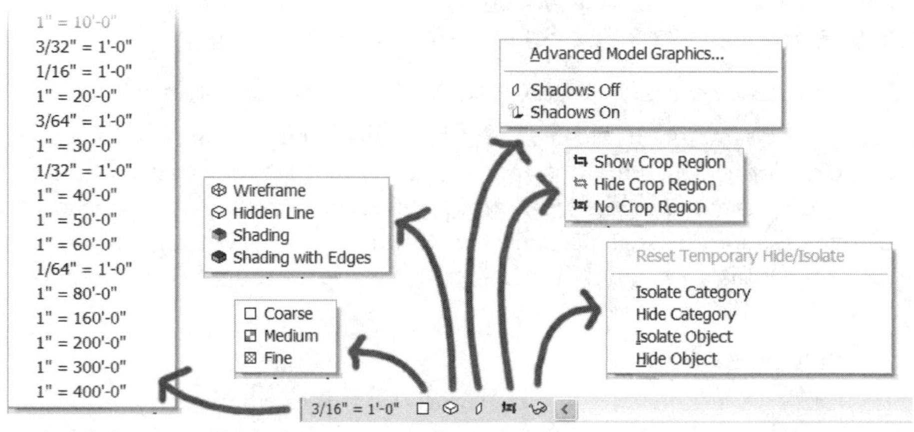

Figure 2–22 *The View Control Bar provides quick access to the most common View settings*

All of the settings, except Hide/Isolate, on the View Control Bar are also accessible from the View's "Element Properties" dialog. To access the View's "Element Properties" dialog, right click in the View window or right-click on its name in the Project Browser and choose **View Properties.**

The last icon on the right: Hide/Isolate, is used to temporarily hide and show elements in your model to make it easier to see things as you work. To use these tools, select an element or elements in the model and then click the Hide/Isolate icon and choose an option. Hide, temporally makes the selected element(s) invisible. Isolate leave the selected elements visible and hides everything else.

The important thing to know about using the Hide/Isolate function in a view, is that the changes it makes to the view window will not be saved outside the current work session. Furthermore, if you were to print or export the current View, Hide/Isolate settings will be ignored. Remember also that while the View Control Bar shows the most common View Properties, right-clicking and choosing the View Properties command will give a dialog with a complete list of properties—many more than on the View Control Bar. If you wish to make the Hide/Isolate settings permanent for instance, you can edit the View Properties and then click the Edit button next to "Visibility." This will call the "View/Visibility Graphics" dialog where you can modify the display of certain Element Categories for that View permanently.

RIGHT-CLICKING

In Autodesk Revit Building, you can right-click on almost anything and receive a context-sensitive menu. In fact, we have already seen examples of this in the previous chapter. These menus are loaded with functionality and will be used extensively.

TIP As a general rule of thumb, "When in doubt, right-click."

The next several figures highlight some of the more common right-click menus you will encounter in Revit Building. Take a moment to experiment with right-clicking in each section of the user interface. You will also discover the typical Windows 2000/XP right-click menus appear in all text fields and other similar contexts (this is used for Cut, Copy, Paste and Select All).

Right-Click on Toolbars

As mentioned above, to load or hide toolbars, right-click in the toolbar area. You simply move your mouse over any visible toolbar icon, and then right-click. This will make a menu appear (see Figure 2–23.)

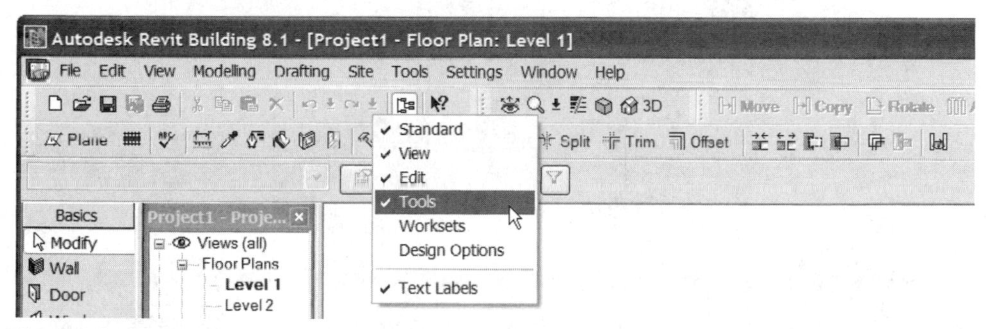

Figure 2–23 *Right-click to load toolbars*

Toolbars that are *currently* displayed on screen will have a check mark next to them. Toolbars that are *not* displayed will have no check mark. Many toolbars also have text labels. These labels provide a text description of the tool name. There is an option to toggle these text labels on and off. Choose an item without a check mark (such as Worksets or Design Options) to display it. The toolbar will appear on screen. Right-click any icon on the newly loaded toolbar and choose it again to hide it.

Right-Click on the Design Bar

The Design Bar was covered in detail above. It is not necessary to display all Design Bar tabs at once. Right-click to decide which tabs you wish to see and those you wish to hide. Just like toolbars, tabs that are displayed will have a checkmark, and those that are hidden will not have a checkmark (see Figure 2–24).

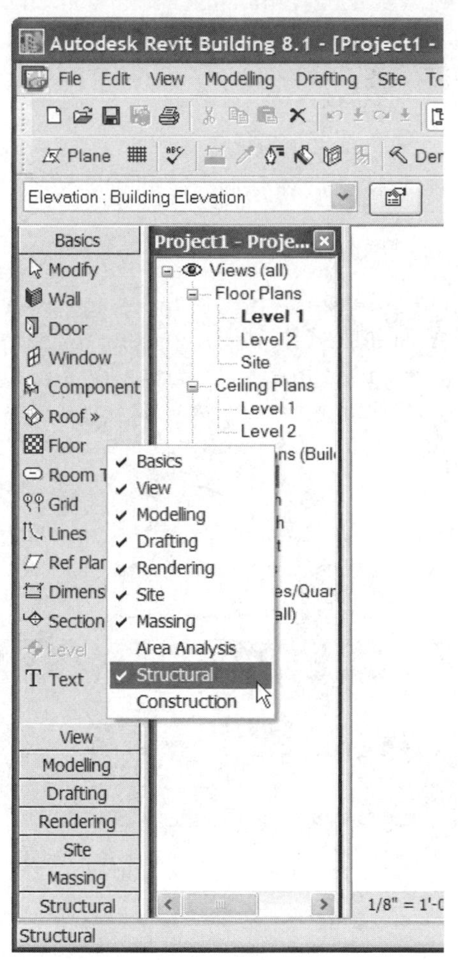

Figure 2–24 *Right-click to display hidden Design Bar tabs*

Right-click on the Design Bar again and choose a checked item to hide that tab. Whatever tabs you choose to display will remain displayed the next time you launch Revit Building.

Right-Click items in Project Browser

Items on Project Browser often have right-click options as well. Context menus are not available for every node of the tree, however you can always right-click directly on a View to receive a context menu of options (see Figure 2–25).

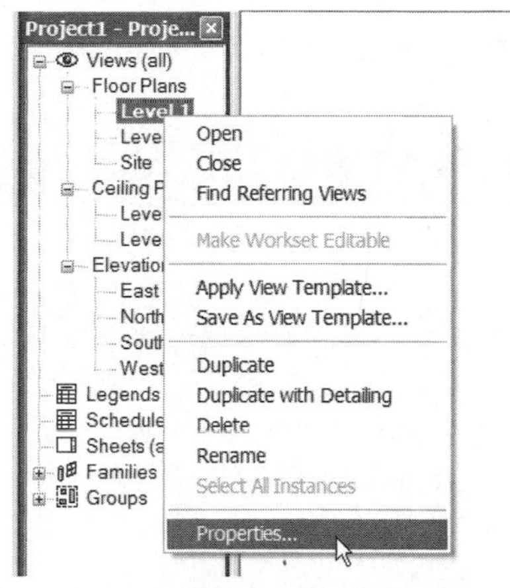

Figure 2–25 *Right-click a View to access options related to working with Views*

A menu will appear with several options. Some of these items have been used already in the previous chapter and above. A particularly useful command that has not been covered yet is the "Find Referring Views" command. When choosing this command, a dialog will appear like the one shown in Figure 2–26.

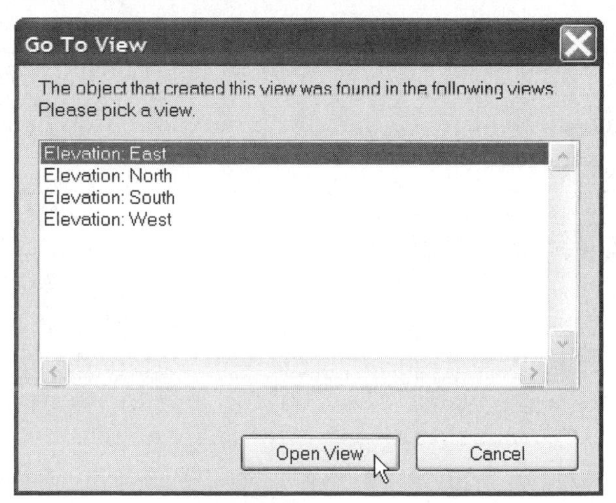

Figure 2–26 *Find Referring Views locates all Views that refer to the plan*

You can choose any elevation View listed in the dialog and then click Open View.

Right-Click in the View Window

You can get a menu similar to the one seen when you right-click a View in Project Browser by right-clicking in the workspace window. To do this, right-click anywhere in the active View window. If there are model elements present in the file, do not right-click the element, right-click the blank space instead. The menu that appears shows some commands that are similar to the menu explored above, but also include many View window navigation commands as well such as Zoom and Scroll (see Figure 2–27).

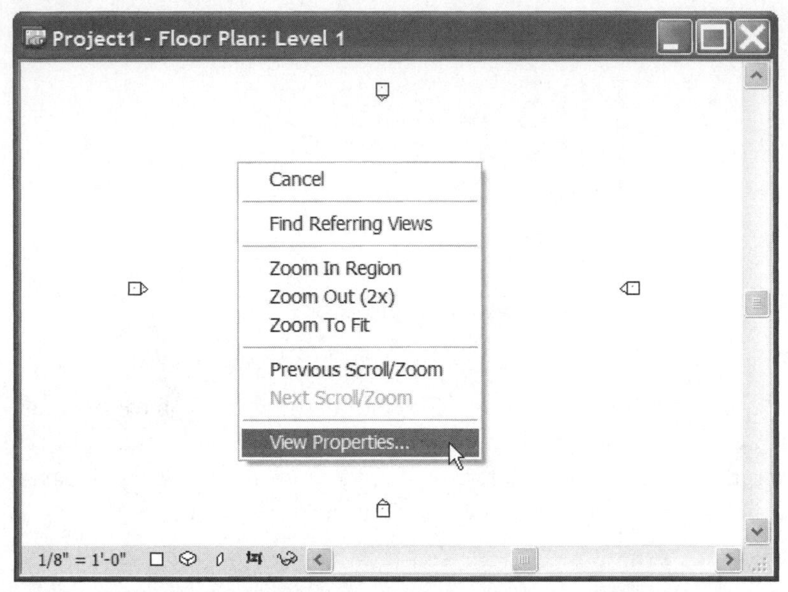

Figure 2–27 *Right-click in a View Window to zoom or access the View Properties*

You can cancel an active command by choosing Cancel from this menu, or by simply pressing the ESC key.

NAVIGATING IN VIEWS

When working with models on a computer screen, you need more than the standard Windows scroll bars to navigate your model. You will need to change the magnification of the model and frequently scroll other parts of the model into view on the limited screen space available. The act of changing the magnification of the screen is referred to as "zooming" and moving the image on screen within the borders of the View window to see parts off screen is referred to as "scrolling."

USING A WHEEL MOUSE TO NAVIGATE

If you have a Windows Intellimouse, (or any third-party mouse with a middle button wheel and the proper driver,) Revit Building provides instant zooming and panning using the wheel! If you don't have a wheel mouse, this might be a good time to get one. This modest investment in hardware will pay for itself in time saved and increased productivity by the end of the first day of usage. Using the wheel you have the following benefits:

- To **Zoom**—Roll the wheel.
- To **Scroll**—Push the wheel down and then drag the mouse.

- To **Dynamic Zoom**—Hold down the CTRL key and then drag the wheel (this is the same as rolling the wheel).
- To **Dynamic Spin**—Hold down the SHIFT key and then drag the wheel (3D Views only).

ZOOM

In addition to the wheel mouse, Revit Building provides 3 basic zooming methods.

Zoom In Region—This command will enlarge an area of the model that you designate by dragging a box around the area on screen.

Zoom Out (2x)—This is also a zoom in (or magnification) command that simply doubles the size of the image on screen.

Zoom To Fit—Most often this is a zoom out (or reduction in magnification) command. The function of this command is to fit the entire extents of model and any annotation objects into the available View window space on screen. A similar command is available on the View menu: **Zoom All to Fit.** This command performs the "Zoom To Fit" command in all of the open View windows rather than just the active one.

You can find these commands in several locations. They are located on the **View > Zoom** menu, on an icon on the View toolbar (both shown in Figure 2–28), the right-click menu (not shown) and by using keyboard shortcuts (as seen next to the zoom commands in Figure 2–28).

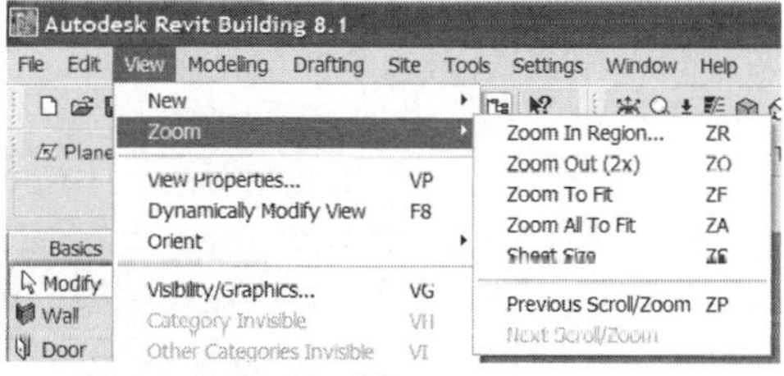

Zoom from the View menu

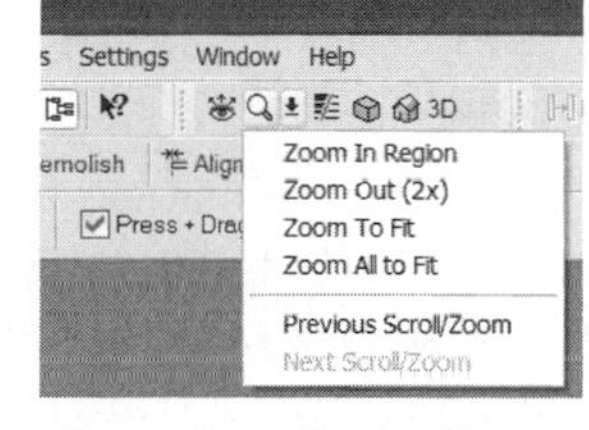

Zoom from the toolbar

Figure 2–28 *Zoom commands are available from various places in the interface*

SCROLL

Scroll bars are located on the bottom and right sides of each View window. You can use these to scroll the window. You can also use the wheel mouse. If you have zoomed or scrolled the window, you can choose the previous and next scroll options from the same Zoom menus shown in Figure 2–28.

SPIN (AND MORE)

If you are working in a three-dimensional View, you can also "spin" the model. This allows you to study the model from all directions and heights. To spin the model, you use the "Dynamically Modify View" icon on the View toolbar (or the

SHIFT + wheel shortcut mentioned above). This command is covered in detail in Chapter 3.

SELECTION METHODS

Before elements can be modified in Revit Building they must be selected. There are various methods used to select elements, some of which are similar to other software. When modifying elements in a model, the basic workflow is as follows:

1. Select the Element(s) to be manipulated with the Modify tool.
2. Issue a command to perform on the selected element(s).
3. Indicate on screen where and how to perform the command.

To summarize this workflow, in Revit Building you typically choose (select) *what* you want to modify and then indicate *how* you want to modify it—to this selection of elements I wish to perform this action.

CREATING A SELECTION SET:

A selection set is one or more selected elements in Revit Building. As you move the Modify tool around a View window in Revit Building you will notice that any elements available for selection will temporarily highlight while under the cursor—this is know as "pre-highlighting." The purpose of pre-highlighting is to preview for you what will be selected if you click the left mouse button. This is particularly helpful when working in a complex model with many elements close together. Use the pre-highlighting (and often the TAB key—see below) as a tool to assist you in accurate selection.

The simplest way to select an element is to click on it. When you select an element in Revit Building it will display in red on screen—this indicates that the element(s) is selected. Once selected, an element will remain selected until you deselect it. You can deselect elements in three ways: selecting another element will automatically deselect the current selection set (unless you hold down the CTRL key) and become the new selection set. You can deselect all elements without creating a new selection set by clicking on a blank portion of the screen (where there are no elements) or by pressing the ESC key.

To create a selection set containing more than one element, use the following techniques:

- Hold the CTRL key down while clicking on another element. The new element will be added to the current selection set (and highlight red as well).
- Hold the SHIFT key down while clicking on a selected element (highlighted red) to remove the element from the current selection set (it will no longer be highlighted in red).

If you accidentally pick an object without holding down the CTRL key you will lose your selection set (which will be replaced with only the one element just picked). You can restore your previous selection set by right-clicking in the View window and then choosing **Select Previous**.

NOTE You can also select previous by holding down the CTRL key and then pressing the LEFT ARROW key.

Either action will restore your previous selection without having to start all over again.

SELECTION BOXES:

Even with the CTRL and SHIFT keys, making selections of multiple elements can be time consuming. The easiest way to create a large selection set is by using a selection box. To create a selection box with the Modify tool, click down with the left mouse button next to an element, hold the button down and drag a rectangular box around the elements you wish to select. Elements will pre-highlight as you make the selection box. You can always combine methods—first create a selection box, and then use the CTRL and SHIFT keys to add and remove from the basic selection.

The direction in which you drag the selection box determines which specific elements are selected. If you create your selection box by dragging the Modify tool from left to right on the screen, the edge of the selection box will appear solid as you drag and only elements *completely* within the box will be selected. If you create your selection box by dragging from right to left on the screen, the edges of the selection box will appear dashed and any element completely or partially included in the box will be selected (see Figure 2–29).

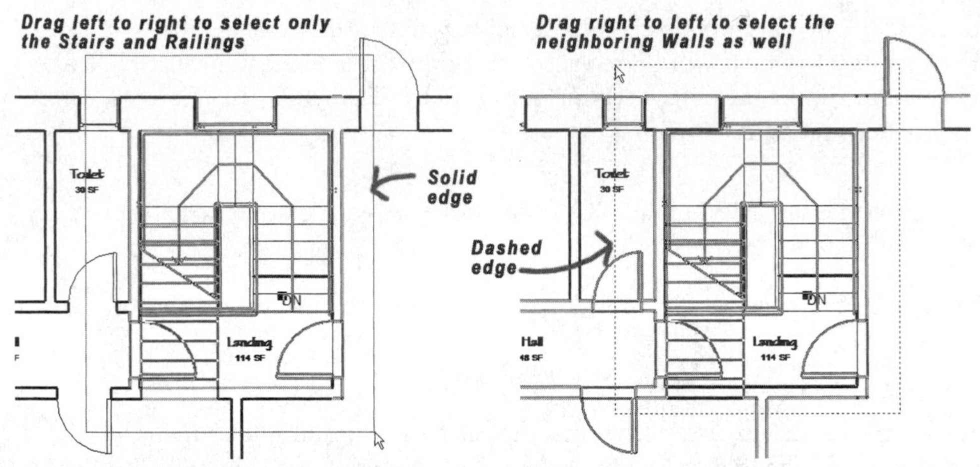

Figure 2–29 *Selection Windows vary with the direction you drag*

FILTER SELECTION

Another approach to building a large selection set is to deliberately select too many elements and then use the Filter function to remove items of particular categories from the selection. The Filter Selection icon is located on the Options Bar and has two major uses:

- Categorizes the selection set by element type.
- Removes complete categories of elements from the current selection set.

Remember, when using the Filter Selection tool, you always start with a selection that includes *more* than the items you want, and you then filter *out* the undesired elements by category (see Figure 2–30).

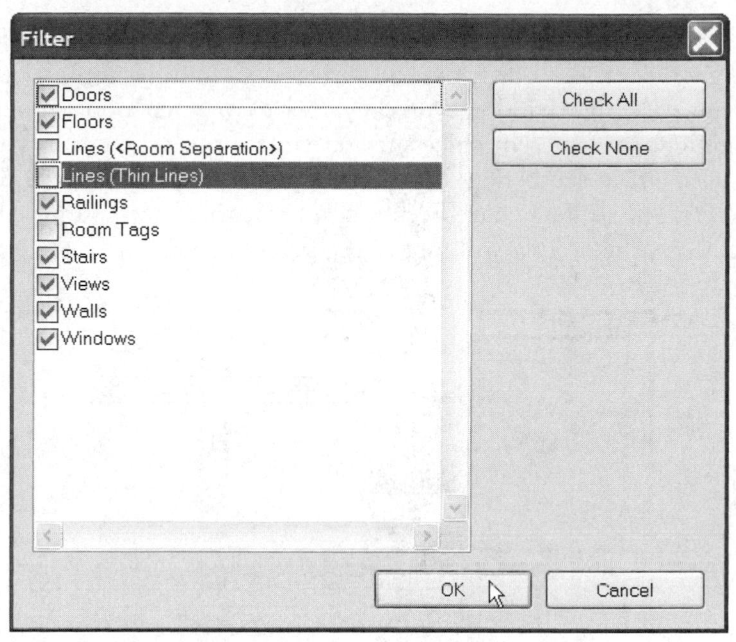

Figure 2–30 *Use the Filter Selection dialog to remove categories of elements from the selection set*

Each of the selection methods will require a little practice to become second nature to you. Take the time to practice each one since selection is such an important and frequent part of nearly all tasks in Revit Building.

THE ALMIGHTY TAB

In Revit Building, you will find that the TAB key is probably the most used and most useful of all the keys on the keyboard. As mentioned in the previous topic, being able to quickly add or remove from a selection set can increase your efficiency during a Revit Building work session. In general, you can think of the TAB key as a "toggle switch," allowing you to cycle through potential selections or other indicators in a View Window. When you attempt to pre-highlight a particular element on screen, sometimes a neighboring element will pre-highlight instead. No matter how subtly you move your mouse in an attempt to capture the desired element, it often proves difficult or impossible to capture the right one. The TAB key provides the solution to this situation. Use it to cycle through adjacent items to highlight the particular element you desire.

While there are dozens of examples of using the TAB key in a Revit Building session, a few examples are presented here that will give you an idea of where using the TAB key proves most handy.

PRE-HIGHLIGHTING ELEMENTS FOR SELECTION

If the tool, whether it is the Modify, Dimension, Split, or another tool, is near two or more Elements in a View window, you can either move your mouse around the view to pre-highlight the different objects for selection, or use the TAB key to pre-highlight each element in succession. Each time you press tab, a different element will pre-highlight until all items have been cycled through. Once you have tabbed through all elements, the cycling will repeat (see Figure 2–31).

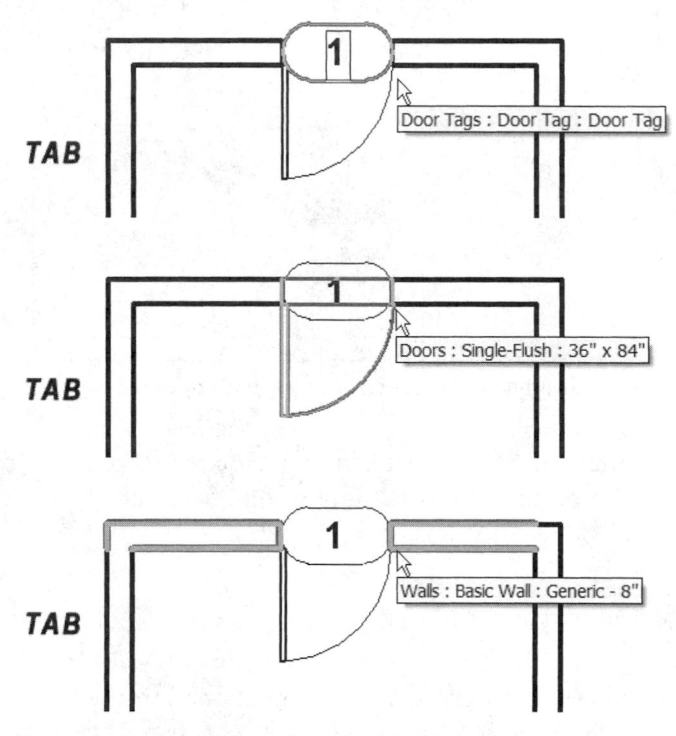

Figure 2–31 *Use the tab to cycle through and pre-highlight nearby elements*

When the element that you wish to select is pre-highlighted, click the mouse as normal to select it.

DURING DIMENSIONING

When you are using existing geometry for reference points such as when adding dimensions, using the align command or tracing a background, you can use the tab key to cycle through possible reference points (see Figure 2–32).

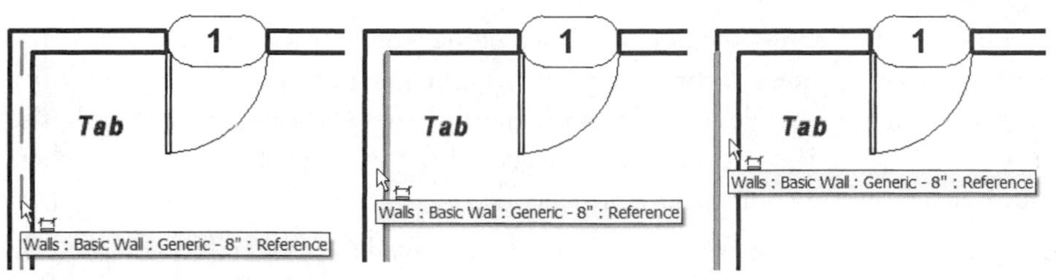

Figure 2–32 *Use the tab to cycle through dimension reference points*

CHAIN SELECTION

Chain selection is a very powerful feature of Revit Building. Chain selection works with Walls or Lines. Like an actual chain, chain selection highlights all of the Walls or Lines that touch one another end to end. To make a chain selection, you first pre-highlight a Wall or Line. Next you press the TAB key and in so doing, any walls or lines that form a chain (i.e. are located end-to-end) with that wall or line are pre-highlighted together. Finally, you click the mouse to make the selection with a single mouse-click (see Figure 2–33).

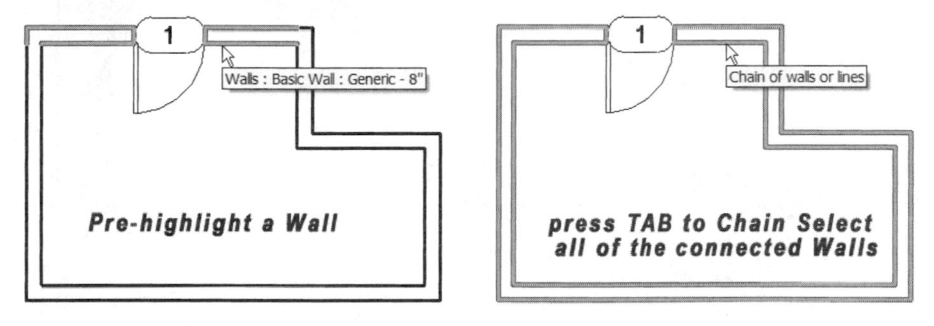

Figure 2–33 *Use the tab key to chain select a collection of Walls*

There are dozens of additional ways to use the TAB key in Revit Building. One of the easiest ways things to do is simply try tabbing in various situations while you work in Revit Building. Other examples will be presented throughout this book in the tutorials that follow. For more discussion on the topic, consult the Autodesk Revit Building online help.

SETTINGS

While not strictly related to interface, this section covers some of the settings you can use to manipulate workspace preferences. In some cases, the settings apply to your Revit Building application in general (in other words, the setting will persist regardless of the project you open) and in other cases the setting is saved with the project and can vary from one project to the next.

Examples of global settings include the User name, the Tooltip Assistance or the Path settings. Any of these settings applies to your installation of Revit Building not to a particular project file. On the other hand, Units settings apply to only the current project file. So if you change the Project Units from Feet-Inches to Inches, this would apply only to the current project file.

While there are many Options and Project Settings available, we will only look at a few of the more common ones here.

THE GENERAL TAB

From the Settings menu, choose the Options command to open the "Options" dialog and view or edit the settings therein. The dialog is divided into five tabs, each containing several related options (see Figure 2–34).

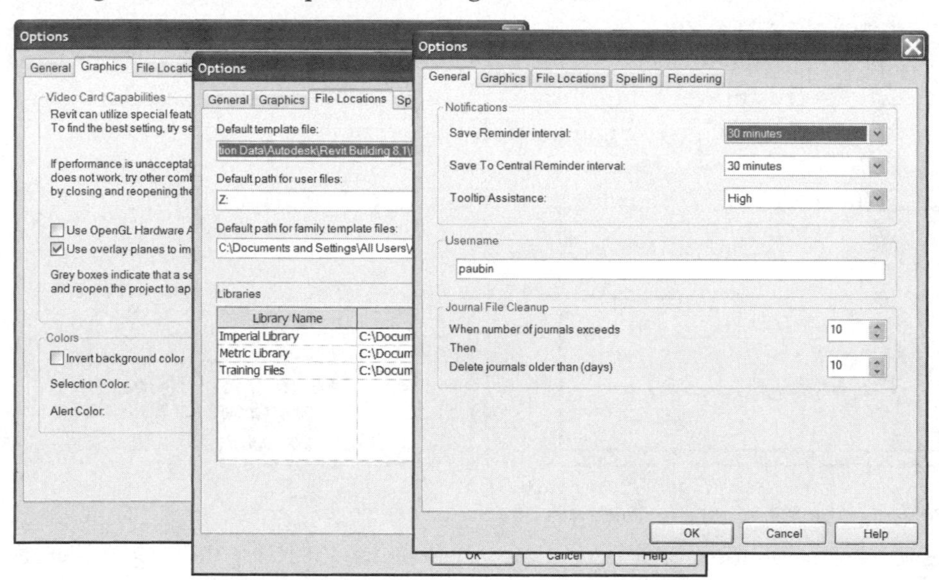

Figure 2–34 *The Options dialog box*

Save Reminder Interval—The default setting is 30 minutes. This means that every 30 minutes during your work session, Revit Building will prompt you to save your file. You can change the increment to suite your preference.

TIP Regardless of your Save Reminder frequency, you should make it a habit to save your work often.

Tooltip Assistance—Use this option to control how much feedback and assistance is given via on screen tooltips.

Username—By default, this is set to your Windows™ login name. The only time this is important is if your are working with Worksets. Worksets enable you to work with teams on the same project. Refer to Appendix A for more information on using Worksets. The important thing to remember about Username is that Revit Building will only allow you to work in a Workset-enabled file if your Username matches the Username that was active when that file was saved.

FILE LOCATIONS

The File Locations tab of the "Options" dialog is where you configure the hard drive and/or network locations of the template and library files used by Revit Building.

Default template file—Project Template files are discussed in detail in Chapter 4. Use the setting here to point to your firm's preferred project template file.

Default path for user files—If you have a specific location on your computer that you typically keep your project files, you can browse to that location here. The folder written here will be used by Revit Building when the Open dialog is used.

Libraries—This is a very handy feature. Each entry in this list will show up as an icon in the Open and Save dialog boxes. In this way, you can jump from one library location to another with a single click.

TIP Add a Library pointer for the folder in which you installed your Mastering Autodesk Revit Building CD ROM files. In this way, you can navigate to this location quickly at the start of each tutorial.

TEMPORARY DIMENSIONS

Although you can toggle the reference point used by temporary dimensions using the TAB key, it is often more efficient to change the default behavior instead. Defaults for Temporary Dimensions can be configured for Walls, Doors and Windows. Depending on your preferences, you can have the Temporary Dimensions default to either the centers (default) or edges of objects. From the Settings menu, choose **Temporary Dimensions** (see Figure 2–35).

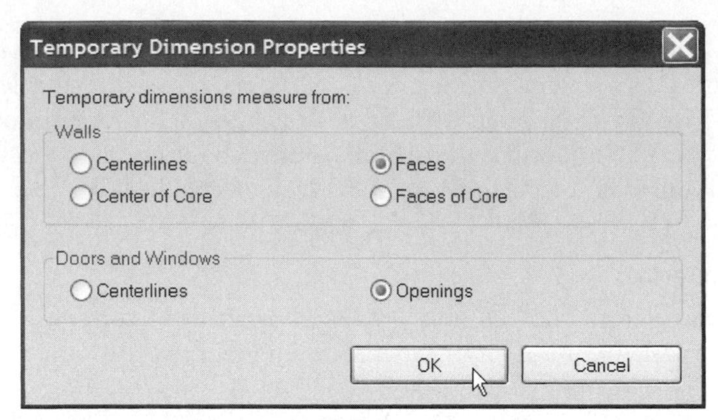

Figure 2–35 *Configure temporary dimension behavior*

Choose your preference for both the "Walls" and "Doors and Windows" settings. Revit Building defaults to centerlines for both. For Walls, you can dimension faces, centerlines and/or Wall Cores. For Doors and Windows choose between centerlines and openings.

SUMMARY

Now that you have completed the Quick Start Tutorial, read the high level overview of building information modeling and Revit Building in Chapter 1 and been guided through an overview of the Revit Building user interface in this chapter, you are now ready to begin creating your first real projects. In the chapters that follow, we will build two projects from scratch: one residential and one commercial. In this chapter you have learned:

The User Interface of Autodesk Revit Building uses many common Windows conventions as well as many unique to Revit Building.

The Design Bar along the left side of the screen contains all of the Tools used in Revit Building organized into Tabs.

The Options Bar which includes the Type Selector and the Properties icon changes to reflect the current tool or command.

One or more Views of the project can be opened at one time in the Workspace area.

The Project Browser, along the left side, contains all of the Views of the project as well as the Families and Groups.

The View Control Bar can be used to manipulate the display settings of the active View.

The Status Bar along the bottom of the screen shows prompts for the current command.

Many commands are available from the context-sensitive right-click menus.

The easiest way to navigate a View is using the wheel of your mouse and the appropriate modifier keys.

Roll the wheel to zoom, drag the wheel to scroll, hold the SHIFT key and drag to spin the 3D model.

Most commands require element selection.

Drag left to right to select all elements within the selection box, right to left to select all those in contact with the box.

The TAB key is used to toggle through potential selections when more than one is possible.

Use the TAB key to "Chain" select Walls or Lines.

Configure settings to suit your personal preferences.

SECTION II

Create the Building Model

The first step toward reaping the benefits of Building Information Modeling is to construct a virtual building model in the software. The tools provided for this purpose in Autodesk Revit Building are many and varied. In this section we will explore the many building modeling tools available such as Walls, Doors, Windows, Columns, Beams, Stairs, Railings, Roofs, Floors, and Curtain Walls. In Chapter 9, we will take a detailed look at Families and the Family Editor.

Section II is organized as follows:

Creating a Building Layout

INTRODUCTION

Autodesk Revit Building can be used successfully in all types of architectural projects and within all project phases, largely due to its robust toolset. Most of our attention in this book is given to the design development and construction documentation phases. The tutorial exercises in this book will explore two building types concurrently, starting at different points in the project cycle, to give a sense of the variety of ways you can approach the design process. Don't become limited to the techniques covered here. The aim of the tutorials is to get you thinking in the right direction. Exploration and play time are highly encouraged.

Architectural projects start in many different ways. You may be commissioned to design a new building from its early conception all the way through construction and beyond, or you may be asked to join a project already in progress as a member of a larger team. Likewise, design data comes in many forms. If the project is new construction, you may have only the site plan and some other contextual data. If the project is a renovation, or if you join a project already in progress, you may use or be given existing design data in many formats from traditional hand-drawn sketches to digitally created images and models in a variety of software formats.

The first two chapters have made you comfortable with the theoretical underpinnings of Autodesk Revit Building. Now that you have the correct mind-set, get ready to roll up your sleeves—it is time to begin creating our first model.

OBJECTIVES

Throughout the course of the following hands-on tutorials, we will lay out the existing conditions for our residential project. In this chapter, we will explore the various techniques for adding and modifying walls, doors, and windows. In addition, we will add plumbing fixtures and other elements to make the floor plan more complete. After completing this chapter you will know how to:

- Add and modify Walls
- Explore Wall properties
- Add and modify Doors and Windows

- Assign Phase parameters to model components
- Add plumbing fixtures
- Add an In-Place Family

WORKING WITH WALLS

Basic object creation in Revit Building involves the simple process of choosing a tool from the Design Bar and then designating the desired settings on the Options Bar. We will begin by working with Walls. You can add Walls point by point or enable the "Chain" option to create them in series (each segment beginning where the previous one ended). Walls will "join" automatically with intersecting Walls at corners. Walls have many parameters such as length, height and type; can have custom shapes and profiles; and can receive Doors and Windows by automatically creating openings for them.

BASIC WALL OPTIONS

Since Walls are the basic building blocks of any building, we will start with them. We will create a new temporary project and sketch some Walls to get comfortable with the various options. Later we will create one of the actual projects that will be used throughout the rest of the book.

CREATE A NEW PROJECT

We will begin our work in a new project created from the Revit Building default template file.

1. Launch Autodesk Revit Building from the icon on your desktop (or from the Windows **Start > All Programs** menu).

Revit Building launches to an empty project based on your default template. Let's be sure that we are starting from the out-of-the-box default template. First we will close the empty project that opened automatically and then create a new one based on the default template file.

2. From the File menu, choose **Close**.
3. From the File menu, choose **New > Project**.
   - In the New Project dialog, in the "Template File" area, be sure that the lower radio button is selected (the one that lists a template file).
   - Verify that the *default.rte* [*DefaultMetric.rte*] template file is listed in the text field.

The default location for the Imperial template file is: *C:\Documents and Settings\All Users\Application Data\Autodesk\Revit Building 8.1\Imperial Templates*.

The default location for the Metric template file is: [*C:\Documents and Settings\All Users\Application Data\Autodesk\Revit Building 8.1\Metric Templates*].

If the template shown on your screen does not match the location and name listed above, click the Browse button to locate it before continuing.

NOTE If you are in a country for which your version of Autodesk Revit Building does not include these template files, they have both been provided with the files from the "Mastering Autodesk Revit Building" CD ROM. Please browse to the*Template* folder to locate them.

- In the "Create New" area, verify that "Project" is selected and then click OK (see Figure 3–1).

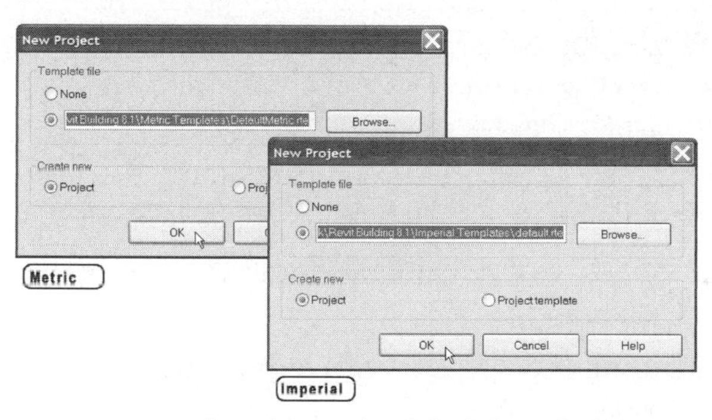

Figure 3–1 *Create a new project based upon the default Revit Building template file*

Depending on your system's settings, the same result can normally be achieved by simply clicking the NEW icon on the Standard toolbar. The NEW icon automatically creates a new project using the default template file. To configure the default template file, choose **Options** from the Settings menu. Click the File Locations tab and then edit the value in the Default template file field.

GETTING STARTED WITH WALLS

Nearly every building component in Revit Building is created following the same basic procedure. Locate the tool for the object you wish to create on the Design Bar (on the left side of the screen, see Chapter 2 for more information). Use the Options bar to configure the settings of the object as you create it. Use your mouse to create the object in the current View in the Workspace.

4. On the Design Bar, click the Basics tab and then click the Wall tool (see Figure 3–2).

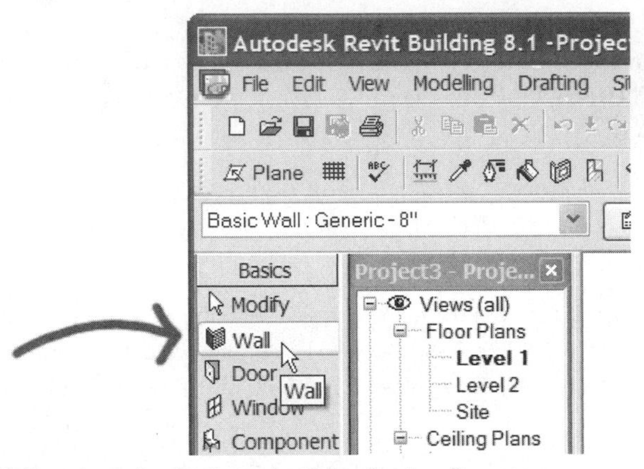

Figure 3–2 *The Wall tool on the Basics tab of the Design Bar*

NOTE If the Basics Design Bar is not showing on your screen, choose Design Bars from the Window menu to retrieve it or right-click on the Design Bar to display it. See Chapter 2 for more information.

The pointer will change to a pencil cursor and the Status Bar prompt will read: "Click to enter wall start point."

- From the Type Selector, choose Basic Wall: Generic - 8" [Basic Wall: Generic – 200mm].

The following list explains the major fields and controls shown on the Options Bar when you are adding a Wall (see Figure 3–3).

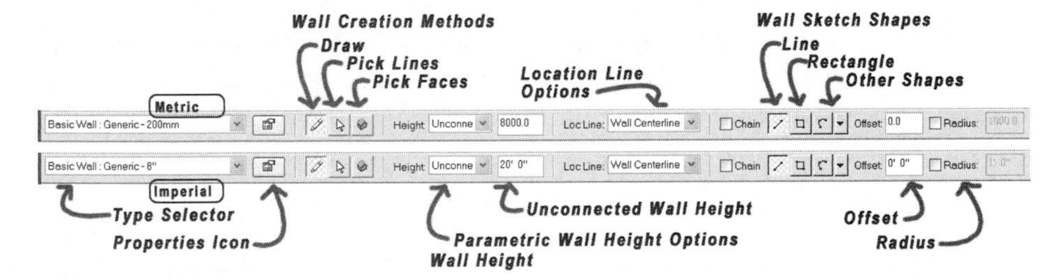

Figure 3–3 *The Options Bar for adding Walls*

- **Type Selector**—Use this drop-down to choose from a list of Wall types in the current file. The specific list is populated by the template we used to create the project. Always choose a type from this list before configuring other options.

NOTE Walls are a System Family. System Families are the elements in Revit Building with the most built-in parameters. (They typically represent building components that are constructed on-site as opposed to delivered and installed) They cannot be edited independently outside of the Project Editor. In order to automatically have customized System Families available to new projects, add and edit them in your project template files. You can create your own project templates by choosing the "Project Template" radio button when creating a new project. You can also transfer Wall Types and other System Families and system settings from existing projects to the current project by opening both projects and then choosing the Transfer Project Standards from the File menu. For more information, search for "Transfer Project Standards" in the online help.

- **Properties icon**—Click this icon to open the Element Properties dialog (see Figure 3–14 below for an example). The Element Properties dialog gives you access to both "Type" and "Instance" parameters. Type parameters apply to all Walls of the same type, Instance parameters apply only to the selected Wall(s).
- **Wall Creation Methods**—Walls can be drawn on-screen or created from existing model components. There are three creation methods, shown by their icons: Draw, Pick Lines and Pick Faces. When you choose Draw, you can create Walls on-screen in a variety of shapes. When you choose the Pick Lines option, the Wall Shapes icons disappear since the shape of the Wall will be determined by the lines picked. Likewise, even fewer options will be available when you choose the Pick Faces option.
- **Height**—There are two options for Height; a drop down list of parametric height options and a text field (available when "Unconnected" height is chosen). In Revit Building, the heights of Walls can be connected to other model components. The most common choice for this is the height of one of the project's levels. When you choose "Unconnected" you are able to simply type in a fixed height for the Wall.
- **Loc Line**—Short for "Location Line." The Location Line is the point within the width of the Wall that will be used as a reference. Choices include: Wall Centerline, Interior and Exterior Finish Faces and several "Core" options. The Core of the Wall will be discussed in detail in later chapters.
- **Wall Sketch Shapes**—These options are only available when you choose the "Draw" icon from the creation methods (see above). There are three basic sketch shape icons in this section: Line, Rectangle and Arc. The Arc icon can be changed to a collection of other options, such as Polygon, Circle and several Arc types.
- **Offset**—Use this field to input a dimension. The Walls you create will be placed parallel to the points you click at a distance equal to this value.

- **Radius**—This option can be used with the arc options to specify a specific radius value rather than clicking the radius on screen.

Now that we have an overview of the available options, let's see some of them in practice.

- Verify that both the "Draw" and the "Line" icons are pushed and then click anywhere on screen to place the first point of the Wall.

Move your mouse to the right, keep it horizontal, but don't click yet (see the top of Figure 3–4).

Notice the dimension on screen as you move the mouse. As you move your pointer, this dimension automatically snaps to whole unit increments. Depending on the size and resolution of your screen, the exact increment may vary. For example, if you are using Imperial units, the increment is likely 1′-0″ and the increment for Metric is likely 100mm.

5. Roll the wheel of your mouse up a few clicks to zoom in a bit and continue to move the pointer left or right (see the bottom of Figure 3–4).

NOTE If you do not have a wheel mouse, you might want to consider purchasing one. You can use the Zoom commands on the View menu instead of the wheel, but the wheel makes quick zooming much easier. Refer to Chapter 2 for more information.

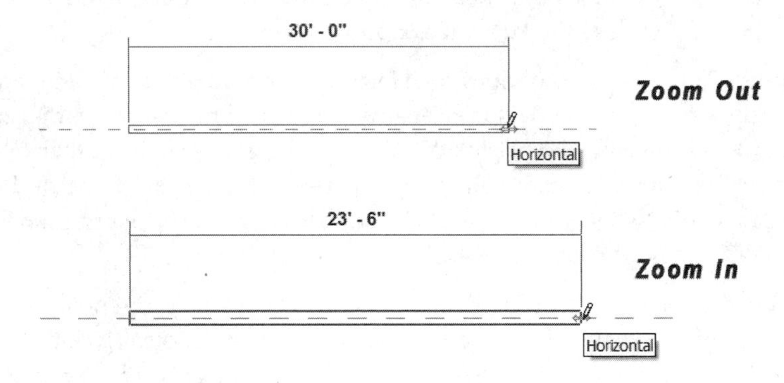

Figure 3–4 *The dimension increment varies with the level of zoom*

As you zoom in, the increment of the dimension will reduce. Depending on the unit type you are using, you may need to zoom in or out to see this. This is easy to do with the wheel. Simply roll the wheel up or down to zoom in or out. If you zoom off screen, keep the command active and hold the wheel in and drag. This will pan the screen. If you do not have a wheel mouse, it is highly recommended that you consider purchasing one. However, you can also access zoom commands on the View and right-click menus.

- Click to set the other point of the Wall.

A single Wall segment will be created.

Often you will want to create more than one Wall segment, each beginning where the previous one ended. You can achieve this with the "Chain" option on the Options Bar.

6. On the Options Bar, place a checkmark in the "Chain" checkbox to enable this option.
   - Click any point on screen to begin placing the next Wall.
   - Click two more points at any locations on screen (see Figure 3–5).

   Notice that the corner where the two Walls meet has formed a clean intersection (called a "Wall Join").

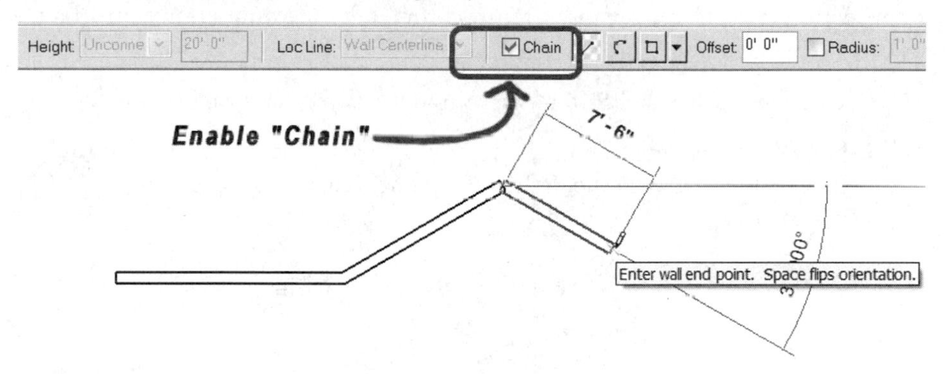

Figure 3–5 *Use the Chain option to create continuous Wall segments*

7. Click the Arc icon.

The default Arc tool first places each of the endpoints and then adds the intermediate point last to define the degree of curvature. Other Arc options are also available if you click the small arrow icon (next to the Arc icon) to reveal the drop down list of Wall shapes available (see the "Wall Shapes" bullet point above).

   - Move the pointer in any direction and click to place the end point of the Arc.
   - Following the cue at the Status Bar, click to place the intermediate point of the Arc.

   Continue to experiment by adding additional Wall segments as desired.

8. To finish adding Walls, click the Modify tool on the Basics Design Bar (or press the ESC key twice).

Think of these few Walls as a simple "warm up" exercise. Drawing them helped us explore some of the basic Wall options. However, these Walls have been placed a bit too randomly to be useful. Let's delete them now and create some new ones a little more deliberately.

USING A CROSSING SELECTION BOX

Before we can delete the existing Walls, we need to select them. There are many ways to select objects in Revit Building. Rather than learn them all at once, let's take them one at a time as our current need dictates. A convenient way to select multiple objects at one time is with the Window and Crossing selection methods. In either technique, you create a box by selecting two opposite corners with your mouse. To create a Window selection, click and drag from left to right. To make a Crossing selection, click and drag the opposite direction—from right to left. A Window selection selects only those items completely surrounded by the box, while a Crossing selects anything touched by or within the box.

9. Click a point below and to the right of the Walls you have on screen.
   - Hold down the mouse button and drag up and to the left far enough to touch all objects with the dashed (Crossing) selection box (see Figure 3–6).

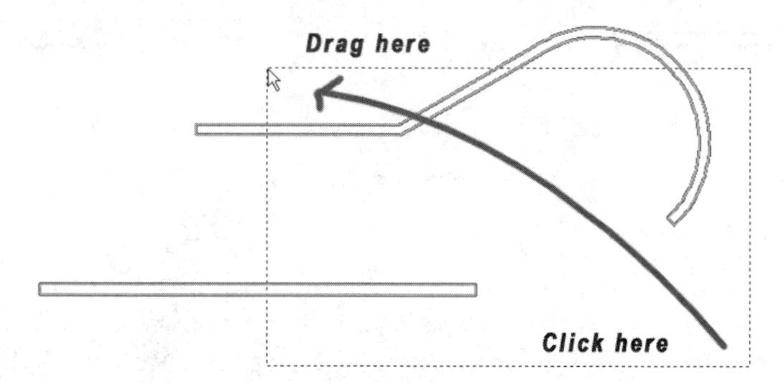

Figure 3–6 *Make a Crossing selection by dragging from right to left*

   - When all of the Walls highlight, release the mouse button.

All of the Walls will turn red to indicate that they are now selected. Once you have a selection of objects, you can manipulate their properties, move, rotate or mirror them or you can delete them.

10. Press the DELETE key on your keyboard to delete the Walls.

ADDING WALLS WITH THE RECTANGLE OPTION

Several of the Wall Shape icons that are available create Walls that form closed geometric shapes like rectangles, circles and polygons. Often using these is the quickest way to layout these more regular shapes. Let's try the rectangle option now.

11. On the Basics Design Bar, click the Wall tool.
    - On the Options Bar, verify that Basic Wall: Generic - 8" [Basic Wall: Generic – 200mm] is chosen from the Type Selector.
    - Verify also that the Draw icon is active. If it is not, push it in now.
    - Accept the defaults for Height and Location Line and then click the Rectangle shape icon (see Figure 3–7).

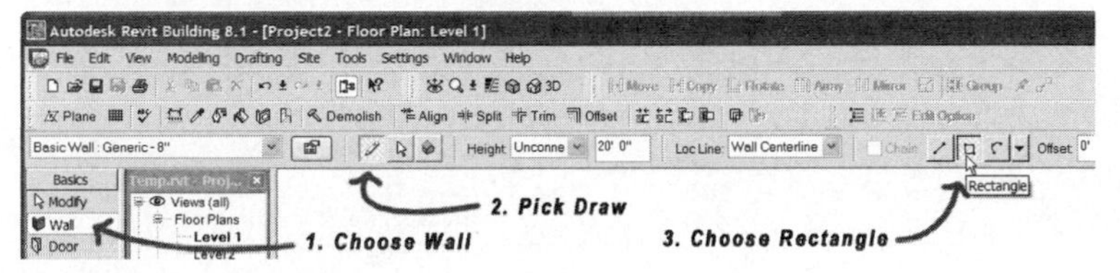

Figure 3–7 *Adding a rectangular sequence of Walls*

In the default Revit Building project that we used as a starting point here, there are four elevation markers. Let's zoom the screen to fit these.

12. On the keyboard, type **ZX**.

NOTE If you prefer, you can also choose **Zoom > Zoom To Fit** from the View menu. However, where available, the keyboard shortcuts like the one suggested in the previous step are usually quicker once you learn them. Menu commands that have keyboard shortcuts list them to the right of the menu pick. Several examples of this are seen on the **View > Zoom** menu.

The mouse pointer will show a small rectangle next to it indicating that we are in rectangle drawing mode.

13. Click a point within the upper left region of the space surrounded by the elevation markers.
    - Move the mouse down and to the right and watch the values of the dimensions.
14. Click a point in the lower right region of the space surrounded by the elevation markers.

Notice that the temporary dimensions continue to display on the Walls just drawn. Furthermore, these dimensions appear in blue. This indicates that their values may be edited dynamically.

15. Click on the blue value of the horizontal dimension.

The value will become an editable text field.

- Type a new value into this field and then press ENTER (The exact value is unimportant— see Figure 3–8).

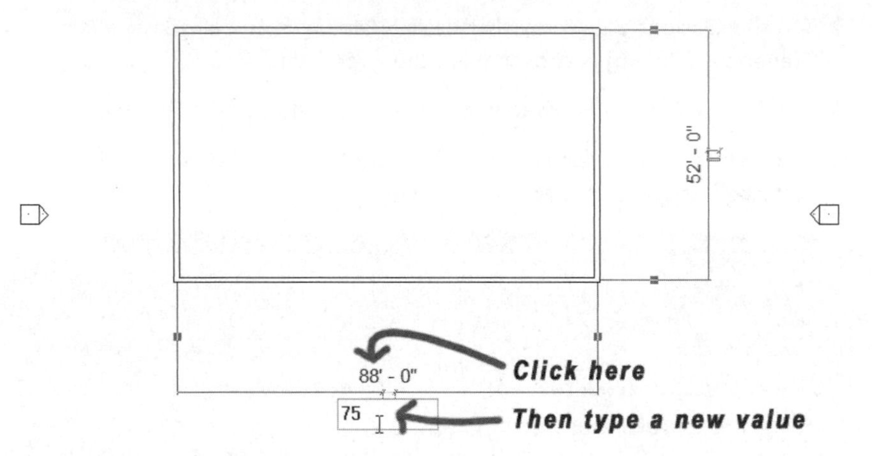

Figure 3–8 *Use the temporary dimensions to edit the locations of the Walls*

16. The Wall command is still active. Using the same technique, draw another rectangle overlapping the first.

Notice that the temporary dimensions now reference points on the first rectangle to points on the new one. You can edit these values in the same way that we edited the ones above. You can also move the witness lines of the dimensions to gain more control over their exact locations. We will explore this below. If you wish, try some of the other sketch shapes like circle or polygon. In the case of the polygon, additional controls will appear on the Options Bar.

17. When you are finished exploring Walls, choose **Close** from the File menu to close the current project. When prompted to save, choose No.

CREATE AN EXISTING CONDITIONS LAYOUT

Now that we have practiced adding a few Walls and seen some of the options available while doing so, let's begin creating an actual model. In this book we will follow two projects from the early schematic phase through to the construction document phase. We will start with the first floor existing conditions for our residential project. The residential project is an 800 SF [75 SM] residential addition. This project will require a little bit of demolition and new construction and will require plans, sections, elevations, details and schedules. In this project we will explore the Phasing tools in Revit Building—Demolition, Existing, and New Construction will be articulated later in the tutorial. This will give us the required separation between construction phases of the project. The completed files for the residential project are available in the *Chapter03\Complete* folder with the files installed from the CD.

CREATE A NEW RESIDENTIAL PROJECT

We will use the same template file that we used above to begin our residential model.

1. Create a new project file using the *default.rte* [*DefaultMetric.rte*] template file as we did in the "Create a New Project" heading above.

   Be sure that the *Level 1* floor plan View is open on screen. You can see this indicated in bold (under Views (all) > Floor Plans) on the Project Browser and in the title bar of the Autodesk Revit Building window.

2. On the Design Bar, click the Basics tab and then click the Wall tool (shown in Figure 3–2 above).

   - From the Type Selector, choose Basic Wall: Generic - 12" [Basic Wall: Generic – 300mm].
   - Verify also that the Draw icon is active. If it is not, push it in now.
   - For Height choose **Unconnected** and set the value to **18'-0"** [**5500**].

Remember: If you are using Imperial units, simply type 18 and then press ENTER. No unit symbol or zero inches is necessary. Also note that to enter a value like 18'-2 1/2" you can type it in several ways. For example you can type 18' 2 1/2" or 18' 2.5" or 18 2 1/2 or 18 2.5 or 18.20833.

   - For the Location Line, choose **Finish Face: Exterior** and then click the Rectangle sketch icon (see Figure 3–9).

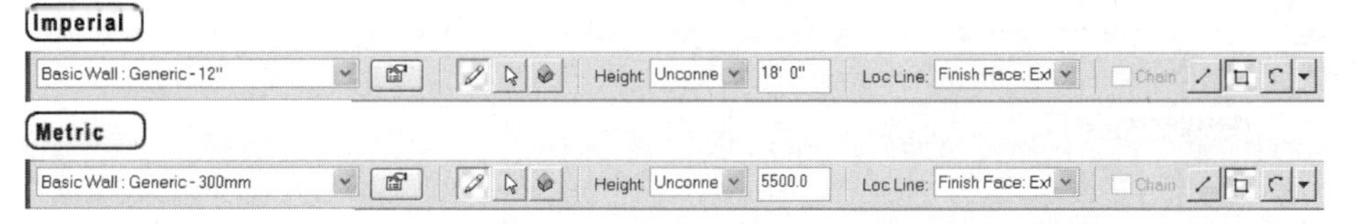

Figure 3–9 *Set the Options for the exterior Walls of the residential project*

3. Click two opposite corners on screen within the space bounded by the elevation markers (the exact size is not important for initial placement).

Notice that even though we have chosen a Location Line of Finish Face: Exterior, that the temporary dimensions still have witness lines at the centerlines of the Walls. This is simply a default behavior independent of the Wall's individual location line setting. If you want to input a value for the dimension based upon the face of the Walls, you can simply move the witness lines.

4. Click the small blue square handle on one of the horizontal dimension's witness lines.

   Notice that the witness line moves to one of the Wall faces. If you click it again, it will move again, this time to the opposite face. One more click returns it to the centerline (see Figure 3–10).

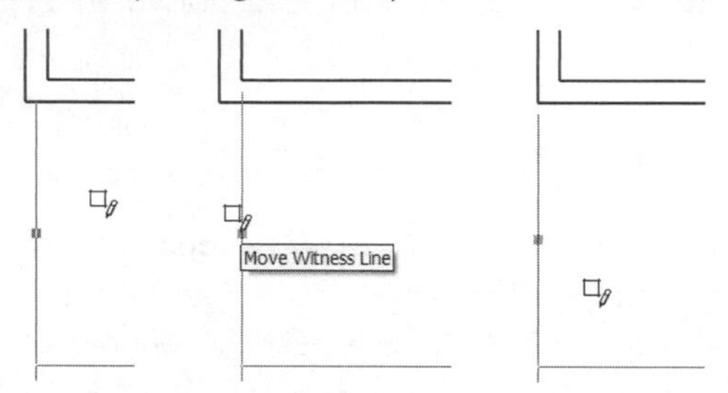

Figure 3–10 *Move the witness lines of the temporary dimension*

   - Click the witness line handle (of the horizontal dimension) until both sides reference the outside edges.
   - Click the blue value of the horizontal dimension and then type: **33′-0″ [10000]** and then press ENTER.

5. Repeat this process on the vertical dimension making the outside face to face dimension equal to: **24′-0″ [73000]**.

TIP If your witness lines are in the wrong locations when you edit the temporary dimension, simply repeat the process to move the witness lines and then repeat the dimension edit process.

CAUTION Be careful not to click the small "permanent" dimension icon when editing the dimension values. Clicking this icon will make the temporary dimension a "permanent dimension" (it will remain in the current View even after the associated Wall is deselected).

6. On the Design Bar, click the Modify tool, or press ESC twice to complete the operation.

You should now have four walls in a rectangular configuration measuring 33′-0″ x 24′-0″ [10000 x 73000] outside dimensions.

7. From the File menu, choose **Save**.
   - In the Save As dialog, navigate to a location on your local system where you wish to save the project.
   - For the File name, type: **03 Residential** and then click the Save button.

OFFSET WALLS

Now that we have the external Walls laid out, let's begin adding the interior partitions. There are several techniques that we could employ. In this sequence, we will use the Offset tool. The Offset tool moves or copies the selected objects parallel to themselves by an amount that you input. This tool is located on the Tools toolbar or the Tools menu. The Tools toolbar is displayed by default and therefore should already be displayed in your Revit Building interface. If you have closed this toolbar, please re-display it now. Refer to Chapter 2 for more information on loading toolbars.

8. On the Tools toolbar, click the Offset tool.

TIP You can also use keyboard shortcuts for many commands if you wish. The shortcut for offset is **OF**. **Offset** is also located on the Tools menu.

The Options Bar for Offset has two modes: Graphical and Numerical. When you choose Graphical, the numeric input field is disabled and you use a temporary dimension on screen to indicate the distance of the offset. When you choose the Numerical option, you input the offset distance first, and then use the pointer on screen to indicate the side of the offset. In both cases, you can enable the "Copy" checkbox which will create a copy of the object as it offsets. If you disable this option, the object you offset will be moved parallel to itself by the amount you indicate.

- On the Options Bar, verify that Numerical is chosen (if it is not, choose it now).
- In the Offset field on the Options Bar, type **12′-5 1/2″ [3794]** and verify that "Copy" is selected.

The mouse pointer will change to an Offset cursor and the Status Bar will prompt for a selection.

- In the workspace, move the Offset cursor over the top edge of bottom horizontal Wall (see Figure 3–11).

A dashed line will appear indicating the location of the Offset. If you move your mouse up and down this line will shift up and down as well.

As in most places in Revit Building, if you press the TAB key, the pre-highlighted selection will cycle to the next available option. In this case, all four Walls will pre-highlight and the offset result would be a concentric ring inside or outside the building. Go ahead and try it if you like. Press the TAB key once, then move the pointer to indicate outside or inside offset and then click. Be sure to undo (Edit menu) after this experimentation.

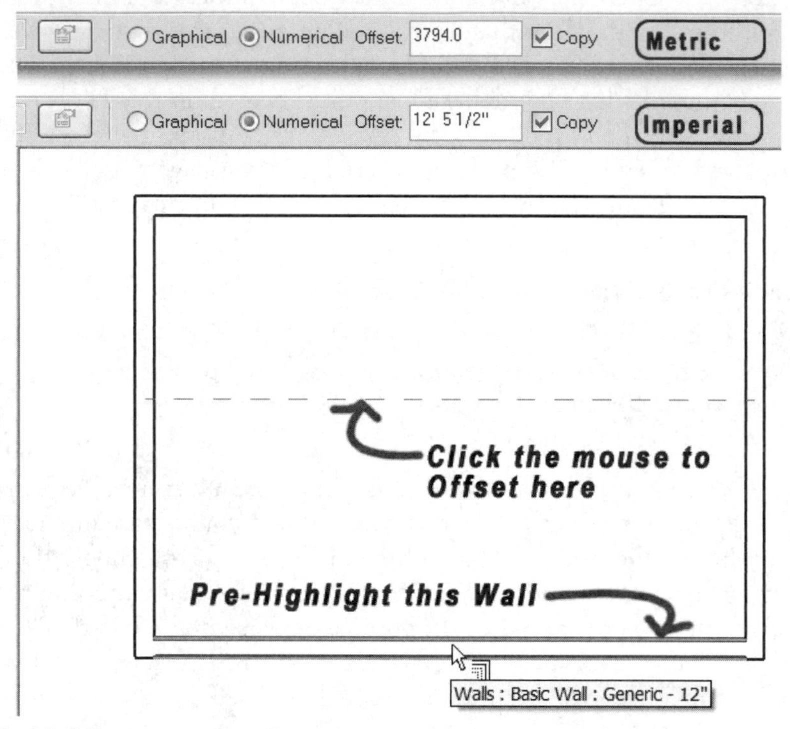

Figure 3–11 *Offsetting a new wall*

- When the dashed line appears above the Wall (inside the house) click the mouse.

A new Wall will appear inside the house.

9. On the Options Bar, type **12′-8 1/2″ [3818]** for the Offset distance this time.

- Move the mouse pointer over the leftmost vertical wall, and then click to offset it inside as well (see Figure 3–12).

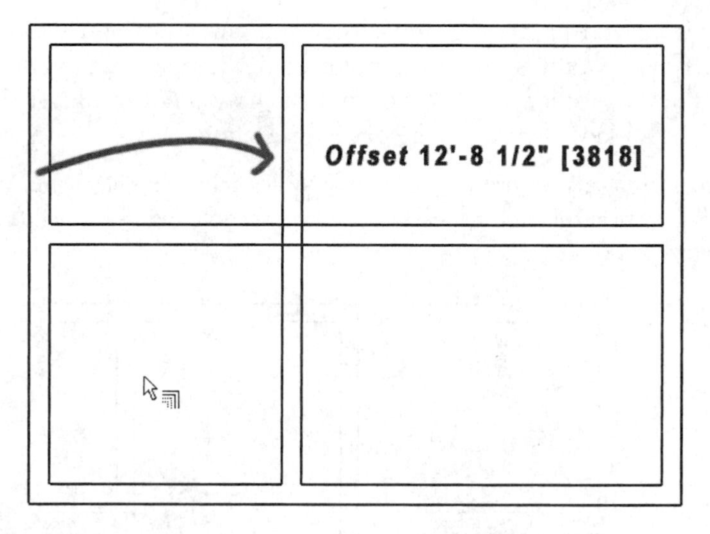

Figure 3–12 *Offset a vertical Wall inside*

NOTE When you Offset the vertical Wall, if you only get half the Wall, simply click the Undo icon and try again.

10. Save your project.

SPLITTING WALLS

Notice that the two new interior Walls overlap one another, but they do not form a clean intersection. In order for the Walls to appear this way, we need to split one of the Walls into two segments. The Split tool is used for this.

11. On the Tools toolbar, click the Split tool.

TIP The keyboard shortcut for Split is **SL**. **Split Walls and Lines** is also located on the Tools menu.

The Split tool has a single option on the Options Bar: "Delete Inner Segment." You would use this option if you split the Wall in two spots and wanted the segment of Wall in between the two splits to be removed. In this case, we simply want to split the one long Wall segment into two shorter segments that touch end to end. Therefore this option is not necessary at this time.

12. Move your mouse pointer over the intersection between the two interior Walls.

Notice that vertical Wall highlights under the Split cursor. If you were to click now, the vertical Wall would split at the point of the intersection. While we do want to split at the intersection, we want to split the horizontal Wall not the vertical Wall. Use the TAB key to cycle to the desired selection.

- Press the TAB key until the horizontal Wall's centerline highlights and a small vertical split line appears at the intersection and then click (see Figure 3–13).

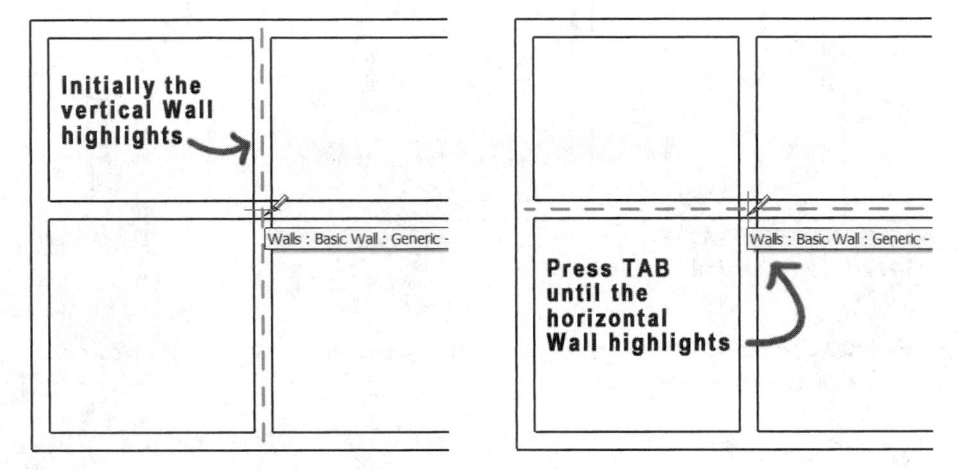

Figure 3–13 *Use the* **TAB** *key to cycle to the desired Wall to Split*

The horizontal Wall will now be split in two at the intersection with the vertical Wall and the intersection will appear properly joined.

13. On the Design Bar, click the Modify tool or press the ESC key twice.

CHANGING A WALL'S PARAMETERS

One of the features that makes Revit Building such a powerful tool is the ability to change an object's parameters at any time, as design needs change. Let's take a look at modifying some of the Walls as we continue with the layout of the first floor existing conditions for the residential project.

14. Select all of the internal Walls created in the previous steps (there are three Walls).

To select all three Walls, you can either click just inside the house near the lower right corner and then drag up and to the left until all three Walls highlight or you can hold down the CTRL key and click each of the three Walls one at a time.

15. On the Options Bar, click the Properties icon (or right-click, and choose **Properties**).

The Element Properties dialog will appear showing both the Type and Instance parameters for the selected Walls.

- In the Element Properties dialog, at the top, choose **Generic – 5″ [Interior - 135mm Partition (2-hr)]** from the Type list.
- In the Instance Parameters area, within the Constraints grouping, choose **Wall Centerline** for the Location Line (see Figure 3–14).

TIP To make this change, click in the Location Line field, and then click the small down arrow that appears to choose from the popup menu, or click in the field and then type "w" to jump to the Wall Centerline option.

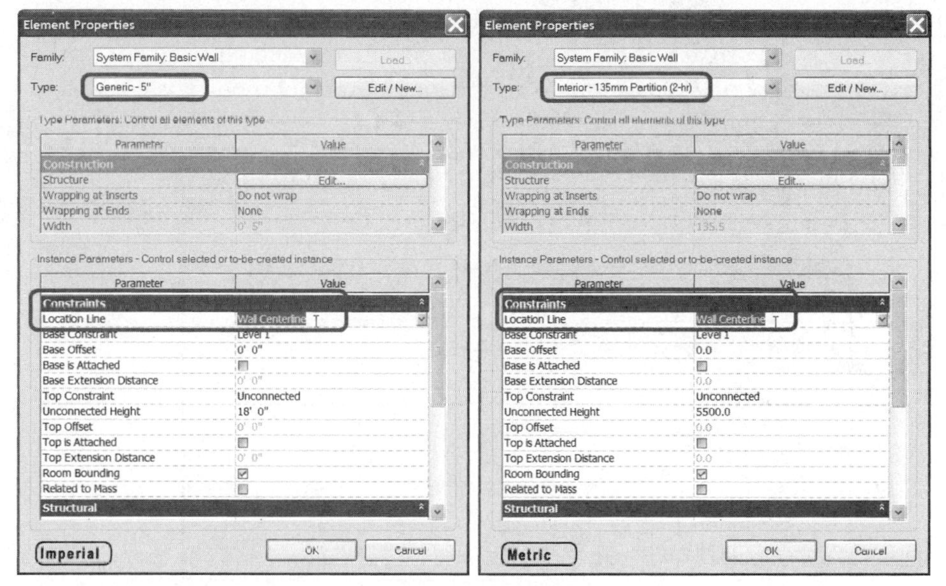

Figure 3–14 *Change the parameters of the Interior Walls*

16. Click OK to dismiss the Element Properties dialog.

    Notice the change to the Walls.

Let's verify that everything is in the correct place. As noted above, we are laying out the existing conditions of our residential project. Therefore, the locations of the Walls that we just placed are based on field dimensions. The room in the lower left corner of the plan is supposed to be 11′-6″ [3500] x 11′-9″ [3658]. We can verify these dimensions and easily make adjustments as necessary.

ADJUST WALL LOCATIONS

In the "Create a New Residential Project" heading above, we learned to use the temporary dimensions and the handles on the witness lines to edit the location of Walls. We can use the same technique to verify our dimensions now.

17. Click on the horizontal Wall to the left.

The temporary dimensions will appear indicating its current location relative to the other Walls. However, depending on the settings of your system, the dimensions may be from the centerlines. The desired size of the room that we are verifying are to the inside faces of the Walls (as you might expect from field dimensions). Therefore, we could use the witness line technique covered above to adjust the dimension points, but this is not persistent and the next time we would need to move the witness lines again. It will be better to change the default behavior of the dimensions from now on. We do this with the **Temporary Dimensions** command from the Settings menu.

18. From the Settings menu, choose **Temporary Dimensions**.

- In the Walls area, click Faces and then click OK (see Figure 3–15).

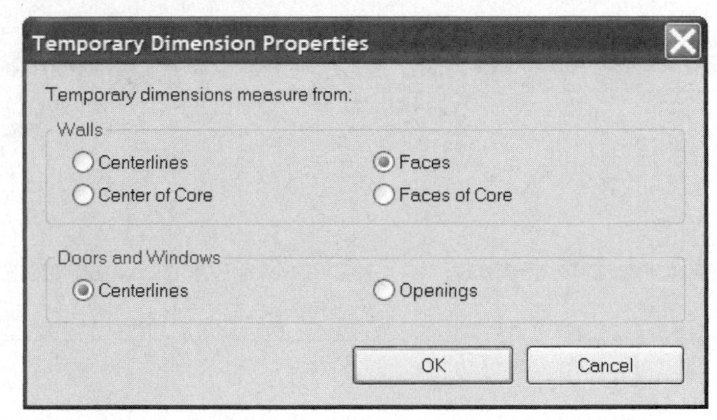

Figure 3–15 *Change the Temporary Dimensions to measure from Wall faces*

19. Click the left horizontal interior Wall again.

The dimension from the inside face of the bottom exterior Wall to the inside face of the selected Wall should be 11′-9″ [3658].

20. Click on the vertical interior Wall.

This time, the dimension from the inside face of the left exterior Wall to the inside face of the selected Wall should be 11′-6″ [3500]. However, this dimension is not correct. Despite best efforts to transfer field measurements to our building model, an error crept into our calculations. Fortunately, errors like this are very easy to correct in Autodesk Revit Building.

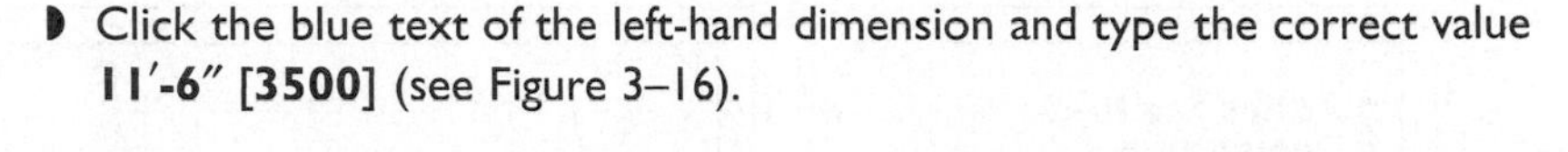

- Click the blue text of the left-hand dimension and type the correct value **11′-6″ [3500]** (see Figure 3–16).

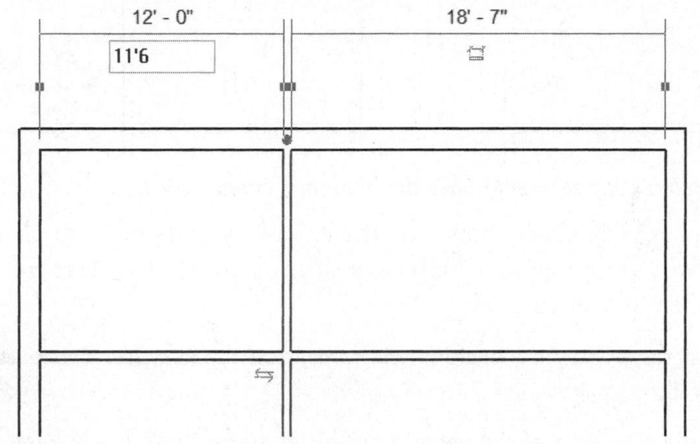

Figure 3–16 *Edit the Temporary Dimension to move the Wall to the correct location*

21. Save the project.

SKETCHING WALLS

The Offset technique covered above is an effective way to add new Walls based upon existing ones. In many cases however, it is easier to sketch the new Walls and then use the temporary dimensions to position them correctly. Let's try that technique next.

22. On the Design Bar, click the Wall tool.

    - From the Type Selector, choose **Generic – 5″ [Interior - 135mm Partition (2-hr)]**.
    - Verify that the Draw icon is selected.
    - For Location Line, choose Wall Centerline.
    - For the Sketch Shape, choose Line and make sure that Chain in *not* selected.

23. Move the mouse next to the exterior vertical Wall on the left.

    Notice the temporary dimension that appears.

    - When the value of the dimension above the interior horizontal Wall is about **4′-0″ [1200]** click to set the first point.
    - Move the pointer horizontally to the right and click to set the second point just past the middle of the plan (see Figure 3–17).

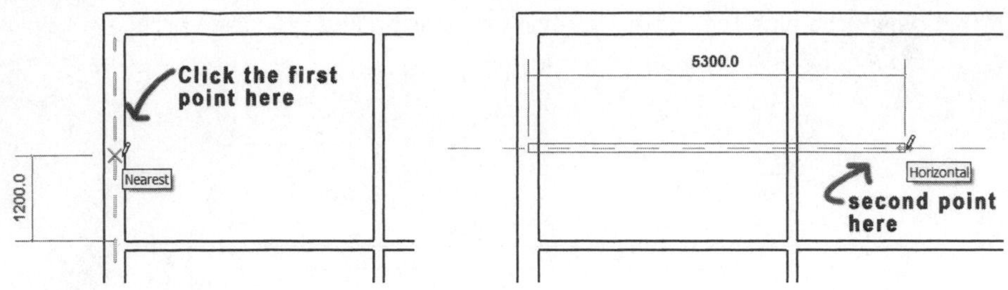

Figure 3–17 *Sketch a Wall segment above the horizontal interior Wall*

The figure shows Metric units, however, the exact locations of both clicks is not critical since we will use temporary dimensions to edit them next. Use the figure to achieve approximate placement.

Several temporary dimensions will remain on screen after the placement of the Wall.

24. Click the blue text of the vertical dimension between the new Wall that you just created and the other horizontal interior Wall.
    - Type **3′-10″ [1170]** and then press ENTER.

NOTE This is a face to face dimension. Once you edited the Temporary Dimension setting above, it should remain persistent.

25. Using the same process, create a vertical Wall approximately **3′-0″ [900]** to the left of the existing interior vertical Wall.
    - Start at the lower exterior Wall and draw up till it intersects with the upper horizontal interior Wall (the one drawn in the last step).
    - Using the temporary dimensions, edit the face to face distance between the two Walls to be **2′-6″ [750]** (see Figure 3–18).

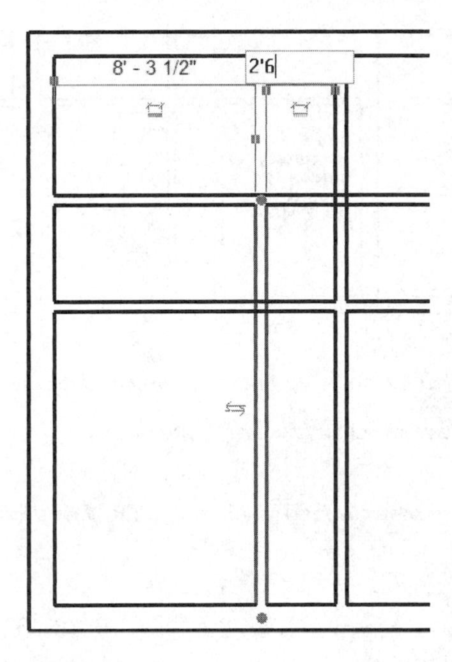

Figure 3–18 *Sketch Walls and use temporary dimensions to fine-tune their placement*

26. On the Design Bar, click the Modify tool or press the ESC key twice.

TRIMMING AND STRETCHING WALLS

The two Walls that we just added frame out a small closet in the left side of the plan. We can use a few techniques to clean up the plan a bit and remove unnecessary Wall segments. We will look at the Trim tool and some control handle editing techniques next.

27. Select the vertical interior Wall that you just completed (the one on the left).

    At each end of the Wall is a small round blue handle.

    - Click and drag the handle at the bottom upward.
    - Using the temporary guidelines that appear, drag up and snap to the horizontal Wall (see Figure 3–19).

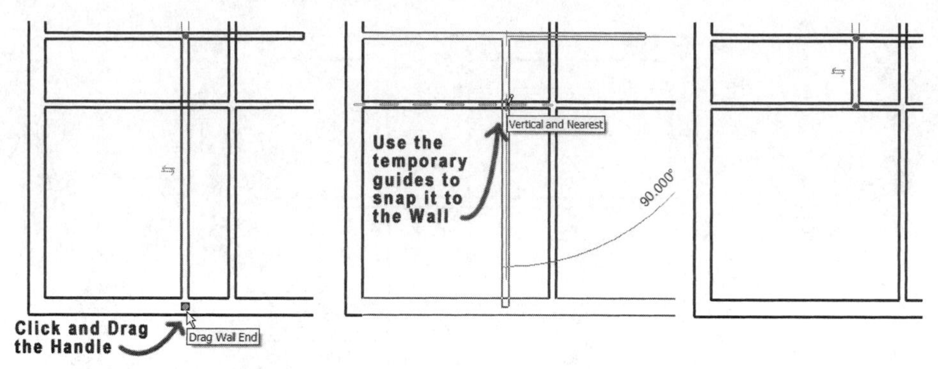

Figure 3–19 *Drag the Wall End Handle up to the intersection with the horizontal Wall*

28. On the Tools toolbar, click the Trim/Extend tool.

TIP The keyboard shortcut for Trim/Extend is **TR**. **Trim/Extend** is also located on the Tools menu.

The Trim/Extend tool has three modes on the Options Bar. The "Trim/Extend to Corner" mode works on two Walls or sketch lines. It will create a clean corner from the two segments by either lengthening or shortening the selected segments. The second mode is "Trim/Extend Single Element." This mode will lengthen or shorten the selected segment to a designated boundary edge. The third option, "Trim/Extend Multiple Elements" works the same way except that multiple Walls or Lines can be lengthened or shortened to the same boundary.

29. Make sure that the first mode is selected (Trim/Extend to Corner) and then click the vertical Wall that was just edited with the handle.

NOTE Pay attention to the prompt at the Status Bar, in this case instructing you to select the first Wall that you want to Trim or Extend. Please note that you are further prompted to select the portion that you wish to keep.

30. Next click the horizontal Wall to the right of the intersection (see Figure 3–20).

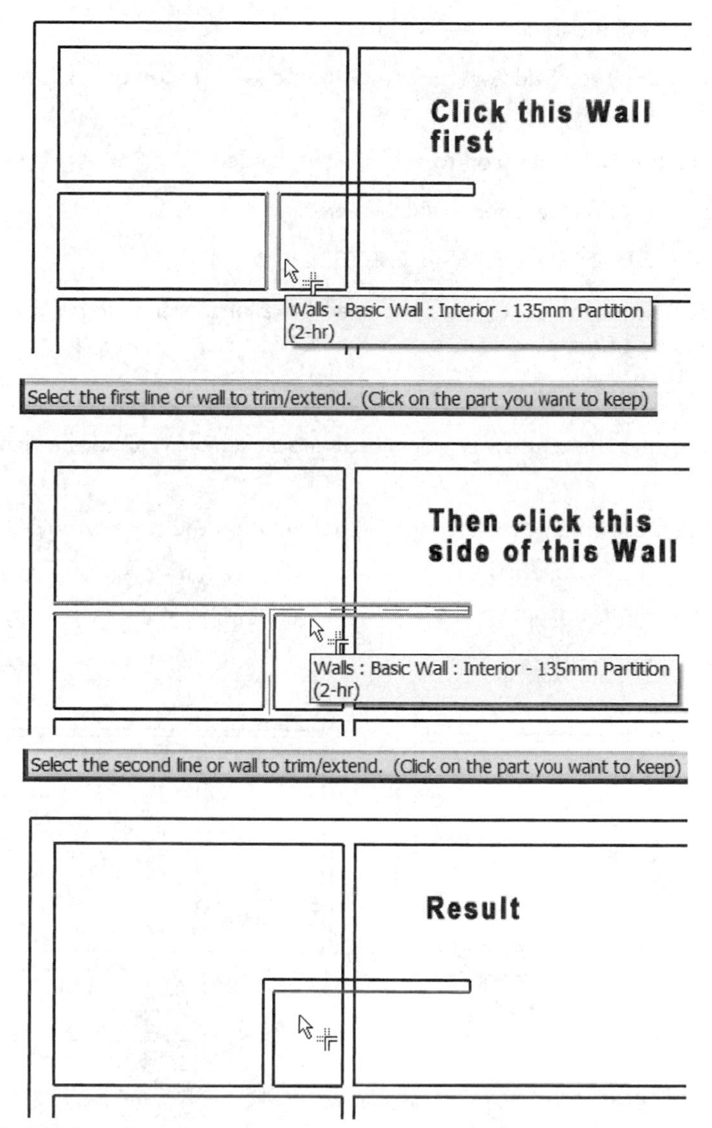

Figure 3–20 *UsingTrim/Extend to create the closet corner*

31. On the Design Bar, click the Modify tool or press the ESC key twice.

ADDING WALLS USING THE PICK LINES MODE

Let's continue adding to our floor plan layout by using a technique that combines features of both the previous techniques. We will add Walls using the Wall tool and its "Pick Lines" mode. We will use this in conjunction with the Offset option on the Options Bar to place the Wall close to where we need it.

32. On the Design Bar, click the Wall tool.
    - From the Type Selector, choose **Generic – 5″ [Interior - 135mm Partition (2-hr)]**.
    - Click the Pick Lines icon to activate this mode.
    - For Location Line, choose Wall Centerline.
    - In the Offset field, type **6′-0″ [1800]**.
33. Move the mouse next to the right side of the long interior vertical Wall in the middle of the plan.

    The right face of the Wall will pre-highlight.
34. Be sure that the temporary guideline indicates a new Wall to the right and then click.
35. On the Design Bar, click the Modify tool or press the ESC key twice.
36. Click on the new Wall, and edit the dimension between this Wall and the one it was offset from to **6′-6″ [1983]** (see Figure 3–21).

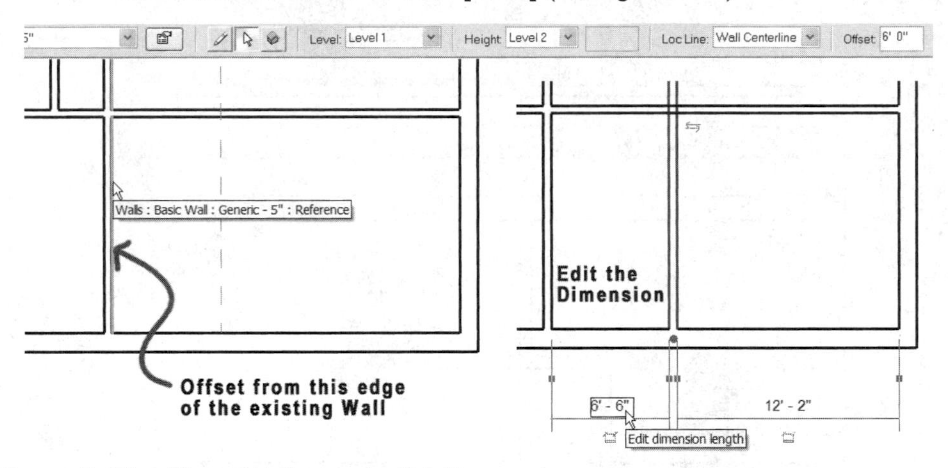

Figure 3–21 *Adding a Wall using the Pick Lines mode*

If you want to calculate the exact offset to use instead of editing the temporary dimension, you can do so. The Offset value will be measured from the line you pick to the Location Line of the new Wall—in this case Wall Centerline. Therefore, half of the Wall's width would have to be added to the dimension. Despite the relative ease of this calculation, it is typically easier to place the Wall close and then edit the dimension to the exact value as we have done here.

TRIM/EXTEND SINGLE ELEMENT MODE

Let's Trim a few more Wall segments.

37. On the Tools toolbar, click the Trim/Extend tool (or type **TR**).
    - On the Options Bar, click the "Trim/Extend Single Element" mode icon.

In this mode you first select the element to use as a boundary and then the click the Wall or line you wish to extend or trim.

TIP Remember to read the prompts on the Status Bar.

- Click the vertical interior Wall on the right as the boundary.
- Click the short horizontal Wall at the top of the closet to extend Figure 3–22).

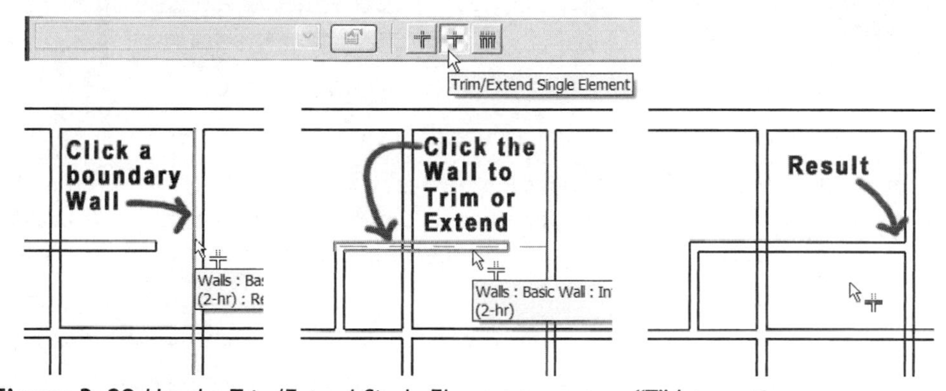

Figure 3–22 *Use the Trim/Extend Single Element to create a "T" Intersection*

38. Use the same technique to trim away the vertical Wall at the top of the closet (Remember to click the piece you want to keep—read the Status Bar).
39. On the Design Bar, click the Modify tool or press the ESC key twice.

USING THE MOVE AND COPY TOOLS

The layout is coming along. Let's use the Move and Copy tools to create the Walls required for the stair hallway.

40. Click the right side of the horizontal Wall (the one we Split above) and then click the Move tool on the Edit toolbar.

TIP The keyboard shortcut for Move is **MV**. **Move** is also located on the Edit menu.

- Snap to the Endpoint at the top of the Wall for the Move Start Point.
- Move the point up approximately **1′-0″ [300]** (see Figure 3–23).

We actually need to move the Wall up 0′-10″ [250]. If you are working in Metric units, you should be able to snap to this value as you move. If you are working in Imperial units, you will need to zoom in to snap to the closest inch. Otherwise, you can use the value indicated here, and then use the temporary dimension to edit it to the correct value.

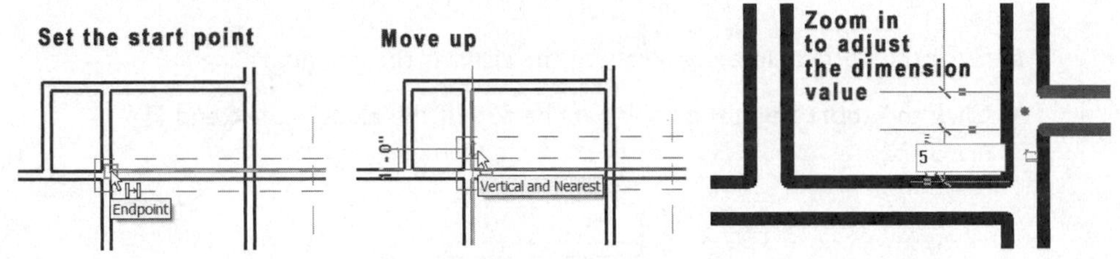

Figure 3–23 *Move the Wall up and then edit the dimension to the correct value*

- Edit the temporary dimension to **0′-5″ [125]**.

You might need to zoom in to edit the value properly. When you are zoomed out, Revit Building might not place the temporary dimensions at useful locations. Zooming in can correct this. If you are working in Imperial units, remember to type the inch (″) symbol when editing the value.

41. Click the long vertical Wall on the left and then click the Copy tool on the Edit toolbar.

TIP The keyboard shortcut for Copy is **CO**. **Copy** is also located on the Edit menu.

- Pick a start point on the selected Wall and then move to the right.
- Move the mouse to the right **3′-6″ [1067]** and click to place the copy.
- Verify that the face to face dimension between the new Wall and the original is **3′-1″ [942]**, if it is not, edit the temporary dimensions (see the left panel of Figure 3–24).

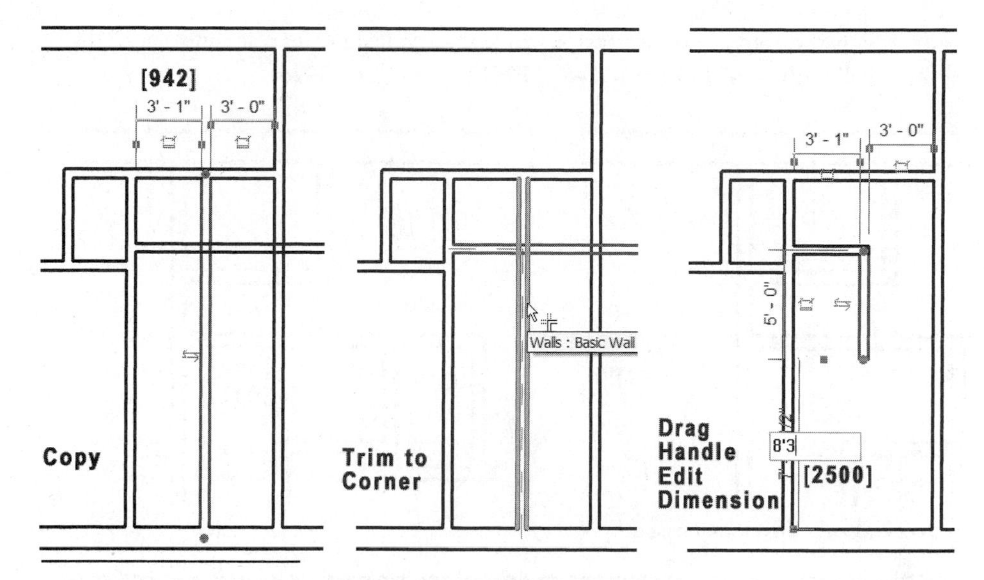

Figure 3–24 *Copy, Trim and Edit Walls for the hallway stairs*

- Using the middle panel of Figure 3–24 as a guide, Trim the two new Walls to form a corner.
- Using the right panel of Figure 3–24 as a guide, drag the handle up and edit the temporary dimension as shown: **8′-3″ [2500]**.

42. Save your Project.

SKETCH THE REMAINING WALLS

At the top middle of the plan is a small half-bath. Let's sketch these Walls in to complete the Wall layout of our first floor existing conditions.

43. On the Design Bar, click the Wall tool.
    - From the Type Selector, choose **Generic – 5″ [Interior - 135mm Partition (2-hr)]**.
    - Click the Draw icon to enable this mode.
    - For Location Line, choose Wall Centerline.
    - For the Sketch Shape, choose Line.
    - Place a checkmark in the Chain checkbox.
    - Verify that Offset is set to zero.
44. Working at the top middle of the plan, sketch two Wall segments to create a small half-bath.

- Edit the temporary dimensions to make the inside clear dimensions of the half-bath **4′-4″ [1300]** x **3′-6″ [1050]** (see Figure 3–25).

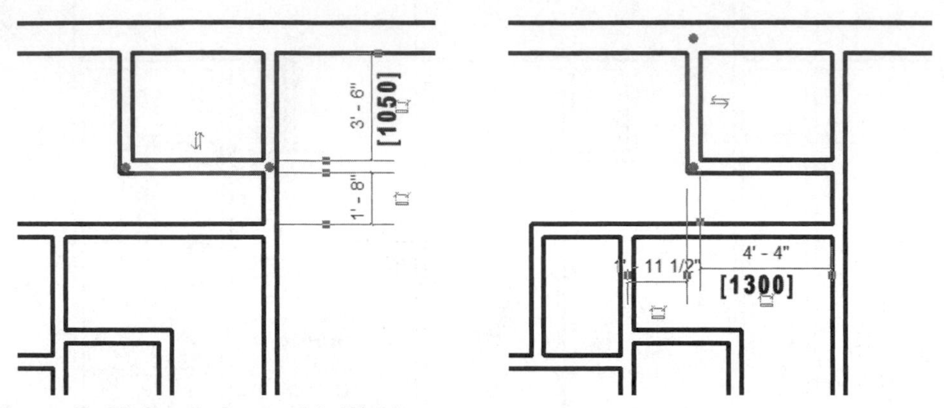

Figure 3–25 *Sketch the small half-bath*

45. Extend the vertical Wall in the middle of the stair hallway up to the half-bath.
    - Use the Split tool with the "Delete Inner Segment" option on the Options Bar to remove the small piece of Wall shown in the middle panel of Figure 3–26.
    - Use the Trim/Extend tool to complete the layout as shown in the right panel of Figure 3–26.

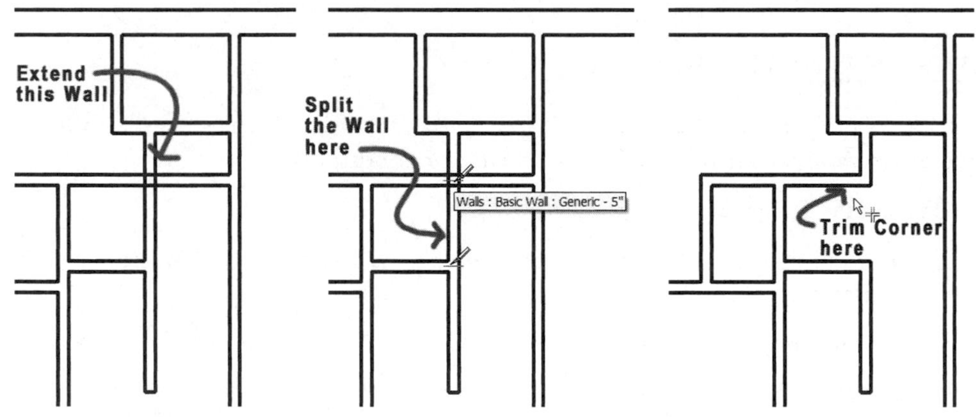

Figure 3–26 *Complete the Wall Layout using Trim/Extend and Split*

46. Save the Project.

WORKING WITH PHASING

We have completed the layout of existing walls on the first floor of the house. Assigning the Walls to a particular construction phase will help us distinguish them as existing construction as the project progresses. Autodesk Revit Building includes robust Phasing tools allowing us to quickly and easily designate these Walls as existing construction. Every element in a Revit Building model is assigned two phasing parameters: "Phase Created" and "Phase Demolished." In this way you can track the life of any element in the model. In the case of the Walls that we have drawn here, all of them are "existing construction" and none will be demolished. We will be demolishing some of the Windows and Doors later on in the chapter.

ASSIGN WALLS TO A CONSTRUCTION PHASE

1. Using a window selection (click and drag to surround all Walls), select all of the Walls in model.
2. On the Options Bar, click the Properties icon (or right-click and choose **Properties**).

   The Element Properties dialog will appear.

   - Beneath the Phasing grouping, for "Phase Created," choose **Existing**.
   - For the "Phase Demolished," verify that **None** is chosen (see Figure 3–27).

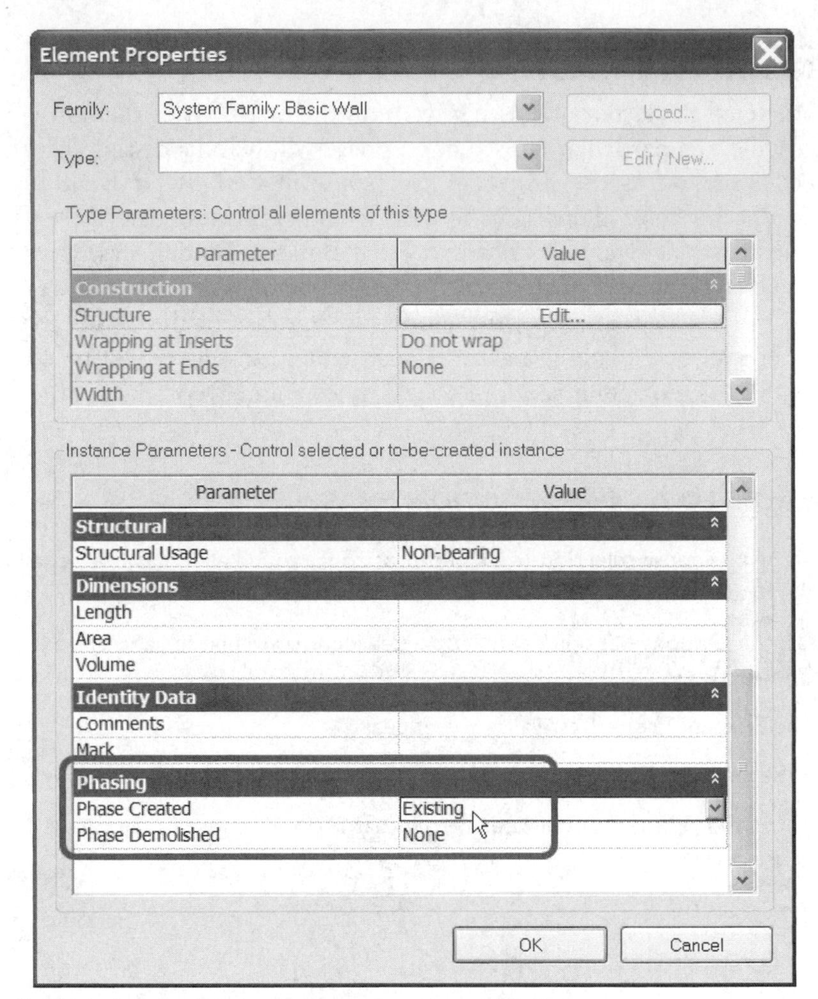

Figure 3–27 *Assign the Existing Phase to the selected Walls*

- Click OK to close the Element Properties dialog.
- Deselect all of the Walls to see the change.

You can deselect the Walls by simply clicking in the white space next to the Walls. You can also click the Modify tool on the Design Bar or press the ESC key twice.

The Walls will display in a lighter lineweight, and will also be colored gray. This indicates that they are existing construction. Assigning a phase to an element automatically assigns a graphic override to it. You can control the settings of these overrides as well as edit and add phases in the Phasing dialog. To open this dialog, choose **Phases** from the Settings menu. Feel free to explore this dialog now if you wish. A good example of a typical task that you could perform in this dialog would be to set up phases of construction in a large project. For example, you could create a "Foundations and Caissons" phase, "Phase 1" and "Phase 2 New Construction" phases. For our residential project, the out-of-the-box phases of "Existing" and "New Construction" are sufficient. Therefore, if you make any changes in the Phases dialog, please do not apply them. **Exploring the View Properties**

At this time, it is appropriate to explore Phasing a bit deeper to get a full understanding of the tools available. Many of the changes we are about to make are not required by our residential project, therefore we will undo the next few steps after we finish exploring.

1. On the Design Bar, click the Wall tool.
   - Accept all the default settings and draw a single Wall segment anywhere in the model, but touching one of the existing Walls.
   - On the Design Bar, click the Modify tool or press the ESC key twice.

Notice that the new Wall came in as New Construction. Notice also that it did join with the existing Wall.

2. Select any Existing (phase) Wall and then click the Properties icon (on the Options Bar).
   - From the Phase Demolished list, choose **New Construction** and then click OK.

Notice that the Wall is now displayed as dashed.

3. Select a different Existing (phase) Wall and then click the Properties icon.
   - From the Phase Demolished list, choose **Existing** and then click OK.

Notice that this time, the Wall has disappeared. In Revit Building, each View has its own display parameters with "Phase" and "Phase Filter" being two of them. In this case, we have the *Level 1* View open on screen. You can verify this on the title bar at the top of the Revit Building window and by noting that *Level 1* is bold in the Project Browser. Let's change the current Phase of this View to Existing to see how the display changes.

4. Make sure that there are no elements selected, right-click in the Workspace and then choose **View Properties**.

TIP The keyboard shortcut for View Properties is **VP**. **View Properties** is also located on the View menu.

The Element Properties dialog will appear.

Notice that the same "Element Properties" dialog appears. However, this time the "element" in question is the View named: "Level 1" rather than a selection of Wall elements. Regardless, to Revit Building both items are "Elements" that have editable properties.

- Beneath the Phasing grouping, for "Phase," choose **Existing**.
- Click OK to see the change.

This change makes the Existing construction phase the current phase for the *Level 1* View. Therefore, the View will now show only items that existed in that phase. This means that no elements assigned to the New Construction phase will show, and the Wall that we specified as demolished in New Construction will now appear solid again. Furthermore, by choosing the Existing phase as the active phase, any elements that we draw will automatically be added to this phase. We will work with this technique below when we add Doors and Windows.

In addition to the current Phase, we can also assign a "Phase Filter" in the View Properties dialog. A Phase Filter does not change the active phase, rather it changes which elements assigned to different phases display and how they appear in the current View. By default this is set to "Show All." Several Phase Filters are included in the default templates from which our project was created. Explore them now to get a sense of how they behave. Please note that the Phase Filter operates on the View's currently active phase. Therefore, choosing "Show Previous Phase," for example, will display only the elements that were created in the previous phase. Like Phases, Phase Filters can be edited in the Phasing dialog. To open this dialog, choose **Phases** from the Settings menu.

5. Return to the View Properties dialog and make the **Existing** Phase active.
   - Select any demolished Walls and return them to **Existing**.
   - Delete any extraneous Walls that you drew.
6. Return to the View Properties dialog and make the **New Construction** Phase active.
   - Select any demolished Walls and return them to **Existing**.
   - Delete any extraneous Walls that you drew.

It may be possible to click the undo icon several times to restore the drawing to the state before we began demolishing and adding extraneous Walls. When you have finished exploring the various View Properties and Phasing options, and restored your model, it should look like Figure 3–28.

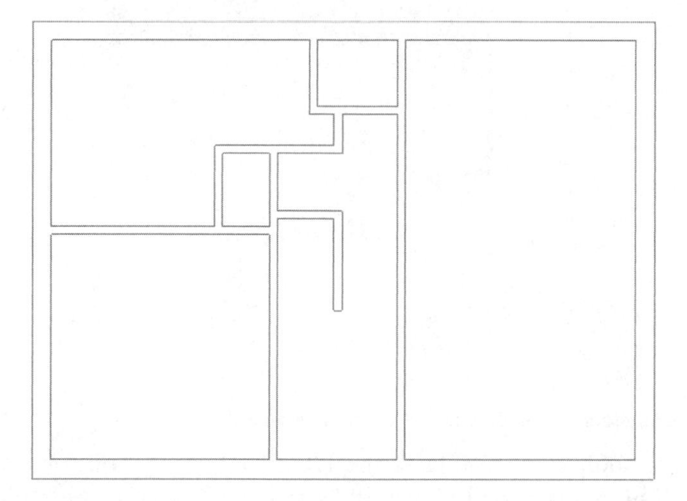

Figure 3–28 *The Residential Level 1 Wall layout*

3. Be sure that **New Construction** is the View's active Phase and then save the project.

WORKING WITH DOORS AND WINDOWS

Continuing with our layout of the residential existing conditions, let's add some Doors and Windows. Doors and Windows automatically interact with Walls when inserted to create the opening and attach themselves to the Wall in a parametric way. We will cover the basics of adding Doors and Windows in Revit Building here.

ADD A DOOR

Doors are Wall-hosted elements. This means that Doors must be placed in Walls. Like Walls, Doors have both Type and Instance parameters. Several Door Types have been included in the out-of-the-box template file from which our residential project was based.

1. Continuing in the *Level 1* View of the Residential project, zoom out to see the entire plan layout.

   You can do this with the mouse wheel or the tools on the **View > Zoom** menu.

2. On the Design Bar, click the Door tool (see Figure 3–29).

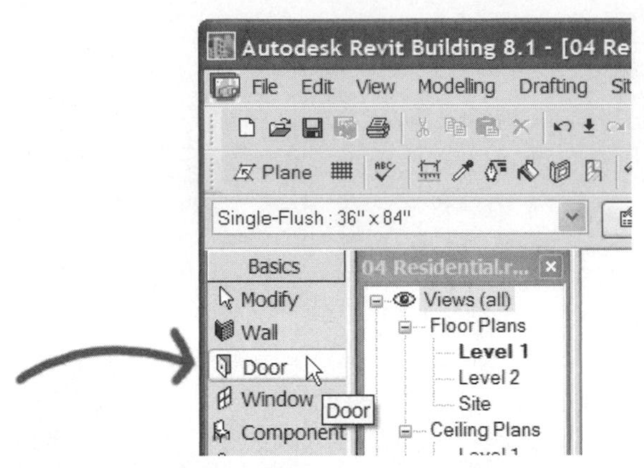

Figure 3–29 *The Door tool on the Basics tab of the Design Bar*

Much like the Wall tool, when you click the Door tool, several options appear on the Options Bar. The pointer for Doors will have a small (+) sign next to the cursor and the Status Bar prompt will read: "Click on a Wall to place Door."

- From the Type Selector, choose Single Flush: 36″ x 80″ [M_Single-Flush : 0915 x 2032mm].

The following list explains the major fields and controls shown on the Options Bar when you are adding a Door (see Figure 3–30).

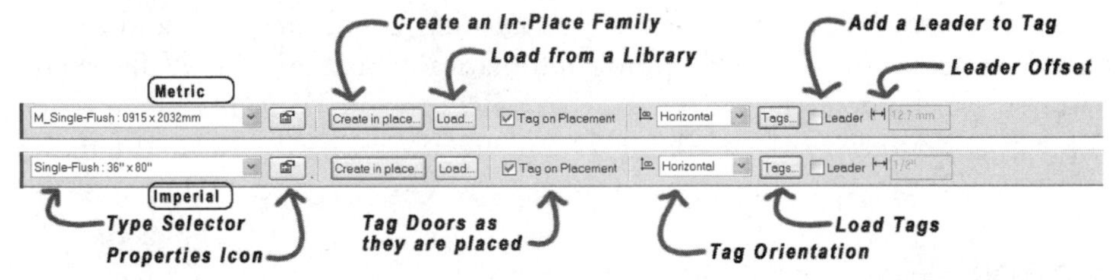

Figure 3–30 *The Options Bar for adding Doors*

- **Type Selector**—Use this drop-down to choose from a list of Door types in the current file. The specific list is populated by the door Families that are already loaded into the template we used to create the project. Always choose a type from this list before configuring other options. You can use the Load button (to the right of the Type Selector) to load additional types from an external library.
- **Properties icon**—Click this icon to open the Element Properties dialog. The Element Properties dialog gives you access to both "Type" and "Instance" parameters. Just like Walls, Type parameters apply to all Doors of the same type, Instance parameters apply only the selected Door(s).

- **Create in place**—Click to create a new in-place Door Family. In Place Families are used to create non-typical "one off" Families. In most cases, it would be rare to use In Place Families for Doors.

NOTE In-place Families are not designed to be moved, copied, rotated, etc. They are meant to be used only once. If you need to use it more than once within this project or in a different project a regular door family should be created in the Family Editor, saved to a library and then loaded into your project as needed. The Family Editor will be explored in Chapter 9.

- **Load**—Noted in the Type Selector above, use this button to access the external libraries and load Door Types not already available in the current project.
- **Tag on Placement**—Check this box to add a Door Tag as Doors are added to the project. Even if you deselect this option, you can still add Tags to Doors later.
- **Tag Orientation**—When "Tag on Placement" is selected, this option sets the orientation of the Tag as either horizontal or vertical.
- **Tags**—Opens the Tag dialog where you can see which Tags are currently loaded in your project and also choose to load additional Tags from a library file.
- **Leader**—Check this to add a Leader to your Tags as they are placed.
- **Leader Offset**—When you enable the Leader option, this field controls the offset of the Leader from the Tag.

As you move your mouse pointer around on screen, a Door will only appear when you move the pointer over a Wall. If you are unhappy with the direction of the door swing, press the SPACE BAR to flip it before you click to place the Door.

3. On the Options Bar, remove the checkmark from the "Tag on Placement" checkbox.
4. Move the mouse to the horizontal exterior Wall at the top right side of the plan.
   - Position the Door roughly in the center of the top exterior Wall of the room on the right and then click the mouse (see Figure 3–31).

   As with Walls, temporary dimensions will guide your placement.

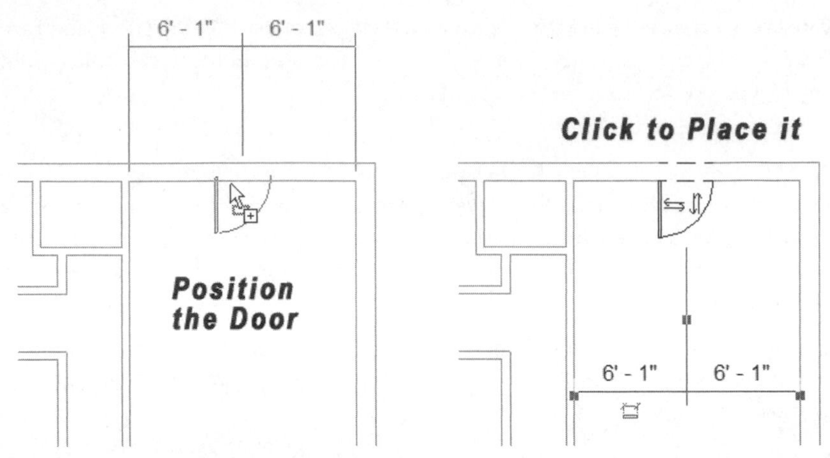

Figure 3–31 *Click to place the first Door*

Notice that the Door appears in the drawing and cuts a hole in the Wall. However, notice that the hole in the Wall is filled with dashed lines. This is because we set the Phase back to "New Construction" above. Therefore, Revit Building is showing this Door as being a new Door placed into an existing Wall. This requires the "hole" to be shown as demo while the Door appears as new.

5. On the Design Bar, click the Modify tool or press the ESC key twice.

CHANGE THE PHASE FILTER

Let's try repeating the Phase Filter exercise above to see the different ways that this condition will display in each phase.

6. Make sure that there are no elements selected, right-click in the Workspace and then choose **View Properties**.

TIP The keyboard shortcut for View Properties is **VP**. **View Properties** is also located on the View menu.

The Element Properties dialog will appear.

- Beneath the Phasing grouping, for "Phase Filter," choose **Show Previous + Demo**.
- Click OK to see the change (see the top of Figure 3–32).

7. Edit the View Properties again.

- Beneath the Phasing grouping, for "Phase Filter," choose **Show Complete**.

- Click OK to see the change (see the bottom of Figure 3–32).

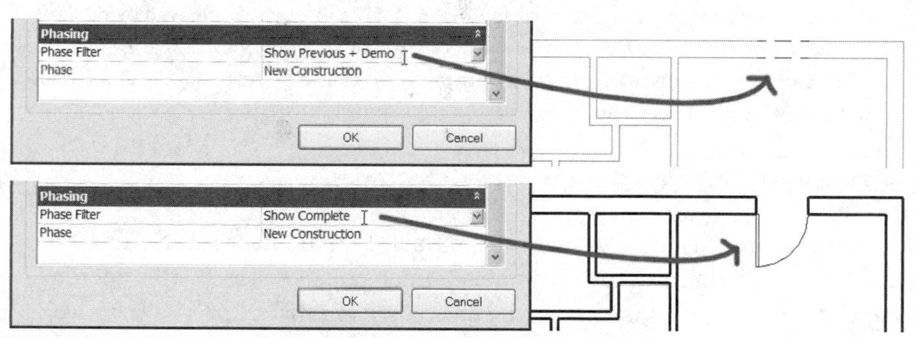

Figure 3–32 *Change the Phase Filter to view the model at different points in time*

The power and potential of the Phasing parameters was seen when we explored these options with just Walls. Now that we have added a Door, we can truly see the full potential of these tools. If this Door truly were a new Door being added to these existing Walls, all of these graphical behaviors would be managed for us automatically by Revit Building simply by assigning the Door to the New Construction Phase parameter as we have done here. It turns out that this Door is actually an existing Door. Therefore, we need to change its Phase parameter to make it display properly.

8. Select the Door in the model and then on the Options Bar, click the Properties icon (you can also right-click and choose **Properties**).

- Beneath the Phasing grouping, for "Phase Created," choose **Existing**.
- Click OK to dismiss the Element Properties dialog.

Since we still have the "Show Complete" Phase Filter active, there will be no apparent change in the model.

9. Return to View Properties dialog (type **VP**) and choose **Show All** for the Phase Filter.

The Door now displays the same as the Wall in which it is inserted and the dashed demolition lines no longer display. This is because the Door and Wall now belong to the same Phase, therefore there is no demo required. Since we are going to add several more existing Doors, let's change the View's active Phase to Existing to save us the trouble of having to edit the Doors (and Windows below) later.

10. Return to View Properties dialog (type **VP**) and choose **Existing** for the Phase.

The Walls will turn bolder to reflect this change.

NOTE A typical set of construction documents requires existing conditions, demolition, and new construction drawings. In Revit Building this is easily achieved by duplicating the Views (plans, sections, and or elevations) and editing the Views' Phase and Phase Filter parameters to display the correct data. Refer to the "Create an Existing Conditions View" heading in Chapter 6 for an example of this.

PLACE A DOOR WITH TEMPORARY DIMENSIONS

Let's add several more existing Doors to our model. For the next several Doors, it will be easier to place them if the temporary dimensions are set to the openings rather than the centers. (This is similar to the change we made for Walls above).

11. On the Settings menu, choose **Temporary Dimensions**.
    - In the Temporary Dimension Properties dialog, in the Doors and Windows area, choose Openings and then click OK.
12. On the Design Bar, click the Door tool.
    - From the Type Selector, choose Single Flush: 30″ x 80″ [M_Single-Flush : 0762 x 2032mm].
    - Verify that "Tag on Placement" is *not* selected.
    - Move the cursor to the upper left corner of the plan and position it so that the Door is being added to the topmost horizontal exterior Wall.
    - When the temporary dimension reads 2′-0″ [600] click the mouse.

    As before, the temporary dimensions will remain on screen until you cancel the command or place another Door.
13. Click the blue value of the temporary dimension on the left and type **2′-4″ [762]** and then press ENTER (see Figure 3–33).

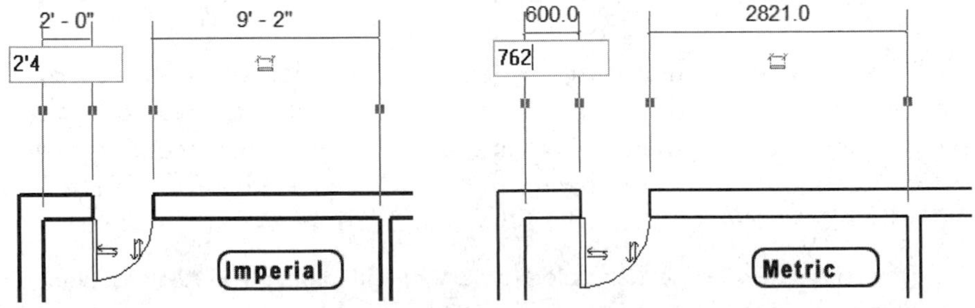

Figure 3–33 *Place the Door in the approximate location, then use the temporary dimensions to fine-tune placement*

The Door will shift the indicated amount.

LOAD A DOOR FAMILY

The next Door that we are going to add is a bi-fold door for the small closet in the middle of the plan. There is no bi-fold type available in the current project. Therefore, we will use the Load function to access the Revit Building library and load some bi-fold Door Types.

You should still be in the Door command. If you have canceled it, click the Door tool on the Design Bar.

As noted, if you open the Type Selector, there are only Single Flush types loaded.

14. On the Options Bar, click the Load button.

- In the Open dialog, click the *Imperial Library* [*Metric Library*] icon on the left side.
- On the right side, double-click the *Doors* folder, choose *Bifold-2 Panel.rfa* [*M_Bifold-2 Panel.rfa*] and then click Open.

There will be a pause while Revit Building loads the Family and its Door Types. If during this process a Save Reminder appears, click Dismiss for now.

- From the Type Selector, choose Bifold-2 Panel : 30″ x 80″ [M_Bifold-2 Panel : 0762 x 2032mm].
- Verify that "Tag on Placement" is *not* selected.

15. Move the pointer over the left vertical Wall of the closet in the middle of the plan.

- Using the temporary dimensions, get the Door centered on the vertical Wall.

Do NOT click the mouse to place the door yet.

- Slowly move the mouse left to the right.

Notice that you can control whether the Door swings into or out of the closet with the mouse, but not which side of the opening (up or down in this case) that it swings. Take note of the Status Bar and in some cases a tip that appears on screen. There it notes that you can use the SPACE BAR to flip the swing.

- Press the SPACE BAR to flip the swing.
- Press it again to flip back. (see Figure 3–34).

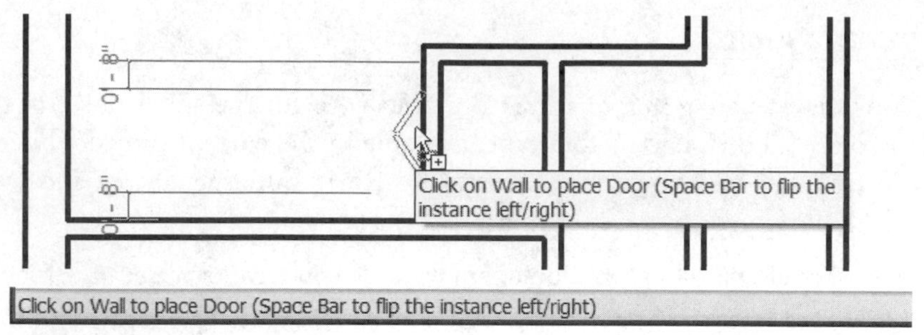
0' - 8"
0' - 8"
Click on Wall to place Door (Space Bar to flip the instance left/right)
Click on Wall to place Door (Space Bar to flip the instance left/right)

Figure 3–34 *Move the mouse to control Door placement, press the* **SPACE BAR** *to flip the Door*

- When the Door opens to the lower portion of the Wall, click the mouse

Don't worry if you added the Door with the wrong swing. We can easily edit this after the Door is placed.

16. On the Design Bar, click the Modify tool or press the ESC key twice.
17. Save you project.

FLIP A DOOR WITH FLIP ARROWS

To flip a Door after it is placed, use its handles.

18. Click on the hinged Door at the top right, (the first Door placed in the previous steps above).

Notice there are two sets of small arrow handles, one group horizontal and the other vertical. Use these handles to flip the Door swing.

19. Click either one of the Flip handles.

    It will pre-highlight and a tip will appear to indicate its function (see Figure 3–35).

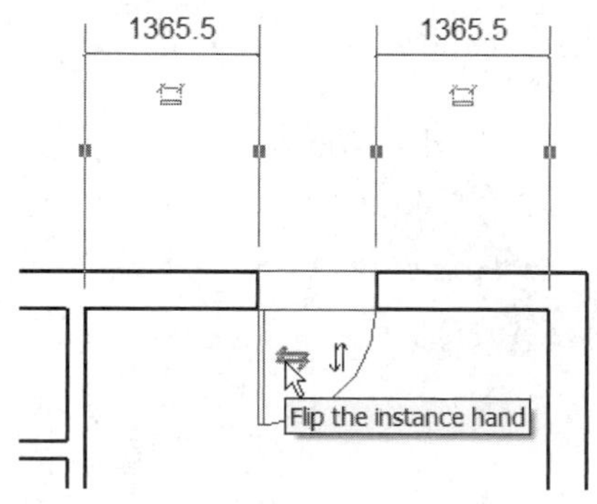
1365.5
1365.5
Flip the instance hand

Figure 3–35 *Flip the instance hand of a Door*

Repeat this process on other Doors if you wish.

CHANGE THE DOOR SIZE

Door size is governed by a Door's Type. If you wish to change the size of Door in your model, simply choose a different type. If you wish to use a size that is not on the list, you must edit the Door Family and create a new type with a different size. This process will be covered below.

20. Select the same Door again—the one in the top right of the plan.
    - On the Options Bar, from the Type Selector, choose a new size, such as Single-Flush : 34" x 80" [M_Single-Flush : 0864 x 2032mm].
21. Click next to the drawing to deselect the Door (or press ESC).

You can edit other Instance or Type parameters for the selected element (Door in this case) by clicking the Properties icon on the Options Bar. For example, we used this technique above to set the Phase Created parameter of our first Door.

EDIT DOOR PLACEMENT WITH DIMENSIONS

We have seen several examples so far of the use of temporary dimensions to control the placement of elements in the model. We can also create permanent dimensions which give us even greater flexibility and control over element placement.

NOTE The following examples would be more appropriate in a New Construction model since it is unlikely that we would need to maintain constraints in an existing layout. These tools are covered here merely for the educational value, not as a recommendation of their usage for existing conditions plans.

22. Select the Door in the top left side of the plan.

    Two temporary dimensions will appear; one on either side—take notice of the small blue handle that looks like a dimension itself (see Figure 3–36).

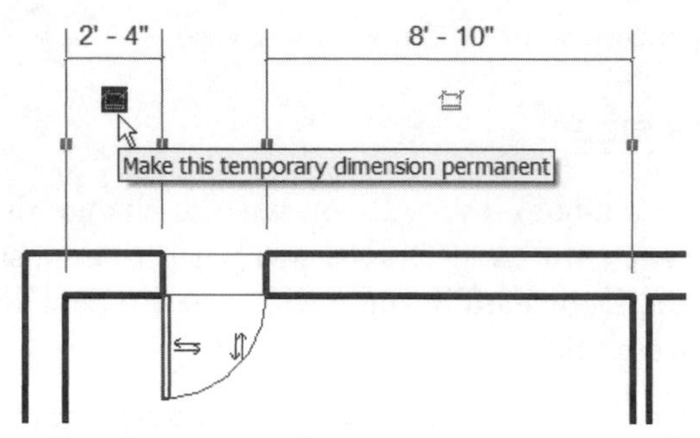

Figure 3–36 *You can convert a temporary dimension to permanent with the handle*

23. On the left side dimension, click this handle to make the dimension permanent.

When you do this, a permanent dimension will appear. It will be colored black. There are four basic colors used in Autodesk Revit Building to indicate various states of selection and/or if the element is editable. Here is a summary of those colors and what they signify:

- **Black**—The element is not selected
- **Gray**—The element is pre-highlighted (caused by the mouse passing over it, or the TAB key being used to cycle to the element).
- **Red**—The element is selected.
- **Blue**—The element is editable (for example, temporary dimensions and handles).

NOTE Selection color can be customized. On the Settings menu, choose **Options**. On the Graphics tab of the Options dialog, change the colors to suit your preference.

When you make the temporary dimension permanent, the new dimension appears in black. The new permanent dimension is a new element in the current View (*Level 1*). Notice that the Door remains selected, and that the temporary dimension also remains but is directly underneath the new permanent dimension. To access the options of the new permanent dimension, we must select it instead of the Door.

NOTE Dimensions are annotation elements. Therefore, they are view-specific—meaning that they appear only in the View in which they are added. Walls and Doors on the other hand are model elements and appear in all views in which they would normally be seen. However, although the display of dimensions are view specific their constraints are not. If the dimension is locked the constraint is enforced throughout the model.

24. Click on the new permanent dimension.

    The Door will deselect (turning black) and the permanent dimension will select (turning red) instead.

    ◗ Click the small padlock icon (see the left side of Figure 3–37).

    The padlock icon will "close" to indicate that the dimension is now constrained.

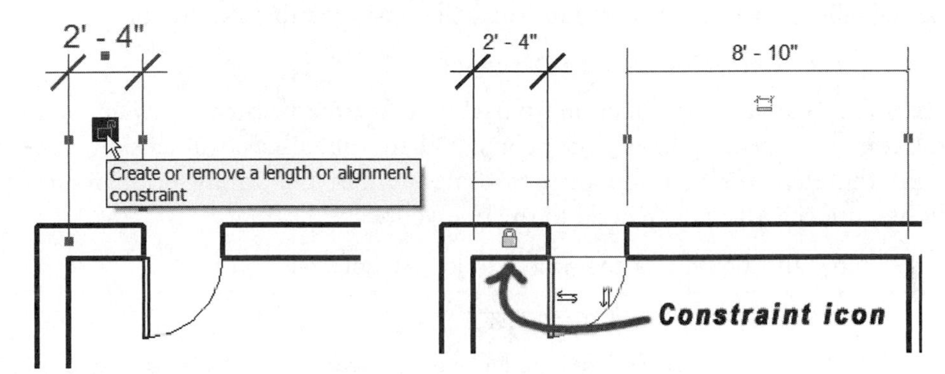

Figure 3–37 *Apply a constraint to the Door position using the padlock on the permanent dimension*

25. Click on the Door to select it.

    The permanent dimension will deselect (turning black) and the Door will select (turning red) instead.

Notice the padlock icon that displays beneath the dimension (see the right side of Figure 3–37). If you click the blue dimension value and attempt to edit it, an error message will display. The only way that you could edit the value would be to first remove the constraint. You can do this by clicking the closed padlock icon. Let's test our new constraint with a quick experiment. We will undo it when we are finished.

26. Click on the left vertical exterior Wall.

    ◗ Move the pointer over the selected Wall (the pointer changes to a small move cursor), click and drag the Wall to the left.

    The exact amount is not important.

Notice that the Wall moved, and the Door moved and maintained its dimension as well.

◗ Undo the change. (Click the Undo icon or press CTRL + Z).

27. Try the same experiment with the right vertical exterior Wall. Move it to the right.

Notice that the Door on that side did not move. It has no constraints applied to it.

- Undo the change.

APPLYING AN EQUALITY CONSTRAINT

Let's apply another kind of constraint to a different Door. The Door that we have on the right should be centered in the room. We can add a permanent dimension with an equality constraint to maintain this relationship automatically for us.

28. On the Design Bar, click the Dimension tool.

We are going to place the dimension ourselves this time because we want more control over the specific dimension points. When you place a dimension, pre-highlight the elements that you wish to dimension. If the wrong element pre-highlights, use the TAB key to cycle to the one you want.

29. Move the mouse over the right exterior vertical Wall.

    The Wall centerline will pre-highlight.

    - Press TAB and repeat until the inner edge of the Wall pre-highlights and then click.
    - Click the center of the Door next.

    The center of the Door should pre-highlight, when it does, click the mouse. Otherwise, use the TAB to cycle to the center first.

    - Set the final point at the inside face of the vertical Wall on the other side of the room—remember the TAB (see Figure 3–38).

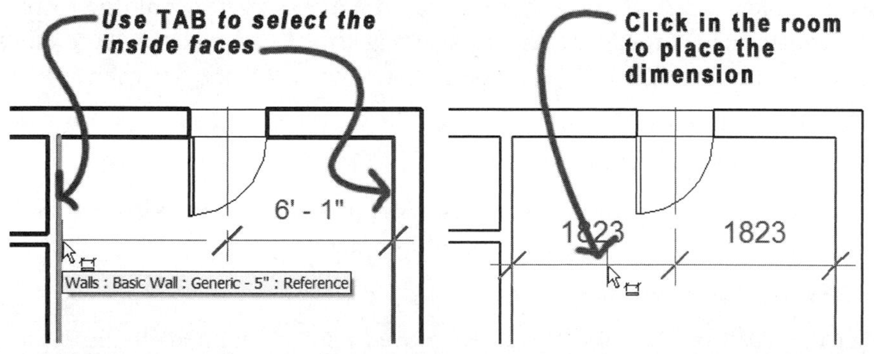

Figure 3–38 *Dimension from the faces of the Walls to the center of the Door using* **TAB** *to cycle to the correct points*

30. To place the dimension, click anywhere in the blank white space of the room.
31. Press ESC twice to exit the dimension command.

32. Click on the new dimension to select it.

Notice the small "EQ" icon with a line though it (see the left side of Figure 3–39). This indicates that the permanent dimension in *not* set to maintain an equal distance. If you click this icon, you enable the equal constraint and force the elements attached to the dimension to remain equally spaced.

- Click the EQ icon to enable the equal constraint.

You will need to click off of the dimension to deselect it in order for the dimension text to properly display. Select the dimension again to see that the Dimension Equality toggle icon no longer has the slash through it (see the right side of Figure 3–39).

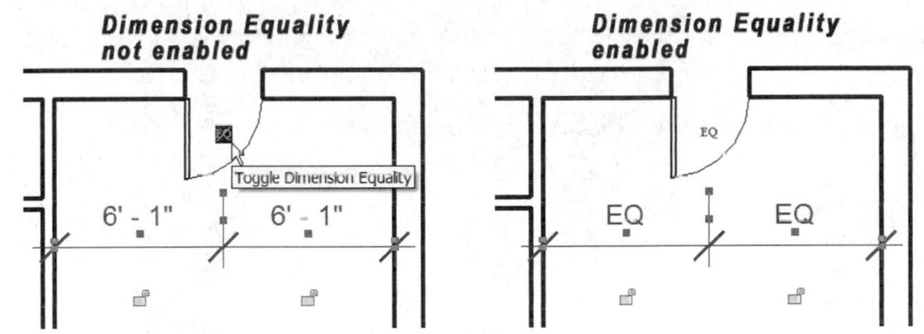

Figure 3–39 *Set the dimension to an equal constraint*

33. Repeat the experiment above, and move the exterior Wall.

Notice that the Door remains constrained to the center of the room.

- Undo the Wall move.

34. Save the Project.

TIP Use CTRL + S to save quickly.

ADD A NEW DOOR SIZE

The Door to the small half-bath at the top of the plan is 2′-4″ [710]. This size is not available in our current list of Door Types. We could try loading the required size from an external library like we did above, but in this case it is just as easy to create a new size. To do this, we need to choose a type that is close to the one that we want and then duplicate and edit it.

35. On the Design Bar, click the Door tool.

- From the Type Selector, choose Single Flush: 30″ x 80″ [M_Single-Flush : 0762 x 2032mm].
- Click the Properties icon.

At the top of the Element Properties dialog, you can see that the Door Family is Single Flush [M_Single-Flush] and the Type is 30″ x 80″ [0762 x 2032mm]. The Family list includes all of the Families loaded in the current project. This currently includes only the single flush and bi-fold Families. Next to the Family list, is the Load button. You can use this button to load additional Families from external libraries as we did above. The Type list includes all of the Types available for the selected Family. If you scan that list, you will not find the size we need which is 28″ x 80″ [0710 x 2032mm]. Next to the Type list is the Edit/New button. You use this button to edit the currently selected Type or to create a new Type by duplicating the current one. Let's duplicate the current Type and modify the size.

36. Next to the Type list, click the Edit/New button.

TIP A shortcut to this is to press ALT + E.

The Type Properties dialog will appear.

At this point, we could edit the Type parameters of the selected Door Type. The change would affect all Doors of this Type in the entire project. It is pretty important to understand this distinction between "Type" and "Instance" parameters. Rather than edit the existing Type, which would give us the size we need but would also reduce the size of the Doors that he have already inserted, let's copy this Type and edit the new copy.

- Next to the Type list, click the Duplicate button.

TIP A shortcut to this is to press ALT + D.

A new Name dialog will appear suggesting the name: Single Flush: 30″ x 80″ (2) [M_Single-Flush : 0762 x 2032mm (2)].

It is a good idea to change this name. Let's follow the convention established with the out-of-the-box Families and name this Type after its size.

- Change the name to Single Flush: 28″ x 80″ [M_Single-Flush : 0710 x 2032mm] and then click OK (see Figure 3–40).

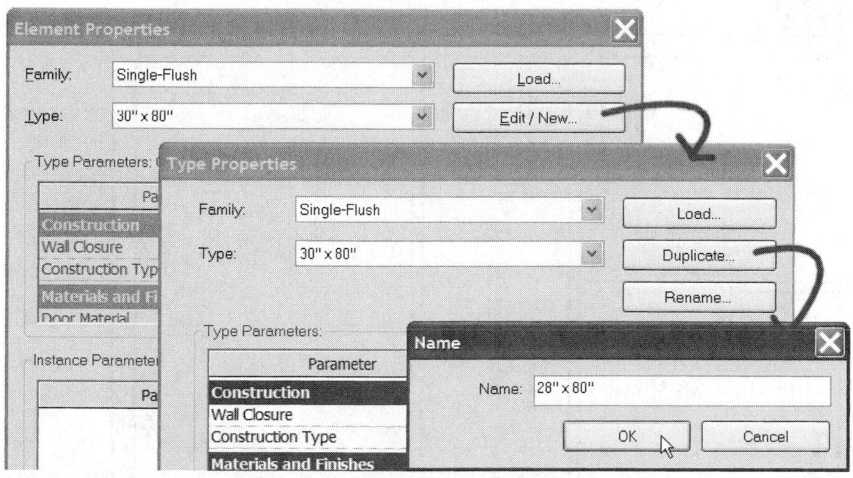

Figure 3–40 *Create a new Type by Duplicating an existing one*

Now that we have created and named a new Type, let's edit it to the size we need.

37. Beneath the Dimensions grouping, for "Width," type **2′-4″** **[762]** and then click OK two times.

We are now ready to add our new Door Type.

38. On the Options Bar, Verify that "Tag on Placement" is *not* selected.
    - Add a Door to the small half-bath at the top of the plan.
    - Use the SPACE BAR to swing the door in to the right (see Figure 3–41).

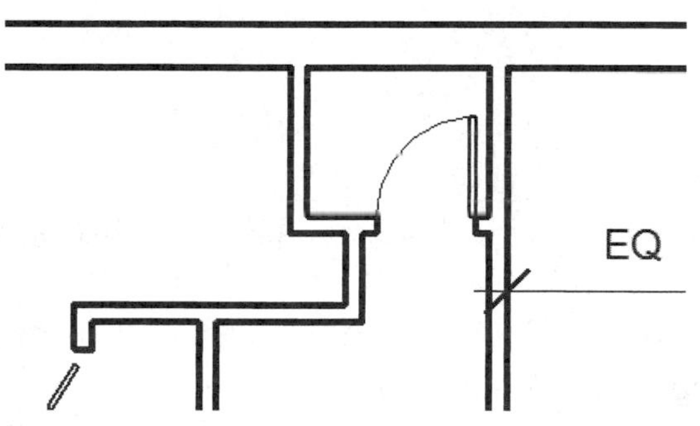

Figure 3–41 *Add a Door in the new Type to the half-bath*

ADD THE REMAINING DOORS

39. Using Figure 3–42 as a guide and the techniques covered above, place the remaining doors in the plan.

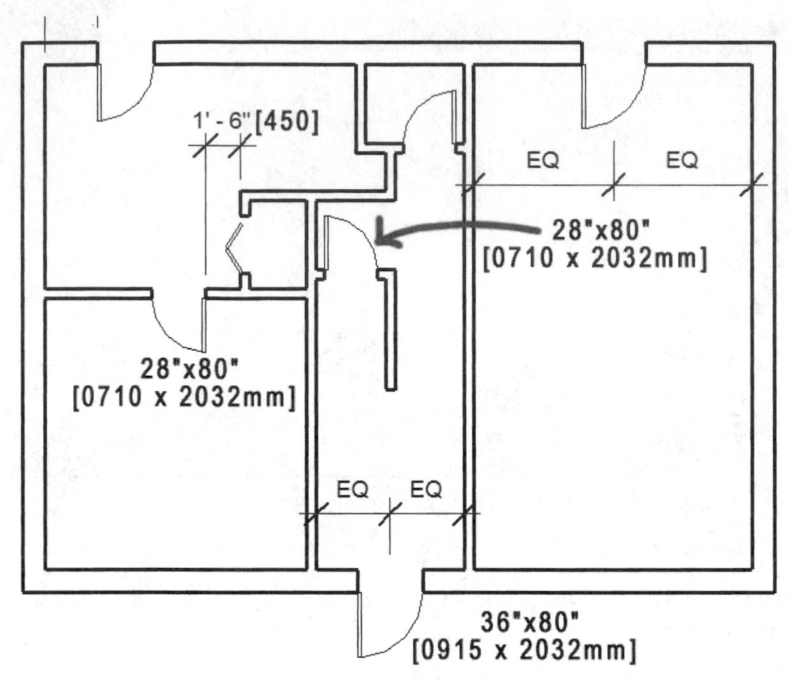

Figure 3–42 *Add the remaining Doors as shown*

Hint: You can select a Door, right-click, and choose **Create Similar** as a way of quickly adding Doors with similar parameters.

40. Save the project.

ADD WINDOWS

Working with Windows is nearly identical to working with Doors. Many of the parameters are the same and placement and manipulation of Windows works the same as with Doors.

41. On the Design Bar, click the Window tool.

All of the options on the Options Bar for Windows match those of Doors. For detailed descriptions of each of these options, see Figure 3–30 above and the accompanying descriptions.

The only Window Family loaded in the current project is the Fixed Family. For this model we need double hung. Let's see what we have in the library.

42. On the Options Bar, click the Load button.

- In the Open dialog, click the *Imperial Library* [*Metric Library*] icon on the left side.
- On the right side, double-click the *Windows* folder, choose *Double Hung.rfa* [*M_Double Hung.rfa*] and then click Open.

The size that we need is not included on the list. So as we did above for the Door, we will create a new type.

- From the Type Selector, choose Double Hung : 36″ x 48″ [M_Double Hung : 0915 x 1220mm].

43. Click the Properties icon.
    - Next to the Type list, click the Edit/New button (or press ALT + E).
    - Next to the Type list, click the Duplicate button (or press ALT + D).
    - Change the name to Double Hung : 36″ x 56″ [M_Double Hung : 0915 x 1422mm] and then click OK.
44. Beneath the Dimensions grouping, for "Height," type **4′-8″** **[1422]** and then click OK two times.
45. Using Figure 3–43 as a guide and the techniques covered above, place the Windows indicated.
    - On the Options Bar, remove the checkmark from the "Tag on Placement" checkbox.
    - Use the new Type that you just created for all of these Windows.

Note that the Window in the lower left corner is placed in at the midpoint of the Wall. This can be done easily by using clicking the control on the temporary dimension to make them permanent dimensions and then clicking the EQ constraint.

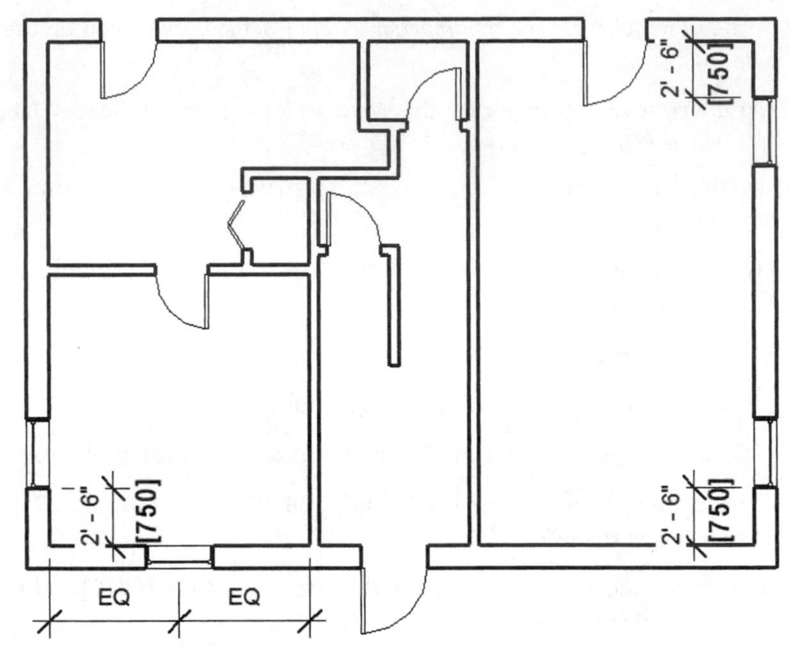

Figure 3–43 *Add several Windows to the plan*

46. Using the process covered here, create two new Window Types:
    - Double Hung : 36″ x 32″ [M_Double Hung : 0915 x 0800mm]
    - Double Hung : 24″ x 32″ [M_Double Hung : 0600 x 0800mm]

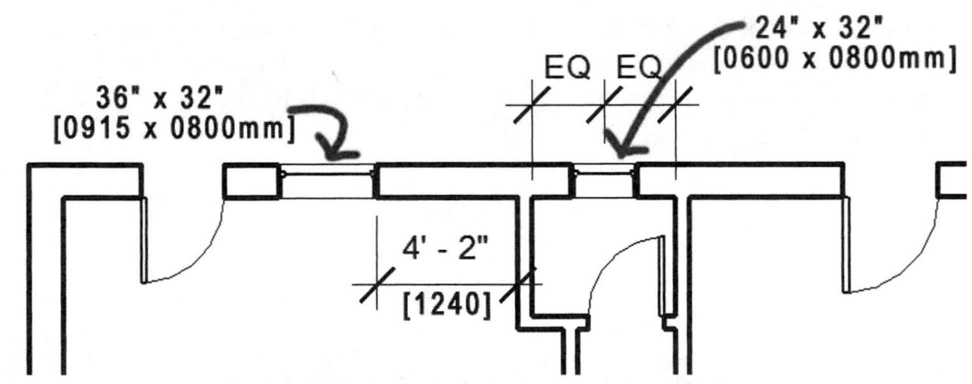

Figure 3–44 *Place the remaining two Windows by creating custom Types*

47. Place the remaining two windows as shown in Figure 3–44.
48. Save the project.

ADD CASED OPENINGS

There are several passages between the hallway and the neighboring rooms that are simply cased openings. In Revit Building you place these the same way as Doors—in fact they are made from Door Families, but they are classified as Components. The element includes the casing and a hole in the Wall, but no Door panel. When the Family for one of these is created, it is not classified as a Door so that it will not appear in Door Schedules.

49. On the Design Bar, click the Basics tab and then click the Component tool.

Components are used to represent many things. In this case, the Cased Opening components that we need are not currently loaded in the model.

- On the Design Bar, click the Load button.
- In the Open dialog, click the *Imperial Library* [*Metric Library*] icon on the left side.
- On the right side, double-click the *Doors* folder, choose *Opening-Cased.rfa* [*M_Opening-Cased.rfa*] and then click Open.

NOTE Remember, the Cased Opening Families are stored in the *Doors* folder, and are essentially Door elements, but they have been classified as Components Instead because typically you would not want cased openings to appear on Door Schedules.

- From the Type Selector, choose Cased Opening : 30″ x 80″ [M_Opening-Cased : 0762 x 2032mm].
- Click on right-most interior vertical Wall (see Figure 3–45)

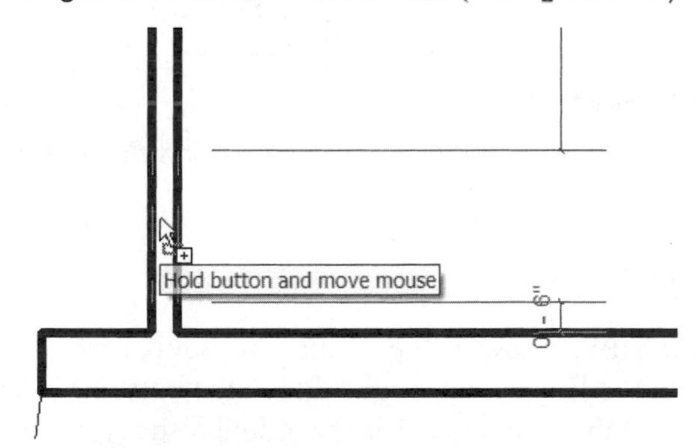

Figure 3–45 *Place a Cased Opening Component in the Wall*

Watch the Status Bar or the tip that appears on screen. You are prompted to select a Wall.

- Using the temporary dimension place it **6″ [150]** from the intersection with the bottom Wall.

TIP You can use temporary dimensions to assist you after placement if you have trouble getting it in the correct spot.

50. Using Figure 3–46 as a guide, place three more Openings of the same size.

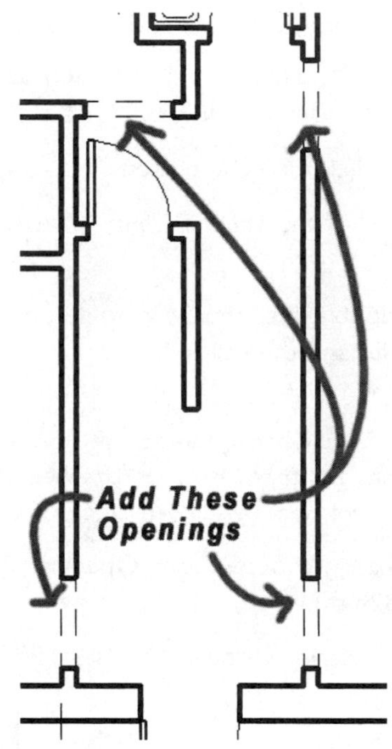

Figure 3–46 *Add Openings to the hallway*

After placement, you can click on a Cased Opening and move it or edit it just like the Doors.

LOAD A CUSTOM FAMILY

There is one additional Window in the existing house that we need to add. In the front of the house in the living room (right side at the bottom) is a picture window comprised of a fixed Window flanked by two double-hung units. We could add this as three Windows simply enough. To save time, this configuration of three Windows has been pre-assembled and saved as a Revit Building Family file and included with the Chapter 3 files from the Mastering Autodesk Revit Building CD ROM. In this sequence, we will load this Family into our current project and add it to the front of the existing house.

51. On the File menu, choose **Load from Library > Load Family**.

   - In the "Open" dialog, choose *Existing Living Room Front Window.rfa* [*Existing Living Room Front Window-Metric.rfa*] and then click Open.

NOTE If necessary, in the "Open" dialog, browse to the *Chapter03* folder to locate the Family file.

The Family will load into the current project. This step simply loaded the Family into the current project. Now that it is loaded, we can use the Window tool to add a Window using this Family to the model. In previous examples, we used the "Load" button on the Options Bar as we were adding the Window to load the Window and place it in a single operation. The same process could be used here. The approach shown is merely an alternative.

52. On the Design Bar, click the Window tool.

The Type Selector should default to the newly loaded Group.

   - From the Type Selector, choose Existing Living Room Front Window [Existing Living Room Front Window-Metric].
   - Add the Window to the bottom horizontal Wall in the room at the right (see Figure 3–47).

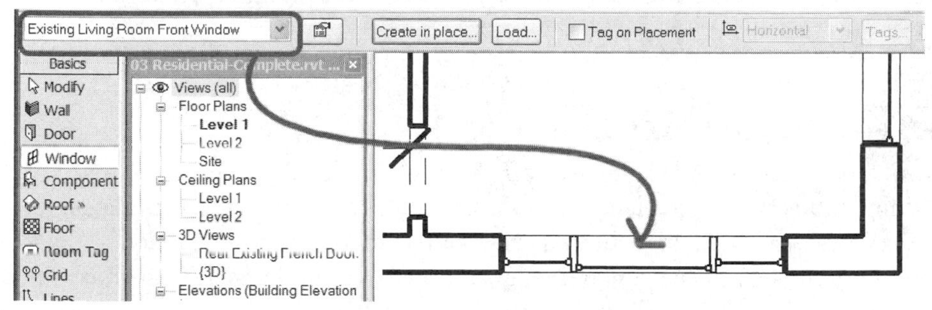

Figure 3–47 *Create an Instance of the imported Family at the front of the house*

This Family is a very simple one that contains three Window elements. This can be thought of as a "nested" Family because the Family file contains reference to other Families. We will explore Families in more detail in Chapter 9.

53. Save the project.

MODIFYING DOOR SIZE AND TYPE

Modifying Doors and Windows is easy. We have already covered most of the basic techniques above. It turns out that the Door in the top right of the plan (the first one we added) is supposed to be a double door. Furthermore, it is actually a French door leading out to an existing patio. Let's make this edit to our model.

54. Select the Single Hinged door in the top right corner of the plan (the first one we added).
    - On the Options Bar, click the Properties icon.
    - Next to the Family list, click the Load button.
    - In the Open dialog, click the *Imperial Library* [*Metric Library*] icon on the left side.
    - On the right side, double-click the *Doors* folder, choose *Double-Glass 2.rfa* [*M_Double-Glass 2.rfa*] and then click Open.
    - From the Type list, choose 68" x 80" [1730 x 2032mm].
    - Click OK.

The new Door Type has replaced the previous one and remains nicely centered since we added that constraint to this Door earlier. In plan View, it is not obvious what affect choosing a glass Door Type had. To see the glass, we will have to view the model in 3D.

NOTE You can use this technique for Windows and other elements as well.

VIEWING THE MODEL IN 3D

We have spent the entire chapter working in a plan View. However, as we have been adding objects in this plan View we are also building a three-dimensional model that can be viewed and edited in elevation, section, and/or 3D views. Revit Building maintains all of the parameters of our building model including those required to view it in 3D, even if we don't explicitly ask it to

OPEN THE DEFAULT THREE-DIMENSIONAL VIEW

Let's have a look at how the model is shaping up in the third dimension.

1. On the View toolbar, click the Default 3D View icon.

TIP If you prefer, you can choose **New > Default 3D View** from the View menu instead.

A new View will appear on screen named: *3D Views : {3D}*. This name is assigned automatically to the default 3D View in Revit Building. The default 3D View is an axonometric of the complete building model looking from the south-east. This View appears in the Project Browser beneath the 3D Views category. From the default View, we can see the Group Window that we added above in the front of the house. Even though the graphics in the Floor Plan View indicated that the middle Window was different than the two flanking Windows, from the 3D vantage point it is now very clear that the middle Window is fixed while the two flanking ones are double-hung.

You can dynamically scroll and or zoom any View as we have already seen. In a 3D View, you can also "spin" the View. When you do this, you can change the angle and height from which we are viewing the model and virtually spin it in all three dimensions.

2. On the View toolbar, click the Dynamically Modify View icon (see Figure 3–48).

TIP The keyboard shortcut for Dynamically Modify View is the F8 key. **Dynamically Modify View** is also located on the View menu.

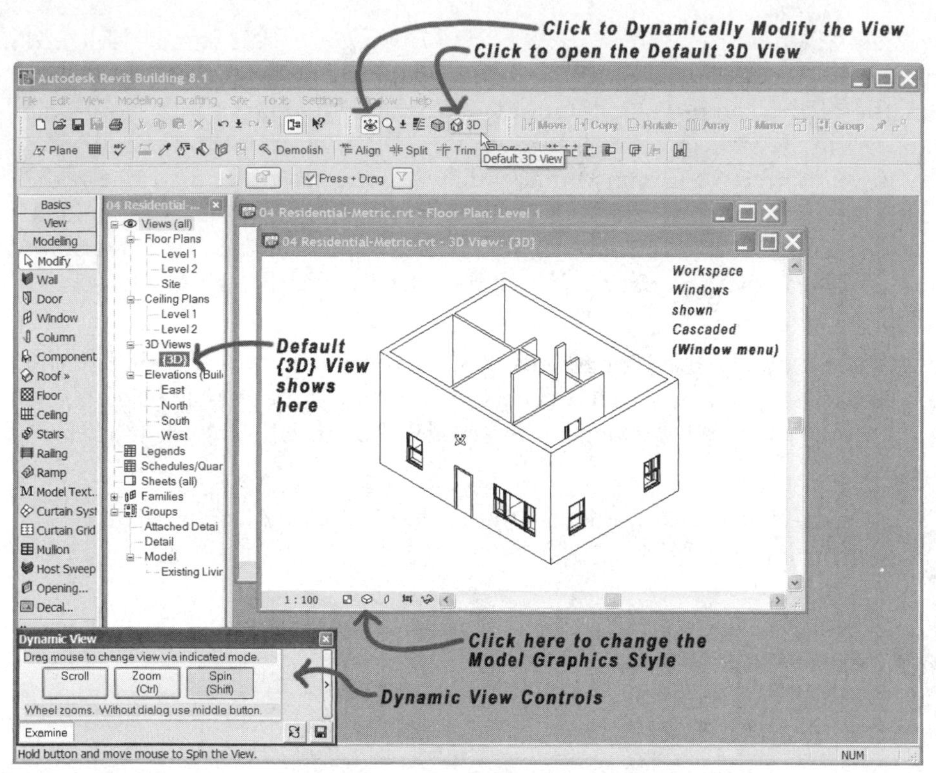

Figure 3–48 *Display the Default 3D View and manipulate it dynamically*

Note the Dynamic View toolbox that appears in the lower left corner of the screen. If you have used this command before and moved the Dynamic View toolbox to another location on screen, Revit Building will restore it to this location instead. In this View, there are three basic controls: Scroll, Zoom and Spin. By now you should already be comfortable with scrolling and zooming. Using the tools in the Dynamic View toolbox behave no differently than they did elsewhere. Spin is only available in a 3D View. In other words, you can open the Dynamic View toolbox from any View such as the Floor Plan View Level 1 that we have worked in throughout the chapter. When you do, the Spin button will be disabled. Let's give Spin a try in our 3D View.

- In the Dynamic View toolbox dialog, click the Spin button.

The pointer will change to a spin icon to indicate that this mode is active.

- Click and drag in the 3D View window to spin the model.
- Drag side to side to rotate the model.
- Drag up or down to tilt the vantage point up or down.
- Spin the model around so that you can see the back of the house.

- Use the Scroll and Zoom controls to fine-tune the View to your liking (see the left side of Figure 3–49).

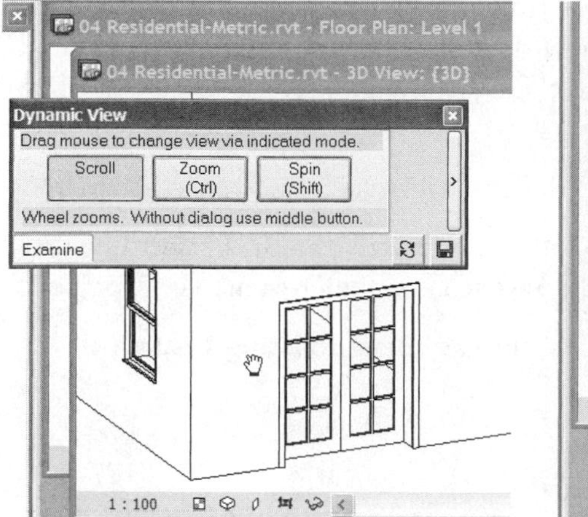

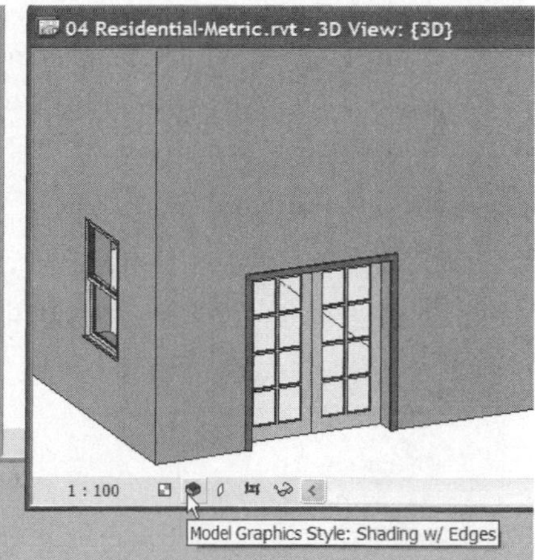

Figure 3–49 *Use the "Dynamic View" toolbox to scroll, zoom or spin the model*

TIP The dynamic viewing functions can also be performed with the CTRL and SHIFT keys as indicated on the buttons in the Dynamic View toolbox. Keep the Scroll button active, and then use the CTRL key to zoom and the SHIFT key to spin. Furthermore, if you have a wheel mouse, you can scroll, zoom and spin anytime using the wheel without having to use the Dynamic View dialog. Drag with the wheel pushed in to scroll, roll the wheel to zoom and hold down the SHIFT key and drag with the wheel to spin. Your specific mouse and mouse driver will determine the exact behavior. Mouse drivers by many manufacturers allow customization of specific button functions.

- On the View Control Bar (bottom left corner of each View window) click the Model Graphic Style icon and choose **Shading w/Edges** from the pop-up menu that appears (see the right side of Figure 3–49).

TIP The keyboard shortcut for Shading with Edges is **SD**. **Shading w/Edges** is also located on the View menu.

The default three dimensional View: *{3D}* is always available. You could certainly use this View exclusively and simply spin, scroll and zoom the View as needed each time you displayed it. However, once you get a 3D View displaying just the way you like, to convey a piece of information about the project in a certain way, you are encouraged to save the View. The new View will be added to the Project Browser beneath the 3D Views category along with any other 3D views your project has. There are two methods for saving the View: you can right-click on *{3D}* beneath 3D Views in the Project Browser and choose **Rename**, or you can reopen the Dynamic View toolbox and click the small Save icon. Either method achieves the same result: a new View will appear in the Project Browser and the name on the title bar will change to reflect this. If you then click the Default 3D View icon on the View toolbar again, a new default *{3D}* View will be created. If you click the 3D View icon and the default *{3D}* View already exists it will open this existing View.

3. Save your modified 3D View.
   - Either right-click the name *{3D}* beneath 3D Views on the Project Browser and choose **Rename**, or use the Save icon on the Dynamic View toolbox.
   - Give the new View a unique name such as "**Rear Existing French Door**".

EDIT IN ANY VIEW

From our explorations in 3D, you may have noticed that many of the interior Walls are the wrong height. We can correct this easily.

4. Close your "Rear Existing French Door" View and then click the Default 3D View icon to create a new *{3D}* View.
   - From the Window menu, choose Tile

TIP The keyboard shortcut for Tile is **WT**.

   - Spin the View down slightly so that you can clearly see the heights of the interior partitions.
   - From the View menu, choose **Zoom > Zoom All To Fit**.

TIP The keyboard shortcut for Zoom All To Fit is **ZA**.

5. Activate the View: *Floor Plan:Level 1* by clicking any where in the View or the View's title bar and then use the Modify tool to create a selection window, click within the exterior walls and drag from right to left to select all interior Walls.

Be sure to select only the interior Walls and not any of the exterior ones. Do not worry about selecting Doors and Openings. We will remove them from the selection next.

All of the selected elements will highlight in red in both views (see Figure 3–50).

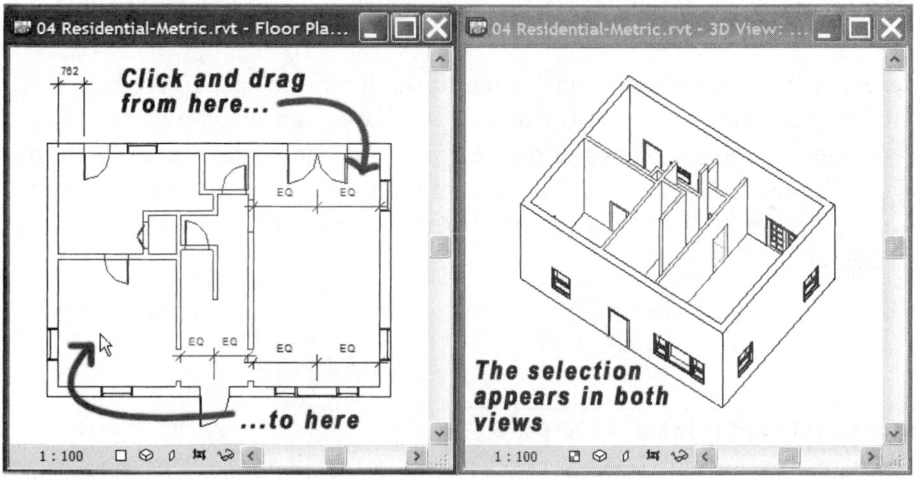

Figure 3–50 *Select all interior Walls*

6. On the Options Bar, click the Filter Selection icon.

   - In the Filter Selection dialog, remove the checkmarks from all boxes *except* "Walls" and then click OK.

   Now only the Walls will be highlighted.

7. On the Options Bar, click the Properties icon.

   The Element Properties dialog will appear.

   - Beneath the Constraints grouping, for "Top Constraint," choose **Up to level: Level 2** and then click OK (see Figure 3–51).

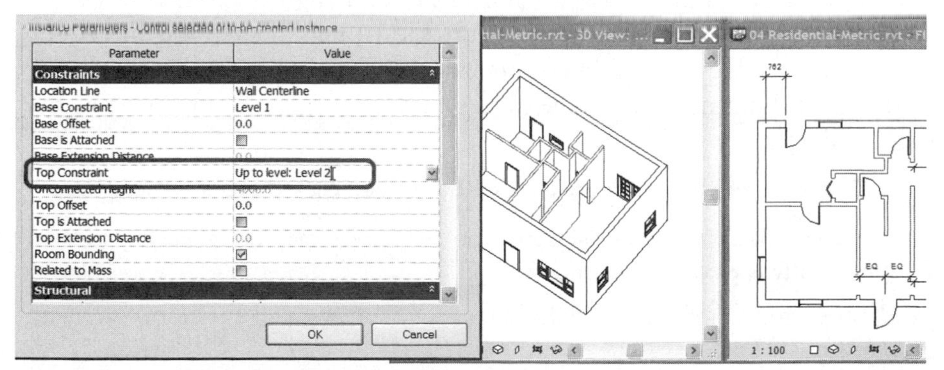

Figure 3–51 *Set the Top Constraint of the interior Walls to Level 2*

Note that even though we made the change for a 2D Floor Plan View, the edit appeared in the 3D View immediately. In Autodesk Revit Building every View is *al-*

ways up to date. There is no need to coordinate or refresh anything. Regardless of the View you are working in, all edits apply directly to the model. You simply choose the most convenient and logical View in which to work and Revit Building takes care of the rest.

In this exercise, we have constrained the top of the interior Walls to the height of the second floor. It turns out that the default template from which we started our project included two levels. However the height of those levels does not match the existing conditions of our residential project. In a future chapter, we will adjust the height of the levels and all of these interior partitions will automatically adjust with the level height.

8. Close the {*3D*} View, maximize the *Floor Plan:Level 1* View and then Save your project.

ADDING PLUMBING FIXTURES

The small half-bath in the top of our plan could use some fixtures. The library of components provided with Revit Building includes a variety of items such as furniture, toilets, trees, parking lot layouts, equipment, electrical fixture symbols, targets, tags, and much more. All of these items are Revit Building Families and many have special behaviors and parameters appropriate to the object that they represent. In this exercise we will simply load the Families we need and insert them in the model. In later chapters we will learn how to manipulate and create our own Families.

ADD COMPONENTS TO THE MODEL

You add Components like plumbing fixtures and furniture in nearly the same way as Walls, Doors and Windows. Each Family may vary slightly depending on the parameters built into it. However, the basic process is the same: choose the Component tool from the Design Bar, choose an existing Type from the Type Selector, or click the Load button to access the library. Then load the Component in the model and make your placements.

1. On the Design Bar, click the Basics tab and then click the Component tool.

   If you open the Type Selector, you will note that no plumbing fixtures currently appear on the list.

As we did above for Doors and Windows, we will simply load the items we need from the library.

2. On the Options Bar, click the Load button.

- In the Open dialog, click the *Imperial Library* [*Metric Library*] icon on the left side.
- On the right side, double-click the *Plumbing Fixtures* folder.

You will note that two Family files exist for domestic toilets, one for 2D and the other for 3D. The reason for this is that in many cases, you will not need to see plumbing fixtures in 3D. By using a 2D symbol in those cases, you can reduce demand on computer resources in the project file. In reality, in a model the size of the one that we are building here, there would be no noticeable difference using either Family. However, let's begin developing good BIM habits right away and choose the 2D Family for this exercise. Like all Families, you can swap out the 2D Component later with a 3D one should project needs dictate.

NOTE The name of the 3D Family here is actually misleading. If you were to choose this Family, you would notice that a 2D symbol would appear in all floor plan views and the 3D would only appear in 3D views. The Family file is composed of both 2D and 3D representations of the toilet and Revit Building knows automatically when to display which one.

NOTE When you create a Building Information Model, it is important to realize that "Model" does not always mean "3D." Nor does "Model" imply that every facet and screw of an item should be painstakingly represented graphically. Always remember to strike a balance between what is conveyed graphically and what is conveyed by other means such as with attached parameters—this is the "Information" part of BIM. In many cases, an "Information" model is much more practical and useful to a project team than a "3D" model. There are many types of models. Statisticians and economists refer to their spreadsheets as "models." As Architects, we tend to only think "3D" when the word model is mentioned. Learning to embrace BIM means understanding that a model is not always 3D, and that the "I" is just as important as, and sometimes more important than the "M."

- Choose *Toilet-Domestic-2D.rfa* [*M_Toilet-Domestic-2D.rfa*] and then click Open.

3. Zoom in on the half-bath at the top of the plan.

- Move the pointer around the four Walls of the half-bath.

Notice how the Toilet automatically orients itself to the Walls (see Figure 3–52). This is because the 2D toilet family was made in a wall-hosted Family template file. It is possible to create a toilet Family that is not hosted by a wall, if it is needed.

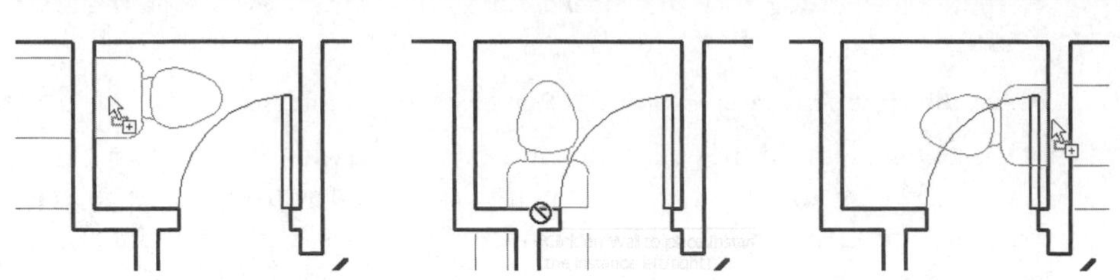

Figure 3–52 *The Toilet Family is wall-hosted, so it automatically attaches to the Walls*

- Click a point on the left vertical Wall to place the toilet in the model.

This toilet room is clearly not up to code! Just place the toilet very close to the intersection with the exterior Wall. We still need room for a sink.

4. On the Options Bar, click the Load button again.

- Return to the same location and Open the *Sink-Single-2D.rfa* [*M_Sink-Single-2D.rfa*] Family this time.

Move the pointer around again and notice that this Component behaves the same way. However, it will be too big for the space that we have available. Sometimes what we find in the field does not meet current building codes. The sink in this existing half-bath is *very* small.

5. On the Options Bar, click the Properties icon.

- Next to the Type list, click the Edit/New button (or press ALT + E).
- Next to the Type list, click the Duplicate button (or press ALT + D).
- Change the name to 18″ x 15″ [450 x 375mm] and then click OK.

6. Beneath the Dimensions grouping, for "Depth," type **1′-3″** [**375**]

TIP Remember, you can simply type **1 3** with a space between the numbers and Revit Building will interpret this as 1′-3″.

- Beneath the Dimensions grouping, for "Width," type **1′-6″** [**450**] and then click OK two times.
- Place the sink next to the toilet. Use the temporary dimensions to fine-tune the placement of both the sink and the toilet.

It will be a tight fit. You will need to set the values of the temporary dimensions to about **1″** [25] between the Walls and the Components to get everything to fit (see Figure 3–53).

TIP You can also use the Nudge tool to move the toilet and or sink. Select the element to nudge and then use the arrow keys to move it. The element will move approximately 2mm on screen. The more you are zoomed into the View the smaller the distance each press of the arrow key will move the object.

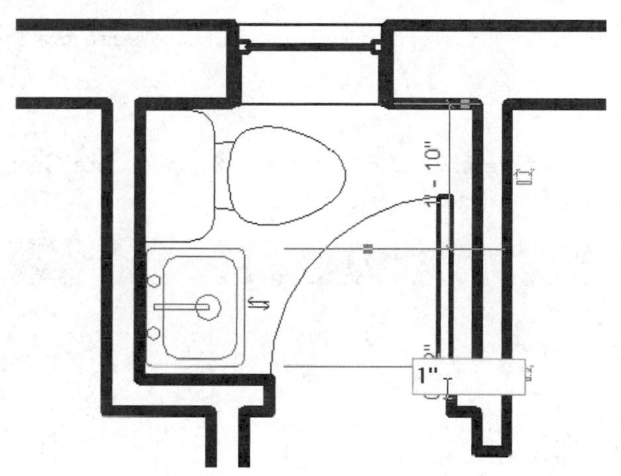

Figure 3–53 *Use the Temporary Dimensions to fine-tune placement of the Components*

7. On the Design Bar, click the Modify tool or press the ESC key twice.
8. Save the project.

CREATE AN IN-PLACE FAMILY

The first floor existing conditions plan is nearly finished. We still need to add the fireplace in the living room. While it would be possible to create a Fireplace Family in our library for use in any project, this would only make sense if we used the same fireplace design often. In this case, we will create the exact fireplace we need for this project directly in-place. In Revit Building, this is called an In-Place Family.

NOTE In-Place Families are not designed to be moved, copied, rotated, etc. They are meant to be used only once. If you need to use it more than once within this project or in a different project a regular door family should be created in the Family Editor, saved to a library and then loaded into your project as needed. The Family Editor will be explored in Chapter 9.

CREATE THE FAMILY AND CHOOSE A CATEGORY

To get started, we will create a new In-Place Family and assign it to a pre-defined category.

1. Zoom in on the middle of the right vertical exterior Wall.

   This is where our fireplace will go.

2. On the Design Bar, click the Modeling tab and then click the Create tool.

TIP Create is short for "Create In-Place Family." **Create** is also located on the Modeling menu.

- In the "Family Category and Parameters" dialog, choose Generic Models and then click OK (see Figure 3–54).

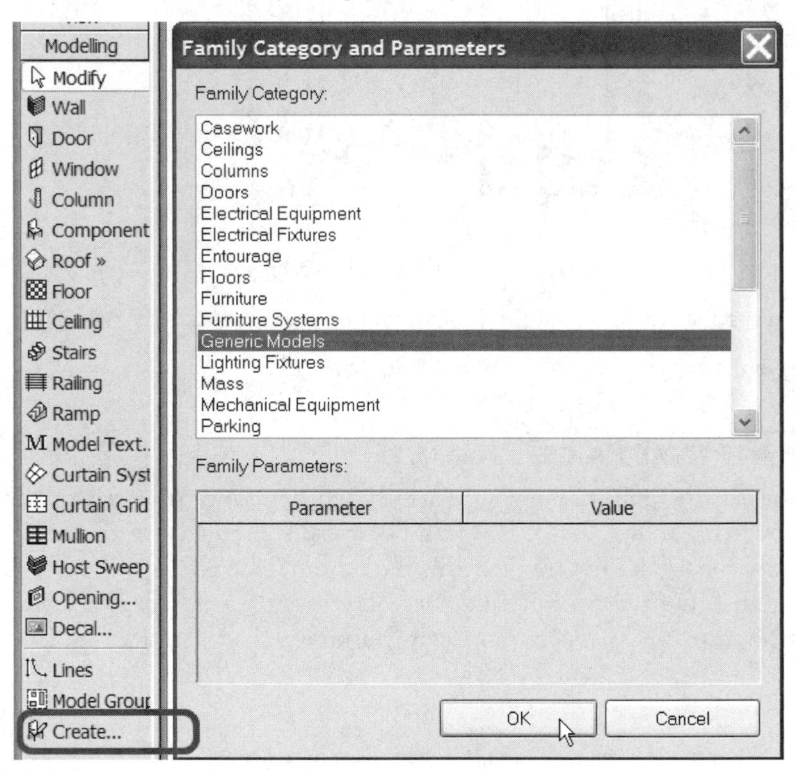

Figure 3–54 *Create an In-Place Family and choose its category*

The Family Category list is a fixed list built into the software. When you create a Family, you must assign it one of these categories. The Family you create will gain the characteristics of the category to which assign it. We chose "Generic Models" here because our fireplace does not fit neatly into any of the other categories. This is sort of a "catch all" category. We will not benefit from many specialized parameters that might be available from others, but our existing fireplace has few specialized needs.

- In the Name dialog, type: **Fireplace** and then click OK.

You are now in "In-Place Family Editing" mode. The Design Bar will change to a collection of In-Place Family Editing tools and all of the other tabs will disappear (see Figure 3–55).

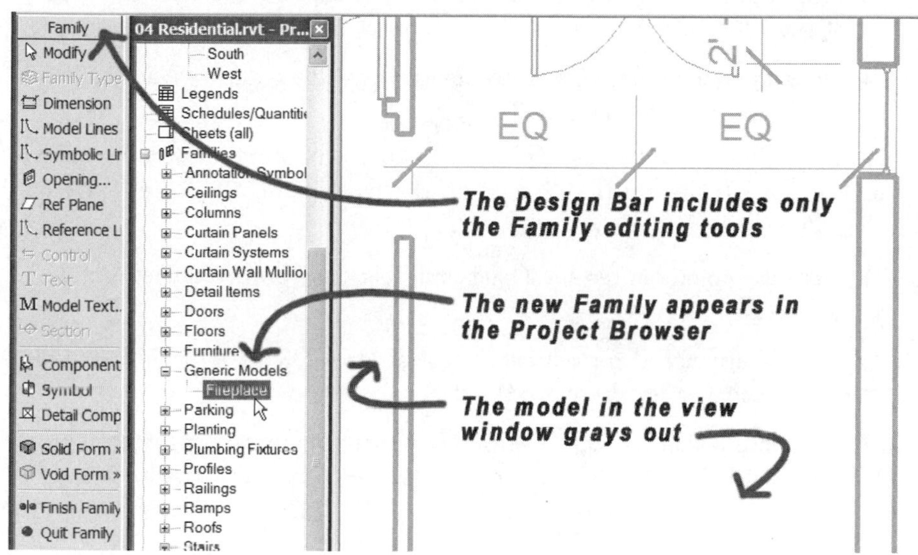

Figure 3–55 *The Family Editor mode is enabled when you create a new Family*

Many common tools can be found on the Design Bar while in Family editing mode. Two tools that we have not seen yet are Solid Form and Void Form. We use these tools to sculpt the form of our Family. You will also note that you can insert Components, Symbols, Dimensions and Detail Components into Families. Some of these items have been explored above, others we will explore in later chapters.

ADDING REFERENCE PLANES

When you construct geometry in Autodesk Revit Building, it is sometimes useful to have guidelines to assist in locating elements. Reference Planes are used for this purpose in Revit Building. You sketch a Reference Plane similar to the way you sketch Walls. They will then appear in any View. You can snap and constrain other elements to Reference Planes, making them useful tools for design layout. You can add Reference Planes anywhere in a Revit Building model, in this example, we will add them within our In-Place Family. When you add them in this way, the Reference Planes will become part of the In-Place Family and will be visible only when editing the In-Place Family.

3. On the Design Bar, click the Ref Plane tool.

4. Move the pointer over the inside corner at the intersection of the right and bottom exterior Walls.

   A small snap marker will appear.

   - Move the pointer up along the Wall until the temporary dimension reads about **7′-0″ [2300]** and then click (see panel "a" in Figure 3–56).
   - Move the pointer horizontally to the right past the exterior Wall and then click (see panel "b" in Figure 3–56).

A small Reference Plane (green dashed line with round blue handles at the ends) will appear. On the left, we started directly on the face of the Wall. Let's stretch the Reference Plane a little longer on that side.

- Place the pointer in the small blue circle, click and drag to the left (inside the house).

The exact distance is not critical. Drag it at least 2′-0″ [600] from the inside face (see panel "c" in Figure 3–56).

- Click in the blue text of the vertical temporary dimension at the bottom, type **7′-11″ [2400]**, and then press ENTER (see panel "d" in Figure 3–56).

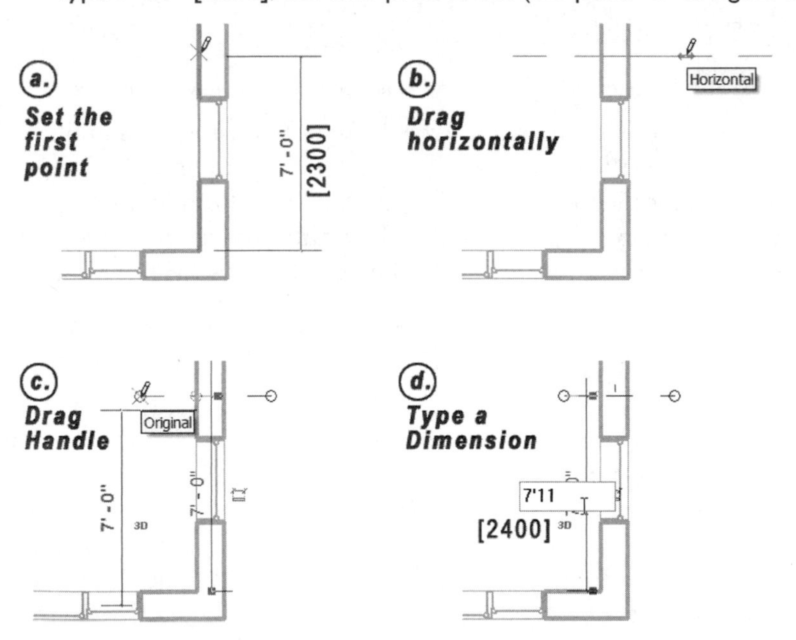

Figure 3–56 *Create a Reference Plane relative to the exterior Wall*

Repeat the process to create three more Reference Planes—one more horizontal and two vertical to frame out the rectangular footprint of the fireplace. When you create the next Reference Plane, Revit Building will automatically snap the end-points to line up with the first one.

5. Line up the end points of the next horizontal Reference Plane with the end points of the first one.

   ◗ Set the vertical temporary dimension at the top to **7′-11″ [2400]** and then press ENTER (see Figure 3–57).

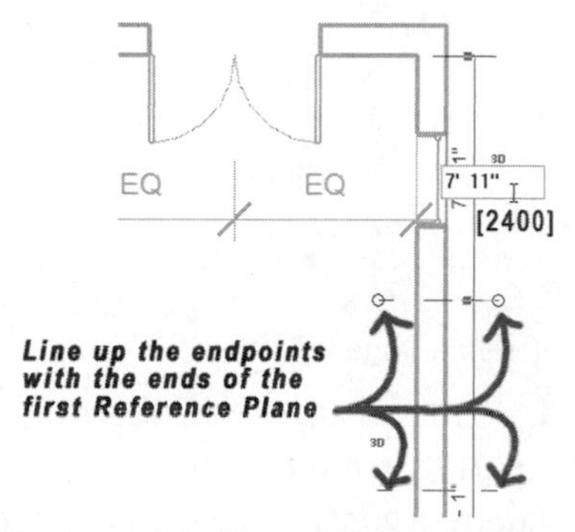

Figure 3–57 *Create a second horizontal Reference Plane aligned with the first*

6. Beginning approximately 6″ [150] to the left of the vertical exterior Wall, draw a Reference Plane and snap it to the top horizontal one.

   ◗ Using Figure 3–58 as a guide, complete the Reference Plane and edit the temporary dimension to **4″ [100]**.

   Refer to Figure 3–59 below for completed Reference Plane layout.

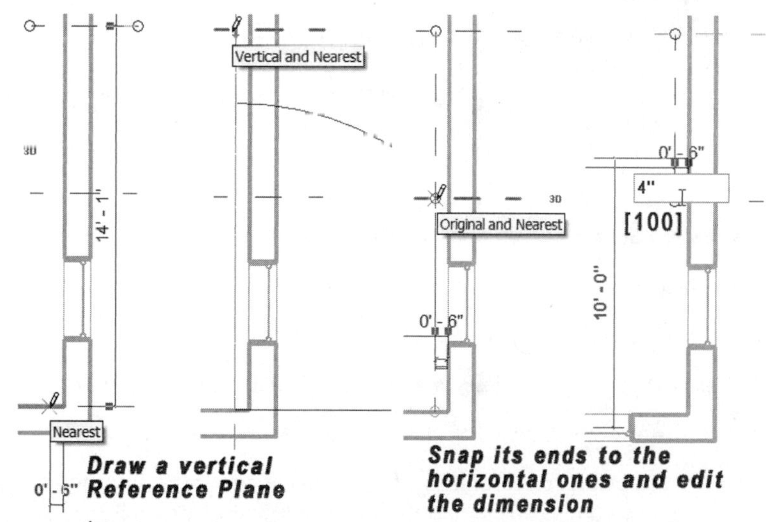

Figure 3–58 *Create vertical Reference Planes relative to the horizontal ones*

7. Create the last vertical one **1′-3″ [380]** to the outside (see Figure 3–59).

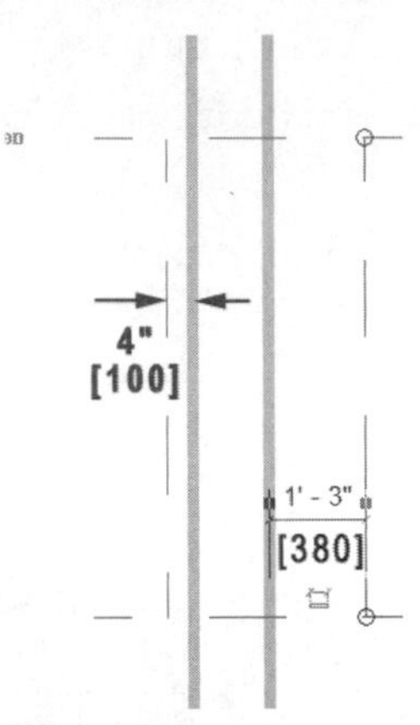

Figure 3–59 *The completed Reference Plane layout*

We now have four Reference Planes that we can use to guide the creation of our fireplace's form. It is not required that you use Reference Planes for this. However, they do make it easier particularly when you begin creating more advanced Families with constraints.

CREATE A SOLID FORM

Using our Reference Planes as a guide, let's create the overall mass of the fireplace.

8. On the Design Bar, click the Solid Form tool, and then from the flyout menu that appears, choose **Solid Extrusion** (see Figure 3–60).

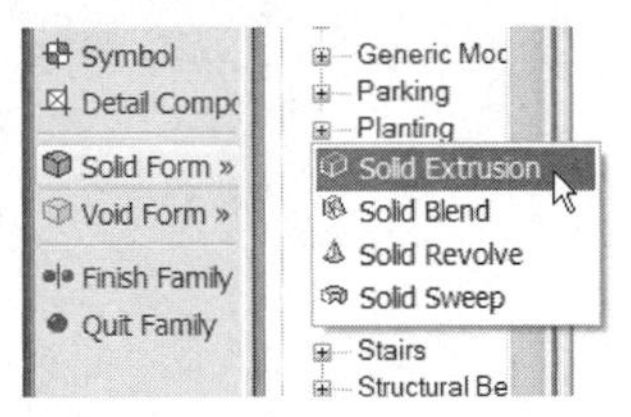

Figure 3–60 *Create a Solid Extrusion from the Design Bar*

This places the Design Bar in Sketch mode. The tools change accordingly.

9. On the Options Bar, in the "Depth" field, type **9′-0″ [2750]**.
   - Verify that the Draw icon is active. If it is not, push it in now.
   - For the sketch shape, click the Rectangle icon.
10. Snap to the intersection of two of the Reference Planes and then snap to an opposite intersection to define the rectangular shape (see Figure 3–61).

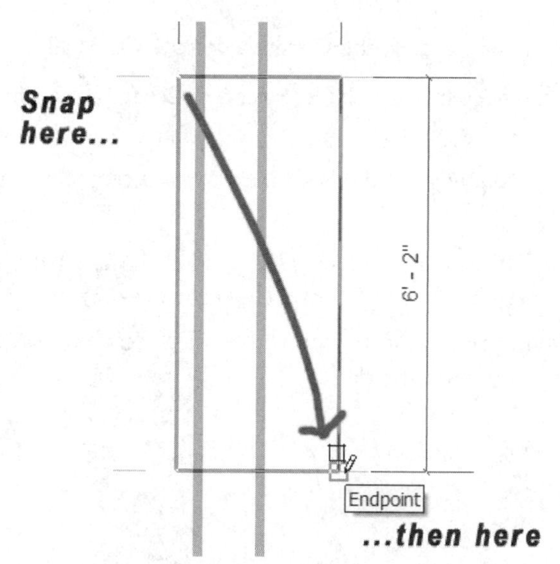

Figure 3–61 *Creating the shape of the fireplace*

Several open padlock icons will appear on each side of the shape. If we wanted to constrain the sides to the Reference Planes, we could click these locks. There is no need for deliberate constraints here so we will not do that for this exercise. Revit Building does interpret the location of this solid as your design intent and internally uses what are called "implied constraints." You do not need to do anything to set implied constraints. It is simply part of the parametric nature of Revit Building. If you add a constraint it will override and remove any implied constraints that may conflict.

11. On the Design Bar, click the Finish Sketch button.

This gives us our basic fireplace mass. We now need to carve out the firebox.

CREATE A VOID FORM

Using the same basic process, we can create a Void form that will carve away from the solid form in our Family giving us the firebox opening.

12. On the Design Bar, click the Void Form tool, and then from the flyout menu that appears, choose **Void Extrusion**.

    This places the Design Bar in Sketch mode again. The tools change accordingly.

13. On the Options Bar, in the "Depth" field, type **4'-0"** **[1200]**.

    - Click the Pick Lines icon.

14. Click the left vertical edge of the Solid Extrusion.

    A magenta sketch line will appear along this edge.

15. On the Options Bar, change the Offset value to **1′-5″ [430]**
    - Click the top edge of the Solid Extrusion to create a magenta sketch line below it.
    - Click the bottom edge of the Solid Extrusion to create a magenta sketch line above it.
    - Change the Offset value to **1′-6″ [350]** and click the right edge to create a magenta sketch line to the left of it (see Figure 3–62).

The side of the offset will pre-highlight before you click so that you can be sure to offset the sketch line to the correct side.

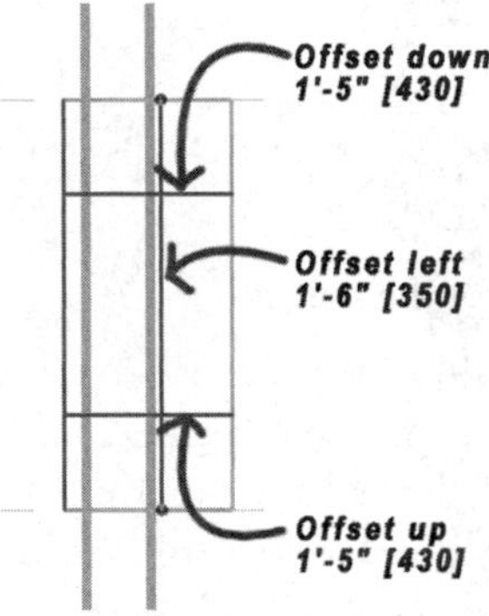

Figure 3–62 *Offset sketch lines to form the firebox shape*

Now we will use the Trim/Extend tool (the same one we used for Walls at the start of the chapter) to cleanup the sketch.

16. On the Tools toolbar, click the Trim/Extend tool (or Type **TR**).
    - On the Options Bar, click the first mode (Trim/Extend to Corner).
    - Click the vertical sketch line on the left toward the middle.
    - Then trim it to the horizontal one at the top.

NOTE Remember, select the portion that you wish to keep.

    - Repeat for the bottom sketch line (see the left side of Figure 3–63).
17. On the Design Bar, click the Modify tool or press the ESC to finish trimming.
18. Click the lower horizontal sketch line and drag the handle on the right up till the temporary dimension reads 20° (see the middle of Figure 3–63).

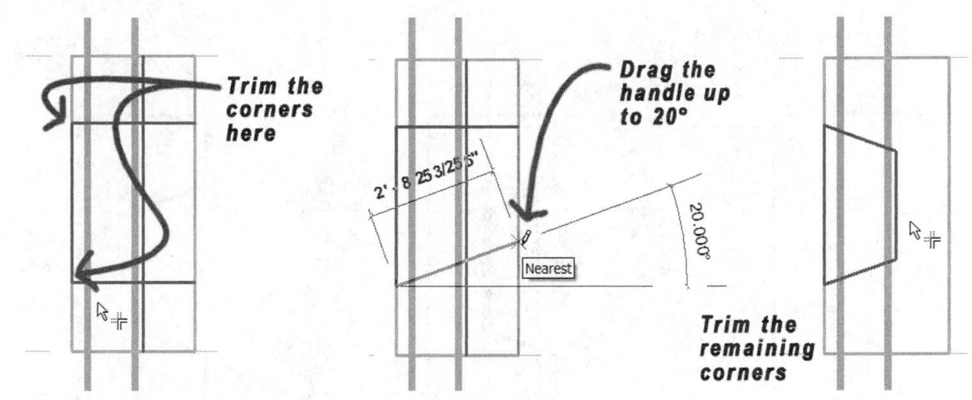

Figure 3–63 *Edit the offset sketch lines to finalize the shape*

- Repeat by stretching the top line down 20°
- Use Trim/Extend once more to cleanup the remaining corners (see the right side of Figure 3–63).

19. On the Design Bar, click the Finish Sketch button.
20. On the Design Bar, click the Finish Family button.

JOIN THE FIREPLACE WITH THE WALL

The Fireplace Family is finished but it is drawn on top of the Wall making both the Wall and the Fireplace hard to read. Let's fix this.

21. On the Tools toolbar, click the Split tool (or type **SL**).
    - On the Options Bar, place a checkmark in the "Delete Inner Segment" checkbox.
    - Split the exterior vertical Wall on both sides of the fireplace (see Figure 3–64).

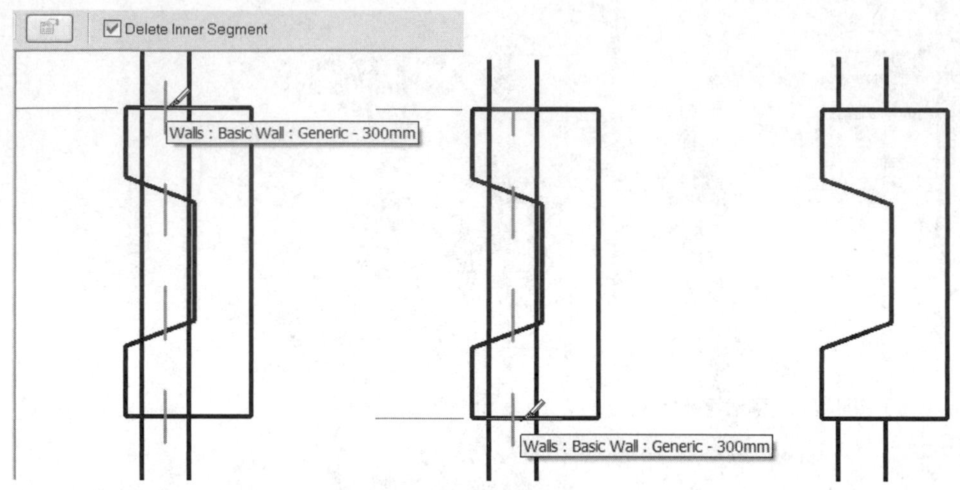

Figure 3–64 *Split the exterior Wall*

- On the Design Bar, click the Modify tool or press the ESC key twice.

This is close to what we want but let's make one more edit.

22. On the Tools toolbar, click the Join Geometry icon (see the top of Figure 3–65).
    - Click one of the exterior Walls.
    - Then click the Fireplace to join them (see Figure 3–65).

TIP Remember to watch the Status Bar for detailed prompts.

- Repeat for the other Wall.

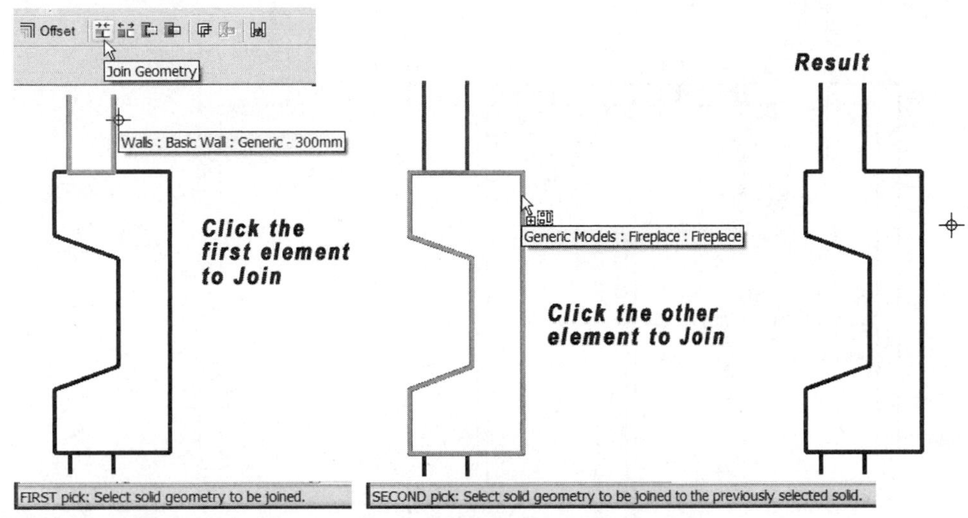

Figure 3–65 *Use Join Geometry to join the Walls to the Fireplace*

23. On the Design Bar, click the Modify tool or press the ESC key twice to cancel the Join command.
24. On the View toolbar, click the Default 3D View icon.
    - Use the techniques covered above and spin the model around so that you can see the Fireplace.

We modeled the fireplace a bit to short. However, for now we will leave this alone. In later chapters we will address the height of the fireplace as well as how it changes on the second floor. The fireplace could also use a mantel and a hearth. However, because there will be no new work done in the living room of this project and therefore no sections or elevations are needed of the fireplace, that extra level of detail is unnecessary for this tutorial. What we have created works well for floor plans. If you wish to try it anyway for the practice, feel free. Select the fireplace, and then on the Options Bar, click the Edit button. This will return you to the In-Place Family editor where you can add these accoutrements.

RESET THE CURRENT PHASE

Congratulations! Our work on residential project first floor existing conditions layout is complete for now (see Figure 3–66). We still need to add the Stairs to this model. However, Stairs will be covered in a dedicated chapter. Therefore, we will save our layout without the Stairs for now.

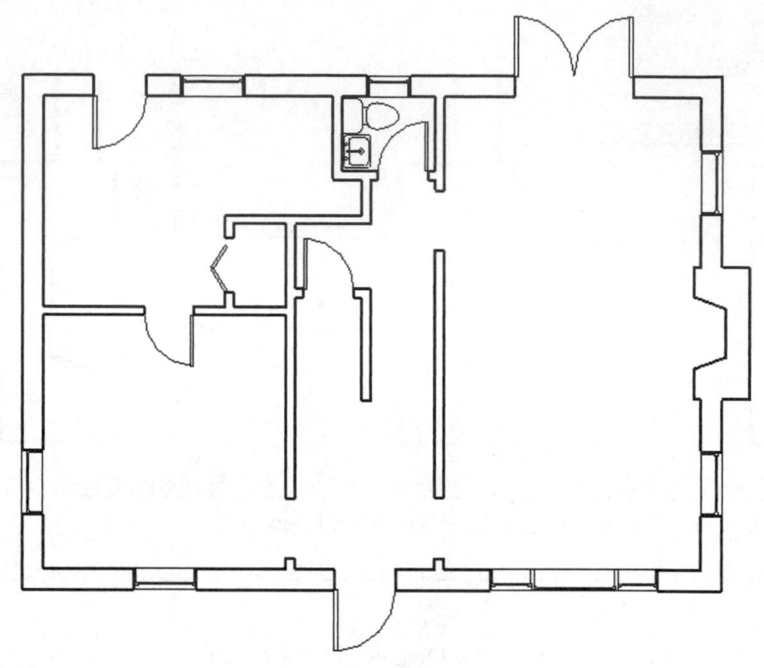

Figure 3–66 *The final first floor existing conditions layout*

25. Make sure that there are no elements selected, right-click in the Workspace and then choose **View Properties** (or type **VP**).
    - Beneath the Phasing grouping, for "Phase," choose **New Construction**.
    - Verify that "Phase Filter" is set to **Show All**.
    - Click OK to see the change.

NOTE Later, in Chapter 6, we will actually duplicate this View and create a permanent Existing Conditions View. For now, we have simply returned the View to its Phase settings at the start of the chapter.

26. Save the project.

SUMMARY

The basic process for adding elements in Revit Building is to choose a tool on the Design Bar, choose a Type from the Type Selector, set additional options on the Options Bar and then click to add the item in the workspace.

Walls can be added one segment at a time or chained to draw them end to end.

Assign elements to construction Phases to show Existing, Demolition and New Construction.

Doors, Windows and Openings automatically "cut" a hole in, and remain attached to, the receiving Wall or Walls.

Add Walls, Doors and Windows quickly, and modify their properties to add detail later.

Use Trim/Extend, Split, Offset, Move and Copy to quickly layout a series of Walls.

Door, Window and Component Types can be included with the project template or loaded as needed from library files.

It is not always necessary to model all elements in 3D. 2D Components can save overhead and computer resources when 3D is not required.

View the model interactively in 3D, plan, section, etc. to study and present the design.

Edit in any View and see the change immediately in all views.

Use In-Place Families to model custom or project-specific components directly in the model where they are required.

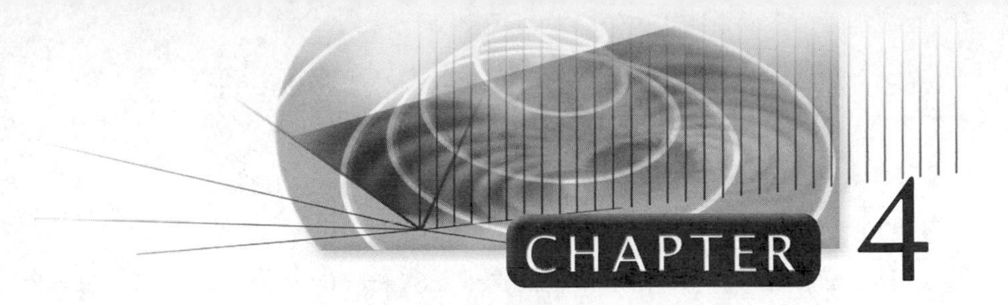

Setting up Project Levels and Views

INTRODUCTION

One of the most important concepts to grasp early in your understanding of Autodesk Revit Building and its implementation of the Building Information Modeling (BIM) concept is the concept of "Views." When you create a Building Information Model, you are creating a single model that represents all aspects of a building project using both graphics (the "M" in BIM) and associated data (the "I" in BIM). The way that you create, manipulate, and extract data from this model is through the use of one or more Views. A wide variety of View types are possible in Revit Building. Among these types are Plan Views, Elevation Views, 3D Views, Sheet Views Schedules, and Legends. Each View presents different vantage point on the building model filtered to a particular need or perspective. For example, a Floor Plan View takes a slice of a building at a particular height and presents it graphically on screen using conventional 2D Architectural abstraction to present the data. On the other hand, a Schedule View presents a collection of common data extracted from the building model presented in tabular format.

Views are not only for viewing data in a Revit Building Information Model; Views are also used to create and edit data and graphics in the building model. When you create a new Revit Building project, the program does not presume to know the type of building you wish to create or the kinds of Views you intend to use to view and edit it. Therefore, you must create the Views in which you wish to work in your projects. Creating Views is simple to do and is done as often as needed throughout the course of the project. However, many typical Views are common to most types of projects. To save time and ensure consistency from one project to the next, Project Templates are available and should be utilized to create new projects. A Project Template is a Revit Building project that has been pre-configured to contain the most common and useful Views, Families and settings needed to quickly start a new project. Autodesk Revit Building ships with several sample Project Templates out-of-the-box and you can customize and save your own templates as well.

OBJECTIVES

In this chapter, we will explore Project Templates and Project Views. First we will explore the many out-of-the-box Project Templates at your disposal. Then we will choose a Project Template from which to begin our second project—the commercial project. As we begin adding some basic geometry to this project, we will use it as a backdrop to explore the many different types of Views available in Revit Building. After completing this chapter you will know how to:

- Open and explore several Project Templates
- Create a new Commercial Project from a Template
- Set up all preliminary Views
- Understand switching between and working with Views
- Work with Sheet Views
- Print a digital cartoon set

UNDERSTANDING PROJECT TEMPLATES

Autodesk Revit Building uses project template files as a means to quickly apply project setup information, enforce company standards and project-specific settings, and save time. This is not unlike other popular Windows software packages such as Microsoft Word. (Please note that although template files apply only at the time of project creation, you can transfer project standards from one project to another at a later time.) A project template is basically a Revit Building project file preconfigured for a particular type of task that has been saved in the template format. Template files have a RTE extension and are available when you choose the **New > Project** command from the File menu. In addition to the time saved when creating projects, templates help to ensure file consistency and office standards by giving all projects the same basic starting point. Revit Building ships with several pre-made template files. These templates are ready to be used "as is." However, because office standards and project-specific needs vary, feel free to modify the default templates as necessary. The exact composition of the template used to create a project is not as important as ensuring consistency in its use to create all new projects. Table 4–1 lists the Revit Building project templates included with the software.

Table 4–1 *Out-of-the-Box Project Templates*

Template File Name	Drawing Units	Rounding	Description
Imperial Templates			
default.rte	Feet and fractional inches	To the nearest 1/32″	The Revit Building default template for Imperial Units.
Commercial-Default.rte	Feet and fractional inches	To the nearest 1/32″	Intended for basic Commercial Projects
Construction-Default.rte	Feet and fractional inches	To the nearest 1/32″	Intended for basic Construction Projects
Residential-Default.rte	Feet and fractional inches	To the nearest 1/32″	Intended for basic Residential Projects
Structural-Default.rte	Feet and fractional inches	To the nearest 1/32″	Intended for basic Structural Projects
Metric Templates			
DefaultMetric.rte	Millimeters	0 decimal places	The Revit Building default template for Metric Units.
DefaultUS-Canada.rte	Millimeters	0 decimal places	The Revit Building default template for Metric Units in the United States and Canada.
Construction-DefaultMetric.rte	Millimeters	0 decimal places	Intended for basic Construction Projects
Construction-DefaultUS-Canada.rte	Millimeters	0 decimal places	Intended for basic Construction Projects in the United States and Canada
Structural-DefaultMetric.rte	Millimeters	0 decimal places	Intended for basic Structural Projects
Structural-DefaultUS-Canada.rte	Millimeters	0 decimal places	Intended for basic Structural Projects in the United States and Canada

CAUTION *Please avoid creating projects without a template, because doing so requires an enormous amount of user configuration before serious work can begin.*

NOTE Revit Building saves virtually all data and configuration within the project file. This includes System Families, Families, settings and often several pre-defined Levels and Views. Therefore, template files provide an excellent tool for promoting and maintaining office standards. Revit Building ships with several sample template files to help you get started. These include project templates tailored to different types of projects such as construction or structural for both Imperial and Metric units. Refer to the table above for more information. Certainly, you will find one of these templates suitable for your firm's needs directly or useable as a good starting point for customization. When you establish and configure office standards, it is highly recommended that these default templates be used as a starting point. Modify them to suit individual project or office-wide needs. Families, title blocks, and other resources can also be stored in separate library files stored on the network or hard drive. The resources can be accessed using the Load button on the Options Bar or within the Element Properties dialog. This approach can help keep the template file size smaller by including only those items needed in all (or most) projects, yet it provides a central repository for additional office standard items. This method also provides additional ongoing flexibility because new items can be easily added to the office standard library. In addition, Revit Building provides the **Transfer Project Standards** command for copying System Family items such as Wall Families and other System Families from one project to another. To use this tool, first open the source project (the one with the standards that you want to copy), and then open the project to which you wish to transfer. Execute this command from the File menu, and check the boxes for the items you wish to transfer in the "Select Items To Copy" dialog.

The best way to demonstrate the importance of using project templates is to create a project with some of them. **Create a New Project with the default Template**

Let's create a new project using the default template and explore some of its settings.

1. Launch Autodesk Revit Building from the icon on your desktop (or from the Windows **Start > All Programs** menu).

Revit Building launches to an empty project based on your default template. Let's be sure that we are starting from the "default" template. First we will close the empty project that opened automatically and then create a new one based on the default template file.

2. From the File menu, choose **Close**.

3. From the File menu, choose **New > Project**.

   - In the New Project dialog, in the "Template File" area, be sure that the lower radio button is selected (the one that lists a template file).

   - Click the Browse button.

   The default location for the Imperial template files is: *C:\Documents and Settings\All Users\Application Data\Autodesk\Revit Building 8.1\Imperial Templates*

   The default location for the Metric template files is: [*C:\Documents and Settings\All Users\Application Data\Autodesk\Revit Building 8.1\Metric Templates*].

   - Browse to this location and then select the *default.rte* [*DefaultMetric.rte*] template file and then click Open (see Figure 4–1).

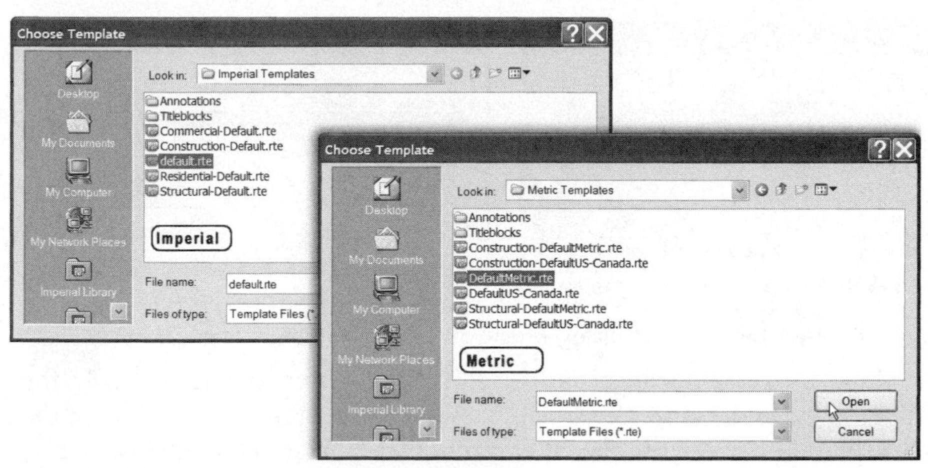

Figure 4–1 *Creating a new Project with no Template*

- In the "Create New" area, verify that Project is selected and then click OK.

You may notice that if you are using Metric units that there is also a *DefaultUS-Canada.rte* template. If you wish, you can use this template instead of the default Metric template. The differences between the two are subtle. Both the *default.rte* Imperial and the *DefaultUS-Canada.rte* Metric templates use the same (square) Elevation Heads. The *DefaultMetric.rte* template uses a round Elevation Head (see Figure 4–2).

Figure 4–2 *The shape of the Elevation Head varies in the North American and non-North American templates*

4. From the Settings menu, choose **Materials** (see Figure 4–3).

Some of the language used in the Materials and other settings varies between the North American and non-North American templates.

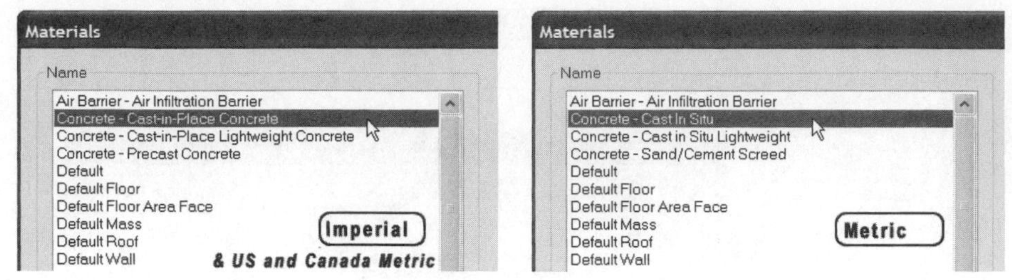

Figure 4–3 *The names of some of the Materials vary*

- Close the Materials dialog when you are finished exploring.

Despite these minor differences, the three "Default" templates are virtually the same. They all have the same Levels and Views pre-configured. They all have the same basic Families loaded and the other settings are similar but with unit-specific or regional vernacular differences (like "Cast-in-Place vs. "Cast In Situ"). On the Settings menu, choose Project Units.

TIP The keyboard shortcut for Project Units is **UN**.

- Explore the various settings without making any permanent changes. Refer to the table above for the default settings of each Template.
- Click Cancel when finished.

5. On the Project Browser, under Views (all), expand Floor Plans, Ceiling Plans and Elevations (Building Elevation).

- There will be three Floor Plan Views: *Level 1*, *Level 2* and *Site*
- There will be two Ceiling Plan Views: *Level 1* and *Level 2*.
- And under Elevations, there are four Elevation Views: *East*, *North*, *South* and *West* (see Figure 4–4).

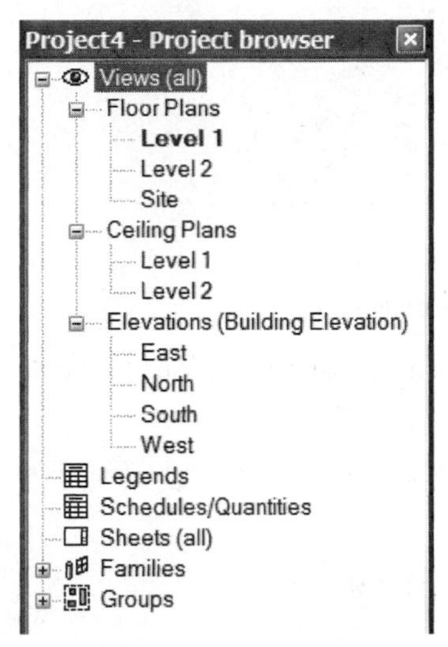

Figure 4–4 *There are several pre-made Plan and Elevation Views*

6. On the Design Bar, click the Basics tab and then click the Wall tool.
   - On the Options Bar, open the Type Selector.
- There will be several Families already in the project (see Figure 4–5).

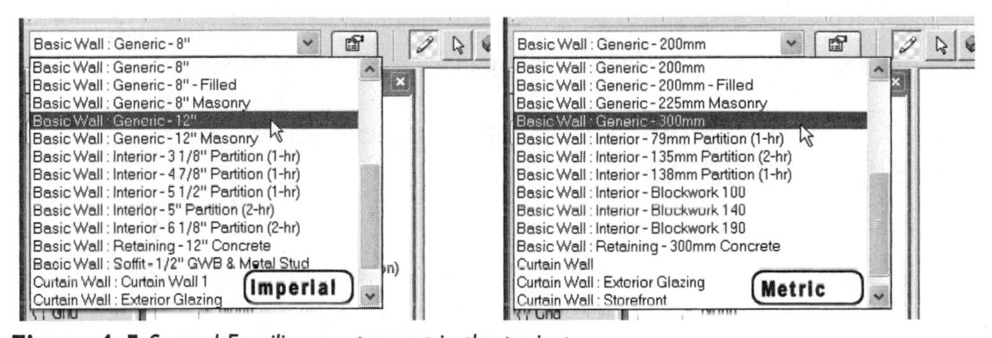

Figure 4–5 *Several Families are present in the project*

- As we saw when opening the Material dialog above, there will already be several Materials in the project (choose **Materials** from the Settings menu to see this).
- Section, Elevation and Level Heads are loaded in the project (however, as noted above, they may vary slightly in Imperial and Metric).

7. On the Design Bar, click the Section tool.
   - Click a point anywhere on screen, and then drag to the right and click again (see Figure 4–6).

Notice that the Section Head and the Section Tail are already configured.

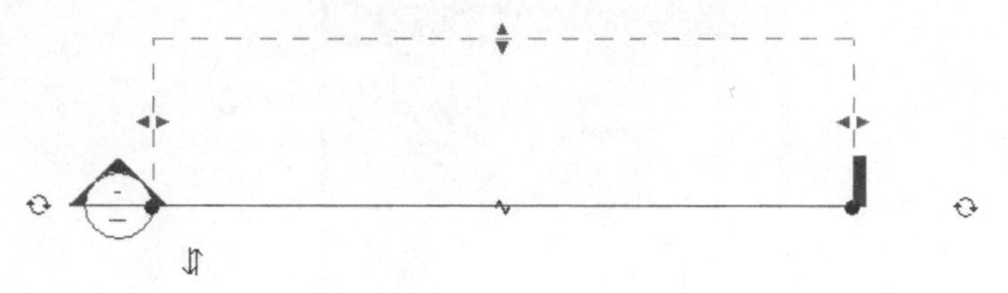

Figure 4–6 *Section Lines are pre-configured*

8. On the Project Browser, expand Elevations (Building Elevation) and then double click *East*.

    - Zoom in on the Level Heads to the right (see Figure 4–7).

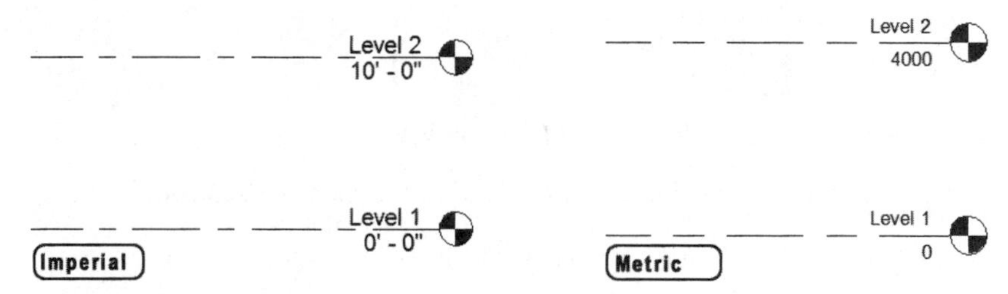

Figure 4–7 *Levels have Level Heads*

You can see what other tags are loaded in the project (from the template) by choosing the options from the **View Tags** menu item on the Settings menu.

CREATE A PROJECT FROM THE CONSTRUCTION AND STRUCTURAL TEMPLATES

Let's repeat the exploration process in some of the other provided project templates.

9. From the File menu, choose **Close**.

    - When prompted to save changes, click No.

10. From the File menu, choose **New > Project**.

    - In the New Project dialog, in the "Template File" area, be sure that the lower radio button is selected (the one that lists a template file).

    - Click the Browse button.

As above, the dialog should open directly to your default template folder. If it does not, browse to the template location listed at the start of the tutorial above.

- Select the *Construction-Default.rte* [*Construction-DefaultMetric.rte*] template file and then click Open.

- In the "Create New" area, verify that Project is selected and then click OK.

The first thing you should notice about this template is that there are several more Views in the Project Browser (see Figure 4–8).

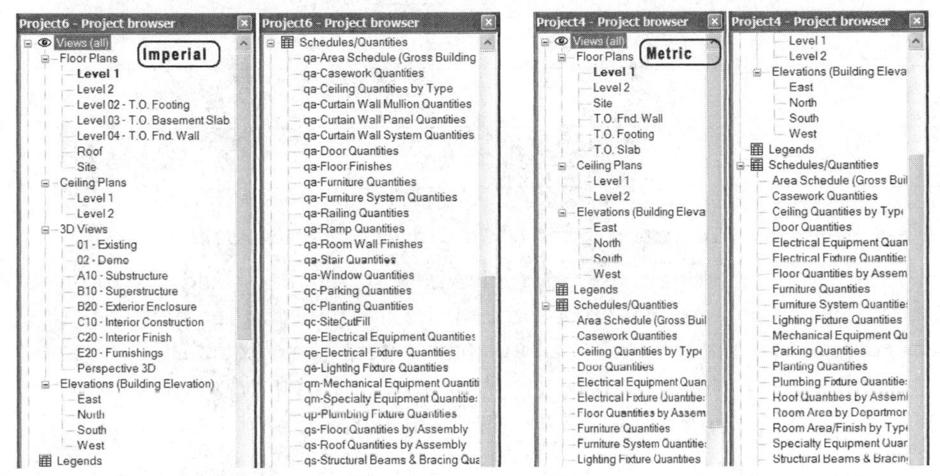

Figure 4–8 *The Construction Template is pre-loaded with dozens of Schedules and Views*

Spend some time in this template exploring the Floor Plan, Ceiling Plan and Elevation Views like we did above. The same four Elevation Views are provided here. There are a few additional Floor Plan Views however, and if you open one of the Elevation Views, you will notice that there are Levels defined for the foundation and footing. The most significant addition to this template that was not included in the Default template is the inclusion of dozens of Schedule Views. As you can see from the names of these Schedules, you can determine quantities for nearly every element in the project.

11. On the Project Browser, double-click the floor plan View: *Level 1* to make it active.

    - Create a Wall anywhere on the screen. Click Modify or press ESC twice when done.

12. On the Project Browser, double-click the *qs-Wall Quantities by Assembly* [*Wall Quantities by Assembly*] Schedule View to open it (see Figure 4–9).

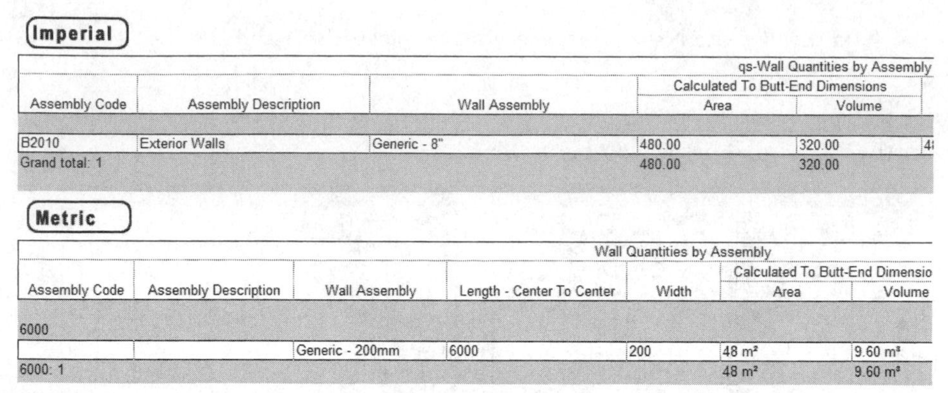

Imperial

qs-Wall Quantities by Assembly					
			Calculated To Butt-End Dimensions		
Assembly Code	Assembly Description	Wall Assembly	Area	Volume	
B2010	Exterior Walls	Generic - 8"	480.00	320.00	4
Grand total: 1			480.00	320.00	

Metric

Wall Quantities by Assembly						
					Calculated To Butt-End Dimensio	
Assembly Code	Assembly Description	Wall Assembly	Length - Center To Center	Width	Area	Volume
6000						
		Generic - 200mm	6000	200	$48\ m^2$	$9.60\ m^3$
6000: 1					$48\ m^2$	$9.60\ m^3$

Figure 4–9 *Schedules will populate automatically as elements are added to the model*

Take a little time exploring these Schedules now if you like. Add a few more Walls, or some Doors and then open some of the Schedules to see how they update. Schedules will be covered in detail in later chapters. The exact list and names of Schedules in the Imperial and Metric templates vary. Therefore, despite your units preference, you may wish to open both templates and look at them.

NOTE The Imperial template provides some Sheets as well. In particular, there is a Sheet named *000 - Temporary Schedule Sheet.* This Sheet has each of the Schedules inserted onto it. There is a piece of text noting "Temporary schedule sheet to allow easy copy and pasting into other projects." This indicates that the Schedules in this template project can be copied from this project and used in any other project.

Let's explore the Structural Template next.

13. Create a new Project based on the *Structural-Default.rte* [*Structural-DefaultMetric.rte*].

    - Explore the Project Browser and other settings as we have above.

The Structural template is much simpler than the others. It really only has one item that distinguishes it from the others. Above we explored the list of Wall Types in the Default Template. This was done by adding a Wall and viewing the list of types on the Options Bar in the Type Selector. You can do the same thing with the Structural elements in this template. You may not, however, see the structural tools on the Design Bar right away. To see the hidden tabs on the Design Bar, right click it (see Chapter 2 for more information). Right-click the Design Bar and choose "Structural" to display the Structural tab. Then choose Column, or Beam or any other structural tool to see its list of Types on the Type Selector. All Families can also be seen on the Project Browser. Expand Families and then expand one of the Structural categories (see Figure 4–10).

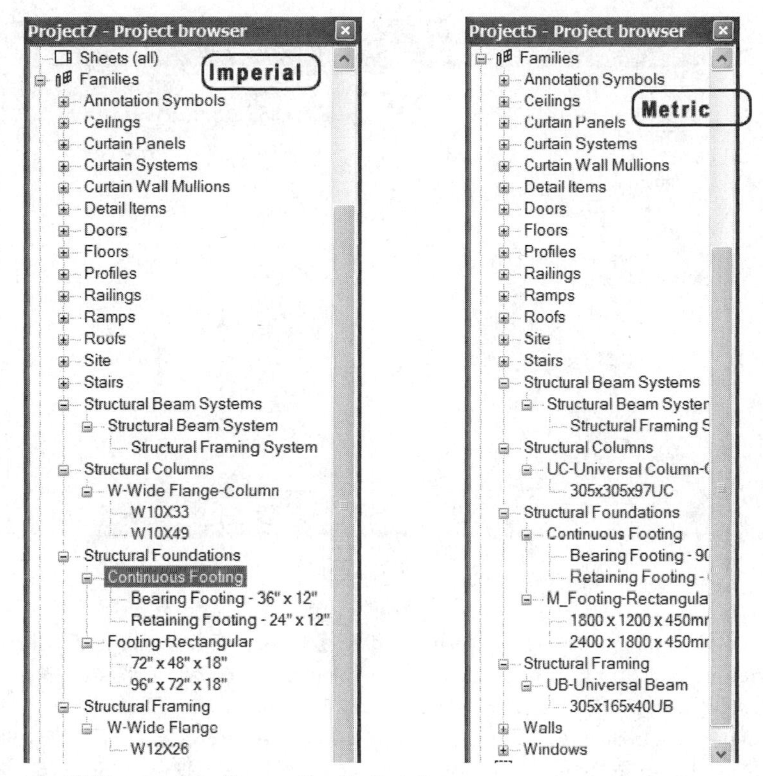

Figure 4–10 *Viewing Families on the Project Browser*

Hopefully, you are beginning to see the benefits to starting new projects with a template. There is really no compelling reason to begin projects any other way. As you work through the exercises in the coming chapters, you will certainly discover areas where the default templates could be enhanced and improved. Make note of these observations as you go. When you are ready, try your hand at creating your own template file. The basic steps are simple:

- From the File menu, choose **New > Project**.
- Load the existing template that is most similar to the one you wish to create.
- Choose the "Project Template" option in the New Project dialog.
- Edit any settings as you see fit. (Change Settings, add or delete Views, load or delete Families, etc.)
- Choose **File > Save As**.
- From the "Save as type" list, choose **Template Files (*.rte)**.
- Be sure to browse to your Template folder, type a name for your new template and then click the Save button (see Figure 4–11).

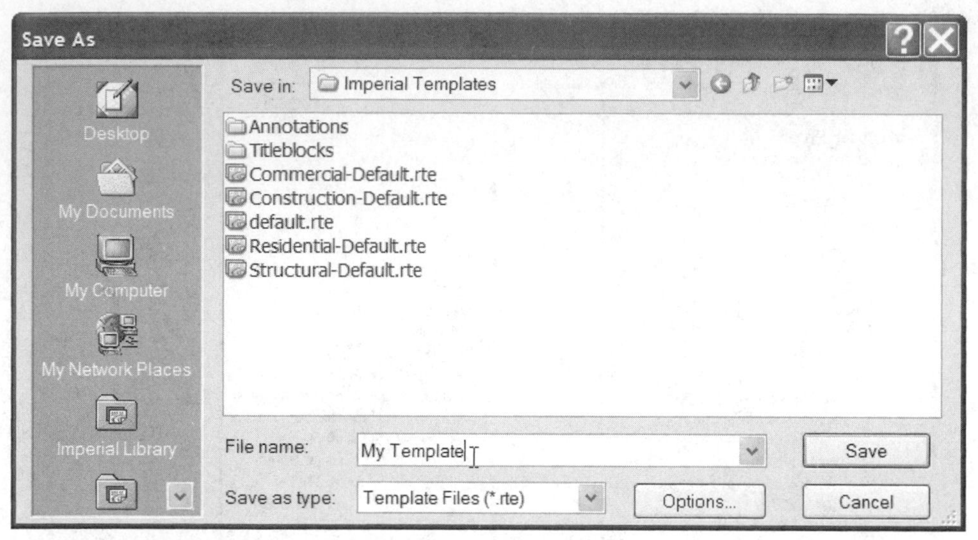

Figure 4–11 *Creating a new Template*

NOTE The templates shipping with Revit Building are ready to use straight out-of-the-box. ,They are excellent starting points for developing your own office-standard project template file(s). It is highly recommended that you become comfortable with the provided offerings and then if necessary, customize them to meet your firm's specific needs. Consider including those settings and Views that people will use most frequently. Once you have a standard template in place, it is imperative that all users be required to use it. In some cases a project will have special needs not addressed by this office standard template. In such cases, project-based derivatives can be created.

STARTING A PROJECT WITHOUT A TEMPLATE (NOT RECOMMENDED)

It is possible to begin a project without a template; doing so is not recommended, however. After exploring several of the provided templates in the exercise above, you should be getting a sense of the variety of settings and elements that are resident in a template. If you start a project without one, you are forced to either configure/create all of these items on your own manually as needed, or import them from other projects or libraries. While it is possible to do this, the amount of extra work that it adds to your project makes it an ill-advised approach to creating a new project. If you have any doubt as to the validity of these claims, try it out for yourself to see.

14. Create a New Project in Revit Building but in the New Project dialog, choose None for "Template file" and then click OK. When prompted for "Initial Units" make your choice of either Imperial or Metric.

15. Repeat the process that was followed above to explore the template.

- Check the Project Browser, check the Families and Types. Check the Materials.
- Add a Section or Elevation and notice that they do not even include Section or Elevation Tags.

16. Close the Project without saving when you are satisfied that you understand the differences between it and a Project created from a template.

SETTING UP A COMMERCIAL PROJECT

Now that we have seen some of the features and benefits of creating projects from templates, let's create an actual project that we will follow (as well as the one started in the last chapter) throughout the remainder of the book. This project is a 30,000 SF [2,800 SM] small commercial office building. The project is mostly core and shell with some build out occurring on one of the tenant floors. The tutorials that follow walk through the startup of the commercial project and the setup of several Views including: plans, elevations, sections and schedules. At the end of the chapter, we will generate sheet Views and print a cartoon set. The completed files for the project are available with the files you installed from the CD in the *Chapter04\Complete* folder on your hard drive.

Getting started with the commercial project gives us a nice practical exercise to further the goals of this chapter: namely the understanding of Project Templates and Revit Building Views. Even though we have explored many templates thus far, it might be difficult to decide exactly which one we should use to get started. In the remainder of this chapter, we will begin modestly with the default template. We will begin adjusting the project levels to suit the needs of our commercial building and then begin creating Views that we will need. In some cases, we will create these items ourselves, and in others we will borrow them from additional templates as appropriate. At the end of the chapter, we will save our work as both our commercial project and as a new template file.

WORKING WITH LEVELS

Before we start our commercial project, close all projects that you currently have open (**File > Close**). If it is easier, you can choose **File > Exit** to quit Autodesk Revit Building and then launch a fresh session. If you do this, be sure to close the empty project that loads automatically upon launch. In either case, be sure that all projects are closed before proceeding. (You do not need to save anything from the previous exercises).

ADDING AND MODIFYING LEVELS

1. From the File menu, choose **New > Project**.

- In the New Project dialog, in the "Template File" area, be sure that the lower radio button is selected (the one that lists a template file).
- Verify that the *default.rte* [*DefaultMetric.rte*] template file is listed in the text field (use the Browse button if necessary to choose it).

NOTE If you are in a country for which your version of Autodesk Revit Building does not include these template files, they have both been provided with the files from the "Mastering Autodesk Revit Building" CD ROM. Please browse to the *Template* folder to locate them.

- In the "Create New" area, verify that Project is selected and then click OK.

By now we have started a few projects this way, and the familiar *Level 1*, *Level 2* and *Site* plan Views will show on the Project Browser. Normally, the names of the Levels in the project will also be used as the names of the Floor Plan Views. As our first task in our new Commercial Project, let's modify and add to the project Levels.

Level—datum elements in Revit Building that are used to represent each of the actual floor levels or other vertical reference points in the building you are modeling. Levels are used to organize projects vertically, in the Z axis.

Typically you will add a Level for each story of the building ("First Floor," "Second Floor," "Roof" etc). In addition Levels can also be used to define other meaningful horizontal planes such as the "Top of Structure" or the "Bottom of Footing." We saw examples of this in several of the sample templates. For our commercial project, we have four stories, a roof and the grade level.

Take a look at the Basics tab of the Design Bar. Notice that the Level tool is grayed out. This is because the currently active View: *Level 1*, is a floor plan View. You cannot add or edit Levels in a floor plan View. To do so, you must switch to an elevation or section View.

2. On the Project Browser, expand Elevations (Building Elevation) and then double click South (see Figure 4–12).

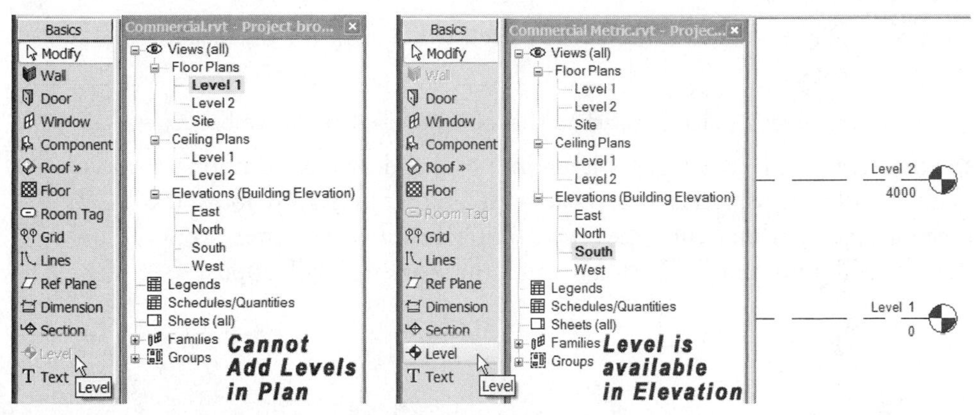

Figure 4–12 *Levels cannot be added or edited in Plan Views*

In this View, you can see two Levels: Level 1 and Level 2.

NOTE Notice that even though there is a "*Site*" Floor Plan View in the Project Browser, that there is *not* a "Site" Level. The *Site* View is actually associated to Level 1. In other words, there can be multiple plan Views associated with the same Level and or Levels that have no plan Views.

3. Zoom in on the Level Heads at the right side of the screen so that you can see them clearly.

Most elements in Revit Building will show a Tool Tip when you pass the mouse over them. Please note that you can turn these tips off in the Options dialog from the Settings menu.

- Move your pointer over the Level line to see a Tool Tip.
- Move your pointer over the Level Head to see a different tip (see Figure 4–13).

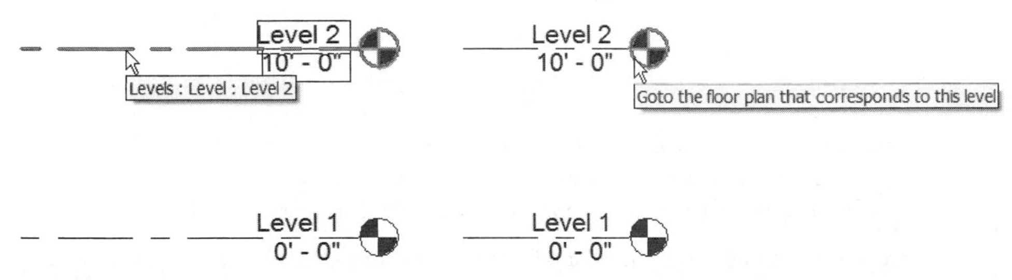

Figure 4–13 *Level Heads can be used to navigate to the associated Plan View*

To select the Level, click the dashed level line as shown on the left side of Figure 4–13. To open the associated floor plan View for the level, double-click the blue Level Head as shown on the right side. This is analogous to clicking on a link on a web page. It opens the associated View. If the Level Head is black that means that there is no associated View. The Level by itself is a datum for documentation and modeling purposes only.

4. Select Level 2 (click the dashed line, not the Level Head symbol).

When you select the Level, it will highlight in red and several additional drag controls and handles appear. As you move your pointer (referred to as the Modify tool) over each one (don't click them yet, simply hover the pointer over each one), the handle will temporarily highlight and the tool tip will appear (see Figure 4–14).

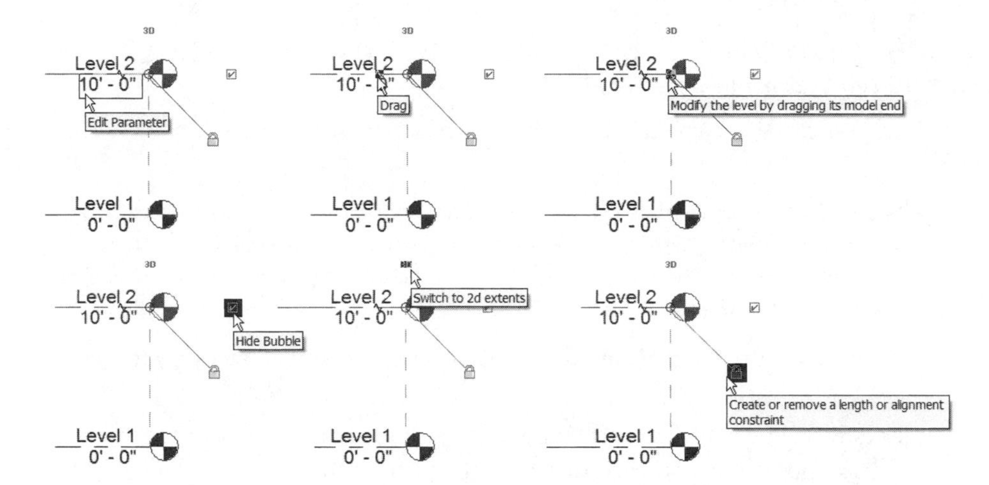

Figure 4–14 *Levels have many control handles and drag points*

Moving left to right in Figure 4–14, the following briefly describes each control.

- **Edit Parameter**—Click the blue text to rename the Level or edit its height. If you rename a Level, you will be prompted to also rename associated plan Views. You can accept or reject this suggestion.

NOTE The height is measured from an "Elevation Base" which is a Type parameter of the Level System Family. The Elevation Base is the zero point from which the Level heights are referenced. This defaults to "Project" which is simply the lowest level of the project. You can also setup a "Shared" Elevation Base (such as height of Sea Level or some other appropriate datum height); to do so, consult the on-line help.

- **Drag**—This small "squiggle-shaped" handle creates an elbow in the Level line. This is useful when the annotation of two Level Heads overlap one another in a particular View. Click this handle to create the elbow, and then drag the resultant drag handles to your liking.
- **Modify the Level Drag Control**—This round handle is used to drag the extent of the Level. If the length or alignment constraint parameter is also ac-

tive, dragging one Level will affect the extent of the other constrained Levels as well.

- **Hide/Show Bubble**—Use this control to hide or show the Level Head bubble at either end of the Level line.
- **2D/3D Extents Control**—When 3D Extents are enabled, editing the extent of the Level line in one View affects all Views in which the Level appears (and is also set to 3D Extents). If you toggle this to 2D Extents, dragging the extent of the Level line affects only the current View.
- **Length and Alignment Constraint**—This padlock icon is used to constrain the length and extents of one Level line to the others nearby. This is useful to keep all of your Level lines lined up with one another.

TIP Another control built into the Level Heads is the ability to quickly jump to the floor plan associated with the Level. To do so, deselect the Level and then double-click the blue Level Head.

5. Click anywhere next to the Level (where there are no objects) to deselect it (or press the ESC key).
6. On the Design Bar, click the Basics tab and then click the Level tool.
   - Zoom out so that you can see the entire length of the Levels.
   - Move your mouse above the left ends of the existing levels.

   A temporary dimension will appear with an alignment vector above the top Level.

   - When the temporary dimension reads 12′-0″ [3600], click the mouse to set the first point (see the left side of Figure 4–15).

TIP Remember, you can zoom in to change the temporary dimension snap increment if necessary. Also, you can simply type the value that you wish while the temporary dimension is active and then press ENTER to apply it. Finally, like other tools that we have already seen in Autodesk Revit Building, you can place the Level at any dimension and then edit the value of the temporary dimension later to the correct value.

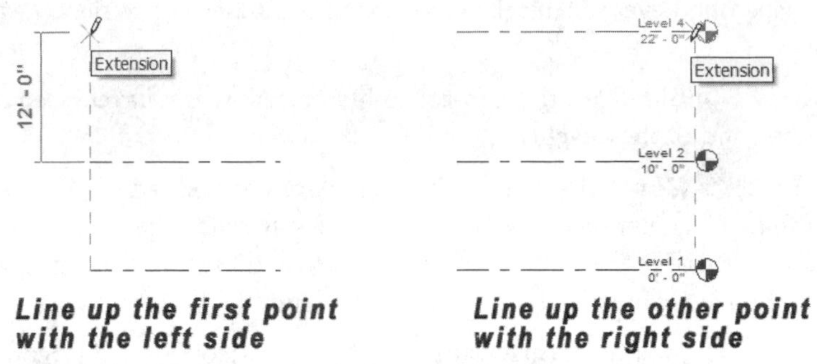

Figure 4–15 *Add a Level aligned with the existing ones*

- Move the pointer over to the right side; when an alignment vector appears above the existing Level Heads, click to set the other point (see the right side of Figure 4–15).

A new Level is added as well as two new plan Views: a *Level 3* Floor Plan View and a *Level 3* Ceiling Plan View (see Figure 4–16). To control whether adding a new Level also creates new plan Views, use the "Make Plan View" checkbox on the Options Bar. Furthermore, if you click the "Plan View Types" button, you can control which type of Plan View(s) are created. By default, a Floor Plan and a Ceiling Plan View are created. Notice also that the "Length and Alignment" constraint is also automatically applied as you add Levels.

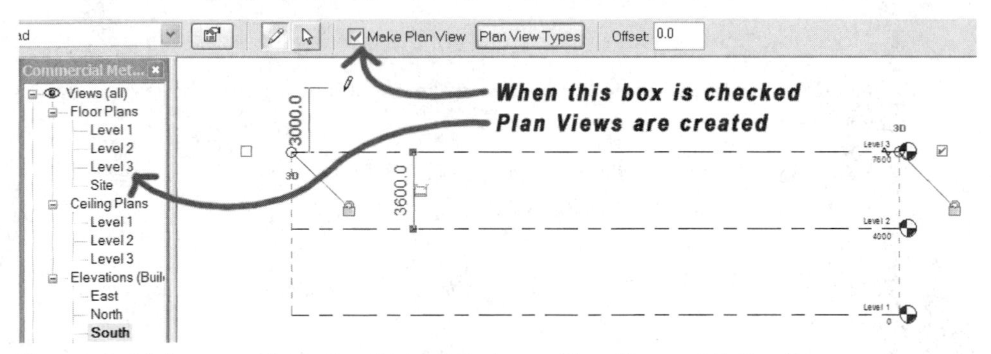

Figure 4–16 *As you add new Levels, by default new Floor Plan and Ceiling Views are created*

The Level command should still be active. If you canceled it, click the Level tool on the Design Bar to restart it.

7. Using the same procedure, to add two more Levels above Level 3, each 12′-0″ [3600] above the previous one.

We need one more Level at the street level of our project. Since there is already a *Site* plan View in the project, we will add the final Level without automatically creating associated Views. Later we will assign our *Site* plan View to this new Level instead of its current assignment (which is Level 1).

8. On the Options Bar, remove the checkmark from the "Make Plan View" checkbox (this checkbox is shown in Figure 4–16 above).
   - Add one more Level 3′-0″ [900] beneath Level 1 (see Figure 4–17).

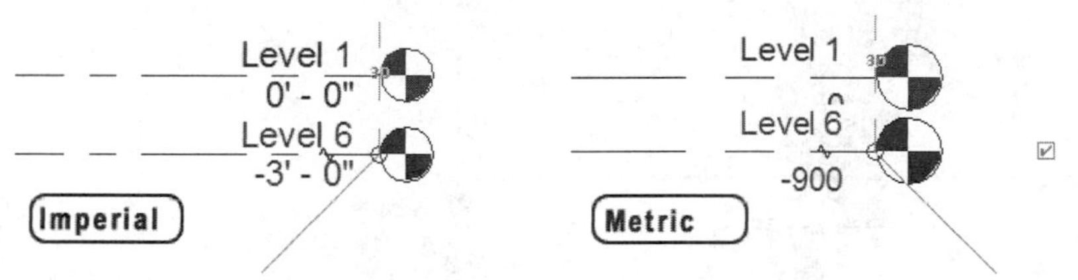

Figure 4–17 *Add the final Level below Level 1 without creating associated plan Views*

- On the Design Bar, click the Modify tool or press the ESC key twice.

Notice that the bottom Level's Level Head is black while the upper Levels are blue. When a tag in Revit Building appears blue it indicates an association to another View. In this case, the blue Level Heads can be double-clicked to open the corresponding floor plan Views. The bottom Level is black because we chose not to have Revit Building create associated plan Views when we added this Level. Level Heads with no associated plan View are often created for Levels that demarcate Top of Steel, or Top of Plate, or Bottom of Footing, etc. New plan Views can be created and associated to any and all existing Levels at any time.

RENAMING LEVELS

We can accept the default level names as they appear, or we can change them to something more suitable for our specific project. The names of Levels can be anything that makes sense to the project team. However, the Project Browser will sort the list alphabetically (notice how "Site" comes after all the "Level" plan Views). Therefore, on larger projects with many levels, it might make more sense to use a naming scheme similar to the default where the floor number is part of the name. In projects with more than 10 stories, you should consider placing a leading zero in the level names so that Level 10 does not inadvertently sort before Level 1. We'll just rename the top and bottom levels in this project to something a bit more descriptive.

9. Select Level 6 (the one we just added at the bottom).
   - Click on the blue text of the name to edit it. Type: **Street Level** for the new name.
   - Repeat this process on Level 5 (the one at the top). Rename it to **Roof**.
   - A dialog will appear asking you if you wish to rename the corresponding plan Views, click Yes (see Figure 4–18).

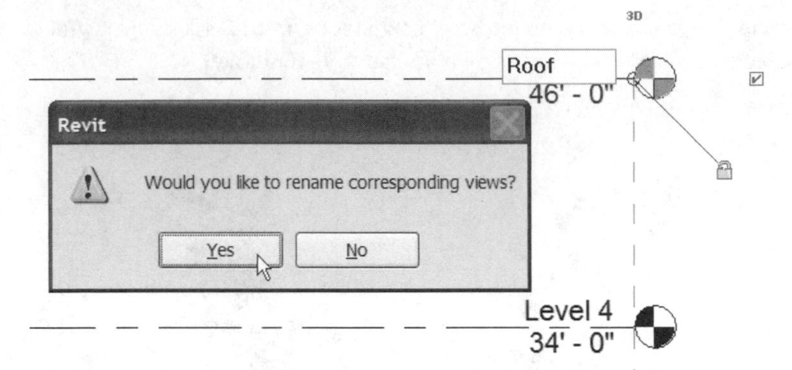

Figure 4–18 *Rename the top Level to Roof and accept the renaming of corresponding Views*

When you choose "Yes" in this dialog, the two associated plan Views: *Level 5* floor plan and *Level 5* ceiling plan, become "*Roof*."

10. From the File menu, choose **Save**.

    - Browse to the *Chapter04* folder, give the project a name such as **MRB-Commercial** and then click Save.

WORKING WITH SITE TOOLS

Now that we have created and named all of our Levels, we can move on to adding some basic building elements to our model. Our building's site plan is a good place to start.

IMPORT A SITE CONDITIONS FILE AND CREATE A NEW VIEW

The default template that we used to start our project included a *Site* plan View. However, the Level association of this View is Level 1. For our project, we need a site plan View that is associated to the Street Level. To do this, we will delete the existing View and create a new one.

1. On the Project Browser, right-click on the *Site* Floor Plan View and choose **Delete**.
2. On the Design Bar, click the View tab and then click the Floor Plan tool.

TIP If you prefer, you can choose **New > Floor Plan** from the View menu.

- In the "New Plan" dialog, select Street Level from the "Floor Plan Views" list.

- From the Scale list, choose **1″=20′-0″** [**1:200**] and then click OK (see Figure 4–19).

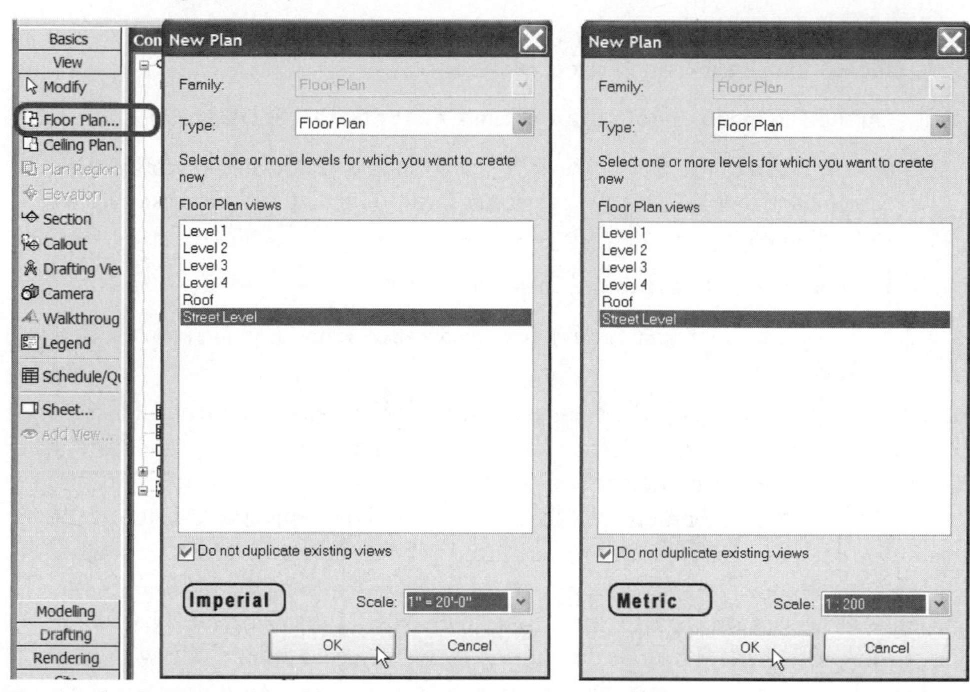

Figure 4–19 *Create a New View associated to the Street Level*

The new View will appear in the Project Browser under Floor Plans and it will open automatically on screen. (If you returned to any elevation View, the Level Head for Street Level will now be blue indicating that there is now a plan View associated to this Level).

NOTE In this sequence we created the Level first (above) without associated Views and then created a new floor plan View. An alternative approach would have been create the Level with "Make Plan Views" checked, but use the "Plan View Types" button to select only a Floor Plan and not a Ceiling Plan.

IMPORT SITE DATA

Frequently you will receive site plan data from outside firms in AutoCAD DWG or Microstation DGN format. Autodesk Revit Building readily imports files saved in either format. The linework in those files can then be used to create site conditions with the Revit Building model. (In order for this to work correctly, the linework in the file has to be drawn at the correct z-height corresponding to the actual contour level). Now that we have created our new *Site* plan View, let's import some contour lines from a DWG file and generate a site model.

Make sure that the Street Level View is open. If you switched to another View above, double-click it on Project Browser now.

3. On the File menu, choose **Import/Link > DWG, DXF, DGN, SAT**.

   In the Import/Link dialog, Revit Building should automatically choose the current folder in which your project is saved (*Chapter04* in this case), if not, you will need to browse there.

   - If necessary, browse to the *Chapter04* folder.
   - Select the *Commercial-Site.dwg* [*Commercial-Site-Metric.dwg*] file (don't click Open yet).

Several options appear beneath the file name and folder section. Imported files can be linked instead of imported. In the "Import or Link" area, if you choose the "Link" option, a live link is maintained to the DWG file so that if it should be changed in its host environment, Revit Building will be able to re-import the changes automatically. Revit Building can interpret the layers or levels contained in the incoming file. If you wish to import only certain layers in the DWG file, you can choose an option next to the Layers list. Most DWG or DGN files are saved in multiple colors. The options under "Layer/Level Colors" allow you to control how this color data is handled on import. If the file you are importing has special scaling needs, you can choose options under Scaling. In most cases you will want to initially leave this set to "Auto-Detect." For "Positioning" there are several options. "Center to Center" is the simplest option. It simply matches the geometric center of the imported file to the geometric center of your active Revit Building View. If the file is a one-time import and you are reasonably certain that you will not need to import additional files, this can be the most convenient option. However, as a matter of best practice "Origin to origin" is usually safer. When you allow Revit Building to align the origin of the DWG or DGN file to the Revit Building model origin, you can later import additional DWG or DGN files based upon the same origin point and be certain that they will automatically align with the existing geometry properly. If neither of these methods meets your needs, "Manually place" is available allowing you to use the mouse pointer to place the imported file in any location that you like. This would not typically be the best option for site plan data, but may be fine for smaller components like furniture or building components designed outside of Revit Building.

- Accept the defaults for the "Import or Link" and "Scaling" options.
- In the "Layer/Level Colors" area, choose "Black and white."

- In the "Positioning" area, choose "Origin to origin" and then click Open (see Figure 4–20).

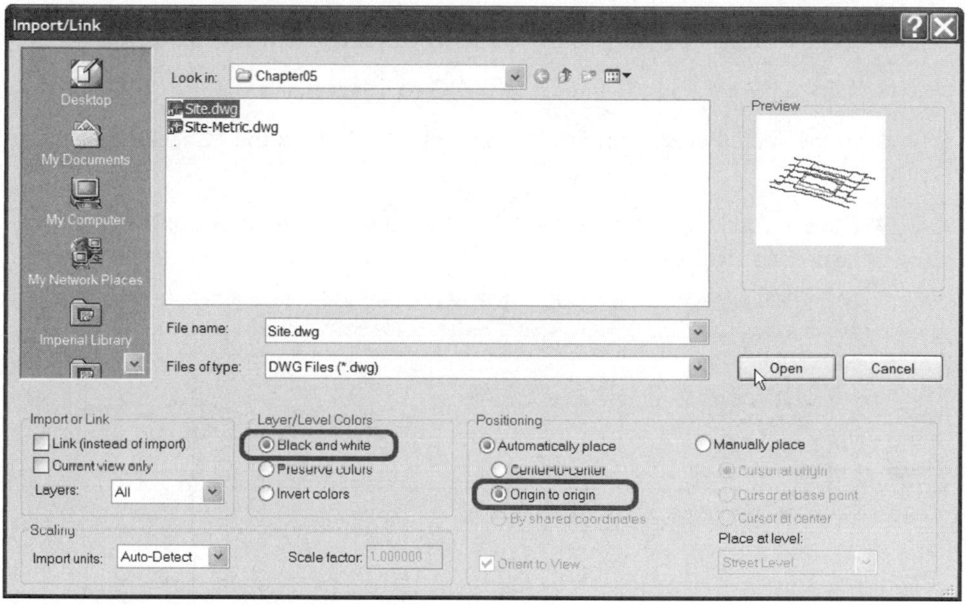

Figure 4–20 *Import the Site data from a DWG file*

4. Type **ZF** on the keyboard, (or choose **Zoom > Zoom to Fit** from the View menu) to zoom to the imported file.

Notice that the file is inserted in the upper right corner of the View, well past the four elevation tags on screen. Despite this, it will still be visible in the elevation Views.

5. On the Project Browser, double-click the *South* elevation to open it.

- Type **ZF** again to fit the View to the screen.

In this View you can see the imported file to the right side. You can also see that it is made up of a series of contour lines set at different heights. It is actually easier to see this in the *East* or *West* View.

- On the Project Browser, double-click the *East* elevation to open it.
- Type **ZF** again to fit the View to the screen.

BUILD A TERRAIN MODEL FROM THE SITE DATA

Now that we have imported the 3D contour line data from the DWG file, we can use the Revit Building Site tools to create a Toposurface from this data.

6. On the Project Browser, double-click the *Street Level* plan to open it again.

7. On the Design Bar, click the Site tab and then click the Toposurface tool.

   The Design Bar changes to show only the Toposurface tools.

   - On the Design Bar, click the → **Use Imported > Import Instance**.
   - Click anywhere on the imported DWG to select it.
   - In the "Add Points from Selected Layers" dialog, click the "Check None" button.
   - Place a checkmark in only the A-Site-Clin checkbox to select only that layer (see Figure 4–21).

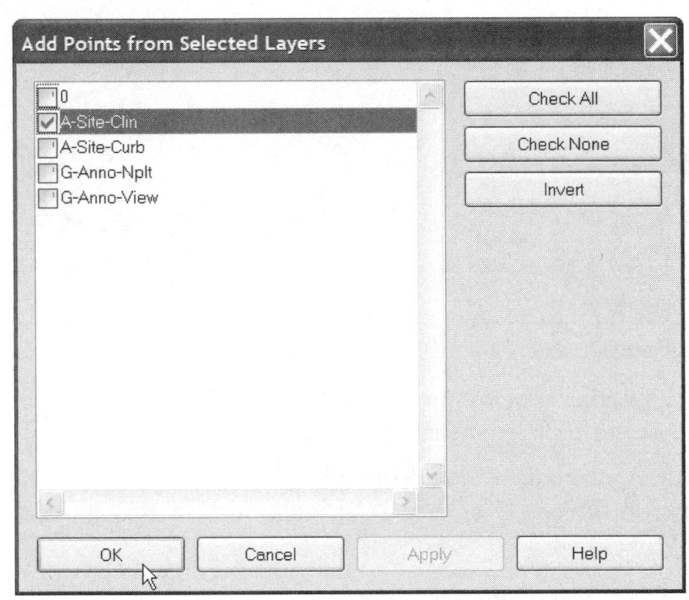

Figure 4–21 *Choose a Layer from which to create the Toposurface*

   - Click OK to create the Toposurface.
   - On the Design Bar, click the Finish Surface button to exit sketch mode and complete the Toposurface.

Several points will be extracted from the geometry on the selected layer and from those points a Toposurface will be created. This is a very simple terrain model and not accurate to the level required by a civil designer, but it will give our building a base upon which to sit.

8. On the View toolbar, click the Default 3D View icon.

   - On the View Control Bar, click the Model Graphics Style icon and choose **Shading** (see Figure 4–22).

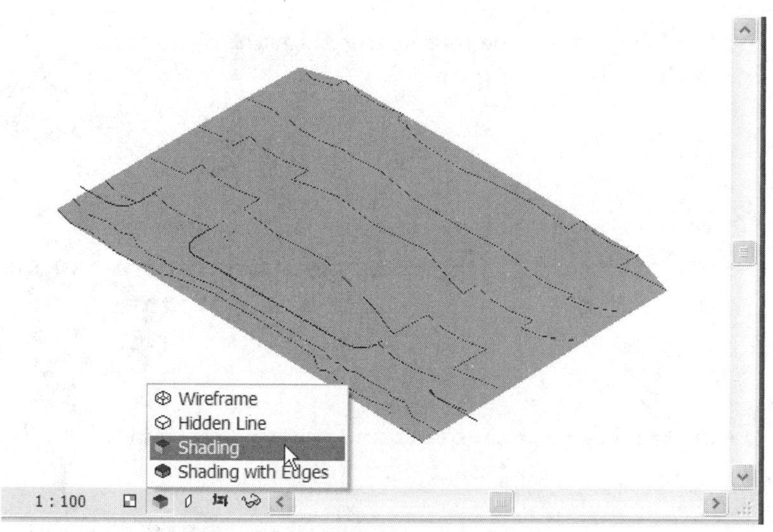

Figure 4–22 *Switch to a 3D View and Shade the model*

- On the View toolbar, click the Dynamically Modify View icon (shown in Figure 3–48 in Chapter 3).
- Use the techniques covered in the "Viewing the Model in 3D" heading in the previous chapter to spin the model around so that you can see the Toposurface from different angles.

ADD A BUILDING PAD

Let's add a Building Pad. A Building Pad adds a simple level surface that cuts into the terrain model as appropriate to give the building model a place upon which to sit. It will be easier to do this without having the model shaded.

9. On the View Control Bar, click the Model Graphics Style icon and choose **Wireframe**.

The site plan data imported from the DWG file includes a rectangle that approximates the rough footprint of the building. We can use this to assist us in sketching the Building Pad.

10. On the Design Bar, click the Pad tool.
    - On the Design Bar (now in Sketch mode) click the Lines tool.
    - On the Options Bar, click the Pick Lines icon.
    - In the View window, position the pointer over one edge of the rectangle in the middle of the site (it should pre-highlight) and then press the TAB key.

- When all four sides of the rectangle pre-highlight, click the mouse to create sketch lines (see Figure 4–23).

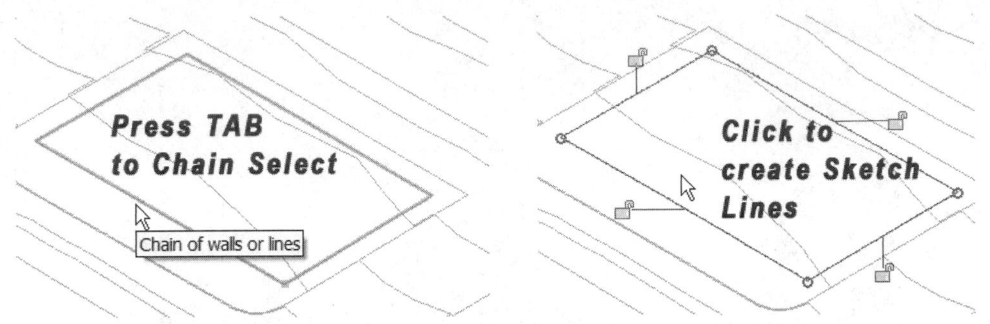

Figure 4–23 *Use the* **TAB key to chain select and create Sketch Lines**

- On the Design Bar, click the Pad Properties button.
- In the Element Properties dialog, for the "Height Offset From Level" parameter, input **2′-0″** **[600]** and then click OK.
- On the Design Bar, click the Finish Sketch button.

11. On the Project Browser, double-click to open the *West* elevation.
    - Zoom as required to see the Pad and its relationship to the Toposurface.
    - Save the Project.

ROUGH OUT THE BUILDING FORM

Programmatic and preliminary design information might be received by the project team in a variety of forms, such as hand-drawn sketches, SketchUp Files, AutoCAD files, or other CAD files. If data is hand drawn but dimensionally accurate, it can be scanned into any popular graphics file format (such as JPG or BMP) and Imported into the Revit Building project. Once you have the scan imported, scale it to the proper size and trace over it with Revit Building objects. If preliminary design data includes field notes or non–dimensionally accurate sketches, transpose the information into Revit Building using objects such as Walls, Doors and Windows (as shown in Chapter 3). It is also common for the site data, toposurface, and building pad to be added later in the design stage. It can be easily done at any point that fits your project and design sequence. Revit Building also contains its own collection of modeling, sketching and preliminary design tools (often referred to as "Building Maker') so that you can also choose to do your preliminary explorations directly within Revit Building (More information can be found in the appendices).

Regardless of the source of preliminary design data, gather all project data together in Autodesk Revit Building early in the project cycle. If you are taking the project through to CDs, also consider developing a digital "cartoon set." A cartoon set is simply a collection of preliminary Sheets that you can print out early and will allow you to quickly assess the quantity and composition of each of the drawings and schedules required in the final document package. Remember, like everything else in Revit Building, the cartoon set will evolve and become more refined as the project progresses. Once you have set up Views and Sheets, they will always update live with the project as the design progresses. The goal is to simply build a rough road map and gain a jump start on production.

CONSTRAINING WALLS TO LEVELS

Now that we have our site contours, a Toposurface and a building Pad, we can add some basic Walls to rough out the overall form of our building on the site. At this early stage of the project, some simple Generic Walls will be sufficient. As the design evolves the wall types can be changed and edited to suit our needs.

1. On the Project Browser, double-click the *Level 1* floor plan View.

We have not looked at this View since before we added the site data. Notice that this View still looks exactly as it did before. This is because by default each floor plan View shows only its own Level. We can however, display any other Level in the building as an underlay to the current View.

2. Right-click in the View window and choose **View Properties**.
   - In the Element Properties dialog, beneath the Graphics grouping, locate the "Underlay" parameter.
   - From the Underlay list (which currently reads: None) choose **Street Level**.

The Street Level View (which contains the Toposurface and Pad from above) will appear in gray on the current View.

   - Zoom so you can see the rectangle in the middle of the site plan.

3. On the Design Bar, click the Basics tab and then click the Wall tool.
   - From the Type Selector, choose Basic Wall : Generic - 12" [Basic Wall : Generic - 300mm]
   - On the Options Bar, click the Pick Lines icon.
   - Verify that Level is set to **Level 1**, and then from the Height list, choose **Roof**.
   - For the Loc Line choose **Finish Face: Exterior** (see Figure 4–24).

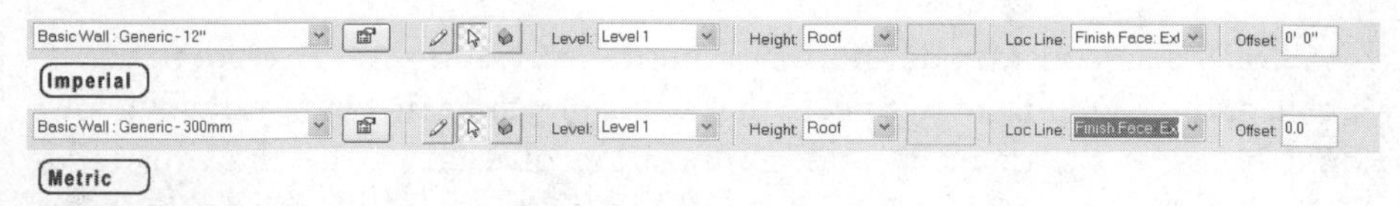

Figure 4–24 *Set the options to create new Walls*

4. In the View window, position the pointer over one edge of the rectangle in the middle of the site (it should pre-highlight as Pads:Pad:Pad 1).

Look carefully at the way that the edge pre-highlights. A dashed green line will appear parallel to the edge on one side or the other. If you move your pointer slightly, the dashed green line will shift to the other side of the edge. This indicates the side on which the Wall will be created.

- When the dashed green line is on the inside of the rectangle, press the TAB key.

  All four edges should pre-highlight with dashed green lines to the inside.

- When all four sides of the rectangle pre-highlight, (and the green dashed lines are on the inside) click the mouse to create Walls.

5. On the Project Browser, double-click to open the *West* elevation View.
   - Zoom in near the bottom of the building
   - Verify that the Walls were created correctly aligned to the edge of the Pad (see Figure 4–25).

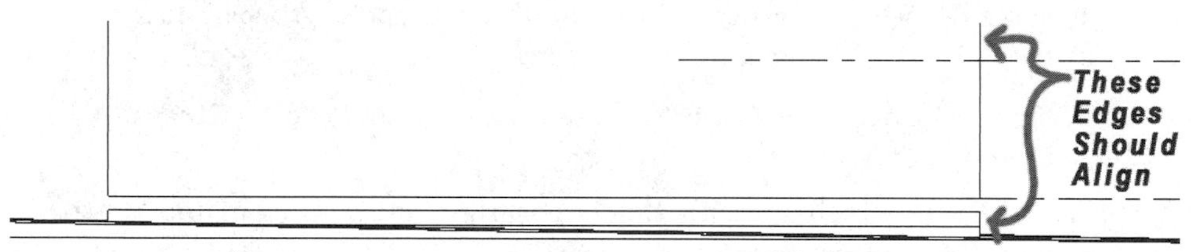

Figure 4–25 *The Exterior Edge of the Walls should align with the Pad edges*

The outside edge of the Walls should be aligned to the outside edge of the Pad. If your Walls were created to the outside of the Pad and are not flush, you can either Undo, or return to the *Level 1* plan View and Change the Wall orientations with the flip handle.

Notice that there is a gap between the bottom edge of the Pad and the Walls. This will eventually be the floor slab of the first floor. We can easily close this gap for now by dropping the bottom edge of the Walls.

6. Pre-highlight one of the Walls, press TAB to pre-highlight the chain of Walls and then click to chain select them.

- On the Options Bar, click the Properties icon (or right click and choose **Properties**).
- In the "Base Offset" field, type **-1′-0″ [-300]** and then click OK.

7. On the Project Browser, beneath 3D Views, double-click on {*3D*} to return to the default 3D View.

- On the View Control Bar, click the Model Graphics Style icon and choose **Shading** (or **Shading with Edges**).

We now have a big hollow box sitting on our terrain. Let's add a simple Roof.

ADD A SIMPLE ROOF

8. On the Design Bar, click the Roof tool and then choose **Roof by Footprint**.

A message will appear requesting that we select the desired Level to build the Roof. This message appears because we are in a 3D View. Naturally we will want to create the Roof at the Roof Level.

- In the "Lowest Level Notice" dialog, choose Roof from the drop-down list and then click Yes (see Figure 4–26).

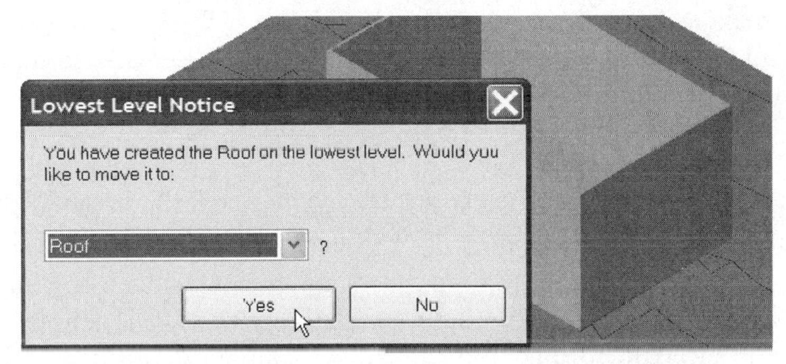

Figure 4–26 *When creating a Roof in 3D, clarify the correct creation Level*

- On the Design Bar (now in Sketch mode) verify that Pick Walls is selected.
- On the Options Bar, clear the checkmark from the "Defines Slope" checkbox.
- Using the TAB key, chain select all four Walls and then click to create sketch lines.
- On the Design Bar, choose Finish Roof.

ADJUST WALL HEIGHT

Our goal at this stage is to create a very rough model that we can use to help setup and understand the Views in our project. However, one quick edit to the Walls is immediately obvious. Let's adjust the Wall height to suggest a parapet at the Roof.

9. Chain select all four Walls and then click the Properties icon (on the Options Bar).
   - For the Top Offset parameter, type **4′-0″ [1200]** and then click OK.
   - On the Design Bar, click the Modify tool or press the ESC key twice.
10. Save the Project.

ADD A MASS

In the coming chapters, the design of both our projects will evolve. At this stage of our commercial project we have little information other than the number of levels and site data that we have already incorporated into the project and few notions of the building's overall form. It is desired that some sort of special treatment be given to the front façade. We don't have the specifics about this treatment yet but at this stage can make a few assumptions. To do this, we will sketch out a simple Massing element on the front façade of our building to suggest that some design element will later occur in this location.

11. On the Project Browser, double-click to open the *Level 1* floor plan View.
12. On the Design Bar, click the Massing tab and then click the Create Mass tool.

A message will appear indicating that the "Show Mass" mode will be enabled. Massing tools are meant as design tools and as such are not visible by default. We can enable their visibility in any View that we wish. This message is simply a courtesy indicating that Massing display can be enabled temporarily for us.

- In the message dialog, click OK after you have read it.
- In the Name dialog, type: **Front Façade** for the name of the Mass that we are creating.
- On the Design Bar (now in Massing mode) click the Solid Form tool and then choose **Solid Extrusion**.
- On the Design Bar (now in Sketch mode) click the Lines tool.
- Sketch the shape indicated in Figure 4–27 using the dimensions indicated.

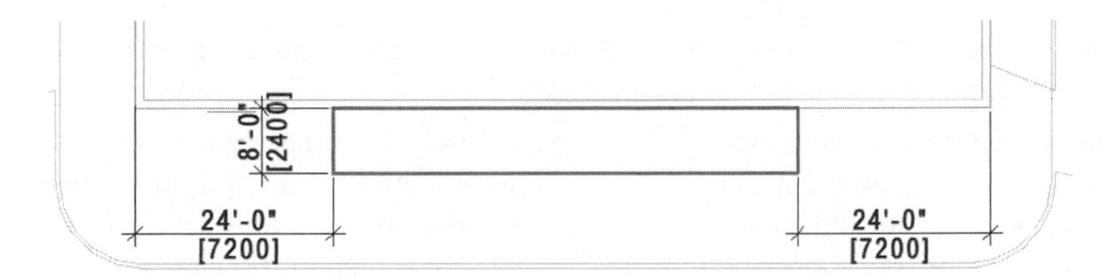

Figure 4–27 *Sketch the footprint of the Mass*

- On the Design Bar, click the Extrusion Properties button.
- In the Element Properties dialog, in the Extrusion End field, type **40′-0″ [12,000]** and then click OK.
- On the Design Bar, click the Finish Sketch button, and then click the Finish Mass button.

13. On the Project Browser, beneath 3D Views, double-click on *{3D}* to return to the default 3D View (see Figure 4–28).

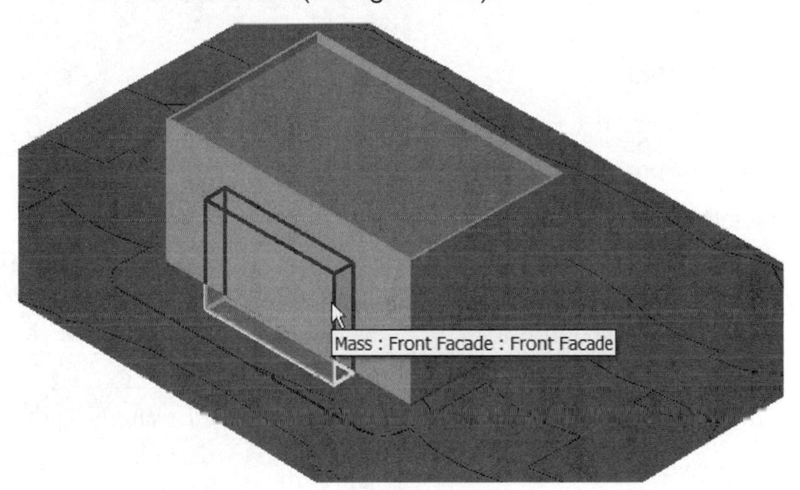

Figure 4–28 *The completed Mass at the front of the building*

Later, we can modify the shape of this Mass and create Walls or Curtain Walls from it. For the time being, we will just keep the Mass as a sort of "stand-in" for later design treatment. We want to be sure that the mass continues to display, so let's edit the Visibility/Graphics of this View.

14. From the View menu, choose **Visibility/Graphics**.

TIP The keyboard shortcut for Visibility/Graphics is **VG**.

- On the Model Categories tab, in the Visibility column, place a checkmark in the box next to Mass and then click OK.

This essentially makes the "Show Mass" toggle from above permanent for this View. If you want to make it permanent in other Views like the elevations, you have to repeat these steps there as well. You do not have to make this change however, as you can always choose **Show Mass** from the View menu if they are not displaying when you reopen the project.

WORKING WITH ELEVATION VIEWS

The default templates that we used to create our project already included four building elevations. As our project progresses, we can work with these elevations, delete them, and/or add additional ones. Elevation Views are vertical slices through the building model cut from a certain point and projected orthographically. Level heads will display automatically in elevation Views and elevation View tags will appear by default in all plan Views to indicate where the elevations are cut. The four default elevation Views are sufficient to our needs at this time. We will make a series of adjustments to them.

MOVING ELEVATIONS

We have added enough geometry to our model to see that it might be beneficial to adjust the locations of our elevation tags and Level heads.

15. On the Project Browser, double-click each elevation View in succession.

Notice that as you open each elevation, some of them show an elevation of the building as you would expect, some might show a section cut through the building while others do not show the building at all. This is due to the locations of the elevation tags in the plan Views.

16. On the Project Browser, double-click to open the *Level 1* plan View.
    - Zoom in on the bottom-most elevation tag.
    - Move the mouse pointer over the tag.

Notice that it is actually made of two pieces. The elevation tag is the square [round] shape, and the arrow (triangle) portion indicates the direction and extent of the associated elevation View (see Figure 4–29). By default one Elevation arrow is active per elevation tag. Each elevation tag can have up to four elevation arrows and corresponding elevation Views.

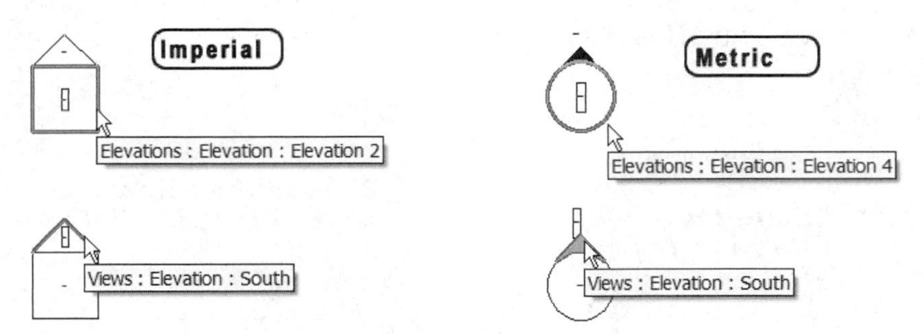

Figure 4–29 *Elevation Tags and Elevation View indicators*

- Select the Elevation Tag (square in Imperial and round in Metric).

Three additional Elevation View Indicator arrows will appear temporarily as long as the Tag is selected. Small checkboxes appear that you can use to add new Elevation Views. If you remove a checkmark, the associated Elevation View will be deleted (see the left side of Figure 4–30). A prompt will appear to warn you of this. To test this out, check one of the empty boxes, then note the addition of a new elevation View on the Project Browser. Remove the checkmark from the same box and a warning will appear indicating that this new View will be deleted. There is also a Rotate handle that appears when the Tag is selected. This will rotate the entire Elevation Tag and all of its associated Views. As the elevation tag is rotated it will snap to 90° increments and perpendicular to adjacent walls and other geometry. Try it out if you like, just be sure to undo before continuing.

- Deselect the Elevation Tag and then select the Elevation View Indicator arrow.

When this arrow is selected, it actually turns blue instead of the customary red. You can double-click the blue arrow to open the associated elevation View. Selecting the elevation arrow also reveals a line perpendicular to the direction of the arrow with a drag control at either end (see the right side of Figure 4–30). This line indicates the extent of the elevation View. Basically, an elevation View is very similar to a Section cut at this line. Anything behind this line will not be seen in the associated View. In order for the extents of the elevation to actually crop the View, you must enable the Crop Box parameter in the associated View.

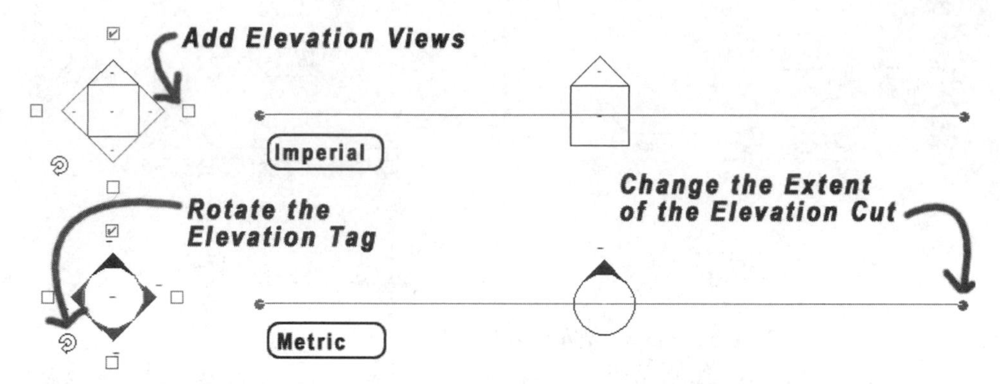

Figure 4–30 *Handles and Drag Controls on the Elevation Tags and Elevation View Indicators*

To get our elevation Views displaying properly, we need to move these Tags and Elevation View arrows. While it is possible to move them separately, it is best to keep them paired up wherever possible.

- Type **ZF** on the keyboard, or choose **Zoom > Zoom to Fit** from the View menu.
- Select both parts of the same elevation Tag (the Tag and the arrow for the bottom-most one pointing up).
- Drag it to just below the building roughly centered on the front façade (see the left side of Figure 4–31).

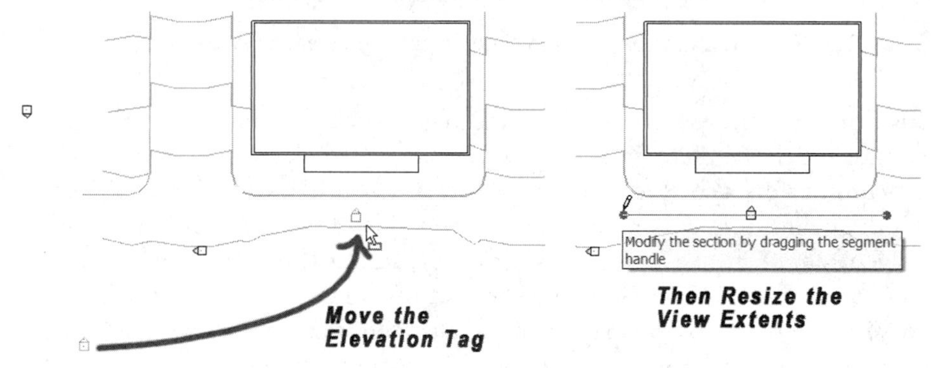

Figure 4–31 *Move the South Elevation Tag and Arrow closer to the building*

- On the Design Bar, click the Modify tool or press the ESC key.
- Click just the Elevation View indicator arrow and then use the drag handles to resize the View Extents to just a bit wider than the model (see the right side of Figure 4–31).
- Repeat this process to move and resize the other three elevations (see Figure 4–32).

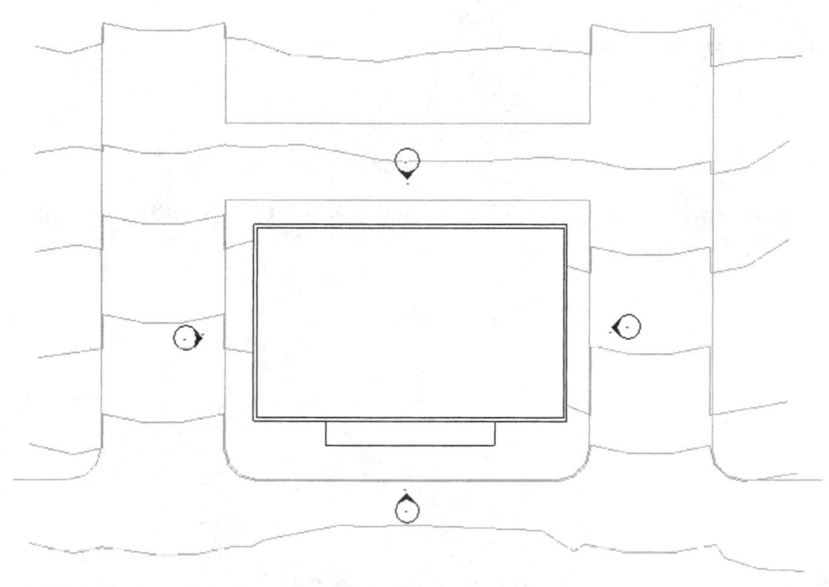

Figure 4–32 *Move the Remaining Elevation Tags and Arrows*

17. Save the Project.

MOVING LEVELS

Now that we have moved and resized the elevations, all of the Views will display correctly. However, the Level Heads still need adjustment.

18. On the Project Browser, double-click each elevation View in succession.

As you can see, the building appears centered in each View now, and they all correctly show as elevations and not sections. However, the Levels are still off-center in all Views. We can select and move them the same way we moved the elevation lines. However, if we use the drag technique, we can accidentally move them vertically as well which would affect our Level heights. To move them precisely left or right only, we will use the Move tool.

19. On the Project Browser, double-click to open the *West* elevation View.

   - Click above and to the right of all the Level Heads and then drag a selection box through them all (down and to the left).

   For this technique to work, the selection box must go from right to left (it will have a dashed edge).

   - On the Edit toolbar, click the Move tool.

TIP The keyboard shortcut for Move is **MV**. **Move** is also located on the Edit menu.

- For the Move Start Point, click anywhere above the model.
- Move the pointer to the left approximately **80′-0″ [16,000]** (see Figure 4–33).

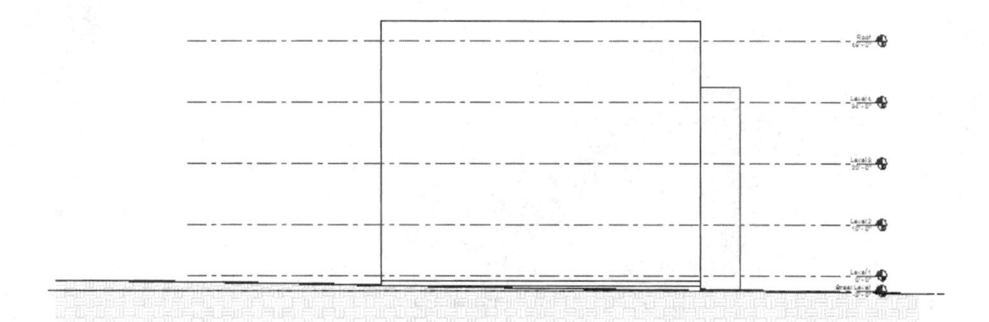

Figure 4–33 *Move the Level Heads to center them on the Elevation*

The precise amount of the move does not have to be exact. However, it is critical that the move be constrained to a horizontal movement. The Move command does this by default. While you can move the pointer and move along an angle, the move command snaps to a horizontal angle automatically.

20. Repeat this process in the *North* elevation View.

Notice that you only need to move the Levels in two elevation Views. Since elevations (like all Revit Building Views) are live Views of the model, changes you made in the *West* View can be seen automatically in the *East* View and likewise the changes in the *North* View appear automatically in the *South*.

21. On the Design Bar, click the Modify tool or press the ESC key.

It may be necessary to adjust the extent of the Levels in either or both of the North/South or East/West directions. To do this, select one of the Level lines, and then drag the handle at the end to lengthen or shorten the Level extent. When you do this to one Level line, all of them will adjust together. This is because they are all constrained to one another (see above). This edit is likely necessary only if you are working in Metric units as the default Metric template from which we began the project has smaller Level extents than the Imperial template.

EDIT LEVEL HEIGHTS

Earlier when we set up all of the Levels, we neglected to adjust the default height between Level 1 and Level 2. By default in the Imperial template it is 10′-0″ and it is 4000mm in the Metric template. However, for this project, the heights of all four Levels should be the same at 12′-0″ [3600]. The change was postponed till now so that we can get a sense of the power of some of the simple constraints that we have already built into our model.

22. On the Project Browser, double-click to open the *West* elevation View.
    - Select Level 2.

Notice the temporary dimensions that appear. You can certainly edit those values to move the Levels to their correct height. However, the problem with that technique, is that you must move the Levels one at a time. A better approach is to use the Move tool like we did in the previous sequence, or use the "Activate Dimensions" Options Bar option for a multiple selection.

23. Select the Level 2, Level 3, Level 4 and the Roof Level.
    - On the Options Bar, click the "Activate Dimensions" button.
    - In the temporary dimension that appears between Level 1 and Level 2, type **12′-0″ [3600]** (see Figure 4–34).
    - Watch the top edge of the Wall as you press ENTER.

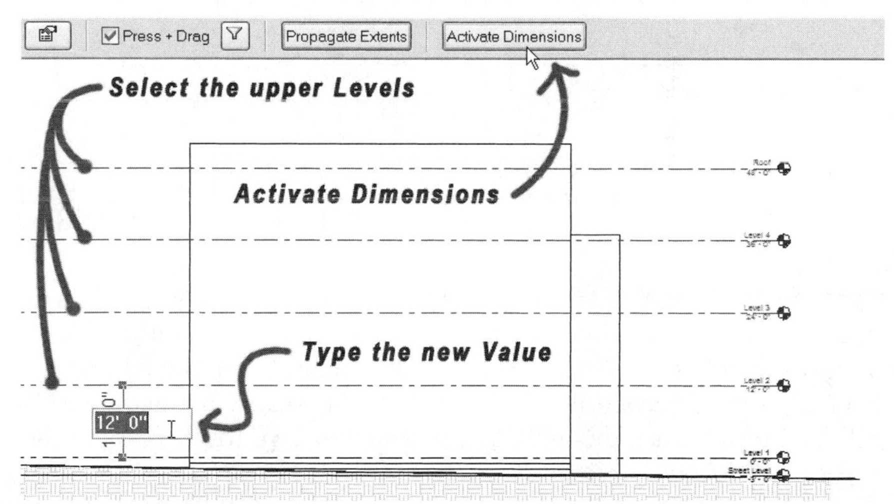

Figure 4–34 *Activate Dimensions on a multiple selection to move them together*

You should notice that the top edge of the Wall moves with the Levels. This is because earlier, we constrained the top of the Wall to the Roof Level. This is another example of the power of the parametric relationships between elements in an Autodesk Revit Building model.

CREATING SECTION VIEWS

Unlike elevations, section Views are not included in the default template. Section Views are also vertical cuts through the building model. They typically show the building in a cutaway fashion in an orthographic View. Level Heads also appear in Sections automatically. Adding Sections to our model is a simple task.

CREATE A SECTION LINE AND ASSOCIATED VIEW

Let's add two building sections, one longitudinal and one transverse.

1. On the Project Browser, double-click to open the *Level 1* floor plan View.
2. On the Design Bar, click the Basics tab and then click the Section tool.
   - Click the first point on the outside left of the building near the middle.
   - Move the pointer horizontally to the right and then click the second point outside the building to the right (see Figure 4–35).

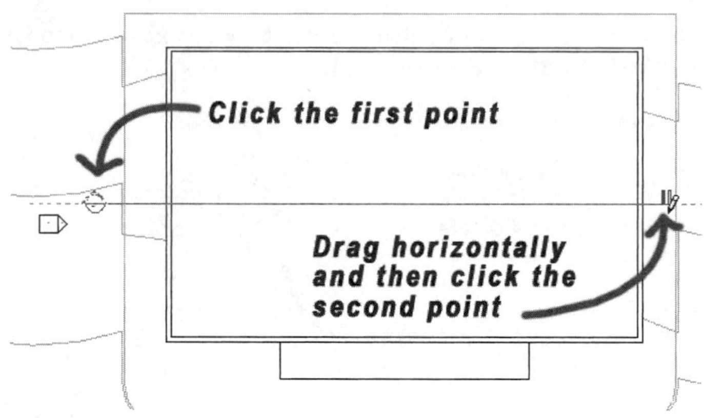

Figure 4–35 *Click two points to create a Section Line*

A section line will appear with several Drag controls and Handles on it. Like the other tags and symbols that we have seen so far, simply hover the Modify tool (mouse pointer) over each control; the handle will temporarily highlight and the tool tip will appear indicating the function (see Figure 4–36). Moving left to right in Figure 4–36, the following briefly describes each control.

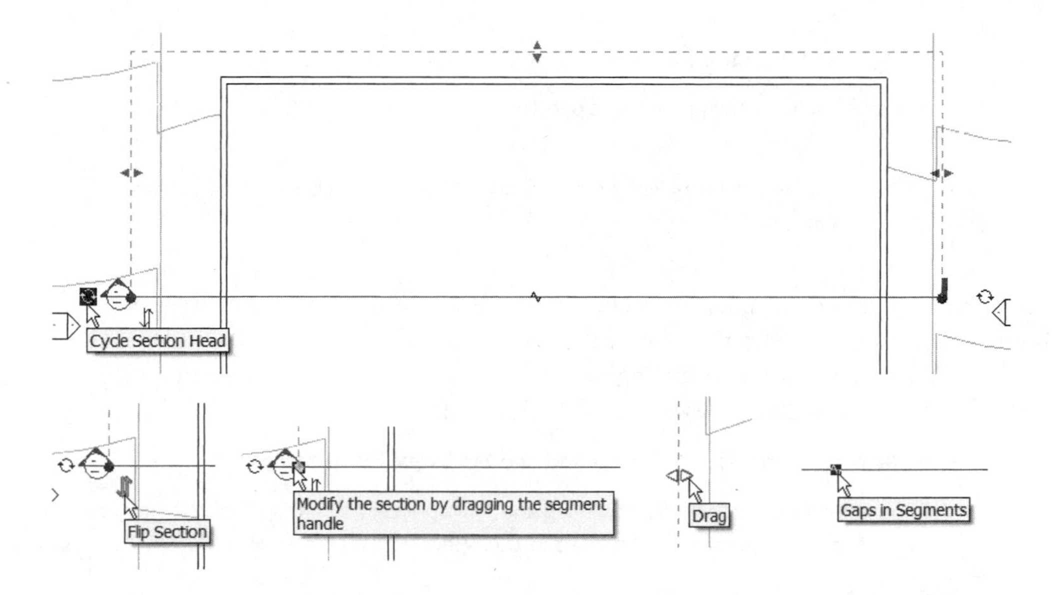

Figure 4–36 *Section Line Drag Controls and Handles*

- **Cycle Section Head**—Each time you click this handle, the Section Head will cycle to a different symbol. You can do this at both ends of the Section Line.
- **Flip Section**—Click this control to flip the Section Line to look the opposite way.
- **Segment Drag Handle**—This handle (round dot) controls the length of the Section Line and move the Section Heads with it. Dragging this handle does not change the size of the Section Crop Box.
- **Drag**—Drag Handles appear on three sides of the Section Crop Box. Use them to define the precise extent of the associated Section View.
- **Gaps in Segments**—This control breaks the Section Line and removes the inner segment. You can then drag the exact length of each end segment.

Notice that when you add the Section marker, a new "Sections" branch appears in the Project Browser and that an associated Section View appears on this branch.

- Using the Drag Handles, make any necessary adjustments to the extent of the section box or the position of the Section Heads.
- If you wish, click the "Gaps in Segments" handle to remove the middle section of the Section Line.

3. On the Project Browser, expand Sections (Building Section).
    - Right click *Section 1* and choose **Rename**.

- Type **Longitudinal** and then click OK.
- Double-click *Longitudinal* to open the View.

TIP You can also double-click the blue Section Head in any plan View to jump to the associated Section View.

Notice the box that surrounds the section in the View. This is the Crop Region for this section and it matches the size of the section box that we created in plan. All graphical Views can have a Crop Region enabled. By default they are turned off in most Views, but in section Views they default to on.

4. Right-click in the View window and choose **View Properties**.
   - Scroll down and notice that both the "Crop Region" and "Crop Region Visible" parameters are enabled in this View.
   - Click Cancel to exit the dialog.
   - Repeat this process in any other View.

Notice that the "Crop Region" and "Crop Region Visible" parameters are turned off typically in plan and elevation Views. We can enable these settings on a per View basis. Crop Regions help you limit the extent of the model objects that are visible in the View to just the area that you wish to study, annotate or print. We will see additional examples of Crop Regions in later chapters.

5. Repeat the steps above to create a vertical section looking to the right.
   - Rename the View to **Transverse**.

Make any additional adjustments that you wish. When you open other Views, the section lines will appear. You can make the same sort of adjustments to them in each View: add gaps to the line, drag the Section Head to a different location. These adjustments are View specific. The same Section marker in the other Views can look different. The location of the Section marker line (perpendicular to its length) is in the exact same place. Or you can adjust the size of the Crop Region from any View. This is not View specific. If you adjust the Crop Region or the location of the Section marker in any View it will move accordingly in all other Views.

6. Save the project.

SCHEDULE VIEWS

We have now created and worked with several types of Views. While floor plans, ceiling plans, sections and elevations may be the most common Views that our projects will contain, Revit Building includes over a dozen distinct View types.

(You can see the complete list on the **View > New** menu). One of the most powerful View type in a Revit Building model is a Schedule View. The Schedule View is not a drawing, but rather a non-graphical View of a collection of related data extracted from the model and presented in tabular format. You do not need to "draw" and manually compile your schedules when you work in Revit Building. You simply create a Schedule View, indicate to Revit Building which pieces of data you wish to include (add Fields) and the data required to populate the Schedule will be extracted from the model and input into the cells of the Schedule View. Essentially, the major concept to understand here is that Views in a Revit Building project can be either graphical (drawings) or non-graphical (schedules and legends).

ADD A SCHEDULE VIEW

In this sequence, we will add a few simple Schedule Views to our project. We will not get into a detailed explanation of scheduling features at this time. Schedules are covered in detail in later chapters.

1. On the Design Bar, click the View tab and then click the Schedule/Quantities tool.

   - In the "New Schedule" dialog, choose Walls from the Category list and then click OK (see Figure 4–37).

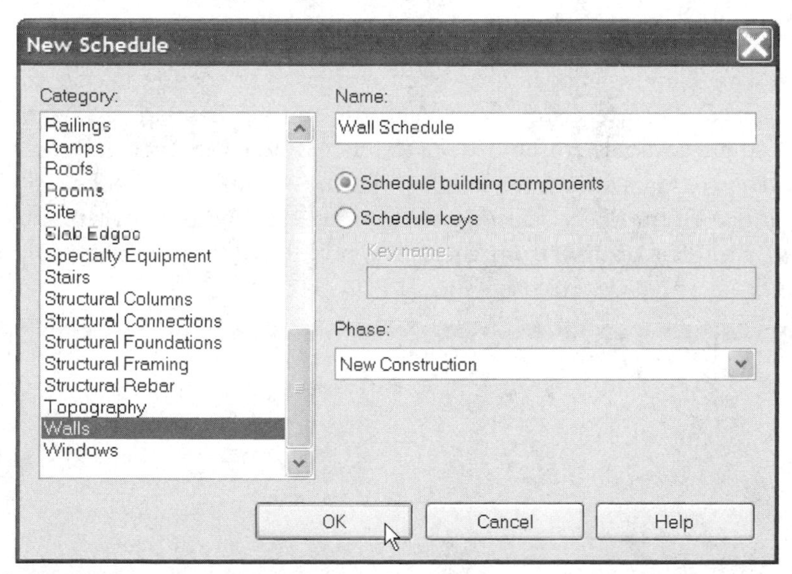

Figure 4–37 *Create a New Wall Schedule*

   - In the "Available fields" column, select "Type Mark" and then click the Add → button.

- Repeat this process for the "Length," "Width," "Family and Type" and "Comments" fields (see Figure 4–38).

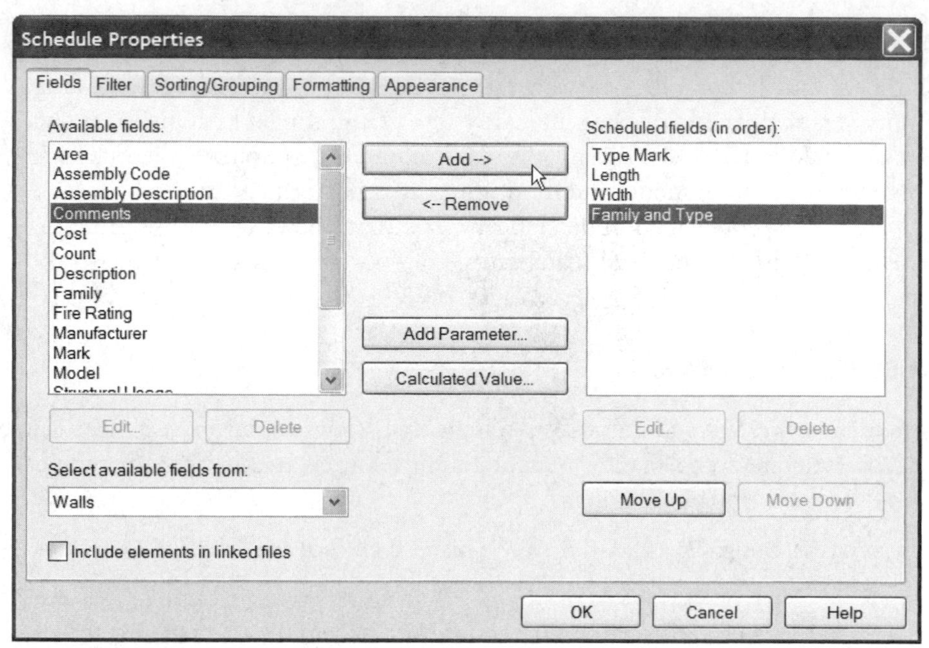

Figure 4–38 *Add Fields to the Schedule*

- Click OK to complete field selection and open the Schedule View.

The *Wall Schedule* View will appear on screen with each of the four Walls that we currently have in our project listed. The Walls do not yet have "Type Marks" or "Comments" so these fields are empty. Later, as we edit the data parameters of these Walls, these fields will update to reflect the latest information. The Schedule View is also added to the "Schedules/Quantities" node of the Project Browser (see Figure 4–39). You can open it from there any time just like the other graphical Views of the project that we explored above.

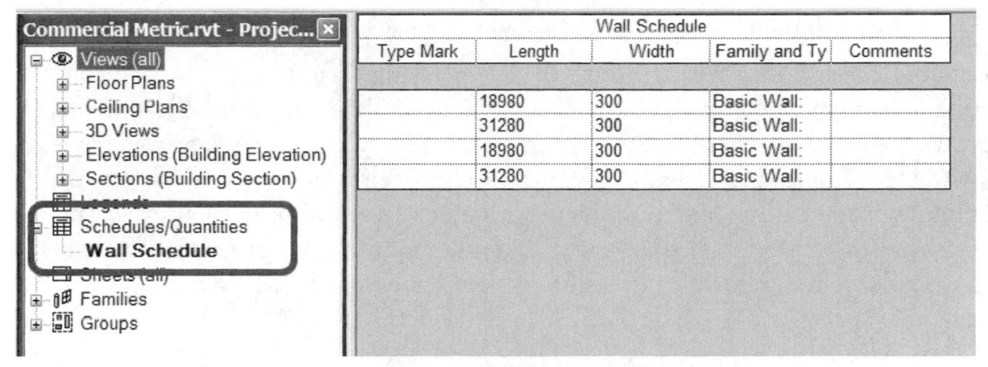

Figure 4–39 *The Wall Schedule appears in the Project Browser and opens on screen*

IMPORTING SCHEDULE VIEWS FROM OTHER PROJECTS

You cannot directly export and import Schedules from one project to another. However, you can copy and paste them between projects which gives the same functionality. When we began this chapter, we opened several out-of-the-box Revit Building template projects. It was noted above that starting a project from a template was the preferred method of beginning a Revit Building project. In some cases, you will begin a project with one template and then later realize that a particular Family, Type or View required already exists in another template file or project. Rather than re-create the item, it is easier to borrow the element from the other project. In this topic, we will create a new project from the *Commercial-Default.rte* template file (installed with the out-of-the-box Imperial templates) and borrow some Schedule Views from that project to use in our own commercial project.

NOTE The *Commercial-Default.rte* template file is installed with the out-of-the-box Imperial templates. If you did not install the Imperial templates, or your version of Revit Building does not give you access to this file, it has been provided for you in the *Out of the Box Templates* folder with the files from the Mastering Revit Building CDROM.

2. On the File menu, choose **New > Project**.

- In the New Project dialog, in the "Template File" area, be sure that the lower radio button is selected (the one that lists a template file) and then click the Browse button.
- Browse to the *Imperial Templates* folder.

NOTE If you do not have access to this folder, you can find a copy of the template in the *Out of the Box Templates* folder instead.

- Select the *Commercial-Default.rte* template file and then click Open.

Take a look around this project. Many of the Views that we have spent time configuring above have been included in this project template. In particular, you will note that the Level structure of this project includes Levels for such datums as bottom and top of footing and framing. Levels such as this are not required, but can be helpful when creating and presenting your model. In addition to the Levels, this project has some useful Schedule Views and a large collection of Sheets. We are going to borrow the Schedules to use in our project and then we'll have a look at some of the Sheets.

3. On the Project Browser, expand Sheets.

   - Double-click to open the *A13 – Schedules* Sheet.

A Sheet is special type of View that is intended to compose and print your document set—they are printing Views. When you create a Sheet, you can add a title block to it, configure the Print Setup to match the paper size and other required printing parameters and add Viewports containing the various Views of your project properly scaled and annotated. We will work with Sheets next, for now, we have opened this existing Sheet because you can copy Schedules from a Sheet in one project and then paste them to a Sheet in another. This method is a bit faster than creating the Schedule View from scratch in the new project.

NOTE If you use the same Schedule Views (or any Views) repeatedly from one project to the next, you should create your own project template that already includes the Views you require.

4. Select all three Schedules on the Sheet. (You can use the CTRL key and click on each one, or drag the mouse right to left to do this).

   - From the Edit menu, choose **Copy to Clipboard**.

TIP The keyboard shortcut for Copy to Clipboard is CTRL + C.

5. On the Window menu, choose **MRB Commercial.rvt – Schedule: Wall Schedule** to return to our commercial project.

6. On the Project Browser, right-click Sheets and choose **New Sheet**.

   - In the "Select a Titleblock" dialog, click OK to accept the default.

This will create a new Sheet using the default titleblock. This is fine for now. Later we can swap in another titleblock Family if we wish.

- From the Edit menu, choose **Paste from Clipboard.**

TIP The keyboard shortcut for Copy to Clipboard is CTRL + V.

If a "Duplicate Types" dialog appears, simply click OK to dismiss it.

- Position the pointer and then click to place the pasted Schedules on the Sheet.

NOTE Depending on your units, and the titleblock that your project uses, the pasted Schedules may or may not fit completely within the border. We can rearrange them later. This process is simply a quick way to borrow the Schedules from a different project and add them to the current project.

- On the Project Browser, expand the Schedules/Quantities branch.

In addition to the three Schedules being added to the new Sheet, there are now three more Schedule Views: "Door Schedule," "Room Schedule" and "Window Schedule" in the Project Browser. You can either zoom in on the Schedule on the Sheet or double-click it on the Project Browser to View its contents.

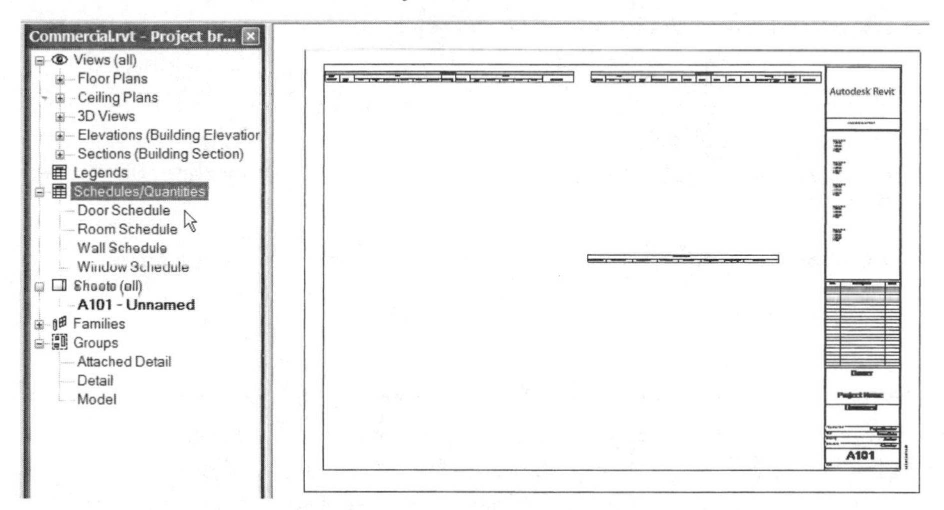

Figure 4–40 *The pasted Schedules appear on the Sheet and in the Project Browser*

7. On the Project Browser, beneath Schedules/Quantities, double-click the *Door Schedule* View to open it.

Notice that this Schedule contains several columns, but lists no Doors. This is because our project does not yet contain any Doors.

ADD A DOOR TO THE PROJECT

When you add elements to the project, they will automatically appear in all appropriate Views—including the Schedule Views. Let's add a Door to our project.

8. On the Project Browser, double-click to open the *Level 1* floor plan View.
   - Zoom in on the middle of the top horizontal Wall in the plan.
   - On the Design Bar, click the Basics tab and then click the Door tool.
   - On the Type Selector, choose Single-Flush : 36″ x 84″ [M_Single-Flush : 0915 x 2134mm] from the list and then add a Door to the middle of the Wall.
   - On the Design Bar, click the Modify tool or press the ESC key twice.
9. On the Project Browser, double-click to open the *Door Schedule* View.

Notice that the Door also appears in the Schedule.

10. Click in the Width field of the Schedule.
    - Change the Width to **4′-0″** **[1200]** and then press enter.
    - In the dialog that appears, confirm the change to the Type by clicking OK.
    - Return to the *Level 1* plan View and confirm the new size.

In Autodesk Revit Building, a change made in one View applies in all associated Views. You are not changing a graphic representation in a drawing. You are actually editing the parameters of an element in your virtual building model. This change must therefore be reflected in any and all Views that show that component of the model.

11. On the Edit menu, choose **Undo Edit Schedule** (or press CTRL + Z).
12. Save the project.

SHEET VIEWS AND THE CARTOON SET

Most architectural projects will need to be printed at some point (and often at several points) in the life of the project. To facilitate this need, Autodesk Revit Building provides Sheet Views. As noted above, a Sheet is a special View containing a titleblock and is intended for printing. There are three approaches to the task of creating Sheet Views. You can create Sheet Views "on the fly" as needed for a particular printing task, you can create a bunch of Sheet Views ahead of time that will automatically update with the project, or you can create your project from a template that already contains a number of preconfigured Sheet Views. However, project submission time is usually a high-stress endeavor characterized by panic as project team members scramble to assemble the required Sheets and print everything required for a particular submission. This frenzy often occurs at the last minute which contributes to the stressful environment. One simple way to help alleviate this situation is to create the Sheet Views early in the project so that they are ready to print at any time.

RENAME A SHEET

Let's first rename our Schedule Sheet.

1. In the Project Browser Right-click the *A101 – Unnamed* Sheet (the one we created above) and choose **Rename**.

   The "Sheet Title" dialog will appear.

   - For the Number, type **A601**.
   - For the Name, type **Schedules**.
   - Click OK to complete the rename (see Figure 4–41).

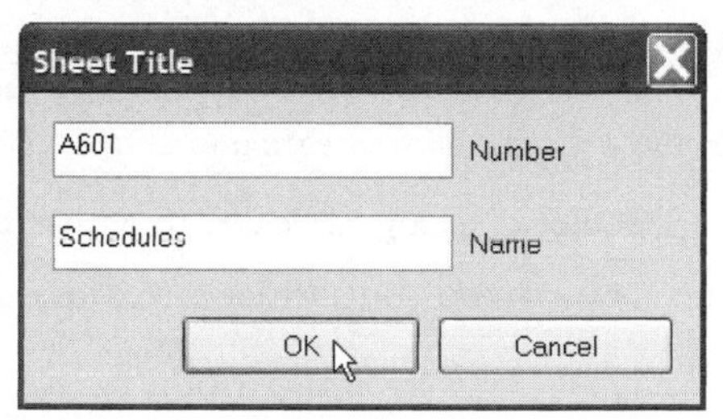

Figure 4–41 *Rename and re-number the Schedule Sheet*

Notice the change on the Project Browser.

2. On the Project Browser, double-click to open the *A601 – Schedules* Sheet.

Notice that the new values are changed on the Sheet too.

- If necessary, re-position any of the existing Schedules on the Sheet to fit within the border.
- From the Project Browser, click the Wall Schedule View and drag and drop it on the Sheet View (see Figure 4–42).

Temporary guidelines will appear to help you align it with the other Schedules.

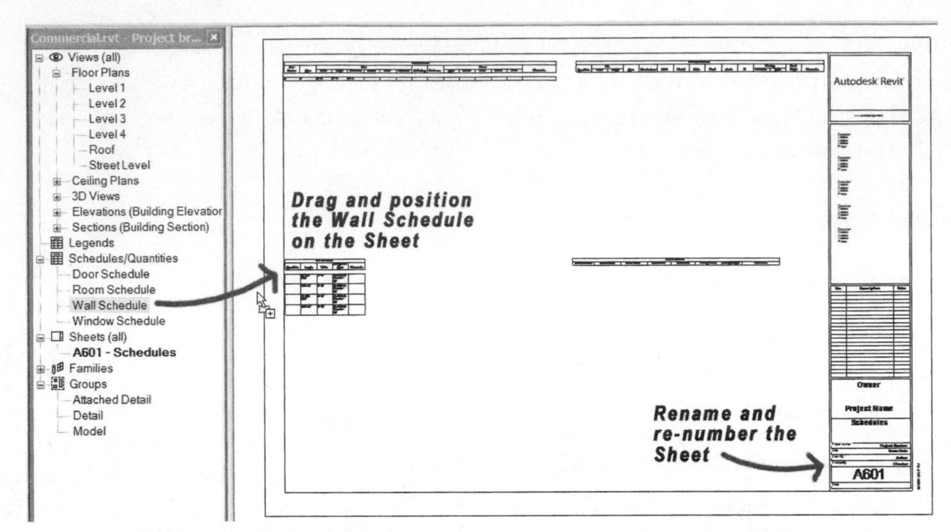

Figure 4–42 *Drag the Wall Schedule to the Sheet and re-position items to fit*

CREATE A FLOOR PLAN SHEET

3. On the Project Browser, right-click Sheets and choose **New Sheet**.
   - In the "Select a Titleblock" dialog, click OK to accept the default.

This new Sheet is also named "Unnamed" but notice that the number automatically enumerated to A602.

4. Repeating the process above, rename the Sheet to **Floor Plans** and number it **A101**.

TIP If you prefer, you can rename and re-number Sheets directly in the titleblock. Click on the titleblock to highlight it and then click the blue text to edit the value. The result will be the same.

Make sure that the *A101 – Floor Plans* Sheet is open.

- From the Project Browser, drag the *Level 1* floor plan View and drop it on the Sheet.
- Zoom in on the View title beneath the floor plan on the Sheet.

Notice that the name of the drawing is automatically "Level 1." While this is logical, and in some cases desirable, chances are we would like the name of this drawing on the printed Sheet to be something else. This is easily changed.

5. On the Project Browser, right-click the *Level 1* View and choose **Properties**.

- In the Element Properties dialog, type **First Floor Plan** in the "Title on Sheet" parameter field (see Figure 4–43).

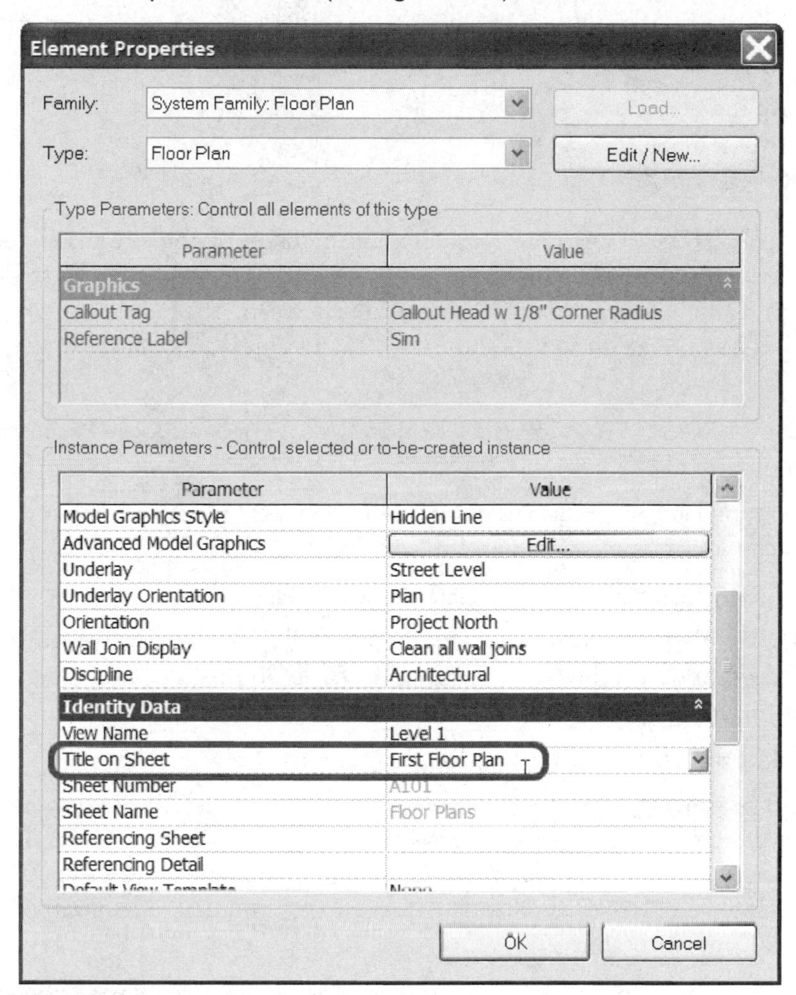

Figure 4–43 *Edit the value of the Title on the Sheet*

- Click OK to complete the change (see Figure 4–44).

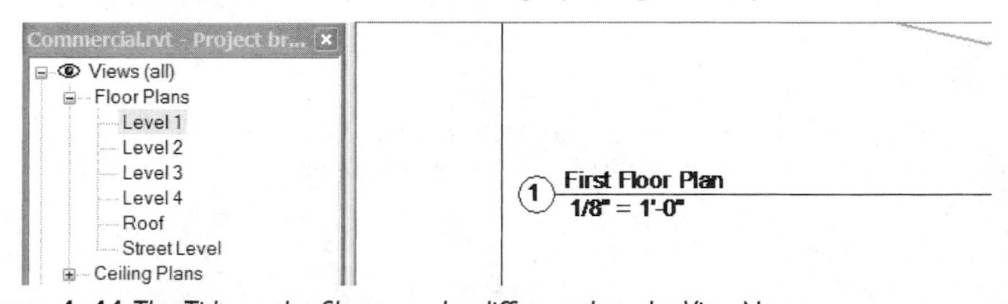

Figure 4–44 *The Title on the Sheet can be different than the View Name*

Notice that the number in the View title tag has automatically filled in with the number 1. This is because this is the first (and currently only) View that we dragged to this Sheet. Revit Building will automatically enumerate the annotation of all Views as you drag drawings onto Sheet Views.

6. Save the project.

EDIT THE CROP BOX

The *Level 1* plan (First Floor Plan) contains an underlay of the site plan (added above). While it might be desirable for some of the site to show on the First Floor Plan, we are not likely to want to show the entire site plan, but rather a small portion of just around the perimeter of the First Floor Plan. To do this, we need to enable the Crop Region of the *Level 1* View.

7. On the Project Browser, right-click the *Level 1* View and choose **Properties** again.
    - In the Element Properties dialog, in the "Extents" grouping, place a checkmark in the "Crop Region" checkbox.
    - Verify that "Crop Region Visible" is also enabled.

A rectangular Crop Region will now surround the plan. To edit it, we can open the *Level 1* View, or edit it directly from the Sheet. To do this, you must "activate" the View on the Sheet.

8. Select the Viewport edge, right-click and choose **Activate View**.

The Viewport is a rectangular region sized a little bit larger than the extents of the model and annotation objects in the View and determines what is seen on the Sheet. When the Viewport has been activated, you can manipulate the View as if it were open directly. When you deactivate the Viewport, you can move it around to compose the layout of the Sheet.

- With the Modify tool (mouse pointer), click on the Crop Region boundary to select it.

Several drag handles will appear on each side. The round ones at the midpoints can be used to edit the size of the Crop Region. We will drag these to make the Crop Region smaller for this View.

- Drag each of the round drag controls in toward the plan until only a small portion of the site plan shows on all sides of the plan (see Figure 4–45).

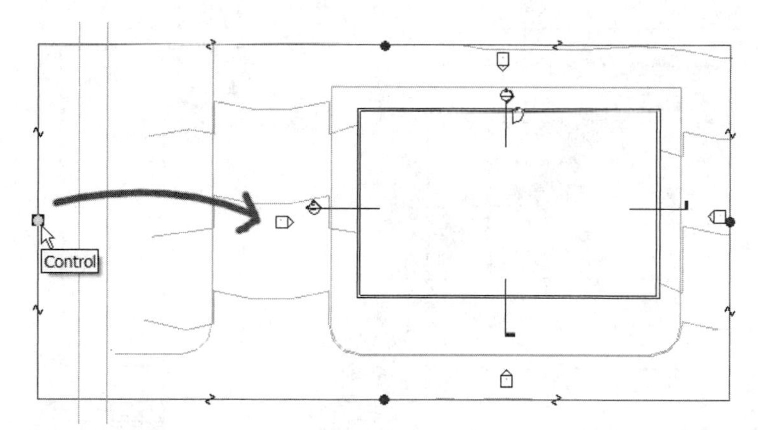

Figure 4–45 *Drag the extents of the Crop Region to show only a small portion of the Site*

- Right-click anywhere inside the Viewport and choose **Deactivate View**.

The View title beneath the plan now needs adjustment. You can move it independently of the Viewport by clicking on it with the Modify tool and dragging it to a new location. To move it with the Viewport, select and move the Viewport and the titlemark automatically moves with it. To resize the View title line, select the Viewport first, then stretch the handles.

- Move and resize the View title as required.
- Move the Viewport to the upper left corner of the Sheet (see Figure 4–46).

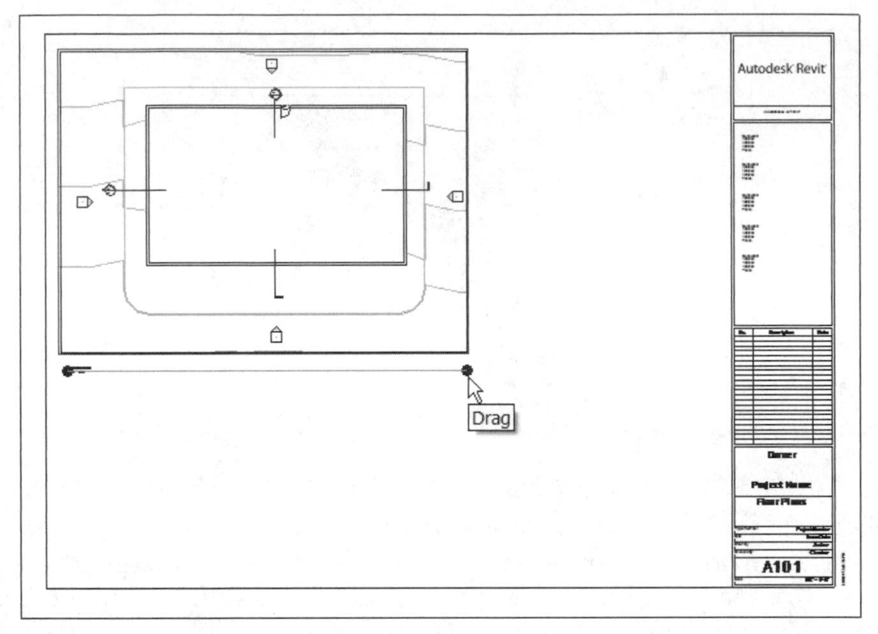

Figure 4–46 *Finalize the size and position of the Viewport and its View title.*

Take note of the Sheet on the Project Browser. When you drag Views onto Sheets, each View will be listed beneath the Sheet. You can double-click on Views from here or the standard location in Project Browser to open them.

CREATE THE REMAINING FLOOR PLAN SHEET FILES

Let's create the remaining floor plan Sheets.

9. Repeating the process outlined above, create Sheet *A102 – Floor Plans.*
   - Create each of the additional Sheets listed in the "Drag to Sheet" column of Table 4–2.
10. On the Project Browser, right-click the *Level 2* floor plan and choose **Properties**.
    - Change the "Title on Sheet" parameter to **Second Floor Plan**.
    - Repeat this process for each View listed in Table 4–2.
    - Drag each of the Views to the appropriate Sheets as indicated in Table 4–2.

None of these Views require that the Crop Region be enabled. Simply edit the "Title on Sheet" parameter and then drag them to the appropriate Sheet.

Table 4–2 *Titles on Sheets for Plan Views*

View Name	Title on Sheet	Drag to Sheet
Floor Plan Views		
Level 2	**Second Floor Plan**	A102 – Floor Plans
Level 3	**Third Floor Plan**	A103 – Floor Plans
Level 4	**Fourth Floor Plan**	A104 – Floor Plans
Roof	**Roof Plan**	A105 – Floor Plans
Street Level	**Site Plan**	A100 – Site Plan
Ceiling Plan Views		
Level 2	**First Floor Reflected Ceiling Plan**	A106 – Reflected Ceiling Plans
Level 2	**Second Floor Reflected Ceiling Plan**	A107 – Reflected Ceiling Plans
Level 3	**Third Floor Reflected Ceiling Plan**	A108 – Reflected Ceiling Plans
Level 4	**Fourth Floor Reflected Ceiling Plan**	A109 – Reflected Ceiling Plans

It turns out that earlier when we added Levels and Views to our project, we added a Roof ceiling plan. It is unlikely that we would need a ceiling plan for the roof. We can simply delete this View.

11. On the Project Browser, beneath Ceiling Plans, right-click the *Roof* View and choose **Delete**.
12. Save the project.

CREATE ELEVATION AND SECTION SHEET VIEWS

We will need a couple of Sheet Views for our Building Elevations and Building Sections.

13. Repeating the process outlined above, create the Sheet Views indicated in Table 4–3.
    - Repeating the process outlined above, edit the "Title on Sheet" parameter of each of the Elevation and Section Views as indicated in Table 4–3.
    - Drag each of the Views to the appropriate Sheets as indicated in Table 4–3.

TIP If your Elevation or Section Views are too wide for the Sheet, you can use the Crop Region handles to resize them. The Crop Region is already on by default for Section Views. For Elevation Views, you must repeat the process outlined above for the *Level 1* View.

Table 4–3 *Titles on Sheets for Plan Views*

View Name	Title on Sheet	Drag to Sheet
Elevation Views		
East	**East Elevation**	A202 – Building Elevations
North	**North Elevation**	A201 – Building Elevations
South	**South Elevation**	A201 – Building Elevations
West	**West Elevation**	A202 – Building Elevations
Section Views		
Longitudinal	**Longitudinal Building Section**	A301 – Building Sections
Transverse	**Transverse Building Section**	A301 – Building Sections

When you have completed these Sheets, your Project Browser should resemble Figure 4–47. Again take note of the elevation and section tags in the various Views. All of them now properly indicate the Sheet references and drawing numbers. All annotations were automatically enumerated in the order in which you dragged them to the Sheets.

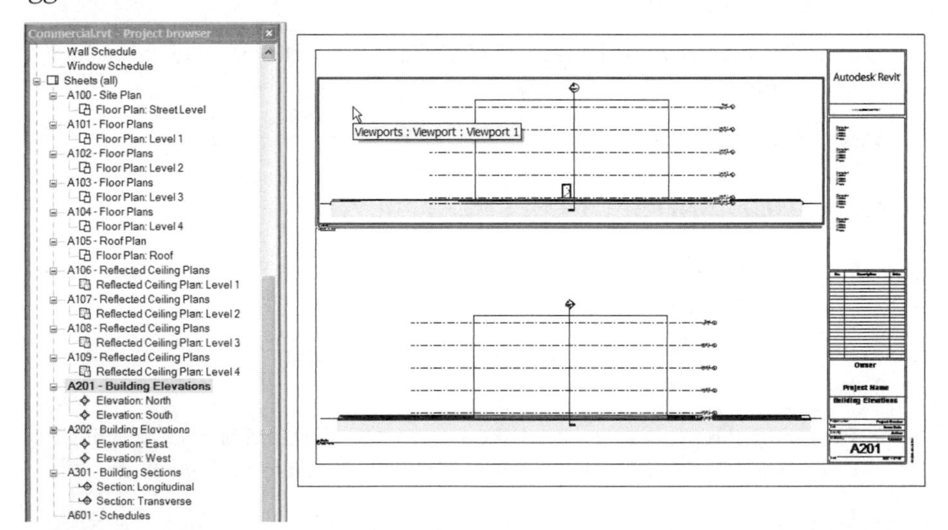

Figure 4–47 *Project Browser showing all Floor Plan, Ceiling Plan, Elevation and Section Sheets*

At this point, every Sheet that we have created is likely still open. You can see which Views and Sheets are open on the Window menu. In a project this size, it is probably OK to have several windows open at once, however, as a matter of best practice, it is a good idea to periodically close unused windows. The easiest way to do this is with the **Close Hidden Windows** command on the Window menu.

14. From the Window menu, choose **Close Hidden Windows**.
15. Save the project.

CREATE A COVER SHEET

The only Sheet we still need at this stage is a cover sheet. To create it, we will load a custom titleblock Family file provided with the files from the Mastering Revit Building CDROM.

16. On the Project Browser, right-click Sheets and choose **New Sheet**.
    - In the "Select a Titleblock" dialog, click the Load button to load a different Family.

The custom Family has been provided in the *Chapter04* folder with the other files for this chapter.

- Browse to the *Chapter04* folder, select the *E1 30 x 42 Cover Sheet.rfa* [*A1 Cover Sheet.rfa*] file and then click Open (see Figure 4–48).

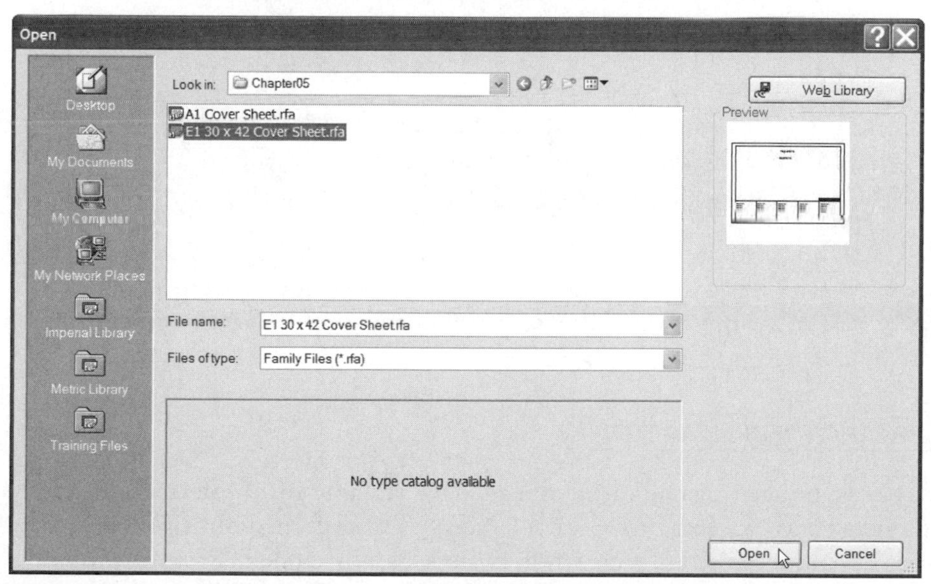

Figure 4–48 *Load the custom Titleblock from the* ***Chapter04folder***

- In the "Select a Titleblock" dialog, click OK to create the new Sheet View Cover Sheet (including our custom titleblock).
- On the Project Browser, right-click the new Sheet (currently *A302 – Unnamed*) and choose **Rename**.
- Re-number it to **G100** and rename it to **Cover Sheet**.

Why don't we add a 3D axonometric of the project to our Cover Sheet? We can use the default *{3D}* View, but it might be better to create a copy of it first.

17. On the Project Browser, expand 3D Views, right-click the *{3D}* View and choose **Duplicate**.
    - Rename the View **Cover Axonometric**.
    - Use the Dynamically Modify View tools to adjust the View to a pleasing vantage point.

TIP If you have a wheel mouse, hold down the SHIFT key and drag with the wheel to quickly spin the model.

18. Drag the *Cover Axonometric* View onto the Cover Sheet and position it.

When you drop the View on the Cover Sheet, it is probably too large for the Sheet. We can easily adjust the scale of a View after it has been placed.

19. On the Project Browser, right-click the *Cover Axonometric* View and choose **Properties**.
    - In the Element Properties dialog, from the View Scale list choose **1/16"=1'-0" [1:200]** and then click OK.
    - Reposition the View as necessary.

TIP If the scale is still too large for the Sheet, you can choose **Custom** from the View Scale list and type in a value. In the Metric files, try a scale 1:250 for example.

EDIT PROJECT INFORMATION

You may have noticed that all of the Sheets have certain bits of information filled in on them such as "Project Name" and "Client." This information is global to the entire project and is stored in a common System Family named "Project Information." We can edit its values at any time and the new values will immediately populate *all* Sheets in the project.

20. On the Settings menu, choose **Project Information**.

   - In the Element Properties dialog, input values for the various fields (see Figure 4–49).

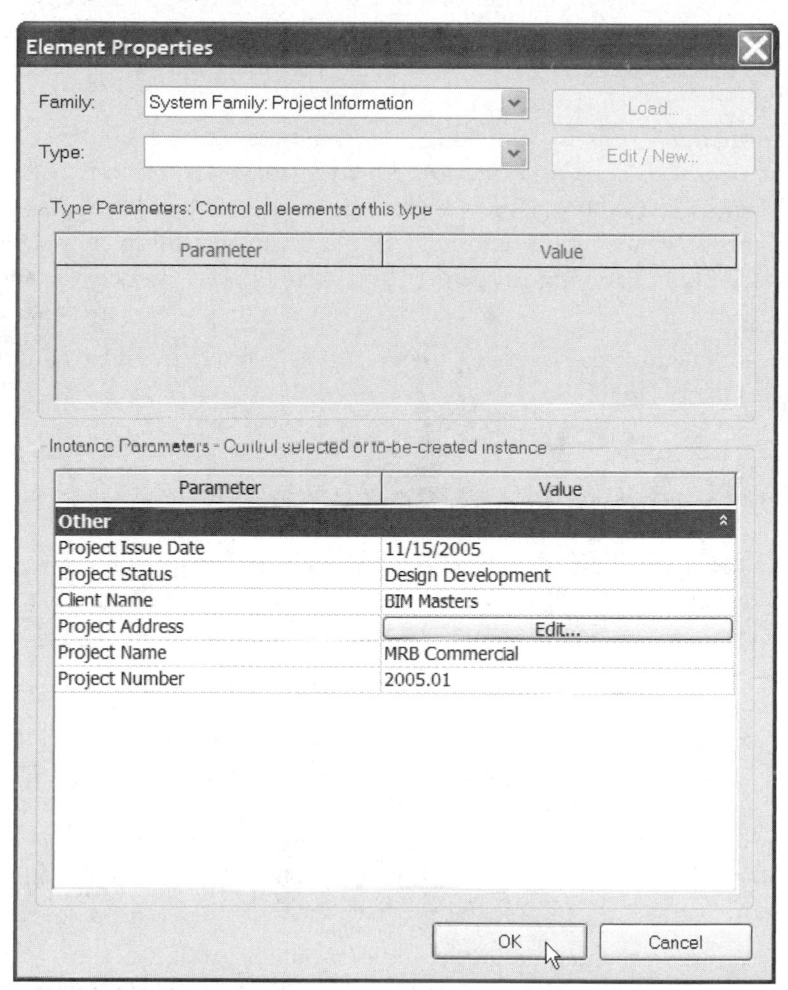

Figure 4–49 *Edit Project Information*

21. Open any Sheet and view the Titleblock fields.

Notice the change to the values. You will also notice that some of the values such as Drawn By did not change. These are sheet specific parameters that need to be edited on a Sheet View by Sheet View basis as is appropriate for this type of information.

22. Save the project.

DRAFTING VIEWS

One other type of View bears mention in this discussion: the Drafting View. A Drafting View is unlike other Views in that it is not necessarily linked back to anything in the model. The purpose of a Drafting View is to create details or other embellishments that are not easily modeled or that you and your project team have deemed not "worth" modeling. Remember that Building Information Modeling is as much "information" as it is "modeling." In some cases, a simple drawing or some sketched embellishment is all that is required to convey a certain bit of information. In these cases, a Drafting View is often appropriate. A Drafting View can have a linked portion of the model as an underlay or it can be a completely separate View comprised only of drafting lines and other non model dependent annotations.

While we are not creating any Drafting Views in this chapter, an entire chapter is devoted to the subject later.

PRINTING A DIGITAL CARTOON SET

Going though the upfront process of creating Sheet Views gives you a few benefits. First, this task can be handled by a single individual early in the project without the pressure of a looming deadline. If one person sets up all the necessary Sheet Views, there is a greater chance for consistency from Sheet to Sheet in terms of naming and drawing placement. Following this process also gives you a digital cartoon set. Just like the traditional cartoon set, the digital version will help make good decisions about project documentation requirements and the impact on budget and personnel considerations. One extra advantage of the digital cartoon set over the traditional one is that it is the real building model and document set! Don't be concerned with the finality that this seems to imply. The documents remain completely flexible and editable (as we have seen).

As the final task in this chapter, go ahead and plot out the project Sheets. We can use the Autodesk DWF Writer printer driver for this task. If you do not have this installed, it is available on the installation CD that comes with Autodesk Revit Building. Please install it now if you do not have it installed already. Please refer to the installation documentation that accompanies the CD for more information. If you prefer, feel free to print hard copy to any printer you have installed on your system.

NOTE Now that we have thoroughly explored Templates and Views and have created a project that contains several Views, we might be looking for ways to save time on the next project we begin. You can take any Revit Building project and save it as a template.

This is very easy to do and allows you to reuse your work on similar projects in the future. To save the current project as a template, simply choose Save As from the File menu. In the Save As dialog, browse to a location to save the template file. The best location is a central location on the network server where all users have ready access. Give the new template file a name, change the "Save as type" to Template Files (*.rte) and then click Save. The template file will be saved to the location you specified and be ready for use.

SUMMARY

Template files provide a means to start new projects with consistent content and setup.

All new projects should be created from an agreed upon office template file.

Levels establish horizontal datums measured vertically in the Z direction for use as floor levels and other reference points.

Walls heights can be constrained to Levels and will thus change height automatically if a Level changes.

Site data can be imported from popular CAD programs and used to generate toposurface elements in Revit Building.

Views are used to study, create, and manipulate model data.

Views can be graphical like plans, sections and elevations or non-graphical like schedules and legends.

Edits made in one View are immediately seen in all appropriate Views.

Sheet Views are used to create drawing sets for printing.

Column Grids and Structural Layout

INTRODUCTION

In this chapter, we will explore the layout of structural components for the commercial project begun in the last chapter. As we have seen, the design is a four-story structure of modest footprint. We will begin with the layout of the column grid lines. We will also add framing members and revisit the residential project to create a foundation plan.

OBJECTIVES

In this chapter we will begin by adding Column Grid Lines and Columns. This grid layout will be added to each level of the project including bubbles and dimensions. We will then add Beams and Joists to complete the framing. After completing this chapter you will know how to:

- Add and modify column Grid lines
- Add and modify Architectural and Structural Columns
- Load Steel Shape Families
- Create Structural Framing elements
- Copy elements to Levels
- Create and manipulate Groups
- Create foundation Wall footings

COLUMN GRID LINES

In Autodesk Revit Building there are several types of datum elements. In the last chapter, we worked with Levels. These are horizontal planes (expressed graphically as lines in Views that show them) that typically define floor levels or stories of the building but that can also be used to define other horizontal datums such as "top of footing" or "bottom of steel." Grids in Revit Building are very similar to Levels in almost every way except that they define vertical planes through the building. Grids are typically used to define the locations of structural columns in the building. Like Levels and section lines, Grids will appear automatically in all appropriate Views such as plans, elevations and sections.

INSTALL THE CD FILES AND OPEN A PROJECT

The lessons that follow require the dataset included on the Mastering Revit Building CD ROM. If you have already installed all of the files from the CD, simply skip down to step 3 below to open the project. If you need to install the CD files, start at step 1.

1. If you have not already done so, install the dataset files located on the Mastering Revit Building CD ROM.

   Refer to "Files Included on the CD ROM" in the Preface for instructions on installing the dataset files included on the CD.

2. Launch Autodesk Revit Building from the icon on your desktop or from the **Autodesk** group in **All Programs** on the Windows Start menu.

   - From the File menu, choose **Close**.

This closes the empty new project that Revit Building creates automatically upon launch.

3. On the Standard toolbar, click the Open icon.

TIP The keyboard shortcut for Open is CTRL + O. **Open** is also located on the File menu.

   - In the "Open" dialog box, click the *My Documents* icon on the left side.
   - Double-click on the *MRB* folder, and then the *Chapter05* folder.

   If you installed the dataset files to a different location than the one listed here, use the "Look in" drop down list to browse to that location instead.

4. Double-click *05 Commercial.rvt* if you wish to work in Imperial units. Double-click *05 Commercial Metric.rvt* if you wish to work in Metric units

   You can also select it and then click the Open button.

The project will open in Revit Building with the last opened View visible on screen.

ADD GRID LINES

Let's begin laying out a column grid in this project.

5. On the Project Browser, double-click to open the *Level 1* floor plan View.

This View shows the first floor plan complete with an underlay of the site plan contour map. The contour map was imported in the previous chapter from a

DWG file. For now, we will hide these contours. There are a few ways to accomplish this; here we will use the Hide/Isolate feature.

6. In the View window, click to select the contour map.

- On the View Control Bar, click the Hide/Isolate pop-up icon and choose **Hide Object** (see Figure 5–1).

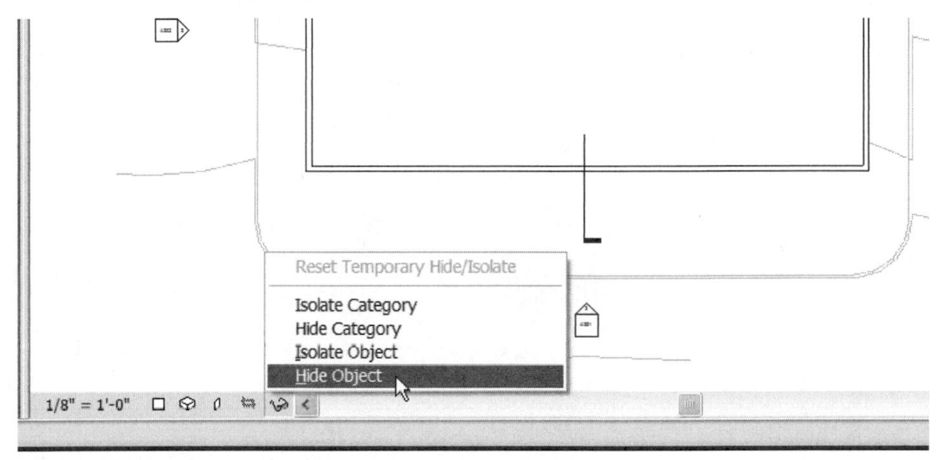

Figure 5–1 *Hide the site contours in the* ***Level 1View***

The Hide/Isolate controls give you a quick way to temporarily hide objects on screen from display. When you choose the Hide commands, the selected elements in the model are temporarily hidden in the current View. When you choose the Isolate options, the selected element(s) remain visible and all other elements are hidden. The Hide/Isolate commands are on screen in the current work session only. In other words, they do not hide and isolate the objects when printed or exported. And the settings are not persistent when a project is closed and reopened.

7. On the Design Bar, click the Basics tab and then click the Grid tool.

TIP The keyboard shortcut for Grid is **GR**. **Grid** is also located on the Drafting menu.

Like many elements in Revit Building, some common tools appear on the Options Bar. In particular, there are two modes: Draw and Pick Lines. Draw is the default and allows you to click any two points to specify the extent of the grid line. Grid lines can be drawn at any angle. You can also specify the grid lines based upon existing elements in the model. To do this, use the Pick Lines mode.

- Click any two points within the building footprint to add a Grid line.
- On the Design Bar, click the Modify tool or press the ESC key twice.

- Click on the Grid line just created.

Notice that many of the same handles and controls appear on the Grid line that appeared on the Level lines in the last chapter (see Figure 4–14 in the previous chapter). Place your pointer over each control to see a screen tip indicating its function (see Figure 5–2). Zoom in as required.

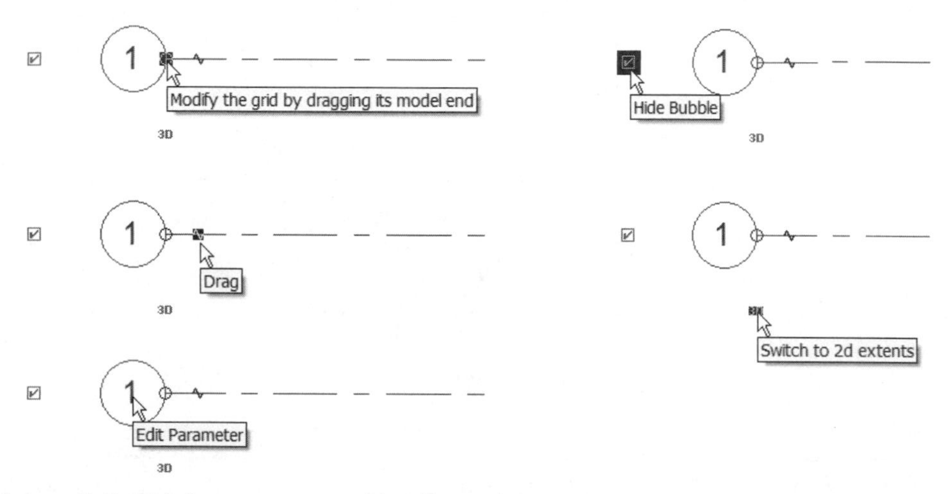

Figure 5–2 *Grids have many control handles and drag points*

Moving left to right in Figure 5–2, the following briefly describes each control.

- **Modify the grid by dragging its model end**—This round handle is used to drag the extent of the Grid. If the length or alignment constraint parameter is also active, dragging one Grid will affect the extent of the other constrained Grid lines as well.
- **Hide/Show Bubble**—Use this control to hide or show the Grid bubble at either end of the Grid line.
- **Drag**—This small "squiggle-shaped" handle creates an elbow in the Grid line. This is useful when the annotation of two Grid lines overlap one another in a particular View. Click this handle to create the elbow, and then drag the resultant drag handles to your liking.
- **2D/3D Extents Control**—When 3D Extents are enabled, editing the extent of the Level line in one View affects all Views in which that Level appears and is also set to 3D Extents. If you toggle this to 2D Extents, dragging the extent of the Level line affects only the current View.
- **Edit Parameter**—Click the blue text to re-number the Grid bubble.
- **Length and Alignment Constraint** (not shown in the figure)—A padlock icon used to constrain the length and extents of one Grid line to the others nearby. This is useful to keep all of your Grid lines lined up with one another.

8. Delete or undo the Grid line.

Sometimes it is easier to use existing geometry to assist in the creation of the Grid lines. This can speed up the creation process and ensure that design relationships are maintained.

9. On the Design Bar, click the Grid tool.
   - On the Options Bar, click the Pick Lines icon.
   - In the Offset field, type **4"** **[100]** and then press ENTER.
   - Position the Modify tool over the inside edge of the bottom horizontal Wall.
   - When the green dashed line appears on the inside of the building, click to add the Grid line (see Figure 5–3).

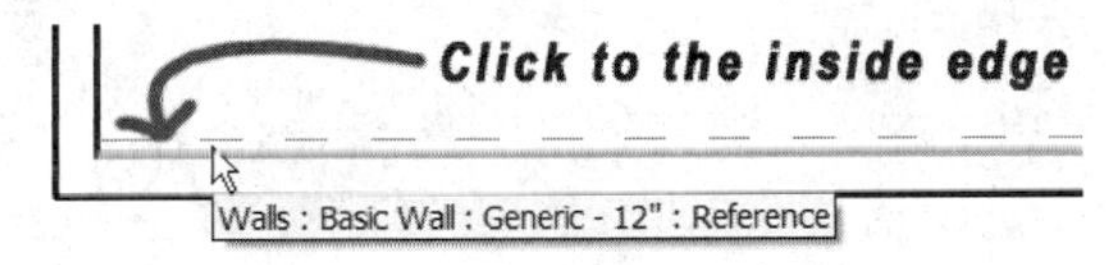

Figure 5–3 *Create a Grid line based on a picked Wall edge*

Depending on whether you deleted the first Grid line or used Undo, the number parameter of this Grid line may be "1" or it may be "2." If it is 2, we can edit it.

- If necessary, click on the Grid bubble text parameter (blue text) and then change the value to **1**.
- Make sure the Grid bubble shows on the left. If it is on the right, Show/Hide Bubble controls on each side to switch it.

You should now have Grid line 1 offset slightly to the inside edge of the bottom horizontal Wall with a bubble on the left. The extent of the Grid line match the length of the Wall. Let's stretch it longer.

10. Click on Grid line 1 and using the model end handles (see above), stretch each end longer away from the outside edges of the model (see Figure 5–4).

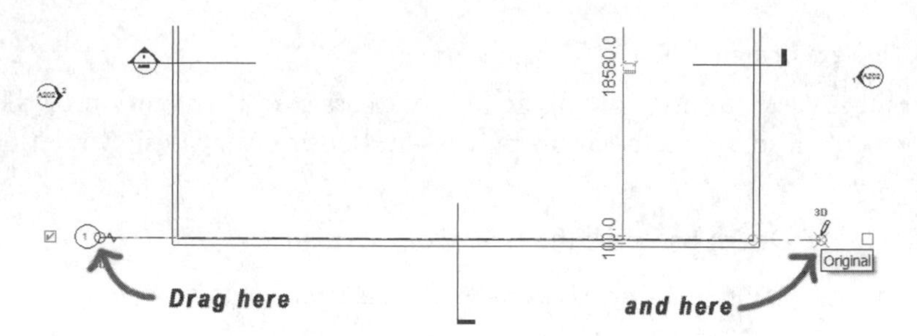

Figure 5–4 *Lengthen the Grid lines by dragging the model end handles*

It is a good idea to get the extent of the first Grid line to your liking before you continue, as the additional Grid lines that you add will snap to this length automatically. However, new Grid lines will automatically constrain to their neighbors, so if you do not stretch it longer first, you can stretch any Grid line later, and they will all lengthen together.

11. On the Design Bar, click the Grid tool.
    - On the Options Bar, verify that the Draw icon is selected.
    - Move the pointer to the right side, just above the previous Grid line.

A guideline will appear aligned with the previous Grid line and a temporary dimension will also appear.

- When the temporary dimension reads approximately 20′-0″ [6000] click to set the first point—make sure it is still aligned.
- Move the pointer to the left and when the alignment guidelines appear, click to place the new bubble aligned to the first (see Figure 5–5).

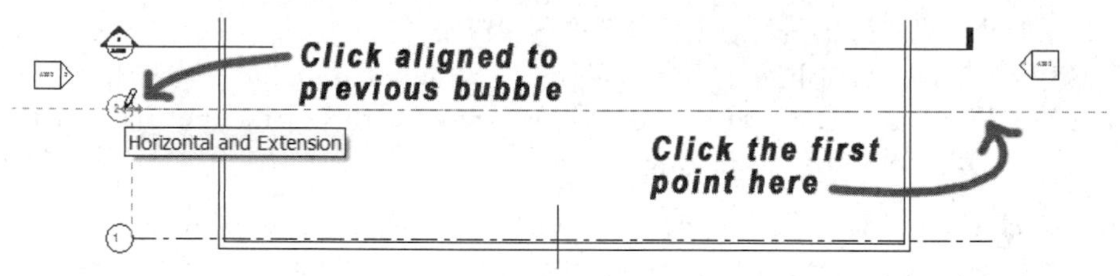

Figure 5–5 *Draw a second Grid line aligned to the first one*

Notice the closed padlock icon that appears connected to the new Grid line. This indicates that the new Grid line is constrained to the previous one. Also take a careful look at the temporary dimensions. The witness lines are actually attached to the Walls, *not* the previous Grid line. We can adjust this and then edit the value to correctly space the two Grid lines.

12. With the Grid line command still active, move the pointer over the lower "Move Witness Line" handle (see panel 1 in Figure 5–6).

- Drag the Witness Line up until it highlights Grid line 1 and then release (see panel 2 in Figure 5–6).
- Edit the temporary dimension value to **20′-0″ [6000]** (see panel 3 in Figure 5–6).

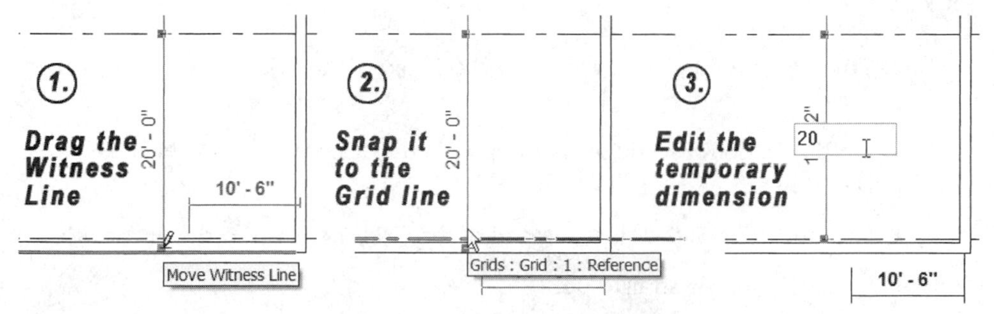

Figure 5–6 *Move the Witness line and then edit the value of the temporary dimension*

At first it might be a bit tricky to do this operation with the command still active. If you prefer, you can press ESC twice to cancel out, select Grid line 2 with the Modify tool, and then perform the steps above. The result will be the same. The motion should be fluid. Move the pointer over the Witness Line handle, click and drag it to Grid line 1 and release. Grid line 1 will highlight as you drag it. When it highlights, you should release the mouse. At this point, we could continue adding Grid lines using the same process. Instead, let's test the alignment constraint, and then explore alternate techniques to create the additional Grid lines.

13. On the Design Bar, click the Modify tool or press the ESC key twice.
14. Select Grid line 2.
    - Using the "Model end" drag handle, drag the bubble horizontally a bit.

Notice that both Grid bubble 1 and 2 move together.

- Position the bubbles where you want them.

15. Save the project.

COPY AND ARRAY GRID LINES

We could draw each additional Grid line as noted above. We can also copy the existing ones using either the Copy tool or the Array tool. The Copy tool is basically a Move command that moves a copy instead of the original. The Array command can create several equally spaced copies in one action. It also has the option to "group and associate" the copies so that the parameters assigned to the array are maintained and remain editable. Let's explore a few options.

16. Select Grid line 2.

- On the Edit toolbar click the Copy tool.

TIP The keyboard shortcut for Copy is **CO**. **Copy** is also located on the Edit menu (Note: choose **Copy**, not **Copy to Clipboard**).

On the Options Bar, there are three checkboxes. The "Copy" box is already selected.

- On the Options Bar, place checkmarks in the "Constrain" and "Multiple" checkboxes.
- For the start point, click at the end of the Grid line (snap to the endpoint).
- Move the pointer straight up.

Notice that with the "Constrain" option enabled, the copies will move only horizontal or vertical.

- Using the temporary dimensions, place two copies above Grid line 2 spaced at **20′-0″ [6000]** (see Figure 5–7).

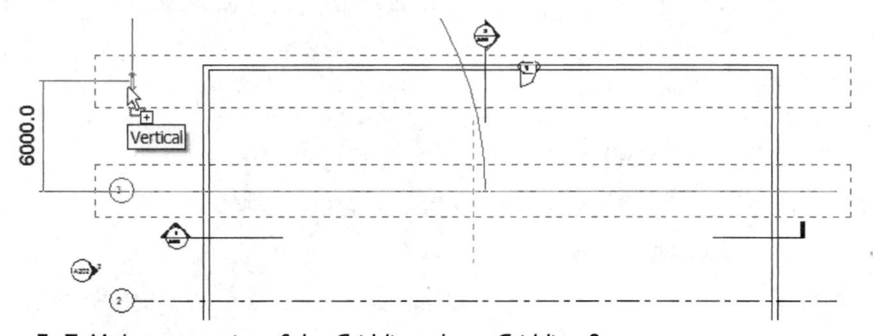

Figure 5–7 *Make two copies of the Grid line above Grid line 2*

- On the Design Bar, click the Modify tool or press the ESC key twice.

Notice that the two new Grid lines numbered automatically as "3" and "4." Revit Building will always number the new Grid lines in sequence after the last one placed. If you were to delete one Grid line 4, and then add another Grid line, it would automatically number to "5." However, if you were to undo the placement of Grid line 4, then the next one would be "4" instead.

17. Click on either of the new Grid lines.

Notice that again, the temporary dimensions reference the Walls and not the other Grid lines. If you wish to move them with the temporary dimensions, you can first move the Witness Lines as we did above. You can also use the Move tool if you prefer. Let's move both Grid lines up slightly to make the middle bay larger.

18. Select Grid lines 3 and 4. (You can use the CTRL key or a Crossing selection to do this—be certain to select only the Grid lines).

   - On the Options Bar, click the Activate Dimensions button.
   - Use the technique outlined above to move the Witness Line of the lower vertical dimension up to Grid line 1.

   When you have moved the Witness Line, the temporary dimension should read 40′-0″ [12,000] (see the left side of Figure 5–8).

   - Edit the value of the lower temporary dimension to **41′-2″ [12,480]** (see the right side of Figure 5–8).

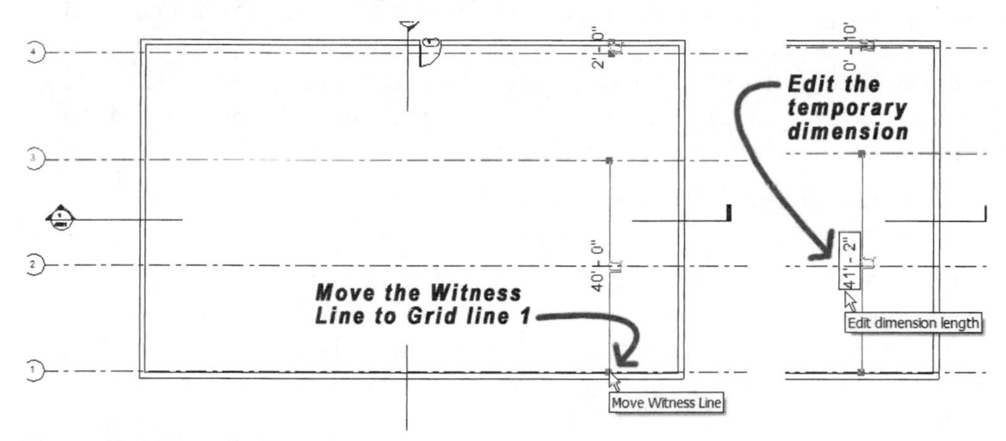

Figure 5–8 *Move Grid lines 3 and 4 using the temporary dimensions*

19. On the Design Bar, click the Grid tool.

   - On the Options Bar, click the Pick Lines icon.
   - In the Offset field, type **4″ [100]** and then press ENTER.
   - Position the pointer so that the green dashed line appears at the inside edge of the left vertical Wall and then click to add the Grid line.

Notice that this Grid line became number 5. For the vertical bays, we want to use letters.

20. Click on the new vertical Grid line.

   - Adjust its length so that it projects above and below the building footprint (like we did above) and then edit the bubble number to **A**.

21. Select Grid line A and then on the Edit toolbar click the Array tool.

TIP The keyboard shortcut for Array is **AR**. **Array** is also located on the Edit menu.

The Array command has many options. First, you can choose between a linear or radial array on the Options Bar. Next there is the "Group and Associate" checkbox. With this option enabled, all of the arrayed items (the original and the copies) will be grouped together. When thus grouped, you can select them later and edit the quantity of items in the array. The remaining items will re-space to match the new quantity either adding or deleting elements in the group. This can be very powerful and useful functionality that has many applications in building design. This being our first opportunity to explore this tool, we will test it out here with our Grid lines. Ultimately, we will want an ungrouped array because the final spacing of our column grid is not equal (as is typical in most buildings), however it will still be educational to explore the "Group and Associate" option of array nonetheless.

- On the Options Bar, verify that the "Linear" icon is chosen and that "Group and Associate" is selected.

Array is much like the move and copy commands in that it requires you to pick two points on screen to indicate the spacing of the array. We have two options for what these two points can represent in the array: they can be the spacing between the each element, or the total spacing of the entire array. To set the spacing between each element, choose the "2nd" option. To set the total spacing, choose the "Last" option.

TIP In this case, the only bad thing about choosing the "Last" option is that the rightmost Grid line will become Grid line B and then the four in the middle lettered sequentially left to right from C to F. You would then need to edit each Grid bubble parameter to fix this. Use the "2nd" option to avoid this problem.

- Choose the "2nd" option.
- For the "Number," type **6** and then place a checkmark in the "Constrain" box.

The number indicates the total quantity of elements in the final array. This quantity includes the original selection. The "Constrain" option works like the same option in the Copy command (see above) by limiting movement to horizontal and vertical only.

- Watch the Status Bar for prompts. Click the first point on the endpoint of the selected Grid line.

- Move the pointer to the right about 20′-0″ [6000] and then click the second point (see Figure 5–9).

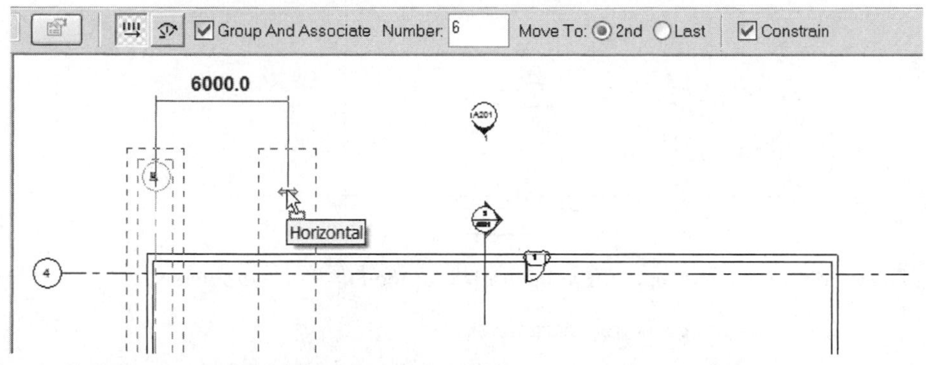

Figure 5–9 *Array a total of 6 Grid lines horizontally*

The last element of the array is close to the inside edge of the right Wall. Using the temporary dimensions like we did above, we can adjust it to the exact location we need. Since the array is grouped, all items will adjust to maintain an equal spacing as we do this.

22. With the array still active, click the last item (Grid line F on the right).

- On the Options Bar, click the Activate Dimensions button.

The last Grid line (labeled F) needs to be 4″ [100] from the inside face of the right Wall (like its counterpart on the left).

- Move Witness Lines and edit the temporary dimension value to **4″ [100]**.

NOTE the Wall is 12″ [300] thick. So you can place the Witness Lines in any convenient location and take this value into account to achieve the correct location.

Notice that all the Grid lines between A and F adjust with this change.

- Make sure that Grid line F is still selected. If you deselected it, click on it again.

Beneath the arrayed items, a temporary dimension will appear with a single text parameter that indicates the number of items in the array (see Figure 5–10).

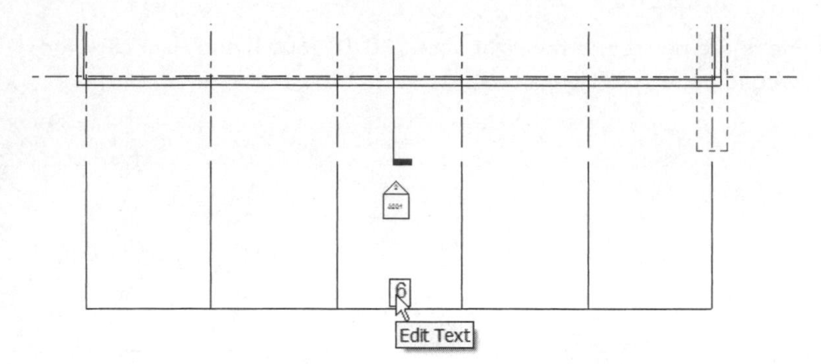

Figure 5–10 *Edit the quantity of items in the array with the Edit Text parameter*

- Click to select this text parameter.
- Type **7** and then press ENTER.

Notice that a new Grid line is added and lettered automatically.

- Repeat the process typing **5** this time.

Notice that two of the Grid lines are deleted.

- Repeat the process one more time setting the value back to **6**. (You can also Undo twice if you prefer).

This is the benefit of using the "Group and Associate" parameter when you create the array. However, as we mentioned above, the spacing of our commercial building's column grid is not actually regular. Therefore, we will actually ungroup this array. Before we do, feel free to experiment further with the array. For example, you can select any one of the Grid lines and move it. All of the others will move accordingly. This can be tricky, as the one you select actually stays stationary and all the others move relative to it. You can also choose the Edit Group button on the Options Bar with one of the elements selected. This will place the Design Bar in an Edit Group mode where you can manipulate the elements within the group. When you choose Finish Group on the Design Bar, the change is applied to all group instances (all of the elements in the array in this case). We will explore Groups in more detail in Chapter 11. Be sure to undo any explorations you made before continuing.

DIMENSION THE GRID LINES

Let's ungroup the array and then adjust the Grid line spacing independently.

23. Select all vertical Grid lines—A through F (use the CTRL key or a crossing selection).

- On the Options Bar, click the Ungroup button.

The Grid lines will now move independently of one another. Try a few moves if you like, but be sure to undo before continuing. We are going to adjust several Grid lines now. This will be easier with some actual dimensions rather than the temporary dimensions.

24. On the Design Bar, click the Dimension tool.
   - Starting with Grid line A, click successively on each lettered Grid line A–F (see Figure 5–11).

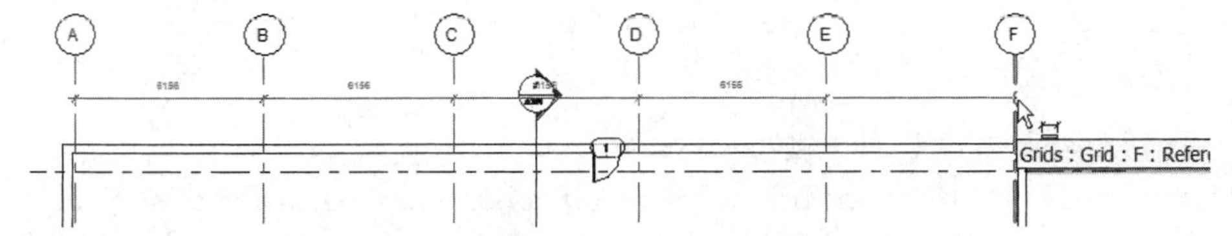

Figure 5–11 *Add a dimension string to the lettered Grid lines*

   - After you click F, move the mouse to a spot where you want to place the dimension string and then click to set the dimension string and end the command.

TIP Make sure that you click in an empty spot to place the dimension, if you click on another model element, it will add a Witness Line instead.

25. On the Design Bar, click the Modify tool or press the ESC key twice.
26. Select Grid line B

Notice that now that we have added a string of dimensions, in addition to the normal temporary dimensions we also can edit the dimension values of the string we just added.

NOTE If the temporary dimensions do not appear on the Options Bar, click the "Activate Dimensions" button.

   - Click on the blue dimension text between Grid line A and B, type **22′-2″ [6740]** and then press ENTER.
   - Select Grid line C next and edit the distance between it and B to **14′-4″ [4400]**.
   - Repeat moving left to right for the remaining vertical Grid lines as shown in Figure 5–12.

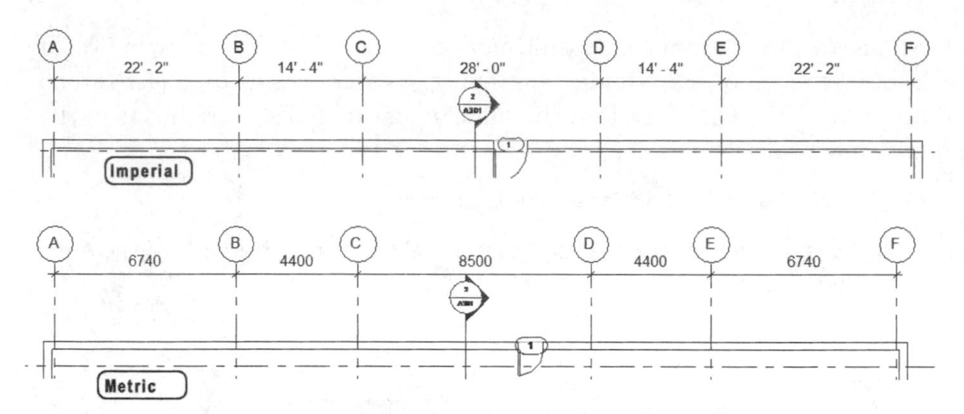

Figure 5–12 *Edit dimension values to move Grid lines*

Our feature façade on the front of the building (we created a Mass for this in the last chapter) will require some structural support. We can add a few additional Grid lines in the front for these.

27. On the Design Bar, click the Grid tool.

- On the Options Bar, verify that the Draw icon is selected.
- Move the pointer to the inside of the building above the lower Wall and click between Grid lines C and D.
- Move down past the building and click again to place the bubble.

The bubble will automatically enumerate as "G."

- Click in the text parameter of the new bubble and change the value to **C.3**.
- Using the temporary dimensions, move the Grid line so that it is **9′-0″** **[2700]** from Grid line C (see Figure 5–13).

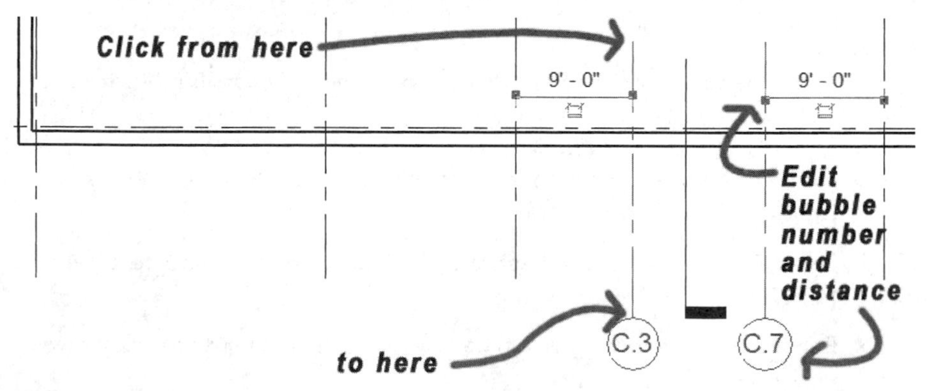

Figure 5–13 *Add Grid lines at the bottom of the plan for support of the front façade*

- Add Grid line C.7 to the other side of the C-D bay (see Figure 5–13).

28. Add a vertical dimension string to the numbered bubble on the left.
29. Save the project.

VIEWING GRID LINES IN OTHER VIEWS

All of the work we have done on Grid lines so far has been in the *Level 1* floor plan View. However like other datums, these grids that we just placed will automatically appear in all orthographic Views.

30. On the Project Browser, double-click to open the *West* elevation View.

Notice how the numbered Grid bubbles appear here running vertically on the elevation.

31. On the Project Browser, double-click to open the *South* elevation View.

This time only the lettered bubbles appear. If you find the bubbles in the middle to be too close together, you can use the drag handles to give them an elbow.

32. Click the "Drag" handle to make an elbow in Grid lines C.3 and C.7. Fine tune the shape as appropriate (see Figure 5–14).

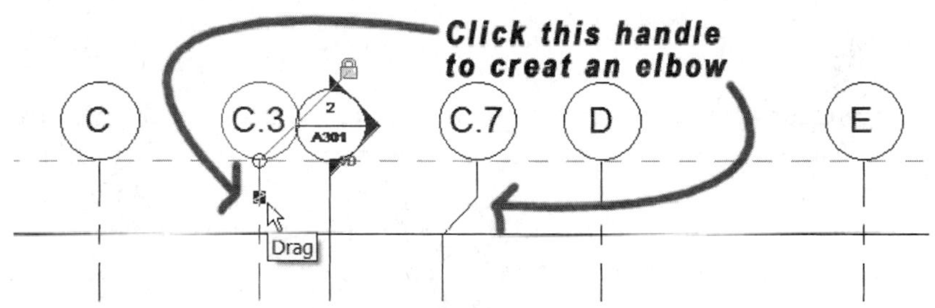

Figure 5–14 *Create elbows in the Grid lines as needed to keep the bubbles from overlapping one another*

33. On the Project Browser, double-click to open the *Longitudinal Section* View.

In this View, Grid lines C.3 and C.7 do not show since the section is looking toward the north. If the section were looking south, they would appear.

34. On the Project Browser, double-click to open the *Level 1* floor plan View.
    - On the Window menu, choose **Close Hidden Windows**.
    - Save the project.

WORKING WITH COLUMNS

A collection of structural tools are included in Autodesk Revit Building. A larger collection of structural tools appears in the Autodesk Revit Structure application. Users of Revit Building can seamlessly share models back and forth between structural consultants using Autodesk Revit Structure. If you are not collaborating with a Structural Engineer who uses Revit Structure, or if it is early in the project and you have not yet selected an engineer or received designs from the consultant, you can use the structural tools provided with Revit Building to begin the layout of Columns and if desired, Beams and Braces as well. Like all Revit Building elements, structural elements remain parametric and editable after they have been added to the model. Therefore, when the structural analysis comes back from the engineer, you can swap in the correct types and sizes to meet the requirements of the design.

Disclaimer:

The shapes used in this book are chosen only for illustration purposes and are not presented as a design solution or to be construed as a recommendation of structural integrity. No structural analysis of any kind has been performed on the designs in this book.

ADD COLUMNS

Revit Building includes two types of columns: Architectural Columns and Structural Columns. An Architectural Column provides the location and finished dimensions (including column wrap, furring, finishes, etc) of a column in an architectural space. The Structural Column is the actual structural material without any enclosure or finish such as the Steel or concrete column. Let's take a look at both kinds.

Continue in the *Level 1* View from above.

1. On the Design Bar, click the Modeling tab and then click the Column tool.

   - From the Type Selector, choose Rectangular Column : 24″ x 24″ [M_Rectangular Column : 610 x 610mm].
   - Move the pointer to the intersection of Grid lines 4 and A.

   The pointer will snap to the intersection automatically (see the left side of Figure 5–15).

   - Click at the intersection to place the Column (see the right side of Figure 5–15).

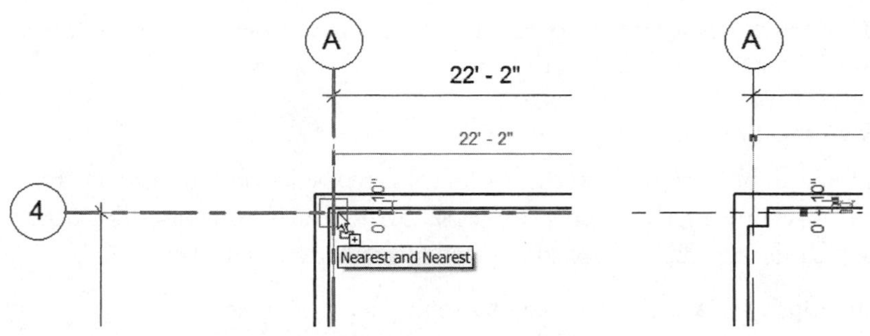

Figure 5–15 *Add an Architectural Column to the A4 Grid intersection*

Notice that the Architectural Column automatically interacts with the Walls at the intersection and graphically displays as an integrated column.

2. Continue placing Columns at all of the intersections around the perimeter of the building (see Figure 5–16).

TIP You can simply click at each intersection or you can use the Copy tool. If you use Copy, be sure to select "Multiple" on the Option Bar, and use the TAB key to select a good start point for the copy.

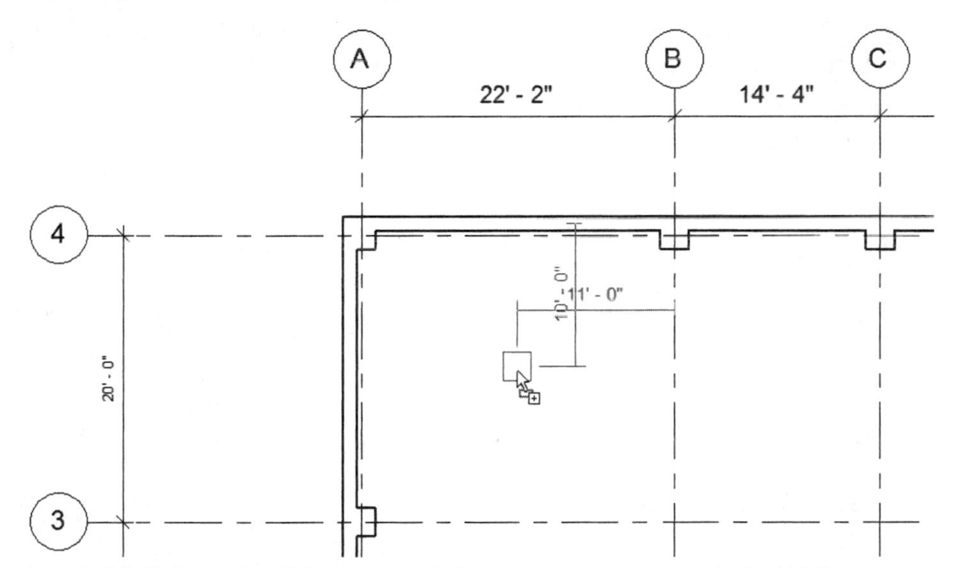

Figure 5–16 *Architectural Columns around the perimeter integrate with the Walls*

- On the Design Bar, click the Modify tool or press the ESC key twice.

Let's try some Structural Columns next.

3. On the Design Bar, click the Structural tab and then click the Structural Column tool.

NOTE If you do not see the Structural tab of the Design Bar, right-click on the Design Bar to display it.

- Open the Type Selector.

Notice that there are only two Structural Column types loaded in the current project (there is only one in the metric project). Structural Columns, like other Families in Revit Building can be loaded as needed from external libraries.

4. On the Options Bar, click the Load button.

If you are working in Imperial units, you should automatically be placed in the *Imperial Library* folder. If you are using Metric units, you should be placed automatically in the *Metric Library* folder. If you have not been placed automatically in one of these locations, browse there now. The easiest way to do this is to click the appropriate shortcut button on the left side of the dialog.

- Double-click the *Structural* folder, then double-click the *Columns* folder and finally double-click the *Steel* folder.
- Click to select (do not double-click) the *W-Wide Flange-Column.rfa* [*M_W-Wide Flange-Column.rfa*].

A list of industry standard steel-shape sizes will appear in a table at the bottom of the dialog. Scroll through the list to see all of the sizes.

- Scroll in the list, locate W12x87 [W310x97] and select it (see Figure 5–17).

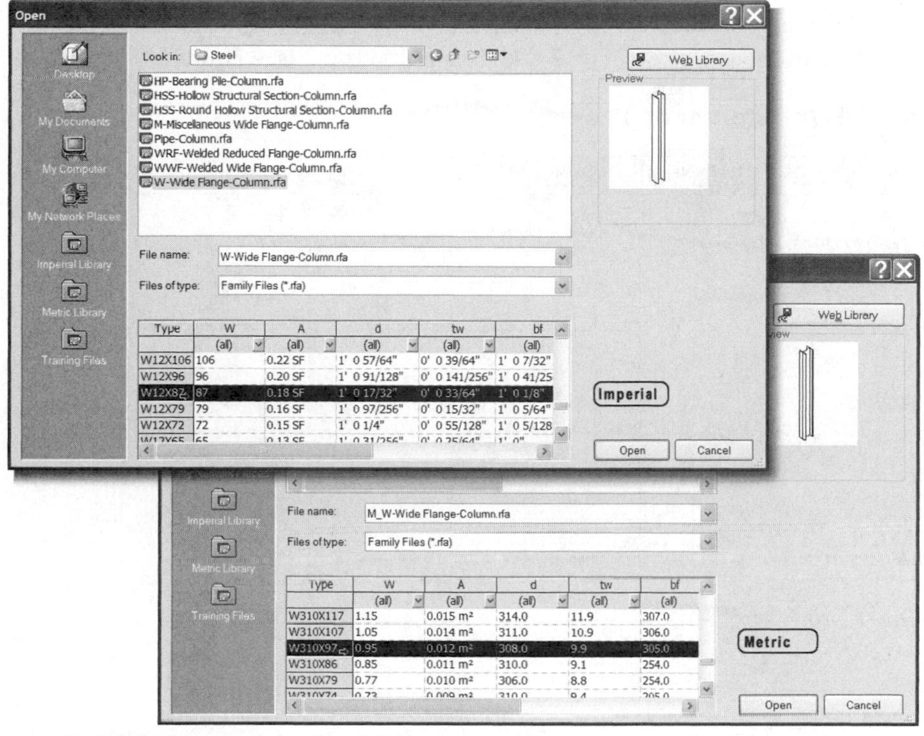

Figure 5–17 *Common Industry Steel Shapes are available to load from the Family file*

- Click Open to load the Family.
- If a "Reload Family" dialog appears, simply click Yes.
- From the Type Selector, choose the newly loaded W12x87 [W310x97] type.

Structural Columns have many of the same parameters as Architectural Columns—and more. If you move the pointer around the screen before you click, you will note that like the Architectural Column, the Structural Column will pre-highlight Grid lines and Walls as possible insertion points. (However if you insert a Structural Column at a Wall, it will not merge with the Wall the same way that the Architectural one did).

In addition to the options that Structural Columns share in common with Architectural Columns, there are several additional options on the Options Bar. For example, two "Place By" options appear: a "Grid Intersection" button which allows you to place several Structural Columns at once on a selection of Grid lines and the "Architectural Column" button which allows you to place a Structural Column at the location of a selection of Architectural Columns. Let's give these a try.

5. On the Options Bar, click the "Grid Intersection" button.

6. Click the pointer above Grid line 3 and to the right of Grid line E and then drag to the left of Grid line B and below Grid line 2 (see Figure 5–18).

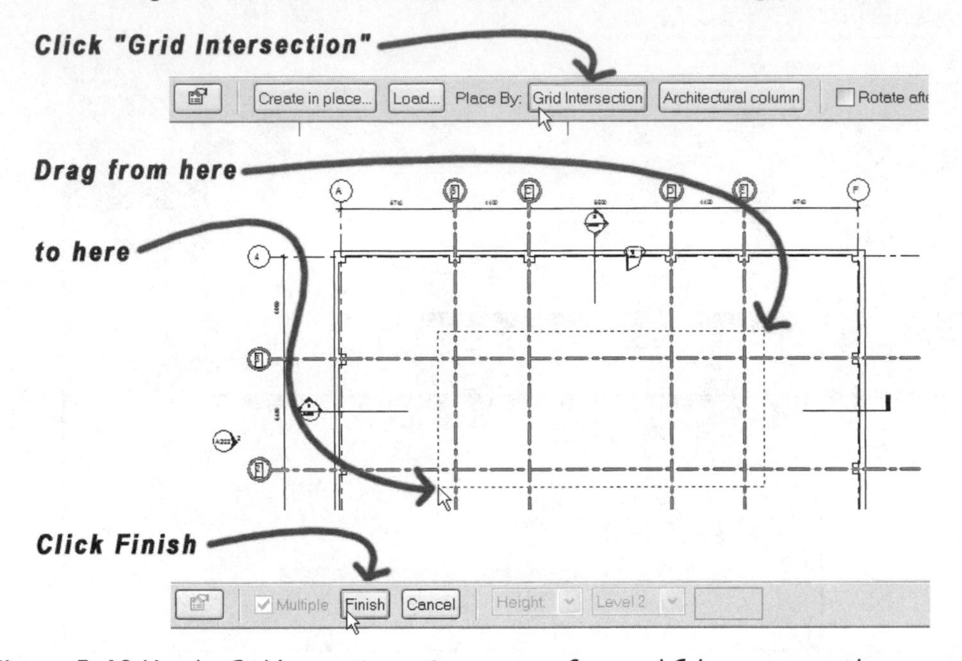

Figure 5–18 *Use the Grid Intersection option to create Structural Columns at several intersections*

The Grid lines will highlight in red and several ghosted columns will appear. If you are satisfied with this selection, you click the Finish button on the Options Bar. Otherwise, you simply select again until you are satisfied. If you decide not to add the Columns, you can click the Cancel button on the Options Bar.

- On the Options Bar, click the Finish button.

You will now have a steel column at each Grid intersection. Let's try the "Architectural Column" option next. In addition, we will also adjust the height of the column as we add it.

7. On the Options Bar, next to Height, choose **Roof** from the list.

   This will create a single continuous Structural Column from Level 1 (the current Level) up to the Roof.

   - On the Options Bar, click the Architectural Column button.
   - Using a crossing or inside selection (click and drag a selection window—either left to right or right to left will work in this case) surround the entire building to select all Architectural Columns.
   - On the Options Bar, click Finish.

- On the Design Bar, click the Modify tool or press the ESC key twice.

EDIT COLUMNS

We now have Structural Columns at all Grid intersections including the ones at the perimeter Walls. However, the ones in the center (not surrounded by Architectural Columns) span from Level 1 to Level 2, while the ones at the perimeter span the complete height of the building. We cannot see this in the current plan View. Let's take a look at the section.

8. On the Project Browser, double-click to open the *Transverse Section* View.

When you switch to this View, you will likely not notice much detail with regard to the Structural Columns. This is because elements in Revit Building display at various levels of detail that automatically simplifies the graphics in smaller scale drawings, and shows more detail in larger scale drawings. You can easily modify the default settings for this behavior, or simply adjust the level of detail display for an individual View. For example, Figure 5–19 shows the variation in the three detail levels when viewing Structural elements such as the Columns we have in our model. As you can see, the Coarse detail level shows simple linework for the steel shape in plan and a single line for it in section (and elevation). This is the currently active display Detail Level in both the floor plan and section Views in our project. If you switch to Medium or Fine Detail Level, the graphics for the steel shape will get progressively more detailed in plan. In elevation and section, the same graphics are used for both these Detail Levels but both are more detailed than Coarse was (see Figure 5–19).

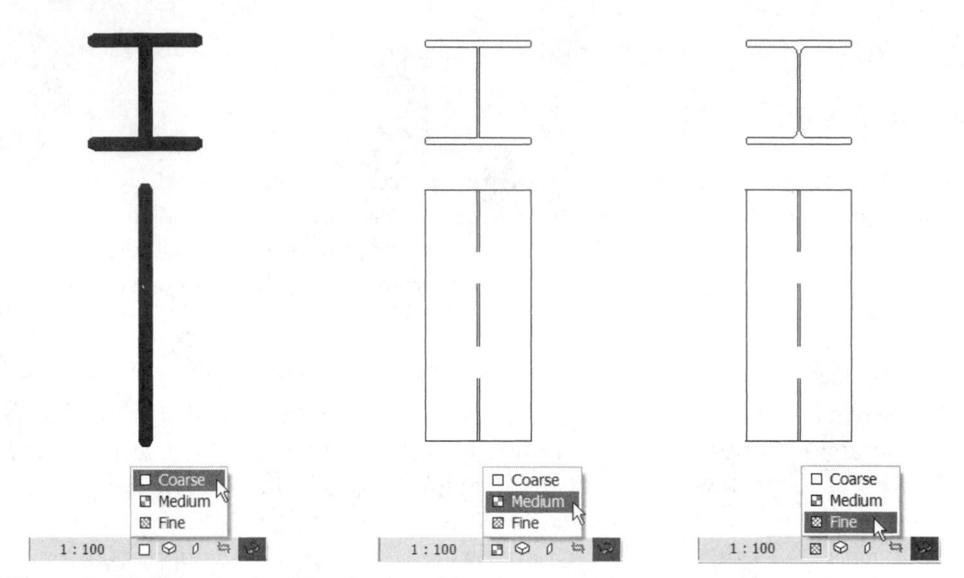

Figure 5–19 *Three levels of Display Detail Level*

9. On the View Control Bar, change the Detail Level to **Medium**.

The Columns will now display with the correct dimensional thickness in the section rather than the diagrammatic single-line display. Zoom around the section as needed to see clearly that the Architectural Columns stop at Level 2 while the Structural ones go all the way to the Roof. Naturally it is likely that these Columns would actually be constructed in sections and not be a single continuous four-story tall Column. However, for design purposes, they function in the building as a single continuous column. Therefore, it is up to you decide in your own projects what the "Height" of the Column represents and build it accordingly. You can make the decision to create the Columns the actual height of the material that will be shipped to the site (two-stories tall for instance); this would be important if you use Revit Building to assist you in deriving construction quantities or when we send the model to our Structural Engineer for structural analysis and design. If you wish to create shorter Columns, simply choose a different Level to end the Columns at, or even choose **Unconnected** and type in an explicit dimension. Then open the plan View for the next Level and copy/paste aligned the Columns there. In this project, we will leave this decision to the Structural Engineers. For now, we will make our Structural Columns span the full height of the building and leave our Architectural Columns (which again represent the finish materials wrapping the Columns) spanning a single Level. Later we will copy these Architectural Columns to the other floors.

10. Zoom in on Grid line 2 or 3 at Level 1.

- Move your pointer over the Columns and note the screen tips that appear (see Figure 5–20).

TIP If you have trouble highlighting the overlapping elements, use the TAB key to cycle through adjacent elements.

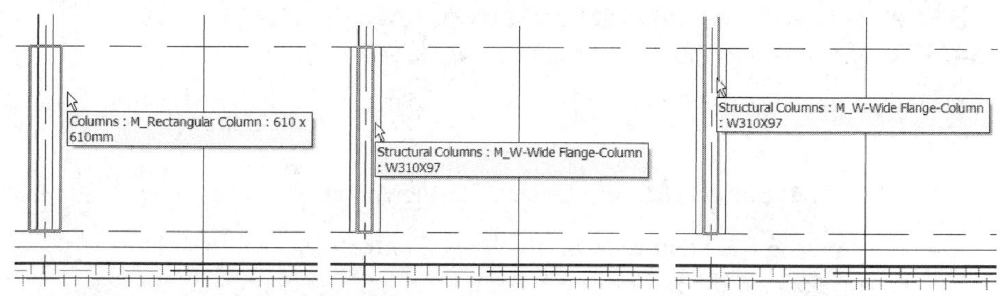

Figure 5–20 *At Level 1, three Columns appear*

From our section View, several of the Columns are in the same plane. We can still select the ones we need to edit in this View, or if you prefer, we can return to the Level 1 plan View and select there. It makes no difference in which View you select them. If you select in section, you can easily avoid the full height Columns with a crossing or inside selection, however, it will be difficult to avoid the Architectural Columns in behind them. Furthermore, some of the Columns would not be selected as they are cropped away in the section View. However, in this case we can make use of another tool to assist in this selection.

11. Select one of the steel Columns (any one, tall or short).
    - Right-click and choose **Select All Instances**.

Be careful with this command. It literarily selects all instances of a Family Type in the entire project—both seen and unseen. In this case, that is exactly what we want, but in many cases it may not be. Keep this in mind.

- On the Options Bar, click the Properties icon.
- In the Element Properties dialog, for Base Level, choose **Street Level**.
- For Top Level, choose **Roof** and then click OK (see Figure 5–21).

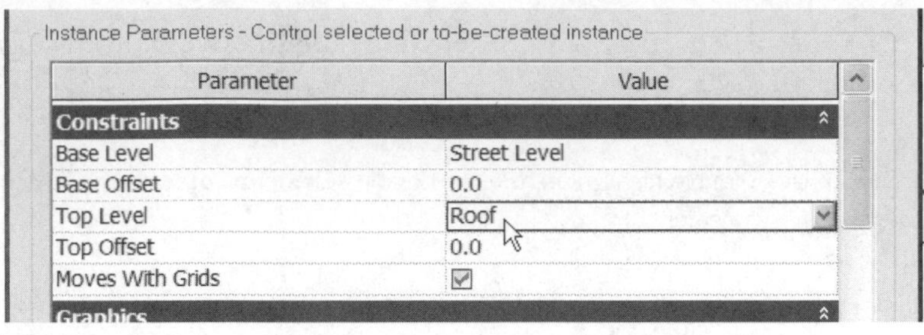

Figure 5–21 *Edit the Base and Top Level parameters of the selected Columns*

Notice the height of all of the steel Columns now spans from below the first floor to the roof.

12. On the View toolbar, click the Default 3D View icon.

- Click on the Roof and then on the View Control Bar, from the Hide/ Isolate pop-up, choose **Hide Object**.

You will notice that all steel Structural Columns were impacted by this change, not just the ones we could see in the section View. This is because as noted above, Select All Instances selects across the entire model, not just the elements visible in the current View.

COPY AND PASTE ALIGNED

Let's copy our Architectural Columns and add them to other floors.

13. On the Project Browser, double-click to open the *Level 1* floor plan View.

We could use the same selection method that we just used to select all of the Architectural Columns, but let's explore another technique this time.

14. Make an inside selection (from left to right) surrounding the entire building (don't include the Grid lines). (See item 1 in Fig- ure 5–22).

This will select all visible elements in the building model.

- On the Options Bar, click the Filter Selection icon (see item 2 in Figure 5–22).
- In the Filter dialog, click the "Check None" button.
- Place a checkmark in the Columns checkbox and then click OK (see item 3 in Figure 5–22).

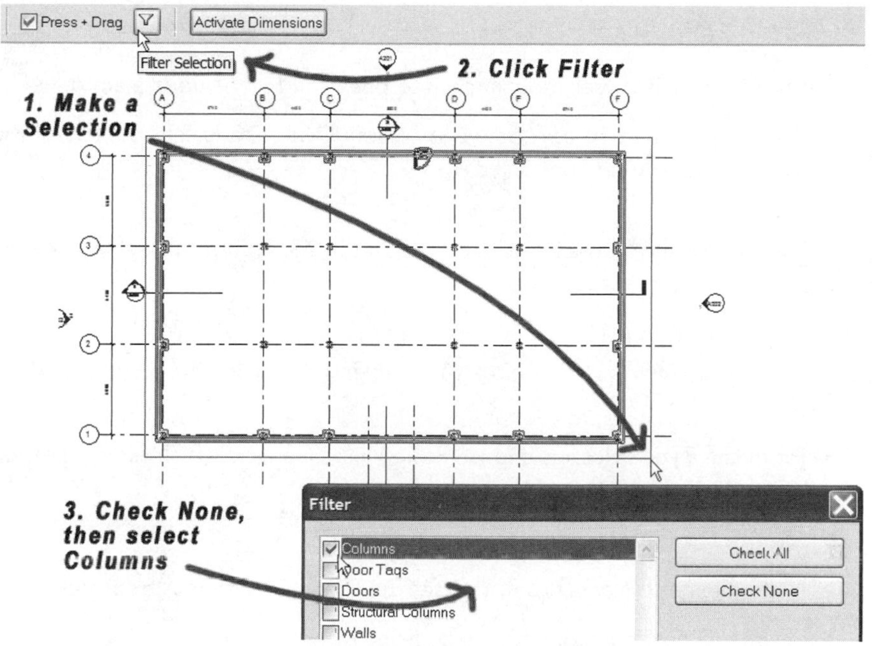

Figure 5–22 *Use Filter Selection to modify a selection and remove unwanted items*

Only the Architectural Columns should now be highlighted in red.

15. From the Edit menu, choose **Copy to Clipboard** or press CTRL + C.
    - From the Edit menu, choose **Paste Aligned > Select Levels by Name**.
    - In the "Select Levels" dialog, use the CTRL key and select Levels 2, 3 and 4 and then click OK.
16. On the Project Browser, double-click to open the *Level 3* floor plan View.

Notice that all of the Columns, both Structural and Architectural show in this View. The Structural Columns show because they intersect this View's cut plane. The Architectural Columns are the copies that we just added. Check other Views if you wish.

17. Save the project.

CORE WALLS

Now that we have a column grid with associated Columns, it is a good time to sketch in the building core. To do this, we will use Structural Walls. Structural Walls are essentially the same as Architectural Walls except that "Structural Usage" parameter is automatically set to **Bearing** for them. Otherwise, we add Structural Walls in the same fashion as architectural Walls.

ADD STRUCTURAL WALLS

1. On the Project Browser, double-click to open the *Level 1* floor plan View.

   ◗ Zoom in on the two bays between column Grids C and E at the top of the plan (between Grid lines 3 and 4).

TIP Right-click and choose Zoom In Region or type **ZR**. **Zoom In Region** is also located on the View > Zoom menu.

2. On the Design Bar, click the Structural tab and then click the Structural Wall tool.

   ◗ From the Type Selector, choose Basic Wall : Generic - 8″ Masonry [Basic Wall : Generic - 225mm Masonry].

   ◗ On the Options Bar, click the Properties icon.

   ◗ In the Element Properties dialog, set the "Base Constraint" to Street Level.

   ◗ Set the "Top Constraint" to Up to level: Roof.

These two settings instruct our Structural Wall to go the full height of the building from the lowest level to the highest. We will also need the core Walls to continue past the Roof to allow stair access to the roof. So let's also add a "Top Offset" parameter. Keep in mind that all of these parameters can be edited later as the project progresses and as design needs dictate.

   ◗ In the "Top Offset" parameter field, type **12′-0″** **[3600]** and then click OK (see Figure 5–23).

Take note of the "Structural Usage" parameter before you close the dialog—note that it is set to "Bearing" automatically.

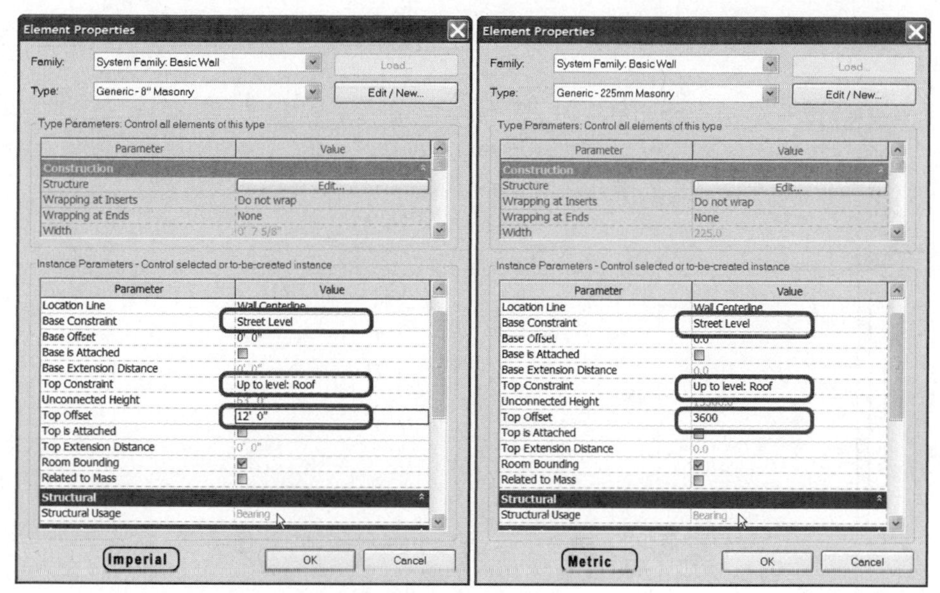

Figure 5–23 *Configure the parameters for Structural Wall the full height of the building*

- On the Options Bar, click the Pick Lines icon.
- Leave the Loc Line set to **Centerline** and in the Offset field, type **1′-2″ [350]** and then press ENTER.

With the Pick Lines option enabled, we will be able to select existing geometry from which to create the Walls. The Walls will be created at a distance of 1′-2″ [350] from the selected geometry.

3. Highlight Grid line C and when the green dashed offset line appears to the right of it, click the mouse to create the Structural Wall.
   - Repeat for Grid line 3 (horizontal) when the offset line appears below the Grid line (see Figure 5–24).

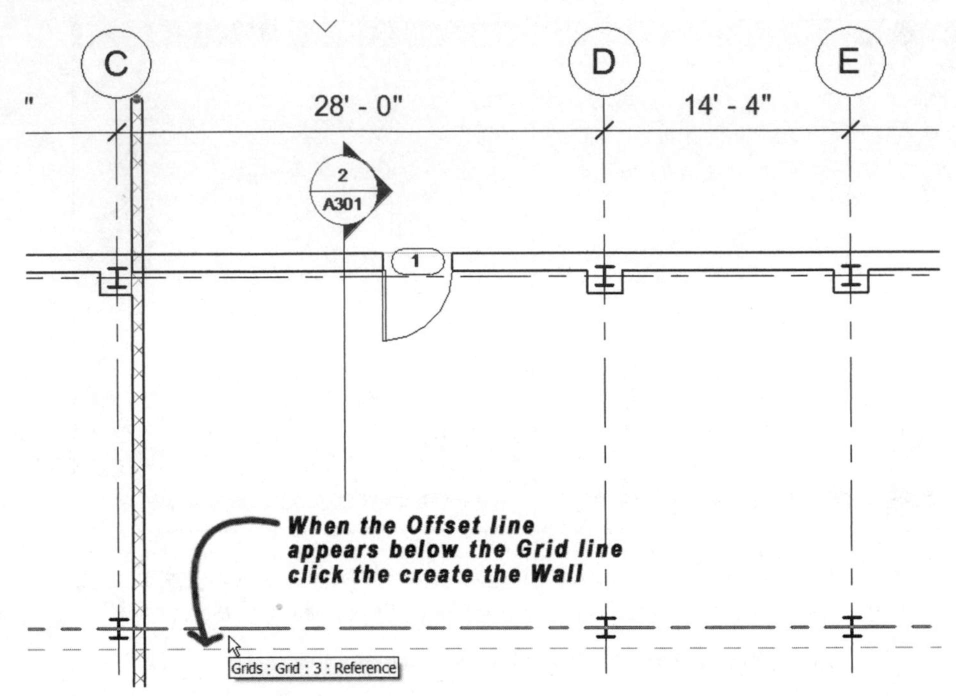

Figure 5–24 *Create Walls from Grid lines using the Pick Lines option with an offset value*

- Change the Offset value on the Options Bar to **2′-8″ [800]** and then create one more Wall to the right of Grid line E.

4. Using the Trim/Extend tool on the Tools toolbar, trim the three Structural Walls to appear as in Figure 5–25.

Use the Trim/Extend to Corner option for the two bottom intersections (remember to select the side of the Wall that you want to keep). Use the Trim/Extend Multiple Elements option for the upper two intersections to make them butt into the inside edge of the exterior horizontal Wall (remember to click the side of the Wall you want to keep). For more information on the Trim/Extend tool, refer to the tutorials in Chapter 3 or search the online Help.

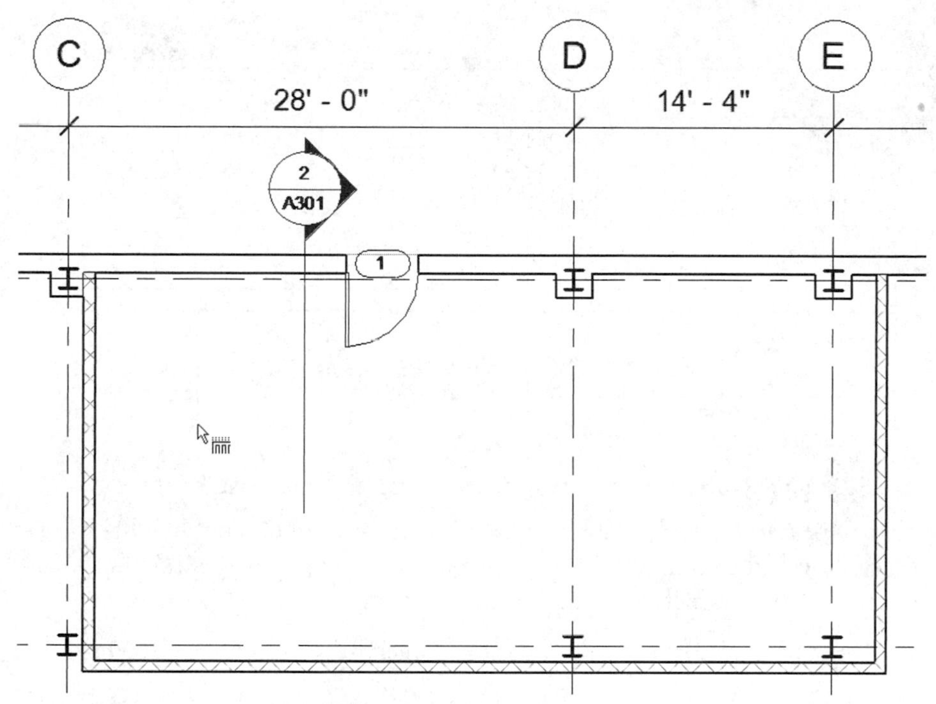

Figure 5–25 *Trim the core Walls to one another*

- On the Design Bar, click the Modify tool or press the ESC key twice.

CONVERT A WALL TO A STRUCTURAL WALL

Let's view the model in section to see what we have so far.

5. Double-click the blue Section Head between Grid lines C and D.

The *Transverse* building section View will open.

6. Click on the Crop Box to select it. Click and drag the small round control handle at the top edge and drag it up enough to see the top of the core Walls (see Figure 5–26).

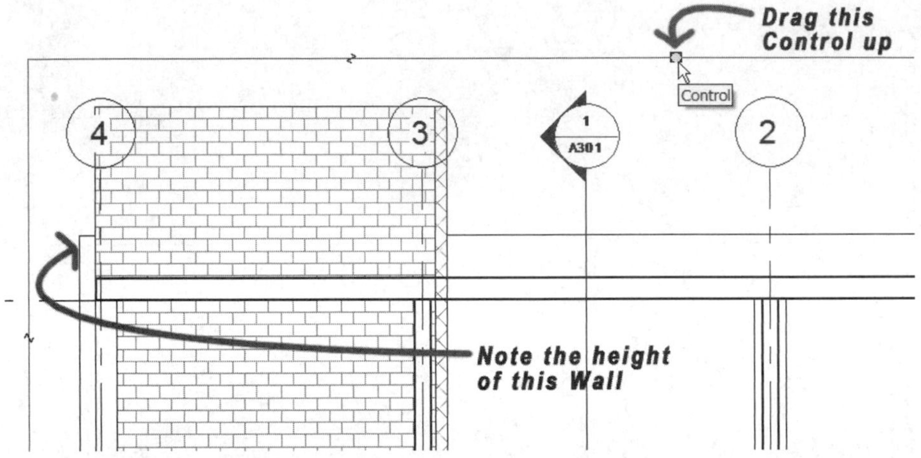

Figure 5–26 *Drag the Crop Box large enough to see the top of the new Walls*

Notice that the exterior Wall at the core only projects to the parapet height. In reality, this Wall would also need to be part of the core. Let's split this existing Wall and then change its parameters to match the rest of the core.

7. On the Project Browser, double-click to open the *Level 1* floor plan View.
8. On the Tools toolbar, click the Split tool.

TIP The keyboard shortcut for Split is **SL**. **Split Walls and Lines** is also located on the Tools menu.

Move the split pointer near the intersection between the vertical core Wall and the horizontal exterior Wall.

- When the pointer snaps to the intersection, click the mouse to split the Wall.
- Repeat on the other intersection.
- On the Design Bar, click the Modify tool or press the ESC key twice.

9. Select the middle horizontal Wall segment (created by the splits) and on the Options Bar, click the Properties icon.
   - From the Type list, choose Generic - 12″ Masonry [Generic - 300mm Masonry].
   - Change the "Base Constraint" to **Street Level** and the "Base Offset" to **0** (zero).
   - Change the "Top Constraint" to **Up to level: Roof** and the "Top Offset" to **12′-0″** [**3600**].

- Beneath the Structural grouping, change the "Structural Usage" parameter to **Bearing** and then click OK.

10. Double-click the blue Section Head between Grid lines C and D again.

Graphically, the Wall should now appear as the other core Walls do (the thickness varies).

11. Save the project.

SLABS

The most obvious omission that you might notice in our current sections is the lack of any floor slabs. Let's add some basic floor slabs.

CREATE A SLAB FROM WALLS

The easiest way to create a Floor slab and locate its boundaries is by using the existing Walls that bound it.

1. On the Project Browser, double-click to open the *Level 1* floor plan View.
2. On the Design Bar, click the Structural tab and then click the Slab tool.

   The Design Bar changes to sketch mode. Pick Walls should be active.

   - If Pick Walls is not active, select it on the Design Bar now.
   - Click on each of the exterior vertical Walls, the bottom horizontal one, and one of the top horizontal ones (see Figure 5–27).

NOTE Be sure to click on the inside edges.

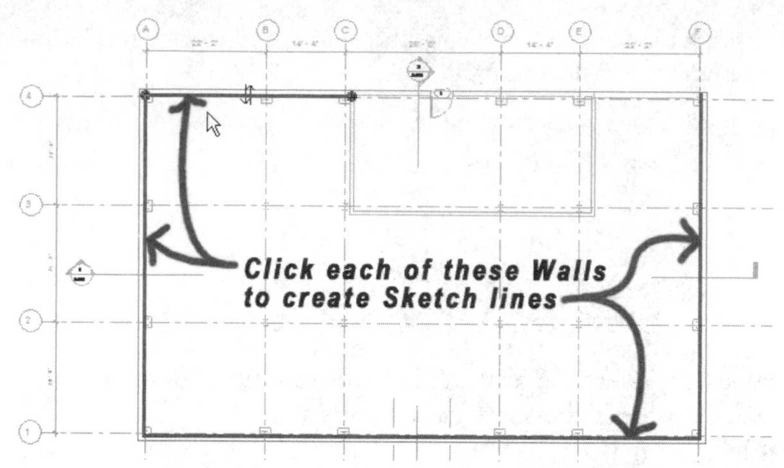

Figure 5–27 *Create sketch lines from Walls*

- Use the Trim/Extend tool to join the two open sketch lines.

NOTE The Trim/Extend tool will remember the last mode you used. Be sure to set it to Trim/Extend to Corner to complete this step.

- On the Design Bar, click the Floor Properties button.
- From the Type list, choose Generic - 12″ [Generic 300mm] and then click OK.
- On the Design Bar, click the Finish Sketch button.
- In the dialog that appears, click Yes (see Figure 5–28).

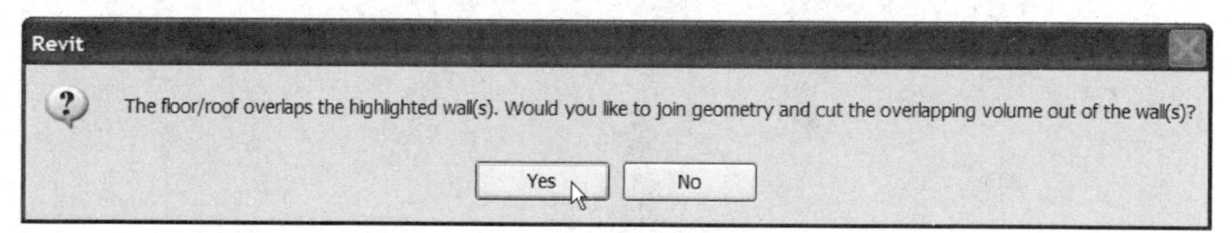

Figure 5–28 *Allow the new Slab to Join with the Walls*

You can always edit the automatic joins later, but in general, it is a good idea to allow Revit Building to apply joins when prompted. This connects the various building elements in logical ways.

COPY FLOOR SLABS TO LEVELS

3. Double-click the blue Section Head between Grid lines C and D again.

The new Floor slab appears at Level 1.

4. Select the new Floor slab, and then press CTRL + C.

Note: If you prefer, choose Copy to Clipboard from the Edit menu.

- From the Edit menu, choose **Paste Aligned > Select Levels by Name**.
- In the Select Levels dialog, select Level 2, Level 3 and Level 4 and then click OK (see Figure 5–29).

TIP use the CTRL key to select more than one Level, or simply drag through the names.

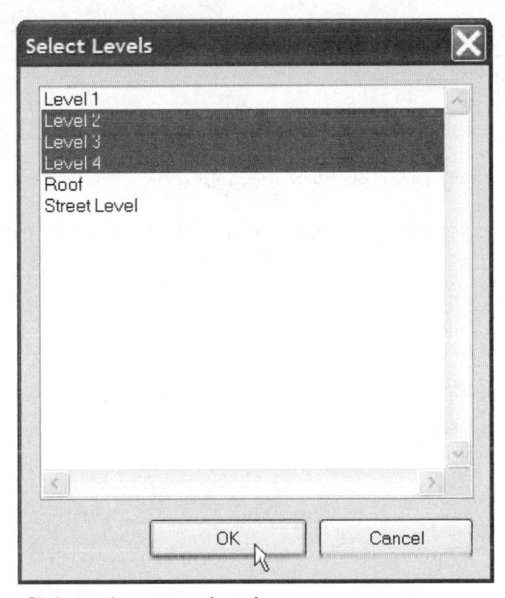

Figure 5–29 *Paste the Slab to the upper Levels*

CREATE A SHAFT

You should now have a Floor slab at each Level and we already had a Roof at the top. However, our building core is now interrupted by these new Floors. We will need a shaft for the elevators and stairs that will come later. Let's add this shaft now to approximate dimensions and then later in Chapter 6 when we add the vertical circulation elements, we can fine tune the size of the shaft to suit the design.

5. On the Project Browser, double-click to open the *Level 1* floor plan View.
6. On the Design Bar, click the Modeling tab and then click the Opening tool.

- In the "Opening Placement Options" dialog, choose "Create shaft opening" and then click OK (see Figure 5–30).

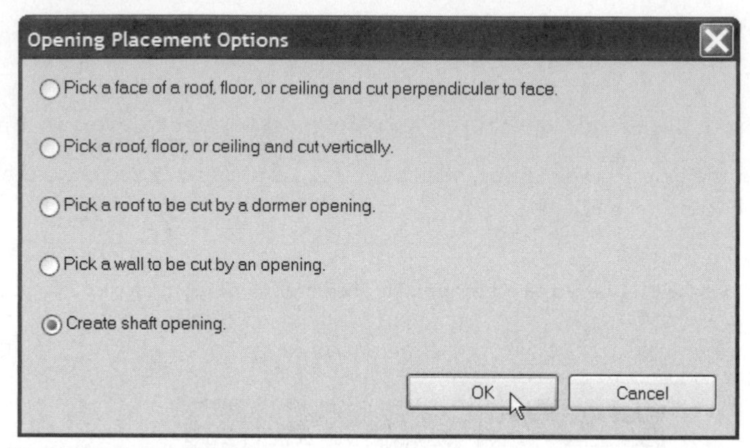

Figure 5–30 *Create a Shaft Opening*

The Design Bar changes to sketch mode. Pick Walls should be active.

- On the Options Bar, click the Rectangle icon.
- Snap to the lower left outside endpoint of the core Walls.
- For the other corner, snap to the inside midpoint of the top exterior Wall (see Figure 5–31).

NOTE If you have difficulty snapping to the Midpoint, snap nearest to the location indicated in the figure instead.

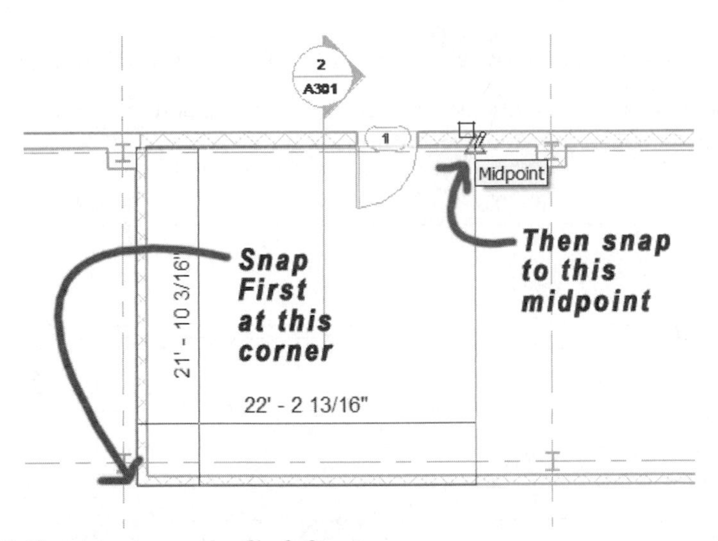

Figure 5–31 *Sketch a rectangular Shaft Opening*

- On the Design Bar, click the Properties button.
- For the "Base Constraint" choose **Level 1** and for the "Top Constraint" choose **Up to level: Roof** and then click OK.
- On the View toolbar, click the Default 3D View icon.
- Using the Spin controls (or hold down the shift key and drag with your wheel button) spin the model around so you can see down into the core.
- On the Design Bar, click Finish Sketch (see Figure 5–32).

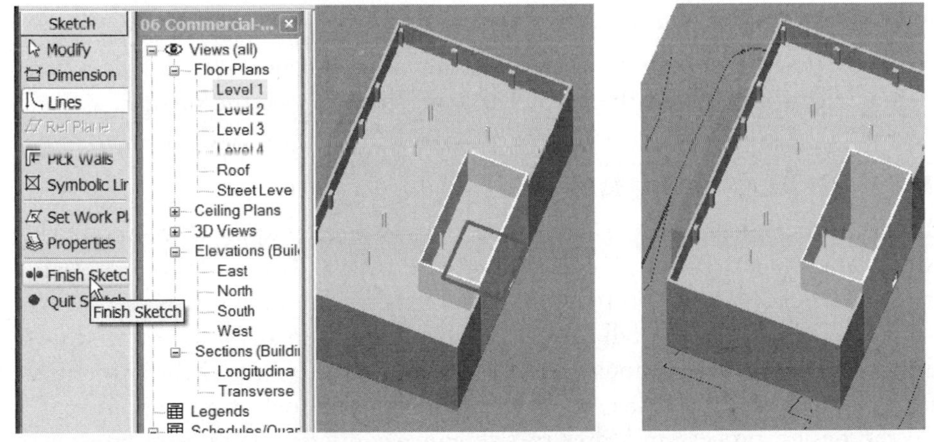

Figure 5–32 *The completed Shaft cuts through all of the Floors*

7. On the Project Browser, double-click to open the *Level 1* floor plan View.
8. On the Design Bar, click the Basics tab and then click the Section tool.

- Create a section (from left to right) through the top portion of the core.
- Click anywhere to deselect the section, then double-click the new Section Head to open the View (see Figure 5–33).

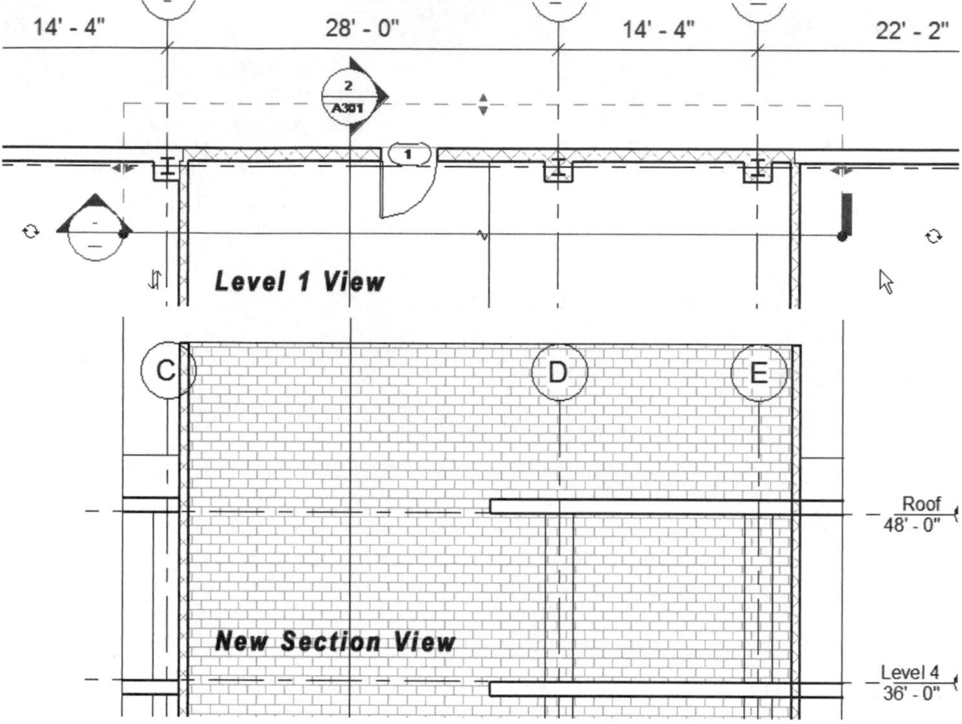

Figure 5–33 *View the Shaft in a new Section View*

Study the way that the shaft has cut the slabs in this and other Views. When you are satisfied with the way that the Shaft has been created, save the project.

STRUCTURAL FRAMING

In addition to the Columns, Grids and Slabs which we have explored so far, Autodesk Revit Building also includes tools to create Beams and Braces. If your primary job function is architectural, you may not be responsible for adding these elements to the model. This task might fall on the structural engineers on your team. However, like the Columns, even though they will ultimately be sized by the Structural Engineer, you can add the basic components to your model at any appropriate point in the design process and later re-size and reconfigure them as appropriate based upon your structural engineer's design and analysis. In this section, we will take a brief look at the tools available. Feel free to explore beyond what is covered here.

WORKING WITH BEAMS

You can add Beams to your model by sketching, or you can create them automatically based upon a column grid. In this exercise, we will use our grid to create Beams.

1. On the Project Browser, double-click to open the *Level 2* floor plan View.
2. On the Design Bar, click the Structural tab and then click the Beam tool.

Like the Columns above, we will first load a Family to use for the Beam shape.

- On the Options Bar, click the Load button.
- Double-click the *Structural* folder, then double-click the *Framing* folder and finally double-click the *Steel* folder.
- Click to select (do not double-click) the *W-Wide Flange.rfa* [*M_W-Wide Flange.rfa*].

A list of industry standard steel-shape sizes will appear in a table at the bottom of the dialog. Scroll through the list to see all of the sizes (similar to Figure 5–17 above).

- Scroll in the "Type Catalog" list area of the "Open" dialog. Select W18x40 [W460x52] and then click Open.

If a "Reload Family" dialog appears, click Yes to accept the reload.

- On the Options Bar, click the Grid button.
- Dragging from right to left, select all Grid lines.
- On the Options Bar, click the Finish button.
- On the Design Bar, click the Modify tool or press the ESC key twice.

After a short pause, it will appear as though the command is finished, but nothing will appear to have been created. This is because the Detail Level of our plan View is set to Coarse.

3. On the View Control Bar, change the Detail Level to **Medium**.

Beams will appear. They have been created at each major Grid line between Columns (see Figure 5–34).

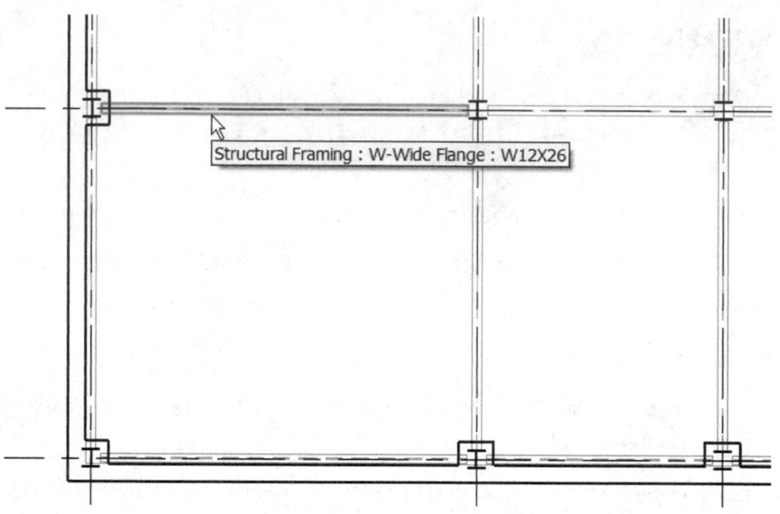

Figure 5–34 *Switch to Medium Detail Level to see the Beams in plan*

4. On the Project Browser, double-click to open the *Longitudinal* section View.
   - On the View Control Bar, change the Detail Level to **Medium.**
   - Zoom in on Level 2 and view one of the Beams.

Notice that they have been added at the same height as the top of the slab.

5. Select one of the Beams.
   - Right-click and choose **Select All Instances.**
   - On the Options Bar, click the Properties icon.
   - Beneath the Constraints grouping, set the Elevation parameter to **-4″ [-100]** (negative values) and then click OK.

The Plane in which the Beams will be created is Level 2. Adding this negative offset moves the Beams below the level enough to allow the topping slab to cover them. (Right now we have a generic slab, but in later chapters, we can change the composition of the slab to show more detail. When we do, the Beams will already be at the correct elevation).

There are other Beam creation methods and options. You can sketch Beams and create them specifically as Girders, Joists, Purlins, etc. When you use the Grid option as we have here, Revit Building determines the usage automatically. In this case we have Girders. You can view the topic on Structural in the online help for more details on this subject. Feel free to experiment further with the various Beam options before continuing.

WORKING WITH BRACES

To complete our preliminary structural layout of our commercial project, let's add a few cross braces at the building core.

6. On the Project Browser, double-click to open the *Level 1* floor plan View.
7. On the Design Bar, click the Structural tab and then click the Brace tool.
   - On the Options Bar, click the Load button.
   - Double-click the *Structural* folder, then double-click the *Framing* folder and finally double-click the *Steel* folder.
   - Click to select (do not double-click) the *L-Angle.rfa* [*M_L-Angle.rfa*].

Once again, a list of industry standard steel-shape sizes will appear in the "Type Catalog" table at the bottom of the dialog. Scroll through the list to see all of the sizes (similar to Figure 5–17 above).

   - Scroll in the list, select L6x6x3/8 [L152x152x9.5] and then click Open.

   If a "Reload Family" dialog appears, click Yes to accept the reload.

8. Verify that L6x6x3/8 [L152x152x9.5] is chosen from the Type Selector.
   - On the Options Bar, from the "Start" list, choose **Level 1**.
   - From the "End" list, choose **Level 2**.
   - Snap the start point to the midpoint of the Column at Grid intersection 3C.
   - Snap the end point to the midpoint of the Column at Grid intersection 4C.
   - Repeat to create another Brace from Grid intersection 3D to 3F.

Feel free to add additional Braces or Beams as desired.

   - On the Design Bar, click the Modify tool or press the ESC key twice.

As with the Beams, Braces will not appear in Coarse Detail Level. You can change this in the *Level 1* plan if you wish. To view either Brace in elevation, you can create additional Section Views as we did above (see Figure 5–35).

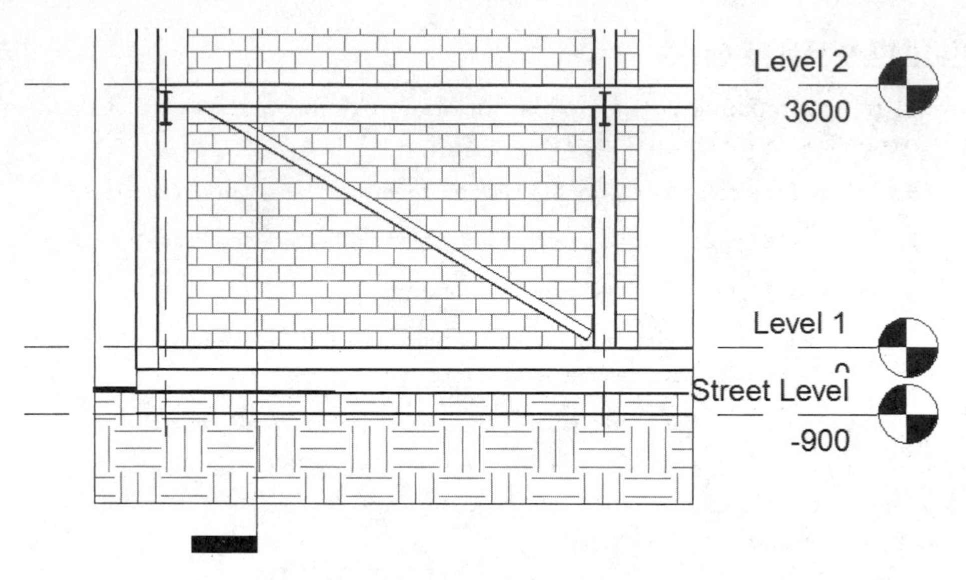

Figure 5–35 *Create a new Section View to see the new Braces*

CREATE A NEW MODEL GROUP

The same Framing layout will occur on all four floors of the project. We could copy it to each Level as we did with the other elements, but let's take advantage of Groups to make it easier to edit the framing later. Anytime that you have a repetitive (or "typical") portion of your building design—such as a typical Stair, Toilet Room or Framing layout as in this case—you can use Groups to manage them. We saw Groups briefly at the start of the chapter with the Array command. In that instance we ungrouped them. In this case, we will create the Group ourselves and leave it grouped. The process is simple. Select the elements, Group them, and then insert instances in the model. Whenever you need to make a change, edit the Group. When you are finished editing, the change will apply to all instances of the Group across the entire project.

9. On the View toolbar, click the Default 3D View icon.
   - Dragging from right to left, select the entire model.
   - On the Options Bar, click the Filter icon.
   - In the "Filter" dialog, click the Check None button.
   - Place a checkmark only in the Structural Framing checkbox and then click OK.

You may not be able to see the selected elements. It depends on the angle of your 3D View. It is not important to see the elements. Choosing "Structural Framing" selected all of the Beams and Braces.

10. On the Tools toolbar, click the Group icon.

- On the Project Browser, expand *Groups* (near the bottom of the list), then expand the *Model* branch.

Notice that there is a "Group 1" listed in the Model category. This is the new Group containing the Structural Framing.

- Right click on Group 1 and choose **Rename** (see Figure 5–36).

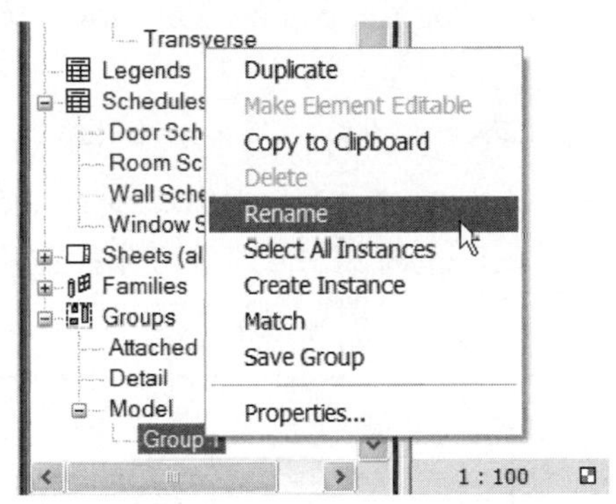

Figure 5–36 *Groups show in the Project Browser. Rename the new Group 1*

- Name the Group **Typical Framing** and then click OK.

11. On the Project Browser, double-click to open the *Level 2* floor plan View.

- Select the Group in the plan (use the TAB key if necessary to select it) right-click and choose **Copy to Clipboard**.
- From the Edit menu, choose **Paste Aligned > Select Levels by Name**.
- In the Select Levels dialog, select Level 3, Level 4 and Roof and then click OK.
- Open any section View to see the result.

EDIT A GROUP

Now that we have a Group of our typical framing, if the design should change, we can edit the Group and the change will propagate across the project to all instances of the Group.

12. On the Project Browser, double-click to open the *Longitudinal* section View.

- Zoom in on Level 1 and view one of the Beams.

- Click on the Beam to select it.

This will select the entire Group.

- On the Options Bar, click the Edit Group button.
- Select the Beam again.

The Design Bar changes to Edit Group mode.

This time, since you are now in the Group edit mode, the Beam will highlight.

13. Using the temporary dimensions, move the Beam by any amount.

An error dialog appears warning you that the elements can no longer be kept joined. This is because you are moving the Beam independently of the elements to which it is connected.

- In the error dialog, click the Unjoin Elements button.
- On the Design Bar, click the Finish Group button.

Notice the change to the other instances of the Group (see Figure 5–37).

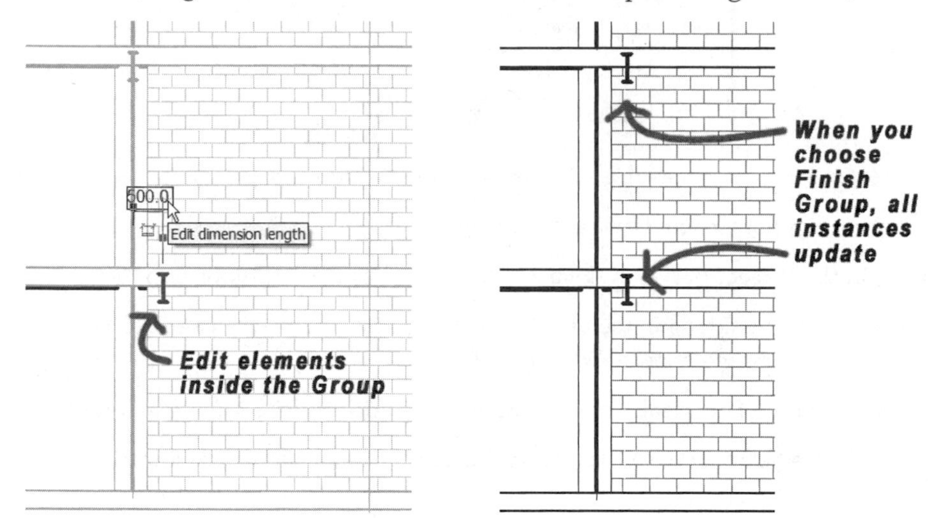

Figure 5–37 *Edit a Group to change all instances at once*

Continue to experiment with the Group if you wish. You can undo the changes when you are finished or you can leave them. It is up to you. This completes the work on the commercial project for this chapter.

As you experiment with the Group, it is important to note that editing the Group and ungrouping it do not yield the same results. When you ungroup a Group, you end up with individual elements that are no longer associated as a Group. Therefore, if you make changes to the ungrouped Elements, such changes will apply only to the Elements themselves and not be propagated back to the other instances of the Group. If you choose to Group the Elements again, you will be creating a new Group, not replacing the existing one.

14. Save and close the project.

FOOTINGS

To complete our brief exploration of structural tools in Autodesk Revit Building we will have a look at the Footing tools provided. To do this, we will switch to the Residential project. Some new Footings are required for the addition on the back of the house.

LOAD THE RESIDENTIAL PROJECT

Be sure that the Commercial Project has been closed and saved.

1. On the Standard toolbar, click the Open icon.

TIP The keyboard shortcut for Open is CTRL + O. **Open** is also located on the File menu.

- In the "Open" dialog box, click the *My Documents* icon on the left side.
- Double-click on the *MRB* folder, and then the *Chapter05* folder.

If you installed the dataset files to a different location than the one listed here, use the "Look in" drop down list to browse to that location instead.

2. Double-click *05 Residential.rvt* if you wish to work in Imperial units. Double-click *05 Residential Metric.rvt* if you wish to work in Metric units

You can also select it and then click the Open button.

The project will open in Revit Building with the last opened View visible on screen.

ADD CONTINUOUS FOOTINGS

You can create two types of Footings in Revit Building: Continuous and Isolated. A Continuous Footing is very easy to add and is associated with a Wall. An Isolated Footing is a Component Family that can be inserted anywhere a Footing is required, such as a pier footing, pile cap, etc. Footing tools are accessed from the Structural tab of the Design Bar.

3. On the Project Browser, double-click to open the *Basement* floor plan View.

4. On the Design Bar, click the Structural tab and then click the Footing > Continuous tool.

The Structural tab should be displayed already from the exercises above. If it is not, right-click on the Design Bar and choose **Structural** to display it.

On the Status Bar, a "Select Wall(s)" prompt will appear.

- Click the vertical Wall on the left of the addition (top of the plan).

A "Warning" message will appear in the lower right corner of the screen. This is considered an "Ignorable" Warning message in Revit Building. It indicates a situation to which you may or may not be aware, but if you choose to ignore it, this condition does not prevent you from continuing your work. In this case, the Footing that we just added is not visible in the current View (see Figure 5–38).

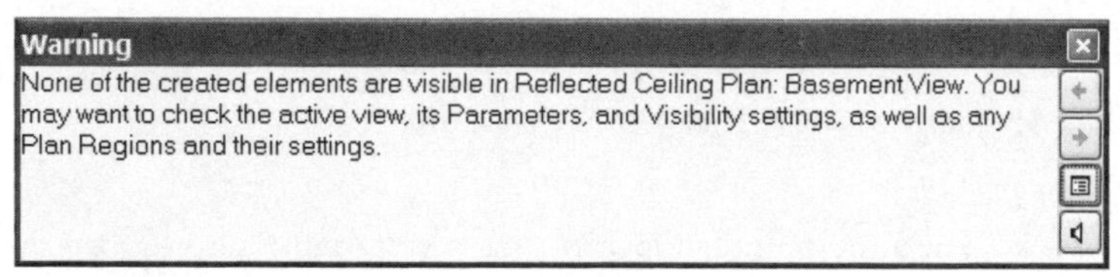

Figure 5–38 *Revit Building "Ignorable" Warning box*

We have to perform a few steps to correct this and display the Footings.

5. Right-click in the *Basement* View window and choose **View Properties** (or just type VP).

- Scroll to the bottom and click the Edit button next to "View Range."
- In the "View Depth" area, type **-2′-0″ [-600]** in the Offset field (see Figure 5–39).

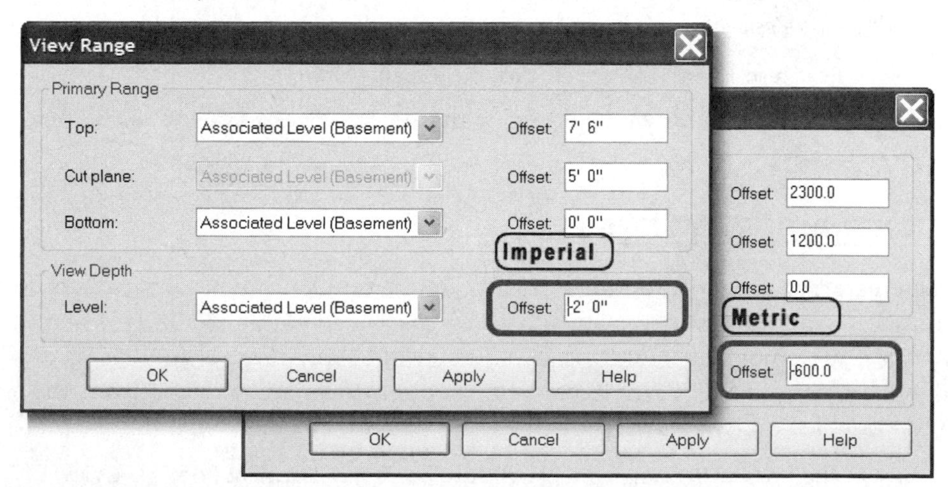

Figure 5–39 *Edit the View Range of the Basement Level Floor Plan*

- Click OK twice to see the results.

If you are working in the Imperial dataset, the Footing displays with a surface pattern. If you want to remove this, select the Footing, click the Properties icon on the Options Bar and then click the Edit/New button. This will edit the Type of the Footing. Next click on the Material, and then the little edit icon next to the Material name. This will load the "Materials" dialog. Finally, Click the edit button next to the Surface Pattern and click the No Pattern button. Click OK three times to return to the model.

We now have the Footings displayed, but they currently display in solid lines. To correct this, we can use the Linework tool to override the display of these lines to dashed in this View.

6. On the Tools Toolbar, click the Linework icon.
   - From the Type Selector, choose **<Hidden>**.
   - Click on each of the Vertical edges of the Footing on screen.

   The lines will display dashed (see Figure 5–40).

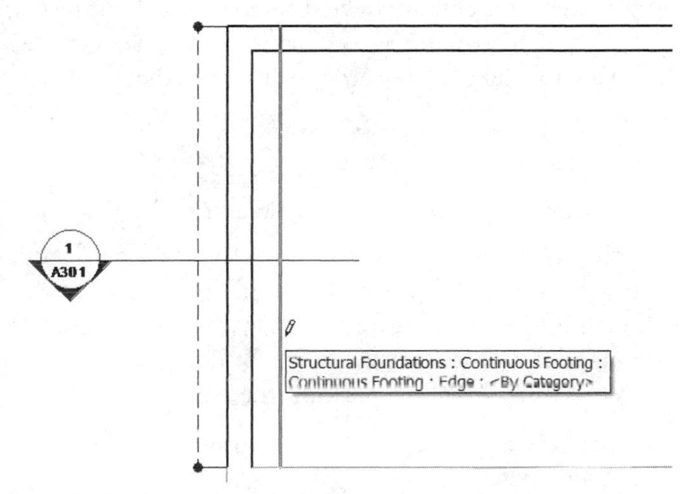

Figure 5–40 *Override the Linework of the Footing edges in this View*

7. On the Design Bar, click the Modify tool or press the ESC key twice.
8. Add two more Continuous Footings to the other two Walls of the addition.
   - Edit the Linework of these in the same way.

You may notice that the small horizontal portion of the Footing where it meets the existing house should also be dashed. If you use the Linework tool, it dashes the entire end of the Footing. However, notice that when you use the Linework tool, there are drag controls on the ends of the linework. You can use these to adjust the portion of the line that receives the override.

9. Use the Linework tool to dash the end of one of the Footings where it meets the existing house.

   - Drag the control to shorten the extent of the linework (see Figure 5–41).

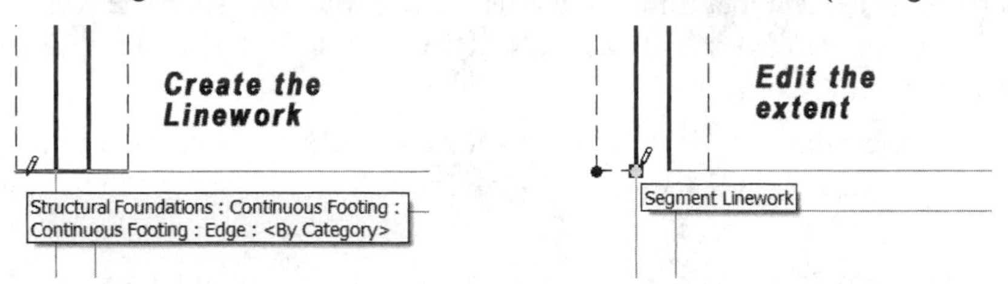

Figure 5–41 *Override the Linework of the Footing ends and use the drag controls to edit the extent*

EDIT THE FOOTING

Like many building model elements in Revit Building, the Footings will interact intelligently with the Walls to which they are associated. For example, if we edit the Profile of the Wall to make a step at the bottom, the Footing will adjust accordingly.

10. Select the horizontal Wall at the top.

    - On the Options Bar, click the Edit Profile button.

The Profile is a sketch that defines the elevation of the Wall. By default, this profile is a simple rectangular shape defined by the length and height of the Wall. We can edit the sketch of this profile to create a Wall with an alternate shape. When you choose this command from a plan View, Revit Building will prompt you select a more appropriate View from which to perform the edit.

- In the "Got To View" dialog, click the *Elevation:North* View and then click the Open View button.

In this View, the terrain model is concealing the foundation Walls and Footings from view. We can use the same technique that we did above to override the display of the linework to make them appear in this View and make them appear as dashed lines. That exercise will be left to the reader. For now, we will simply hide the Terrain so that we can work with the Footings and Foundations Walls.

11. Select the Toposurface, and then on the View Control Bar, choose **Hide Object** from the Hide/Isolate popup.

You should now be able to see the Footings and Walls. Remember that Hide Objects is limited to this work session only.

- Using the Offset tool, offset the bottom edge down **1′-6″** [**450**].
- Draw a vertical sketch line (click the Lines tool on the Design Bar) at **11′-0″** [**3300**] from the right side.
- Using the Trim/Extend tool, complete the sketch as shown in Figure 5–42.

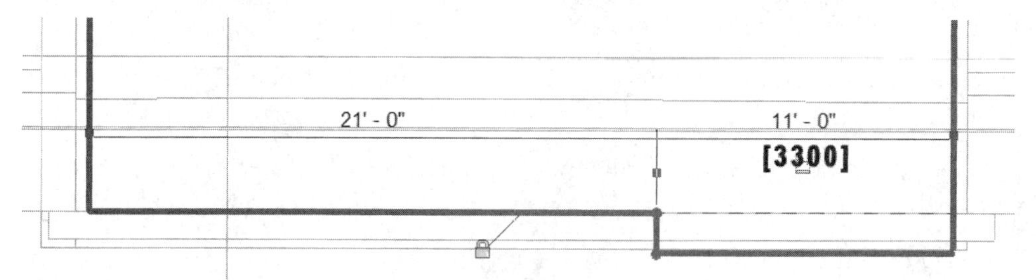

Figure 5–42 *Create a step in the Footing (sketch lines enhanced in the figure for clarity)*

12. On the Design Bar, click the Finish Sketch button.

The Footing will automatically adjust to match the new configuration of the Wall.

13. Perform similar Steps on the adjoining Wall (right side in the current elevation).

Study the model in various Views. Remember that you will need to hide the Toposurface temporarily in other Views to see the foundations (see Figure 5–43).

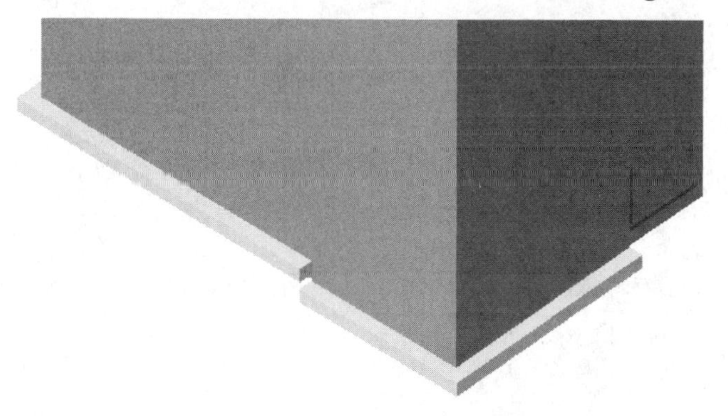

Figure 5–43 *The completed foundation in the* ***{3D}*****View**

There is plenty more that we could do to finish up the foundation of the residential project. We could add existing Footings (remember to work in that phase), we could add an Isolated Footing to the chimney and the columns in the basement. We could also add isolated Footings to the Commercial project. When you choose the Isolated Footing tool, you will be prompted to load a Family for its use. This is left to the reader as an additional exercise. Feel free to experiment in both projects with these tools.

SUMMARY

Column Grid Lines are datums that establish our Column Grid intersections.

Column Grid Lines share many features with Levels and appear automatically in all orthographic Views.

Structural Tools in Revit Building are accessible from the Structural tab of the Design Bar.

Revit Building includes many Structural Tools useful to Architects. For complete Structural Tools including interface with analysis packages, look into Revit Structure.

Structural Columns represent the actual structural support members of the building

Architectural Columns are typically used to represent the finished Column as it would be seen in the building. Use them for Column wraps, etc.

Structural Walls are Revit Building Walls that have their Structural Usage parameter set to Bearing.

Floor Slabs can be created from bounding Walls.

You can create Framing layouts and plans using Beams and Braces

Group a collection of elements and reuse it elsewhere in the model. If the Group changes, all instances will update.

A Continuous Footing is associated to Walls.

An Isolated Footing is a Component Family that you place in your model.

Use the Linework tool to override the display of elements in a particular View.

Vertical Circulation

INTRODUCTION

In this chapter, we will look at Stairs and Railings. We will explore these elements in both of our projects. The Residential Project contains existing Stairs on the interior and an existing exterior Stair at the front entrance. The core plan of the commercial building will include Stairs, elevators, and toilet room layouts. The exterior entrance plaza leading up to the commercial building calls for Stairs, a Ramp, and Railings.

OBJECTIVES

We will add Stairs for the existing conditions of the residential project and lay out the core for the commercial project. Our exploration will include coverage of the Stairs, Railings, Ramps and Elevators. After completing this chapter you will know how to:

- Add and modify Stairs
- Add and modify Railings
- Add and modify Floors and Shafts
- Add and modify Ramps
- Add Elevators

STAIRS AND RAILINGS

Stairs in Autodesk Revit Building are "sketch-based" objects. Sketch-based objects require that you sketch out their basic form with simple 2D sketch lines. This linework is then used to generate the 2D and/or 3D form of the final object. We have seen other examples of sketch-based objects such as the Roof and floor Slabs in the previous chapters. Stairs (like Walls) are System Families. This means that the Stair Types that we will have available to us must be part of the original template from which our project was built. If we want to use a Stair Family that is not part of this original template, we have to either create a duplicate of it from one of the existing ones within the project, or use the Transfer Project Standards command to borrow the Family from another project (Copy and Paste can also be used).

Stairs, like most Revit Building elements, have both instance and type parameters. Type parameters include riser and tread relationships, stringer settings, and basic display settings. Width and height parameters and clearances belong directly to the Stair object (instance parameters). Stairs automatically create Railings by default. For situations where the Railing is not required, we simply delete it after we build the Stair.

INSTALL THE CD FILES AND OPEN A PROJECT

The lessons that follow require the dataset included on the Mastering Revit Building CD ROM. If you have already installed all of the files from the CD, simply skip down to step 3 below to open the project. If you need to install the CD files, start at step 1.

1. If you have not already done so, install the dataset files located on the Mastering Revit Building CD ROM.

   Refer to "Files Included on the CD ROM" in the Preface for instructions on installing the dataset files included on the CD.

2. Launch Autodesk Revit Building from the icon on your desktop or from the **Autodesk** group in **All Programs** on the Windows Start menu.

   - From the File menu, choose **Close**.

This closes the empty new project that Revit Building creates automatically upon launch.

3. On the Standard toolbar, click the Open icon.

TIP The keyboard shortcut for Open is CTRL + O. **Open** is also located on the File menu.

   - In the "Open" dialog box, click the *My Documents* icon on the left side.
   - Double-click on the *MRB* folder, and then the *Chapter06* folder.

   If you installed the dataset files to a different location than the one listed here, use the "Look in" drop down list to browse to that location instead.

4. Double-click *06 Residential.rvt* if you wish to work in Imperial units. Double-click *06 Residential Metric.rvt* if you wish to work in Metric units

   You can also select it and then click the Open button.

The project will open in Revit Building with the last opened View visible on screen.

ADD A STAIR TO THE RESIDENTIAL PLAN

This is the Residential project that was begun in Chapter 3. (We added some footings to it in the previous chapter, and if you did the additional exercises at the end of Chapter 3, you also laid out the second floor existing conditions). In Chapter 3, you may recall that we completed the existing conditions without adding any Stairs. Our first exploration of Revit Building Stairs will be to add those to the project now.

CREATE A STAIR

Adding Stairs is much like adding any other Revit Building sketch-based object. Use the Stair tool on the Design Bar to begin creating a Stair. Since it is a sketch-based object, the Design Bar immediately changes to Sketch mode. Sketch the required Stair components and then finish the sketch to create the Stair and its associated Railings.

5. On the Project Browser, double-click to open the *First Floor* plan View.

NOTE An additional View named *First Floor (Chapter 3)* of the first floor also appears in the Project Browser. This View contains the dimensions that were used in Chapter 3 to set the locations of the Walls. While you can delete the dimensions without removing the associated constraints, it is more difficult later to modify those constraints if required. The recommended approach (used here) is to duplicate the View and apply dimensions to the copied View. Think of this as a "working View" as opposed to the original from which it was copied. The original is more likely to be presented and printed on Sheets, the "working" View simply provides a convenient place for us to work. Remember, in Revit Building Views are simply live snapshots of the total model. Edits you make to model geometry in one View affect all other appropriate Views automatically. So even though the dimensions appear in only one of the floor plan Views, the constraints apply to the elements in the model in every View.

We'll start with a very simple Stair—the existing front entrance stairs.

6. Right-click and choose **Zoom In Region**.

TIP The keyboard shortcut for Zoom In Region is **ZR**. **Zoom In Region** is also located on the Views > Zoom menu.

- Click and drag a region around the front door at the bottom of the plan (see Figure 6–1).

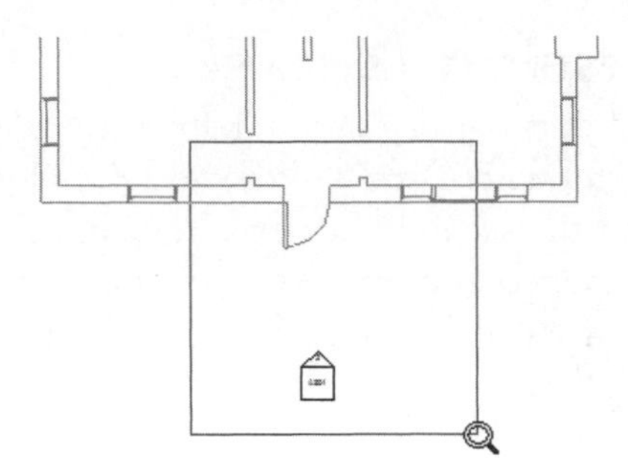

Figure 6–1 *Zoom in on the front entrance*

7. On the Design Bar, click the Modeling tab and then click the Stairs tool.

   The Design Bar changes to sketch mode and shows only the Stair tools.

The Design Bar will go into sketch mode showing only the tools available in that mode. The common "Modify," "Dimension," "Ref Plane," "Finish Sketch" and "Quit Sketch" tools will appear. In addition, Stairs have two sketch creation modes: "Boundary and Riser" and "Run." (Note that Boundary and Riser are two different buttons that are used together as parts of the same sketch). You basically sketch out the plan view of the Stair in simple 2D linework and when you finish the sketch, Revit Building creates a Stair element from it that includes all of the correct parameters. A Stair sketch requires at least two boundary lines and several riser lines. If you use the "Run" option, you simply draw the path of the Stair, and Revit Building will automatically create the required Boundary and Riser lines. If you prefer, you can choose either the "Riser" or "Boundary" mode to draw these items manually. In general, you can create most common stair configurations including straight runs, U-shaped, L-shaped, etc. with the "Run" option. Use the other modes when you wish to modify the automatically created graphics to add a special feature to the Stair or when you want to build a sculptural Stair. We will explore these options later in the tutorial. For now, let's stick with the "Run" option.

Also like other sketch-based Revit Building objects, the Stair has a "Stair Properties" button. You use this button to access the instance and type parameters of the Stair that you are sketching. You cannot use the Properties icon on the Options Bar for this. This is because the Options Bar shows options for the current element. When you are sketch mode, the current element is the Model Lines that you are sketching, *not* the object (Stair in this case) that they will become. An additional button also appears for Stairs. This is the Railing Type button. When you create a Stair, Revit Building automatically adds a Railing, which is a separate object. Most stairs require railings, so Revit Building adds them automatically. Use this control to choose the Railing you wish it to add, or if you prefer, choose "None" to instruct it not to add a Railing (see Figure 6–2).

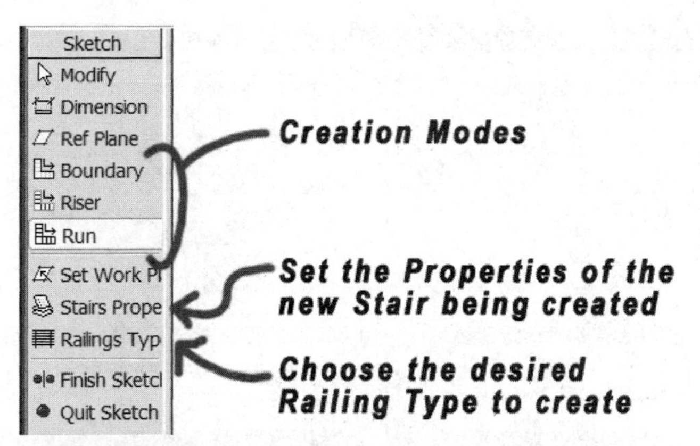

Figure 6–2 *The Sketch-mode Design Bar for Stairs contains some unique tools*

- On the Design Bar, verify that the "Run" option is selected.
- On the Design Bar, click the Stair Properties button.
- In the Element Properties dialog, from the Type list, choose **Monolithic Stair**.
- Beneath the "Constraints" grouping, for the "Base Level" choose **Site**, and for the "Top Level" choose **First Floor**.
- Beneath the "Graphics" grouping, uncheck all the boxes.

These boxes control where and when arrows and annotation will appear indicating the direction of a Stair in plans. In this case, this is a simple existing entrance stair. It does not require any graphics in plan.

- Beneath the "Dimensions" grouping, set the "Width" to **4′-0″ [1200]**.
- Beneath the "Phasing" grouping, from the "Phase Created" list, choose **Existing** and then click OK (see Figure 6–3).

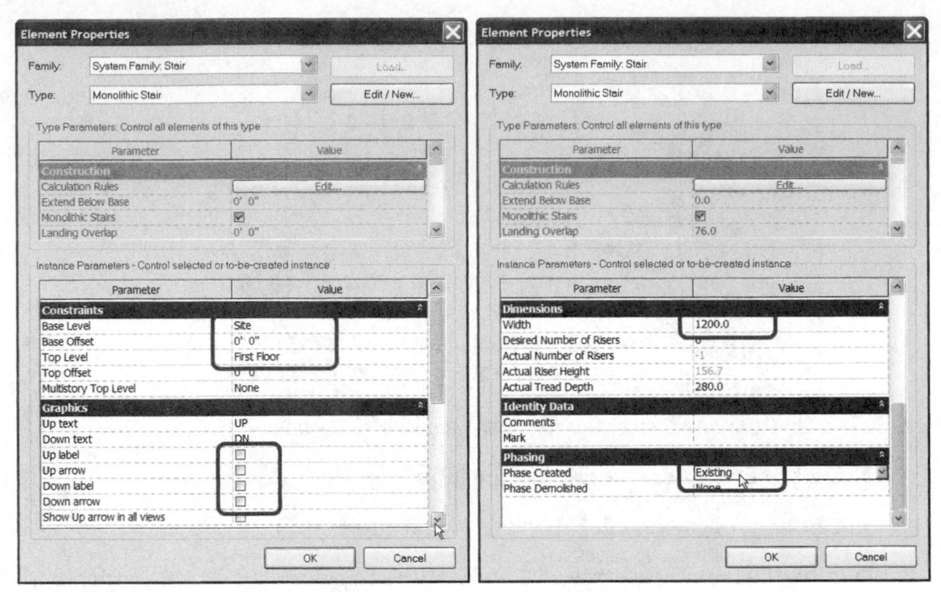

Figure 6–3 *Configure the Stair parameters*

8. Move your pointer to the middle of the front Door.

   - When the guideline appears at the midpoint, click the mouse to set the first point of the Stair run (see the left side of Figure 6–4).

When you create the Stair, the first point is at the bottom of the run and the last point you click is at the top. If you click just two points, you get a straight run. As you drag the mouse (after the first pick) a note in grey text will appear indicating how many risers you have placed. If you click again before you use up all the risers, you create a landing. In this case, we only have 6 risers and want a straight run, so we will click only one more point. However, because we go from bottom to top, we are going to click the second point "up" (inside the house) and then move the resultant Stair sketch to the proper location outside the house. As an alternative to this method, we could have created Reference Planes first to place the Stair in the precise location. We will try the first approach below.

   - Move the pointer straight up and when the number of risers reads "0 Remaining" click the second point (see the middle of Figure 6–4).

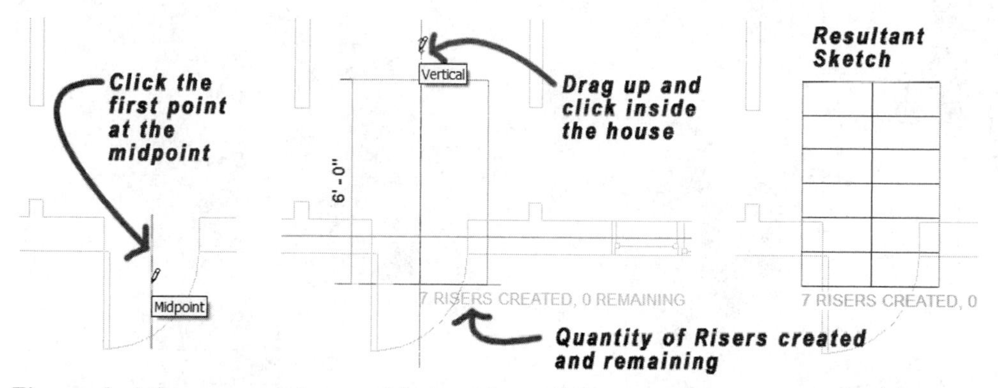

Figure 6–4 *Create a straight run of Stairs with two clicks*

A sketch of blue and green lines will appear indicating the Stair (see the right side of Figure 6–4). The green lines are the boundary lines of the Stair, and the blue lines are the risers. You can edit this sketch in any way before clicking Finish Sketch on the Design Bar. In this case, the only edit we need to do is move the sketch to the correct location.

9. Dragging from left to right, surround the entire Stair sketch.

NOTE Don't worry about selecting the neighboring elements. When you are in sketch mode, you can only select elements of the sketch.

All the sketch lines will highlight red.

10. On the Tools toolbar, click the Move tool.
    - Snap to the Endpoint at the top of the sketch (see the left side of Figure 6–5).
    - Snap to the Intersection of the sketch and the outside edge of the house (see the middle of Figure 6–5).

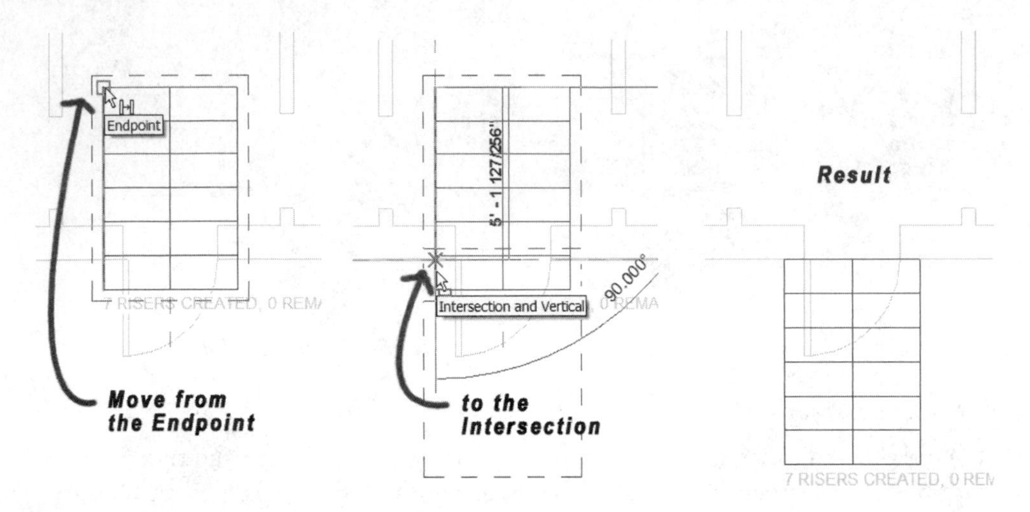

Figure 6–5 *Move the Stair Sketch and snap it to the outside of the house*

The new location of the sketch should match the image on the right side of Figure 6–5.

7. On the Design Bar, click the Railings Type button.
   - In the Railings type dialog, choose Handrail - Rectangular [900mm] and then click OK.
8. On the Design Bar, click the Finish Sketch button.

You will exit the sketch mode and the Stair will appear in the plan. You can view it in other Views as well, to see that it was created properly. Try any section, elevation or 3D View for this. To get the best look at it, we are going to view it in elevation perpendicular to the direction of travel.

MODIFY A STAIR

Sometimes when you create an element, it looks fine in plan, but then you look at it in another View and notice that something is not correct. This is the case here.

9. On the Project Browser, double-click to open the *East* elevation View.
   - Use Zoom In Region again, to zoom in on the Stair (see Figure 6–6).

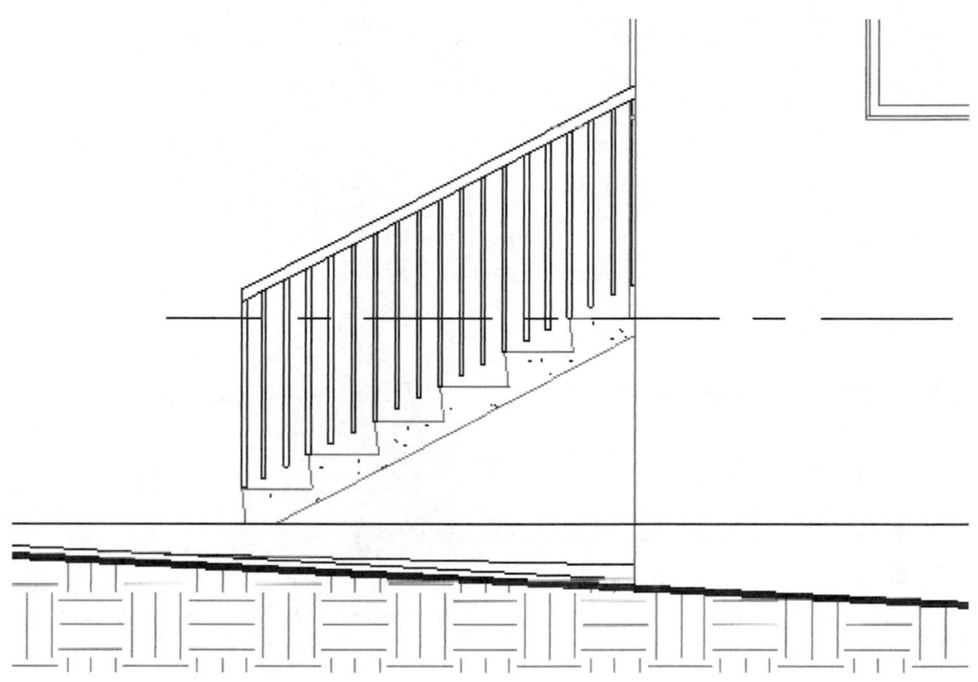

Figure 6–6 *Study the Stair in an elevation View*

The Stair itself looks fine, but since we set the bottom edge at the Site Level, and since the terrain slopes away from the Level height, the Stair appears to float. Let's measure the distance of the Stair above the terrain, and then adjust the Stair accordingly.

10. On the Tools toolbar, click the Tape Measure tool.
    - Snap to the bottom corner of the Stair for the first point.
    - Snap to the terrain directly below the Stair for the second point (see Figure 6–7).

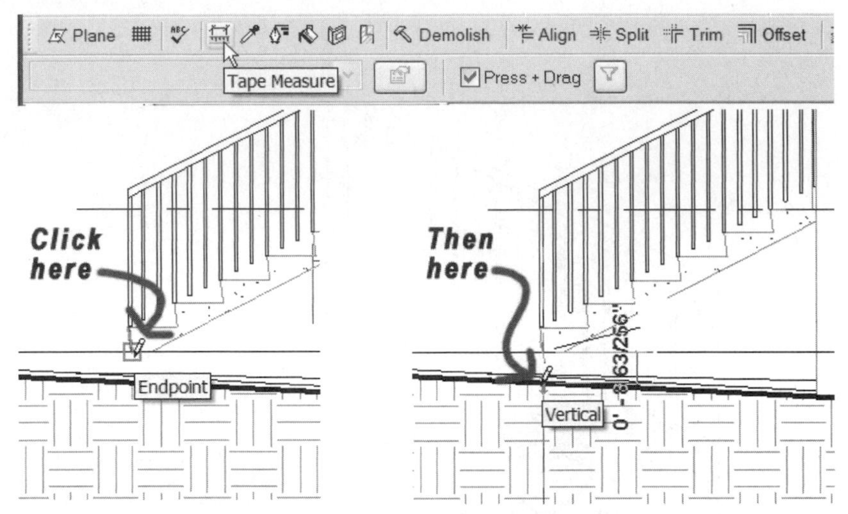

Figure 6–7 *Measure a distance with the* **Tape Measure tool**

TIP Use your wheel mouse to zoom in closer if necessary to click the right point or read the result.

The distance rounded off should be 8″ [200]. We will use this value to adjust the Stair which will add a Riser to it.

11. Select the Stair, and then on the Options Bar, click the Edit button.

A dialog will appear indicating that you need to choose a View better suited to the edit than the currently active elevation View. Revit Building does this when you choose to edit an element in a View that would be difficult or impossible to achieve.

12. From the "Go To View" dialog, choose *Floor Plan: First Floor* and then click Open View.
    - On the Design Bar, click the Stair Properties button.
    - In the Base Offset field, type **-8″** **[-200]** and then click OK.

In the workspace window, the gray note beneath the Stair will now read: "7 Risers Created, 1 Remaining." By making this change, we now need an additional Riser. We can simply add one to the sketch. We can use the sketch tools on the Options Bar to draw the new Riser, or we can copy it from the bottom one instead. The exact technique is not important.

13. Select the bottom Riser.
    - On the Tools toolbar, click the Copy tool.

- Pick a base point and then drag straight down 1′-0″ [280].
- Use the Trim/Extend tool to join the two Boundary lines to the new Riser line (see Figure 6–8).

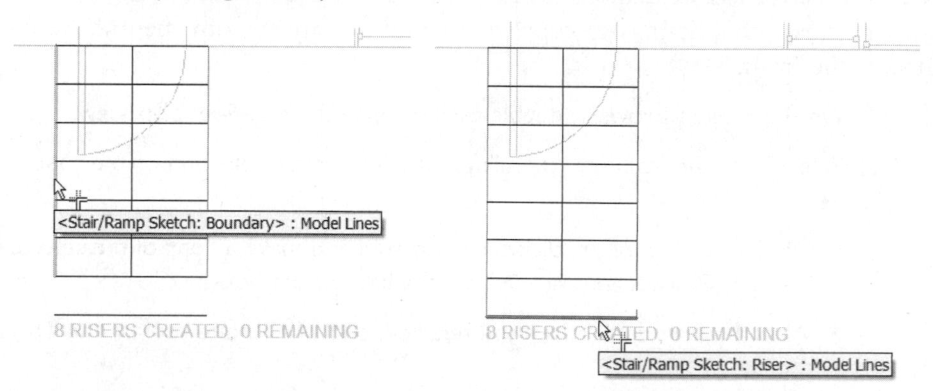

Figure 6–8 *Copy the Riser and then Trim/Extend it to the Boundary Lines*

14. On the Design Bar, click the Finish Sketch button.
15. On the Project Browser, double-click to open the *East* elevation View.

Zoom in again if necessary. Notice that the Stair is now touching the ground instead of floating in space (see Figure 6–9).

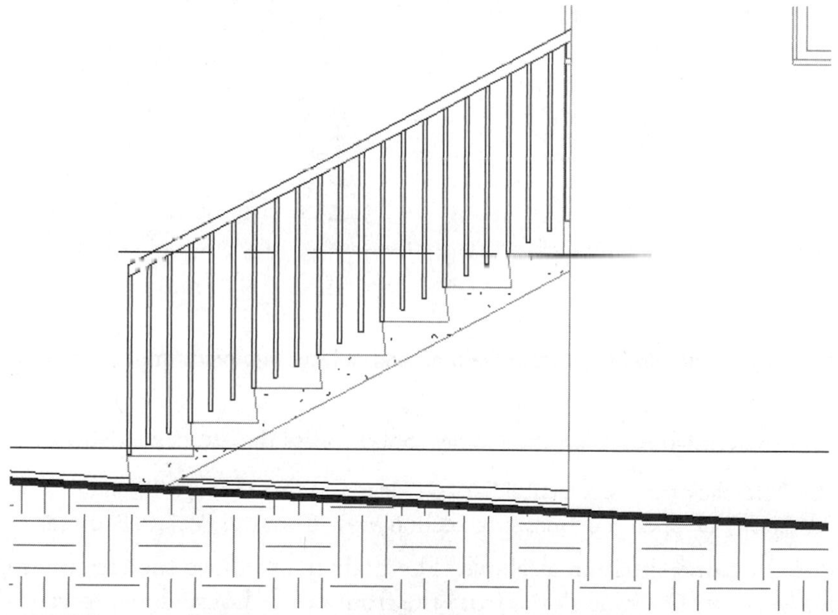

Figure 6–9 *After the edit, the Stair sits properly on the grade*

16. Save the project.

COPY AND DEMOLISH A STAIR

At the back door of the existing house is another Stair like the one in the front. However, this one will be demolished to make way for the new addition. The easiest way to create this Stair is to copy, rotate and modify it from the one we just created at the front.

17. On the Project Browser, double-click to open the *First Floor* plan View.
18. Select both the Stair and its Railings. (use a marquee selection or the CTRL key).
    - On the Tools toolbar, click the Copy tool and make a copy of the selected Stair and Railings and place it near the back of the house.
    - With the Stair and Railings still selected, click the Rotate tool on the Tools toolbar.
    - On the Options Bar, type **180** in the Angle field and then press ENTER.
    - Move the Stair and Railings so that it touches the house centered on the existing back Door (see Figure 6–10).

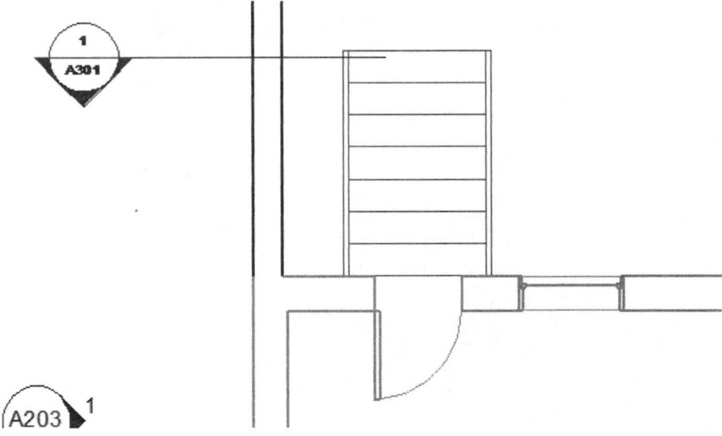

Figure 6–10 *Copy, Rotate and then Move the Stair and Railings into place at the back Door of the existing house*

19. On the Design Bar, click the Modify tool or press the ESC key twice.
20. Double-click the blue vertical Section Head (cutting through the new addition) or open the *Transverse* section View from the Project Browser.

You will note the same situation as above. The Stair floats above the terrain at the back. This is because the grade slopes from the front of the house down toward the back. So by the time we reach the back of the house, the Stair requires yet another Riser.

21. Repeat the Edit process above.
    - Measure the required "Base Offset" for the Stair.
    - Copy an additional Riser to the sketch.
    - Trim/Extend the Boundary Lines and then Finish the Sketch.
22. Select both the Stair and its Railings. (use a marquee selection or the CTRL key).
    - On the Options Bar, click the Properties icon.
    - In "Element Properties" dialog, beneath the "Phasing" grouping, choose **Existing** for the "Phase Created" parameter.
    - Choose **New Construction** for the "Phase Demolished" parameter (see Figure 6–11).

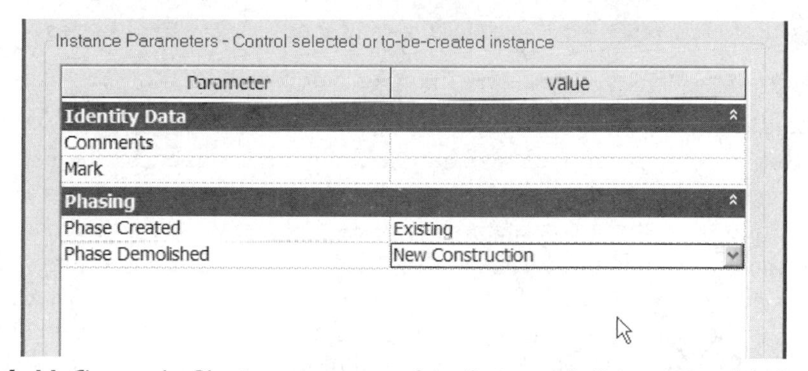

Figure 6–11 *Change the Phasing parameters of the Stair and Railings to demolish them*

NOTE There is also a Demolish tool on the Tools toolbar. However, it only changes the "Phase Demolished" parameter, not the "Phase Created" so you will still need to edit the Element Properties to change both.

The Stair and Railings will turn dashed to indicate that they are to be demolished. If you wish, you can repeat the experimentation that we did in Chapter 3 to edit the parameters of the *First Floor* plan's View Properties and choose different Phase Filters to see how these elements change graphics and display over the course of time. Be sure that you return the "Phase Filter" to **Show All** and the "Phase" to **New Construction** for the *First Floor* View when you are finished experimenting.

CREATE AN EXISTING CONDITIONS VIEW

The next Stair that we will build is the main stair in the existing house. Before we do, let's explore a technique to make working with phases a bit simpler. In Chapter 3, we temporarily changed the current Phase of the *First Floor* plan View (named

Level 1 in that chapter) to "Existing" to add all of the existing construction. At the end of the lesson, we set the Phase back to "New Construction" We could repeat that process here. As an alternative approach, you can maintain two first floor plan Views—one set to the Existing phase, and the other set to the New Construction phase. In this way, you can simply open the View for the phase in which you wish to work.

23. On the Project Browser, right-click the *First Floor* plan View and choose **Duplicate**.

- Right-click the new *Copy of First Floor* View and choose **Rename**.
- For the new Name type **Existing First Floor** and then click OK.

The new Existing First Floor View should have opened automatically. If it did not, double-click to open it now.

- Right-click in the workspace window and choose **View Properties**.
- From the "Phase" list, choose **Existing** and then click OK.

This will set the current Phase to Existing so that any elements we add while working in this View from now on will automatically be assigned to the existing construction phase. We did not *need* to create this View, but sometimes it is easier than remembering to set the Phase parameter of each object we create or changing the Phase of the View back and forth. With Existing set as the Phase, notice that the rear Stair is no longer dashed. It is still set to be demolished, but since we are now viewing the plan as it looks during the Existing Phase, these Stairs are not yet demolished at this time. Also, the Walls of the addition have disappeared. Again, at this point in time, those Walls do not yet exist. If you open the original *First Floor* View it will still show all of the new construction and the demolished Stair.

CREATE A STRAIGHT INTERIOR STAIR

The existing interior Stair is also a simple straight-run stair going from the first to the second floor. However, there are a few custom details on the first few treads that will make the Stair a bit more interesting.

24. Working in our newly created *Existing First Floor* plan View, Zoom and Pan to the middle of the existing house layout.
25. On the Design Bar, click the Modeling tab and then click the Stairs tool.

    The Design Bar again changes to sketch mode and shows only the Stair tools.

26. On the Design Bar, click the Ref Plane tool.

Sometimes it is difficult to set the precise location of the element that you are placing. In the case of Stairs, the Run option adds the Stairs from the midpoint at the bottom of the run. We do not currently have a convenient point at that location that we can snap to. A Reference Plane can help. Reference Planes are simple datum objects in Revit Building. You can add them anywhere that you need assistance in achieving alignments or reference points. Think of them as the blue pencil lines in hand drafting.

NOTE The way in which Reference Planes are presented here is an acceptable use for them. However they are also part of the constraint mechanism in Revit Building. This means that they are very similar to Levels & Structural Grids (see previous chapters). If an element somewhere else in the model is constrained to a Reference Plane and then later that Reference Plane is deleted, important constraints will be lost. As an alternative to the method shown here, we can create a line type with special properties such as thin light blue and dashed. Then, instead of Reference Planes for this situation, we could add Detail Lines to the View as "Construction Lines" and choose the Type we defined from the Type Selector. Again, either approach can work. Just be certain that all team members are informed of the preferred approach.

- On the Options Bar, click the Pick Lines icon.
- In the Offset field, type **1′-6 1/2″ [471]**. (This is half the width of the Stair).
- Highlight the right side of the vertical Wall on the left of the stair space.
- When the green guideline appears in the middle of the stair space, click to create the Reference Plane (see Figure 6–12).

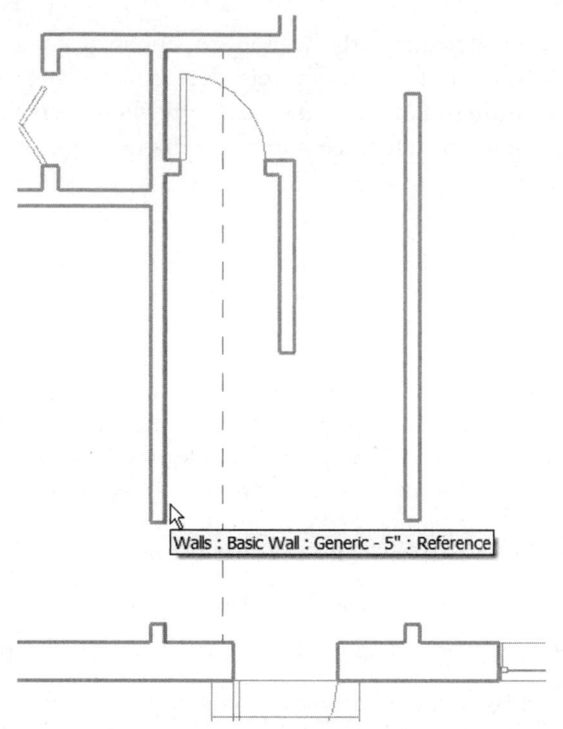

Figure 6–12 *Offset a Reference Plane to the right of the left vertical Wall*

- On the Options Bar, change the Offset parameter to **3′-6″** [**1100**].
- Offset a Reference Plane up from the inside face of the bottom exterior Wall (see Figure 6–13).

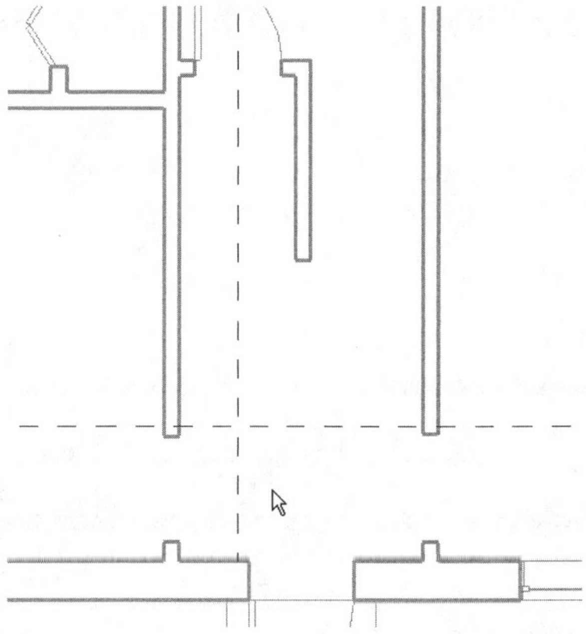

Figure 6–13 *Offset a Reference Plane up from the bottom exterior Wall*

27. On the Design Bar, click the Modify tool or press the ESC key twice.

28. On the Design Bar, click the Run tool.

   - On the Design Bar, click the Stair Properties button.
   - In the Element Properties dialog, from the Type list, choose 7″ max riser 11″ tread [190mm max riser 250mm going].
   - Verify that the "Base Level" is set to **First Floor** and the "Top Level" is **Second Floor**.
   - Beneath the "Graphics" grouping, place checkmarks in only the "Up label" and "Up arrow" boxes; be sure the others are cleared.
   - In the "Dimensions" grouping, change the Width to **3′-1″** **[942]** and then click OK (see Figure 6–14).

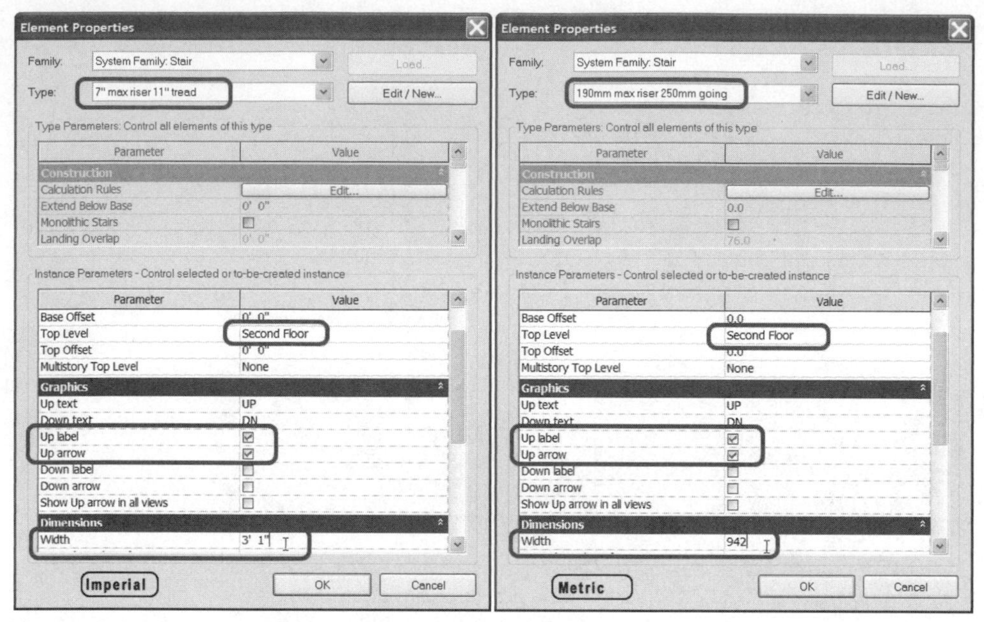

Figure 6–14 *Set up the parameters for the Stair*

29. Click the first point of the Stair run at the intersection of the two Reference Planes.
    - Move the pointer straight up far enough to place all of the risers in a single run and then click.

The sketch created will be too long for the space. The parameters configured above were all "instance" parameters. To adjust our Stair to fit the space, we need to modify the "type" parameters. We will create a new Type and adjust its parameters in the next topic.

CREATE A NEW STAIR TYPE

In this sequence, we will edit the parameters of the Stair Type and make some minor modifications to help it conform to the existing structure. This will involve applying less stringent Tread and Riser rules on our Stair. Since the stairs existing in this house were built before current code requirements were in place, an existing Stair Type needs to be created allowing more freedom.

30. On the Edit menu, choose **Undo Run** (or press CTRL + Z).

This will remove the sketch so that we can adjust the parameters and re-create it.

31. On the Design Bar, click the Stair Properties button.
    - At the top of the dialog, click the Edit/New button.

- Click the Duplicate button.

TIP To do this quickly, press ALT + E and then immediately press ALT + D.

- In the "Name" dialog, type **Existing House Stair** and then click OK.

Here you can enter a range of values for minimum and maximum tread and riser. You can also enter building code values for your jurisdiction in a rule-based calculator (click the Edit button next to "Calculation Rules"). Values assigned to these rules will constrain the parameters of the Stair as it is being sketched. Again, because the Stair we are building is an existing Stair, we will set the limits very broadly so that we can enter actual values without restriction.

- Beneath the "Treads" grouping, type **8″ [200]** for the "Minimum Tread Depth."
- From the "Nosing Profile" list, choose **Default**.
- Beneath the "Risers" grouping, **12″ [300]** for the "Maximum Riser Height" (see Figure 6–15).

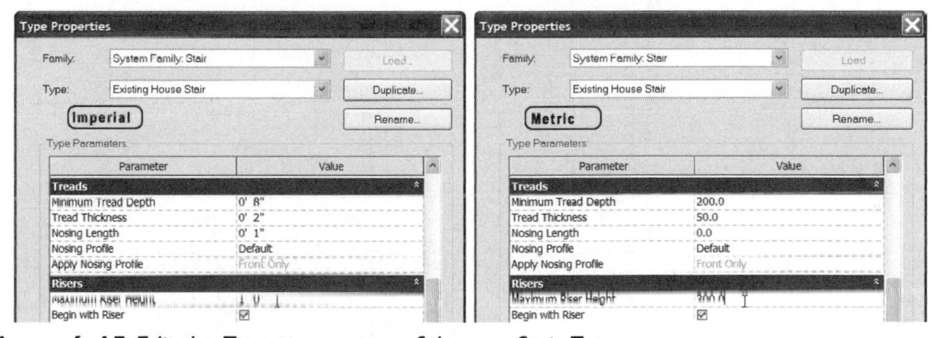

Figure 6–15 *Edit the Type parameters of the new Stair Type*

32. Click OK to accept the values and return to the "Element Properties" dialog.

- Beneath the "Dimensions" grouping, set the "Actual Tread Depth" value to **8″ [250]**.
- Set the "Desired Number of Risers" parameter to **14** and then click OK (see Figure 6–16).

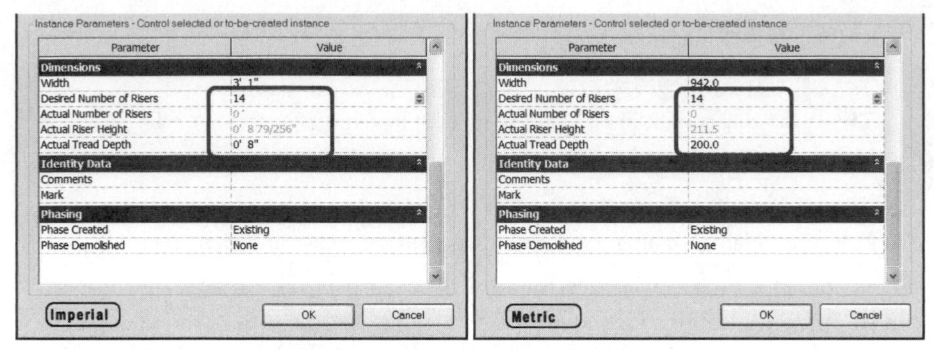

Figure 6–16 *Set the actual Tread and Riser settings of the existing house Stair*

33. On the Design Bar, click the Run tool.

    - Click the first point of the Stair run at the intersection of the two Reference Planes.

    - Move the pointer straight up far enough to place all of the risers in a single run and then click.

Notice that the Stair sketch now fits the available space (see Figure 6–17).

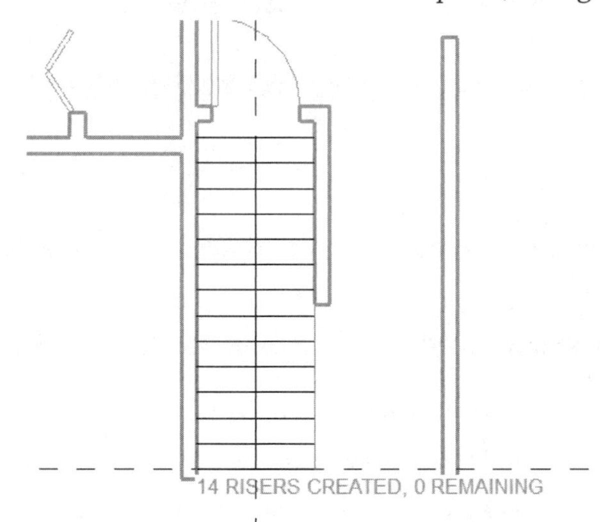

Figure 6–17 *The Stair sketch based on the newly created Stair Type properly fits the space*

34. On the Design Bar, click the Finish Sketch button.

The Stair object will appear with its railings. Notice that a cut line appears automatically. Also, since we asked only for "up" annotation, we only get an arrow and label pointing up, not up and down. We will add the Stair down to the basement below.

EDIT THE STAIR TYPE

When you look at the Stair we created, it appears to be too wide—part of it overlaps the neighboring Walls. This overlap is actually the Stringer of the Stair and the Railing that sits on top of it. We will need to make some adjustments to the Stair Type to address this.

35. Select the Stair.
    - On the Options Bar, click the Properties icon.
    - Click the Edit/New button (or press ALT + E).
    - Beneath the "Stringers" grouping, choose **Open** from both the "Left Stringer" and the "Right Stringer" lists.
    - From the "Stringer Top" list, choose **Match Level**.

An "Open" stringer is notched in the shape of the treads and risers and supports the treads from underneath. The "Closed" stringer occurs at the edges of the Stair with the treads and risers spanning in between. A typical wooden Stair would use Open, while a steel pan Stair would be Closed. To create a typical concrete Stair, choose Closed and then also select "monolithic" which makes a single continuous stringer across the bottom of the Stair. The tops of the stringers can be cut to match the Level or they can be left uncut. Several of stringer settings are shown in Figure 6–18.

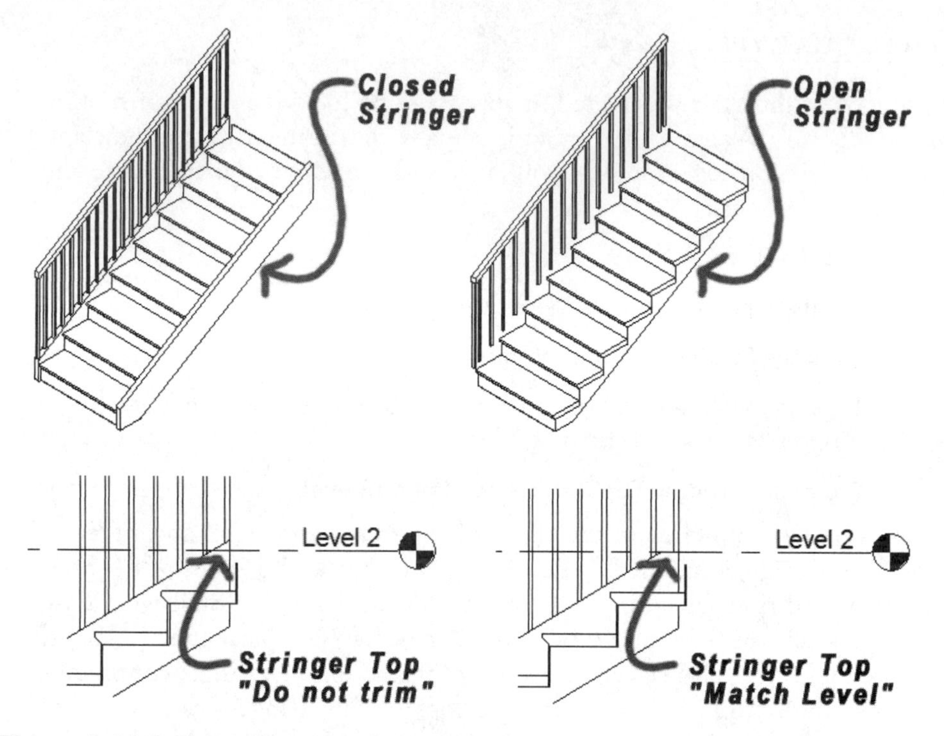

Figure 6–18 *Exploring the various Stringer settings*

In addition to these settings, we can also edit the dimensions of the stringer material. In this case we will leave those set as is.

36. Click OK twice to apply the changes and return to the model.

It appears as though nothing has changed in our plan View. However, if you slowly pass your mouse over the Stair and its edge, you will note that when the Stair pre-highlights, it has indeed reduced in width. The overlapping portion that we still see is actually the Railing.

37. Click on one of the Railings.

- Click the small "Flip Railing Direction" control to flip the Railing to the other side of the stringer.
- Repeat on the other side (see Figure 6–19).

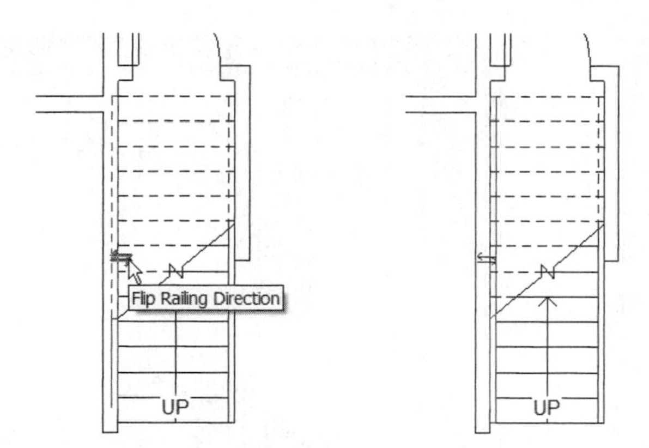

Figure 6–19 *Flip the Railings to the inside edges of the Stair*

CREATE THE "DOWN" STAIR

There is also an existing Stair going down to the basement. The simplest way to create this one is with Copy and Paste.

38. On the Design Bar, click the Basics tab and then click the Section tool.
   - Create a section line running vertically through the Stair.
   - Deselect the Section Line (click away from it) and then double-click the Section Head to open the associated View.

In this View, you can see very clearly that we have a single Stair spanning from first to second floor.

39. Select the Stair.
   - From the Edit menu, choose **Copy to Clipboard** (or press CTRL + C).
   - From the Edit menu, choose **Paste Aligned > Select Levels by Name**.
   - In the "Select Levels" dialog, choose Basement and then click OK.
40. With the new Stair still selected, on the Options Bar, click the Properties icon.
   - Verify that the "Base Level" is Basement and that the "Top Level" is First Floor.
   - Remove the "Top Offset" by typing a value of **0** (zero) and then click OK (see Figure 6–20).

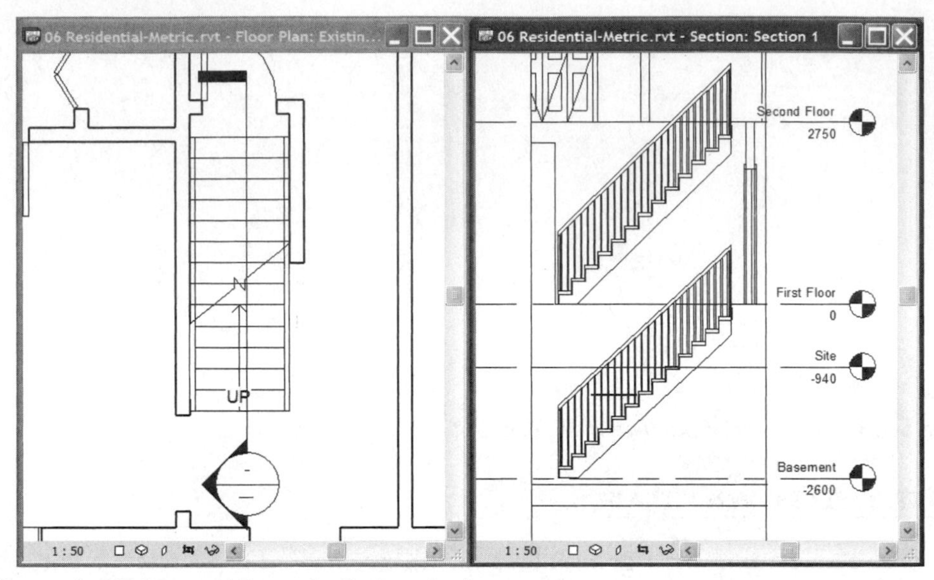

Figure 6–20 *Copy and Paste the Stair to the Basement*

41. On the Project Browser, double-click to open the *Existing First Floor* plan View.
42. Select the basement Stair (the one just pasted—you may need to use the TAB key to select the right one).
    - On the Options Bar, click the Properties icon.
    - Beneath the "Graphics" grouping, clear the "Up arrow" and "Up label" checkboxes.
    - Place checkmarks in the "Down arrow" and "Down label" checkboxes and then click OK.
43. On the Project Browser, double-click to open the *Basement* floor plan View.

Notice only the basement Stairs appear here. If you open the Second Floor plan View as well, you will note that only the upper Stairs appear. However, since we have not yet added any Floors to our residential model, they will show even beneath the closet at the front of the house. We'll adjust this in later chapters.

EDIT THE STAIR SKETCH

So far we have used the "Run" option to automatically create very simple Stair sketches based upon our riser and tread dimensions. However, you can edit these simple sketches to add architectural detailing to your Stairs. You can also completely create your own custom sketch to create very complex Stair configurations.

In this example, we will widen the lower portion of the Stair and add a bull nose to the bottom two treads.

44. On the Project Browser, double-click to open the *Existing First Floor* plan View.

45. From the Window menu, choose **Close Hidden Windows.**

46. Select the "Up" Stair (use the TAB key if necessary to assist in selection).

- On the Options Bar, click the Edit button.

This returns you to sketch mode and reveals the sketch lines created when we added the Stair. By adjusting these sketch lines, we can change the shape of the Stair.

- Position your pointer over one of the blue Riser lines to pre-highlight it.

Notice that each Riser line pre-highlights individually; also, note that a screen tip appears indicating its category and type. The sketch lines of the Stair are actually "Model Lines." There are two types of drafted line elements in Revit Building: they are Model Lines and Drafting Lines. A Model line, while two-dimensional, is a part of the building model. Therefore, like other elements of the model, it will appear in *all* appropriate Views. A Drafting line is also a two-dimensional piece of linework, but it appears *only* in the View in which it is drawn. Model Lines are used to represent real things, while Drafting Lines are like annotation; used to embellish a particular View to convey intent.

- Pre-highlight one of the green Boundary lines (see Figure 6–21).

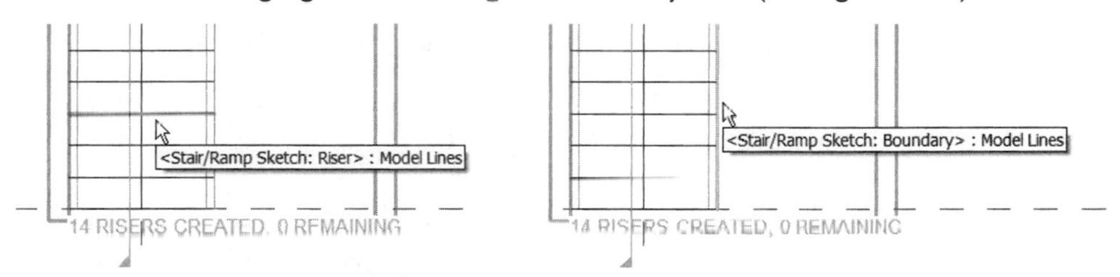

Figure 6–21 *Pre-highlight the sketch lines of the Stair*

The Boundary lines are also Model Lines. Model Lines are used in the Stair sketch because they indicate that actual form of the Stair in the model. If we want to customize the shape of the Stair, we simply edit these model lines in sketch mode. The Design Bar will change to Sketch Mode again. When you are working with a Stair in Sketch Mode, there are three sketching methods on the Design Bar: Boundary, Riser and Run. We have seen Run already. When you wish to edit an existing Stair, or when you wish to draw a Stair in free-form mode, you use the Riser and Boundary options. Let's start by editing the boundary.

47. On the Design Bar, click the Boundary tool.

- On the Options Bar, select the Draw and Line icons and place a checkmark in the "Chain" checkbox.
- Click the first point of the new Boundary Line at the intersection of the right-hand Boundary Line and the bottom of the existing Wall (see "Point 1" in Figure 6–22).
- Click the next point at the other side of the Wall's width (see "Point 2" in Figure 6–22).

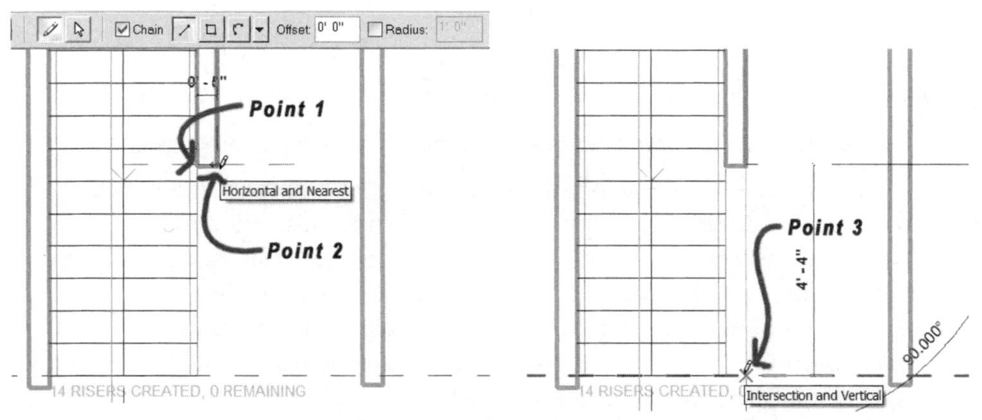

Figure 6–22 *Sketch two new Boundary Lines snapping to the existing Wall*

- Place the last point aligned with the original Boundary Line's bottom end and the Wall's right face (see "Point 3" in Figure 6–22).
- On the Design Bar, click the Modify tool or press the ESC key twice.
- Use the Trim/Extend tool (with the "Trim/Extend to Corner" option) to remove the unneeded segment (see Figure 6–23).

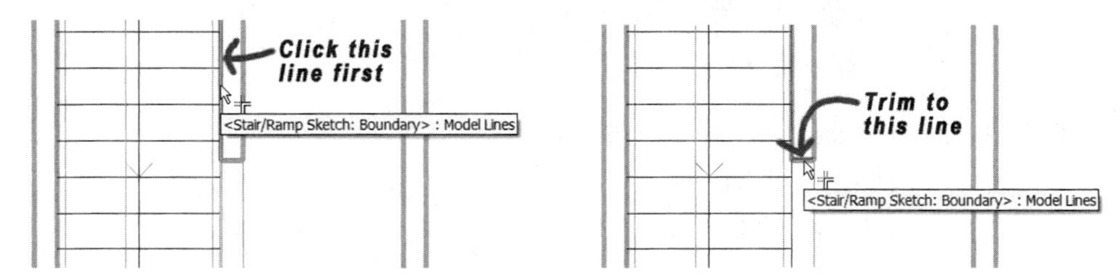

Figure 6–23 *Trim the existing Boundary Line to join it to the new sketch*

- With Trim/Extend still active, on the Options Bar, choose the "Trim/Extend Multiple Elements" icon.
- For the Trim/Extend Boundary, click the newly drawn Boundary Line (vertical one).

- Click each of the Riser Lines to extend them to this Boundary (see Figure 6–24).

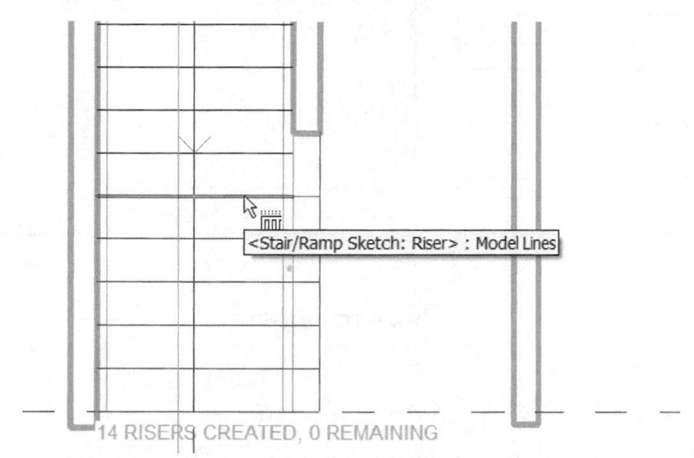

Figure 6–24 *Extend the Riser Lines to the new Boundary Line*

- On the Design Bar, click the Modify tool or press the ESC key twice.

On the lower portion of the Stair, the risers will now come out flush with the Wall. Before we add the bull nose to the lower two steps we need to stretch the Boundary Line back a bit. Otherwise, the stringer will follow the shape of the bull nose.

48. Click the right Boundary Line.
    - Using the drag handle at the bottom, drag the endpoint up and snap it to the end of the third Riser Line.
    - When the error dialog appears, click the Unjoin Elements button (see Figure 6–25).

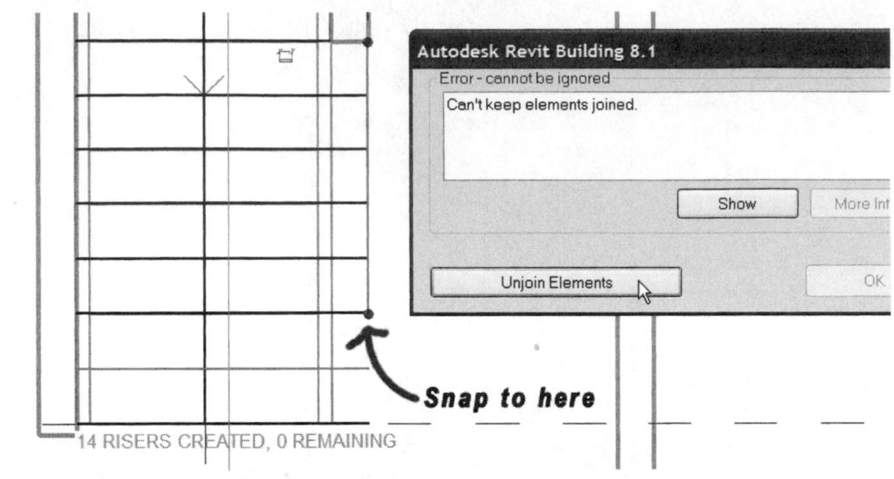

Figure 6–25 *Reduce the length of the Boundary Line by two treads*

The Boundary Line is joined to each of the Riser lines. This is required to create a valid Stair. When we stretch the Boundary Line as indicated above, it breaks this relationship with the lower Risers. We will edit the shape of the Riser lines below, to join them back to the Boundary.

ADD A BULL NOSE RISER

To add the bull nose Riser, we simply sketch them in using the Riser sketch method on the Design Bar.

49. On the Design Bar, click the Riser tool.
    - On the Options Bar, clear the "Chain" checkbox.
    - Click the "Arc passing through three points" icon.
    - Place the first point of the arc at the right endpoint of the bottom Riser Line (see "Point 1" in Figure 6–26).
    - Place the second point at the endpoint of the third Riser line from the bottom (see "Point 2" in Figure 6–26).

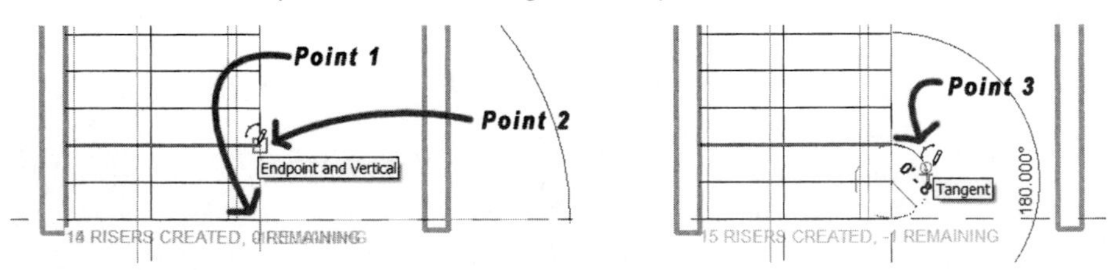

Figure 6–26 *Add an Arc to the lower Riser*

- Move the pointer to the right and when the shape snaps to a half-circle, click to place the last point (see "Point 3" in Figure 6–26).
- Repeat the process to create a smaller Arc (one tread deep) between the second and third risers (see Figure 6–27).

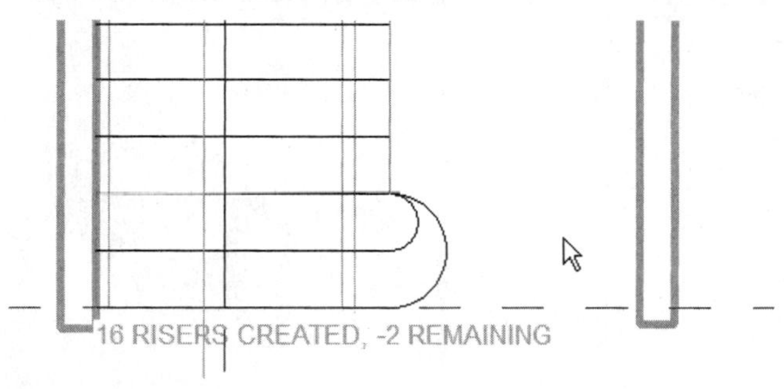

Figure 6–27 *Add an additional Riser Arc*

50. On the Design Bar, click the Finish Sketch button.

VIEW THE STAIR IN 3D

At this point, it might be easiest to understand what we have created if we view the edited Stair in 3D. We have used the "Default 3D View" toolbox several times already. However, this tool will show the entire model in 3D. With all of the exterior Walls visible, it will be impossible to see the inside Stairs. There are a few tricks we can use to quickly view just the Stairs in 3D.

51. On the Window menu, choose **Close Hidden Windows.**

NOTE If this command is not available, then you already have only the current Window open.

- On the View toolbar, click the Default 3D View icon.
- From the Window menu, choose **Tile**.

TIP The keyboard shortcut for Tile is **WT**.

52. Click in the *First Floor* View window to make it active.

- Select the Stair, Railings and their immediately adjacent Walls (see the left side of Figure 6–28).

TIP Use the CTRL key to click on each object that you wish to add to the selection. Try to avoid selecting the basement Stair.

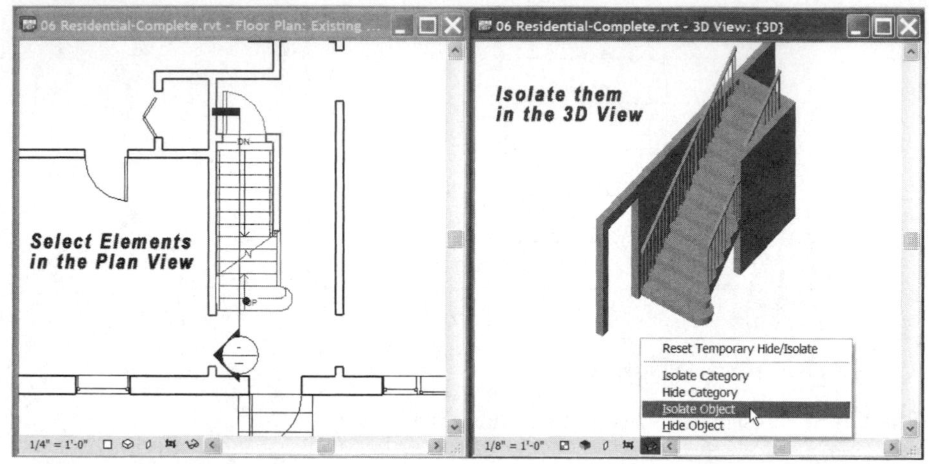

Figure 6–28 *Select elements in the plan View, isolate them in the 3D View*

TIP The keyboard shortcut for Isolate Object is **HI**. **Isolate Object** is also located on the View > Temporary Hide\Isolate menu.

- From the View Control Bar in the 3D View window, choose **Isolate Object** from the Hide/Isolate pop-up menu (see the right side of Figure 6–28).

Feel free to use the "Dynamically Modify View" controls to spin the isolated selection around to an alternative viewing angle. Notice the way the two treads at the bottom of the Stair have the bull nose applied to them. Notice also the way that the stringer stops before these two treads. These are direct effects of the way that we sketched the Stair above.

ADJUST A RAILING SKETCH

Having completed the edit of the Stair above, it is now apparent that the Railing could use a bit of adjustment as well. You edit Railings the same way as Stairs—edit the sketch.

53. Zoom in on the Railing on the right side.

As you can see, this Railing has matched the shape of the Boundary Line that we sketched above and make a slight jog as it goes up the run of Stairs. Let's eliminate the portion of the Railing above the jog (adjacent to the Wall).

54. Select the Railing on the right side (use TAB if necessary to select it. Select in either View).
    - On the Options Bar, click the Edit button.
    - Delete the top two sketch lines, (the vertical one and the short horizontal one that are adjacent to the Wall).
    - On the Design Bar, click Finish Sketch (see Figure 6–29).

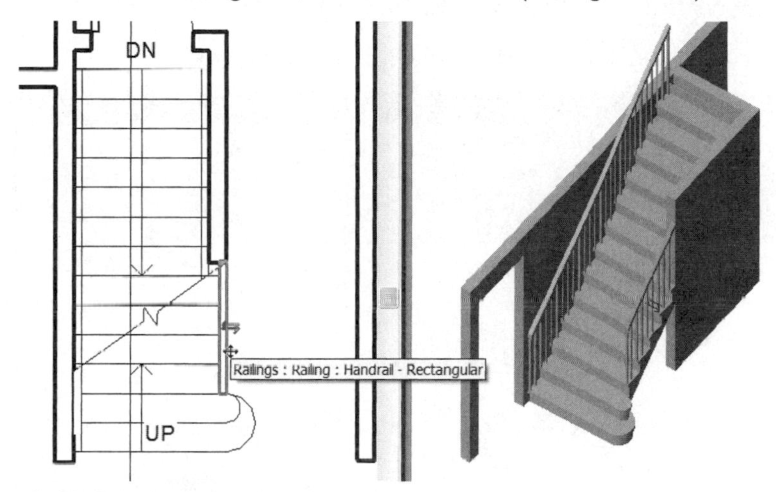

Figure 6–29 *Edit the Railing Sketch*

CREATE A NEW RAILING TYPE

The Railing on the other side of the Stair does not need balusters or posts. It is attached to the Wall. To represent this correctly, let's duplicate the existing Railing Type and then modify it.

55. Select the long Railing on the left side of the Stair.
    - On the Options Bar, click the Properties icon.
    - At the top of the dialog, click the Edit/New button.
    - Click the Duplicate button.

TIP To do this quickly, press ALT + E and then immediately press ALT + D.

- In the "Name" dialog, type **Existing House Railing – No Balusters** and then click OK.

- In the Type Properties dialog, beneath the "Construction" grouping, click the Edit button next to "Baluster Placement."
- In the "Edit Baluster Placement" dialog, in the "Main Pattern" area (at the top), choose **None** from the Baluster Family list for element 2 (see the top of Figure 6–30).
- In the "Posts" area (at the bottom), choose **None** for all three components (see the bottom of Figure 6–30).
- Clear the "Use Baluster Per Tread On Stairs" checkbox.

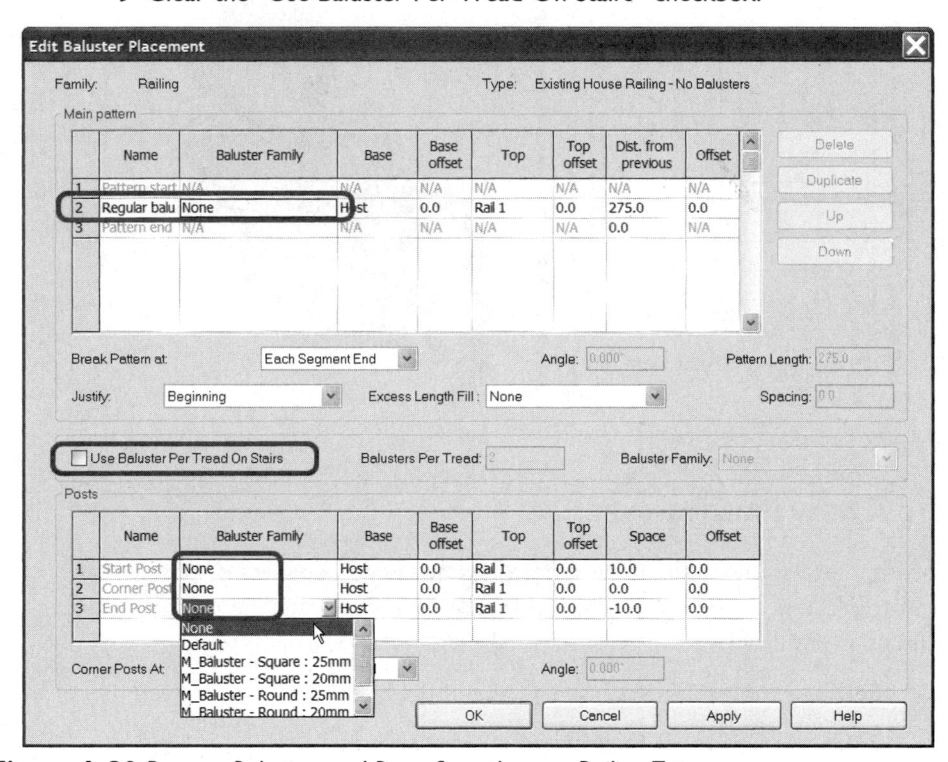

Figure 6–30 *Remove Balusters and Posts from the new Railing Type*

- Click OK three times to return to the model window.

ADJUST WALL PROFILE

Although our Stair looks fine, the 3D View reveals that the Wall on the right does not continue under the Stair and the Basement Door is too tall and intersects the Stair.

56. In the 3D View window, on the View Control Bar, choose **Hidden Line** from the Model Graphics popup.

TIP The keyboard shortcut for Hidden Line is **HL**. **Hidden Line** is also located on the View menu.

- From the View menu, choose **Orient > East**.

The options on the View > Orient menu allow you to match the 3D View to several preset vantage points or to match to another existing View, such as a section. If you wish, you can also save the View as a new View in the project.

57. Select the Wall to the right of the Stair.

    It is to the right in plan, it is in front of the Stairs in the 3D View.

    - On the Options Bar, click the Edit Profile button.
    - Edit the sketch lines as indicated in Figure 6–31.

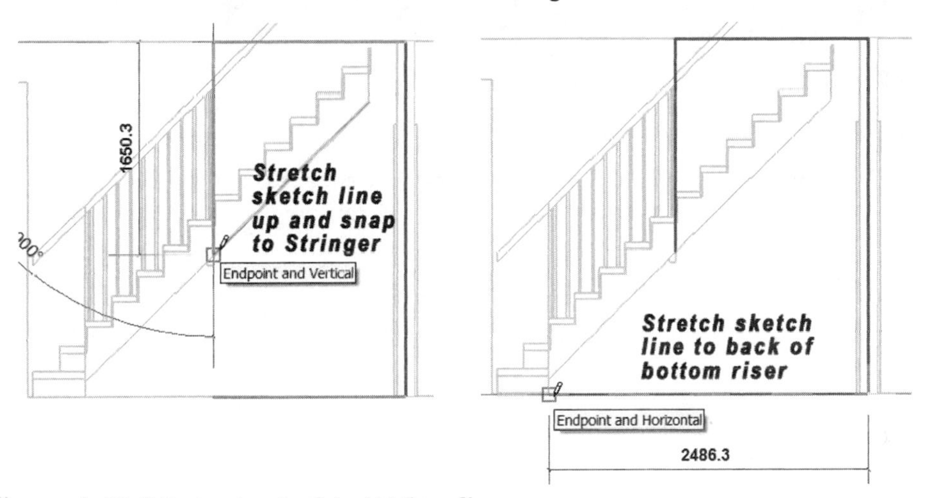

Figure 6–31 *Edit the sketch of the Wall profile*

- On the Design Bar, click the Lines tool.
- On the options Bar, place a checkmark in the "Chain" checkbox.
- Draw two sketch lines to close the Wall profile shape (see Figure 6–32).

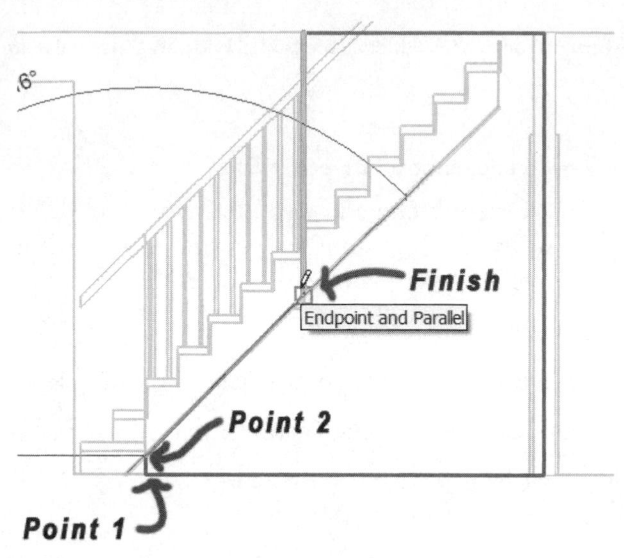

Figure 6–32 *Complete the Wall Profile Sketch*

58. On the Design Bar, click the Finish Sketch button.

   - From the View menu, choose **Orient > Southeast** (see Figure 6–33).

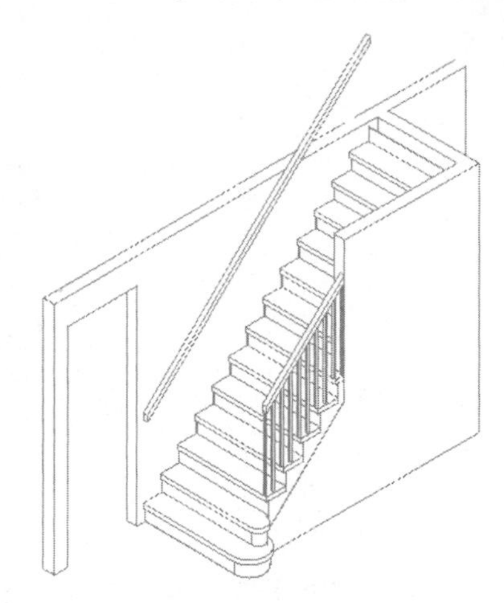

Figure 6–33 *The completed residential Stair hallway*

59. Save and Close the Residential project.

COMMERCIAL CORE PLAN

While we have done much with the Stair tools in the residential project, there is still more we can explore. To continue our exploration of vertical circulation, we will switch to the Commercial Project. In this project, we will add a multi-story

Stair and some elevators in the building core and we will add some ramps to the front entrance of the building.

LOAD THE COMMERCIAL PROJECT

Be sure that the Residential Project is saved and closed.

1. On the Standard toolbar, click the Open icon.

TIP The keyboard shortcut for Open is CTRL + O. **Open** is also located on the File menu.

- In the "Open" dialog box, click the *My Documents* icon on the left side.
- Double-click on the *MRB* folder, and then the *Chapter06* folder.

If you installed the dataset files to a different location than the one listed here, use the "Look in" drop down list to browse to that location instead.

2. Double-click *06 Commercial.rvt* if you wish to work in Imperial units. Double-click *06 Commercial Metric.rvt* if you wish to work in Metric units

You can also select it and then click the Open button.

The project will open in Revit Building with the last opened View visible on screen.

UNDERSTANDING LINKED PROJECTS

The project is in much the same state as we left it at the end of the previous chapter. However, some important changes have been made since we closed it there. For this reason, be certain that you use the new dataset provided for Chapter 6 and do not attempt to continue in your own files from the previous chapter. The geometry has been broken out into three separate Revit Building projects, each in its own *RVT* project file. The project that we have just loaded *06 Commercial.rvt* [*06 Commercial Metric.rvt*] contains the architectural elements of our commercial building; the Walls, Doors, Windows, Column enclosures (Architectural Columns), Floors and Roof. Two additional Revit Building project files have been provided in the *Chapter06* folder. They are *06 Commercial-Site.rvt* [*06 Commercial-Site Metric.rvt*] which contains the terrain model (toposurface), the linked DWG file and the Building Pad and *06 Commercial-Structure.rvt* [*06 Commercial-Structure Metric.rvt*] which contains the Columns, Beams and Braces. All three projects contain the Levels and Grids.

When the architectural model *06 Commercial.rvt* [*06 Commercial Metric.rvt*] opened above, you will notice that the toposurface does appear in the file, but that none of the structural Columns, Beams and Braces do. This is because a link has already been established between the *06 Commercial-Site.rvt* [*06 Commercial-Site Metric.rvt*] file and the current *06 Commercial.rvt* [*06 Commercial Metric.rvt*] project file. Let's take a look at this.

3. From the File menu, choose **Manage Links**.
   - In the "Manage Links" dialog, click the RVT tab.
   - Click on the *06 Commercial-Site.rvt* [*06 Commercial-Site Metric.rvt*] entry at the left (see Figure 6–34).

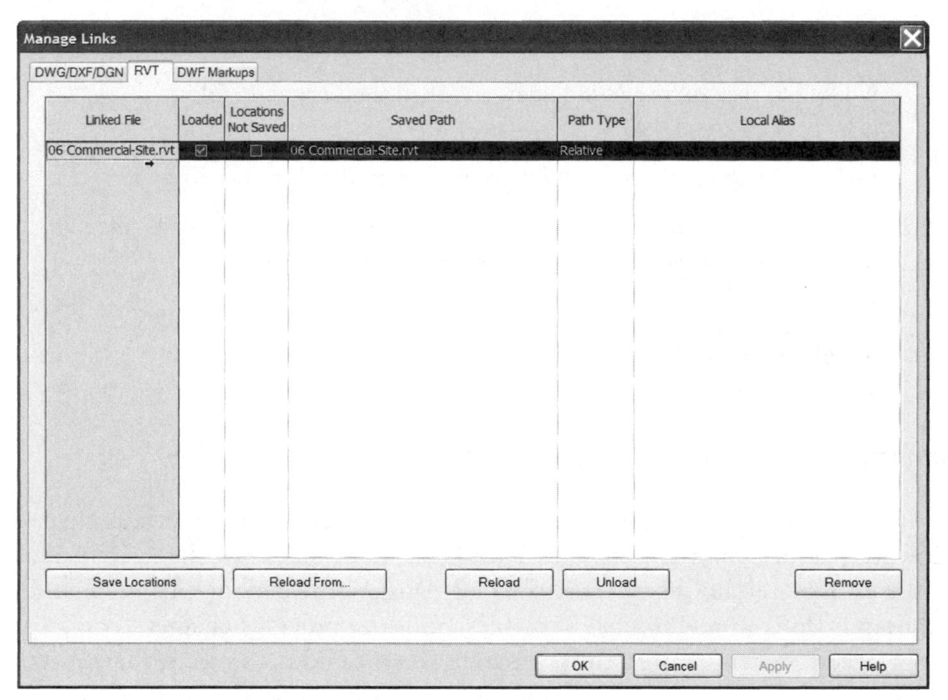

Figure 6–34 *The Manage Links dialog*

The linked file will highlight in the dialog. Several buttons at the bottom of the dialog will become available when the file is selected. With "Save Locations" you can save the linked file to a new location on your hard drive or network. If the file has moved and cannot be found by Revit Building, you can use the "Reload From" button to locate the file. "Reload" and "Unload" can be used to refresh and "turn off" the linked file respectively. If you no longer want the file to be linked, you can click "Remove" to break the link and remove the instance from the current project.

Notice that there is also a DWG/DXF/DGN tab and a DWF Markup tab. In the previous chapter we linked a DWG file for the site plan data. That file has been moved to the *06 Commercial-Site.rvt* [*06 Commercial-Site Metric.rvt*] project which is why it does not show up here. If you were to open the *Site* project, you could use the "Manage Links" dialog in that project to see and edit the DWG link.

NOTE If you try to open a project that is actively linked by the currently open project, Revit Building will display a warning asking you to close the current project first or launch a separate Revit Building session. You cannot have both projects open at the same time in the same Autodesk Revit Building session. This limitation does not prevent two different users from working simultaneously in each of the projects since each team member will be working on a different system. However, if both users are actively changing their respective models, you should use the "Manage Links" dialog periodically to reload the linked project(s).

- Click Cancel to dismiss the dialog without making any changes.

Let's link the structural file into the architectural so that the Columns and framing that we built in the previous chapter will be visible here again.

4. From the File menu, choose **Import/Link > RVT**.

   The "Add Link" dialog should open to the Chapter06 folder automatically. If it does not, browse there now.

   - Select the *06 Commercial-Structure.rvt* [*06 Commercial-Structure Metric.rvt*].
   - In the "Positioning" area (bottom right corner of the dialog) choose "Origin to origin" and then click Open.

A warning dialog will appear informing you that models linked to the file we are linking will not appear in this project. The structural file also has a link to the site file. Since we already have a link to that file, it would not be desirable that it show here anyhow (see Figure 6–35).

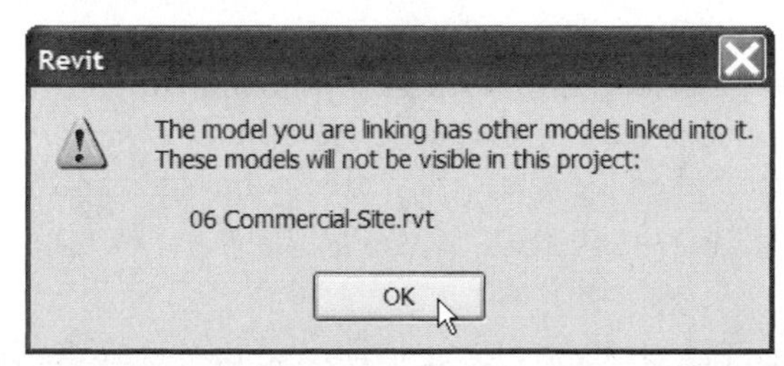

Figure 6–35 *Only the selected model is linked, not any nested models linked within it*

- Click OK to dismiss the warning.

Notice that the core walls and the Columns and framing re-appear in exactly the same locations in which we built them in the previous chapter.

5. Save the project.

ADD A NEW STAIR TO THE COMMERCIAL PLAN CORE

Now that we have the structural model linked into the architectural one, let's take a look at the progress in the building core and add a new egress Stair.

6. On the Project Browser, double-click to open the *Level 1* floor plan View.

   - Right-click in the workspace and choose **Zoom In Region**.

TIP The keyboard shortcut for Zoom In Region is **ZR**. **Zoom In Region** is also located on the View > Zoom menu.

   - Drag a box around the core to zoom in on it (see Figure 6–36).

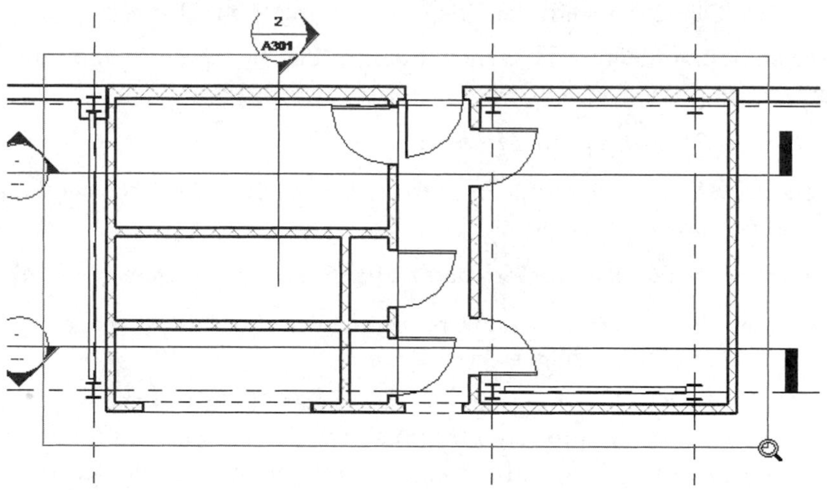

Figure 6–36 *Zoom in on the core and note the new Walls and Doors added since the previous chapter*

Notice that some Walls, Doors and Openings have been added to the core. These items are part of the *06 Commercial-Structure.rvt* [*06 Commercial-Structure Metric.rvt*] file that we just linked because they are in bearing and shear structural walls. We will add a U-shaped egress Stair to the space in the upper left corner of the core.

7. On the Design Bar, click the Modeling tab and then click the Stairs tool.

   The Design Bar changes to sketch mode and shows only the Stairs tools.

Above, we used the Run tool to create a straight Stair. To create a Stair with a landing, you simply click short of the complete run. When you do this, a landing will be created in your sketch automatically. We will use this approach to create a U-shaped Stair.

- On the Design Bar, click the Stair Properties button.
- Verify that the Type is 7″ max riser 11″ tread [190mm max riser 250mm going].
- In the "Instance Parameters" area, beneath the "Dimensions" grouping, change the Width to **3′-8″** [**1100**].
- Set the "Desired Number of Risers" to **21** and then click OK.

8. Locate the first point on the inside edge of the right vertical Wall of the Stair core.
   - Drag the mouse horizontally to the left. When the message reads "11 Risers Created, 10 Remaining" click the mouse to set the next point (see Figure 6–37).

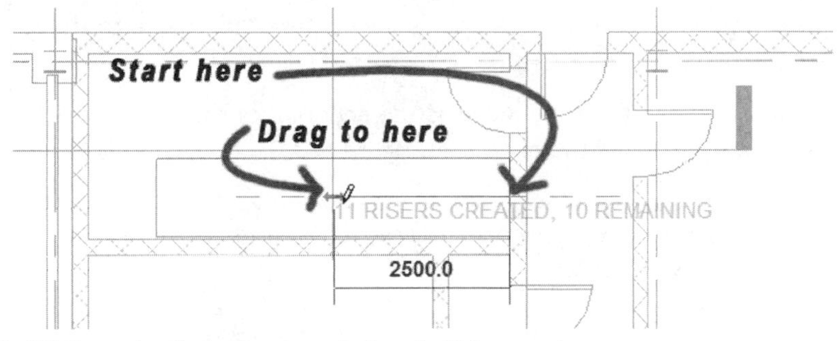

Figure 6–37 *Start the Stair sketch and place half the treads*

- Move the pointer above the point just clicked, and then click again (see Figure 6–38).

The point is approximately 2′-0″ [600] above the previous one, but the exact location is not important right now and will be adjusted below.

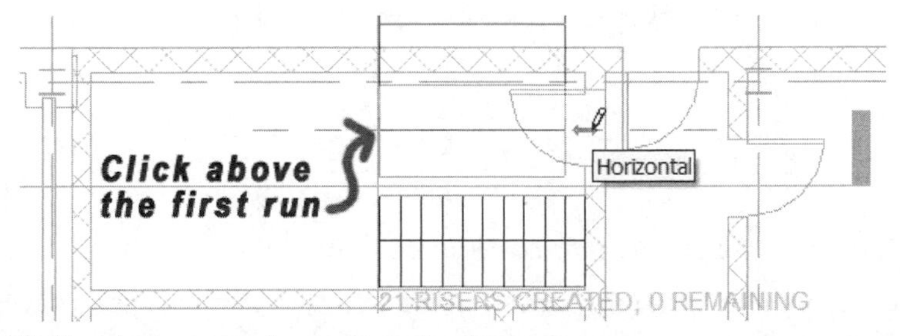

Figure 6–38 *Sketch the second run above the first*

- Drag horizontally to the right and click to place the remaining risers.

A sketch of the U-shaped Stair will appear. The size is not exactly correct and it is not in the right spot. We can adjust both of these issues easily.

9. Select the green vertical sketch line on the left.

   This is part of the Stair Boundary.

   - Click the Move tool on the Edit toolbar.
   - Move the Boundary line to the left **1′-0″ [300]**.

Notice that the other Boundary lines remain attached to the moved line.

   - Click off of the line to deselect it, or press ESC.

10. Using a marquee selection, select the entire Stair sketch.

NOTE Don't worry about selecting the neighboring elements. When you are in sketch mode, you can only select elements of the sketch.

   - Using the Move tool again, snap the start point of the move to the lower left endpoint.
   - Snap the end point of the move to the endpoint of the Wall at the lower left inside corner of the room (see Figure 6–39).

TIP If you have trouble getting Revit Building to automatically present the endpoint, press the TAB key to cycle to it.

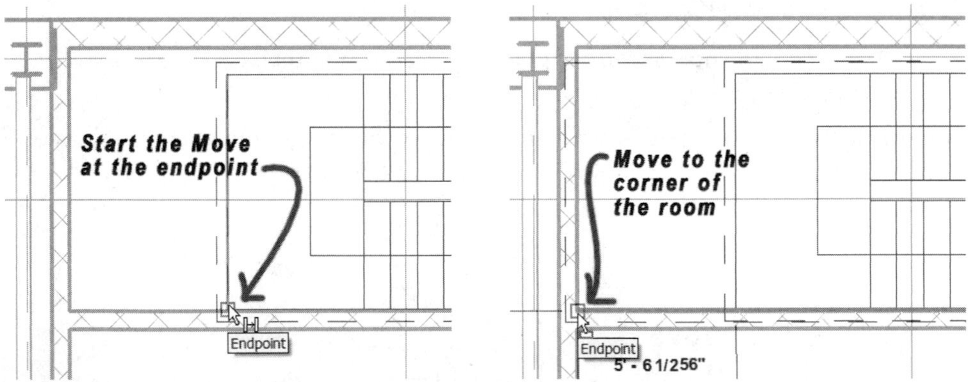

Figure 6–39 *Move the entire sketch to the corner of the room*

This positions the Stair in the right location. However, the Stair Type that we are using here uses Closed stringers instead of open ones like the Stair above. Therefore, we need to take the thickness of our stringers into account when establishing our final position.

11. Using Move again, move the entire sketch up vertically **2″ [50]** and right horizontally **2″ [50]** (see Figure 6–40).

TIP Simply start the Move, click any start point, move the mouse horizontally or vertically in the desired direction and then type the move value (**2″ [50]** in this case) on the keyboard followed by an ENTER.

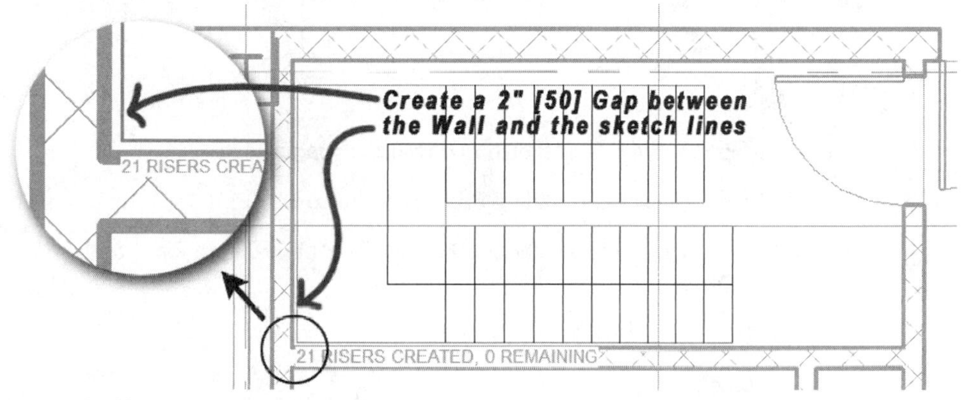

Figure 6–40 *Move the entire sketch to create a gap for the stringer*

12. Using a marquee selection, surround just the top run of the Stair sketch (see Figure 6–41).

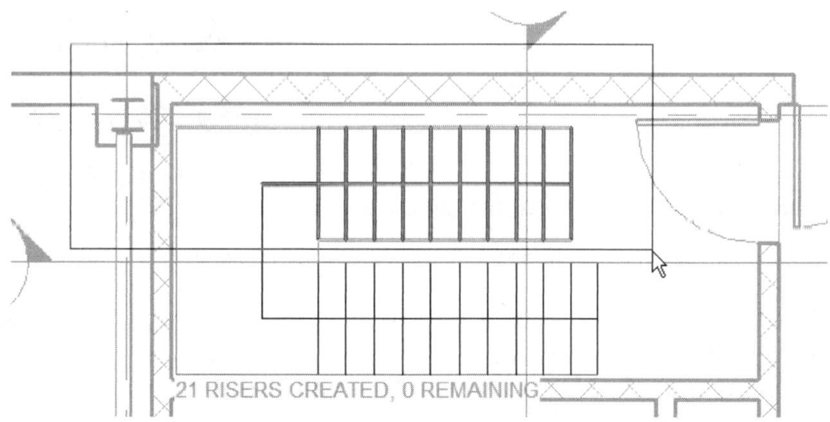

Figure 6–41 *Select the top run of the Stair sketch*

- Using the same techniques, move it from the top left endpoint, to the inside horizontal edge of the room.
- With the same sketch lines still selected, move them down **2″ [50]** (see Figure 6–42).

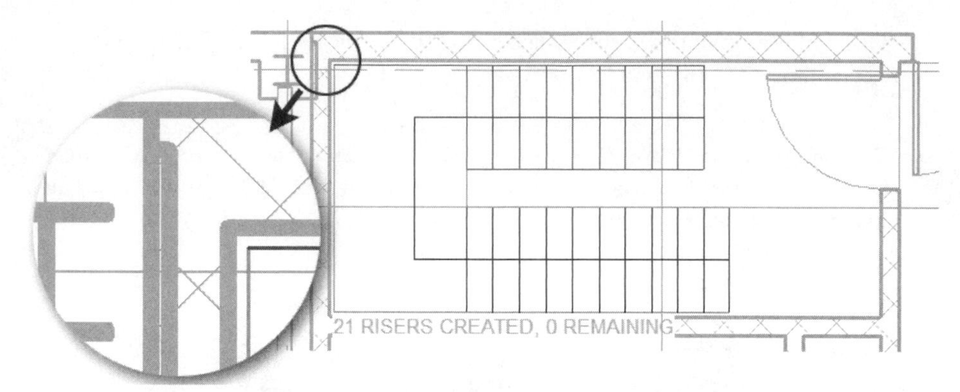

Figure 6–42 *Move the top run of the Stair sketch and create a gap for the stringer*

13. On the Design Bar, click the Railing Properties button.
    - In the "Railing Type" dialog, choose Handrail – Pipe [900mm Pipe] and then click OK.
    - On the Design Bar, click Finish Sketch.
14. Save the project.

CREATE A MULTI-STORY STAIR

Let's take a look at our Stair in a section View.

15. A section line runs horizontally through the Stair. Double click the blue section head to go to that View.

As you can see, the Stair that we built only occupies the first floor. We can edit the properties of the Stair to make it apply to multiple stories.

16. Select the Stair object in the current Section View.

    Be sure to select the Stair and not the Railing for this step.

    - On the Options Bar, click the Properties icon.
    - In the "Instance Parameters" area, beneath the "Constraints" grouping, choose **Roof** from the "Multistory Top Level" list.
    - Click OK to apply the change (see Figure 6–43).

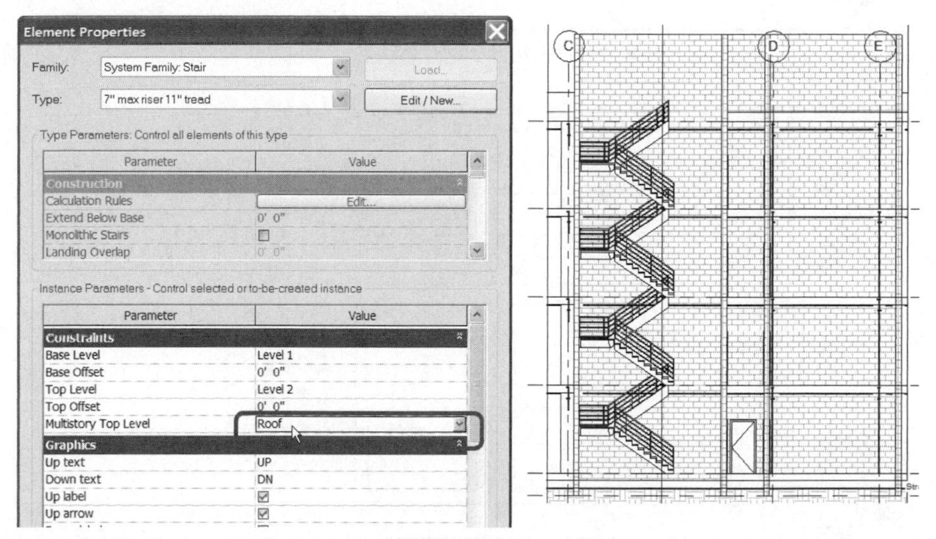

Figure 6–43 *Change the Stair to a multistory Stair*

17. On the Project Browser, double-click to open the *Level 1* floor plan View.

Notice that only the UP Stair annotation displays on the first floor and that the Stair above the cut plane is shown dashed.

18. On the Project Browser, double-click to open the *Level 3* floor plan View.

Notice that in this View (and in *Level 2* and *Level 4* as well) that both the UP and DOWN Stair annotations display.

19. On the Project Browser, double-click to open the *Roof* plan View (see Figure 6–44).

Notice that here, only the DOWN annotation shows.

TIP The UP and DOWN annotations can be moved. First, select the Stair. A blue control dot appears next to the text. Click and drag the text to a better location.

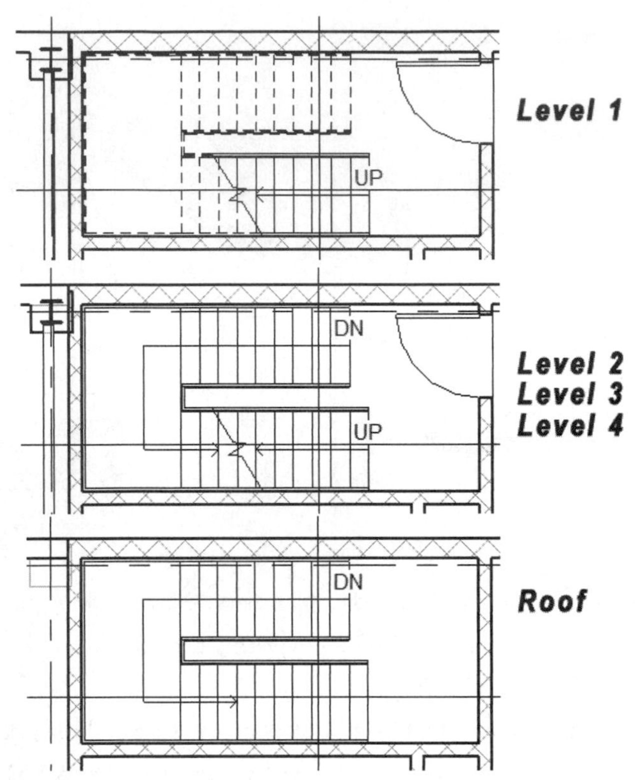

Figure 6–44 *Stairs automatically display directional annotations appropriate to the level*

20. Save the project.

FLOORS, LANDINGS AND SHAFTS

In the previous chapter, we created some simple Floors and a Roof for the project. We used a Shaft element to cut these horizontal surfaces through all floors at the stair and elevator shafts. This is a perfectly valid approach to this situation. There are however, other approaches that we could take in this type of situation as well. Another glance at the section in our current model will reveal that there are no landings on the entrance side (right) of the Stairs. We have a single continuous Floor slab across the entire floor plate with a single Shaft cutting out both the Stair and Elevator spaces.

At this point, to accommodate the Stair elements that we have added, and add landings on the entrance side to meet the Stairs, we need to adjust our existing Shaft. We could edit the shape of the Shaft, to simply reveal more of the slabs in the shape of these landings, or we could remove the Shaft, edit the Slab sketch and/or build the landings as separate Floor elements. The exact approach that we take really depends on the type of model we wish to create and what level of detail we wish to include in the element geometry. On the one hand, we might take a subtractive approach: create a Floor, and then cut away the void with the Shaft. On the other hand, we could take an additive approach and construct each piece of Floor construction in its final constructed shape.

In order to see both approaches, we'll use the additive approach on the Stair tower and the subtractive approach on the Elevator core. To do this, we need to first adjust the size of the Shaft element that we added in the last chapter to the size of only the elevator core. Then we need to edit the existing Floor elements to accommodate the Stair tower and finally add new Floors for the landings.

EDIT THE SHAFT

Editing the Shaft is easy. We simply select it and then edit the sketch.

1. On the Project Browser, double-click to open the *Level 2* floor plan View.
2. Move the pointer near the outside edge of the Stair or Elevator spaces.
   - When the screen tip reads "Shaft Opening: Opening Cut" and the Shaft pre-highlights, click the mouse to select it.

   The Shaft will turn opaque and block the other elements while it is selected.

   - On the Options Bar, click the Edit button.
   - Click the top horizontal sketch line and move it down to the top elevator Wall.
   - Move the bottom line up and the right side over to the left to match the shape of the elevator core (see Figure 6–45).

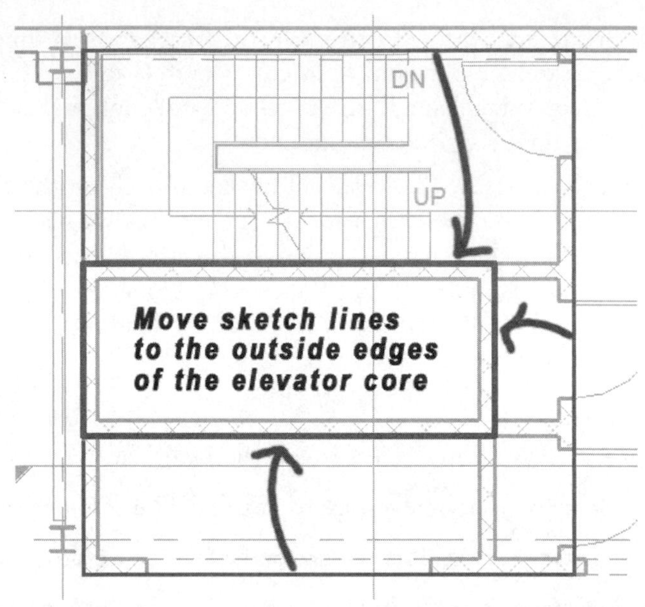

Figure 6–45 *Move sketch lines of the Shaft to fit the elevator core (sketch lines enhanced for clarity)*

TIP You can make your sketch lines appear bolder on screen as you work by editing their lineweight in the Line Styles dialog (Settings menu).

3. On the Design Bar, click the Finish Sketch button.

The effect of this change will not be very apparent in plan Views. If you wish, you can go to the default *{3D}* View, hold your SHIFT key down and drag with the wheel on your mouse to spin the model around and see the effect. If you don't have a wheel, or if you prefer, you can also use the Dynamically Modify View (F8) icon instead. Spinning the model in this way will reveal that the Shaft now only cuts the Floor slabs at the elevator shaft.

EDIT THE FLOORS

If you spun the model in 3D above, you noticed that the previous edit has now restored our Floor slabs in the Stair tower. This will make it difficult for us to climb these stairs. As was noted above, a Shaft provides a quick and convenient way to cut several Floors at once, however the alternative to a Shaft is to edit the sketch of the Floors themselves to model it in its true shape (additive) rather than carve a hole from it later (subtractive). It should be noted again that we are showing both approaches here for the education value. In your own projects, either approach can be employed based upon your team's preferences.

If you switched to another View such as *{3D}* return to *Level 2* now.

4. Continuing in the *Level 2* View, select the Floor.

TIP to do this in a plan View, place your pointer at the edge of the outer Wall where the Floor edge ought to be and then press TAB until the screen tip reveals the Floor element.

- On the Options Bar, click the Edit button.
- On the Design Bar, click the Lines tool.
- On the Options Bar, click the Draw icon.
- Place a checkmark in the "Chain" checkbox.

5. Create three sketch lines by tracing the inside edges of the stair tower space.
6. Use the Split tool (with the "Delete Inner Segment" option chosen) to remove the segment at the top of the Stair (see Figure 6–46).

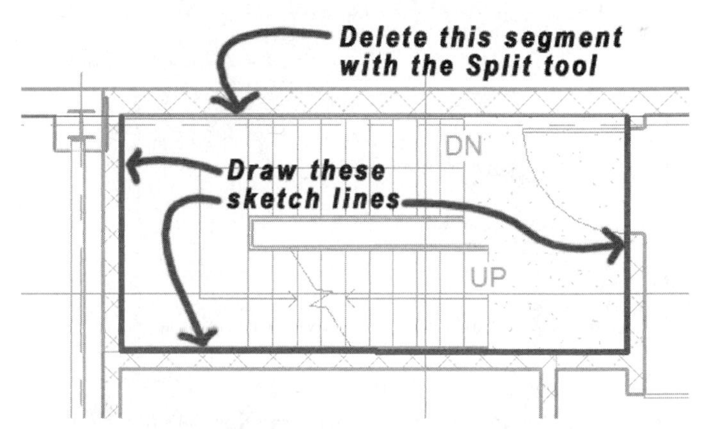

Figure 6–46 *Add sketch lines around the Stair tower space (sketch lines enhanced for clarity)*

- On the Design Bar, click the Finish Sketch button.
- In the Alert dialog that appears, click Yes to accept the joining of geometry.

7. Repeat the edit Floor sketch on the *Level 3*, *Level 4* and *Roof* Views.
8. Check the model in the *{3D}* and/or one of the Section Views.

TIP If you open the *{3D}* View, you can display it in hidden line, wireframe or shaded. The Shaded with Edges option often offers the best contrast and readability.

9. Save the project.

ADD A LANDING (FLOOR)

Our commercial Stair Tower is nearly complete. We have completed the work we needed on the Floors and Roof. We are now ready to add some new Floor elements for the Stair Landings. The first floor is fine as is, so let's start with the second floor.

10. On the Project Browser, double-click to open the *Level 2* floor plan View.
11. On the Design Bar, click the Modeling tab and then click the Floor tool.

    The Design Bar changes to sketch mode and shows only the Floor tools.

    - On the Design Bar, click the Lines tool.
    - On the Options Bar, click the Pick Lines icon.
    - Click the inside edge of the right vertical Wall and each of the inside edges of the two horizontal Walls (see Figure 6–47).

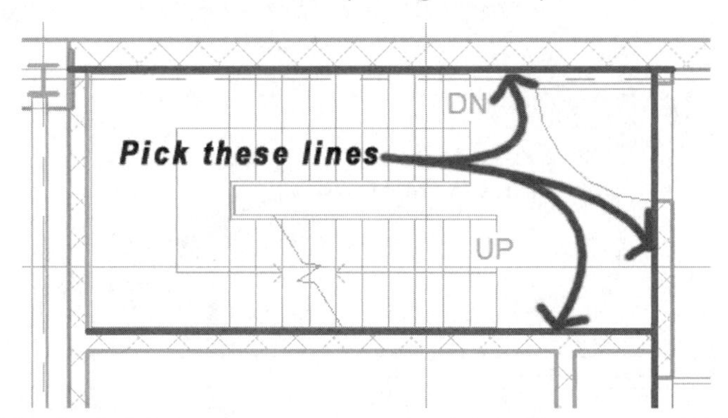

Figure 6–47 *Use the Pick Lines option to create sketch lines from existing Wall edges (sketch lines enhanced for clarity)*

- Continuing with Pick Lines, click each of the tread lines of the Stair on the right side (see the left side of Figure 6–48).
- Use the Trim/Extend tool with the "Trim/Extend to Corner" option to finish the sketch (see the right side of Figure 6–48).

TIP Remember to pick the side of the lines that you wish to keep.

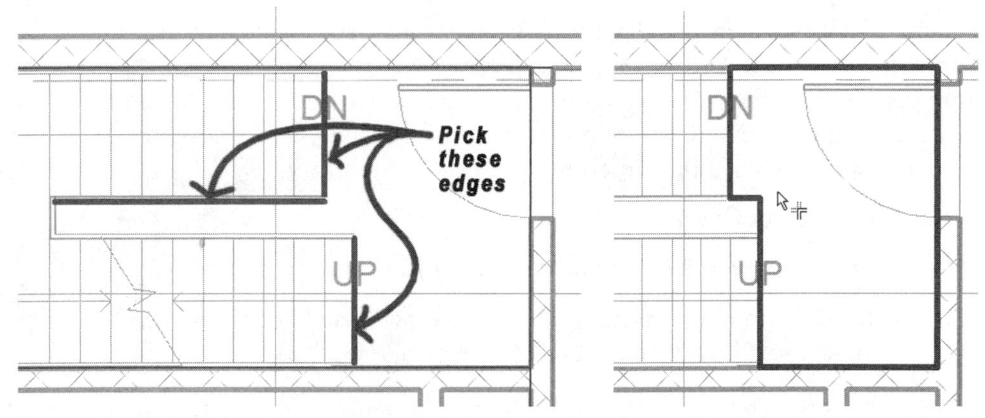

Figure 6–48 *Pick the remaining lines and then Trim the corners (sketch lines enhanced for clarity)*

12. On the Design Bar, click the Floor Properties tool.
    - From the Type list, choose LW Concrete on Metal Deck [Floor : Insitu Concrete 225mm] and then click OK.
    - On the Design Bar, click the Finish Sketch button.

COPY THE LANDING TO LEVELS

13. Double-click the blue section head running horizontally though the Stair to go to that View.

You should be able to see the new Floor landing at the second floor. We could use landings on the other floors as well, so let's copy and paste this one to the other levels.

14. Select the Floor landing, and from the edit menu choose **Copy to Clipboard** (or press CTRL + C).
    - From the Edit menu choose **Paste Aligned > Select Levels by Name**.
    - In the "Select Levels" dialog, hold down the CTRL key and select *Level 3*, *Level 4* and *Roof*.
    - Click OK to complete the operation.

Landings will be added to the other levels.

15. Zoom in on the top of the Stair tower at the Roof Level.

Notice that the Stair is a bit too short and that the landing is also too low. The vertical position of the Roof element is different than that of the Floor slabs. Notice that the Roof sits on top of the Beams, while the Floors sit in plane with the Beams. The construction of both the Floors and the Roof includes steel framing joists, metal deck and concrete topping. The Roof will have a membrane roofing layer on top as well. At this stage of the project, we have only a simple generic structure in place to represent this structure. In later chapters, we can edit the Floor and Roof Types to add the appropriate layers. For now, just getting the Roof positioned properly relative the structure is all that is necessary.

16. Select the Roof element.
    - On the Options Bar, click the Properties icon.
    - Beneath the "Constraints" grouping, edit the "Base Offset From Level" to **-1′-0″** **[-400]** and then click OK (see Figure 6–49).

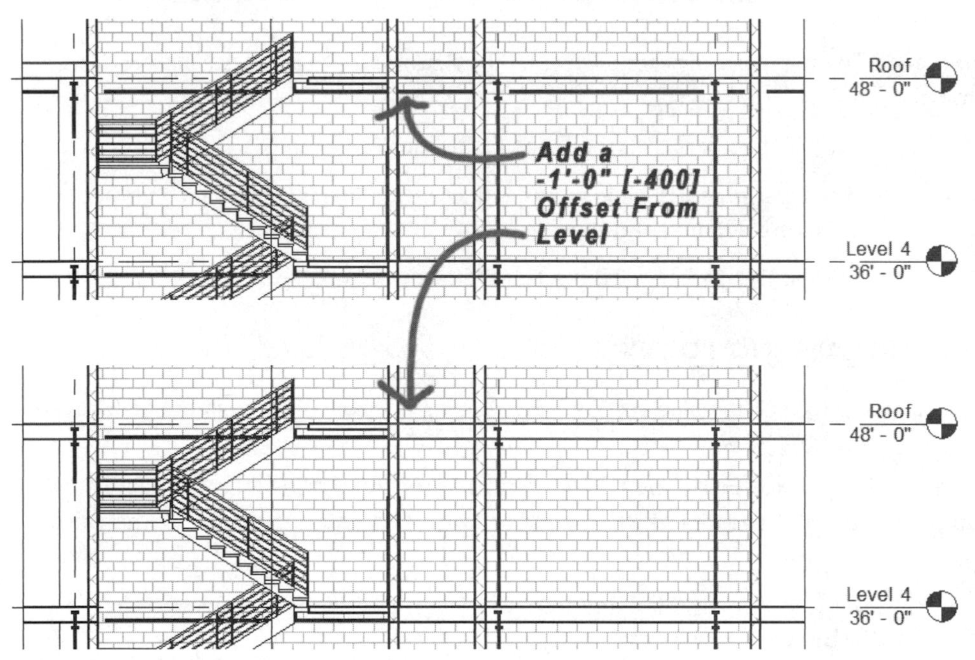

Figure 6–49 *Adjust the Roof offset from level to match the Floors*

EDIT A MATERIAL (IMPERIAL ONLY)

If you are working in the Imperial dataset, the Type chosen for the landing Floor includes a stipple pattern in the plan Views. This pattern is really not necessary for our plan Views. Removing it is easy. If you are working in the metric dataset, the following steps are not required, but feel free to follow along to understand the process if you wish.

17. On the Project Browser, double-click to open the *Level 2* floor plan View.
18. Select the Landing.

- On the Options Bar, click the Properties icon.
- Next to the Type list, click the Edit/New button (or press ALT + E).
- In the "Type Parameters" area, beneath the "Construction" grouping, click the Edit button next to Structure.

The "Edit Assembly" dialog will appear revealing the structure of the selected Floor type. We are editing at the type level right now, so any change we make will automatically apply to all elements that use this Type in the entire project. This means that *all* of our landings will receive the change when we are finished editing.

- Next to Layer 2, click in the Material cell.

A small expand (down pointing arrow) icon will appear in the right side of the field. This icon is used to open a browse window to choose a material for the layer (see Figure 6–50).

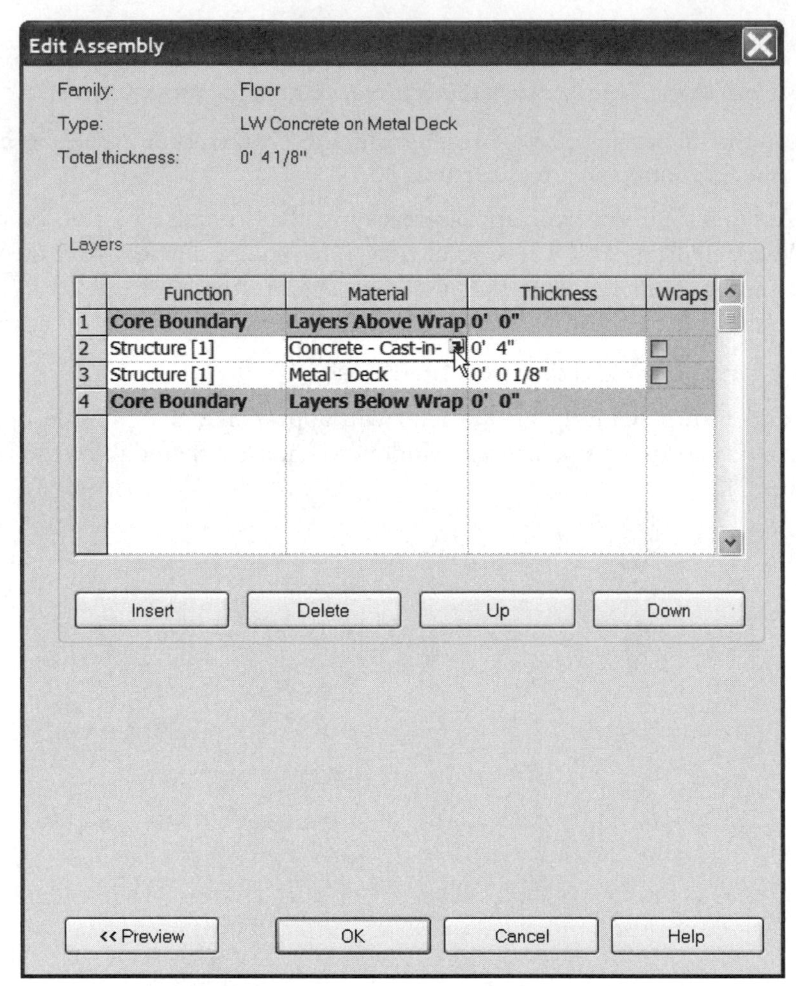

Figure 6–50 *Each Layer of the Floor's Structure is assigned a Material*

- Click the expand icon to open the "Materials" dialog.

Notice that the currently selected Material is highlighted in a long list of available Materials.

19. At the bottom of the dialog, click the Duplicate button.

NOTE We could edit the existing Material directly, but it is good practice to get in the habit of renaming (or duplicating) any Families, Types or Styles that you edit. This way, you can always quickly locate the elements that you have modified. In an office environment, establishing a good logical naming scheme for such elements is also highly recommended.

- Name the new Material **MRB Landing Concrete Topping** and then click OK.

TIP If you want your custom elements to sort automatically to the top of the list, add an underscore [_] to the front of the name. **_MRB Landing Concrete Topping** in this case.

- On the right side of the "Materials" dialog, in the "Surface Pattern" area, click the expand icon for Pattern.
- In the "Fill Patterns" dialog, click the No Pattern button (see Figure 6–51).

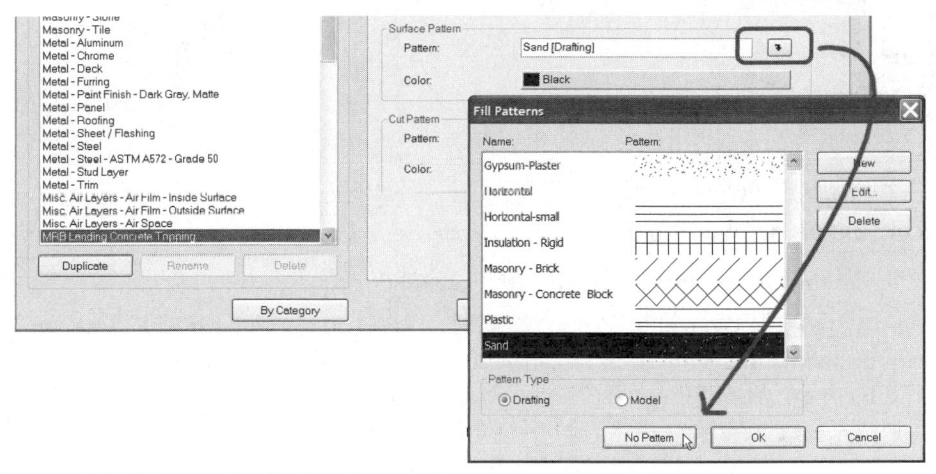

Figure 6–51 *Remove the Surface Pattern from the new Material*

20. Click OK four times to return to the model and view the results.
21. Save the project.

EDIT RAILINGS

One last finishing touch remains for our Stair tower. The Railings should wrap around at the landings. This can be accomplished by editing the Railing sketch. When you create a Stair, the Railings are added automatically. By default they match the same shape as the Stair. However, like the Stairs, they are controlled by a sketch that we can edit.

22. On the Project Browser, double-click to open the *Level 2* floor plan View.
23. Select the inside Railing.

    Make sure you select the Railing and not the Stair, use the TAB key if necessary.

    - On the Options Bar, click the Edit button.

The Design Bar changes to sketch mode and shows only the Railing tools.

- On the Design Bar, click the Lines tool.
- On the Options Bar, place a checkmark in the "Chain" checkbox.
- Draw the two segments indicated in Figure 6–52.

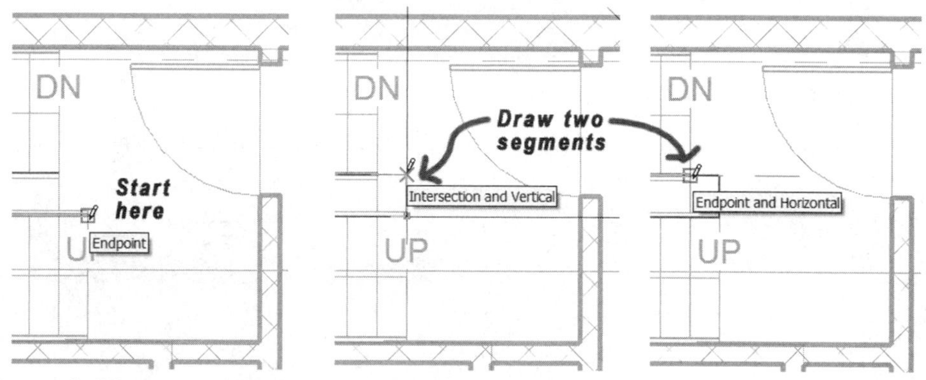

Figure 6–52 *Draw two segments in the Railing sketch*

24. On the Design Bar, click the Finish Sketch button.

The best way to view the result is in a *3D* View. If we use the default 3D View, we will have to isolate the Stair (like we did above in the residential project) in order to see the Railing. Another option is to create a new *3D* View based upon another View. In this case, creating a *3D* View from the *Section at Stair Tower* View would be the best choice. Since we already have a default *3D* View, our best option is to copy this View, rename it and then adjust its orientation to match the section.

25. On the Project Browser, right-click the {*3D*} View and choose **Duplicate**.
    - Right-click the *Copy of* {*3D*} View and choose **Rename**.
    - In the "Rename View" dialog change the name to **Stair 3D Section** and then click OK.
26. Click in the workspace window to shift focus from the Project Browser to the model window.
    - From the View menu, choose **Orient > To Other View**.
    - In the "Orient To Other View" dialog, select the *Section at Stair Tower* View and then click OK (see Figure 6–53).

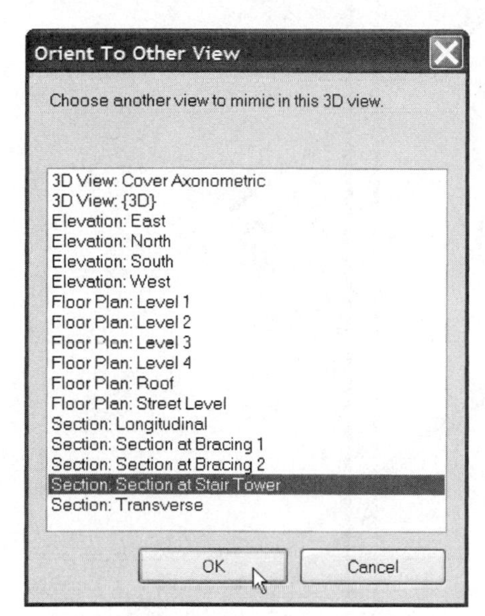

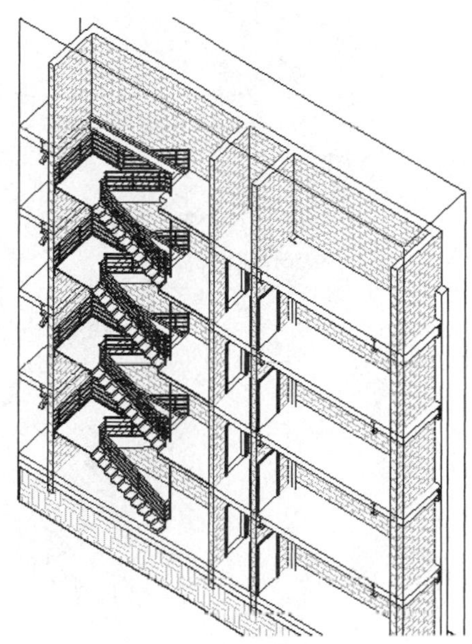

Figure 6–53 *Orient the View to the Section at Stair Tower then spin it in 3D*

27. Hold your SHIFT key down and drag with the wheel on your mouse to spin the model around and see the effect.

NOTE If you don't have a wheel, or if you prefer, you can also use the Dynamically Modify View (F8) icon instead.

28. Zoom in on the second floor at the point where we edited the Railing.
    - Pan to other floors as well.

Notice that the Railing now wraps around the gap between the two runs and this change has occurred on all Levels (see Figure 6–54).

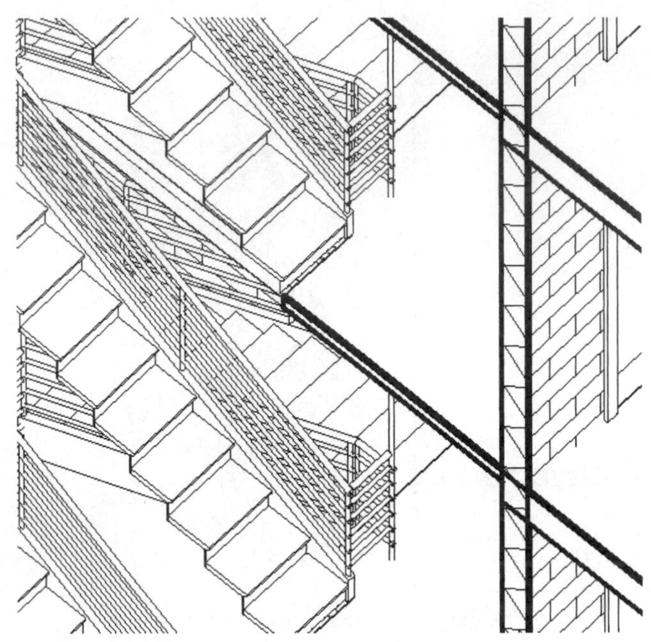

Figure 6–54 *The Railing edit we made is visible on all Levels*

For the top and the bottom of the Stair, you can add additional Railings if you like to complete the layout. The easiest way to do this is to select the Railing, right-click and choose **Create Similar.** Sketch the new Railing and use the Set Host tool on the Design Bar to attach it to the Landing.

TIP If you witness a strange artifact for the balusters in the 3D View, you can edit your sketch and pull the vertical sketch line back slightly from the horizontal segment. Leave the vertical segment about a inch [few millimeters] short.

ADJUST OUTSIDE RAILING

Above in the "Create a New Railing Type" of the residential project, we created a new Railing Type that did not have any balusters. A Railing Type similar to this has been provided in the dataset that we can use for the Railing along the Stair core Walls.

29. Select the outside Railing (the one adjacent to the Wall).
    - From the Type Selector on the Options Bar, choose **MRB Handrail Only** for the Type.
30. Save the project.

RAMPS AND ELEVATORS

In this section, we will add some ramps and elevators. Ramps and elevators are handled quite differently from one another. Creating a ramp in Revit Building is much like adding a Stair. Elevators are simply Families that insert in our models like other Components.

ADD A RAMP

A Ramp tool is provided on the Modeling tab of the Design Bar. A ramp is added in much the same way as a Stair. To save a bit of time and effort, the dataset includes some Reference Planes at the front of the building. These have been placed to make it easy for us to the sketch the ramp. An entrance patio is also provided as an In-Place Family.

1. On the Project Browser, double-click to open the *Street Level* plan View.
   - Zoom in on the lower right corner at the front of the building.

You will notice four reference planes in this location. We will use these to help us sketch the ramp.

2. On the Design Bar, click the Modeling tab and then click the Ramp tool.

Notice that the Design Bar changes to a series of tools that are identical to the Stair Design Bar we saw above.

- On the Design Bar, click the Ramp Properties tool.
- In the "Instance Parameters" area, beneath the "Dimensions" grouping set the Width to **3′-0″** [**900**].
- Verify that the "Base Level" is set to **Street Level** and the "Top Level" is set to **Level 1** and then click OK.
- Starting at the lower left endpoint of the inclined Reference Plane, click on each Reference Plane endpoint moving first left to right, then up, then right to left (see Figure 6–55).

NOTE Like when we created the U-shaped Stair above, you only click the start and end of the runs. The landings fill in automatically.

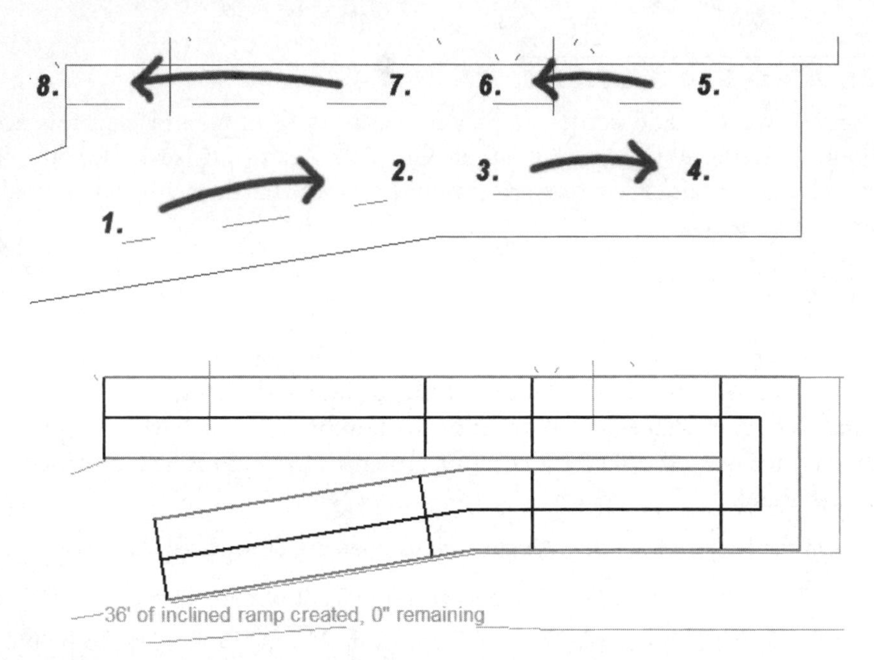

Figure 6–55 *Sketch the ramp using the provided Reference Planes*

If you wish, you can edit the shape of the green outline on the right side at the landing to make the ramp match the building Pad in the linked Site model. Try moving the vertical green boundary line to the right about 1′-0″ [300] to widen the landing, or even add a curve to it.

NOTE While we have provided the Reference Planes in this exercise to make the process of sketching the ramp go smoothly, you can use any of the sketching techniques that we covered above on Stairs when adding ramps in your own projects. Remember that you can sketch the Run, or edit the Risers or Boundaries directly with the appropriate tools on the Design Bar.

3. On the Design Bar, click the Finish Sketch button.
   - View the ramp in various Views to see the results (see Figure 6–56).

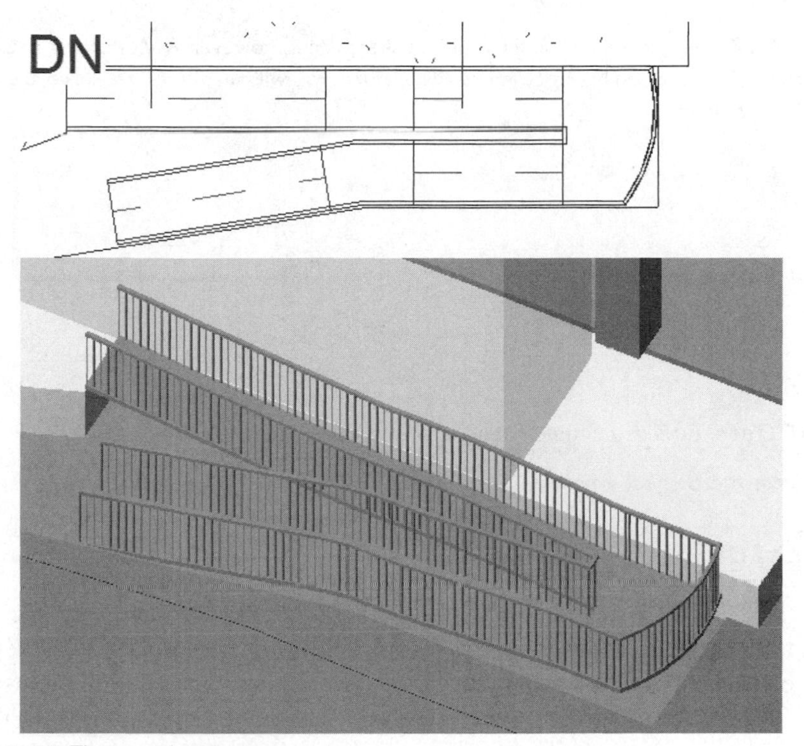

Figure 6–56 *The completed ramp*

4. Return to the *Street Level* View and then Mirror the ramp to the other side.
   - Use the Draw option to pick a midpoint about which to mirror.
5. Save the project.

ADD RAILINGS TO THE PATIO

Let's add guardrails to the Patio.

6. On the Project Browser, double-click to open the *Level 1* floor plan View.
7. On the Design Bar, click the Modeling tab and then click the Railing tool.
   - On the Design Bar, be sure that the Lines tool is chosen.
   - On the Options Bar, click the Pick Lines icon.
   - Set the "Offset" value to **4″** **[100]** and then press ENTER.
8. Using Figure 6–57 as a guide, place Railing sketch lines around the patio edges.

NOTE The figure shows all the required sketch lines, however, Revit Building does not allow disconnected Railing segments. Therefore, you will need to create three separate Railing elements.

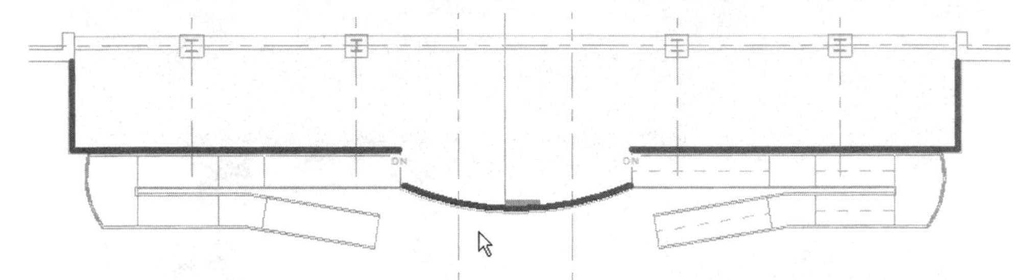

Figure 6–57 *Create Railings from the Patio surface (sketch lines enhanced for clarity)*

9. Save the project when finished.

ADD ELEVATORS

An elevator in Revit Building is placed using the Component tool in the same way that we added plumbing fixtures in Chapter 3. Elevator Families can be created from scratch or derived from manufacturers drawing files. Some sample elevator Families are provided on the Autodesk Web Site. We can access this Web Library directly from the Autodesk Revit Building interface. If you do not have Internet access, the Family files have been provided with the files installed from the Mastering Revit Building CDROM in a folder named *Autodesk Web Library*. Feel free to access the following content from that location instead of directly from the Web library as indicated in the following steps.

10. On the Project Browser, double-click to open the *Level 1* floor plan View.
11. On the Design Bar, click the Basics tab and then click the Component tool.
    - On the Options Bar, click the Load button.
    - In the "Open" dialog, click the Web Library button at the top right corner (see Figure 6–58).

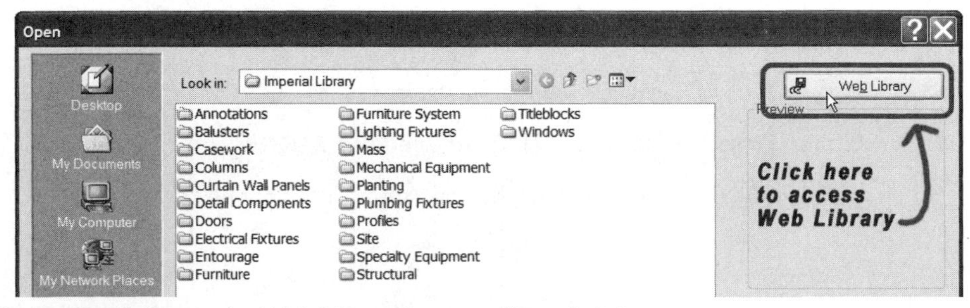

Figure 6–58 *Access the Web Library from any "Open" dialog*

The Web Library will open in your default web browser. If it does not appear as the active window, look down on your Windows Task Bar for it. Click the icon on the Task Bar to bring it to the front. This is a typical web page with navigation links on the left side and a main preview pane on the right. Autodesk provides a library for both Revit Building and Revit Structure accessed by the buttons at the bottom of the page. On the left are links to several libraries, Revit Building 8.1, and various manufacturer libraries. These libraries are provided by building component manufacturers or organizations and give access to Revit Families that you can download and load into your projects. Libraries from previous releases of Revit Building can be found under "Archived Libraries" (see Figure 6–59). For this exercise, we will go directly to the component we need for our project, but please feel free to spend some time exploring in this web site and download and add any of the components you wish to the current project. A new version of the dataset will be provided in the next chapter, so you do not have to worry about straying from the exercise. Play time is highly encouraged and is often a great way to learn Revit Building!

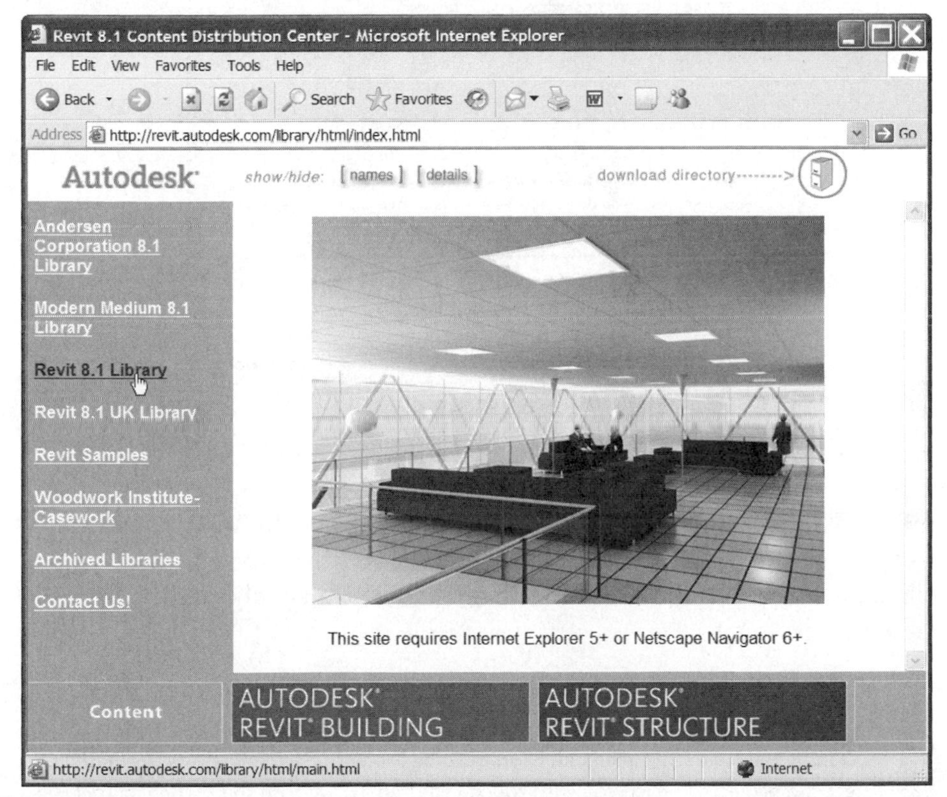

Figure 6–59 *The Autodesk Revit Building Web Library Home page in Internet Explorer*

12. On the left, click the *Revit 8.1 Library* link.

This will expand an index beneath it.

- Click on *Specialty Equipment*, and then *Conveying Systems*.
- On the right, click the *Elevator-Hydraulic* link to access this file.

A "File Download" dialog will appear.

- In the "File Download" dialog click Open (see Figure 6–60).

NOTE If you do not have Internet access, open the file from the *Autodesk Web Library* folder of the dataset installed from the Mastering Revit Building CDROM instead.

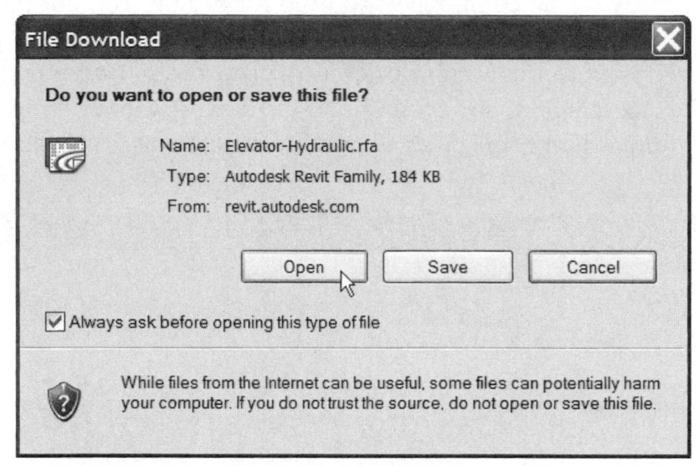

Figure 6–60 *Open the Elevator-Hydraulic Family file*

NOTE If a message appears about another user having the file open, simply click OK.

When you open the file, it will open directly in the Revit Building workspace in "Family Edit" mode. Family Edit mode is a special mode for editing Revit Building Families. The tools on the Design Bar will be unique to this mode, but otherwise most of the interface will remain the same. Here you could edit the Family file before adding it to your project. We will not edit it, but simply save the file and add it to our project.

13. On the Design Bar, click the Load into Project button.

NOTE If you have more than one Revit Building project open, you will be prompted to select the project in which you wish to load the Family. Choose the *06 Commercial* [*06 Commercial Metric*] project and then click OK.

- Back in the commercial project, on the Design Bar, click the Component tool.

The Elevator should now be loaded and already selected on the Type Selector.

- Place two elevators in the elevator core space.

Notice that they will snap automatically to the nearby Walls. Snap them to the lower Wall of the core space (see Figure 6–61).

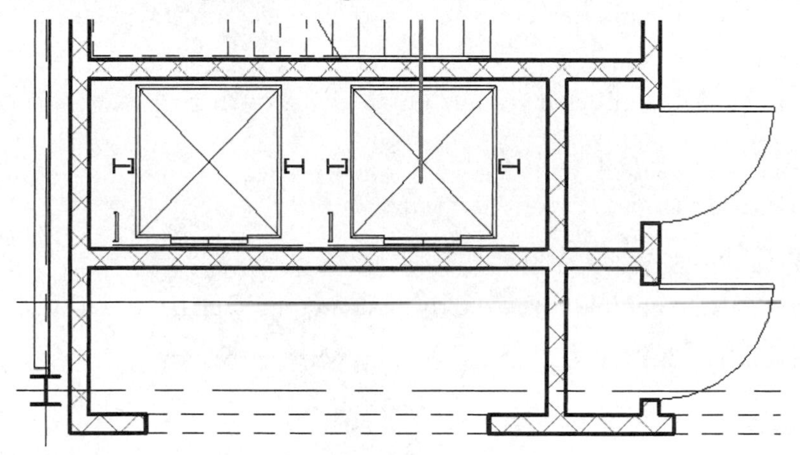

Figure 6–61 *Place two elevators in the core*

- On the Design Bar, click the Modify tool or press the ESC key twice.
- Open each of the other floor plan Views to see the elevators.

14. Save and close the project.

SUMMARY

Stairs and Railings offer flexible configuration with Type-based parameters and Element-based variations.

Stairs are "sketch-based" objects—create them with a 2D footprint sketch that includes risers and boundary edges.

Like other Revit Building elements, Stairs and Railings can be demolished or added to any phase.

Edit a Stair Type to apply changes to all Stair elements sharing that Type.

Edit the Stair Sketch to make customizations such as bullnosed treads.

Edit the Railing Sketch to modify the extent of the Railing.

Adjust Wall Profiles to cut away the space where they intersect Stairs.

Stairs can be made multiple stories tall by assigning the height constraints.

You can use Floor elements to model landings where necessary.

Ramps are similar to Stairs and are also sketch-based.

Pre-made elevator component families can be loaded into the project and placed in the model.

7

Horizontal Surfaces

INTRODUCTION

In this chapter we will focus on Roofs and Floors in both our residential and commercial projects. Both Roofs and Floors in Revit Building are sketch based objects. We will begin in the residential project with gable Roofs. We will understand how to make Walls attach to Roofs, edit Roof structure, apply edge conditions and gutters. The commercial project will give us the opportunity to explore flat Roofs and create Floors.

OBJECTIVES

In Revit Building you can construct Roofs in a variety of ways. Roofs can interact with the Walls of the building and we can apply custom treatment to their edges. After completing this chapter you will know how to:

- Build Roofs
- Create a Custom Roof Type
- Add and modify Floors
- Work with Roof Edges and Gutters
- Understand and utilize Slope Arrows
- Attach Walls and Join Roofs

CREATING ROOFS

Since we are working concurrently on two different projects in this book, we will get an opportunity to look at both traditional sloped residential roofs and "flat" commercial roofs. The only real difference between the two in Revit Building is the slope parameters that we assign and the way that we treat the edges. Let's start with the residential project.

INSTALL THE CD FILES AND OPEN A PROJECT

The lessons that follow require the dataset included on the Mastering Revit Building CD ROM. If you have already installed all of the files from the CD, simply skip down to step 3 below to open the project. If you need to install the CD files, start at step 1.

1. If you have not already done so, install the dataset files located on the Mastering Revit Building CD ROM.

   Refer to "Files Included on the CD ROM" in the Preface for instructions on installing the dataset files included on the CD.

2. Launch Autodesk Revit Building from the icon on your desktop or from the **Autodesk** group in **All Programs** on the Windows Start menu.

   - From the File menu, choose **Close**.

This closes the empty new project that Revit Building creates automatically upon launch.

3. On the Standard toolbar, click the Open icon.

TIP The keyboard shortcut for Open is CTRL + O. **Open** is also located on the File menu.

   - In the "Open" dialog box, click the *My Documents* icon on the left side.
   - Double-click on the *MRB* folder, and then the *Chapter07* folder.

   If you installed the dataset files to a different location than the one listed here, use the "Look in" drop down list to browse to that location instead.

4. Double-click *07 Residential.rvt* if you wish to work in Imperial units. Double-click *07 Residential Metric.rvt* if you wish to work in Metric units

   You can also select it and then click the Open button.

The project will open in Revit Building with the last opened View visible on screen.

CREATE AN EXISTING ROOF PLAN VIEW

In the previous chapter, we introduced the concept of creating a separate View for the existing construction and setting its parameters accordingly. We did this while adding Stairs in the first floor. Let's use the same technique now to create an "Existing Conditions Roof Plan" View.

5. On the Project Browser, right-click the *Roof* plan View and choose **Duplicate**.

   - Right-click on *Copy of Roof* and choose **Rename**.

- Name the new View **Existing Roof** and then click OK.
- On the Project Browser, right-click on *Existing Roof* and choose **Properties**.
- In the Element Properties dialog, beneath "Instance Parameters" choose **Existing** for the "Phase" and then click OK.

The new construction will disappear and only the existing second floor will show in gray as an underlay to the current View. Recall that in Chapter 3 we assigned the Second Floor as an underlay to this View. Copying the View here also copies this underlay parameter in our *Existing Roof* View.

ADD THE EXISTING ROOF

You may note that the temporary Roof provided in previous chapters has not been included for this chapter project. There is valuable experience to be gained in creating the Roof from scratch as we will do in the tutorial that follows. We will begin with a simple gable Roof on the existing house and then add a slightly more complex double gable on the new addition.

6. On the Design Bar, click the Basics tab and then click the Roof tool and then from the flyout, choose Roof by Footprint.

   The Design Bar changes to sketch mode and shows only the Roof tools.

   - On the Design Bar, verify that the Pick Walls tool is selected.
   - On the Options Bar, verify that the "Defines slope" checkbox is selected.
   - In the Overhang field, type **6″ [150]**.
   - Verify that "Extend to wall core" is not selected.
   - Move the pointer over the topmost horizontal Wall and when the green dashed line appears above and to the outside, click the mouse (see Figure 7–1).

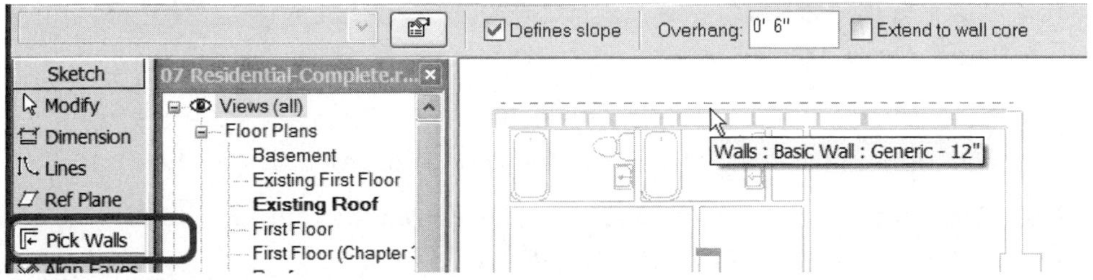

Figure 7–1 *Set the parameters for the first edge of the Roof on the Options Bar*

- Repeat this on the lower horizontal Wall to create another sketch line below and to the outside of that Wall.

7. On the Options Bar, clear the "Defines slope" checkbox and then click the outside vertical Wall on the left.

- Also click to the outside of vertical Walls as shown in Figure 7–2.

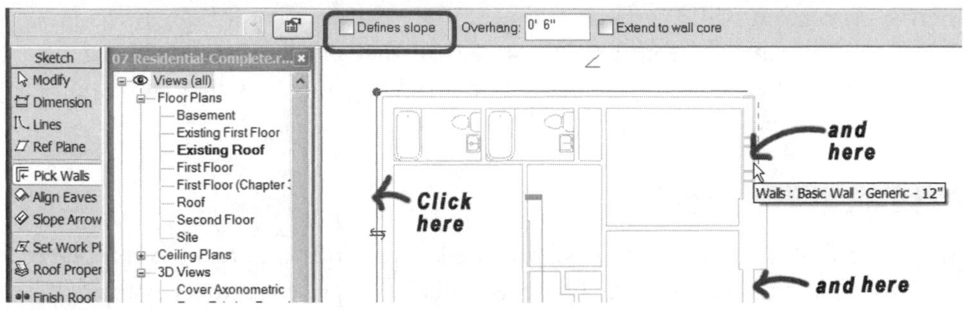

Figure 7–2 *Create the gable ends by clearing "Defines slope"*

Since we have the chimney on the right side, we need to make the Roof cut around it. This can be accomplished by drawing the remaining sketch lines relative to the chimney.

8. On the Design Bar, click the Lines tool.

- On the Options Bar, place a checkmark in the "Chain" checkbox.
- Using Figure 7–3 as a guide, sketch the remaining three sketch lines.

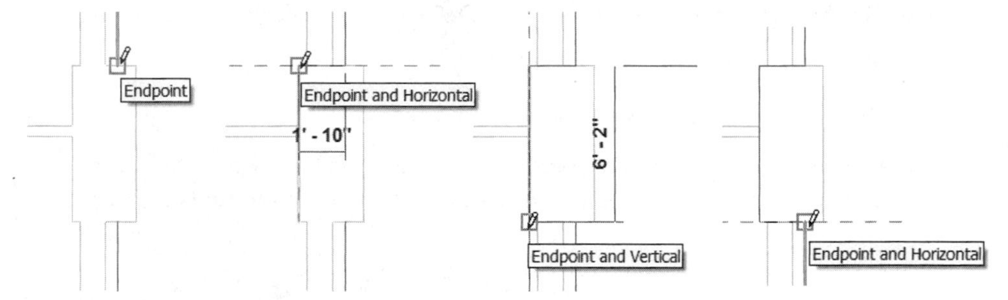

Figure 7–3 *Sketch around the chimney*

- On the Design Bar, click the Modify tool or press the ESC key twice.

Before we complete the sketch, let's adjust the slope and properties of the roof.

9. Click on the top horizontal sketch line, hold down the CTRL key and then click the bottom horizontal line as well.

- On the Options Bar, click the Properties icon.

- In the "Element Properties" dialog, beneath the "Dimensions" grouping, change the Roof slope. If you are working in Imperial units, type **6″** for the Rise/12″ parameter. If you are working in Metric, set the Slope Angle to **26.57**.
- Click OK to dismiss the "Element Properties" dialog.

TIP As an alternative, you can select the sketch line and then edit the slope directly with the temporary dimension that appears next to the slope indicator or in the properties of the slope defining line.

10. On the Design Bar, click the Finish Roof button.
11. On the View toolbar, click the Default 3D View icon (or open the {*3D*} View from the Project Browser).

Notice that the two Roof edges that we designated as "Defines slope" have a pitch sloping up to a single gable ridge down the middle of the existing house. Feel free to use the "Dynamically Modify View" tools to spin the model around and see it from different angles.

ADJUST THE CHIMNEY

Notice also that the Roof has a cutout for the space of the chimney. (However, at the moment the chimney is too short). We can adjust the height of the chimney by simply dragging the control handles. To do this with a bit of accuracy, let's first lay down a reference plane.

12. On the Project Browser, double-click to open the *East* elevation View.
13. On the Design Bar, click the Basics tab and then click the Ref Plane tool.
    - Click the first point above the roof to the left of the chimney.
    - Click the other point in a horizontal line above the top of the Roof to the right of the chimney.

An "ignorable warning message" will most likely appear when you place the second point of the Reference Plane (see Figure 7–4). An ignorable warning message is simply an informational message that alerts you of a situation that may not otherwise be apparent. This type of dialog appears in the lower right corner of the screen, requires no user action and can be safely ignored. Simply continuing to draw or picking other tools closes this warning dialog. You do not need to click the close button to close it. They are tinted in color that can be customized in the Options dialog.

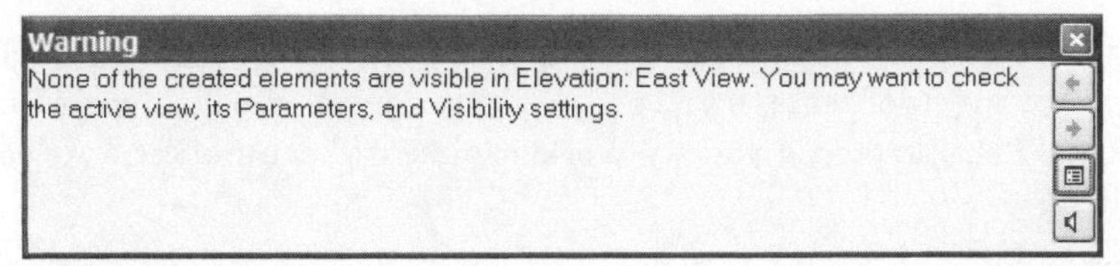

Figure 7–4 *"Ignorable Warning" dialogs report information that may be useful, but can be safely ignored*

This message appeared because the elevation View has a crop boundary enabled. This crop region is visible by default in the section Views, but invisible in elevation Views. We could turn on the visibility of Crop Region in this View so that the Reference Plane would become visible. However, it is likely that we sketched it too high above the building anyway. Let's use the temporary dimension to move it first and see if it comes into view before we take any action on the Crop Region. Incidentally, the display of the Crop Region is easily toggled on or off from the icon on the View Control Bar.

14. With the Reference Plane (the one you just drew) still active, click in the blue text of the temporary dimension.

    - Type a value of **8′-6″ [2600]** and then press ENTER (see Figure 7–5).

NOTE Be certain that the temporary dimension measures from the Roof Level to the Reference Plane. If it does not, adjust the witness lines first.

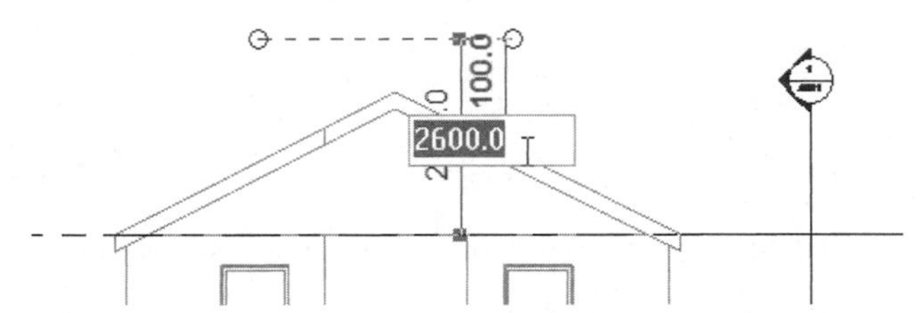

Figure 7–5 *Edit the temporary dimension to move the Reference Plane*

- On the Design Bar, click the Modify tool or press the ESC key twice.

If you deselected the Reference Plane already, you can either undo the last command, or you can use the Crop Region icon on the View Control Bar to toggle on the Crop Region, then use the control handle to make it larger. This will bring more of the model into view and the Reference Plane will appear above the Roof. Click to select it and then adjust the temporary dimension as noted above. You can toggle the display of the Crop Region off again when finished.

15. Click on the chimney to select it.

You may recall that the chimney was constructed in Chapter 3 from an In-Place Family.

- Drag the control handle at the top and snap it to the Reference Plane.
- Click the small padlock icon (close the lock) to constrain the top edge of the chimney to the Reference Plane (see Figure 7–6).

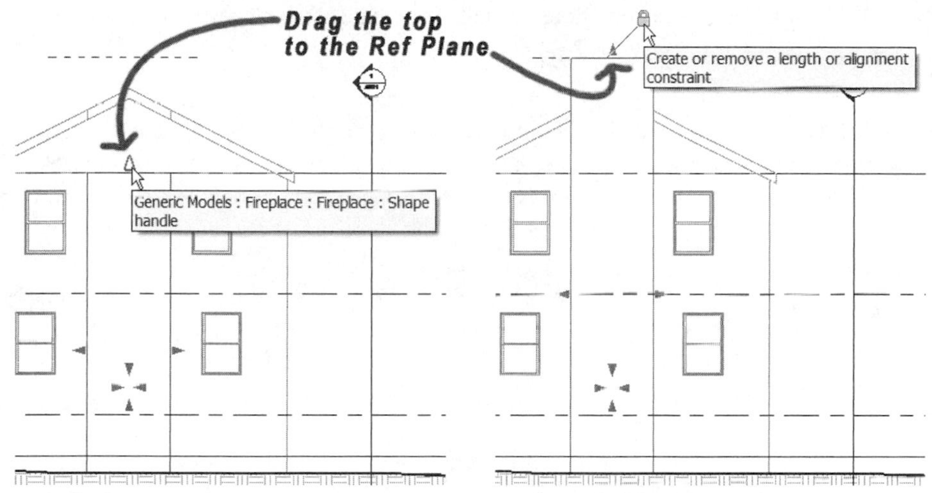

Figure 7–6 *Drag the top of the chimney and lock it to the Reference Plane*

Constraining the geometry to the Reference Plane is not required, but it makes it easier to adjust the height of the chimney later if required. Simply move the Reference Plane and the top of the chimney will follow. Try it now if you like.

16. Save the project.

ATTACH WALLS TO THE ROOF

It may not be apparent from the elevation View, but the Walls do not project all the way to the Roof. In some cases when you draw a Roof, you will be prompted to automatically attach the Walls to the Roof. Usually for this to occur, the Walls must intersect the Roof. In this case, the Walls stopped beneath the Roof. However, we can still manually attach them to the Roof. It is a good idea to get in the habit of checking for this condition.

17. From the Window menu, choose **Close Hidden Windows**.
    - On the View toolbar, click the Default 3D View icon (or open the *{3D}* View from the Project Browser).
    - From the Window menu, choose **Window Tile**.

TIP The keyboard shortcut for Tile is **WT**.

18. Pre-highlight one of the exterior Walls of the existing house.
    - Press the tab key to highlight a chain of connected Walls.
    - With all of the existing house Walls pre-highlighted, click the mouse to select them (see Figure 7–7).

NOTE If you have trouble chain-selecting the existing Walls, you can use other selection methods such as the CTRL key, or you can even duplicate the {3D} View and make an "Existing Conditions 3D" View from which to perform these steps like we did previously for the plan.

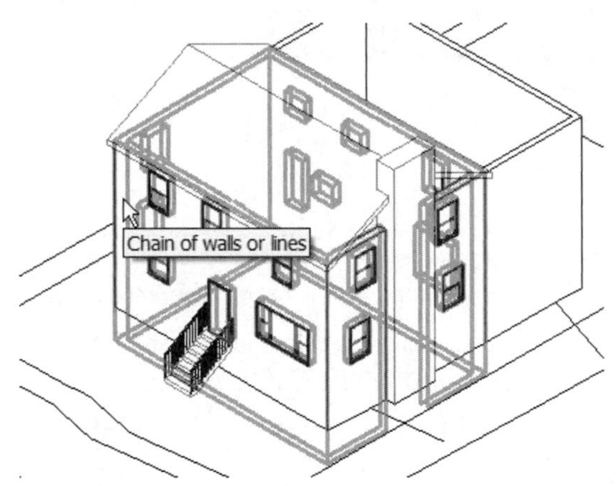

Figure 7–7 *Select a chain of Walls*

19. On the Options Bar, click the Attach button.
    - On the Options Bar, verify that "Top" is selected and then click the Roof (see Figure 7–8).

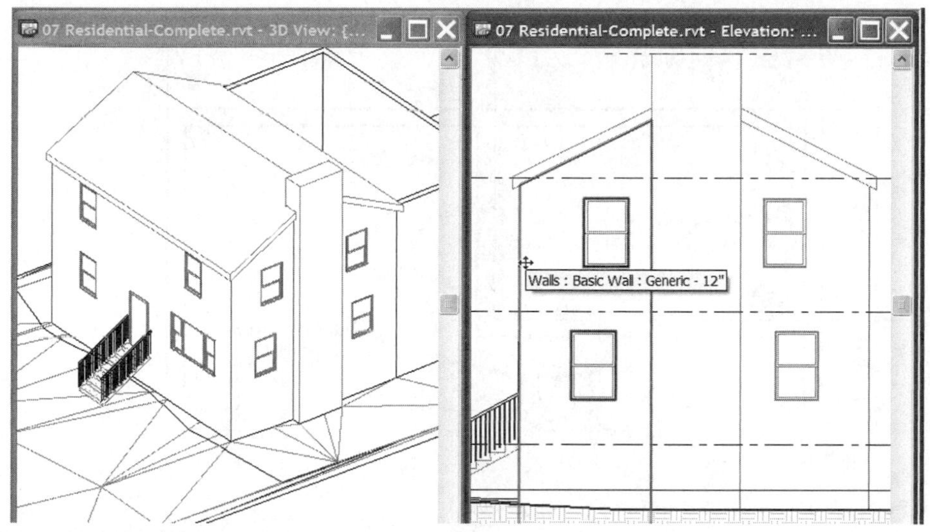

Figure 7–8 *Attach the Walls to the Roof*

- Deselect the Walls.

EDIT THE SECOND FLOOR WALL LAYOUT

There are many other Roof options to explore. The best place to do this in this project is on the new construction in the addition rather than the existing construction. The second floor of the addition will extend the two existing bathrooms and have an outdoor patio on the left side of the plan. On the north and west sides of the patio, we will have low height parapet-type Walls. The Roof will cover the other portions of the addition, but the patio will be uncovered. Before we add the Roof, we need to modify the Walls a bit to reflect these design features.

20. On the Project Browser, double-click to open the *Second Floor* plan View.
21. Using techniques covered in Chapter 3, add two Walls as shown in Figure 7–9. Use the same Type as the other three Walls in the addition.
    - Set the Location Line to **Finish Face:Exterior**.
    - Set the "Base Constraint" to **Second Floor** and the "Top Constraint" to **Up to level: Roof**.

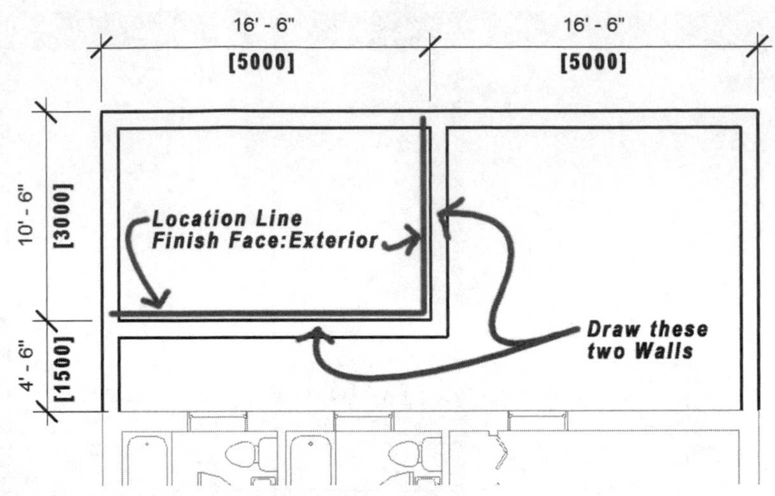

Figure 7–9 *Add two new Walls to the Second Floor*

The space that we have just described in the top left corner is the outdoor patio. As noted above, it will have low height Walls on two sides. Now let's edit the original exterior Walls to reflect this condition. To do this, we are going to edit the Profile of the Walls.

22. Select the top horizontal Wall.

    - On the Options Bar, click the Edit Profile button.

    The "Go To View" dialog will appear.

    - In the "Go To View" dialog, choose *Elevation:North* and then click Open View (see Figure 7–10).

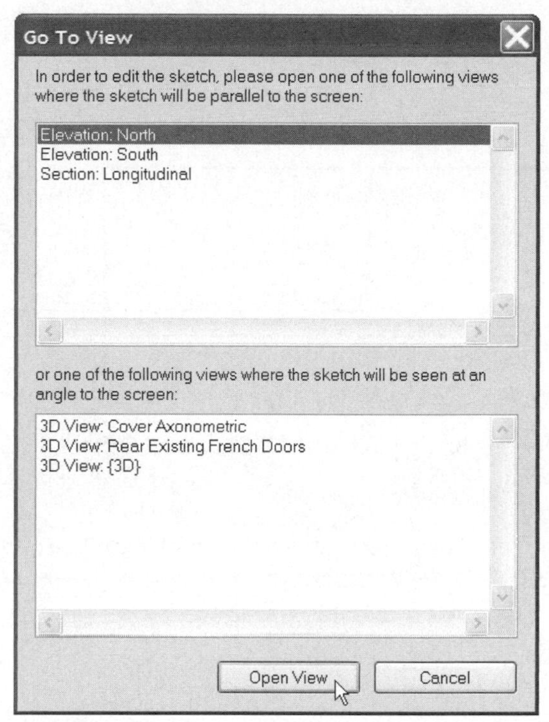

Figure 7–10 *Certain edits prompt for a more appropriate View in which to perform the action*

When you begin an operation that cannot easily be performed in the current View, Revit Building will prompt you to open a more appropriate View. The "Go To View" dialog suggests all appropriate Views in the current project in which to perform the operation. You should now be looking at the back of the house with the selected Wall in sketch mode. In this mode we can edit the shape of the Wall to sculpt it to meet the needs of the design.

23. On the Design Bar, click the Lines tool.
    - On the Options Bar, click the Pick Lines icon.
    - Click on the right vertical edge of the intersecting Wall in the middle of the second floor (see Figure 7–11).

NOTE This is the outside edge of the Wall we drew above.

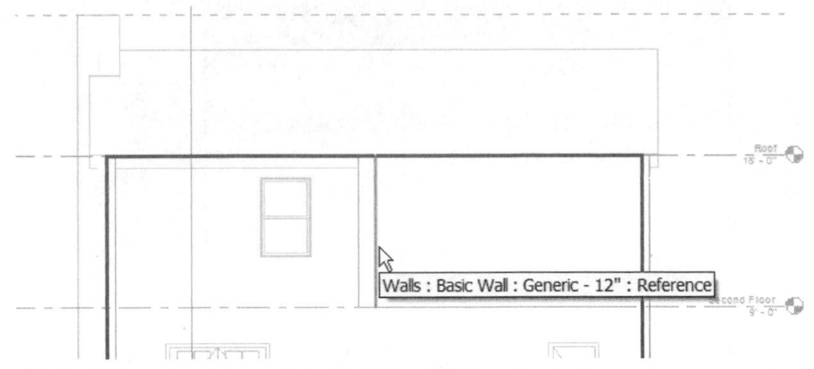

Figure 7–11 *Add a sketch line to the Wall Profile Sketch relative to the intersecting Wall*

- On the Options Bar, type **3′-6″** [**1050**] in the "Offset" field.
- Highlight the Second Floor Level line.
- When the green line appears above the Level line, click to create the sketch line (see Figure 7–12).

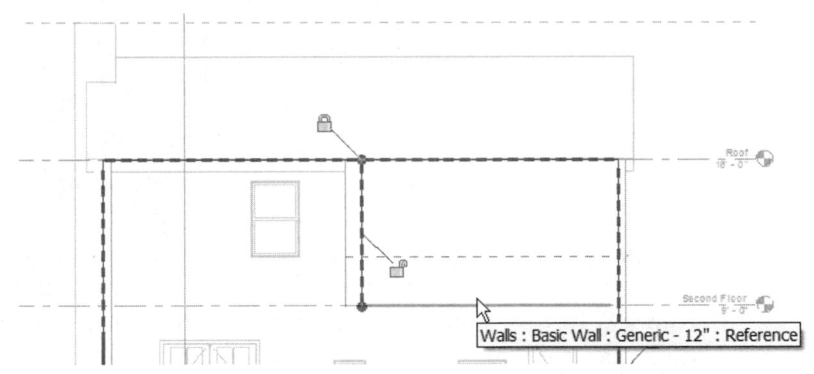

Figure 7–12 *Add another sketch line relative to the Second Floor Level Line*

24. Use the Trim/Extend tool to cleanup the sketch and close all of the corners (see Figure 7–13).
    - Use the Trim/Extend to Corner option on the Options Bar.

NOTE Remember to click the side of the line that you wish to keep.

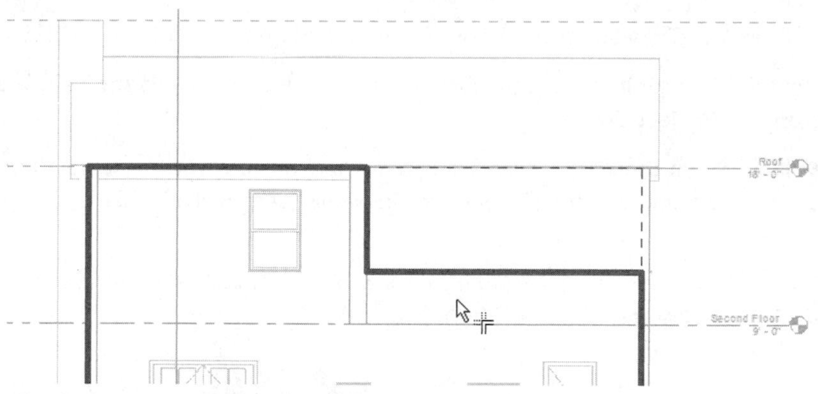

Figure 7–13 *Complete the Wall Profile Sketch (sketch lines enhanced for clarity)*

25. On the Design Bar, click the Finish Sketch button.
26. Select the other Wall (the vertical one on the right in the current View—on the left in the plan View) and repeat the process.
    - When prompted, open the *Elevation:West* View.
    - Again use the outside edge of the Wall we drew above to create the vertical sketch line and then offset a line up from the Second Floor Level Line as before.
    - Trim/Extend to complete the sketch (see Figure 7–14).
    - On the Design Bar, click the Finish Sketch button.

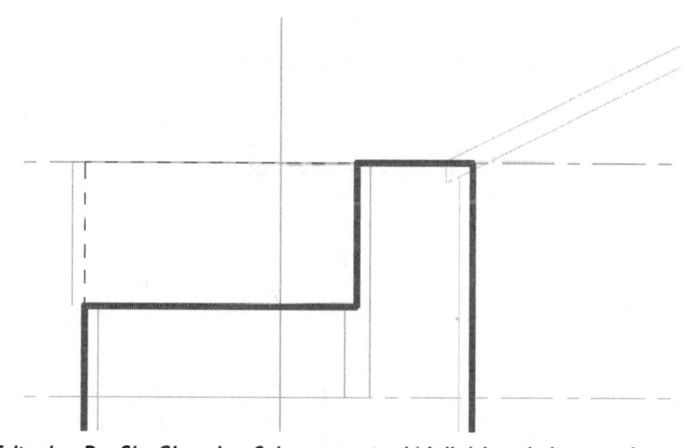

Figure 7–14 *Edit the Profile Sketch of the opposite Wall (sketch lines enhanced for clarity)*

The new addition will be brick veneer on a stud wall backup. We can apply a Wall Type that represents this type of construction to our new construction Walls.

27. Select all of the Walls in the addition (the three original ones and the two new ones we just added).

- From the Type Selector, choose Basic Wall : Exterior - Brick on Mtl. Stud.

28. On the View toolbar, click the Default 3D View icon (or open the *{3D}* View from the Project Browser).
    - Hold down the SHIFT key and drag the wheel on your mouse to spin the model around and see the edits to the Walls (see Figure 7–15).

NOTE If you do not have a wheel mouse, use the "Dynamically Modify View" tool on the View toolbar instead.

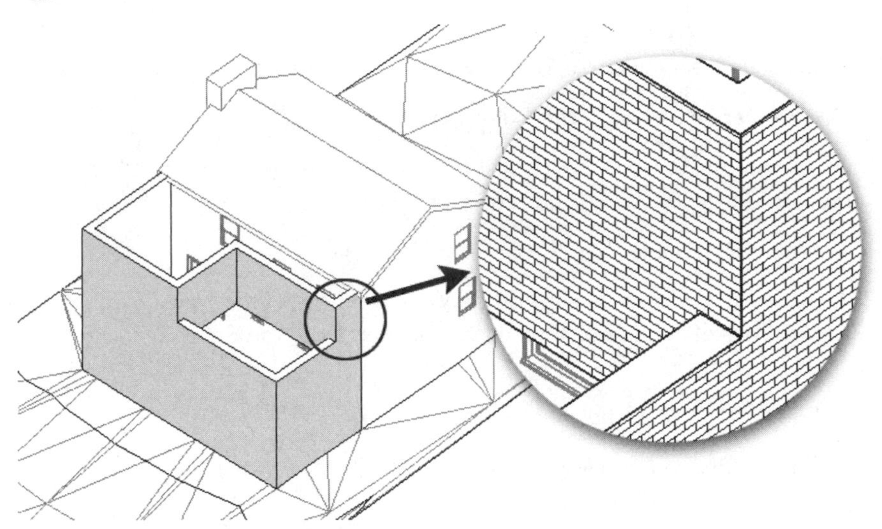

Figure 7–15 *View the model in 3D to see the completed Wall edits*

29. On the Project Browser, right-click the *{3D}* View and choose **Duplicate**.
    - Name the new View **New Addition Axon**.
30. Save the project.

ADD THE NEW ROOF

Now that we have prepared the second floor Wall layout, we are ready to begin roofing the addition.

31. On the Project Browser, double-click to open the *Roof* plan View.

NOTE Be sure to open "*Roof*" this time, not "*Existing Roof.*"

32. On the Design Bar, click the Basics tab and then click the Roof tool and then from the flyout, choose Roof by Footprint.
    - On the Options Bar, place a checkmark in the "Defines slope" checkbox.
    - In the Overhang field, type **6″** [**150**].
    - Place the two sketch lines indicated in Figure 7–16.

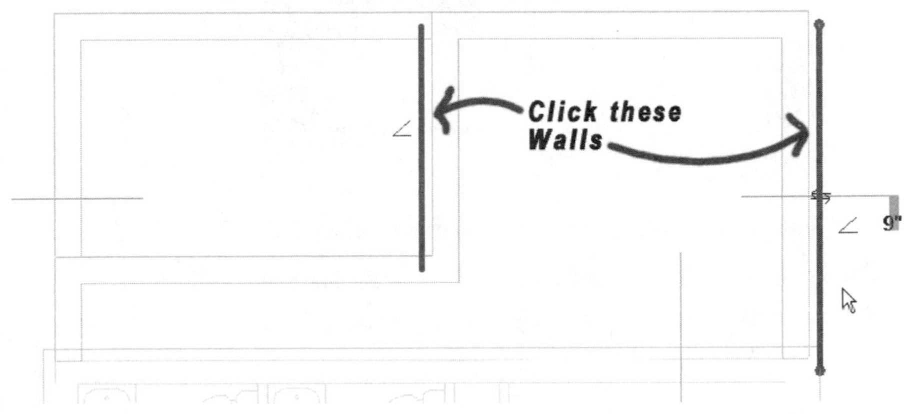

Figure 7–16 *Create the sloped sketch lines*

33. On the Design Bar, click the Modify tool or press the ESC key twice.
34. Select both of these sketch lines.
    - On the Options Bar, click the Properties icon.
    - In the "Element Properties" dialog, beneath the "Dimensions" grouping, change the Roof slope. If you are working in Imperial units, type **6″** for the Rise/12″ parameter. If you are working in Metric, set the Slope Angle to **26.57**.
    - Click OK to dismiss the "Element Properties" dialog and deselect the sketch lines.

TIP As an alternative, you can select the sketch line and then edit the slope directly with the temporary dimension that appears next to the slope indicator.

35. On the Design Bar, click the Lines tool.
    - Clear both the "Defines slope" and "Chain" checkboxes.
    - Draw a horizontal line aligned to the edge of the existing house Roof and the width of the two sketch lines we already have.

- Use the Pick Lines option and a **6″ [150]** offset to create the top sketch line (see Figure 7–17).

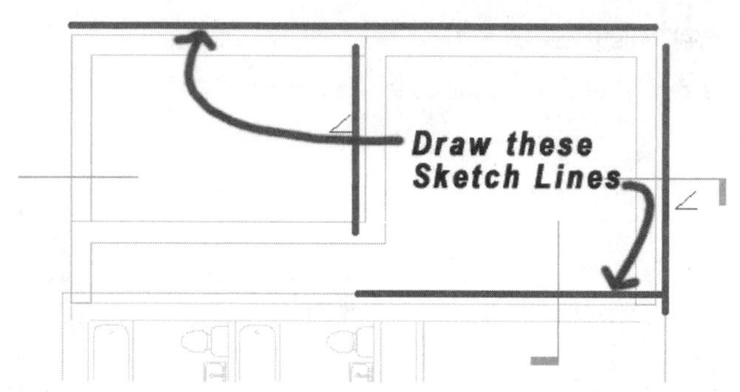

Figure 7–17 *Create two horizontal sketch lines that do not define slope*

36. Using the Trim/Extend tool, cleanup the rectangular sketch shape.

- On the Design Bar, click the Finish Roof button.
- On the Project Browser, double-click to open the *New Addition Axon* 3D View.
- Spin the model to see the interaction of the two Roofs clearly.

JOIN ROOFS

As you can see in the 3D view, the new Roof does not intersect with the existing one. This is easily corrected.

37. On the Tools toolbar, click the Join Roofs icon.

TIP Join/Unjoin Roof is also located on the Tools menu.

Take a look at the Status Bar (lower left corner of the screen) and notice the message. You are prompted to "Select an edge at the end of the roof that you wish to join or unjoin." We want to select one of the edges of the new construction Roof. Then we will be prompted to select a face to which to join. In that case, we will select the face of the existing Roof.

- Click on the edge of the new construction Roof as indicated on the left side of Figure 7–18.
- Click on the face of the existing construction Roof as indicated on the right side of Figure 7–18.

NOTE To pick a Roof face of the existing Roof, move the Modify tool over any of the edges of the existing Roof. The perimeter of the existing Roof will pre-highlight. Use the TAB key to cycle the selection, and click when the desired Roof face is pre-highlighted to select it.

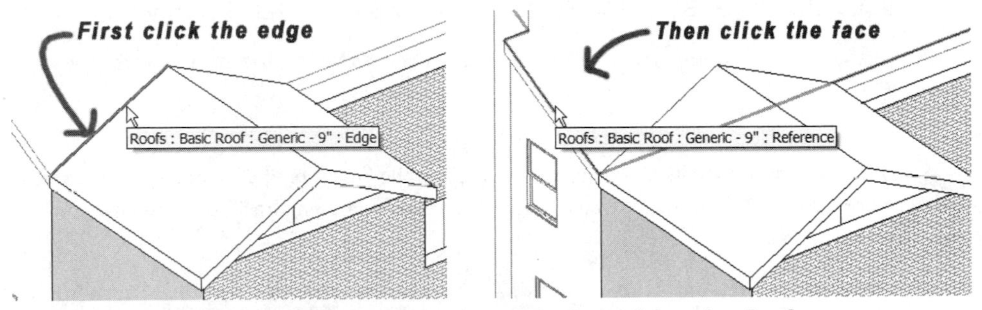

Figure 7–18 *Select an edge of the Roof to join to the face of the other Roof*

The new construction Roof will now extend over the existing Roof and form nicely mitered intersections. The same tool can be used in reverse if you ever need to unjoin a Roof.

38. Repeat the entire process to create the shorter new Roof on the other side of the addition (see Figure 7–19).

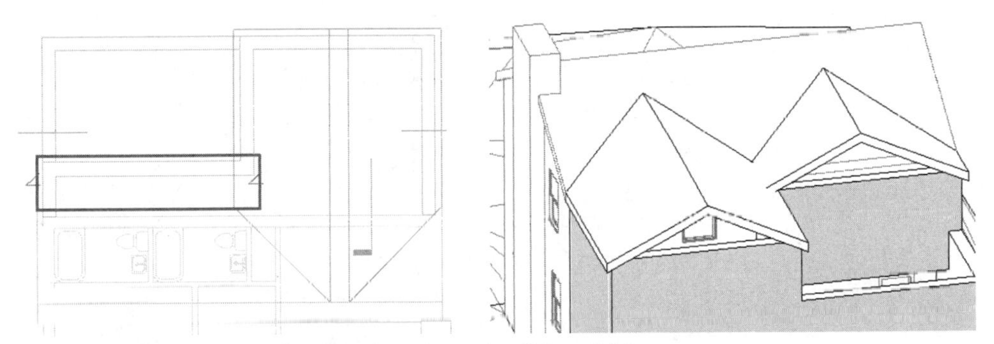

Figure 7–19 *Create the Roof on the other side of the addition*

Join Roofs works when the two Roofs meet perpendicular to one another. We will use a different technique to join the two new Roofs together. For that we will use the more generic Join Geometry tool.

39. On the Tools toolbar, click the Join Geometry icon.

TIP Join Geometry is also located on the Tools menu.

Again, watch the Status Bar prompts.

- Click on the first new construction Roof, then click the other to join them.
- On the Design Bar, click the Modify tool or press the ESC key twice.

40. Using the technique covered above, Attach the Walls to the new Roofs (see Figure 7–20).

TIP If you have trouble attaching the tall Wall on the East side of the house to the new Roof, attach it first to the existing construction Roof and then attach it to the new construction Roof.

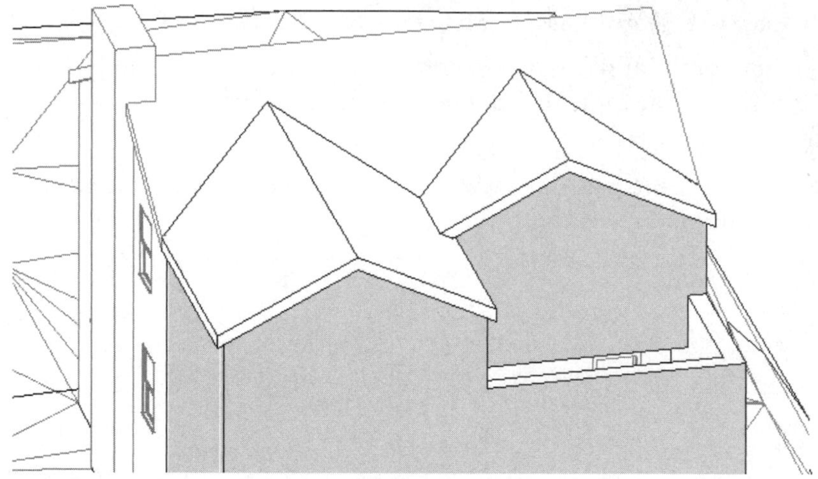

Figure 7–20 *Join the two new construction Roofs with Join Geometry*

41. Save the project.

EDITING ROOFS

Now that we have created the Roofs for the Residential project, let's turn our attention to editing Roofs that already exist in a Revit Building model. In this case, they will be the ones that we have just added. In your own projects, you will work with Roofs in much the same way as other Revit Building elements—start by laying them out with simple generic parameters and then over the course of the project, layer in additional details and parameters as design and project needs dictate. In this sequence, we'll explore some of the possibilities available for editing Roof elements in Autodesk Revit Building.

MODIFY THE ROOF PLAN VIEW

You may have noticed that the *Roof* plan View looks a bit odd. The Roof elements are being cut in plan which shows us only part of the sloped surface. While appropriate for an attic space, for an actual Roof Plan, this is not typically the way we would want to represent it.

Continuing in the Residential Project.

1. On the Project Browser, double-click to open the *Roof* plan View.
   - Right-click in the workspace and choose **View Properties**.
   - In the "Element Properties" dialog, beneath the "Extents" grouping, click the Edit button next to "View Range."
   - In the "Primary Range" area, change the "Top" Offset to **15′-0″** [**4500**].
   - Change the "Cut Plane" Offset to **10′-0″** [**3000**] and then click OK (see Figure 7–21).

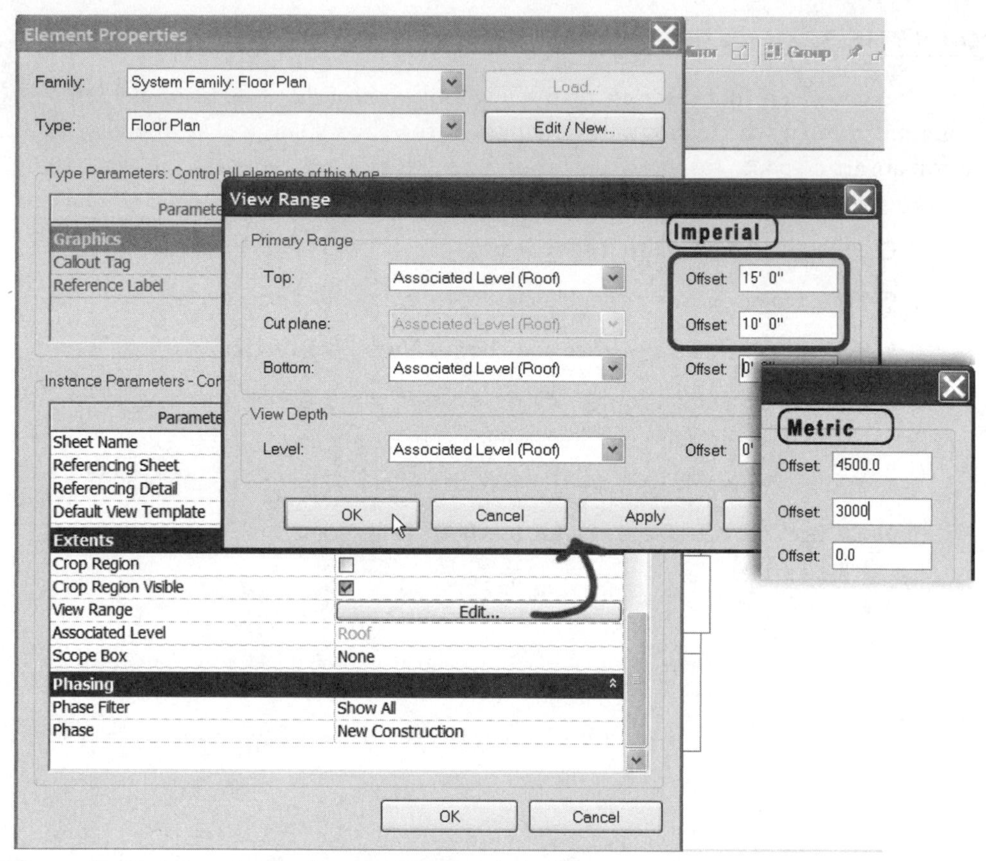

Figure 7–21 *Adjust the Cut Range of the* Roof *plan View*

2. Click OK twice to return to the "Element Properties" dialog.

   - Beneath the "Graphics" grouping, choose **None** for the "Underlay."
   - Click OK to return to the *Roof* plan View and see the results (see Figure 7–22).

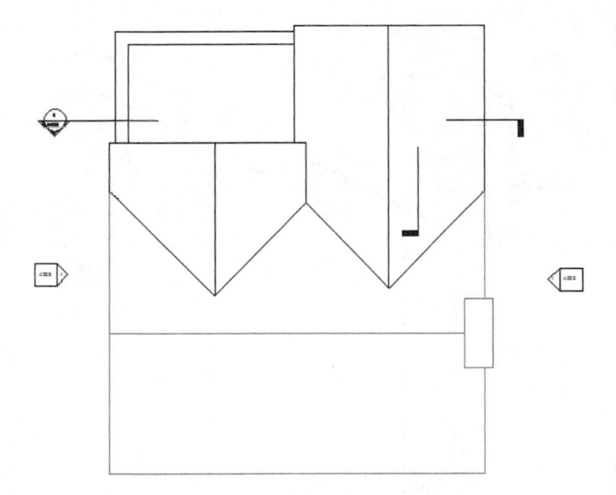

Figure 7–22 *Turn off the Underlay of the Second Floor and view the* Roof *plan*

UNDERSTAND ROOF OPTIONS

The Roof element has several options that we have not yet explored. Let's take a look at some of them now.

3. On the Project Browser, double-click to open the *Longitudinal* section View.
   - Zoom in Region around the eave at the left side of the section (see Figure 7–23).

Notice the way the Roof intersects the attached Wall.

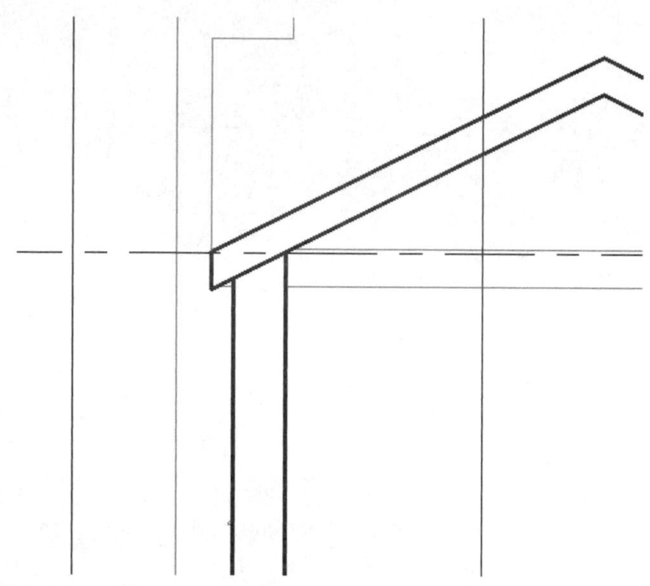

Figure 7–23 *Zoom in on the eave condition*

Above we swapped in a more detailed Wall Type for these Walls. However, since we are currently viewing the model in "Coarse" Display mode, we do not see any difference in the Wall structure. To see the more detailed Wall structure, we need to adjust the detail display level of the current View.

4. On the View Control Bar, click the Detail Level icon and choose **Medium** (see Figure 7–24).

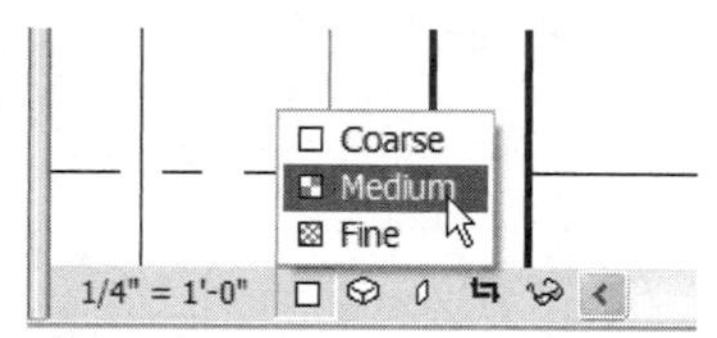

Figure 7–24 *Change the Detail Level to Medium*

The Layers that makeup the Wall Type's structure will appear including brick, a stud layer and air gap. This Detail Level will make it a little easier to understand the various Roof options that we are about to explore.

5. Select the Roof.

- On the Options Bar, click the Properties icon.
- Beneath the "Construction" grouping, change the "Rafter Cut" to **Two Cut – Plumb** and then click OK.

The results will probably not be satisfactory.

- Return to the Roof Properties and set the "Fascia Depth" to **6″** [**150**].

- Repeat the process and choose **Two Cut – Square** (see Figure 7–25).

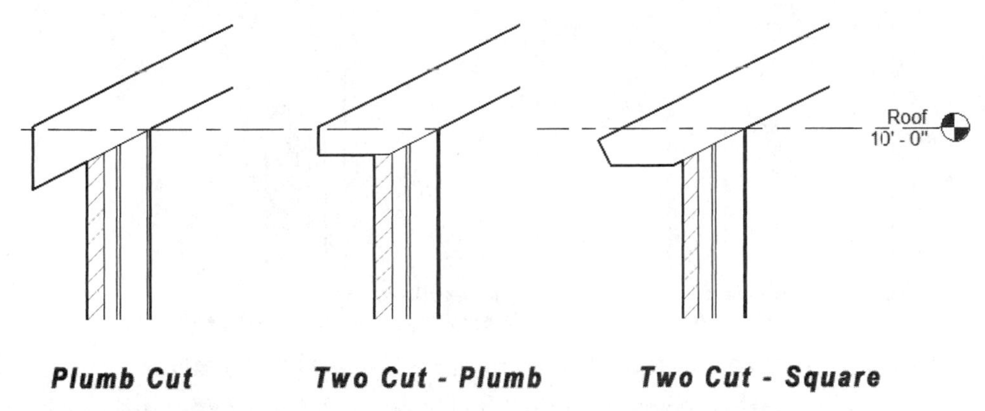

Figure 7–25 *Comparing the Rafter Cut options (Illustrations simplified for clarity)*

The section shown in the figure has been simplified to remove any of the beyond components (the Far Clip Offset of the section was adjusted to achieve this). However, the cut components of your model should match the figure fairly closely. The panel on the left shows the original condition—which is the default. The middle and right panels show the conditions that we just tried.

6. Return to the Roof "Element Properties" dialog once more and change the "Rafter Cut" to **Two Cut – Plumb** and then click OK.

When we were adding the Roof above, you may have noticed the "Extend to wall core" checkbox on the Options Bar. We did not use this option when creating the Roof. You can use this setting when building the Roof on top of a complex Wall Type (like the one we now have) that includes more than one layer. When you choose this option, the Overhang setting will be measured relative to face of the core layer rather than the finish face of the Wall. Figure 7–26 shows this option used with and without attaching the Walls to the Roof. (The Core layer of the Wall has been shaded gray in the figure for clarity). If you wish, you can experiment with these options. Remember, if you choose to edit the Roof while the section View is open, you will be prompted to open a more appropriate View—choose *Floor Plan:Roof.* Cancel or undo any edits made it you do experiment with this setting.

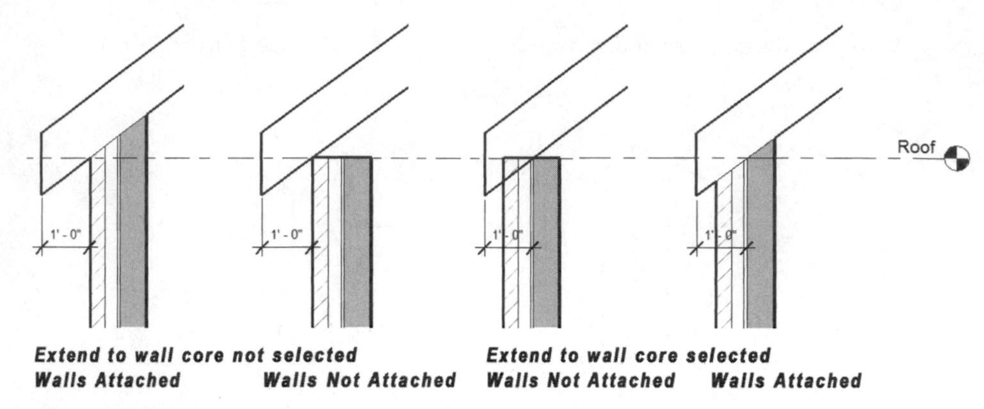

Figure 7–26 *Understanding the effect of the "extend to wall core" and Attach options*

There is one other setting worth exploring at this point. When you create your Roof, Revit Building offers two modes of construction: Rafter or Truss. The difference between these two settings is simply the point that it uses as the spring point for the Roof.

7. Select the Roof.
   - On the Options Bar, click the Properties icon.
   - Beneath the "Construction" grouping, change the "Rafter or Truss" to **Truss** and then click OK.

Notice how the entire Roof appears to move up. Rafter measures the plate of the Roof from the inside edge of the Wall. Truss measures from the outside edge. This option is only available for Roofs created using the "Pick Walls" option.

8. Return to the Roof "Element Properties," change the "Rafter or Truss" to **Rafter** and then click OK.
9. Select the other two Roofs (one new construction, one existing) and then edit the Properties.
   - Change the "Rafter Cut" to **Two Cut – Plumb**.
   - Set the "Fascia Depth" to **6″** **[150]** and then click OK.
10. Save the project.

CREATE A COMPLEX ROOF TYPE

Much like Walls, Roofs can be composed of several Layers of structure. You can create your own Roof Type that contains the structure you need, or transfer an appropriate Type from another project using the Transfer Project Standards feature. In this example, we will build a new Roof Type from scratch.

11. Select all of the Roofs.
    - On the Options Bar, click the Properties icon.
12. Next to the Type list, click the Edit/New button.

TIP A shortcut for this is to press ALT + E.

The Type Properties dialog will appear.

- Next to the Type list, click the Duplicate button.

TIP A shortcut for this is to press ALT + D.

A new Name dialog will appear. By default "(2)" has been appended to the existing name.

- Change the name to **MRB Wood Rafters with Asphalt Shingles** and then click OK (see Figure 3–40 in Chapter 3 for an example).

13. In the "Type Properties" dialog, next the "Structure" click the edit button.

In the "Edit Assembly" dialog, you can see that the Roof Type currently contains only a single generic Layer. This is because we duplicated it from the generic Roof Type that we started with. You can add, edit and delete Layers to the Roof structure in this dialog. We will keep the structure of our Roof Type fairly simple. We need a structure layer, which will be wood rafters, a plywood substrate and asphalt shingles for the finish layer. When you build a new Type in Autodesk Revit Building, there are several things to consider. We can add as much or as little detail to the structure of the Roof Type as we wish. In some cases it will prove valuable to accurately represent each piece of the Roof's actual construction. However, also consider the potential negative impact that highly detailed Types can have on drawing legibility. As a general rule of thumb, you should seek to build your models as accurately as possible while remembering that any architectural drawing includes a certain degree of abstraction as a matter of industry convention or simply a matter of clarity. All of these points hold true in other areas of Revit Building such as creating and editing Wall Types as well. We will see more on this in coming chapters. For this example, we will abstract our Roof construction to just the three Layers noted above. For now, we will not include building paper, insulation or interior finish. These items can be added to the Roof Type later (which will automatically apply to all Roof elements that reference the Type) or we can apply these items graphically as drafting embellishment in a Detail View (see Chapter 10 for more information on Detail Views).

EDIT ROOF STRUCTURE

The first step will be to edit the existing Layer to meet our needs. After that, we will insert the additional Layers. The "Edit Assembly" dialog lists each Layer of the Roof Type in a list with a numeric index number next to each one. There are four columns next to each item.

- **Function**—Click in this field for a list of pre-defined functions. The functions include "Structure," "Substrate," "Finish" and others. The number next to the function name indicates the priority of the Layer with regard to material "wrapping" joins. In this way, the Structure Layer of one Roof will attempt to join with the Structure Layer of another. One [1] is the highest priority while five [5] is the lowest. "Membrane Layers" have zero thickness and thus do not have priority nor do they join.
- **Material**—Materials determine the graphical characteristics of each Layer. Material properties include patterns, render material and shading.
- **Thickness**—This is the dimensional thickness of the Layer.
- **Wraps**—Controls if the Layer wraps around corners at the ends or at openings. If this is not selected, the Layer simply cuts perpendicular at the ends and openings.

14. In the "Edit Assembly" dialog click in the Material cell for the Structure Layer (Layer number 2).

    A small expand (downward pointing arrow) icon will appear at the right side of the cell

    - Click the small expand icon to open the "Materials" dialog.
    - From the "Name" list, select Structure - Wood Joist/Rafter Layer [Structure - Timber Joist/Rafter Layer] and then click OK.
    - In the Thickness field, type **7 1/2″ [190]** and then press ENTER (see Figure 7–27).

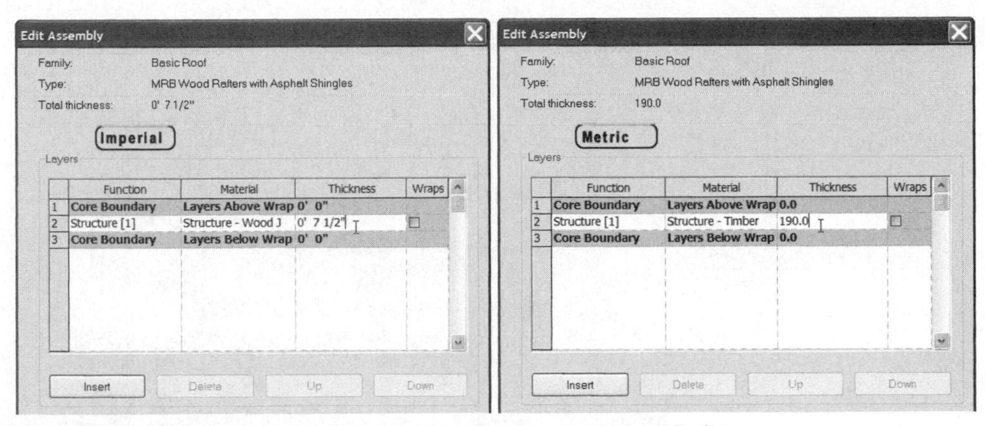

Figure 7–27 *Edit the existing Structure Layer to become Wood Rafters*

15. Beneath the list of Layers, click the Insert button.

    A new zero thickness Structure Layer will appear.

Notice that the new Layer appears within the "Core Boundary." Roofs and Walls have a Core that contains the structural Layers. You can have additional Layers on either side or both sides of the Core. If a new Layer that you insert does not appear in the desired location, select it and then use the Up and Down buttons to adjust its position in the structure of the Roof.

- Change the Function of the new Layer to **Substrate [2]**.
- Change the Material to Wood - Sheathing – plywood.
- Set the Thickness to **5/8″ [16]**.

16. Click the Insert button again.

- With the new Layer highlighted, click the Up button to place it above the Core Boundary.
- Change the Function of the new Layer to **Finish [4]**.
- Change the Material to Roofing – Asphalt [Roofing – Asphalt Shingle].
- Set the Thickness to 1/4″ **[6]**.

If you wish to see how the Type looks so far, click the Preview button at the bottom left corner of the dialog (see Figure 7–28).

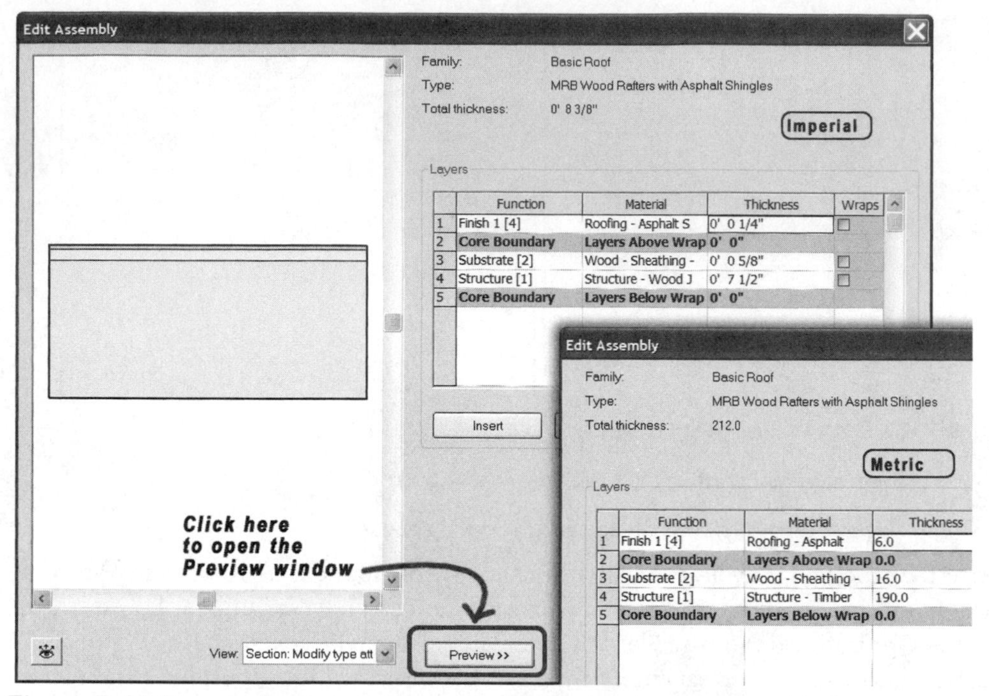

Figure 7–28 *Open the Preview window to see the completed structure graphically*

17. Click OK three times to return to the workspace window.

Notice that the new Roof Layers appear in the section View—if you switched to a different View, please return to the *Longitudinal* section View now. Also note that the new Layers will only appear if you left the section View in the Medium Display mode from above. If you set it back to Course, the graphics will simplify to show the outer edge of the Roof only. If you have difficulty seeing the sheathing and asphalt shingle Layers, try toggle the "Thin Lines" setting. You can find an icon for this on the View toolbar. Thin Lines is also available from the View menu (see Figure 7–29).

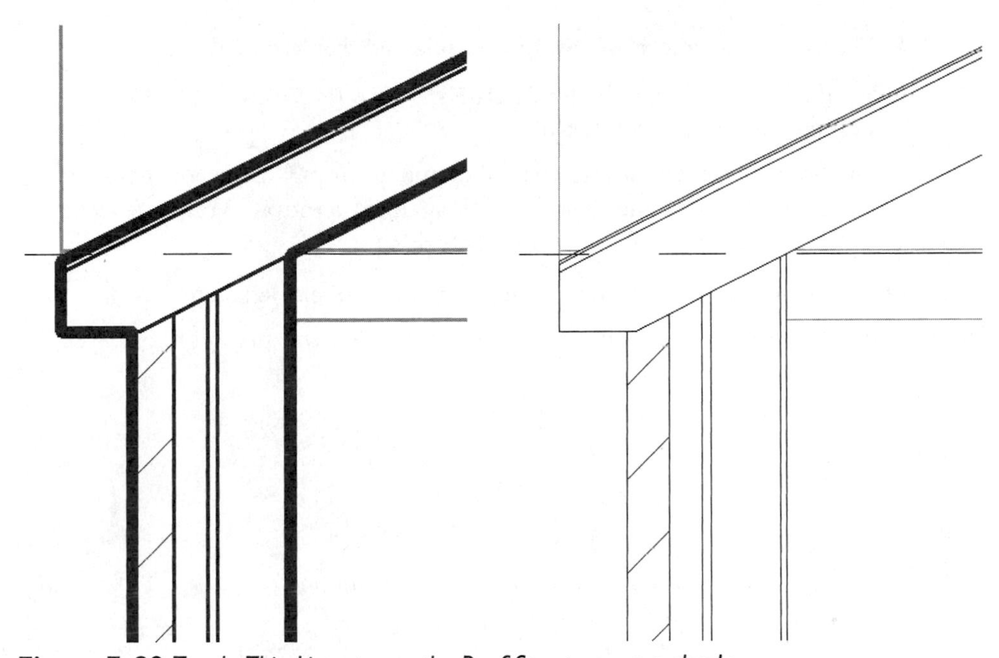

Figure 7–29 *Toggle Thin Lines to see the Roof Structure more clearly*

Toggle Thin Lines back off after you are satisfied with the Roof structure.

APPLY A HOST SWEEP TO ROOF EDGES

Creating the new Type and editing its structure provides a satisfactory representation of the overall Roof construction. However, the edges of the Roof could use some embellishment. A Host Sweep allows us to apply a detailed profile condition the edges of a Roof or along the length of a Wall. In this example, we will explore the use of Host Sweeps to apply fascia boards and gutters to our Roofs.

18. On the Project Browser, double-click to open the *New Addition Axon* View.
19. On the Design Bar, click the Modeling tab and then click the Host Sweep tool.
    - From the flyout menu that appears, choose **Roof Fascia**.

There are some Fascia Types already resident in the project. However, as with the Roof Type above, we will create our own. A Fascia is a simple Revit Building element. Its primary parameter is the assignment of a Profile shape. We have a 6″ [150] Roof edge as defined above. Let's choose a Profile close to this depth.

20. On the Options Bar, click the Properties icon.
    - Next to the Type list, click the Edit/New button.
    - Next to the Type list, click the Duplicate button.

- Change the name to **MRB Simple Fascia** and then click OK.
- From the "Profile" list, choose Fascia-Flat : 1 x 6 [M_Fascia-Flat : 19 x 140mm] and then click OK twice.

A glance at the Status Bar will reveal the following prompt: "Click on edge of Roof, Soffit, Fascia, or Model Line to add. Click again to remove." With the *New Addition Axon* View open, it is easy to accomplish this.

21. Move the pointer over the various edges of the Roof elements on screen.

Notice that both the top and bottom edges of any give edge will pre-highlight (see Figure 7–30).

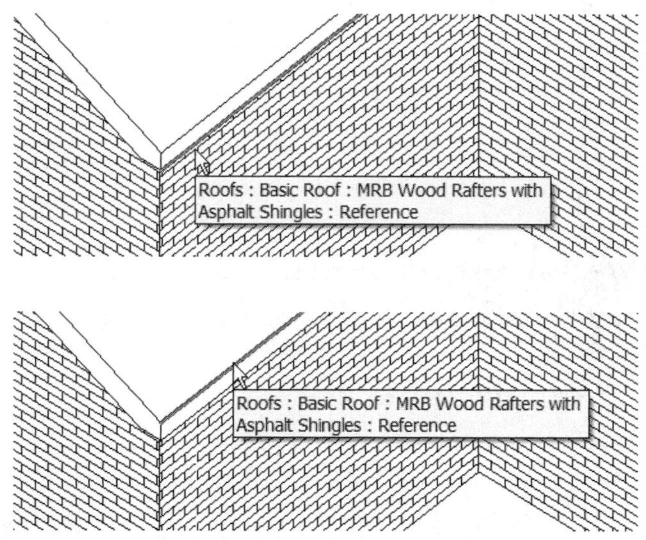

Figure 7–30 *You can Pre-Highlight either the top or bottom edge*

- Click on an edge to apply the Fascia.
- Place Fascia boards on the top edges of all of the horizontal (fascia) edges of the new Roof and all edges of the existing Roof as shown in Figure 7–31.

NOTE Be sure to click the top edge of each Roof edge, not the bottom.

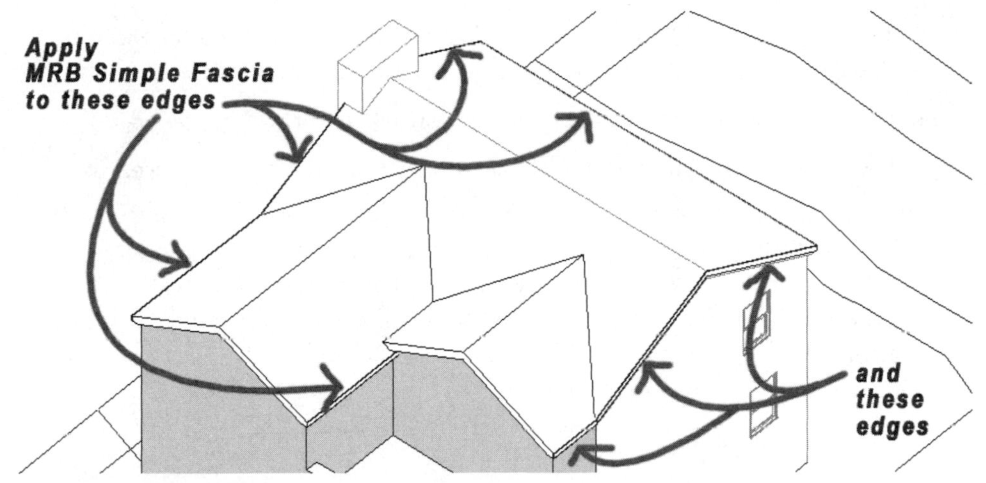
Apply
MRB Simple Fascia
to these edges
and
these
edges

Figure 7–31 *Adjust Gutter Horizontal Offset*

- On the Design Bar, click the Modify tool or press the ESC key twice.

We did not apply a Fascia to the sloped (rake) edges of the new construction Roof yet because we are going to use a different Type for these. However, before we create the Type, let's load a more complex profile from the library.

22. On the File menu, choose **Load from Library > Load Family**.

You should automatically be placed in either the *Imperial Library* or *Metric Library* folder. If this has not occurred, you can use the shortcut icons on the left side of the dialog to jump to those locations. If you do not have these icons, the appropriate files have been provided with the files installed from the Mastering Revit Building CDROM in the *Autodesk Library* folder. You can navigate manually to that location and access the Families from there.

- Double-click on the *Profiles* folder and then double-click the *Roofs* folder.
- Select the *Fascia-Built-Up.rfa* [*M_Fascia-Built-Up.rfa*] file and then click Open.

This loads the new Profile Family into the current project. We now need to repeat the steps above to create a new Fascia Type using this Profile.

23. On the Design Bar, click the Modeling tab and then click the Host Sweep tool.

- From the flyout menu that appears, choose **Roof Fascia**.
- On the Options Bar, click the Properties icon.
- Next to the Type list, click the Edit/New button and then the Duplicate button.
- Change the name to **MRB Built-up Fascia** and then click OK.

- From the "Profile" list, choose Fascia-Built-Up : 1 x 8 w 1 x 6 [M_Fascia-Built-Up : 38 x 140mm x 38 x 89mm] and then click OK twice.

24. Click each of the remaining Roof edges (remember to click the top edge).

    - On the Design Bar, click the Modify tool or press the ESC key twice.

Zoom in on one of the intersections between the Rake and Fascia conditions. If you select one of the Fascia boards, you will notice a drag control at the ends. You can stretch this control to modify the way the two Fascia boards intersect. Also, with one of the Fascia or Rake boards selected, you will notice a "Change Mitering Option" button on the Options Bar. The options are "Horizontal," "Vertical" and "Perpendicular." Try them out if you wish to see how each option behaves.

25. Save the project.

ADD GUTTERS

Another type of Host Sweep available to Roofs is a Gutter. These are conceptually the same as Fascia boards. They use a Profile to determine the cross-section shape and sweep it along the path of the Roof edge(s).

26. Zoom out to see the whole Roof.
27. On the Design Bar, click the Modeling tab and then click the Host Sweep tool.

    - From the flyout menu that appears, choose **Roof Gutter**.
    - Click one of the horizontal edges of the new construction Roof.

If the gutter is not visible, it needs to be flipped. There are flip controls on the selected Gutter.

- Click the Flip control to flip the gutter to the outside.

28. Add Gutters (and flip as required) to the remaining horizontal new construction Roof edges (see Figure 7–32).

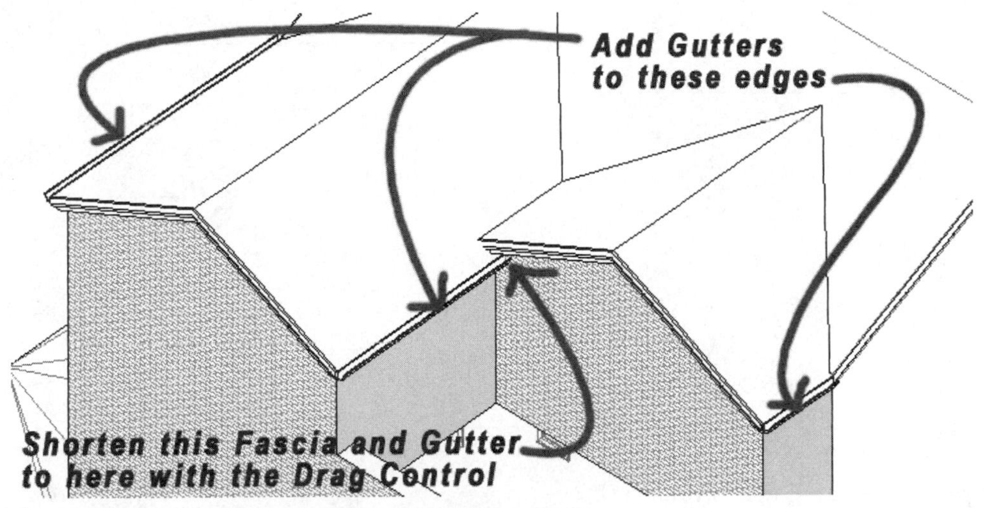
Add Gutters
to these edges
Shorten this Fascia and Gutter
to here with the Drag Control

Figure 7–32 *Add Gutters to the new construction Roof*

After you place the Gutters, pre-highlight them and notice the one in the valley between the two new Roofs is too long. You can use the Drag control on the Fascia board (and the Gutter if necessary) to shorten it.

29. On the Project Browser, double-click to open the *Longitudinal* section View.
30. Zoom in on one of the Gutters (turn on Thin Lines if necessary).

Notice that the Gutter overlaps the Fascia board. This is because they share the same reference point. We can apply an offset to the Gutter equal to the thickness of the Fascia profile to compensate for this.

31. Select all of the Gutters.
    - On the Options Bar, click the Properties icon.
    - In the "Element Properties" dialog, change the "Horizontal Profile Offset" to 3/4" [19] and then click OK (see Figure 7–33).

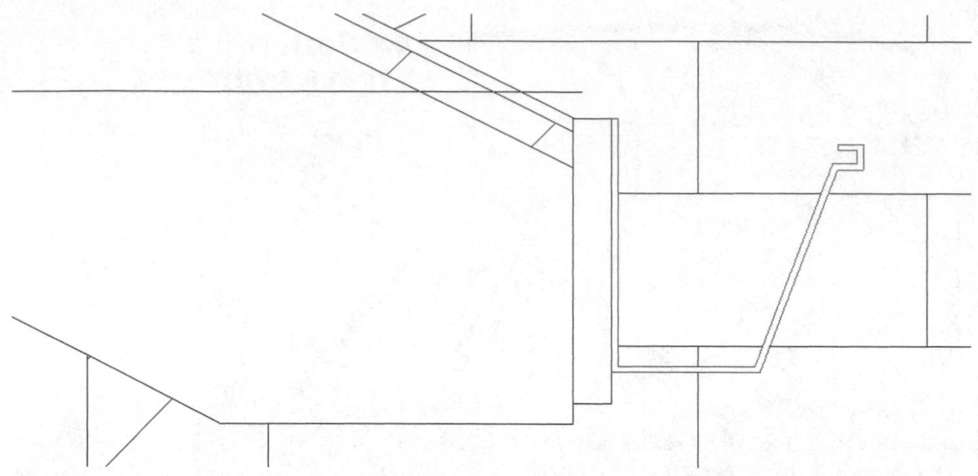

Figure 7–33 *Adjust Gutter Horizontal Offset*

SKYLIGHTS

As a finishing touch to the residential Roof, let's add a skylight in one of the new Roofs. To do this, we must load another Family into the project.

32. On the Project Browser, double-click to open the *Roof* plan View.
33. On the File menu, choose **Load from Library > Load Family**.
    - Double-click on the *Windows* folder.
    - Select the *Skylight.rfa* [*M_Skylight.rfa*] file and then click Open.
34. On the Design Bar, click the Basics tab and then click the Window tool.
    - From the Type Selector, choose Skylight : 28″ x 38″ [M_Skylight : 0711 x 0965mm]
35. Place it approximately as indicated in Figure 7–34.

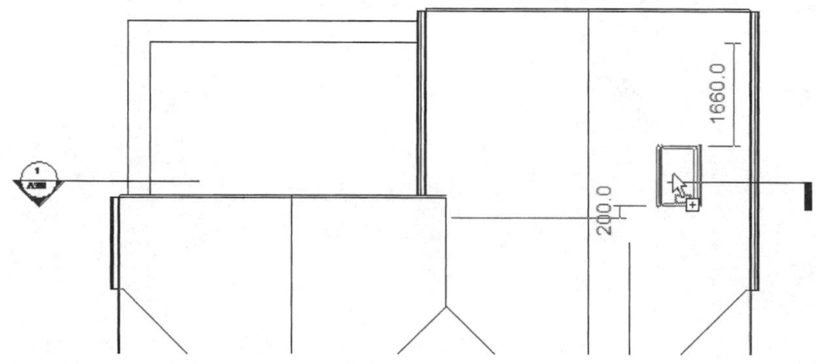

Figure 7–34 *Place a skylight in the Roof*

A vertical temporary dimension will remain after you place the skylight. Its witness lines should be between the outside horizontal Wall and the top edge of the skylight.

- Adjust the witness lines if necessary and edit the value of the temporary dimension to **5′-0″ [1500]**.

36. On the Project Browser, double-click to open the *Longitudinal* section View.

The skylight should appear cutting through the Roof.

37. Save the project.

CREATING FLOORS

Use Floor elements in Revit Building to model the actual floor surfaces and construction in your projects. Most often the Floor element is a simple flat structure that you create via a closed sketch. In some cases, the Floor may slope like in parking garages or theaters. In this topic, we will add some and modify some Floor elements.

ADD FLOORS

We have already worked with Floors a little in our commercial project; so you have probably noticed that our residential project currently has no Floors.

1. On the Project Browser, double-click to open the *First Floor* plan View.
2. On the Design Bar, click the Basics tab and then click the Floor tool.
   - On the Design Bar, click the Pick Walls tool.
   - On the Options Bar, place a checkmark in the "Extend into wall (to core)" checkbox.
3. Pre-highlight and then click to select the vertical Wall of the addition on the left.
   - Using the CTRL key, select the other two exterior Walls of the new addition.
   - Using the CTRL key again, select the existing Wall between the house and the addition.

Sketch lines will appear at all four Walls. The one for the existing Wall is currently on the inside edge, but we need it to be on the exterior side adjacent to the new construction.

- On the Design Bar, click the Modify tool or press the ESC key twice.

4. Drag the horizontal sketch line between the existing house and the addition to the other side of the existing Wall (see Figure 7–35).

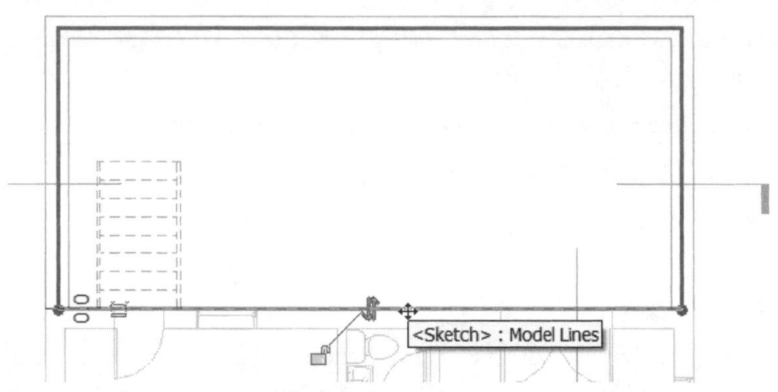

Figure 7–35 *Modify the sketch lines*

5. On the Design Bar, click the Floor Properties button.
   - From the Type list, choose Wood Joist 10" - Wood Finish [Standard Timber-Wood Finish] and then click OK.

The structure of the Floor is similar to the structure of the Roof that we built above. Feel free to edit the Type Parameters to see the structure but be sure to not save any changes at this time.

6. On the Design Bar, click the Finish Sketch button.
   - In the dialog that appears, click Yes to join the geometry (see Figure 7–36).

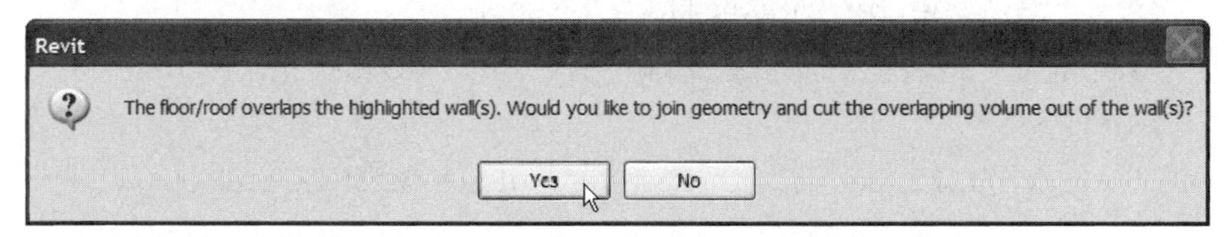

Figure 7–36 *Allow Revit Building to join the Floor to the neighboring Walls*

7. Repeat the entire process on the second floor.

Create the sketch the same way initially, however, modify the sketch to conform to the "L" shape (excluding the outdoor patio) of the interior space of the addition. You can use an additional "Pick Walls" action to do this (see Figure 7–37). Remember to edit the Floor Properties and choose the same Type as the first floor.

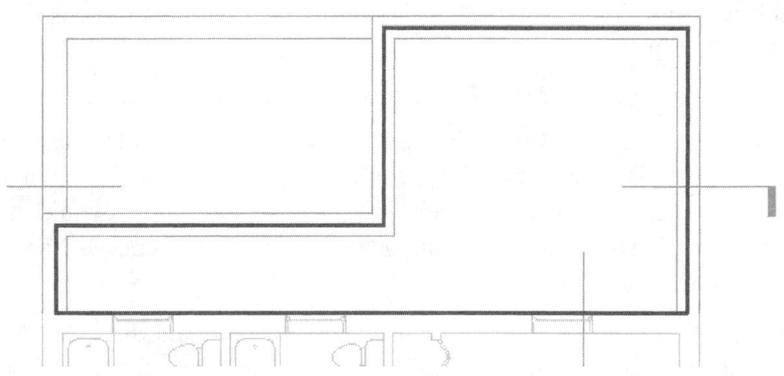

Figure 7–37 *Create the sketch for the second floor*

CREATE A FLOOR TYPE

The last Floor that we need to make is the one for the patio space on the second floor. The process is similar to the above.

8. On the Design Bar, click the Basics tab and then click the Floor tool.
   - On the Design Bar, click the Pick Walls tool.
   - On the Options Bar, place a checkmark in the "Extend into wall (to core)" checkbox.
9. Create sketch lines for each of the four Walls that make up the patio and use the Trim/Extend tool to complete the shape (see Figure 7–38).

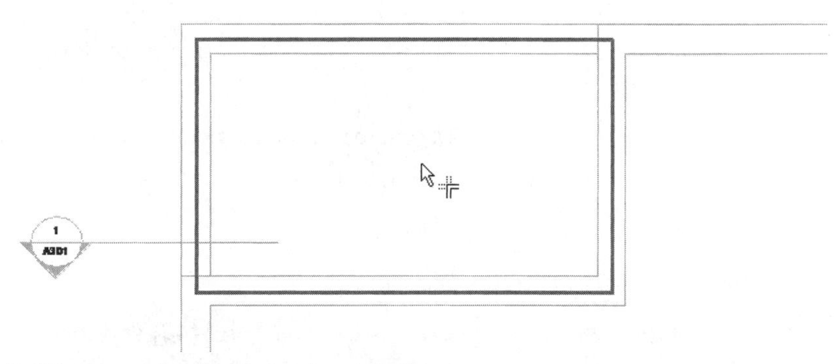

Figure 7–38 *Create the sketch for the patio floor*

10. On the Design Bar, click the Floor Properties button.
    - From the Type list, choose Wood Joist 10″ - Wood Finish [Standard Timber-Wood Finish].

- Next to the Type list, click the Edit/New button and then the Duplicate button.
- Change the name to **MRB Patio Floor** and then click OK.

The patio will have a wood deck built up on top of sleepers to provide drainage below. Therefore, we can use nearly the same component makeup as the other Floor system and simply insert the sleeper Layer. While this Floor Type would have a membrane Layer and flashing, we will not include those in the model but rather save those components for details.

11. Click the Edit button next to Structure, select Layer 1, and then click the Insert button to add a new Layer.
    - Move it to down to beneath the Finish Layer (but above the Core Boundary).
    - Set the new Layer Function to **Substrate [2]**.
    - Change the thickness of the new Layer to **1 1/2″ [40]** and change the Material to Wood - Stud Layer.
    - Click OK three times to return to the sketch.
    - On the Design Bar, click the Finish Sketch button.
    - In the dialog that appears, click Yes to accept joining with the Walls.

ADJUST FLOOR POSITION AND JOINS

A couple of issues remain with the two Floors on the second floor. First, the Walls that highlighted automatically for join did not include the two Walls we added at the start of the chapter. Second, the patio Floor is too low relative to the other one. We can see these issues best in section.

12. On the Project Browser, double-click to open the *Longitudinal* section View.

Examine the section to see both conditions noted here.

13. Select the patio Floor (it is on the right in the section).
    - On the Options Bar, click the Properties icon.
    - For the "Height Offset From Level" value, type **1 1/2″ [40]** and then click OK.

This is the same amount as the thickness of the sleepers.

14. On the Tools toolbar, click the Join Geometry icon.
    - For the first pick, click the patio Floor.

- For the second pick, click the vertical Wall (see Figure 7–39).

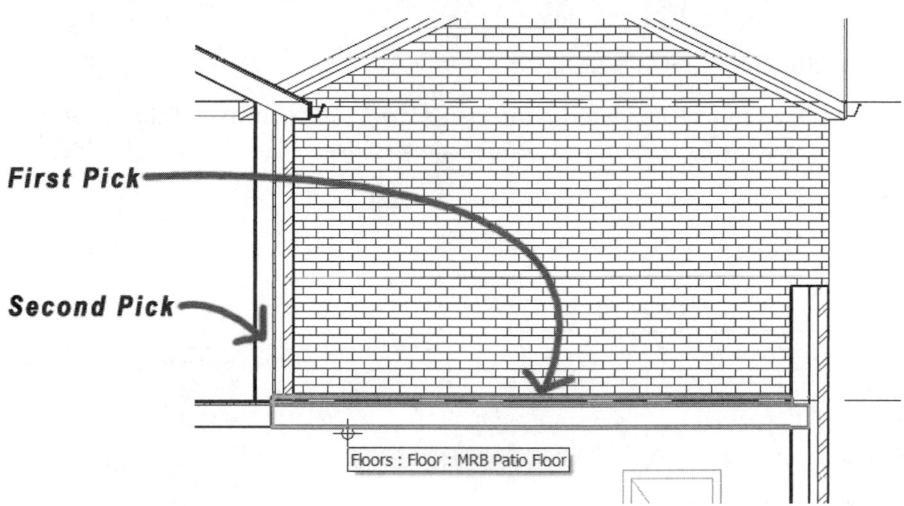

Figure 7–39 *Join the Floor to the Wall*

The Join command will remain active. You can perform several Join operations in a row.

15. Join the same Wall to the other Floor.
16. Join the two Floors to one another (see Figure 7–40).

TIP Note the "Multiple Join" option on the Options Bar. Place a checkmark in this box to Join several elements in one operation.

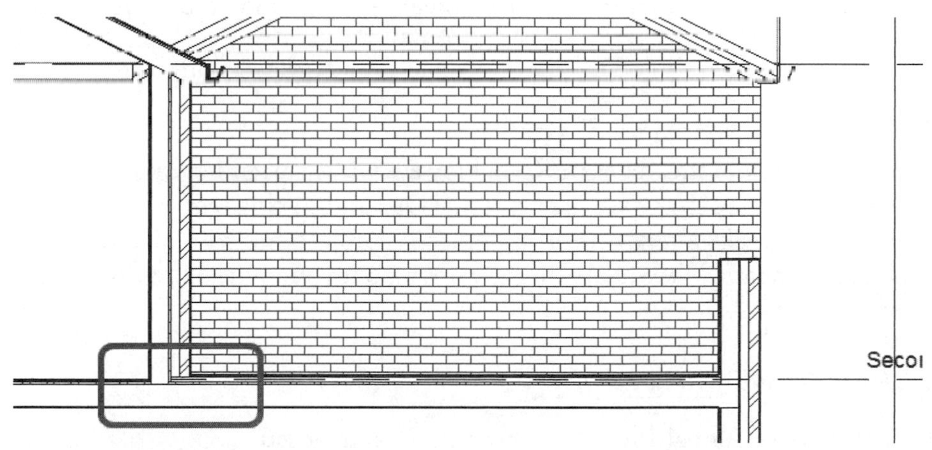

Figure 7–40 *Join the Floors and the Wall*

17. On the Project Browser, double-click to open the *Second Floor* plan View.

- Create a section cutting vertically (parallel to *Transverse*) through the patio (see Figure 7–41, left side).
- Rename it **Patio Section**.
- Open this section, change the Detail Level to **Medium** and then repeat the steps here to join the Floors and Wall (see Figure 7–41, right side).

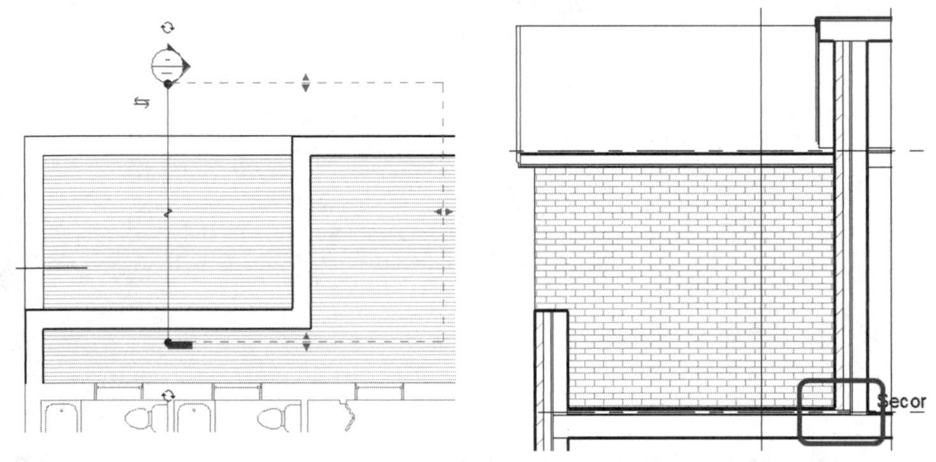

Figure 7–41 *Cut a section to assist in joining the remaining Floors and the Wall*

18. On the Design Bar, click the Modify tool or press the ESC key twice.
19. Save and close the project.

Above we mentioned the priorities of the Layers within the Roof (also Wall and Floor) structure. In these two sections, you can see this interaction very clearly. Notice the way that the structural Layer of the Floor cuts into the Core of the Walls.

COMMERCIAL PROJECT ROOF

Our commercial project already has a Roof. However, it is currently just a flat slab. Also, the Stair tower does not yet have a Roof. Our aim in this section will be to refine the Roof element already in the commercial project and to add additional required Roof elements. We will also begin work on our commercial project's Roof Plan View.

LOAD THE COMMERCIAL PROJECT

Be sure that the Residential Project has been saved and closed.

1. On the Standard toolbar, click the Open icon.

TIP The keyboard shortcut for Open is CTRL + O. **Open** is also located on the File menu.

- In the "Open" dialog box, click the *My Documents* icon on the left side.
- Double-click on the *MRB* folder, and then the *Chapter07* folder.

If you installed the dataset files to a different location than the one listed here, use the "Look in" drop down list to browse to that location instead.

2. Double-click *07 Commercial.rvt* if you wish to work in Imperial units. Double-click *07 Commercial Metric.rvt* if you wish to work in Metric units

You can also select it and then click the Open button.

The project will open in Revit Building with the last opened View visible on screen.

CREATE A ROOF BY EXTRUSION

Till now we have created our Roofs with the footprint option. It is also possible to create a Roof by sketching the profile of it in section and extruding this profile to form the Roof. In the Quick Start chapter we looked briefly at creating a Roof by Extrusion, let's look at that process in detail for the Roof at the top of our Stair tower.

3. On the Project Browser, double-click to open the *South* elevation View.

To assist us in placing the Roof, we will add some new Levels. You may recall from Chapter 4 that a Level can be created with automatically associated plan Views, or it can be created without them and simply used for reference. We do not need a separate plan for the Roof of the Stair Tower, so the Levels we add here will not have associated plan Views.

4. Select the Roof Level line.

- On the Tools toolbar, click the Copy tool.
- Click anywhere to set the start point.
- Move the mouse straight up, type **8′-0″** **[2400]** and then press ENTER.
- Repeat the process and create another copy **4′-0″** **[1200]** above the previous copy (or **12′-0″** **[3600]** above the Roof Level).

Notice that both copies have black-colored Level heads. A Level head will be blue if it has an associated floor plan View, and black if it does not. Take a look at the Floor Plans on the Project Browser to confirm that no new floor plan Views have been created.

5. Click on the blue text of the new Level heads and rename the lower one to: **Stair Roof Low** and the upper one to: **Stair Roof High** (see Figure 7–42).

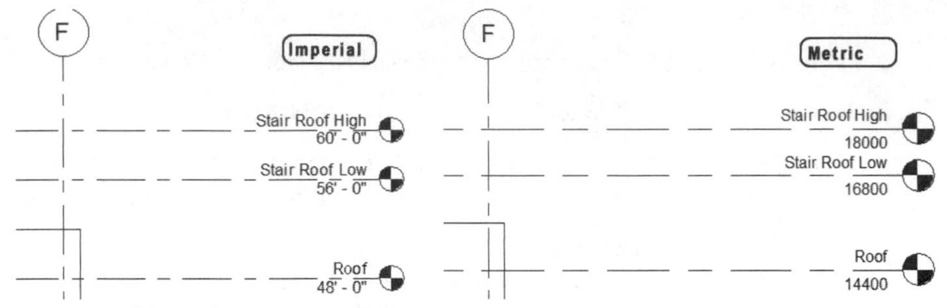

Figure 7–42 *Copy two new Levels without Associated Floor Plan Views*

6. On the Design Bar, click the Basics tab and then click the Roof tool and then from the flyout, choose Roof by Extrusion.

A "Work Plane" dialog will appear. The Work Plane is the plane in which we will sketch. In this case, because we are making a Roof by extrusion, an effective Work Plane will be perpendicular to the Roof Levels. Any of our numbered Grid Lines can serve this purpose—the numbered Grid Lines form planes parallel to the screen in the current elevation View. Once we have chosen a plane, we will be able to sketch the shape of our Roof. When we finish the sketch, it will extrude perpendicular to the selected plane.

- In the Roof "Work Plane" dialog, choose Grid 4 (from the "Name" list) and then click OK (see Figure 7–43).

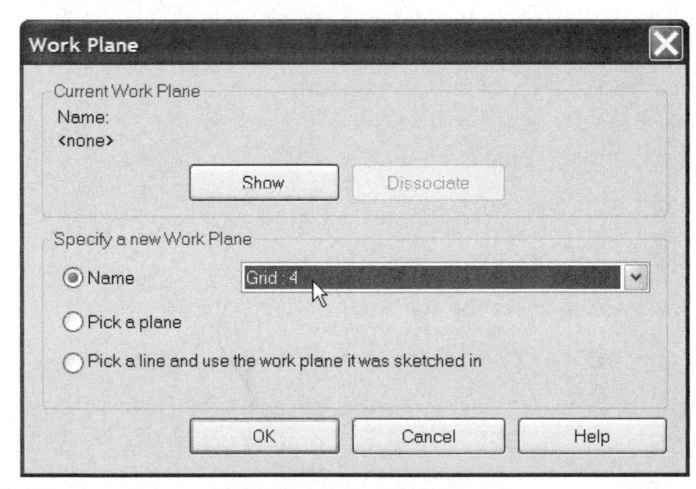

Figure 7–43 *Choose a Column Grid Line as the Work Plane*

As you can see in the figure, it is also possible to pick the face of some geometry such as a Wall in the model to set the Work Plane. In this case the named Plane associated with the Grid line works best.

- In the "Roof Reference Level and Offset" dialog that appears, select **Stair Roof Low** and then click OK.

The Roof must be associated with a Level to determine its height. As you can see in the dialog, you can choose to create the Roof at any Level and also apply an offset above or below the Level if necessary. In this case, we specifically created Stair Roof Low for this purpose, so no offset is required.

The Design Bar changes to sketch mode and shows only the Roof tools.

7. On the Options Bar, click the "Arc passing through three points" icon.
   - For the Arc start point, click the intersection of the Stair Roof Low Level and the left edge of the core Wall (see the top panel of Figure 7–44).
   - For the Arc end point, click the intersection of the Stair Roof High Level and the right edge of the core Wall (see the middle panel of Figure 7–44).
   - For the Arc intermediate point, click somewhere near the middle of the top edge of the core Wall (see the bottom panel of Figure 7–44).

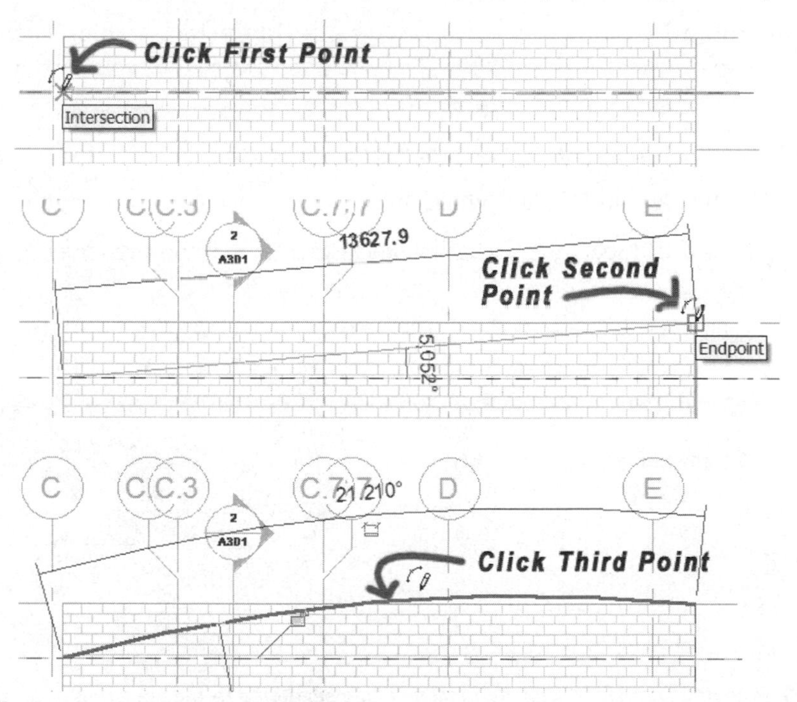

Figure 7–44 *Sketch an Arc Profile for the Extruded Roof*

You can click on the sketch line and use its drag controls to fine tune the shape to your liking. When you draw a Roof by extrusion, you create an open shape, not a closed shape. The thickness of the Roof material will be determined by the Roof Properties and the Type assigned to it—just like the other Roofs.

- On the Design Bar, click the Modify tool or press the ESC key twice.

8. On the Design Bar, click the Properties button.

- In the "Element Properties" dialog, from the Type list, choose Steel Truss - Insulation on Metal Deck – EPDM [Steel Bar Joist - Steel Deck - EPDM Membrane].

In the "Instance Parameters" area, notice that the Work Plane parameter is set to Grid 4 and that this is unavailable for edit. To change the Work Plane after creation, Select the Roof; then click the "Edit Work Plane" button on the Options Bar. This opens the Work Plane dialog that allows you to pick a new work plane. The geometry then moves appropriately.

- For the "Extrusion Start" type **-2′-0″ [-600]**.
- For the "Extrusion End" type **17′-0″ [5100]**.
- Click OK to return to the sketch.

You probably noticed that there were no overhang parameters in the "Element Properties" dialog or on the Options Bar. To create an overhang, you simply edit the sketch line, or add additional segments.

9. On the Design Bar, verify that **Lines** is still selected.

- On the Options Bar, click the Line icon.
- Add a 2′-0″ [600] long horizontal line at each end of the Arc (see Figure 7–45).

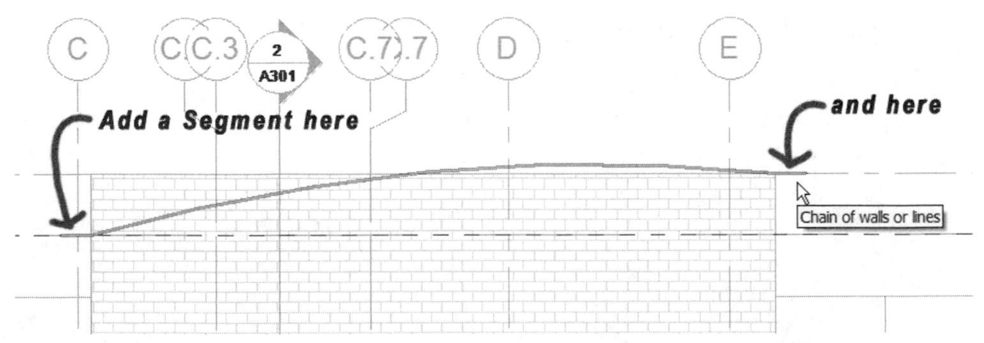

Figure 7–45 *Sketch overhangs by adding additional Line segments to the sketch*

10. On the Design Bar, click the Finish Sketch button.

The Roof should appear with its thickness determined by the Type that we chose.

11. On the Project Browser, double-click to open the *{3D}* 3D View (see Figure 7–46).

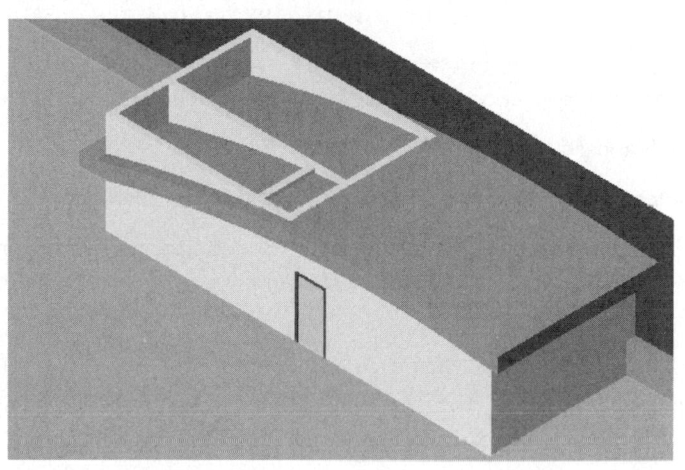

Figure 7–46 *The new Roof intersects the core Walls*

It appears as though our Walls could use some adjustment. The only problem is that the Walls reside in the linked structural RVT file. Therefore, we will need to come up with a way to temporarily copy the Roof over to that file. The simplest way would be to copy and paste. However, there are a few complications in the current situation. First, when you copy and paste, Revit Building will require the pasted element to come in relative to a Level. This would normally not be an issue except that in this case our Roof element references a Level that we just added and therefore does not exist in the structural file. The other issue is that Revit Building will not allow us to open a linked file at the same time as the host within the same Revit Building session. This will make copy and paste difficult. Our options include copying the Roof to a new intermediate file, and then copy and pasting from there, or to create a Group from the Roof and save it out to a Group file. In either case, we are basically copying the Roof off to a temporary file where we can then paste or import it into the structural model.

12. Select the new Roof.
    - On the Edit toolbar, click the Group icon.

If you wish, you can expand the Model Groups category on the Project Browser and rename the Group. However, since it will only be temporary for our purposes, we can safely skip that step.

- From the File menu, choose **Save to Library > Save Group**.

- In the "Save Group" dialog, choose Group 1 from the "Group To Save" list and then click Save.

The "Save Group" dialog should default to the same folder in which your project is located. This is a good place to save the Group, however, feel free to save it in any location you wish.

13. From the file menu, choose **Close**. When prompted to save, choose **Yes**.
14. On the Standard toolbar, click the Open icon.

    The "Open" dialog box should already be pointing to the *Chapter07* folder.

15. Double-click *07 Commercial-Structure.rvt* if you wish to work in Imperial units. Double-click *07 Commercial-Structure Metric.rvt* if you wish to work in Metric units

    You can also select it and then click the Open button.

The project will open in Revit Building with the last opened View visible on screen.

USING THE COPY/MONITOR TOOL

We have saved the extruded Roof to a Group file and it can now be added to the structural model, however, the Roof references the new Levels that we created in the architectural model. Those Levels do not yet exist in the structural file. Let's use this opportunity to be introduced to the "Copy/Monitor" tool in Autodesk Revit Building. With this tool, you can copy Levels, Grids and Columns between projects. Once copied, they will continue to be monitored. In this way, if the host element should change, you can use the Copy/Monitor tool to update the copy.

16. On the Project Browser, double-click to open the *South* elevation View.

While the structural project does host a link to the Site project, it does not currently host the architectural model. Let's first establish a link to the architectural file, and then use the Copy/Monitor to copy the new Levels over.

17. From the file menu, **Import/Link > RVT**.

    - In the "Add Link" dialog box, select *07 Commercial.rvt* [*07 Commercial Metric.rvt*].

    Do not double-click it.

    - In the "Positioning" area, choose the "Origin to origin" option and then click Open.
    - In the warning dialog that appears, click OK.

The architectural model will appear on screen.

18. On the Tools toolbar, click the Copy/Monitor icon.

    - From the pop-up that appears, choose **Select Link**.

TIP Copy/Monitor > Select Link is also located on the Tools menu.

19. Click on the architectural model to select it. (The easiest way to select it is to click on one of the Level lines added above).

    The Design Bar changes to Copy/Monitor mode and shows only the Copy/ Monitor tools.

    - On the Design Bar, click the Copy tool.
    - Click on each of the two new Level lines (see Figure 7–47).

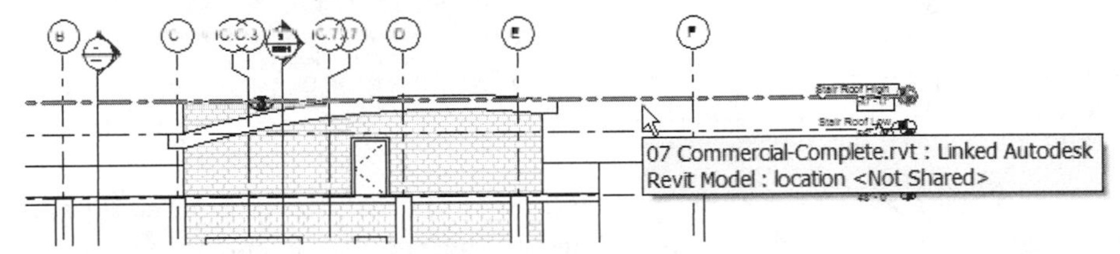

Figure 7–47 *Select the Levels using the Copy tool*

- On the Design Bar, click the Modify tool or press the ESC key twice.

Notice the small eyeball icon that appears next to the elements after you click them. This icon indicates that the copied element is being monitored.

20. On the Design Bar, click the Finish Mode button.

With monitoring enabled for these elements, if the host elements change in the linked file, these changes will be flagged in a Coordination Review. (see Figure 7–48).

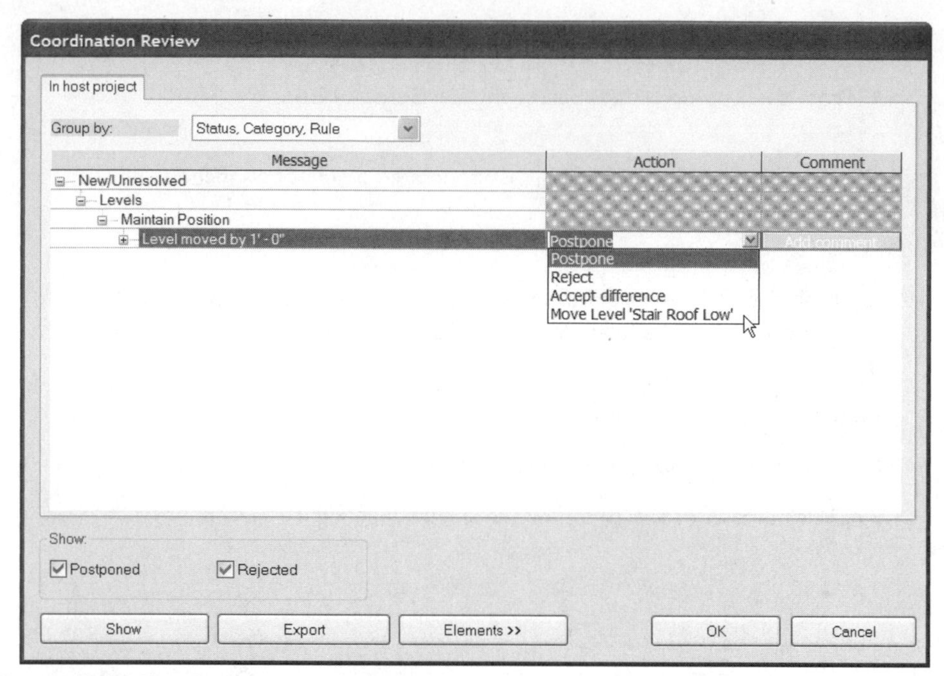

Figure 7–48 *The Coordination Review dialog*

To see this you will need to close the structural project, open the architectural one and make a change to one of the monitor Levels. Then you close and save the architectural model and return to the structural. There you enable the Copy/Monitor tool and click the Coordination button. In the "Coordination Review" dialog that appears, you can see the items that have changed. You then can choose to postpone the change, reject the change, accept the difference—which basically means to allow the two projects to differ. The final option is to accept the change (in Figure 7–48 this is listed as Move Level "Stair Roof Low'). If you wish to experiment with this tool, feel free. You will either need to close and open each project in between changes (since the projects are linked to each other) or open two concurrent Revit Building sessions.

IMPORT THE ROOF GROUP

Now that we have the new Levels added to our structural model, we can import the Group that we created above. Remember, we used a Group in the previous exercise simply as a convenient way to export and import the geometry.

21. From the File menu, choose **Load from Library > Load Group**.

    The "Load Group" dialog should open to the same folder as your current project. If it does not, browse there now.

- In the "Load Group" dialog, choose *Group 1* and then click Open.

22. On the Project Browser, expand the *Groups* category and then the *Model* category.

- Right click *Group 1* and choose **Create Instance** (see Figure 7–49).

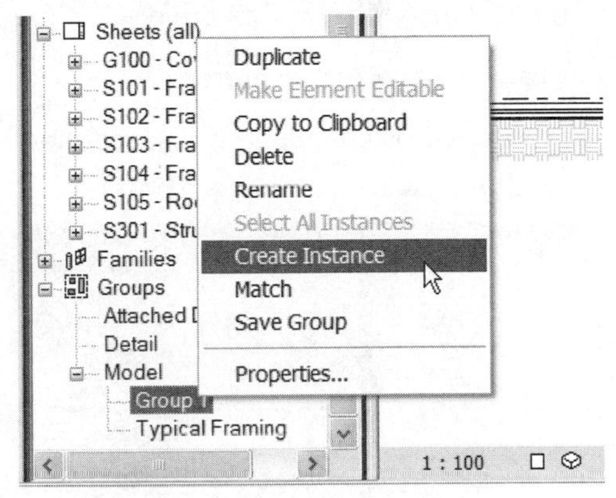

Figure 7–49 *Create an Instance of the Imported Group*

The Group will appear in its original location and a move operation will automatically be initiated. We want the Group to appear in the same location that it was in before, therefore, we need to "move" it by a zero offset.

23. Type **0** and then press ENTER.

The Group should appear in the original location and remain selected. You are still in the Group placement mode. Notice the buttons on the Options Bar.

- On the Options Bar, click the Finish button.

The Group will deselect and Group placement will be complete.

24. Select the Group instance in the model again.

- On the Options Bar, click the Ungroup button.

If you move the pointer over the imported Roof, you will notice that now it pre-highlights just the Roof, rather than the rectangular bounding box of the Group that contained it.

25. On the Project Browser, double-click to open the *Roof* plan View.

- Dragging from left to right, surround the entire core.

Both Walls and Doors will be selected by this action.

- On the Options Bar, click the Filter Selection icon.
- In the Filter dialog, clear the "Doors" checkbox and then click OK.

This will remove Doors from the selection leaving only the core Walls on the Roof level selected.

26. On the Project Browser, double-click to open the *South* elevation View.

If you prefer, you can tile the two windows with the command on the Window menu instead. The selection of Walls will remain active.

- On the Options Bar, click the Attach button.
- Verify that Top is selected on the Options Bar and then click the Roof (see Figure 7–50).

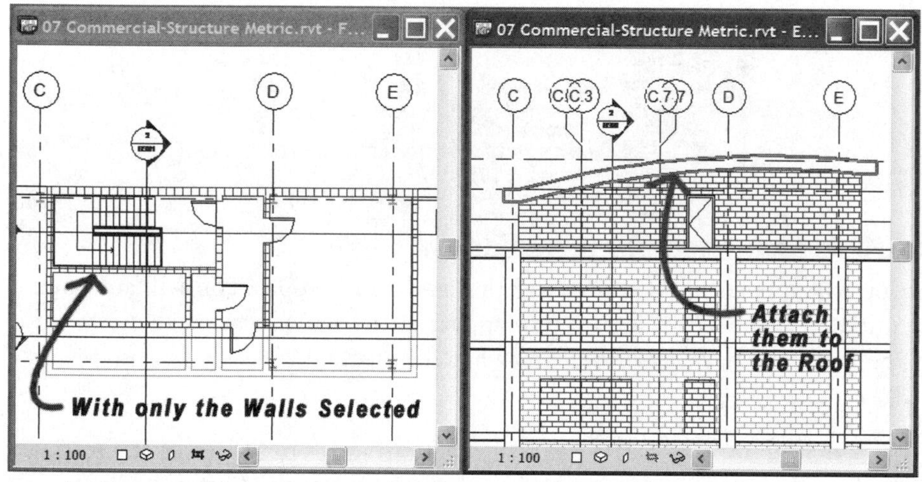

Figure 7–50 *Attach the core Walls to the imported Roof*

NOTE Using this technique to cut the top edge of the Walls requires the Roof to remain in the structural model. If you were to delete the Roof, the Walls would revert to their original shape. Since the Copy/Monitor only works on Levels, Grids and Columns, we will have to coordinate this redundancy in our geometry manually. (Should the architectural model change, we would need to repeat the process above to re-create the exported Group and then re-import it back in here). It is manageable however since it is a single element. If you don't want the Roof to show in the structural model, you can edit the View Visibility in each of the Views in which you want to disappear and turn off the Roofs category. An alternative to this approach would be to stop the shear Walls in the structural model at the top of the flat roof and then add the "penthouse" Walls in the architectural model as separate elements. If we took that approach, we would simple need to keep the position of the architectural penthouse Walls coordinated with the structural shear Walls should they move as the building design evolves.

27. From the File menu, choose **Manage Links**.

- On the RVT tab, select *07 Commercial.rvt* [*07 Commercial Metric.rvt*] and then click the Unload button.
- Click OK to exit the dialog.
- Edit the Visibility/Graphics in any Views where you do not wish to see the Roof (like elevations, sections and 3D Views) and then clear the "Roofs" checkmark.

TIP The keyboard shortcut for Visibility/Graphics is VG. **Visibility/Graphics** is also located on the View menu.

28. From the file menu, choose **Close**. When prompted to save, choose **Yes**.
29. On the Standard toolbar, click the Open icon.

    The "Open" dialog box should already be pointing to the *Chapter07* folder.

30. Double-click *07 Commercial.rvt* if you wish to work in Imperial units. Double-click *07 Commercial Metric.rvt* if you wish to work in Metric units

    You can also select it and then click the Open button.

The project will open in Revit Building with the last opened View visible on screen. This should be the *South* elevation View since we last worked in this above. If this View does not open automatically, open it now. You should now see that the roofline of the Walls (from the linked structural file) is properly cut to follow the Roof element. Feel free to study this in other Views like the *{3D}* View as well.

WORKING WITH SLOPE ARROWS

For the final exercise in this chapter, we will explore the creation of a Roof using Slope Arrows. Slope Arrows provide an alternative way to designate the slope of a Roof. With a Slope Arrow, you specify the elevation height at the high point and the low point rather than indicating the rise over run or slope angle.

CUT A ROOF INTO TWO PIECES

The first thing that we want to do is basically cut the Roof as it currently exists in the model into two pieces. In other words, we want to create a ridge between column lines C and D. To do this, we need to edit the Roof sketch and eliminate half the Roof. However, before we do that, let's make a copy of the Roof in place so that we can use the existing sketch to form each of the two new Roofs.

1. On the Project Browser, double-click to open the *Roof* plan View.
2. Select the Roof element and press CTRL + C.

- Press CTRL + V to paste the Roof.

The Roof will appear in its original location and a move operation will automatically be initiated. We want the Roof to paste in the same location that it was in before, therefore, we need to "move" it by a zero offset (like we did with the Group above).

- Type **0** and then press ENTER.
- On the Options Bar, click the Finish button.

3. Select one of the Roofs (there are now two in the same spot).
   - On the Options Bar, click the Edit button.
   - Edit the sketch as shown in Figure 7–51.

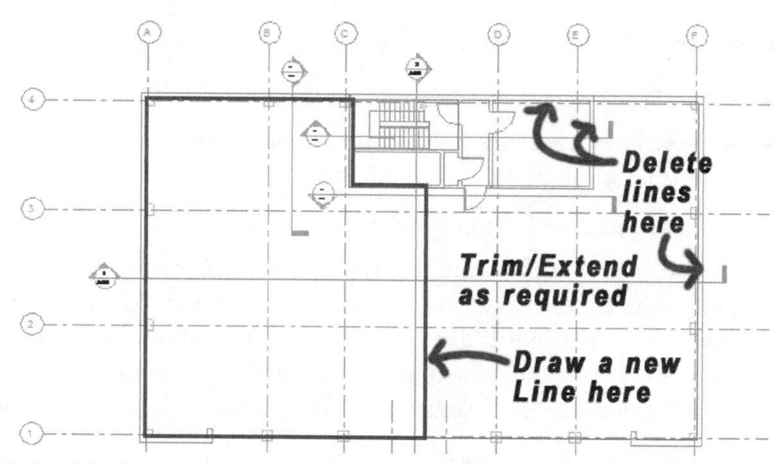

Figure 7–51 *Edit the sketch to remove half of the Roof (sketch lines enhanced for clarity)*

NOTE To draw the new sketch line indicated in the middle of the figure, click the "Lines" tool on the Design Bar and draw it point-to-point.

4. On the Design Bar, click the Finish Roof button.
5. Select the other Roof and then click the Edit button on the Options Bar.
   - Edit the sketch to match to opposite shape (keep the portion removed on the other Roof, remove the portion kept on the other) (see Figure 7–52).

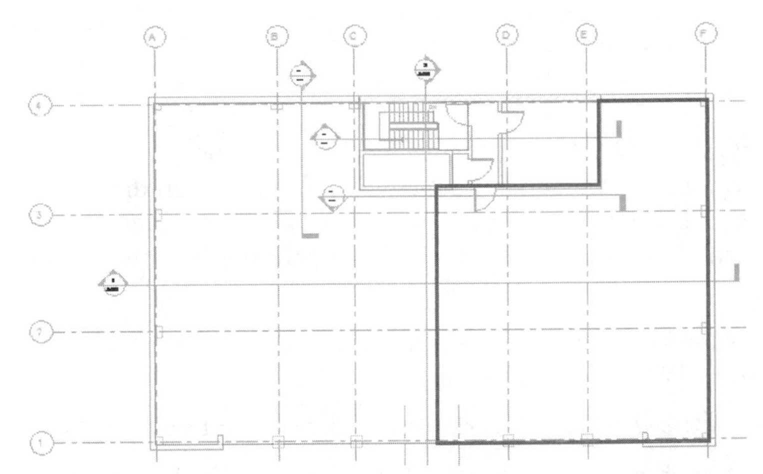

Figure 7–52 *Edit the sketch to remove the other half of the Roof (sketch lines enhanced for clarity)*

ADD SLOPE ARROWS

Now that we have our two Roofs, the high points will be at the edges of the two Roofs which will mean that the Roofs will slope down to low points in the middle of the plan. Drain pipes will occur at the columns along column lines B and E. The point where the two Roofs meet will be a high point. Since this is a commercial Roof, it is mostly flat. The slope must be at least 1/4″ per foot. A convincing argument could be made that it is unnecessary to model such a shallow slope. This relates to the various discussions made throughout this book regarding Building Information Modeling and when to rely on the "information" part over literal modeling. It is the opinion of this text that sketching in the slope of this flat commercial Roof as Detail lines directly on top of the *Roof* View would be sufficient for most applications. The true slope of the Roof could be handled with notes and details elsewhere in the project. However, an equally compelling argument could be made that on larger projects with very expansive Roofs, that the modeling of even small slopes can help discover coordination errors with structure and systems long before the project enters construction. The Roof is being modeled with its slope here as an exercise to showcase the use of Slope Arrows. This is by no means the only application of Slope Arrows. It simply fits conveniently with the rest of the dataset in this context. In the additional exercises (Appendix A) is another example of Slope Arrows in the residential project and the Autodesk Revit Building tutorials (available from the Help menu) gives another useful example. Ultimately the choice of whether to model very shallow slopes accurately or not in actual projects will be left to the reader. More information on these types of BIM issues can be found in Chapter 1.

6. Select the Roof on the left.

   - On the Options Bar, click the Edit button.
   - On the Design Bar, click the Slope Arrow tool.

When you draw a Slope Arrow, you first click its Tail and then drag to its Head. In this case, we will sketch four Slope Arrows that have their Tails snapped to the outer edges of the Roof sketch and their Heads pointing into the center of the Roof footprint.

- For the start point, click the midpoint of the vertical Roof sketch line in the middle of the plan.
- For the end point, drag horizontally and then snap to column line B.
- Create three more as shown in Figure 7–53.

Be sure to snap each start point at the outer edge of the Roof sketch and snap each end point to an appropriate Grid line as shown.

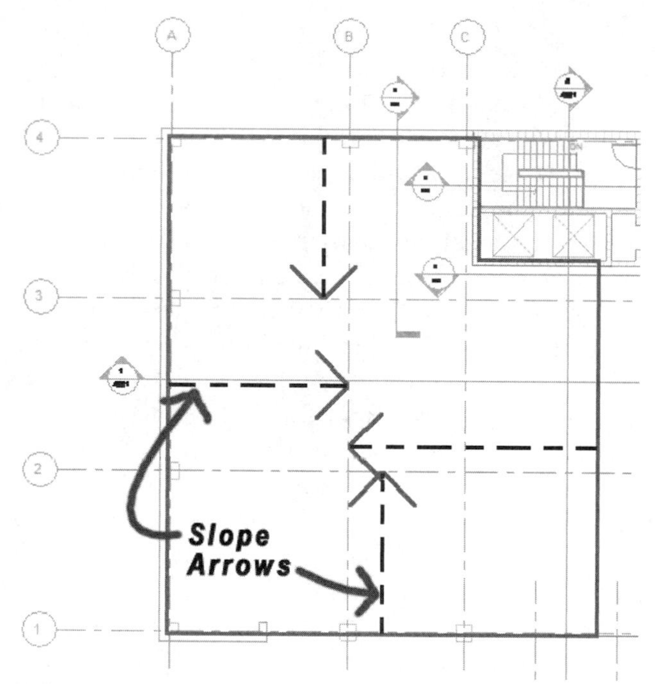

Figure 7–53 *Add Slope Lines to the Roof sketch (sketch lines enhanced for clarity)*

7. Select all four Slope Arrows.

   - On the Options Bar, click the Properties icon.

In the "Constraints" area, the "Specify" parameter allows you to set the slope of the Slope Arrow by specifying the height of the Head and Tail or by indicating the actual Slope. If you choose "Height Offset at Tail" you indicate the offset of the Head and the Tail relative to a base Level. If you choose "Slope" you indicate slope in the traditional fashion as we did for Slope Defining Roof Edges above. We will leave the default setting of "Height Offset at Tail" for this exercise. This will allow us to set the high and low points of the Roof rather than the slope. In this way, we determine exactly where the ridges and valleys occur as we have sketched them in plan. If you choose Slope, the valleys might shift.

- Beneath "Constraints" change the "Height Offset at Tail" to **8″** **[200]**.
- Change the "Height Offset at Head" to **0** (see Figure 7–54).

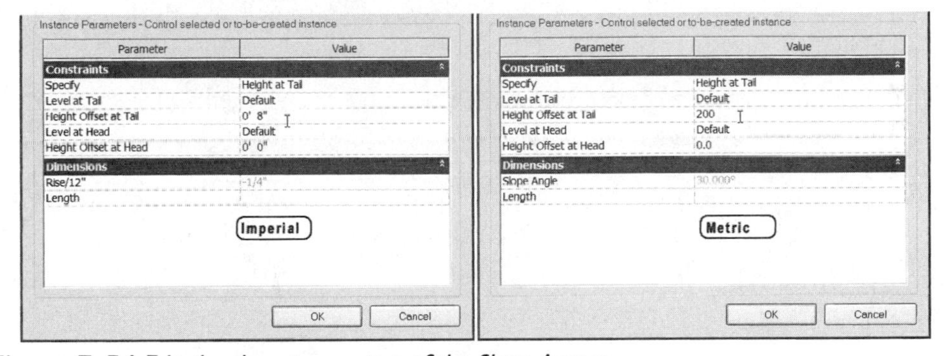

Figure 7–54 *Edit the slope parameters of the Slope Arrows*

- Click OK to accept the parameter changes and dismiss the dialog.

8. On the Design Bar, click the Finish Roof button.

You may recall in the previous chapter that we adjusted the offset from the Level of the Roof to compensate for its position in section. Now that we have modeled a true slope to the Roof, this offset is no longer necessary. The best way to see this is in the *{3D}* View with "Shading with Edges" display mode enabled.

9. Select both Roofs and on the Options Bar, click the Properties icon.

   - Change the "Base Offset From Level" to **0** and then click OK.

10. Repeat the process of adding Slope Arrows to the other Roof.

    For the vertical Slope Arrow at the top, you may need to go lower than Grid line 3. Try making it 24′-0″ [8000] long using the temporary dimension.

    - Use the same values for the Head and Tail offsets.
    - Study the model in section and 3D when finished (see Figure 7–55).

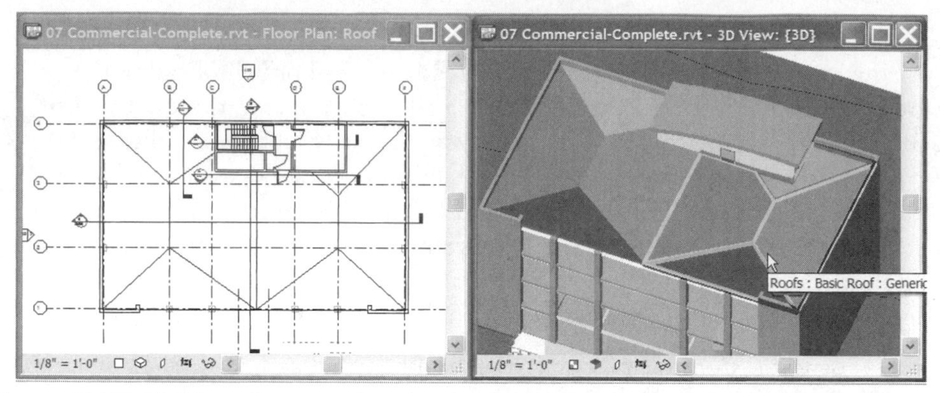

Figure 7–55 *Add Slope Arrows to the other Roof and study the model in 3D*

Feel free to study the model in other Views and experiment further with the Roofs and Floor slabs.

11. Save and Close the commercial project file.

SUMMARY

Roofs are sketch-based elements that can be generated from existing Walls or manually drawn sketch lines.

Walls can be attached to Roofs and remain attached as the model is modified.

Join Roofs together to resolve the intersection of complex Roof planes.

Roofs have many options for their construction and how they interact with neighboring elements.

You can apply edge conditions to each Roof Fascia and Rake.

Gutters and other sweep profiles and Host Sweeps can be applied to Roof edges.

Skylights interact with and cut holes in Roofs in the same way Windows interact with Walls.

Create complex Roof Types that include layers of structure that share many features with complex Wall Types.

Floors are added and modified via sketch mode similar to Roofs.

Floors may also have complex structure like Roofs and Walls.

Edit a Roof shape at any time by editing the sketch of the Roof.

Use Slope Lines to create complex Roof sloping patterns.

CHAPTER 8

Developing the Exterior Skin

INTRODUCTION

In this chapter, we will enclose our commercial project with a building skin. The skin will be comprised of a masonry enclosure on three sides, with various curtain wall elements on the front and sides of the building. The front façade curtain wall begins on the second floor and spans the height of the third and fourth floors. In Chapter 4, we created a massing element to suggest this design element. We will now replace this Mass with a Curtain Wall.

OBJECTIVES

In order to complete the shell of the commercial project we will apply a more detailed Wall Type to the skin Walls already in the project. We will also build Curtain Walls and Curtain Systems for the front and side façades. Upon completion of this chapter you will be able to:

- Swap Wall Types
- Create a vertically compound Wall Type
- Add Curtain Walls to the model
- Build a Curtain Wall Type
- Modify a Curtain Wall

CREATING THE MASONRY SHELL

With the exception of the large portion of the front façade that was cut away in previous chapters, the majority of the skin of the commercial building is comprised of masonry Walls. We will perforate portions of this masonry skin with Curtain Systems later in the chapter, but we will begin by swapping out the simple generic Wall Types used in early chapters with a more detailed Wall Type appropriate to the design at this stage.

INSTALL THE CD FILES AND OPEN A PROJECT

The lessons that follow require the dataset included on the Mastering Revit Building CD ROM. If you have already installed all of the files from the CD, simply skip down to step 3 below to open the project. If you need to install the CD files, start at step 1.

1. If you have not already done so, install the dataset files located on the Mastering Revit Building CD ROM.

   Refer to "Files Included on the CD ROM" in the Preface for instructions on installing the dataset files included on the CD.

2. Launch Autodesk Revit Building from the icon on your desktop or from the **Autodesk** group in **All Programs** on the Windows Start menu.

   - From the File menu, choose **Close**.

This closes the empty new project that Revit Building creates automatically upon launch.

3. On the Standard toolbar, click the Open icon.

TIP The keyboard shortcut for Open is CTRL + O. **Open** is also located on the File menu.

   - In the Open dialog box, click the *My Documents* icon on the left side.
   - Double-click on the *MRB* folder, and then the *Chapter08* folder.

   If you installed the dataset files to a different location than the one listed here, use the "Look in" drop down list to browse to that location instead.

4. Double-click *08 Commercial.rvt* if you wish to work in Imperial units. Double-click *08 Commercial Metric.rvt* if you wish to work in Metric units

   You can also select it and then click the Open button.

The project will open in Revit Building with the last opened View visible on screen.

CREATING A WALL TYPE

As you can see, the project is largely unchanged from the previous chapter. We still have the very simple generic Wall Type used for the building skin. We'll start here.

5. On the Project Browser, double-click to open the *Level 1* floor plan View.

6. Pre-highlight one of the Walls (try the vertical one on the left).
   - Press the TAB key once to pre-highlight a chain of Walls.
   - Click the mouse to select the chain of Walls.
   - Move the pointer to the other side of the plan and pre-highlight the vertical Wall on the right.
   - Press the TAB key once to pre-highlight a chain of Walls. *Do not click yet!*
   - Hold down the CTRL key and then click to add the second chain of Walls to the selection (see Figure 8–1).

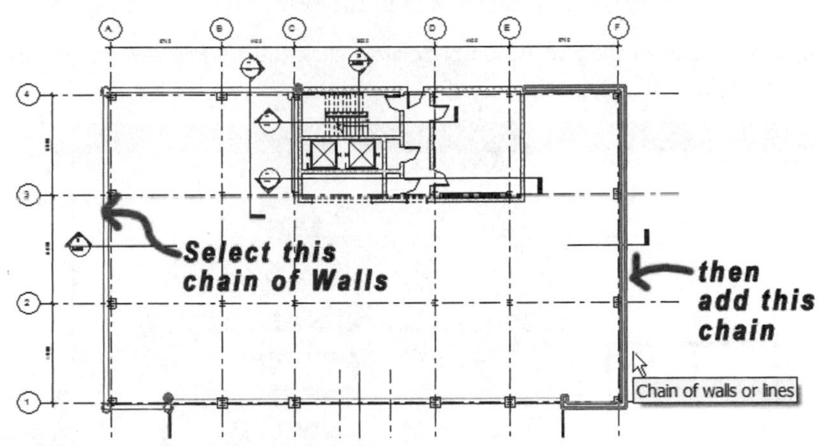

Figure 8–1 *Select a chain of Walls and then use the* **CTRL key to add a second chain to the selection**

All of the exterior Walls should now be selected.

7. On the Options Bar, click the Properties icon.
   - In the Element Properties dialog, click the Edit/New button.
   - Click the Duplicate button.

TIP If you prefer, you can hold down the ALT key and the press E followed by D.

   - For the name type **Exterior Shell** and then press OK.

EDIT WALL TYPE STRUCTURE

8. In the Type Parameters dialog next to "Structure" click the Edit button.

If you completed the previous chapter and worked through the roof tutorials, you saw a very similar dialog there. The structure of Wall Types and Roof Types have much in common. Wall Types however have more parameters available.

9. Highlight Layer 2 (Structure [1]).
   - In the Thickness column, type **7 5/8″ [190]**.
   - In the Material column, click the small expand (down pointing arrow) icon to choose a Material.
   - In the Materials dialog, choose Masonry - Concrete Masonry Units [Masonry - Concrete Blocks] and then click OK.
10. At the bottom of the Edit Assembly dialog, click the << Preview button (see Figure 8–2).

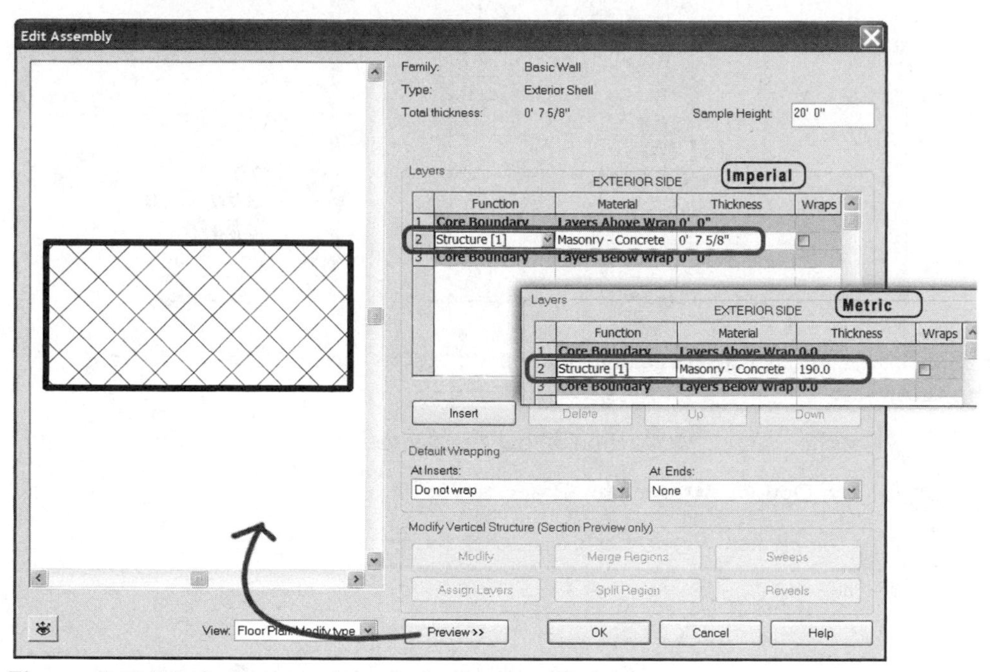

Figure 8–2 *Edit the Structure Layer to be Concrete Block*

11. Beneath the Layers list, click the Insert button.
    - Move the new Layer up above the Core Boundary.
    - From the Function list, choose **Thermal/Air Layer [3]**.

Functions determine the way that Walls join with one another. A pre-defined list of Functions is included with Revit Building. For more information, refer to the online Help.

- Change the Material to Air Barrier - Air Infiltration Barrier.
- Change the Thickness to **2″ [50]**.

12. Insert one more Layer above the air gap Layer.
    - Set its Function to **Finish 1 [4]**.
    - Set its Material to Masonry – Brick.
    - And set its Thickness to **3 5/8″ [90]** (see Figure 8–3).

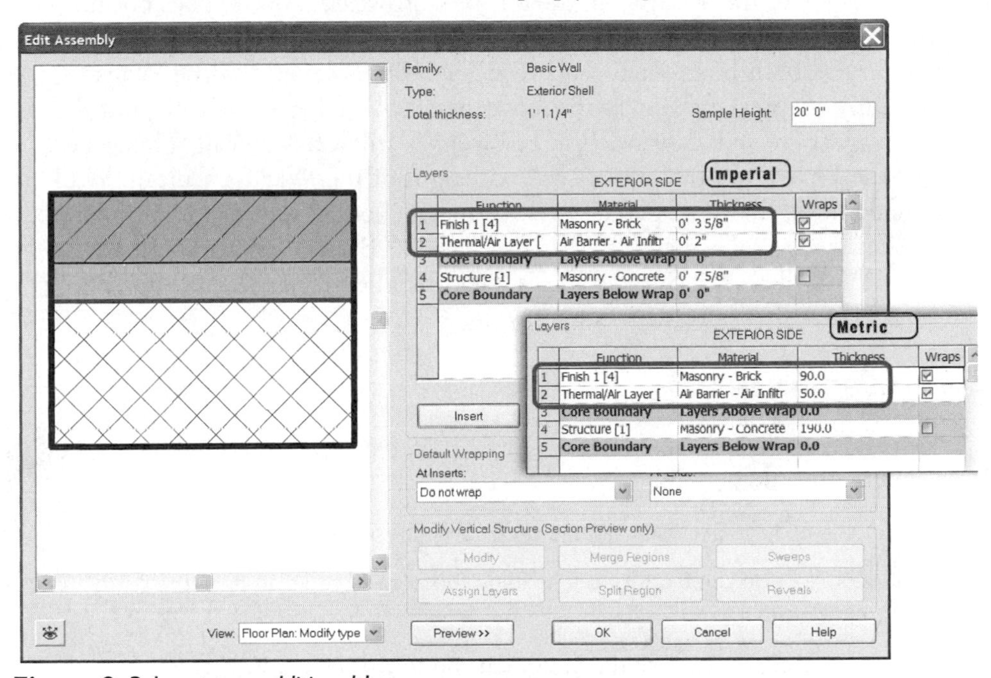

Figure 8–3 *Insert two additional Layers*

13. Click OK three times to return to the workspace window
    - Study the results in the floor plan View.
    - Open the *{3D}* View and study the results there as well.

WORKING WITH CURTAIN WALLS

Curtain Walls in Revit Building, like their real-life counterparts, are panelized wall systems. They are drawn the same way as Walls and using the same tool. To create a Curtain Wall, you simply choose the appropriate Type from the Type Selector as you draw the Wall. A Curtain Wall Type sets up a panel modulation along the length and/or height of the Wall. Each panel can be assigned specific Family elements such as glass or stone panels or even other Wall Types. The edges between the panels are mullions. You can decide which edges should receive mullion elements and which type of mullion each should have. You can even control the way that the mullions intersect with one another.

DRAWING CURTAIN WALLS

The simplest ways to create a Curtain Wall is to draw it using the Wall tool. The Wall tool allows you to create three types of Wall: Basic Wall, Curtain Wall and Stacked Wall. All of the Walls that we have created so far are Basic Walls. Basic Walls can have one or more Layers in their Type's Structure. "Basic" does not necessarily mean simple or generic. A Basic Wall is defined by having a single continuous set of Layers both horizontally and vertically. The "Generic" Type Walls we used in the early chapters are among the most "basic" Wall Types available, but the Wall Type we just created above with three Layers is still a Basic Wall. This is because all of the Layers run the full length and height of the Wall. A Curtain Wall as mentioned above, defines a panel system along the length and/or height of the Wall and a Stacked Wall actually "stacks" two or more Basic Wall Types on top of one another. We will look at Stacked Walls later on. The major focus of this chapter will be on Curtain Walls.

DRAW A CURTAIN WALL

Using the Wall tool, let's create our first Curtain Wall at the front entrance to the building on the first floor.

1. On the Project Browser, double-click to open the *Level 1* floor plan View.
   - Zoom in on the front of the building (bottom of the plan near the patio and Ramps).
2. On the Design Bar, click the Basics tab and then click the Wall tool.
   - From the Type Selector, choose Curtain Wall:Curtain Wall 1 [Curtain Wall].
   - On the Options Bar, verify that Draw is selected.
   - From the "Height" list, choose **Level 2**.

   Do *not* select "Chain."
3. Click the first point of the Curtain Wall at the midpoint of the short vertical masonry Wall at the left (see Figure 8–4).

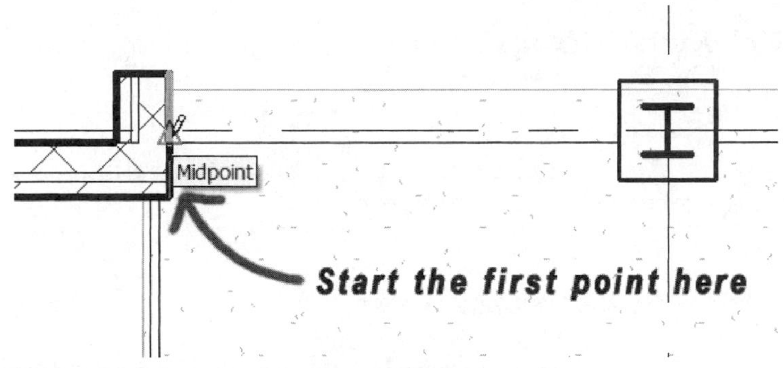

Figure 8–4 *Place the first point of the Curtain Wall*

4. Click the end point of the Curtain Wall at the midpoint of the short vertical masonry Wall on the other side (see Figure 8–5).

TIP While the command is active, hold down the wheel on your mouse and drag to pan to the other side.

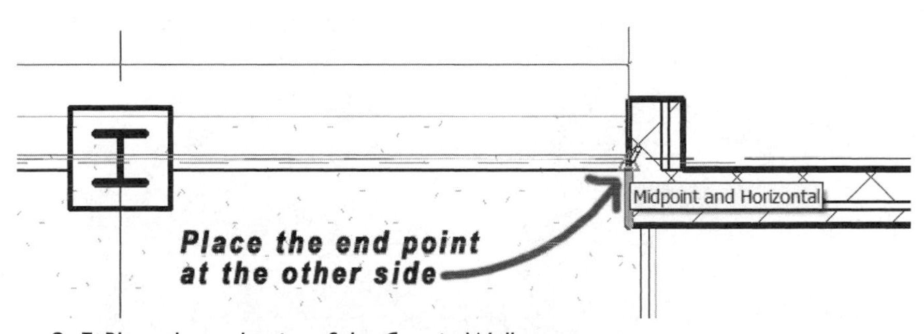

Figure 8–5 *Place the end point of the Curtain Wall*

5. Using the "Move Witness Line" drag control, move the witness line to the outside edge of the horizontal masonry Wall.

   - Change the value of the temporary dimension to **6″** **[150]** (see Figure 8–6).

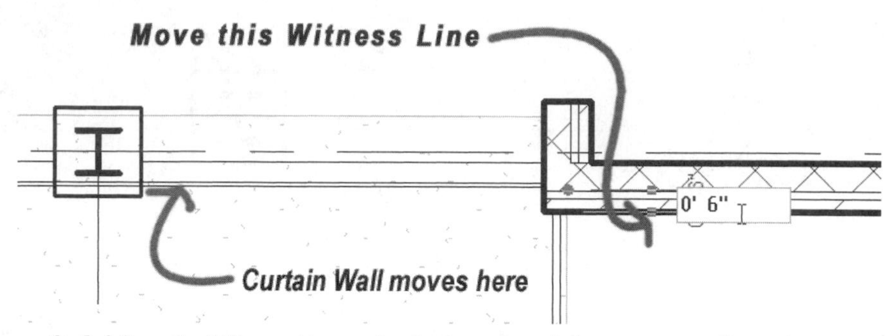

Figure 8–6 *Adjust the Witness Line and edit the value of the temporary dimension*

- On the Design Bar, click the Modify tool or press the ESC key twice.

EDIT THE CURTAIN WALL

Let's take a look at what we have and make some adjustments.

6. On the View toolbar, click the Default 3D View icon (or open the {*3D*} View from the Project Browser).
   - If necessary, spin the model around to the front so that you can see the new Curtain Wall (see Figure 8–7).

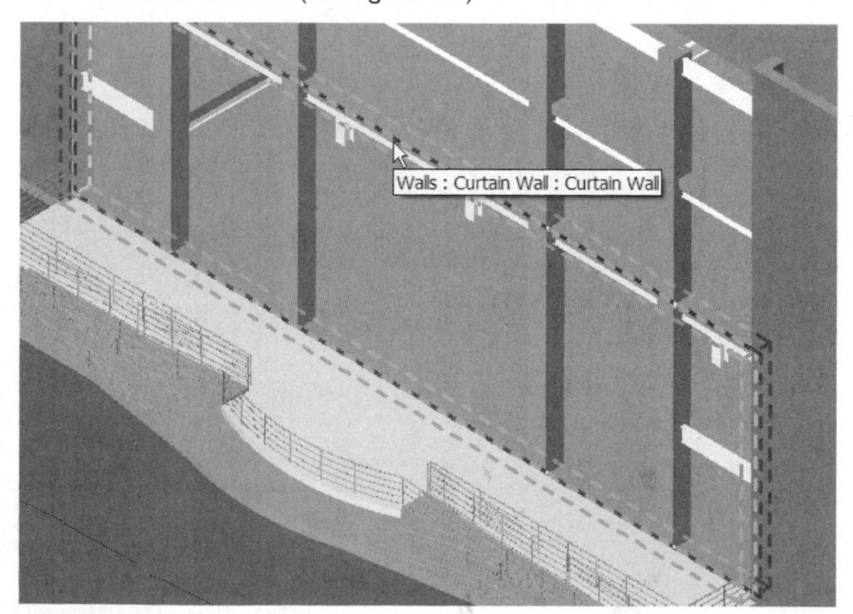

Figure 8–7 *Study the model and the new Curtain Wall in the* {3D} *View*

The first thing that you are likely to notice is that the Curtain Wall is two stories tall even though we set the Top Constraint to Level 2. It turns out that a height offset was applied to the Curtain Wall by default as we were drawing it. Let's remove that now.

7. Select the Curtain Wall and on the Options Bar, click the properties icon.

   - In the "Element Properties" dialog, beneath the "Constraints" grouping, change the "Top Offset" to **0** (zero) and then click OK.

The Curtain Wall should now match the height of the first floor only.

A Curtain Wall is comprised of a Grid Pattern in both the horizontal and vertical directions. Within each of the cells defined by these grids, is a panel. Panel Families can be made to represent anything from glass to louvers to Doors. The number of grid divisions in the pattern is either a parameter of the Curtain Wall Type or can be defined on the Curtain Wall instance directly. You can select the grid lines and panels independently. By default, the Curtain Wall itself pre-highlights first. To select the internal elements, you use the TAB key.

8. Place the pointer over the Curtain Wall to pre-highlight it.

   - Press the TAB key once, then again, and again.

Each time you TAB, a different portion of the Curtain Wall will pre-highlight. As is consistent throughout Revit Building, you can then click the mouse to select that element.

CREATE A WORKING VIEW

For this particular Curtain Wall, we have one large panel of glazing. Therefore the tabbing does not yield very interesting results yet. Naturally, it would be unlikely to have a single continuous panel of glass across the entire front of the building. We will learn how to sub-divide the Curtain Wall next. We can sub-divide the Curtain Wall in any View. It will be easiest to create a new View for this purpose. The reasons are two-fold: first, it will be easier to see and work on the Curtain Wall if we isolate it from the rest of the model, and second creating a View specifically for this purpose can essentially "save" our isolated View of the model where a tool like Isolate Objects would last only for the current Revit Building session.

9. On the Project Browser, double-click to open the *Level 1* floor plan View.

   - On the Design Bar, click the Basics tab and then click the Section tool.
   - Draw a Section line in front of (below in plan) the Curtain Wall just a little wider both left and right.
   - Using the Control Handle, drag the "Far Clip Offset" back to just the inside of the Curtain Wall (see Figure 8–8).

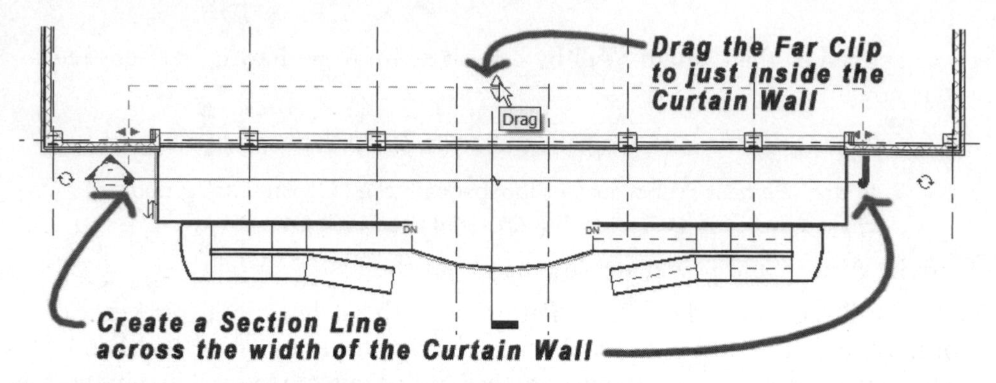

Figure 8–8 *Create a Section around just the Curtain Wall*

10. Click off of the Section line and then double-click the blue section head to open the section View.
    - Drag the top of the crop box down to just above the top edge of the Curtain Wall.
11. On the Project Browser, right-click *Section 1* and choose **Rename**.
    - Change the name to: **Entry Curtain Wall Section** and then click OK.
12. On the Project Browser, right-click the *{3D}* View and choose **Duplicate**.
    - Right-click *Copy of {3D}* and choose **Rename**.
    - Change the name to: **Entry Curtain Wall 3D** and then click OK.
13. On the View menu, choose **Orient > To Other View**.

TIP If this command is grayed out, click in the *Entry Curtain Wall {3D}* View to make it active.

- In the Orient To Other View dialog, choose *Section: Entry Curtain Wall Section* and then click OK.

14. On the View Control Bar, change the Model Graphics Style to **Hidden Line** (see Figure 8–9).

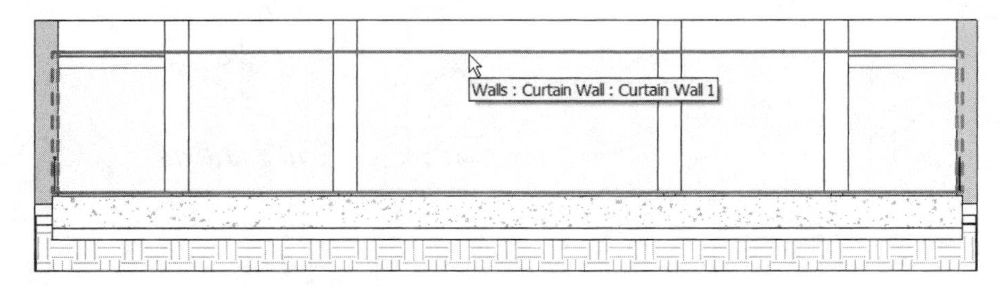

Figure 8–9 *Orient the copied 3D View to the Section at Curtain Wall and view it as Hidden Line*

We now have a section and a 3D View that are cropped to show only the Curtain Wall across the front entrance. In the 3D View, if you hold down the SHIFT key and drag the model to spin the View, you will notice that the View is cropped to match the section in all directions. To return to the head-on (elevation) view point, choose **Orient > To Other View** from the View menu again. This View will make it easier to work on the Curtain Wall. The 3D View is good for checking your progress, but you will get some additional editing functionality in the section View. We will use both below as we work.

ADD CURTAIN GRIDS

Now that we can see the Curtain Wall clearly without the rest of the model cluttering our view, let's sub-divide the Curtain Wall with Curtain Grids.

15. On the Project Browser, double-click to open the *Entry Curtain Wall Section* View.
16. On the Design Bar, click the Modeling tab and then click the Curtain Grid tool.

- Move the pointer near the top edge of the Curtain Wall.

A Grid line and temporary dimensions will appear.

- Click to create a Grid line in the middle bay near one of the columns.
- Click to create another in the same bay near the other column (see Figure 8–10).

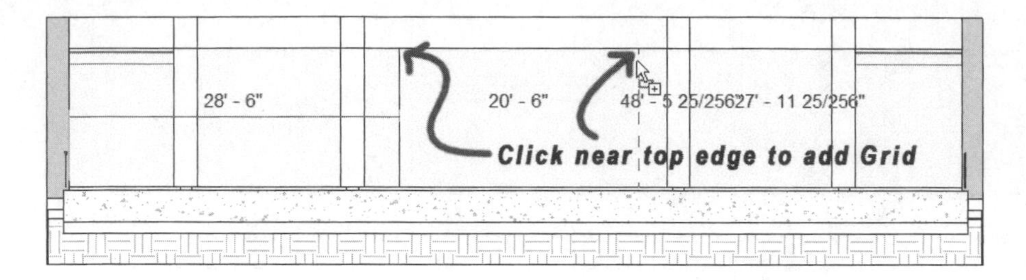

Figure 8–10 *Add two Grid lines near the middle columns*

- On the Design Bar, click the Modify tool or press the ESC key twice.

17. Select the new Grid Line on the left (you should be able to just click it, but use the TAB key if necessary).

    Beneath the temporary dimension a small dimension icon will appear.

    - Click on the dimension icon beneath the temporary dimension on the left (see Figure 8–11).

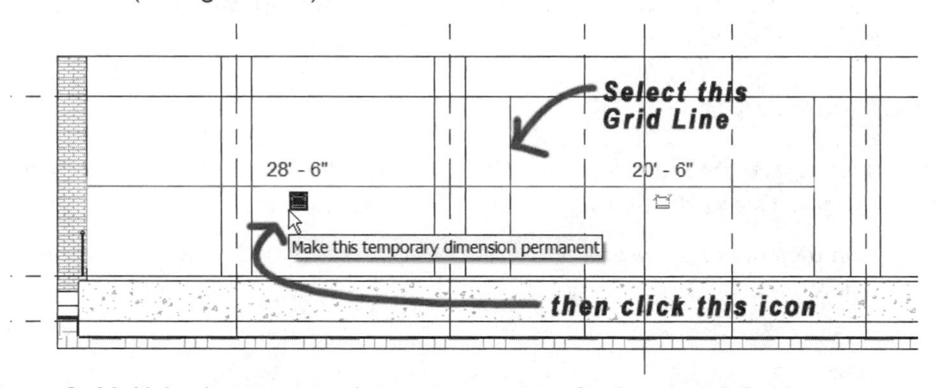

Figure 8–11 *Make the temporary dimension permanent for the selected Grid Line*

- Repeat by selecting the Grid line on the right and then making the temporary dimension on the right permanent as well.

    Do not make the temporary dimension in the middle permanent.

18. Click away from the selected element to deselect it (or on the Design Bar, click the Modify tool or press the ESC key twice).

    - Drag the leftmost witness line of the dimension on the left to the right face of the column.

    - Drag the rightmost witness line of the dimension on the right to the left face of the column (see Figure 8–12).

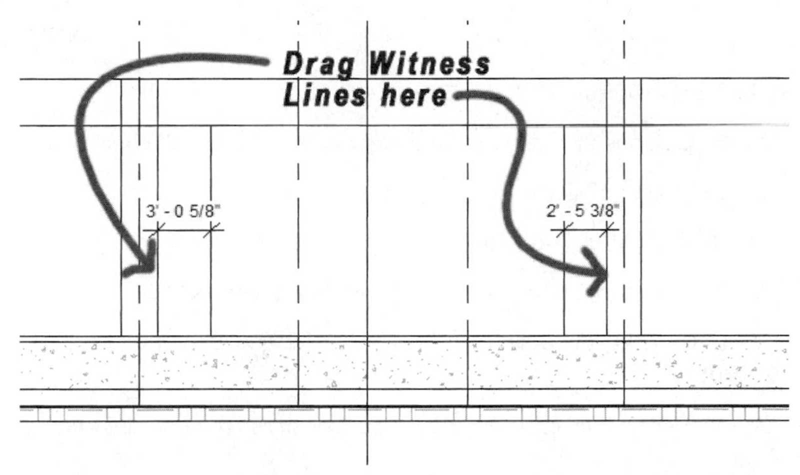

Figure 8–12 *Move the Witness Lines of the Dimensions to the faces of the Columns*

19. Select the new Grid Line on the left (you should be able to just click it, but use the TAB key if necessary).

Notice how the temporary dimensions now include the dimension that we just created and modified. We can now edit the value of this dimension to place the Grid Lines where we would like relative to the Columns.

- Click on the temporary dimension between the Grid Line and the Column (the one edited above).
- Input a value of **1′-0″ [345]**.
- Repeat on the other side.

You should now have a middle bay that has a Grid Line 1′-0″ [345] away from the Columns on each side. The distance between the Grid Lines should be 24′-0″ [7200]. If this is not the case, make any required adjustments (see Figure 8–13).

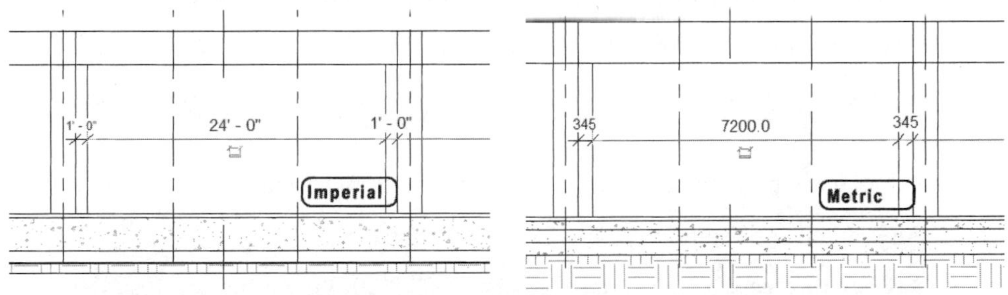

Figure 8–13 *Move both Grid Lines relative the neighboring Columns*

We can add additional Curtain Grid Lines by sketching them like the ones above, or we can copy them from the existing ones. Let's copy the ones we have to create three wide bays and two small bays for the entrance to our building.

20. Select the Grid Line on the left.
21. On the Tools toolbar, click the Copy tool.
    - On the Options Bar, place a checkmark in the "Multiple" checkbox.
    - Pick any start point and then drag to the right.
    - Type **6′-0″ [1800]** and then press ENTER to create the first copy.
    - For the next copy, type **3′-0″ [900]** and then press ENTER.
    - Create another one at an offset of **6′-0″ [1800]** and then one more at **3′-0″ [900]**.
    - On the Design Bar, click the Modify tool or press the ESC key twice (see Figure 8–14).

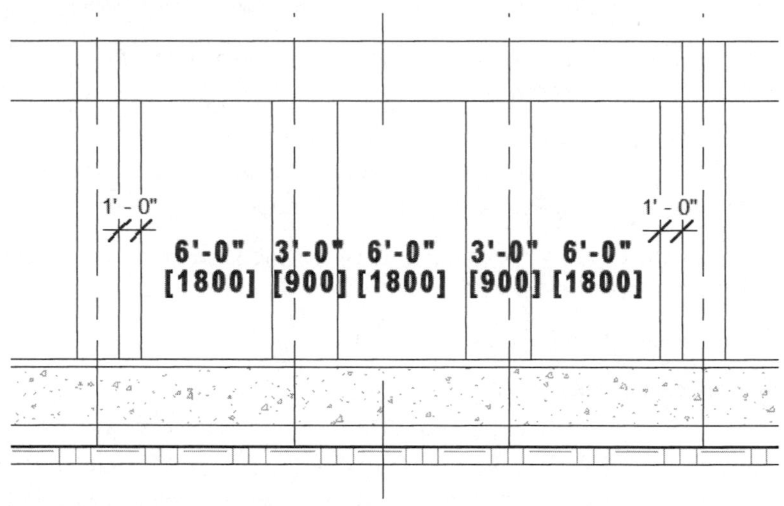

Figure 8–14 *Create several copies of the Grid Line making an A B A B A rhythm*

22. Zoom out so you can see the entire Curtain Wall.
23. On the Design Bar, click the Modeling tab and then click the Curtain Grid tool.
    - Move the pointer near the left edge of the Curtain Wall.

    A Grid line and temporary dimensions will appear.

    - Click to create a horizontal Grid Line 4′-0″ [1200] from the top (use the temporary dimensions to position it if necessary).

At this point we have created several Grid Lines in our Curtain Wall. We can now use the tabbing technique mentioned above to see that this procedure has also cut the one large Curtain Panel into several smaller ones at each division we created.

We can select any or all of these panels and assign them to other Types. Some of the bays could be made solid construction like stone or brick while some could remain glass. We can also insert Curtain Wall Doors into the Curtain Wall the same way. A Curtain Wall Door is actually a Panel Family that looks like and schedules like a Door. The Grid Lines while they show on screen do not yet have any physical materials assigned to them yet. We can use a tool on the Design Bar to assign Mullions to these Grid Lines. We will explore each of these techniques next.

24. Save your project.

ASSIGN CURTAIN PANEL TYPES

Now that we have several Curtain Grid Lines dividing our Curtain Wall into individual panels, let's assign some panel types to these bays. Our building needs an entrance. The three bays that we have roughed out in the front need some Doors.

25. Place your Modify tool (mouse pointer) over the edge of one of the wider bays created above.
    - Press tab until the panel in this bay pre-highlights—when it does, click to select it (see Figure 8–15).

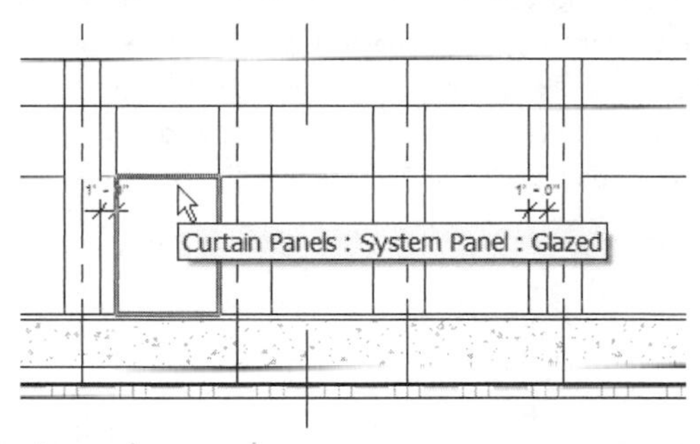

Figure 8–15 *Select the larger bays at the entrance*

- Pre-highlight the next wide panel using the same technique.
- Hold down the CTRL key and click to select it.

You should have two wide panels selected after this action.

- Repeat once more to select the remaining wide panel.

26. With the three panels selected, click the Properties icon on the Options Bar.
    - In the "Element Properties" dialog, click the Load button.

The Open dialog should default to the location of your installed Revit Building libraries. By default, the US version opens to the *Imperial Library* folder. If you installed both Imperial and Metric content, you should also have Imperial Library and Metric Library icons on the outlook bar at the left. If you do not have these icons, the Families that are referenced below have been provided in the *Autodesk Web Library* folder with the files installed from the Mastering Autodesk Revit Building CD ROM. Feel free to access the required content from that location instead.

- In the Open dialog, double-click the *Doors* folder.

If you do not see a *Doors* folder, browse first to the location where you installed the Mastering Autodesk Revit Building CD ROM files and then double-click the *Autodesk Web Library* folder, then the *Doors* folder.

- Select the *Curtain Wall-Store Front-Dbl.rfa [M_Curtain Wall-Store Front-Dbl.rfa]* file and then click Open.

NOTE If you do not have this folder, you can find the Families noted above in the *Autodesk Web Library\Doors* folder with the files you installed from the Mastering Autodesk Revit Building CD ROM.

NOTE Please note that even though the files referenced above are stored in a *Doors* folder, they are actually Curtain Panel Families and not Door Families. While they function like Doors in the building, since they are inserted into a Curtain Wall, they are actually Curtain Panel Families. If you wish to include these "Doors" on your Door Schedule, you will need to be sure to include them in the Schedule filtering later or to choose a Panel Family that behaves like a regular Door.

27. In the "Element Properties" dialog, click OK to see the results (see Figure 8–16).

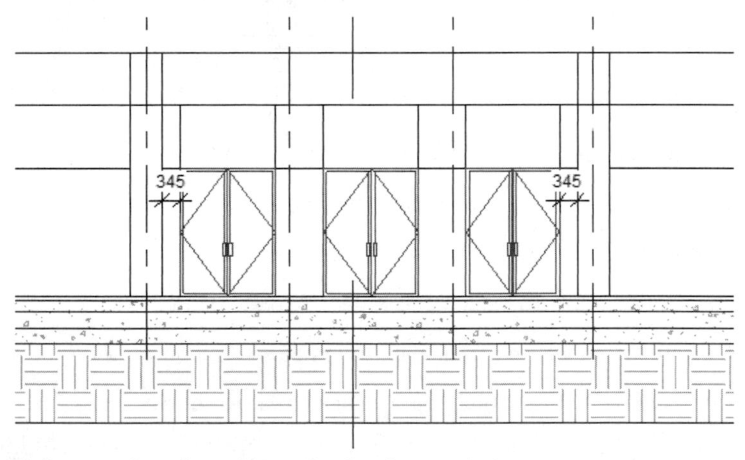

Figure 8–16 *Swap in Storefront Doors for the three wide bays*

ASSIGN MULLIONS

Our next task is to apply some mullions to the Curtain Grid Lines.

28. On the Design Bar, click the Modeling tab and then click the Mullion tool.

   - From the Type Selector, verify that Rectangular Mullion : 2.5″ x 5″ rectangular [Rectangular Mullion : 50 x 150mm] is selected.

On the Options Bar, there are three methods to create Mullions: a single Grid Line Segment, an Entire Grid Line, or All Empty Segments. Entire Grid Line is the default choice.

   - With the "Entire Grid Line" option selected, click on the horizontal Grid Line (the one 4′-0″ [1200] from the top).

If you are zoomed in close enough, you should notice that the Doors adjust in size to accommodate the mullion.

   - Try the "Grid Line Segment" option next on any Grid segment.
   - Use the "All Empty Segments" option to complete the process of adding mullions (see Figure 8–17).

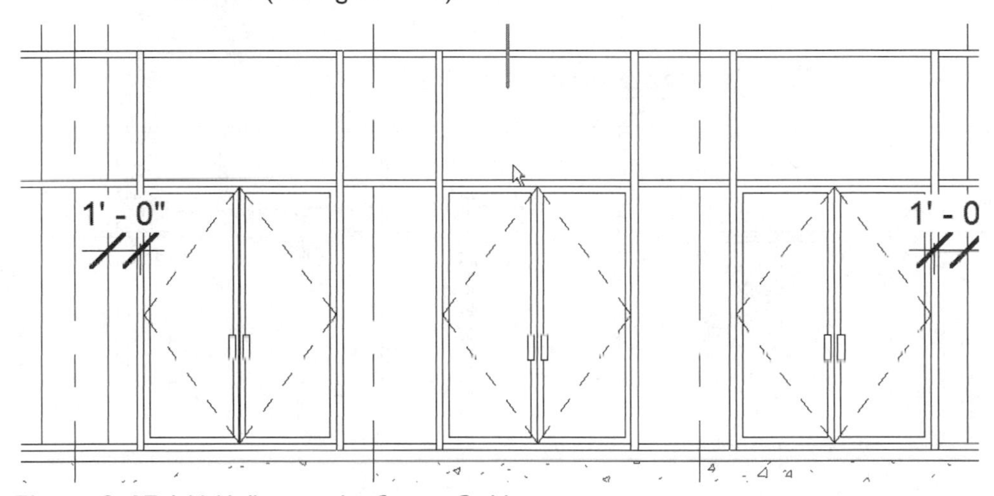

Figure 8–17 *Add Mullions to the Curtain Grid Lines*

EDIT MULLION JOINS

You will notice that the vertical mullions have been given priority over the horizontal ones. If you prefer to have the horizontal mullions continuous and have the verticals stop and start at each intersection, you can toggle the Mullion Join behavior for each join.

29. Select one of the vertical mullions at the Doors.

Notice the Join icons that appear at each end of the Mullion (see Figure 8–18).

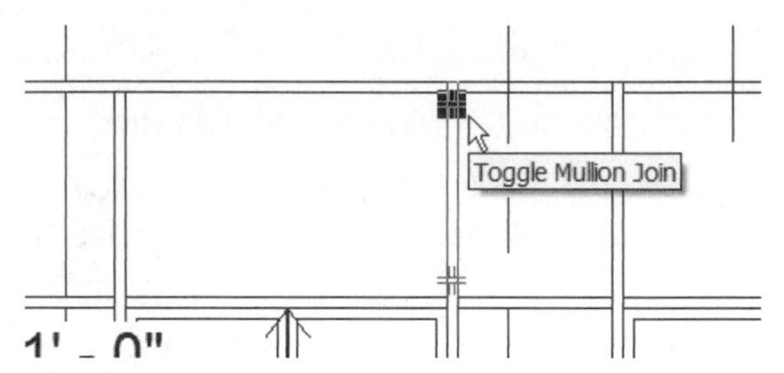

Figure 8–18 *Edit Mullion Joins using the control*

- Click the Toggle Mullion Joins icon to model the join to your liking.

CREATE ADDITIONAL GRID LINES AND MULLIONS

The middle entrance bay is complete, but the other bays could use some more embellishment.

30. Using the techniques covered above, create a new Grid Line **1′-0″ [345]** from edges of the two middle columns as shown in Figure 8–19.

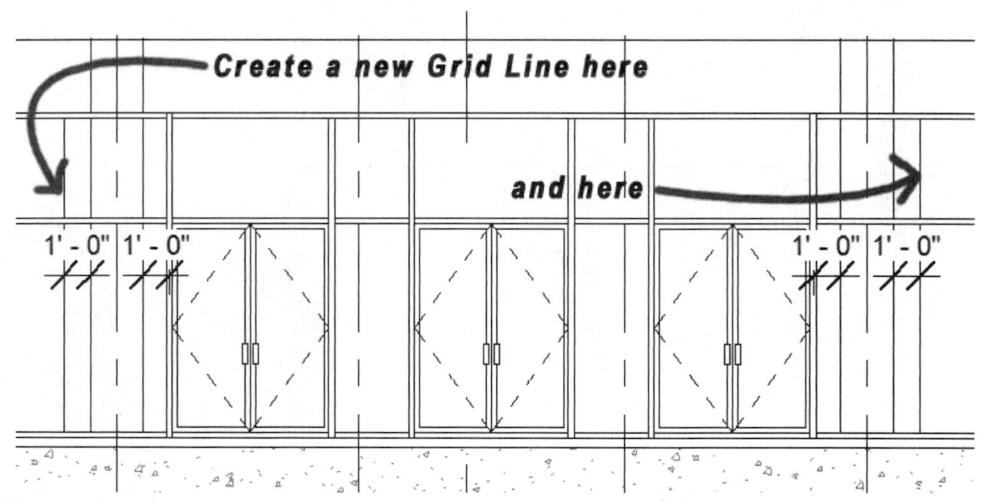

Figure 8–19 *Add more Grid Lines*

31. Add Mullions to each of these new Grid Lines.
32. Repeat the entire process to create Grid Lines and Mullions on each side of the remaining Columns (see Figure 8–20).

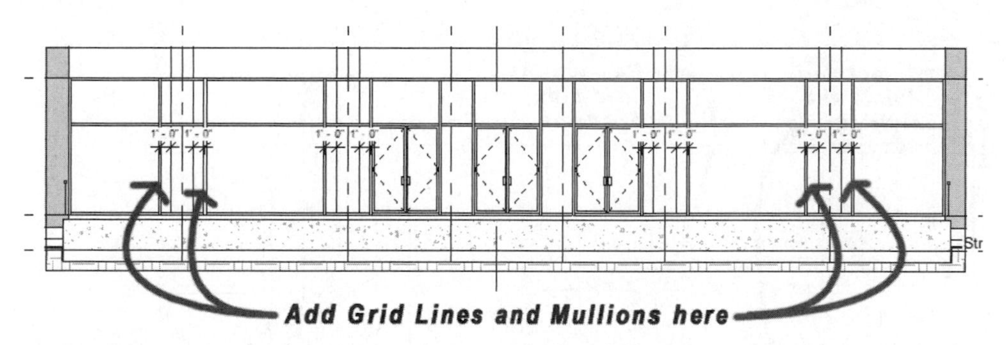
Add Grid Lines and Mullions here

Figure 8–20 *Add more Grid Lines and Mullions at the remaining Columns*

USING A WALL TYPE FOR A PANEL

At each of the Columns, we want to remove the horizontal Mullions and swap out the glazing for a solid Wall panel.

33. Zoom in on one of the Columns.

- Using the CTRL key, select all three of the horizontal mullions crossing through the Column.

TIP If you have trouble with the selection, remember your TAB key.

You only want the piece of Mullion between the two vertical Mullions adjacent to each column. If you edited the Mullion Joins so that the entire horizontal Mullion is selected, edit the joins again so that you can select just the small piece.

- Delete the three selected Mullions (press the DELETE key) (see the left panel of Figure 8–21).
- Select the horizontal Grid Line.
- On the Options Bar, click the Add or Remove Segments button.
- Click the Grid Line in-between the two vertical Mullions (see the middle panel of Figure 8–21).

The Grid Line will turn dashed between these Mullions to indicate that a portion of it has been removed.

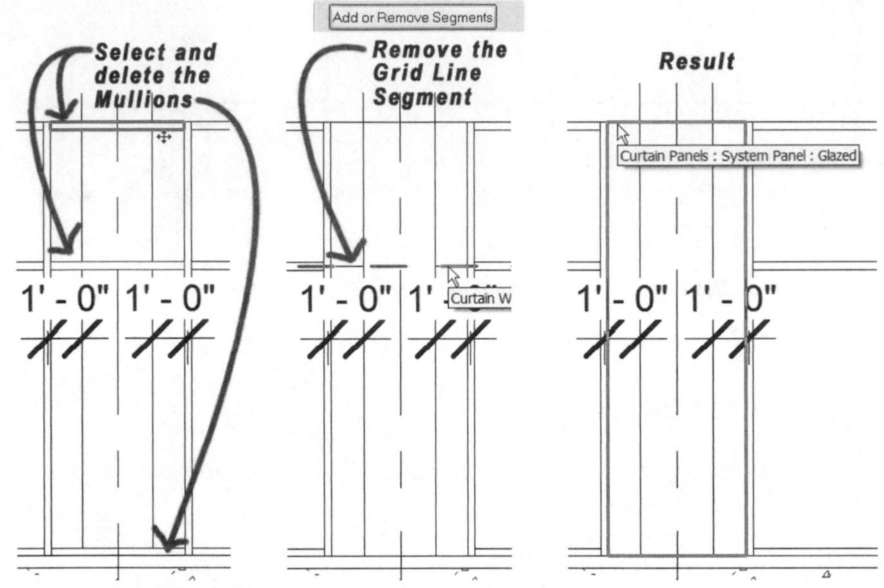

Figure 8–21 *Delete Mullions and Remove a Grid Line*

You will now have a continuous vertical panel at the Column as shown in the right panel of Figure 8–21. You may need to use the TAB key to pre-highlight it to see this.

34. Using the TAB key, pre-highlight and then select the new full height panel at the Column.
    - From the Type Selector, choose Basic Wall : Generic - 8″ [Basic Wall : Generic - 200mm].
35. Repeat the entire process for the other three Columns on the front façade.
    - On the Project Browser, double-click to open the *Level 1* floor plan View.
    - Zoom in on the Columns at the front of the façade to see the result (see Figure 8–22).

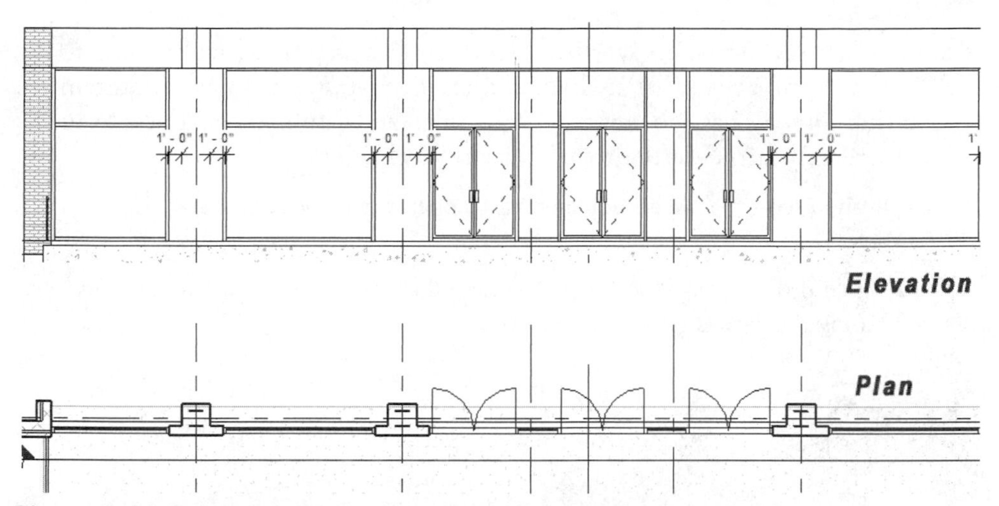

Figure 8–22 *Swap in a Wall Type in place of the Curtain Panels*

36. On the Project Browser, double-click to return to the *Entry Curtain Wall Section* View.

37. Using the TAB and CTRL keys, select each of the four lower panels (two to the right of the entrance, two to the left).

   - From the Type Selector, choose Curtain Wall : Storefront (see Figure 8–23).

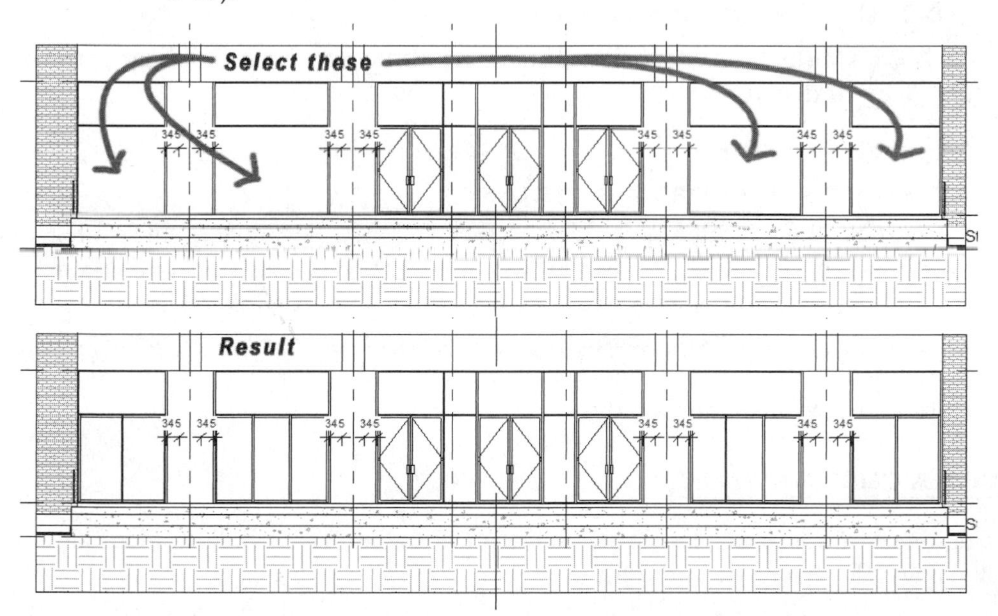

Figure 8–23 *Swap in a Curtain Wall Type for the selected Panels*

There are two interesting points worth mention on this step: first, we are able to use one Curtain Wall Type as a Panel on another. Second, we can have a Curtain Wall Type that already has sub-divided Grid Lines and Mullions assigned to it. We will explore this further in the remainder of the chapter.

38. On the Project Browser, double-click to open the *Entry Curtain Wall 3D* View.

   - Hold down the SHIFT key and then drag the mouse to spin the model and study the results (see Figure 8–24).

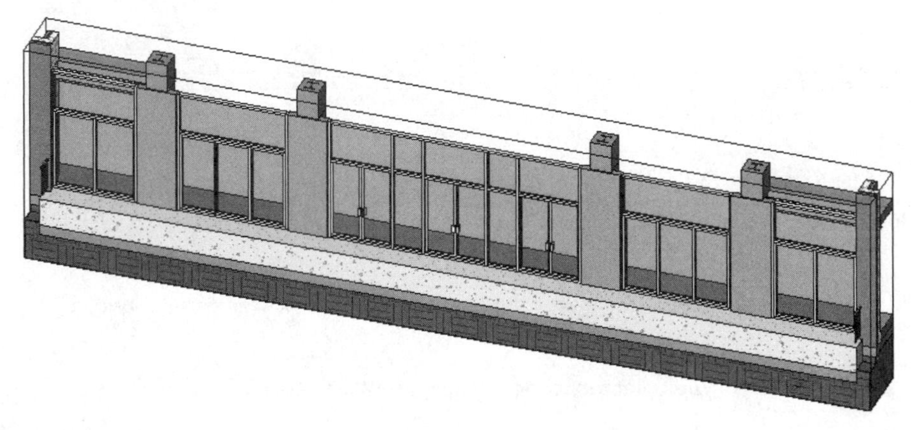

Figure 8–24 *Spin the model to study the results*

If you wish to see the Curtain Wall in the context of the rest of the building, open the *{3D}* View and study it there.

39. Save the project.

CREATING HOSTED CURTAIN WALLS

The Curtain Wall we created here filled a space that had no previous enclosure. We drew it just like any other Wall and then edited it. You can also create Curtain Walls from existing Walls or even have them cut away from existing Walls. In this way the Wall will "host" the Curtain Wall. In this sequence we will use one of the Types provided in the current file and draw a Curtain Wall that will be hosted in the exterior shell Walls of the building.

DRAW A CURTAIN WALL

The first part of this exercise is similar to the previous one. We will draw a Curtain Wall using a Type already in the file.

1. On the Project Browser, double-click to open the *Level 2* floor plan View.
2. On the Design Bar, click the Basics tab and then click the Wall tool.

- From the Type Selector, choose Curtain Wall : Storefront.
- On the Options Bar, verify that Draw is selected.
- From the "Height" list, choose **Level 4**.

Recall that above, the default is for a Top Offset to be applied to the Curtain Wall as it is drawn. In this case (as we did above), we will remove this offset.

- On the Options Bar, click the Properties icon.
- In the "Element Properties" dialog set the "Top Offset" to **0** (zero) and then click OK.
- On the Options Bar, place a checkmark in the "Chain" checkbox.

3. Click the first point of the Curtain Wall at the midpoint of the brick face of the short horizontal masonry Wall at the front of the building on the right side (see Figure 8–25).

   Use the TAB as necessary to assist you in selection of the proper point.

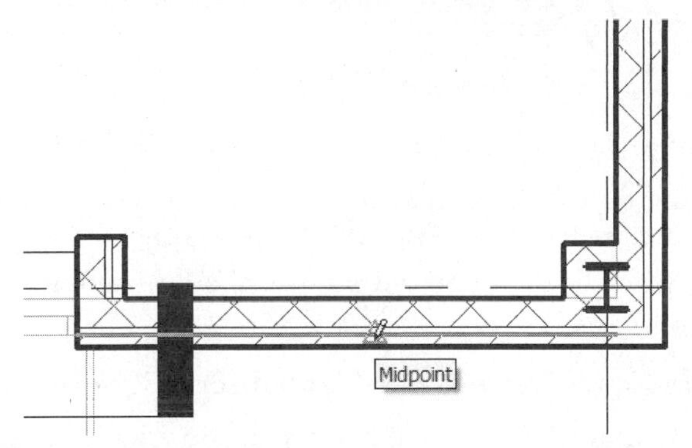

Figure 8–25 *Draw a new Curtain Wall on the second floor directly on top of the existing Walls*

- Create two Curtain Wall segments, one horizontal, one vertical along the inside face of the brick.
- Stop the vertical Curtain Wall when it is **54′-0″ [16,200]** long (about half way between Column Line 3 and 4) (see Figure 8–26).

Ignore any warning dialogs that may appear.

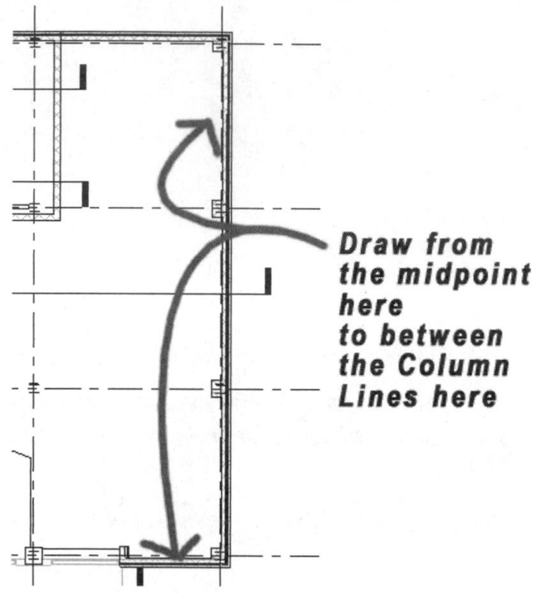

Figure 8–26 *Draw two Curtain Wall segments*

- On the Design Bar, click the Modify tool or press the ESC key twice.

In some cases, depending on the Type of Curtain Wall you select on the Options Bar, the new Curtain Wall will cut the host Wall automatically. This is what has occurred here. In cases where it does not, we can use the Cut tool to manually cut the Wall with the Curtain Wall.

4. Zoom in on the lower area with the short horizontal Curtain Wall.

Because we chose the Curtain Wall : Storefront Curtain Wall Type above, this Curtain Wall already has mullions assigned to the Grid Lines. However, the corner where the two segments meet needs some work. Also, the glazing intersects the Columns.

5. Select the short horizontal Curtain Wall at the front of the building.

- Click the "Flip Wall Orientation" control.
- On the Options Bar, click the "Activate Dimensions" button.
- Edit the temporary dimension in the thickness of the Wall to move the Curtain Wall out (away from the interior).

Experiment to find the right value. Try about **6"** [**150**].

- Repeat both steps on the vertical Curtain Wall (see Figure 8–27).

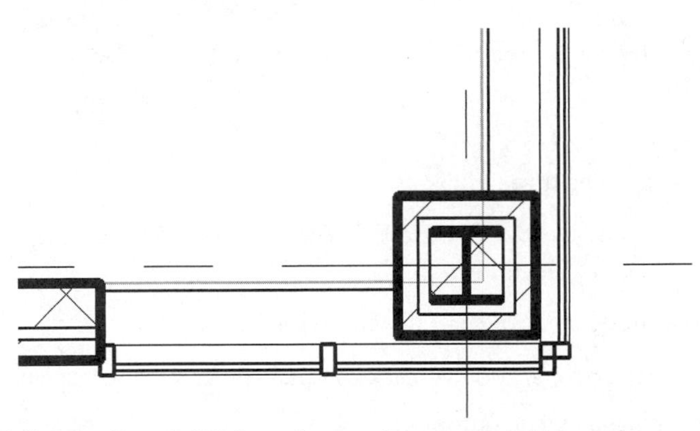

Figure 8–27 *Flip the Curtain Walls and move them toward the outside*

6. On the Project Browser, double-click to open the *{3D}* View.
   - Spin the model as needed to gain a clear view of the two Curtain Wall segments.
7. Select all of the Mullions at the corner (six total).

You can zoom in and use the CTRL key to select them, or you can use a marquee selection and then use the Filter Selection icon to remove everything but Mullions from the selection. Notice the small pushpin icons attached to each Mullion. You may need to zoom to see them all—there is one attached to each of the six Mul lions. Since the Curtain Wall Type that we used here assigned the Mullions automatically, these Mullion definitions are pinned to the Type. If you look at the Type Selector, you will notice that you cannot change the Type. To delete or change a Mullion, you must unpin the Mullion first.

8. Click each of the six pushpins to unpin them (see Figure 8–28).

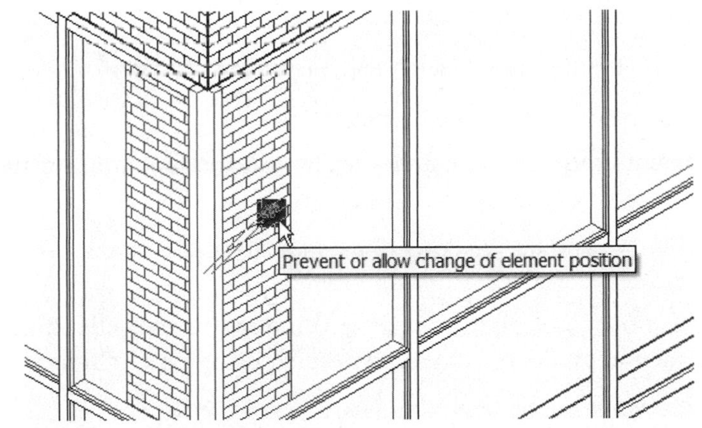
Prevent or allow change of element position

Figure 8–28 *Unpin the selected Mullion elements*

9. Delete one of each pair of Mullions (three total).

TIP You can try this in floor plan Views instead if you find it easier.

10. Select the remaining three Mullions.
    - From the Type Selector, choose L Corner Mullion : 5″ x 5″ Corner [Quad Corner Mullion : Quad Mullion 1].
    - Study the model in both the {*3D*} and *Level 2* plan Views.
11. On the Project Browser, double-click to open the *Level 2* floor plan View.
12. Select both Curtain Wall segments.

Keep in mind that the Curtain Wall segment is made up of many subparts. The whole curtain wall is represented by a green dashed that will appear when the Modify tool is passed over the top of the wall.

13. On the Tools toolbar, click the Mirror tool.

TIP The keyboard shortcut for Mirror is **MM**. **Mirror** is also located on the Edit menu.

- On the Options Bar, verify that the "Copy" checkbox is selected.
- Click the Draw option icon.
- Snap to the midpoint of the Floor slab curve (in the center of the plan) and then drag straight down and click again.

The two Curtain Walls will mirror over to the other side of the plan and automatically cut the Walls.

14. On the Project Browser, double-click to open the {*3D*} View.
    - Spin the model and study the results.

If any of the segments did not cut automatically, use the Cut tool on the Tools toolbar to cut them manually using the Curtain Wall.

15. Add ordinary punched Windows to the masonry Walls on the first and forth floors above and below the Curtain Walls.

Simply open the *Level 1* floor plan, add some Windows to one side, mirror them to the other and then copy them to the *Level 4* plan using paste aligned.

16. Study the results in the {*3D*} View (see Figure 8–29).

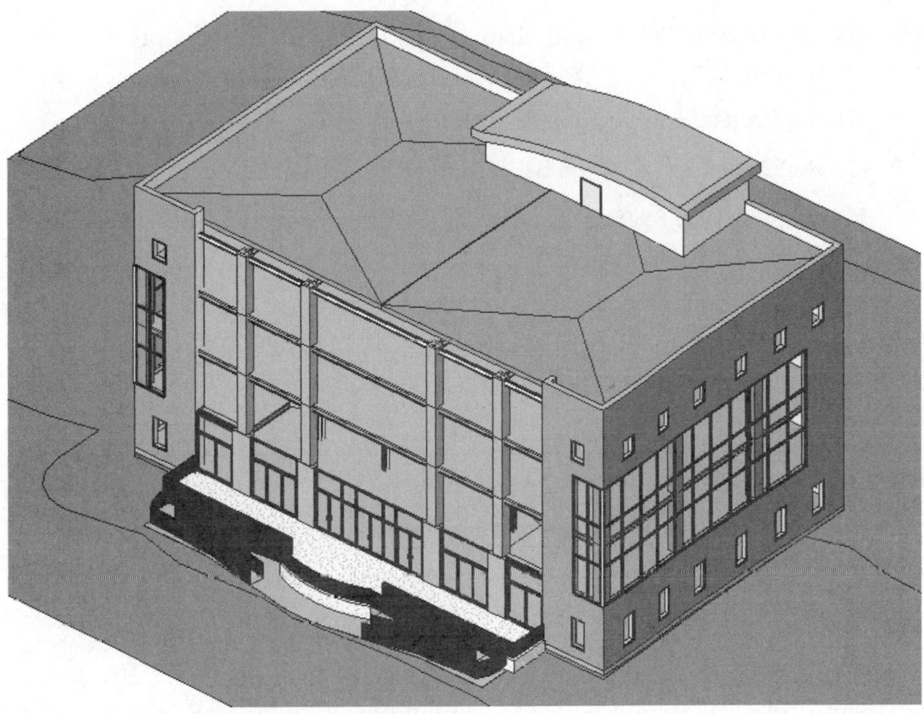

Figure 8–29 *Add Windows to the other floors*

17. Save the project.

CREATE A CUSTOM CURTAIN WALL TYPE

We have now worked with a simple Curtain Wall with no divisions (in which we added all of the Curtain Grid Lines and Mullions manually) and a Curtain Wall Type that included Grid spacing and Mullions within its Type. In this sequence we will combine techniques starting with the creation of our own Curtain Wall Type. The goal is to try to include as many of the divisions and Mullions within the Type so that they occur automatically. We can then finish the design using the manual Grid line and Mullion techniques already covered.

CREATE WALLS BY PICKING LINEWORK

The second floor of our building includes a Floor slab with a curved cutout for a two-story atrium. We can use this linework to create a series of Walls of similar shape for the front façade.

Curtain Walls are a type of Wall. There are a few ways that we can go about adding and customizing our Curtain Wall Type in our project. In this example, we will start with some basic Walls and then swap them with our custom Curtain Wall Type as we build it.

1. On the Project Browser, double-click to open the *Level 2* floor plan View.
2. On the Design Bar, click the Basics tab and then click the Wall tool.
   - From the Type Selector, choose Basic Wall : Generic - 5″ [Basic Wall : Generic - 125mm].
   - On the Options Bar, click the Pick Lines icon.
   - Verify that on the Options bar the "Level" parameter reads **Level 2** and set the "Height" parameter, to **Roof** (see Figure 8–30).

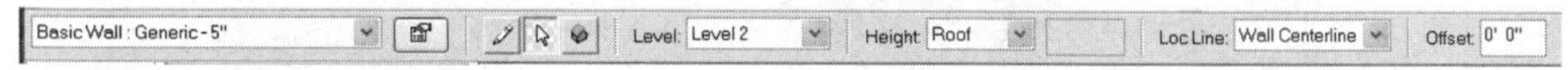

Figure 8–30 *Set the Options for creating the new Walls*

   - Click the Properties button and verify that the "Top Offset" is set to **0** (zero) and then click OK.

With these settings, we are going to create a Curtain Wall from Level 2 to the Roof using the Generic 5″ [Generic - 125mm] Type by picking lines from existing geometry. At the front of the building you can see the edge of the Floor slab. It has a curved atrium cut away from it between column lines B and E. We are going to use the lines here to create the Wall.

3. Using Figure 8–31 as a guide, click the lines indicated to create the Walls (5 total).
   - On the Design Bar, click the Modify tool or press the ESC key twice.

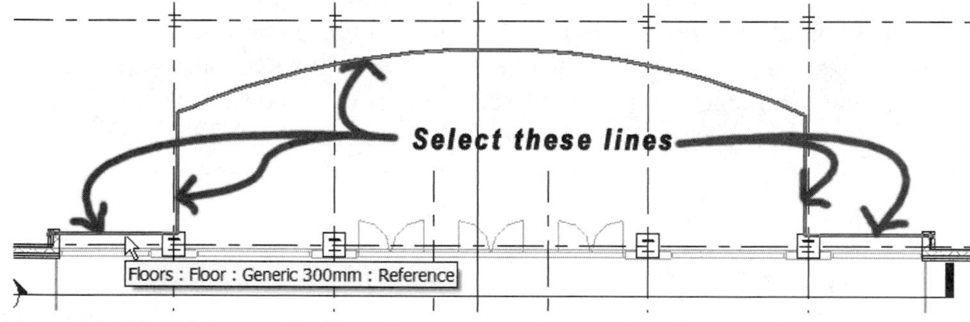

Figure 8–31 *Pick lines in the Floor slab to create Curtain Wall segments*

We now have the basic shape of the façade only on the inside rather than the outside. Let's make a few adjustments and then mirror it to the outside.

4. Select the small vertical Wall segment on the left.

- Using the Move tool on the Edit toolbar, using any start point, Move it to the left **1′-4″** [**400**].
- Repeat on the other side, moving the vertical Wall on the right to the right **1′-4″** [**400**].
- Move the curved Curtain Wall down **6′-0″** [**1800**] (see Figure 8–32).

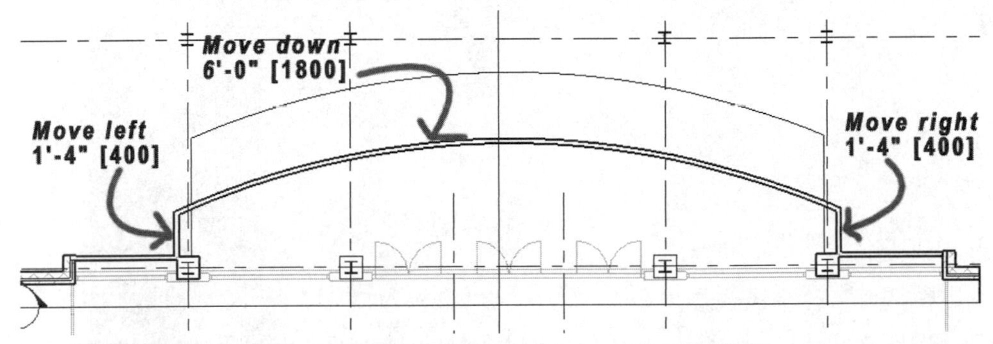

Figure 8–32 *Make adjustments to the positions of the new façade Walls*

Noticc that with each move, the adjacent Walls adjust accordingly and stay attached to the one that is moving.

5. Pre-highlight one of the façade Walls, press the TAB key to pre-highlight the chain of Curtain Walls, and then click to select them.

NOTE If necessary, tab more than once to pre-highlight them all. You can also use CTRL to add to the selection and SHIFT to remove. Select just the 5 façade Walls.

6. On the Edit toolbar, click the Mirror tool.
   - On the Options Bar, verify that Pick Lines is selected.
   - Clear the checkmark from the "Copy" checkbox.
   - Click on Column Line 1 to use it as the mirror line.
7. Study the results in the *{3D}* View (see Figure 8–33).

Figure 8–33 *Study the new Curtain Walls in 3D*

8. Save the project.

DUPLICATE AND EDIT A CURTAIN WALL TYPE

So now that we have some Wall elements in the basic shape of our façade, we can begin working on a Curtain Wall Type for them. Like most Family Types, we will begin with an existing Type and duplicate it. To do this, we will work from the Family tree in the Project Browser.

9. On the Project Browser, locate the *Families* node, and click the plus (+) sign icon next to it to expand it.

   - Click to expand *Walls* next, then *Curtain Wall* (see Figure 8–34).

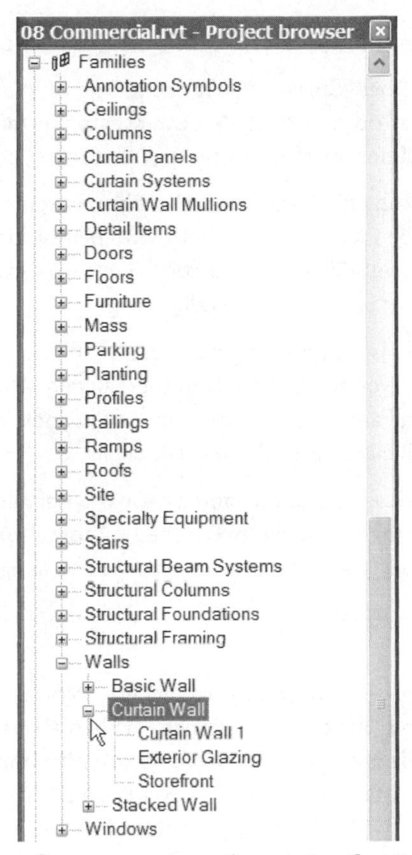

Figure 8–34 *Use the Project Browser to access the existing Curtain Wall Types in the current project*

Each of the Types currently resident in the existing project will appear in the tree listing. You can add an instance of a Type to the model or edit them directly from here as well as duplicate them to create new Types.

- Right-click on Storefront and choose **Duplicate**.

A new Type named Storefront (2) will appear in the list.

- Right-click Storefront (2) and then choose **Rename**. Name the new Curtain Wall Type: **MRB Front Façade** and then press ENTER.

- Right-click MRB Front Façade and choose **Properties**.

This takes you directly to the "Type Properties" dialog for this new Type. This dialog is much like any other "Type Properties" dialog. However, there are more parameters available than many other objects in Revit Building. Following is a brief description of some of the important parameters.

Under the "Construction" grouping, are the following parameters:

- **Wall Function**—Since Curtain Walls are a type of Wall, they have the same list of "Functions" as other Walls. We can create Curtain Wall Types that fulfill a variety of functions such as "Exterior," "Interior" and "Foundation."
- **Automatically Embed**—When this checkbox is selected, the Curtain Wall will attempt to embed itself automatically within another Wall like the one that we created above. When this function is off, you can still use the Cut tool to embed a Curtain Wall in a Wall.
- **Curtain Panel**—This parameter gives a list of all possible panel Types. Choose the default type for all panels in the Curtain Wall Type here. You can always override the Panel Type on one or more Panels in the specific instance of the Curtain Wall element in the model.
- **Join Condition**—Several options appear for the default way that Mullions will join to one another. You can make the horizontal or the vertical continuous, make the border continuous or choose not to assign an option.

Beneath the "Vertical Grid Pattern" and the "Horizontal Grid Pattern" groupings, are three parameters each:

- **Layout** and **Spacing**—Layout choices include "None," "Fixed Distance," "Fixed Number" and "Maximum Distance." When None is chosen the Curtain Wall will include a single panel across the entire horizontal or vertical dimension as appropriate. Fixed Distance will create Curtain Grid Lines at the spacing indicated in the "Spacing" parameter. There may be panel space left over when using this option. Maximum Distance is similar in that it specifies the *largest* that the spacing can become but will size all bays equally up to the size indicated. The Spacing parameter is used for both of these settings.
- **Adjust for Mullion Size**—This parameter adjusts the location of Curtain Grid lines to maintain equal Panels even when Mullion sizes vary.

Beneath the "Vertical Mullions" and "Horizontal Mullions" groupings, you can choose from a list of available Mullion Families to use for the Borders and the Interior Mullions. These Mullions will be created automatically and will be pinned to the Curtain Wall as it is created (see above). You can unpin any Mullion as we did in the exercise above and assign an alternate Type as design needs dictate.

Now that we have explored the settings overall, let's configure the first set of parameters for the Curtain Wall design.

10. Beneath the "Construction" grouping, change the "Join Condition" to **Horizontal Grid Continuous**.
11. Change the parameters of the "Vertical Grid Pattern."
    - Change the "Layout" type to **Fixed Distance**.

- Set the "Spacing" to **11′-0″** [**3300**].

12. Change the parameters of the "Horizontal Grid Pattern."
    - Verify that the "Layout" type is set to **Fixed Distance**.
    - Set the "Spacing" to **12′-0″** [**3600**] (see Figure 8–35).

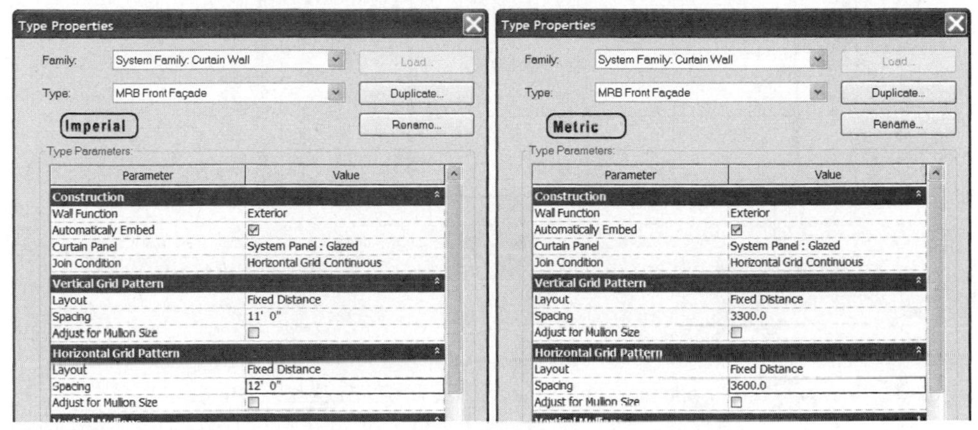

Figure 8–35 *Configure the Type Parameters for the Curtain Wall*

13. Click OK to accept and apply the changes.

Let's apply the new Type to the façade Walls in our model. The *{3D}* View of the model should still be open on screen. If it is not, re-open it now.

14. Select the curved Wall in the middle of the façade.
    - From the Type Selector (on the Options Bar), choose Curtain Wall : MRB Front Façade.

As we can see, the new spacing has been applied to the Curtain Wall in both directions. However, also note that the spacing begins from one end of the Curtain Wall façade and is not centered. In some designs, this may be the desired result, but in this one, we want the pattern centered. Unlike the spacing parameters themselves, the Justification of the pattern is an instance parameter.

15. With the Curtain Wall still selected, on the Options Bar, click the Properties icon.
    - Beneath the "Vertical Grid Pattern" grouping, set the "Justification" to **Center** and then click OK.

    The change to the justification will be easier to see in elevation.

    - On the Project Browser, double-click to open the *South* elevation View (see Figure 8–36).

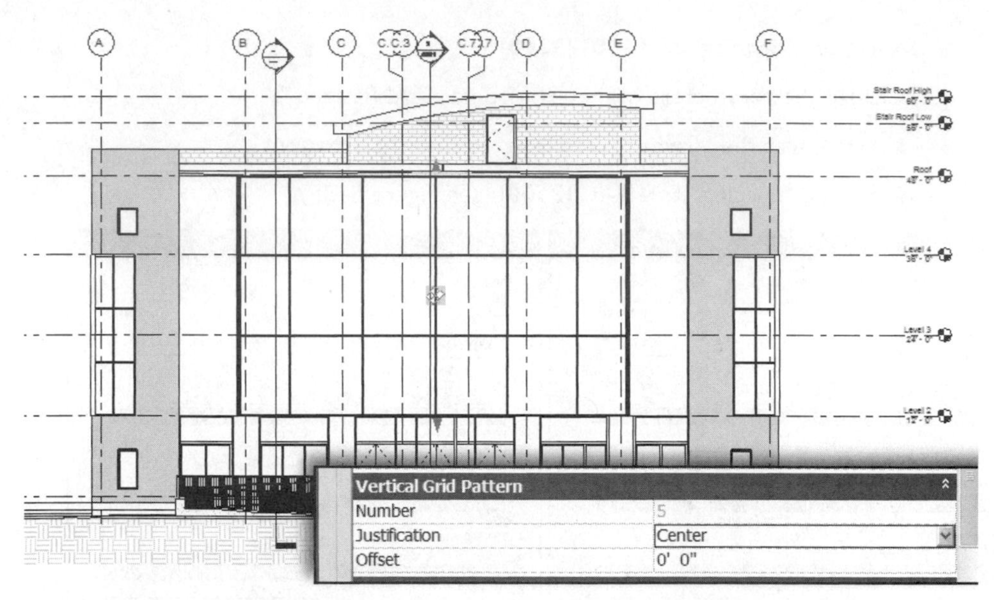

Figure 8–36 *Changing the Justification of the Vertical Grid Pattern centers the bays*

EDIT THE HORIZONTAL MULLIONS

We now have the overall grid pattern established with the major horizontal and vertical spacing in place. The next thing to address is the shape and size of the mullions. In this sequence, we will edit the horizontal mullions to represent spandrels and the vertical Mullions to represent piers. We did not edit the Mullion assignments in the sequence above. Therefore, a simple rectangular mullion has been applied to all of the Curtain Grid Lines of our design. (This was a parameter of the original Storefront Type from which we created our Type). We need to define a few new Mullion Families and then apply them to the horizontal and vertical orientations of the Curtain Wall. We can define these families on the Project Browser the same way we did the Curtain Wall Type.

16. On the Project Browser, locate the *Families* node again, It should still be expanded from above—if not, click the plus (+) sign icon next to it to expand it.
    - Click to expand *Curtain Wall Mullions* next, then *Rectangular Mullion.*
17. Right-click on 2.5″ x 5″ rectangular [50 x 150mm] and choose **Duplicate**.
    - For the name type **Spandrel** and then press ENTER.
    - Right-click on Spandrel and then choose **Properties**.

There are several parameters in the "Type Parameters" dialog for Mullions. Let's take a look at several of them. Under the "Constraints" grouping, are the following parameters:

- **Angle**—Use this constraint to rotate the Mullion relative to the Curtain Wall (see top left of Figure 8–37).
- **Offset**—Input a positive or negative value here to shift the position of the Mullion in or out relative to the Curtain Wall (see top right of Figure 8–37).

Under the "Construction" grouping, are the following parameters:

- **Profile**—By default there are two basic shapes for Mullion profiles: rectangular and circular. You can create custom profile Families that can be loaded into your project and then will appear in the list here (see middle left of Figure 8–37).
- **Position**—Two options exist for Position: Perpendicular to Face is the default condition. This orientation, as its name implies, sets the mullion profile relative to the Curtain Wall. Parallel to Ground rotates the mullion profile and is useful in conditions such as in a sloped glazing or a sloped curtain system (Not shown).
- **Corner Mullion**—Toggle this checkbox to make the mullion a corner condition (Not shown).
- **Thickness**—This is the depth of the Mullion as measured along its axis perpendicular to the Curtain Wall (Not shown).

Under the "Materials and Finishes" grouping, you can choose a Material for the Mullion from the list of Materials available in the project. Unlike the "Thickness" parameter listed above, width is divided into two separate width parameters under the "Dimensions" grouping:

- **Width on side 1** and **Width on side 2**—If you set both of these parameters to the same value, then your Mullion will be centered on the Curtain Grid line (see bottom left of Figure 8–37). If you specify different settings, you will essentially shift the Mullion relative to the Grid line (see bottom right of Figure 8–37). In either case, the overall width will be the total of both settings.

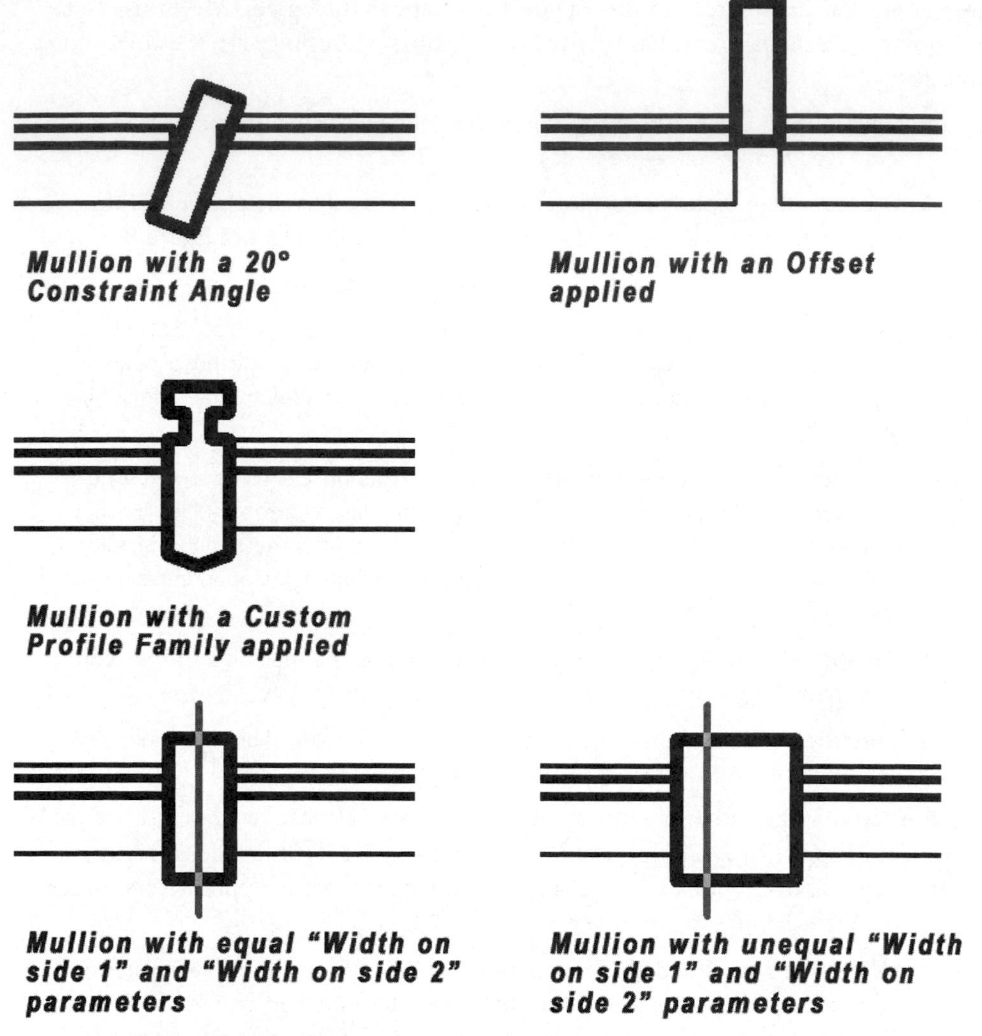

Figure 8–37 *Understanding Mullion Type parameters*

- In the "Type Properties" dialog, type **1′-0″ [300]** for the "Thickness," "Width on side 1" and "Width on side 2."

Remember that the total width is the sum of both the "Width on side1" and "Width on side 2" parameters, so in this case, this makes the Mullion 2′-0″ [600] wide.

- From the "Material" list, choose Masonry – Stone and then click OK.
- Click OK again dismiss the "Type Properties" dialog and return to the model.

The Mullion you just created will not appear in the model yet. All we did was create a new Mullion Type. We have not used it in the actual model yet.

18. On the View toolbar, click the Default 3D View icon (or open the *{3D}* View from the Project Browser).

    If you are not looking at the front façade, spin the model to see it.

19. On the Project Browser, beneath the Families node, right-click on the MRB Front Façade Curtain Wall Type and choose **Properties**.

At this point, since we have used the Curtain Wall in the model, you could use the alternative process of selecting the large curved Curtain Wall element, (Be sure to select the Curtain Wall and not the mullions or Grid lines—use the TAB if necessary) on the Options Bar, click the Properties icon and then click the Edit/New button (or press ALT + E) to edit the Type properties.

- Beneath the "Horizontal Mullions" grouping, choose **Rectangular Mullion : Spandrel** for each of the Mullion conditions: "Interior Type,""Border 1 Type" "Border 2 Type" (see Figure 8–38).

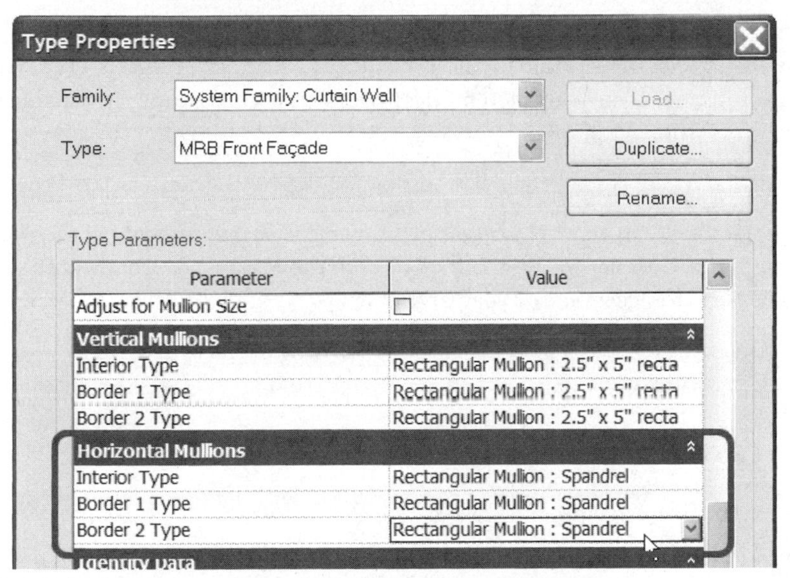

Figure 8–38 *Assign the new Mullion Type to the Horizontal Mullions*

20. Click OK to see the result (see Figure 8–39).

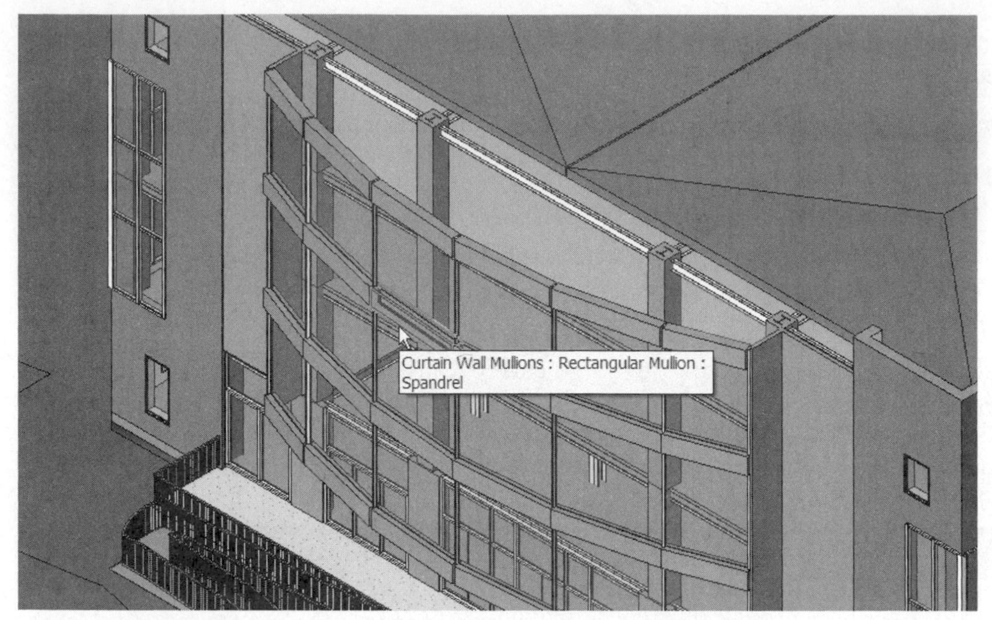

Figure 8–39 *Result of the Spandrel Mullions being assigned to the model*

While the spandrels in the middle of the Curtain Wall are continuous horizontally as we indicated in the Type Parameters, it appears that this is not the case for the top and bottom border Mullions.

21. Return to the "Type Properties" dialog for the MRB Front Façade Type.

    Remember, you can right-click on the Project Browser beneath Families, or select the Curtain Wall in the model, click the Properties icon, and then click Edit/New to access the "Type Parameters."

    - Change the "Join Condition" to **Border and Horizontal Grid Continuous**.

EDIT THE VERTICAL MULLIONS

Repeating nearly the same process, we can edit the Vertical Mullions. We also need a new Type here.

22. On the Project Browser, expand the *Families > Curtain Wall Mullions > Rectangular Mullion* node again.
23. Right-click on 2.5″ x 5″ rectangular [50 x 150mm] and choose **Duplicate**.
    - For the name type **Pier Mullion** and then press ENTER.
    - Right-click on Pier Mullion and then choose **Properties**.

Configure it the same way as the Spandrel Mullion. Refer to the process above in the "Edit the Horizontal Mullions" topic.

- Set the "Thickness" to **1′-0″ [300]**.
- Set both the "Width" parameters to **6″ [150]**.

This will make a 1′-0″ [300] square pier Mullion.

- Choose the same Material as above: Masonry – Stone and then click OK.

24. Return to the "Type Properties" dialog for the MRB Front Façade Type again.
    - Beneath the "Vertical Mullions" grouping, for "Interior Type," choose **Rectangular Mullion : Pier Mullion.**
    - For the "Border 1 Type" and "Border 2 Type" Choose **None** (see Figure 8–40).
    - Click OK to accept the changes and close the dialog.

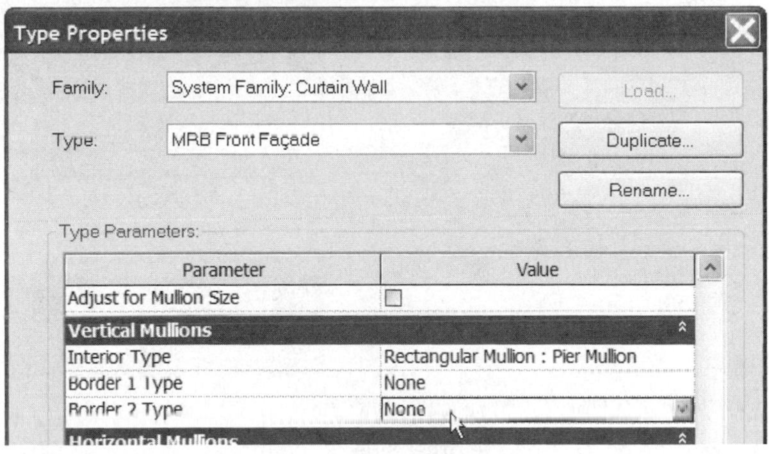

Figure 8–40 *Remove Mullions from the Vertical Borders*

We chose "None" for the two "Border Types" because our design calls for butt glazing at these locations. However, if you study the model in 3D, you will notice that there are still Mullions in these locations. This is because it is possible to have Revit Building automatically apply a new Mullion Type to Curtain Grids or existing Mullions via the Curtain Wall Type, but it will not delete existing Mullions automatically. If you draw a new Curtain Wall using this Type as it is currently configured, it would *not* have Mullions at the start and end vertical borders. (Try it out if you like, but be sure to erase any Curtain Walls that you draw when you are finished exploring). The one we have on screen however, still has the small rectangular Mullions inherited from the Storefront Type duplicated above.

There are two approaches that we could take to rectify the situation. First, we could simply delete the extraneous Mullions. In this model, that would be fairly easy to do. However, in larger projects this might become tedious. The other approach is a work around method that is quite effective. We can change the Type of the Curtain Wall back to a Basic Wall Type and then switch it right back to a Curtain Wall. When you change a Curtain Wall to a Wall, all Mullions and Grids (whether manual or Type-driven) will be deleted. This is because these items are not supported on Basic Walls. Let's explore both approaches briefly.

25. Using the TAB key, select one of the vertical edge Mullions.
    - Right-click and choose **Select Mullions > Select Mullions on Gridline**.
    - Press the DELETE key to delete the selected Mullions (see Figure 8–41).

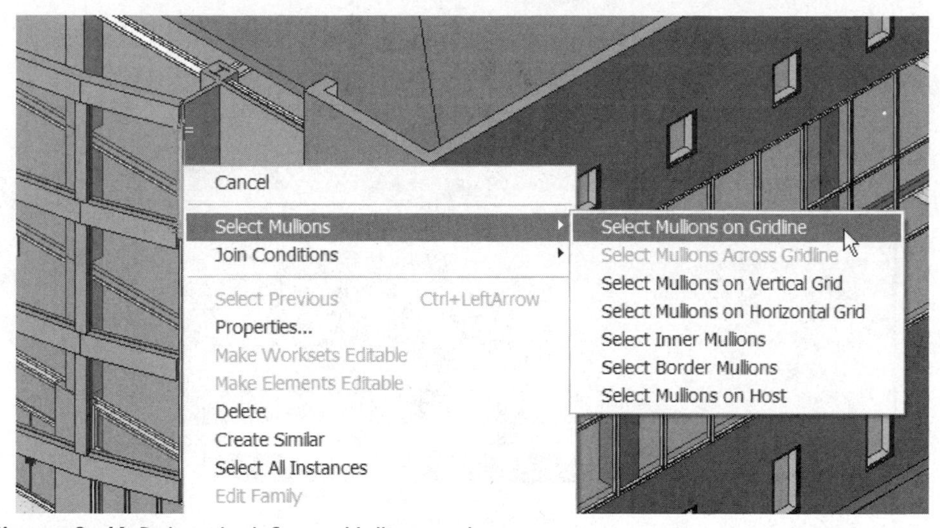

Figure 8–41 *Delete the left over Mullions at the corners*

    - Repeat the process on the other side of the model, or

26. Select the curved Curtain Wall.
    - From the Type Selector choose Basic Wall : Generic - 5" [Basic Wall : Generic - 125mm].

    An "Ignorable Warning" will appear explaining that Grids and Mullions were deleted. You can simply keep working and this dialog will close on its own.

    - With the Curtain Wall still selected, from the Type Selector choose MRB Front Façade.

Regardless of the process you choose, the vertical Mullions at the edges should no longer be there.

27. Save the model.

SUB-DIVIDE CURTAIN WALL BAYS

We can use virtually any of the techniques that we have covered so far to sub-divide the Curtain Wall that we have into smaller bays. We can add Curtain Grid lines and then apply Mullions to them and we can even nest other Curtain Wall designs in as Panels. Let's open the *South* elevation View and experiment with a few of these techniques.

28. On the Project Browser, double-click to open the *South* elevation View.

Our design currently has six bays—four equal ones in the center and two slightly smaller ones at the ends. Remember, the fact that the pattern is centered is a function of the Element's instance properties and not something that we can set in the Type. The spacing of the bays was determined by the Type and is at this point rather large. We can sub-divide these panels into smaller ones by simply adding Grid lines as we did above.

29. On the Design Bar, click the Modeling tab and then click the Curtain Grid tool.

- Move the pointer near the left edge of the Curtain Wall.

A Grid line and temporary dimensions will appear.

- Using the temporary dimensions, click to create a horizontal grid line at the two-thirds mark (see Figure 8–42).

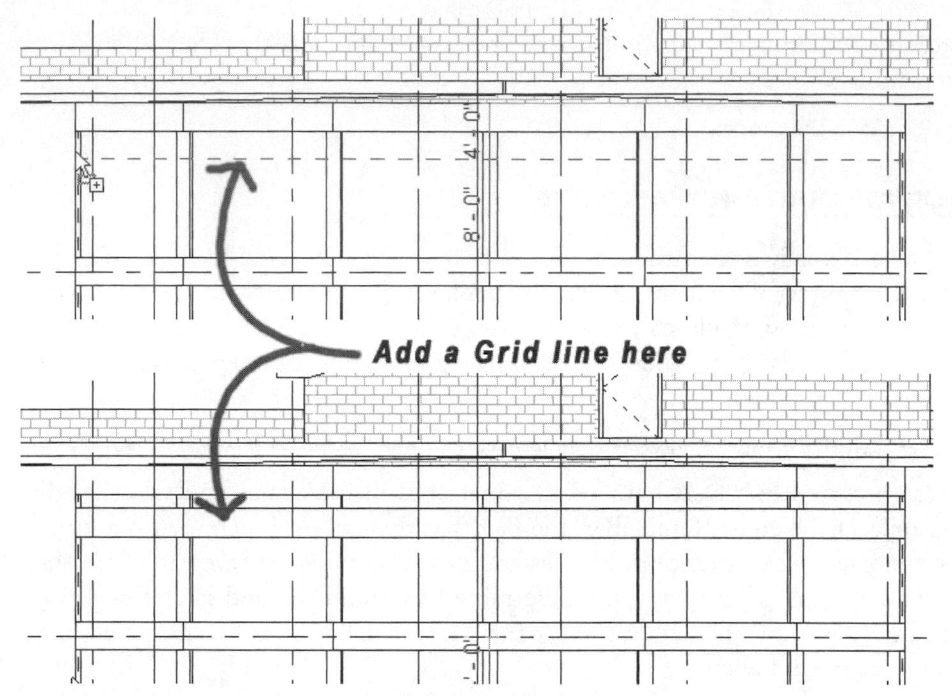

Figure 8–42 *Create a Grid line that divides the bay at the two-thirds mark*

Notice that the new Grid line automatically creates Mullions in the Spandrel Type. It might be nice to use something a little less "heavy" for these intermediate Mullions. In fact, let's make a new Mullion Type for these horizontal bands.

30. On the Project Browser, expand the *Families > Curtain Wall Mullions > Rectangular Mullion* node (as we did above).
31. Right-click on 2.5″ x 5″ rectangular [50 x 150mm] and choose **Duplicate**.
    - For the name type **Horizontal Band Mullion** and then press ENTER.
    - Right-click on Horizontal Band Mullion and then choose **Properties**.
    - Set the "Thickness" to **1′-6″ [450]**.
    - Set both the "Width" parameters to **2″ [50]**.

This will make a 4″ x 18″ [100 x 450] band Mullion.

- Leave the Material set to: Metal – Aluminum and then click OK to exit the dialog.

32. Pre-highlight one of the new Mullions (added automatically above to the new Grid line), right-click and choose **Select Mullions on Gridline**.

This is a handy way to select an entire row of Mullions very quickly. Notice that when these Mullions highlight, that they are all pinned to the Curtain Wall (see

Figure 8–43). Before we can assign the new Horizontal Band Mullion to them, we must unpin them. You can click each of the pushpin icons individually, but this would be very tedious. Instead, use the command on the Edit menu.

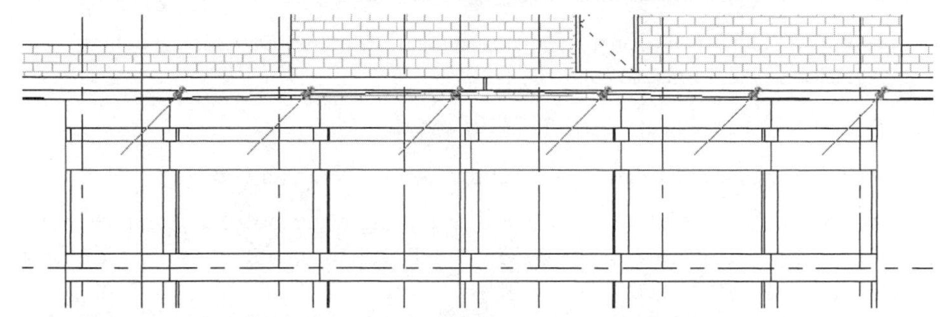

Figure 8–43 *The selected Mullions are all pinned to the Curtain Wall Type*

33. From the Edit menu, choose **Unpin Position**.

TIP The keyboard shortcut for Unpin Position is UP.

- From the Type Selector, choose Horizontal Band Mullion.

34. Repeat this process to add horizontal Grid lines and Mullions to the other two floors as well.
    - On the Design Bar, click the Modeling tab and then click the Curtain Grid tool.
    - Add a horizontal Grid line to each of the remaining floors at two-thirds the height.
    - Select and unpin the Mullions and then assign the Horizontal Band Mullion Type.

ADJUST MULLION POSITION

Take a close look at the Curtain Wall in the South elevation View and you will note that the size of the horizontal bays is not equal.

35. On the Project Browser, double-click to open the *South* elevation View.

Notice that the Level lines for Level 3 and Level 4 pass through the middle of the Spandrel Mullions, but at Level 2 and the Roof they do not. It takes two steps to adjust this.

36. On the Project Browser, double-click to open the *Level 2* floor plan View.

37. Select the Curtain Wall and each of the other four Walls (created at the start of the sequence) that make up the front façade.

   - On the Options Bar, click the Properties icon.
   - In the "Element Properties" dialog, beneath the "Constraints" grouping, set the "Base Offset" to **-1′-0″ [-300]**.
   - Set the "Top Offset" to **1′-0″ [300]** and then click OK.
   - If an error message appears, click the Delete Elements button.

If you return to the *South* elevation, you will notice that the bottom Spandrel has shifted down so that it is now centered on the Level line, but all of the others shifted down as well.

38. Select the Curtain Wall.

In the center of the Curtain Wall a "Configure Grid Layout" control appears.

   - Click the "Configure Grid Layout" control to activate it (see Figure 8–44).

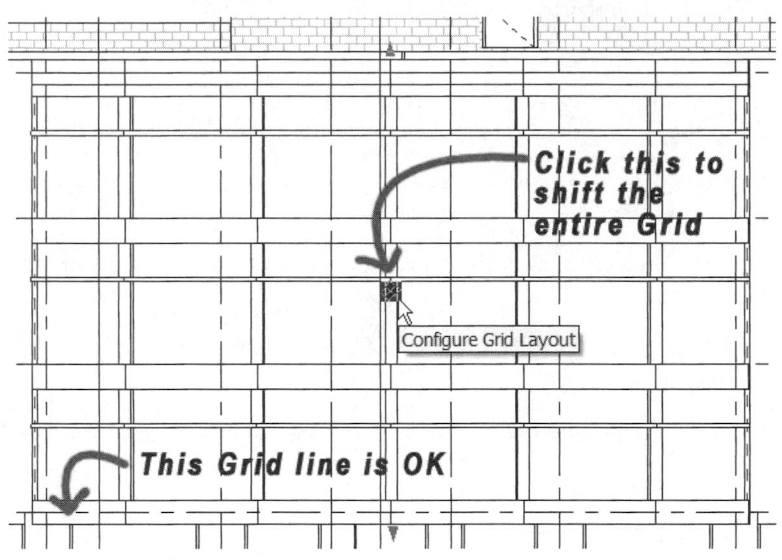

Figure 8–44 *Apply an Offset to the Curtain Wall to shift the start point of the bays*

Two temporary dimensions will appear: one at the top prefixed by the number 1, and one at the left prefixed by 2. Number 1 (at the top) shifts the Grid vertically, while number 2 (at the left) shifts it horizontally.

   - On the left side, click in temporary dimension number 2, type **1′-0″ [300]** and then press ENTER.

Another Autodesk Revit Building error dialog will appear. The messages appear in this case because each time we shift the Grid with the techniques covered here, it forces Revit Building to create new Mullions based on the Curtain Wall Type. Some of the elements that it is creating or replacing need to be deleted and the error message is confirming that situation.

- In the error message box, click Delete Elements.

All of the Spandrels should now be centered on their respective Level lines (see Figure 8–45).

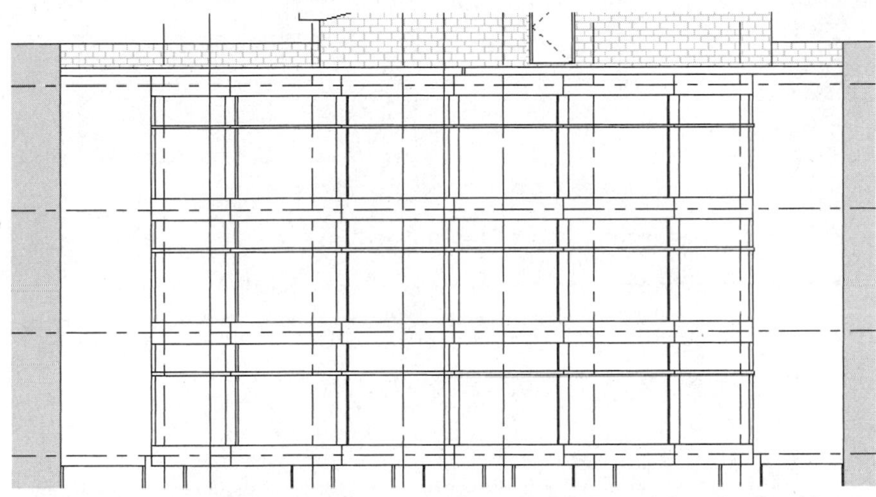

Figure 8–45 *With the Offsets applied, the Spandrels are all centered on the Level lines*

SWAP PANEL TYPES

Using a process nearly identical to the above, we can select groups of Panels and swap them for a different Type as well.

39. On the View toolbar, click the Default 3D View icon (or open the {*3D*} View from the Project Browser).
40. Pre-highlight one of the small horizontal panels at the top of the Curtain Wall (see Figure 8–46).

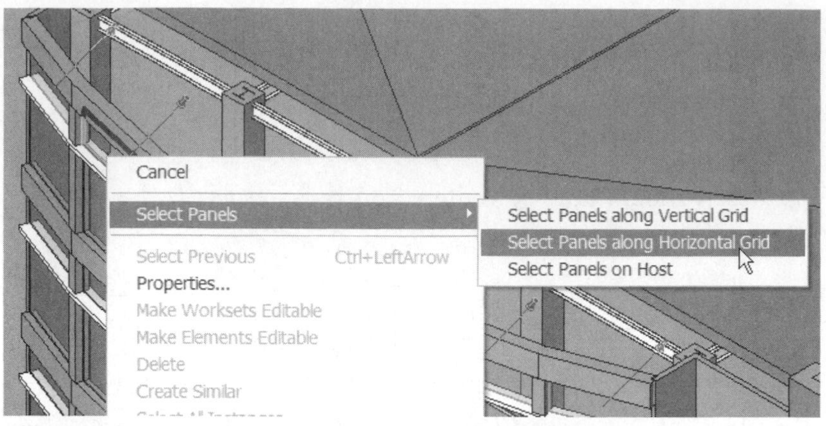

Figure 8–46 *Select one Panel then right-click to select the entire row*

- From the Edit menu, choose **Unpin Position** (or simply type UP).
- From the Type Selector, choose System Panel : Solid.
- Repeat for the thin top bay on each of the other floors (see Figure 8–47).

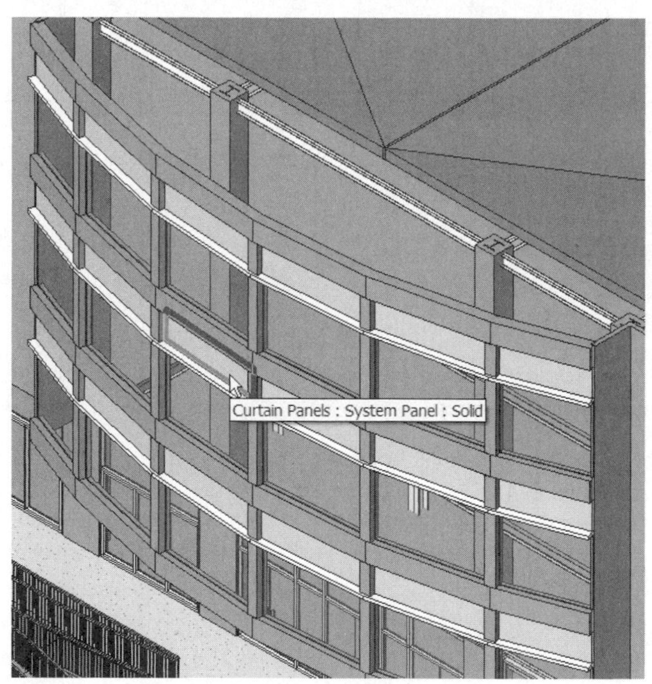

Figure 8–47 *Apply solid panels to all of the top bays*

Plenty more refinements remain to finalize the front façade of our commercial building. We could sub-divide the bays further in the vertical direction and either apply an existing Mullion Type to these Grid lines, or create another new one.

The Other Walls on the front façade could use some punched Windows or we could even choose a Curtain Wall Type for them instead. The Panels of the Curtain Walls can even be substituted for another Curtain Wall Type. Be careful when doing this however as Revit Building tends to be very fussy about what it allows here. In some cases you can nest your Curtain Wall Types to make a very complex design. In other cases, Revit Building will display an error and not allow you to complete the task. In these cases, use the manual Grid line technique covered here instead. While it is a bit more work to do it this way, Revit Building rarely generates an error when doing so. Whatever explorations you wish to continue on to in this dataset, the specifics are left to you as an exercise. We will explore one more topic before ending this chapter—Stacked Walls.

WORKING WITH STACKED WALLS

At the start of the chapter, we build a custom Wall Type and applied it to the exterior Walls. We can go even further and create a special kind of Wall Type called a "Stacked Wall." A Stacked Wall simply uses two or more Revit Building Wall Types stacked on top of one another to form a more complex design.

CREATE A STACKED WALL TYPE

In the simple example that follows, we will create a Stacked Wall that combines the Wall Type we created at the start of the chapter with another that is already resident in the dataset to form a base condition at the first floor.

1. On the Project Browser, expand the *Families* > *Walls* > *Stacked Wall* node.
2. Right-click on Exterior - Brick Over CMU w Metal Stud [Exterior - Brick Over Block w Metal Stud] and choose **Duplicate**.
    - Right-click on Exterior - Brick Over CMU w Metal Stud (2) [Exterior - Brick Over Block w Metal Stud (2)] and choose **Rename**.
    - For the name type **Brick Exterior with Stone Base** and then press ENTER.
    - Right-click on Brick Exterior with Stone Base and then choose **Properties**.
    - Beneath "Type Parameters" click the Edit button next to "Structure."

The Structure of a Stacked Wall is very simple. You designate each Type to use and how tall it should be. You can insert additional Types by clicking the Insert button. Use the Up and Down buttons to move the various Wall Types relative to one another. Each Type will be assigned an explicit height except for one (use the

"Variable" button to choose which is variable). This one will adjust to the actual height of the individual Wall instance in the model.

3. Click next to Type 1.
   - From the pop-up menu, choose Exterior Shell (this is the Type we created at the start of the chapter).
   - For Type 2, choose Exterior Base.
   - Change the height of Exterior Base to **6′-0″** [**1800**].
   - In the Offset field next to each Type, input **4″** [**100**].

This shifts the Wall Type overall to compensate for the additional thickness that the base material adds. In this way, the Wall will remain in the same relative position in the model.

- Click the Preview >> button (see Figure 8–48).

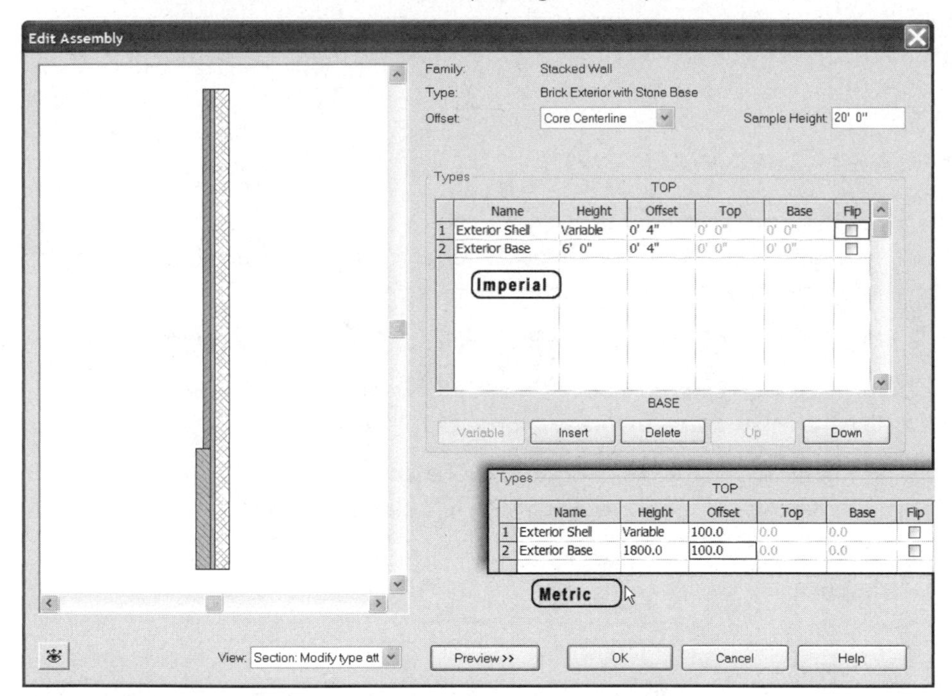

Figure 8–48 *Configure the Stacked Wall Type*

4. When you are done studying the settings, click OK twice to dismiss the dialogs.
5. Select one of the Exterior Shell Walls.
   - Right-click and choose **Select All Instances**.

- From the Type Selector, choose Stacked Wall: Brick Exterior with Stone Base (see Figure 8–49).

You might get an Ignorable Warning. You may also need to re-cut the Wall Geometry with the Curtain Walls on the sides of the building. Make these and any other final adjustments that you wish.

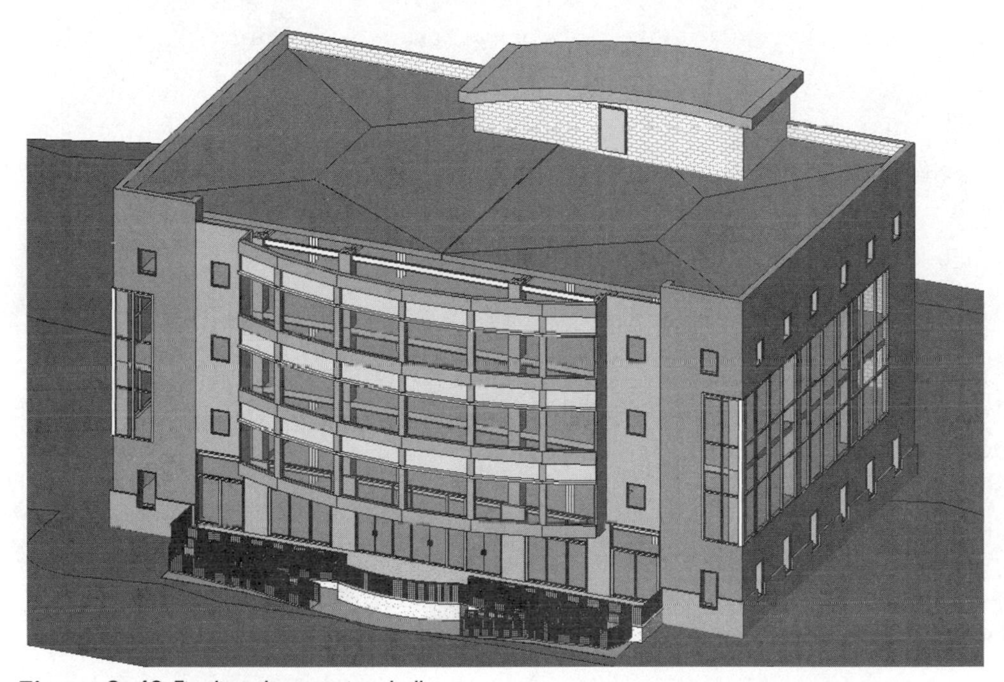

Figure 8–49 *Finalize the exterior shell*

6. Save the model.

SUMMARY

Revit Building provides three types of Walls: Basic Wall, Curtain Wall and Stacked Wall.

You can create a custom Wall Type and edit it to add and delete Layers.

Curtain Walls can be drawn like Walls or even placed within other Walls.

Use Cut Geometry to cut the mass of the Curtain Wall as a hole from the Wall.

Curtain Wall Grids can be sketched manually or defined within the Curtain Wall Type.

Mullions can be added to Grid lines.

Mullion Joins can be edited to suit your preferences.

Walls can be used as panels for Curtain Walls.

A Curtain Wall Type can contain a pre-defined pattern of horizontal and vertical Grid lines.

A Stacked Wall contains one or more Basic Walls stacked on top of each other.

CHAPTER 9

Working with Families

All elements that you create and use in an Autodesk Revit Building project belong to a Family. As noted in Chapter 1, a Family is an object that has a specific collection of parameters and behaviors. Within the limits established by the Family and its parameters, an endless number of "Types" can be spawned. Understanding Families and how to manipulate them is an important part of learning Autodesk Revit Building. Families are the cornerstone of the Revit Building parametric change engine.

OBJECTIVES

While there are several kinds of Families in Revit Building, the main focus of this chapter will be "Model Families." Model Families are simply Families that describe some physical component in the model. Conceptually however, topics discussed in this chapter apply to any kind of Revit Building Family, including annotation Families such as tags and title blocks. After completing this chapter you will know how to:

- Explore the Families contained in the current project
- Insert instances of Families in your model
- Manipulate Family Types
- Create custom Families
- Create custom Parametric Families

KINDS OF FAMILIES

In the "Autodesk Revit Building Elements" heading of Chapter 1, a detailed discussion is given of the various kinds of elements available in Autodesk Revit Building. They include Hosts (System Families), Components (Families), In-Place Families, Datums, Views and Groups. Figure 1–1 gives a summary of this.

System Families are simply more intelligent elements that have more pre-programmed behaviors than Component Families. An example is the ability to host other objects and automatically adjust depending on what is being hosted—such as a Wall (System Family) that hosts a Window (Family). The Wall knows how to automatically interact with the window. System Families can not be created or edited outside of the Project Editor. There must always be at least one type of every System Family in a project (Wall, Level, Roof, Text notes, etc.). This allows you to duplicate the one Type and make unlimited additional Types—but always in the Project Editor. Most, if not all of their behaviors and parameters for such elements are pre-programmed without the need for user interaction.

Families are created externally (in the Family Editor) and are loaded into projects. The user can create nearly unlimited parameters and behaviors for such a Family. Since they are created outside of the project editor context, they can be loaded into any project.

As has already been noted in the introduction, many of the concepts covered in this chapter might apply equally to both System Families and Families. However, for the purposes of the following discussions and tutorials, we will limit our discussion to the use and manipulation of Families—in particular, Model Families. Should you need to modify System Families many of the same techniques can be employed.

FAMILY LIBRARIES

Although a well-conceived template project is a critical component in successful Revit Building implementation (refer to Chapter 4 for more information), a well-stocked library is just as important. In Revit Building, all Families are part of the project file. Therefore, it can be tempting to include *all* required Families directly in the project template file so that each new project starts with *everything*. However, each Family that is included in the template increases the size of every project file created from that template even when there are no instances of those Families in the model. This can add up quickly to bloated and unwieldy project file sizes. Furthermore, it makes it difficult to manage your library of components if they all live in and are being copied to every active project in the office. For these reasons, it is important to consider carefully those items that are better included in a readily accessible office standard library instead of the template file.

UNDERSTANDING WHAT IS PROVIDED "OUT OF THE BOX"

The "Library" is nothing more than a series of folders on your hard drive. Revit Building ships with a large library of ready-to-use items. Some of these are included in sample template files (see Chapter 4) and others are provided in Family files (RFA) in library folders. For example, the default United States installation installs the "Imperial Library" located by default in the *C:\Documents and Settings\All Users\Application Data\Autodesk\Revit Building 8.1\Imperial Library* folder (see Figure 9–1). (Please note that the "*Application Data*" folder is hidden by Windows™

by default. To make it visible, choose **Folder Options** from the Tools menu in Windows Explorer and turn on Hidden Files)

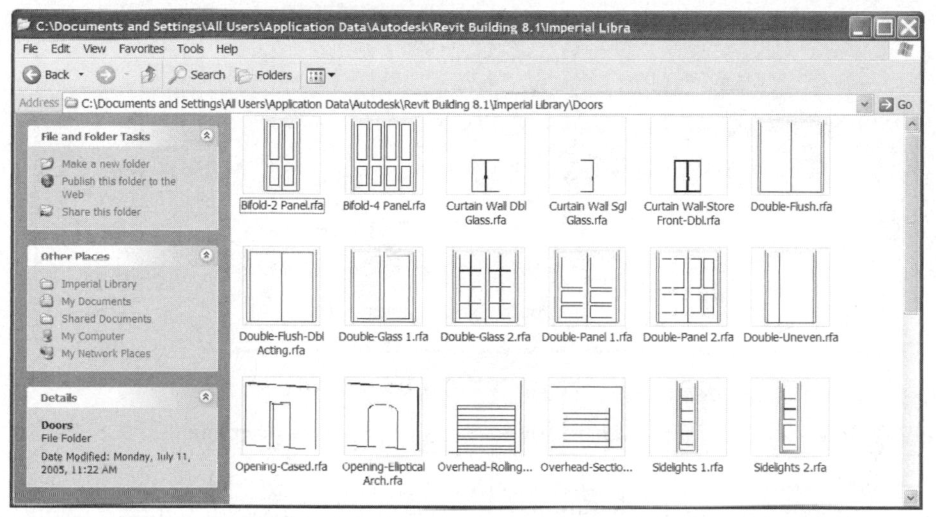

Figure 9–1 *An example of one of the folders in the Revit Building Imperial Library*

NOTE The specific location of your library files will vary with your locality and particular version of Revit Building. Check the documentation that came with your product for the specific location.

In Figure 9–1 the Doors folder is shown in Windows Explorer in "Thumbnail" view. This is a convenient way to quickly assess the contents of each library folder and subfolder. There are many items in the library. For example, the Imperial library contains nearly 700 Family files. As we have seen in previous chapters and as we will discuss in more detail below, a Family can contain one or several "Types." Therefore, the nearly 700 imperial Family files represent potentially several thousand more than that in readily available component elements that can be added to our projects. It is a good idea to become as familiar as you can with the provided content. The reason is simple; it is always easier to use or modify something that exists than it is to create it from scratch. You can browse through the library in either Windows Explorer as already noted or directly from Revit Building. If you are in Revit Building, simply choose **Open** from the File menu and then click the shortcut icon on the left (i.e. *Imperial Library* or *Metric Library*) to open your library. Double-click a subfolder to view its contents. You can select a file (single-click) to see a preview on the right, or if you wish, you can use the View Menu icon at the top of the dialog to browse by Thumbnails.

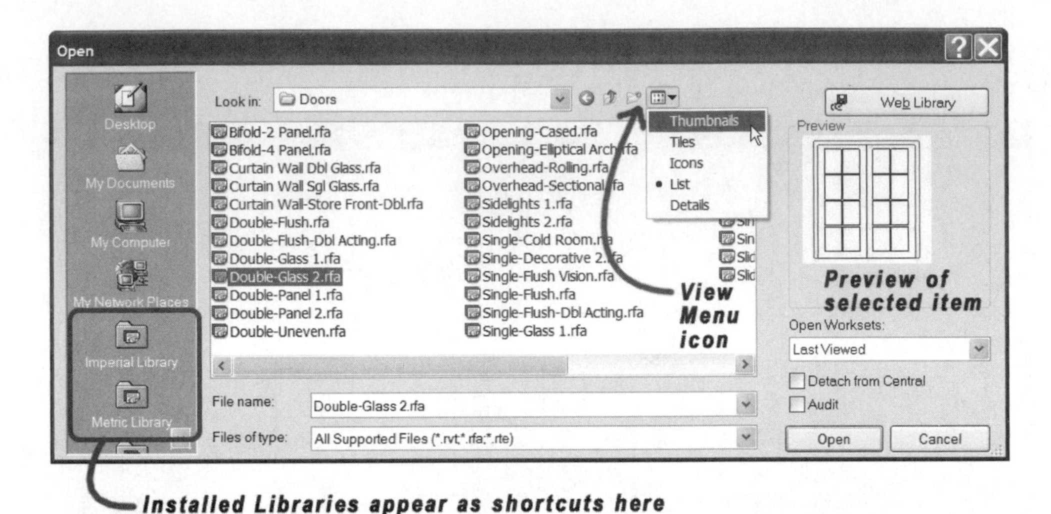

Figure 9–2 *Access a Library from within Revit Building using the Open command*

No matter what method you use to browse your library, you can always open a Family file directly into the Revit Building interface. In Windows Explorer, simply double-click on the file you wish to edit. From the open command, double-click the file or select it and then click the Open button. The Family file will open into the Revit Building interface. The interface when a Family file is loaded is slightly different than when a project file is loaded. In such a state, the interface is often referred to as the "Family Editor." While in the Family editor, the Design Bar will show only one tab with Family-specific tools. Some of the other functions will appear differently as well. You will still have a Project Browser, but it typically includes only a few Views that have been developed for editing the current Family. These Views are only seen while in the Family Editor and do not appear in projects to which the Family is loaded (see Figure 9–3).

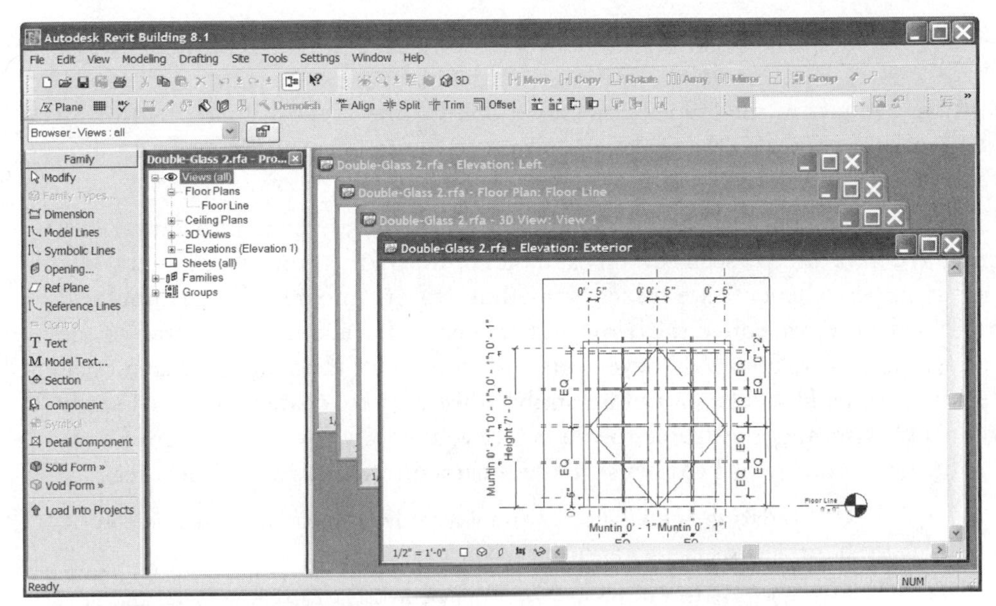

Figure 9–3 *The Family Editor*

If you are following along in this passage within Revit Building, feel free to view and open as many Family files as you wish. However, please do not make any edits or save any changes to the out-of-the-box Family files. Treat this as an exploratory exercise. In the lessons that follow later in this chapter, we will have the opportunity to load, edit and build our own Families. The purpose of the current discussion is to give you familiarity with what has been provided with your software.

FAMILY STRATEGIES

Do take the time to explore your Family library. As you analyze each of the library items, many of the items will seem useful and there are likely those that you will deem not useful. Still others will prove useful after they have been modified in some fashion. If you take the time to perform this analysis of the included library content before committing any time and resources to customizing existing items and/or building new items from scratch you can save yourself a great deal of effort and frustration. The other benefit to this process is that you will undoubtedly discover ways of performing certain tasks or representing certain items that you had not considered on your own. In other words, reverse engineering existing Families can prove to be a tremendous learning experience. Just remember not to save changes to existing files. Always make a copy or "Save As" first.

In addition to any libraries that have been installed with your product on your local system you may also have access to other libraries maintained by your firm's IT personnel. Libraries of this type are typically stored on the company network and accessible via your local or wide area network. Furthermore, Autodesk maintains a Web Library of Revit Building and Revit Structure content. It is accessible from within the product in the "Open" dialogs. Look for the "Web Library" button in the upper right corner of the dialog (an example can be seen in Figure 9–2). The Web Library will be covered in more detail in the tutorials below.

Nearly everything you do in Revit Building involves the use or manipulation of a Family. We have already seen several examples of this in the previous chapters. We "used" Families whenever we added something to our model. We manipulated Families whenever we edited the Type and or clicked the "Duplicate" button in the "Type Properties" dialog. Whenever you wish to add an element to your model, the element you add will be part of a Family. Always try to locate and use a Family that already exists and is available to you before you modify or create custom Families. The basic strategy of Family usage is to maintain the following priorities:

- 1st—**Use Existing** See if what you want already exists somewhere (in the current project or in a library) and simply use it.
- 2nd—**Modify Existing** If the precise Family you want does not exist, find a Family that is close to what you need, duplicate and modify it.
- 3rd—**Create New** If necessary; create a new Family to represent exactly what you need.

This basic approach is somewhat obvious and logical. However, it is surprising how often users will either resort immediately to building custom components without first checking the libraries, or do the opposite and settle for some less than ideal component in their models. Follow this simple three-step guideline, and you will always have the right Family for the job. If you follow the recommendations here and become familiar with the libraries available to you, it is likely that you will only need to "Use" or "Modify" a Family rather than "Create" a new one. The remainder of this chapter is devoted to tutorials that will illustrate this.

ACCESSING FAMILIES IN A PROJECT

If you wish to use an existing Family in your project, the first step is to explore what is available. The first place you should look is the current project. You can quickly see all of the Families available in the current project in the Project Browser.

INSTALL THE CD FILES AND OPEN A PROJECT

The lessons that follow require the dataset included on the Mastering Revit Building CD ROM. If you have already installed all of the files from the CD, simply skip down to step 3 below to open the project. If you need to install the CD files, start at step 1.

1. If you have not already done so, install the dataset files located on the Mastering Revit Building CD ROM.

   Refer to "Files Included on the CD ROM" in the Preface for instructions on installing the dataset files included on the CD.

2. Launch Autodesk Revit Building from the icon on your desktop or from the **Autodesk** group in **All Programs** on the Windows Start menu.

   - From the File menu, choose **Close**.

This closes the empty new project that Revit Building creates automatically upon launch.

3. On the Standard toolbar, click the Open icon.

TIP The keyboard shortcut for Open is CTRL + O. **Open** is also located on the File menu.

   - In the "Open" dialog box, click the *My Documents* icon on the left side.
   - Double-click on the *MRB* folder, and then the *Chapter09* folder.

   If you installed the dataset files to a different location than the one listed here, use the "Look in" drop down list to browse to that location instead.

4. Double-click *09 Commercial.rvt* if you wish to work in Imperial units. Double-click *09 Commercial Metric.rvt* if you wish to work in Metric units

   You can also select it and then click the Open button.

The project will open in Revit Building with the last opened View visible on screen.

ACCESSING FAMILIES FROM PROJECT BROWSER

There are a few ways to see which Families are already loaded into a project. Whenever you choose a tool from the Design Bar, the list of Family/Type combinations will appear in the Type Selector. The format is **Family Name : Type Name.** We have seen this already in several earlier chapters. While this method is effective for a particular class of element such as a Door, or a Window, the easiest way to see a list of loaded Families is via the Project Browser. The Project Browser has several major nodes or branches. The first branch is the Views branch. We have spent nearly all of our time in previous chapters on this node. Beneath this one are specialized View types such as Legends, Schedules and Sheets. The Families and Groups nodes are at the bottom. If you expand the Families node, you will see each element category currently loaded in the project. Expand any category, such as Doors, and you will see each Family of that category. Finally, expanding one step further reveals all of the Types for each Family.

5. On the Project Browser, double-click to open the *Level 3* floor plan View.

NOTE To make the contents of the Project Browser easier to read, drag the edge of the project Browser window to widen it.

6. On the Project Browser, collapse the *Views (all)* and *Sheets* nodes (click the minus (-) sign icon).
   - Expand the *Families* Node, and then expand *Doors.*
   - Expand Single-Flush [M_Single-Flush] (see Figure 9–4).

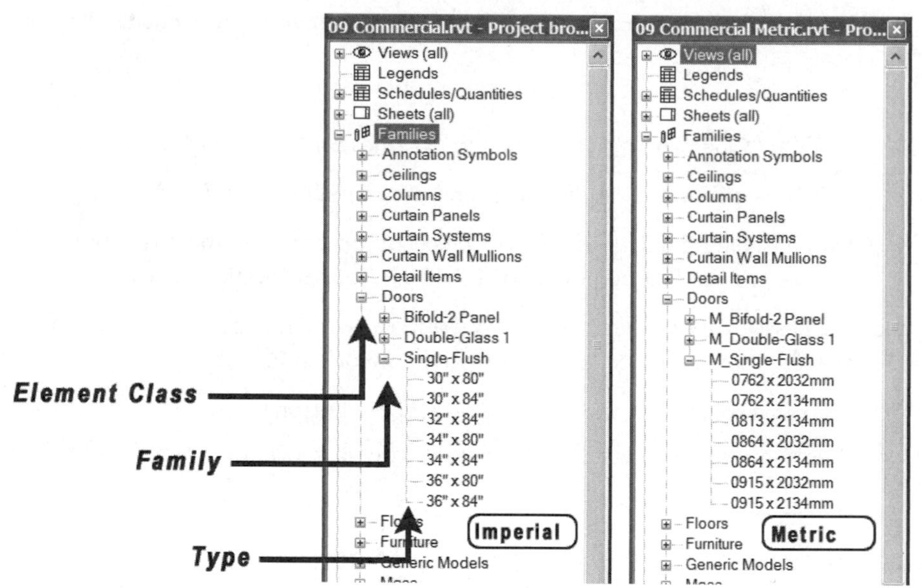

Figure 9–4 *Expanding the Families branch of the Project Browser*

7. On the Design Bar, click the Basics tab and then click the Door tool.
   - Open the Type Selector and compare the list of Family : Types to the list shown in Figure 9–4.
   - On the Design Bar, click the Modify tool or press the ESC key twice.

As noted above, all of the Family/Type combinations for Doors will be listed in the Type Selector. In addition to simply taking inventory of Families and Types in the project, you can also use the list on Project Browser to manipulate and interact with Families. This is achieved via the right-click menu. You can right-click on the *Family* and *Type* nodes of the Project Browser tree. (Right-clicking the *Families* or *Element* class nodes will not produce a menu).

8. Right-click on the Type 36″ x 84″ [0915 x 2134mm] and take note of the menu that appears (see Figure 9–5).

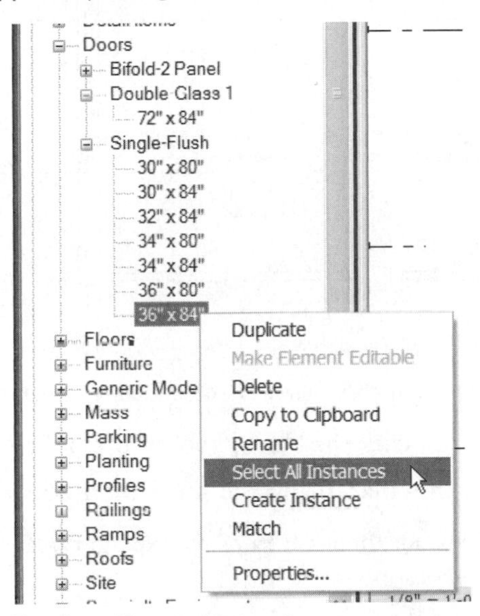

Figure 9–5 *Right-click a Family or Type in Project Browser to access menu options*

- From the right-click menu, choose **Select All Instances**.

All of the Doors in the model that use this Type will be selected. Be careful with this particular command as it selects *all* instances in the model, not just those visible in the current View. There are many other commands available on the right click menu as well. You can Duplicate the current Type, Rename it, or edit its Properties. You can also create an Instance directly from this menu, rather than first executing the tool on the Design Bar and choosing the Type from the Type Selector. In some cases this can be quicker; the end result is the same, it is simply a matter of personal preference.

- Press the ESC key to deselect the Doors.

9. Right-click on the Type 36″ x 84″ [0915 x 2134mm] and choose **Create Instance**.

- On the Options Bar, remove the checkmark from the "Tag on Placement" checkbox.

- Add a Door to the conference room in the bottom middle of the plan (see Figure 9–6).

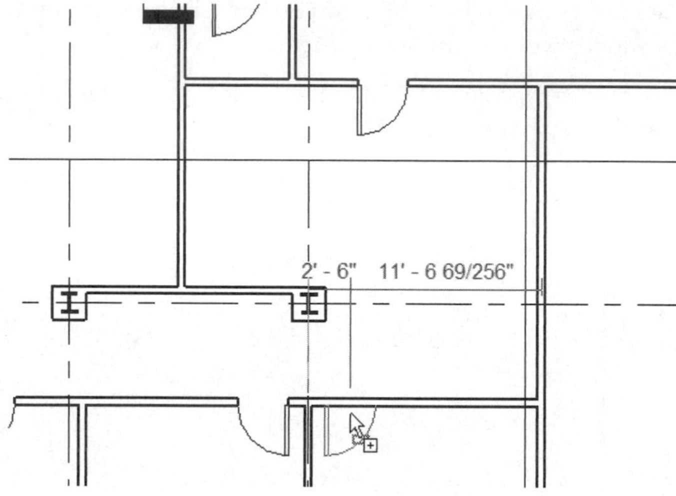

Figure 9–6 *Use the Family Tree right-click menu to add a Door to the Conference Room*

- On the Design Bar, click the Modify tool or press the ESC key twice.

Let's add some furniture to this plan using the same method.

10. On the Project Browser, beneath Families, expand the *Furniture* node and then the *Desk* [*M_Desk*] node.
    - Right-click on 72" x 36" [1830 x 915mm] and choose **Create Instance**.
    - Place the new Desk in one of the offices.
    - Repeat in other offices. Press the SPACEBAR to rotate the Desk before placement (see Figure 9–7).

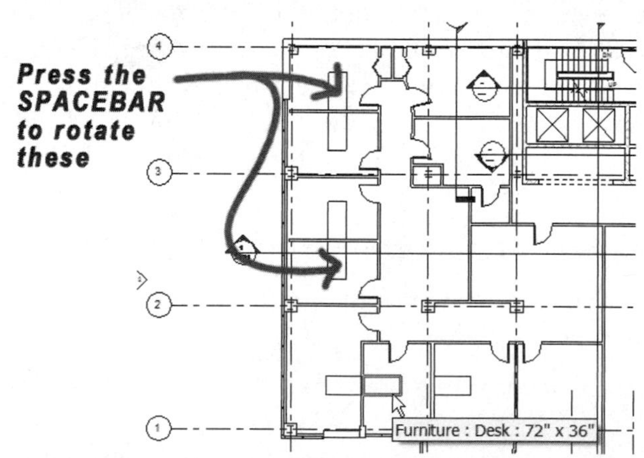

Figure 9–7 *Place Desk Families into the model from the Project Browser*

- On the Design Bar, click the Modify tool or press the ESC key twice.

11. Save the model.

MATCH TYPE PROPERTIES

One of the useful functions on the Type's right-click menu is "Match." With this command, you can apply the Type parameters of the item highlighted in Project Browser to an element already in the model. This is a quick way to "paint" a Type's properties onto existing element. We can try this out on the entrance to the suite on Level 3. Currently there is a single Door there. Let's make it a double glass Door.

12. On the Project Browser beneath *Families*, expand *Doors* then *Double-Glass 1* [*M_Double-Glass 1*].

- Right-click on 72″ x 84″ [1830 x 2134mm] and choose **Match**.

When the cursor is moved into the plan view it will change to an eyedropper shape.

- Click on the Door at the entrance to the suite (see Figure 9–8).

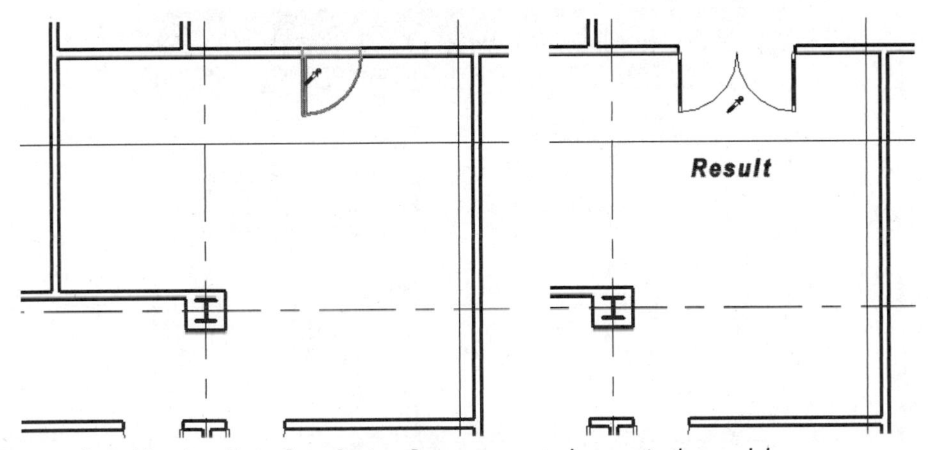

Figure 9–8 *Match a Type from Project Browser to an element in the model*

The Door will change to a double glass Door.

- On the Design Bar, click the Modify tool or press the ESC key twice.

LIBRARIES

So far we have limited our exploration to Families that are already part of our current project. As was mentioned however at the start of the project, you can find potentially limitless Families in external libraries. You have both the out-of-the-box library installed with your version of the software and also remote libraries accessible via the World Wide Web. The process of using such resources is nearly identical to that already covered with the additional step of first locating and loading the Family in the appropriate library.

As we have already mentioned, a library is simply a collection of files and folders stored on your local system or a remote server. To access and place a Family in a project, you must first load it. There are several ways to do this. While you are adding many elements to the model, a Load button is often presented on the Options Bar and is also available within the "Element Properties" dialog. Click this button to open a dialog and browse to an appropriate library folder and Family file. You can also load a Family without first executing a particular model element tool. The command is on the File menu.

LOAD FAMILIES FROM "PREFERRED" LIBRARIES

Continue in the same dataset as the previous topic.

1. From the File menu, choose **Load from Library > Load Family** (see Figure 9–9).

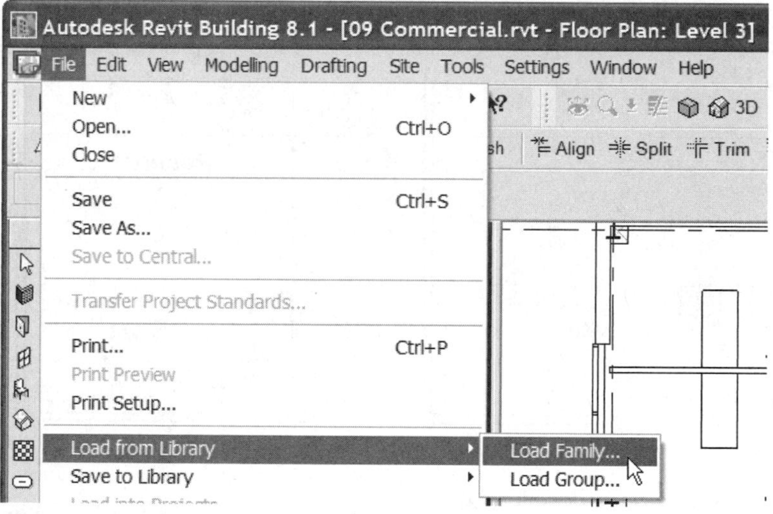

Figure 9–9 *Load a Family from a Library*

An "Open" dialog will appear starting in the default library folder as indicated in your "Options" dialog (Settings menu). Depending on the options you chose during installation, there may be additional libraries available to you. Each library that you have installed or added will appear as a shortcut on the left of the dialog. To add shortcuts, choose Options from the Settings menu, and then click the "File Locations" tab (see Figure 9–10). The small bank of icons in the "Libraries" area on the right side allow you to add a Library shortcut, delete a shortcut and move items up and down in the list.

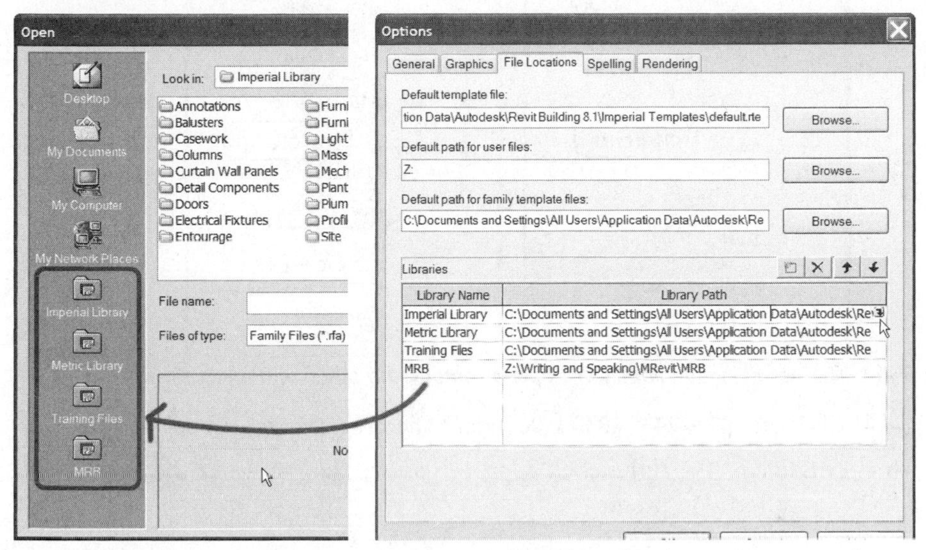

Figure 9–10 *Libraries show as shortcuts on the left of "Open" dialogs, add more in Options*

If your preferred library did not open automatically, use the shortcut icons on the left or the drop-down menu at the top to locate your desired library. The library folder will typically contain several sub-folders. Each folder can contain additional folders or Revit Building Family (RFA) files.

2. On the left side, click the *Imperial Library* [*Metric Library*] shortcut icon.

- At the top of the dialog, click the View Menu icon and choose **Thumbnails.**

This is not necessary, but it makes it easier to quickly locate the Family you need.

- Double-click the *Furniture* folder, select the *Credenza.rfa* [*M_Credenza.rfa*] file and then click Open.

NOTE If your version of Autodesk Revit Building does not include either of the libraries mentioned herein, both Family files have been provided with the files from the Mastering Autodesk Revit Building CD ROM. Look for them in the *Autodesk Web Library\Furniture* folder wherever you installed the dataset files on your local system.

- On the Project Browser, beneath Families, expand *Furniture* to see the new Credenza Family in the list (see Figure 9–11).

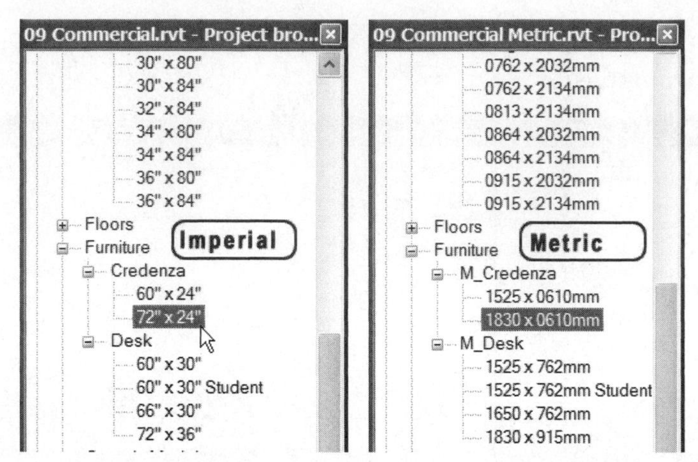

Figure 9–11 *The newly loaded Family will appear among the others in Project Browser*

- Click on 72″ x 24″ [1830 x 0610mm] and drag it to the model window.

This is an alternative to the right-click approach above that achieves the same end. Use whichever method you prefer.

- Place a Credenza in each office.

Remember to press the SPACEBAR to rotate them as you place them.

ACCESS WEB LIBRARIES

In addition to the libraries installed with the product several Web sites are available containing Autodesk Revit Building content and Families. Accessible from directly within the product it's the Autodesk Web Library. You can access it via any "Open" dialog using the "Web Library" button in the upper right corner of the dialog.

3. From the File menu, choose **Load from Library > Load Family**.

Let's give our occupant in the corner office a more fitting desk.

- In the top right corner of the dialog, click the "Web Library" button (see Figure 9–12).

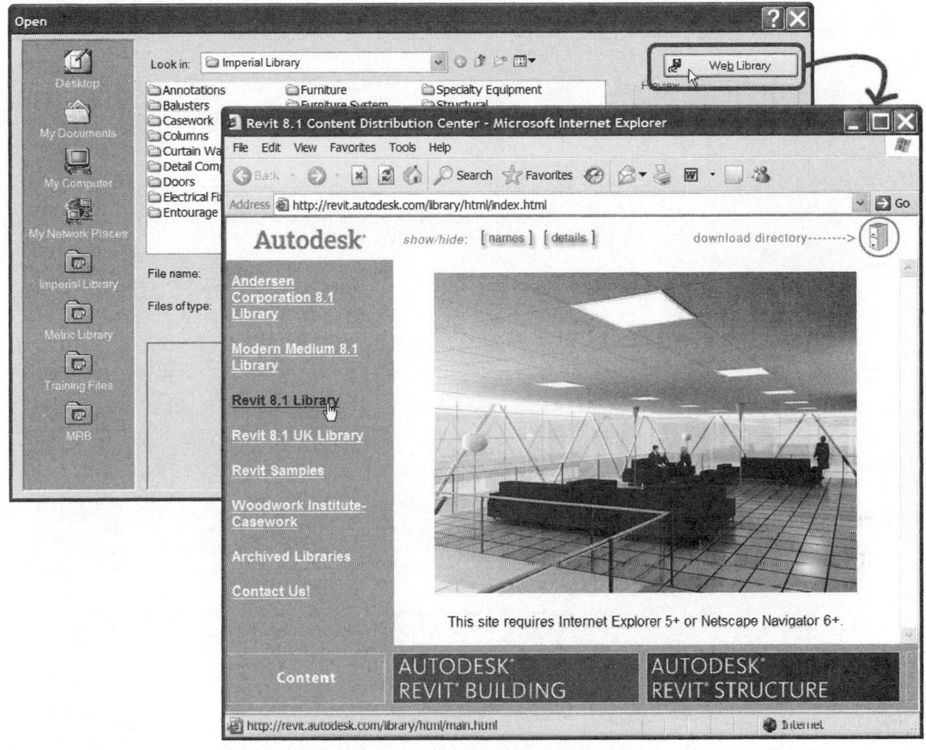

Figure 9–12 *Use the "Web Library" button to access the Autodesk Web Library*

This will open your Internet browser to the Autodesk Revit Building Library web page where you can view and download available content. The web browser application may have opened behind Revit Building. Click on the browser's icon from the Windows Task Bar to make it the active window.

- On the left side of the page, click the "Revit 8.1 Library" link.

This will expand the library.

- Click the Furniture link.

This will reveal a page full of furniture Families on the right. Each Family will have an image preview.

- Scroll as necessary and click the link beneath "Desk-Executive" to download this Family file.
- In the "File Download" dialog that appears, click the Open button (see Figure 9–13).

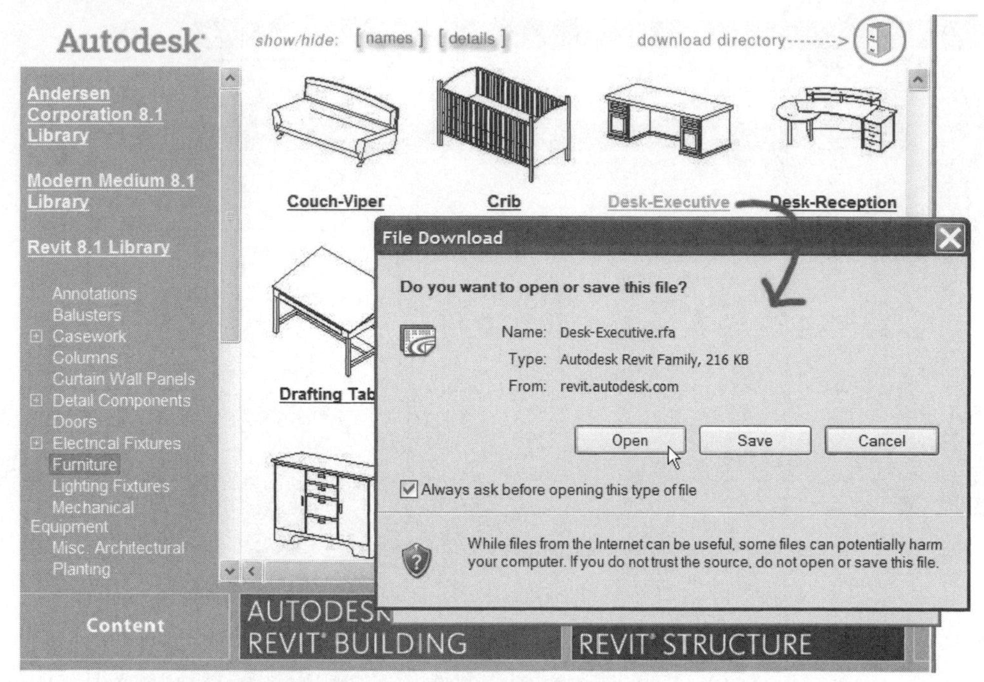

Figure 9–13 *Download a Family from the Web Library and Open it in Revit Building*

This action will download the Family file and open it in your current Revit Building session. We discussed the Family editor above. The Desk-Executive Family file is open in the Family Editor. If you wish, you can study each View, but please do not make any edits to this file.

NOTE In order to access this library using this method, you must have an Internet connection. If you do not have access to the Internet, both Family files have been provided with the files from the Mastering Autodesk Revit Building CD ROM. Look for them in the *Autodesk Web Library\Furniture* folder wherever you installed the dataset files on your local system.

4. On the Design Bar, click the Load into Projects button.

This will load the Family into the commercial project and make its View window active again. (Please note that the Family file is still open in the Family Editor in the background). When convenient you can switch from the project to Family Editor and close the Family file.

5. On the Project Browser, beneath Families, expand *Furniture* to see the new Desk-Executive Family in the list.

- Expand Desk-Executive to reveal that the only Type in this Family is also named Desk-Executive.

- Right-click Desk-Executive and choose **Match**.
- Click the Desk in the corner office to change it to the new Desk-Executive Family.
- On the Design Bar, click the Modify tool or press the ESC key twice.
- Move the Desk as required.

6. Save the model.

There are many websites devoted to the distribution of Revit Building Families and general tech support and discussions. See Appendix A for a further list of websites offering such resources.

EDIT AND CREATE FAMILY TYPES

Till now, we have only worked with existing Families and Types. In many cases, you will need to edit existing Types and create your own Types within existing Families. You should try both of these approaches before resorting to creating a completely new Family. There will certainly be situations where it is appropriate to create new Families. This will be covered below. Let's look at what we can do with Types first.

VIEW OR EDIT TYPE PROPERTIES

In some cases, you will want to understand more about the particular Family or Type you are selecting before you place it in the model. You can view the Type Properties or edit the Family directly from Project Browser. This is also accomplished via the right-click menu.

1. On the Project Browser, beneath *Families*, expand *Furniture* then *Desk [M_Desk]*

- Right-click on 72″ x 36″ [1830 x 915mm] and choose **Properties**.

This method takes you directly to the "Type Properties" dialog. We have seen enough examples of this dialog in previous chapters to know that if we make any edits here, they would apply at the Type level and therefore affect *all* Desks currently in the model (since they all currently share this Type).

- In the "Type Properties" dialog box, click on the "<< Preview" button at the bottom left corner of the dialog.

This will expand the width of the dialog to include an interactive viewing pane on the left side.

- At the bottom of the viewing pane, click the "View" pop-up menu and choose **3D View:View 1** (see Figure 9–14).

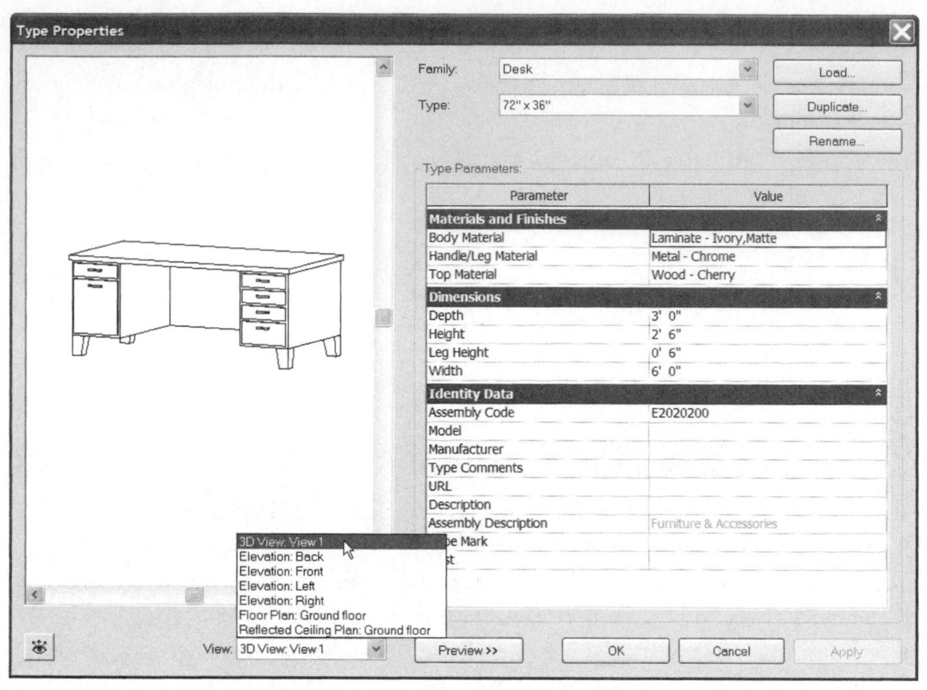

Figure 9–14 *Open the View Pane and choose a 3D view point*

There are several other view options, such as plan, elevation and ceiling views. Try them all out if you wish and then return to the 3D view when finished. If you need to get a better look in any view, you can use standard navigation techniques. Right-click in the viewer to get a menu of standard zoom and scroll commands. There is also a "Dynamically Modify View" icon.

2. At the lower left corner of the dialog, click the "Dynamically Modify View" icon.

- Click the "Spin" button and then drag the mouse in the viewer to spin the Desk model.
- Try other modes if you wish and then close the "Dynamic View" toolbox when finished.

Take a close look at the parameters available at the right of the dialog. First, you can choose any Type from this Family from the Type list at the top and the viewer will react immediately. Try choosing each of the other two Types one at a time to see this. Return to the Type: 72″ x 36″ [1830 x 915mm] when finished. Notice that all three Types look essentially the same. This is critically important to understand. Study the various parameters listed in the tables below the Family and Type lists. These are the "Type Parameters." The shape of this desk, with its two pedestals, four legs, closed back and top are all characteristics of the Family called:

"Desk [M_Desk]." The overall width, height and even the height of the legs on the other hand are parameters of the each Type. As you choose a different Type from the list, one or more of these values will change accordingly. Then it is like choosing between several models of a certain make car. A Honda Civic comes in several models, some have higher quality wheels, or better sound systems, but they are all *Civics*. In this analogy, "Civic" is the Family and "LX" or "EX" is the Type. Returning to our Desk on screen, what this means is that if you simply want a different width or height Desk, or perhaps wanted brass hardware rather than the default chrome, you would simply create a new Type. If however, you wanted a different "make" of desk, with only a right or left pedestal for instance, you would need to create a new Family. Let's explore both scenarios.

- Click Cancel in the "Type Properties" dialog to dismiss it without making any changes.

CREATE A NEW TYPE FROM PROJECT BROWSER

For the group of offices running vertically along the left, let's create a new Type in this Desk [M_Desk] Family that is a slightly smaller size, yet a bit larger than the other Types currently available.

3. Right-click on 72″ x 36″ [1830 x 915mm] and choose **Duplicate**.

This will create a new Type named 72″ x 36″ (2) [1830 x 915mm (2)] with the name highlighted and ready to be renamed. (Alternately, if you prefer, you can right-click the Family name and choose New Type, rename it, and the result will be the same).

- Name the new Type: **66″ x 30″ [1650 x 762mm]** and then press ENTER.

You can give the Type any name you want, but usually it is helpful to include in the name what makes this Type different from others, in this case the size. This is standard procedure in all of the provided Revit Building content and a good practice to follow.

4. Right-click on 66″ x 30″ [1650 x 762mm] and choose **Properties**.

- In the "Type Properties" dialog, change the Depth to **2′-6″ [762]**.

Notice that when you click in this field, a temporary dimension appears in the viewer to indicate how this parameter will apply (see Figure 9–15).

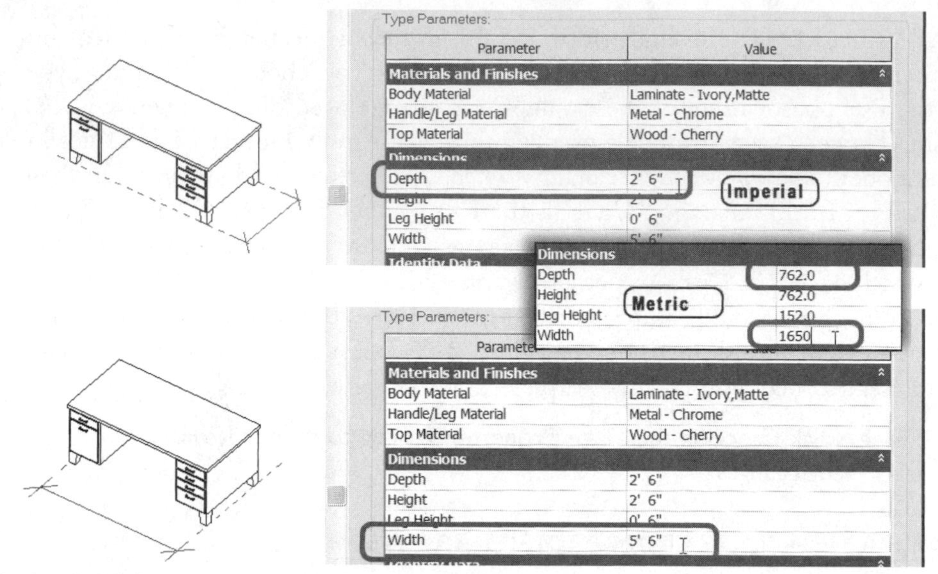

Figure 9–15 *Edit the Width and Depth of the new Type*

- In the "Type Properties" dialog, change the Width to **5′-6″** [**1650**] and then click OK.

5. Right-click on the new Type, and choose **Match**.

- Click on each of the Desks in the offices along the left side of the plan to change them to the new Type.
- Move the Desks if necessary to align with the Walls.

TIP You can use the Align tool here and then apply a lock constraint to keep the Desk aligned to a particular Wall. Be careful not to "over" constrain the model however.

6. Save the model.

Because we can see that there is more than one Type in this family, with different names, we have a good indication that this family is parametric, and new types with different parameters, for example sizes or materials, can be created.

If there is only one Type, and it is the same name as the family, that typically means that particular family is not parametric, and it is not possible to make multiple types from that family.

CUSTOMIZING FAMILIES

While a great deal of content is available to us in the installed and online libraries, we often need components in our projects for which a suitable Family is not readily available. In these situations, we can create our own custom Family. Custom Families can be simple or complex. In this topic, we will look at various examples of creating and using custom Families in our projects.

DUPLICATE AN EXISTING FAMILY

When you decide to build your own Family, always try to start with an existing Family that is close to the Family you wish to create. Doing so will require you only to edit the existing Family which is more expedient than starting from scratch. For example, flanking the corridor of our tenant suite to the right are two secretarial spaces. Perhaps the client would like desks in this area that have a CPU cubby rather than the two drawer pedestals in the Family we currently have loaded. We can duplicate the existing Family, and then modify it to make this change. This is much more efficient than modeling the entire desk over again.

7. On the Project Browser, beneath *Families*, expand *Furniture* and then right-click on *Desk [M_Desk]*.

   - Choose **Edit** from the right-click menu (see Figure 9–16).

TIP As an alternative, select one of the desks in the model, right-click and choose **Edit Family** (or click the Edit Family button on the Options Bar).

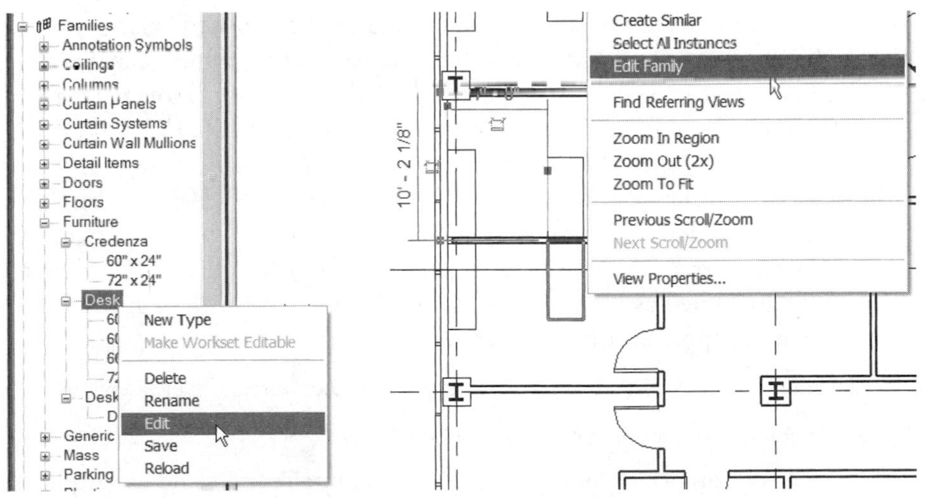

Figure 9–16 *Edit a Family from Project Browser or the model*

- In the message dialog that appears, click Yes to confirm opening the Family file.

The *Desk* [*M_Desk*] Family will open into the Family editor with a 3D View active. (You are no longer in the commercial building project file. You are now in the Desk Family file and can edit it directly).

8. Move your Modify tool over each part of the desk and pause for the tool tip to appear (see Figure 9–17).

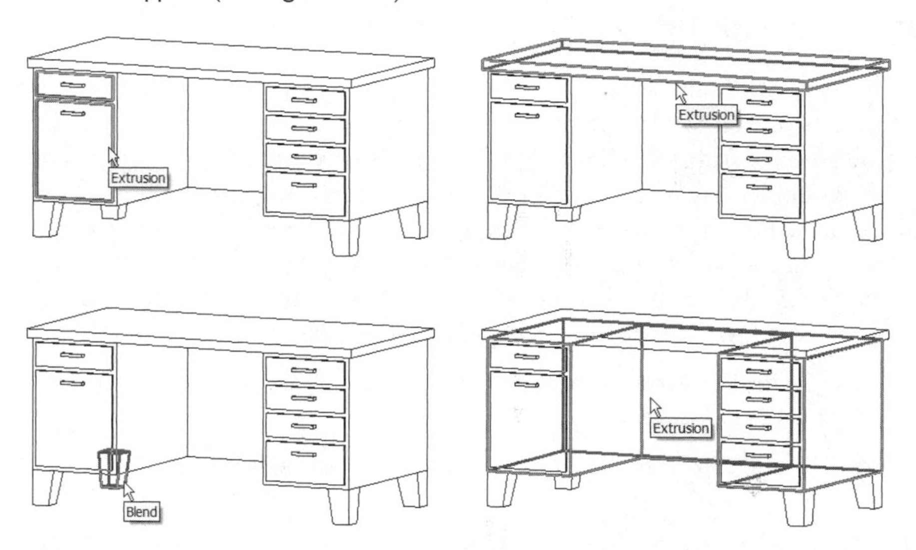

Figure 9–17 *The Desk is made from various Solid Forms*

The desk is made from a collection of Solid Extrusions, Sweeps and Blends. An Extrusion is a 3D form created by a 2D closed shape that is "pushed" along a perpendicular path. A Sweep is an extrusion that follows a path, which can be any shape. A Blend is form that starts with one closed shape and then "morphs" into a second closed shape. There is one last type of solid (not used in this Family) called a Revolve. This shape is derived from rotating a 2D closed shape along an axis. Each of these four basic shapes can be made as solids or voids. Solids represent actual materials in the model, while voids carve away from the solid form. In addition to this "raw material" a Family can also contain other Families. These "nested" Families are added as Components to the Family within the Family Editor. We will see an example of a nested Family below.

9. On the Project Browser, expand *Views (All)* and then expand *Elevations*.

- Double-click to open the *Front* elevation View.

As you can see, navigating the Views in the Family editor is the same as within a project.

10. Select the Extrusion on the left (that represents the drawer fronts) and then hold down the CTRL key and select the handles as well (see Figure 9–18).

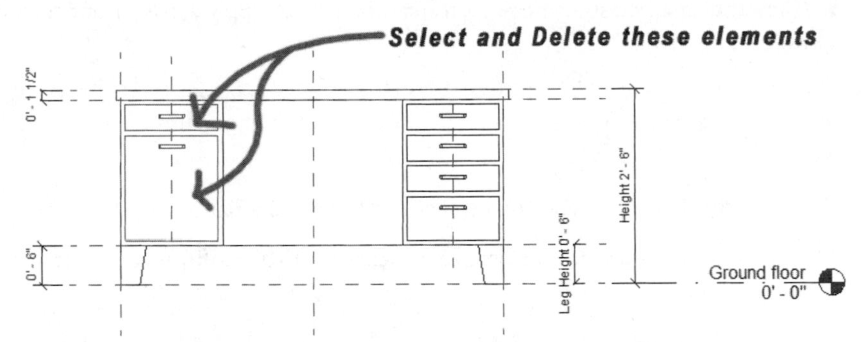

Figure 9–18 *Select and Delete the Drawer Fronts and Handles on the left*

- Press the DELETE key to erase both elements.

The author of this Family chose to create a single extrusion that contained two shapes—one for the top drawer and another for the bottom. While this is a perfectly valid approach, it forces us to edit both drawers together. In your own Families think about such issues carefully. This approach may ultimately be the best, but in some cases, you may decide to instead model each drawer separately.

The pedestal on the left will become a cubby for a CPU. This is why we have deleted the drawer fronts. We now need to make the cubby a bit narrower and then cut a void from it to complete the cubby.

11. Click on the Extrusion that comprises the major form of the desk.

Notice all of the shape handles that appear. We can simply drag one of these to make the cubby narrower (see Figure 9–19).

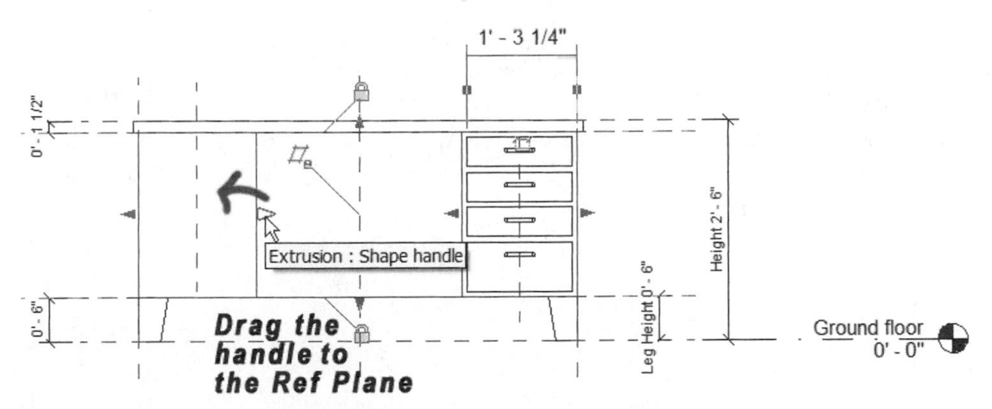

Figure 9–19 *Use the Shape Handles to resize the Cubby Pedestal*

12. Drag the shape handle (indicated in Figure 9–19) to the left and snap it to the Reference Plane (green dashed vertical line in the middle of the pedestal).
    - Click the small open padlock icon that appears to apply a constraint to this Reference Plane.

This action "locks" this side of the extrusion to this Reference Plane. If the Reference Plane later moves, the edge of the extrusion will move accordingly. Let's try it out.

13. Deselect the Extrusion and then select the Reference Plane.
    - Edit the temporary dimension that appears to **1′-8″ [500]** (see Figure 9–20).

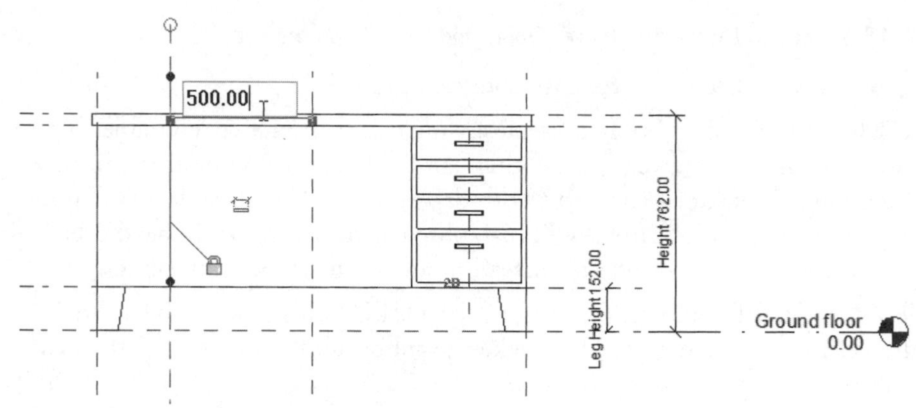

Figure 9–20 *Move the Reference Plane to move the constrained Extrusion edge*

Let's add the void next.

14. On the Design Bar, click the Void Form tool and then choose Void Extrusion from the flyout.

Since we are working with solids and voids which are three-dimensional forms, we need to indicate to Revit Building our preferred working plane. A Working Plane establishes a 2D surface in the model in which we can sketch our 2D shapes. In this case, since we are working on the Front elevation, a Work Plane parallel to Front makes the most sense (see Figure 9–21).

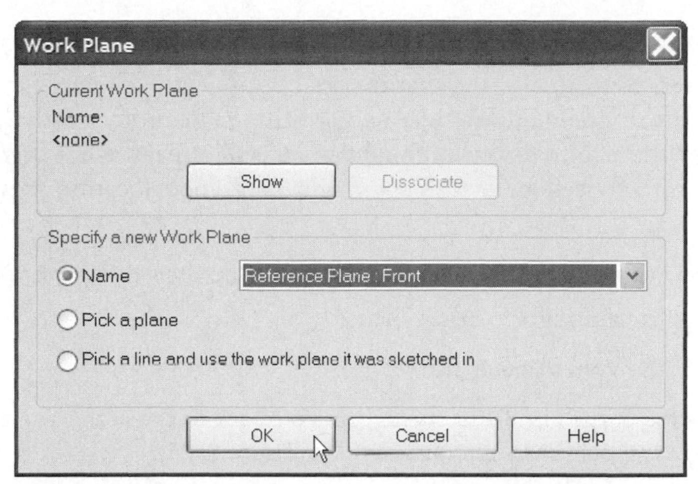

Figure 9–21 *Choose a Reference Plane in which to sketch the Void shape*

- In the "Work Plane" dialog, choose **Reference Plane : Front** from the "Name" list and then click OK.
- On the Options Bar, click the Rectangle icon and in the "Offset" field, type -3/4" [-19] (Note the value is negative).
- Using Figure 9–22 as a guide, snap to opposite corners of the CPU pedestal.
- On the Options Bar set the Depth to **1'-0"** [**300**].

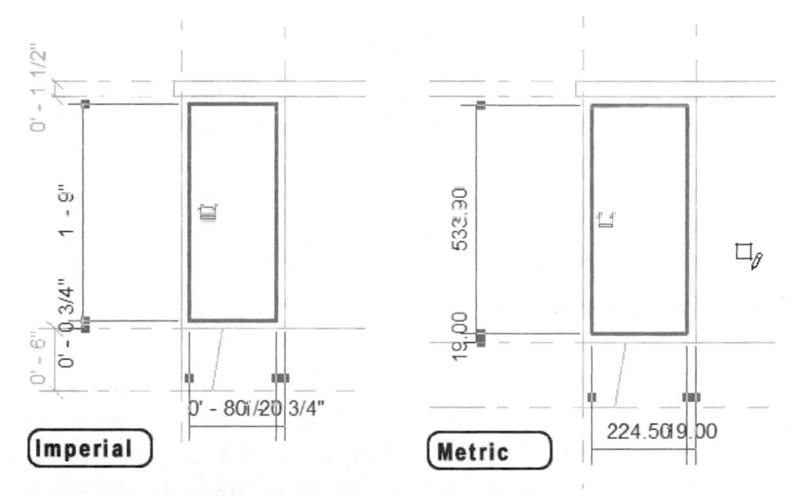

Figure 9–22 *Sketch the Void Extrusion shape*

- On the Design Bar, click the Finish Sketch button.

15. On the Project Browser, double-click to open the *View 1* 3D View.

From this View you can see that the Void is cutting away from the overall form of the desk. However, you can also see that it does not project back very far into the desk's depth. We can easily adjust this. (To see the Void more clearly, pass the Modify tool over the cubby to pre-highlight it—it will appear as an orange outlined Void element). The floor plan View might be a good location to work for this edit.

16. On the Project Browser, double-click to open the *Ground Floor* plan View.

The Void will appear among the other shapes.

- Click the Void shape in plan.
- Use the Shape Handle to stretch the Void back to the rear plane of the desk (it should snap automatically) (see Figure 9–23).

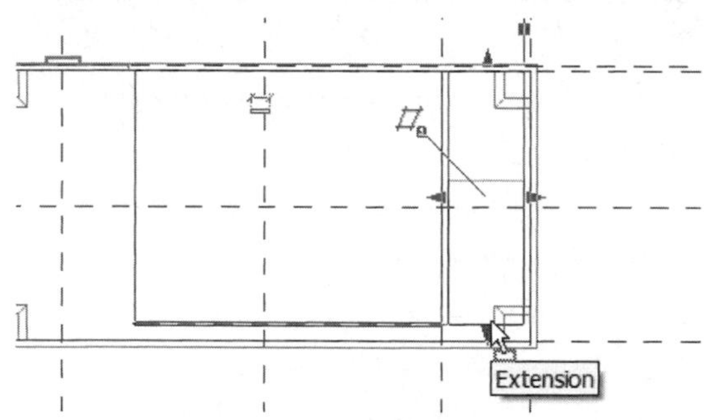

Figure 9–23 *Stretch the depth of the Void using Shape Handles*

17. Return to the 3D View to see the results.

With the Void still selected note the Depth field on the Options bar shows the current dimension. The Depth can also be edited here as well.

SAVE A NEW FAMILY

We are ready to save the results and load our new Family into the project.

18. From the File menu, choose **Save As**.

Autodesk Revit Building imposes no limitations on where you can save Family files, however the location where you save Family files is a very important consideration particularly in team environments. Check with your IT support personnel regarding the preferred location for saving Family files. Sometime firms have a "check-in" process for newly created content. Follow whatever guidelines or practices are in place in your company. For this exercise, we will simply save the Family file (RFA) to our *Chapter09* folder. If you decide later that you wish to use this Family in real projects, you can copy it from this location to a suitable location on your company network.

- In the "Save As" dialog, browse to the *Chapter09* folder in the location where you installed the Mastering Autodesk Revit Building CD ROM files.
- In the "File name" filed, type: **Desk-Secretary.rfa** for the name and then click Save.

It is very important to use Save As rather than Save. If you simply save the Family, it will overwrite the existing *Desk [M_Desk]* Family from which we started. Before we load this Family back into the project, let's edit its Types. Remember, the Family controls the available parameters and physical form of the element. However, the Type(s) can have specific values for the established Family parameters. Following the lead of the Family from which we created this one, our Types will have predefined values for the sizes.

19. On the Design Bar, click the Family Types tool.

The "Family Types" dialog will appear. At the top is a drop-down list showing each of the existing Types inherited from the *Desk [M_Desk]* Family. Some of these are no longer necessary.

20. From the "Name" list, choose 60″ x 30″ Student [1525 x 762mm Student] and then click the Delete button.

- Repeat for the 72″ x 36″ [1830 x 915mm] Type (see Figure 9–24).

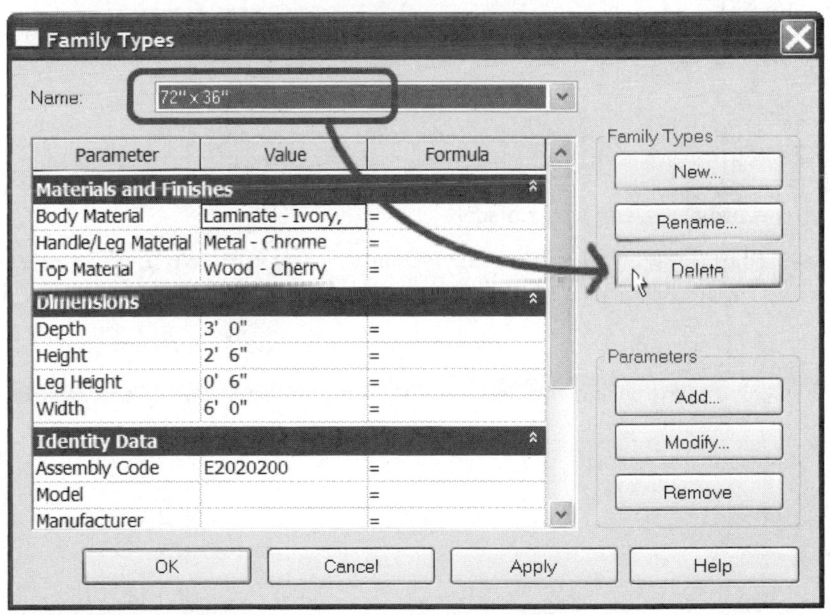

Figure 9–24 *Delete unneeded Types*

The remaining Types can stay for this Family.

21. Click OK to dismiss the dialog.
    - On the Toolbar click the Save icon (or choose **Save** from the File menu).

LOAD THE FAMILY INTO THE PROJECT

22. On the Design Bar, click the Load into Projects button.

As we saw above, this will return the commercial project to the front and add this Family to the Project Browser ready to be placed in the model.

23. Add two of the new desks to the model as shown in Figure 9–25.

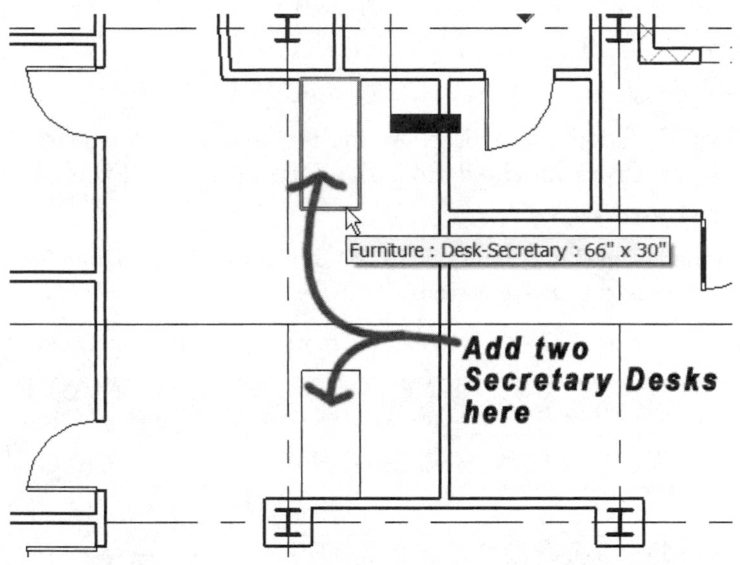

Figure 9–25 *Add two new Desks to the model*

From the floor plan View, we cannot really see any of the edits we made. Let's add a Camera View to the project so we can get a look at our new Component Family.

24. On the Design Bar, click the View tab and then click the Camera tool.
    - Click a point behind the desks and then drag toward one of the desks.
    - Click again in the center of one of the desks (see Figure 9–26).

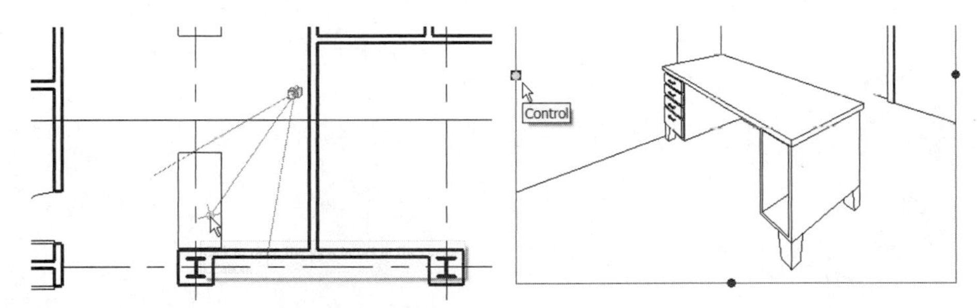

Figure 9–26 *Add a Camera View from the plan*

- Use the Control handles to adjust the size of the 3D perspective View so you can see the whole desk.

If your desk is facing the wrong way, return to the *Level 3* View and mirror them. Be sure to remove the checkmark from the "Copy" option on the Options Bar.

EDIT A FAMILY

Since this desk is now a computer-type desk, we should probably add a keyboard shelf in the middle. To do this, we simply repeat the process above and return to the Family Editor. We can model the shelf with a simple Solid Extrusion. We have to pay attention to where we sketch the extrusion. As with the void above, the shape of the keyboard shelf will be associated with a Work Plane. We could simple use the existing ground plane, and then adjust the extrusion parameters to make it sit at the correct height in elevation and 3d, or we can create a new "Keyboard Shelf" Reference Plane. Using the Reference Plane will give us more control.

25. On the Project Browser, double-click to open the *Level 3* floor plan View.
    - Select one of the Secretary Desks, right-click and choose **Edit Family**.
    - In the dialog, click Yes.
26. On the Project Browser, double-click to open the *Ground floor* plan View.
27. On the Design Bar, click the Solid Form > Solid Extrusion tool.
    - Using the rectangle option on the Options Bar, sketch a rectangular shape for the keyboard shelf in plan (see Figure 9–27).

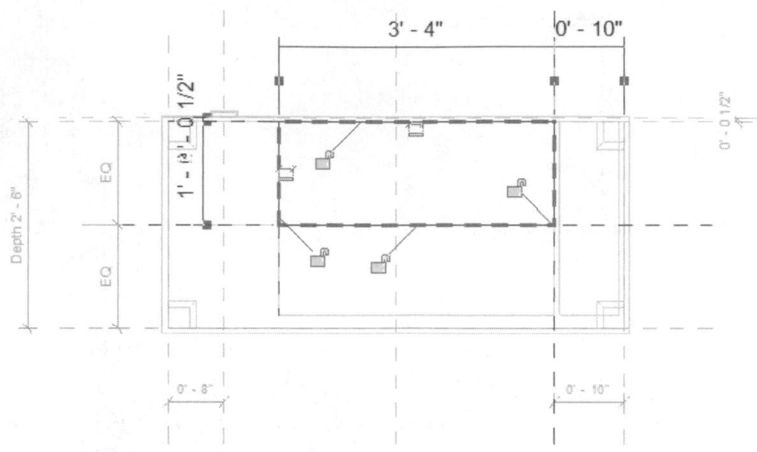

Figure 9–27 *Sketch the Keyboard Shelf*

- On the Design Bar, click the Extrusion Properties button.
- In the "Element Properties" dialog, set the "Extrusion Start" to **2′-1″** **[625]** and the "Extrusion End" to **2′-1 3/4″** **[644]** and then click OK.
- On the Design Bar, click the Finish Sketch button.

28. Return to the 3D View (see Figure 9–28).

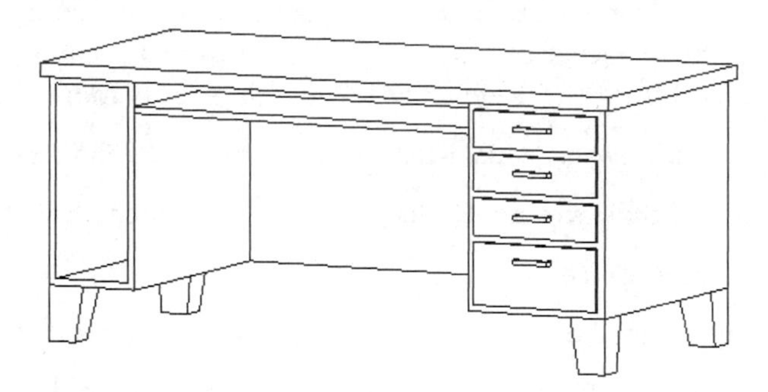

Figure 9–28 *The Secretary Desk with the Keyboard Shelf added*

29. Save the Family, click Yes to overwrite the existing family, and then click the Load into Projects button.

- In the "Reload Family" dialog, click Yes to confirm reloading the Family and updating the project.
- On the Project Browser, double-click to open the *3D View* / 3D View.

Notice how the update to the desk now appears in the project. If you wish, you can rename this camera View to something more descriptive.

30. Save the project.

BUILDING FAMILIES

In most cases, the preceding process will enable you to produce the Family you need by leveraging your existing library content. However, in some cases it will be necessary or even in some cases easier to start from scratch. In this case, you will simply create a new Family file and begin modeling the element you require. All Families are based upon templates. Autodesk Revit Building ships with a large collection of pre-made Family templates from which to choose. It is important to select the Family template which best corresponds to the kind of Family you wish to create. This is because the Family is categorized for use in projects based on the template you choose. It is difficult to change the Host Element of a Family once it has been created, so choose your template carefully.

CREATE A NEW FAMILY FILE

Keeping with the furniture layout on the third floor of the commercial project a little longer, let's create a custom reception desk for the lobby to the suite.

1. From the file menu, choose **New > Family**.

This will open the "New Family" dialog box to the folder which contains all the available Family templates. As we noted the importance of selecting the proper template, it is a good idea to familiarize yourself with this list (see Figure 9–29).

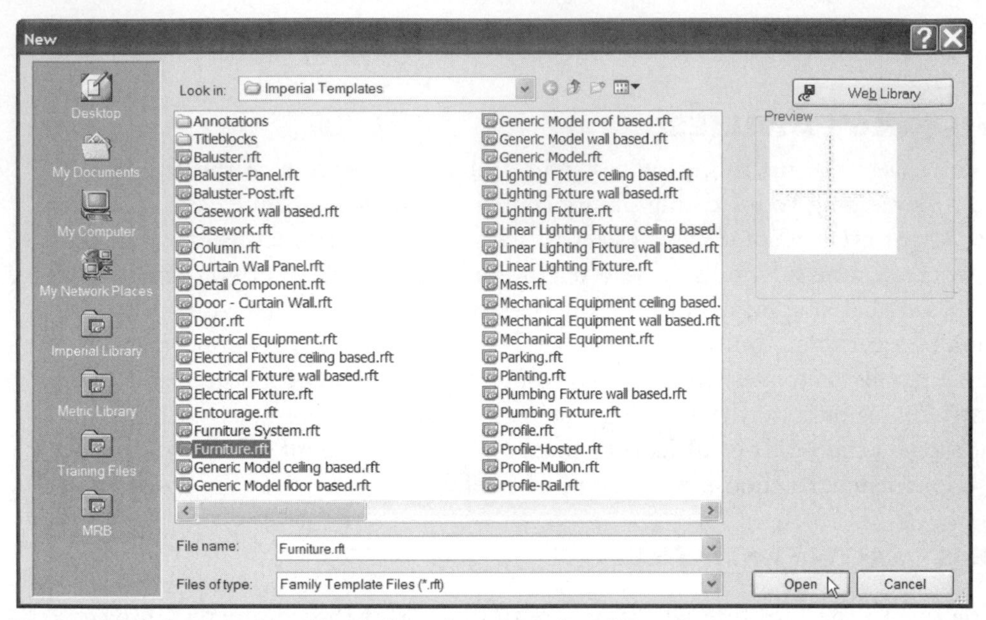

Figure 9–29 *Create a New Family from the existing list of Templates*

Our reception desk is best categorized as furniture. Therefore, we will select the *Furniture.rft* template.

- Browse to you Templates folder if necessary.
- Select *Furniture.rft* [*Metric Furniture.rft*] and then click Open.

NOTE if your version of Autodesk Revit Building does not contain either of these templates, they have been provided in the *Out of the Box Templates* folder with the files installed from the Mastering Autodesk Revit Building CD ROM.

As with the previous exercise, we are now in the Family Editor. The difference here is that the current View contains only two Reference Planes and no geometry. It is often easier when building Families to work with tiled windows. Most Family templates open several Views at once—a plan, a 3D and two elevations. However, if we were to tile the windows now, in addition to these four Views of the Family, our screen would also tile all of the Views we have open from the commercial project and the Desk-Secretary Family file. Therefore, for now let's close the project and other files and leave only the new Family file open.

2. From the Window menu, choose **Desk-Secretary.rfa - 3D View: View 1**.
   - From the File menu, choose **Close**.
3. From the Window menu, choose **09 Commercial.rvt - Floor Plan: Level 3 [09 Commercial-Metric.rvt - Floor Plan: Level 3]**.

➧ From the File menu, choose **Close**.

If prompted to save, choose Yes.

If you open your Window menu again, only the four Views mentioned above will be listed (see Figure 9–30).

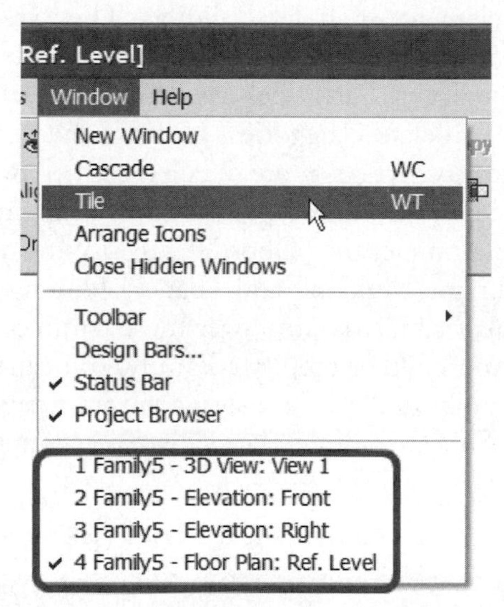

Figure 9–30 *Close other projects and Families so that only the current Family file's Views show in the Window menu*

4. From the Window menu, choose **Tile**.
5. From the View menu, choose **Zoom > Zoom All To Fit**.

TIP The keyboard shortcut for Zoom All To Fit is **ZA**.

While it is not necessary to close the other projects and tile the Views, it will be easier to work on the Family in this environment. As we make changes in any View, we will see the results immediately in the others. Since we will be making many three-dimensional edits, this will be very helpful.

6. From the File menu, choose **Save**.

➧ In the "Save as" dialog, type **Reception Desk** for the Name and then click Save.

CREATING SUB-CATEGORIES

Revit Building classifies elements in a fixed hierarchy starting with Categories, then Families, then Types and finally Instances. We have been working with Families, Types and Instances throughout this chapter. Categories however are the broadest classification of elements in Revit Building. They are sub-divided into Model, Annotation and Imported object groupings for convenience. Each Family belongs to a Category. The list of Categories available is predetermined by Revit Building—we cannot add or delete Categories. However, we can add and delete "Subcategories." A Subcategory gives us more detailed control over the visibility of Model and Annotation elements by allowing parts of an element to appear differently from the whole. For example, the "Doors" Category includes Subcategories for "Elevation Swing," "Frame/Mullion" and "Panel" (among others). Each of these Subcategories gives us visibility control over these common sub-components of Door elements. When you build or modify a Family, you can assign the various components within the Family to any of the available Subcategories. To do this, you must first know what Subcategories are available. You can see a complete list by opening the "Object Styles" dialog.

7. From the Settings menu, choose **Object Styles** (see Figure 9–31).

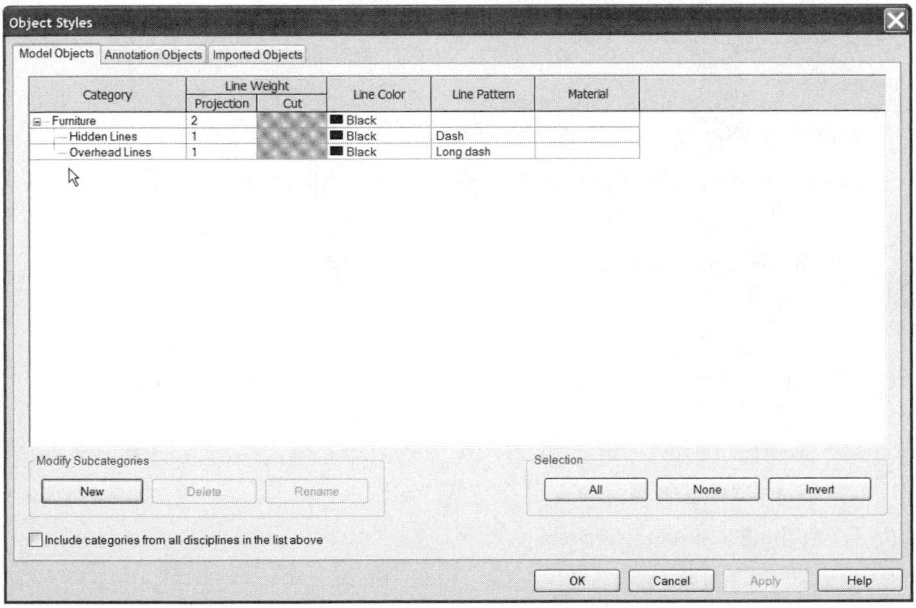

Figure 9–31 *Object Styles shows the available Categories and Subcategories*

Since we are currently in a Family file, the list of available Categories includes only those relating to the Family template in which we are editing—in this case Furniture. (If you wish to see a more complete list, you will need to repeat this command in a project file, however, if you do this, when you close the project file, you will need to re-tile your Family Windows on screen). When working on a Family, this is the way you can check what kind of Family you have if you are unsure. Notice that Furniture contains two Subcategories: "Hidden Lines" and "Overhead Lines." While you can create or edit Subcategories in this dialog at any time in your Family editing process, adding some Subcategories now will make editing easier since they will be available during the creation process.

We will add three new Subcategories to our Family each one representing one of the materials that we intend to use in the design: frame, hardware and top.

8. At the bottom of the dialog, in the "Modify Subcategories" grouping, click the "New" button.

    - Name this new Subcategory **Screen** (see Figure 9–32).

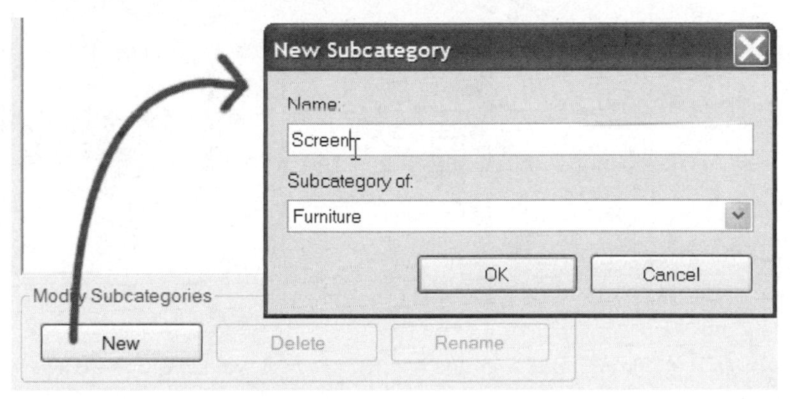

Figure 9–32 *Create a new Subcategory of Furniture*

- Repeat the process to create **Frame** and **Worksurface**.

Now that we have created our three Subcategories we can adjust their properties. For these Subcategories, we will assign materials to them.

9. Select the "Frame" Subcategory and then click in the Material field.

    A small expand (down pointing arrow) icon will appear at the right side of the cell

    - Click the small expand icon to open the "Materials" dialog.
    - From the "Name" list, select Default and then click the "Duplicate" button
    - In the "New Material" dialog, Type **MRB - Natural Birch Wood**.
    - On the right side, in the "AccuRender" area, click the expand arrow icon next to "Texture."

- In the "Material Library" dialog, on the left side, expand the *_accurender* folder, then the *Wood* folder and click on the *Birch* folder.
- In the middle pane, click on Natural:Medium Gloss and then click OK (see Figure 9–33).

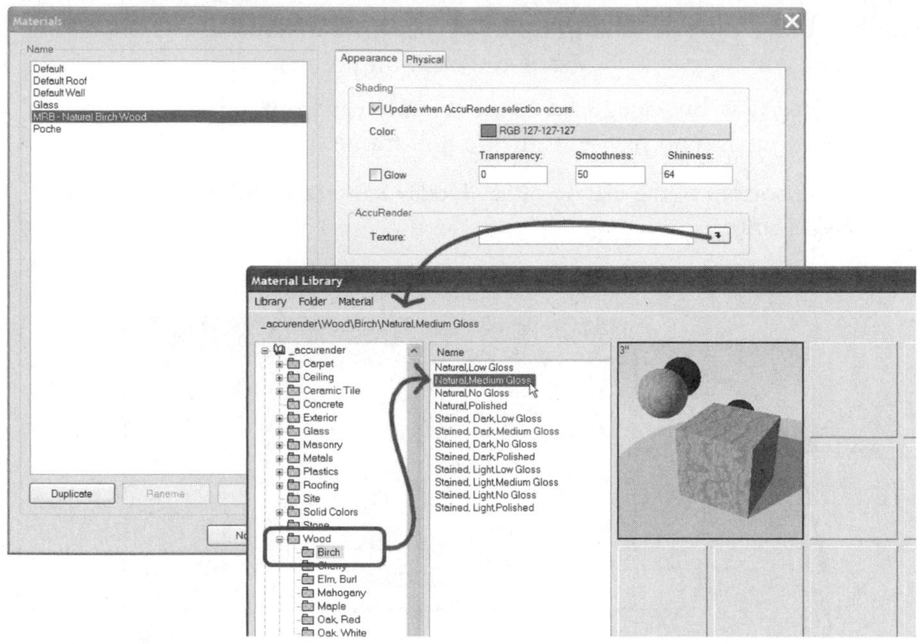

Figure 9–33 *Assign a Wood Material from the AccuRender Material Library to the Frame Subcategory*

Notice how this also automatically fills in the "Shading" parameters with a color to approximate the AccuRender material we have chosen. You can override this color without changing the AccuRender assignment if you wish, but for this example we will accept the default designation.

- Click OK to return to the "Object Styles" dialog.

10. Select the "Worksurface" Subcategory and then click expand icon in the Material field.

    - Repeat the process above to create another new material named: **MRB - Natural Walnut Wood**.
    - From the *_accurender\Wood\Walnut* folder, choose Natural:Medium Gloss and then click OK twice.

11. Select the "Screen" Subcategory this time and assign it to the MRB - Natural Birch Wood material created above (see Figure 9–34).

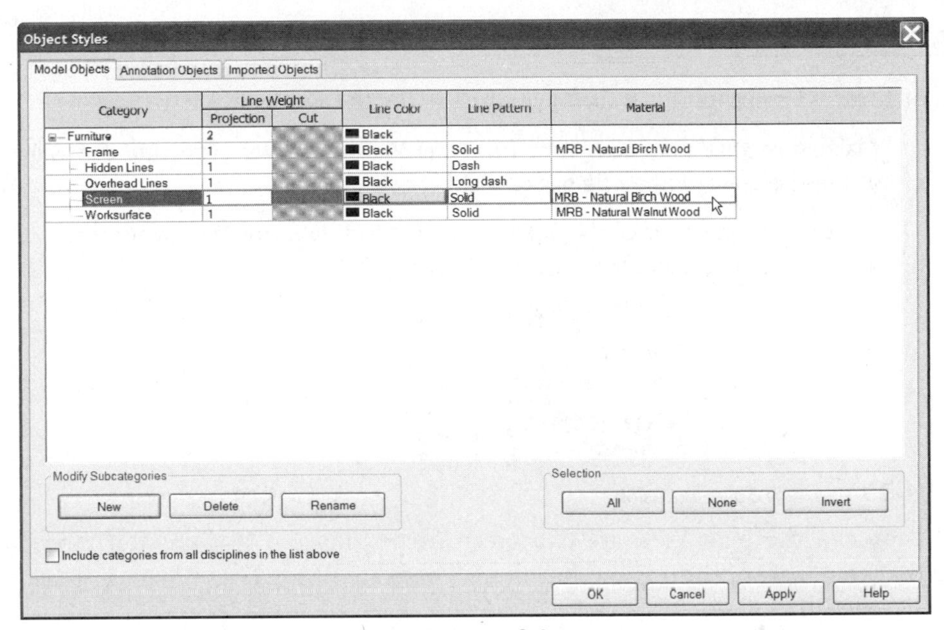

Figure 9–34 *Assign Materials to the each custom Subcategory*

Note that there are other parameters that we could configure here for each Category and Subcategory. Among them are "Line Weight," "Line Color" and "Line Pattern." Notice that the Line Weights and Line Patterns for some of the items do vary. If you click the "Annotation Objects" tab, you will also note that the Line Color of the elements listed there also varies. We can accept the defaults for all of these values.

12. Click OK in the "Object Styles" dialog.

As we already noted, adding Subcategories is optional and can be performed before adding any geometry as we have done here, or later in the modeling process as well. We did it first to assist us in differentiating elements as we build the Family geometry.

BUILD FAMILY GEOMETRY

As we have already seen, there are two methods to create objects in our Family: adding Solid Forms and adding Void forms. Solid Forms represent the actual physical materials from which our Family object is created. Void forms simple carve away from Solid forms to help us create more complex shapes. Both Solid Forms and Void Forms include four methods—Extrusion, Blend, Revolve and Sweep. Some of these have already been covered. We will see examples of each of these in this exercise.

CREATE A SOLID EXTRUSION

The first form we will build is a simple extrusion for the worksurface.

Make sure that the *Floor Plan : Ref. Level* View is active. (The title bar will appear bold—click the title bar to make it active).

13. On the Design Bar, click the Solid Form tool and then from the flyout that appears, choose **Solid Extrusion** (see Figure 9–35).

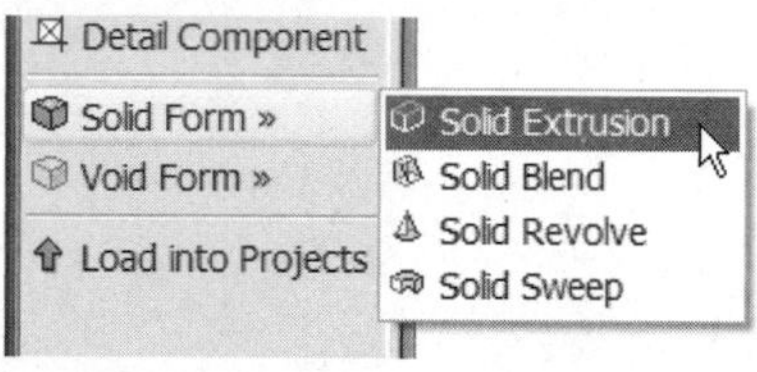

Figure 9–35 *Create a Solid Extrusion*

In the center of the plan View are two green dashed lines. These are Reference Planes that were part of the original Furniture Template. We will use them to center our geometry.

14. On the Design Bar, be sure that the Lines tool is selected.
    - On the Options Bar, type **-0′ 3/4″ [-19]** in the "Depth" field.

Be sure to use a negative value for the "Depth." We will discuss the reason below.

- Click on the Rectangle icon and draw a rectangle that is roughly centered on the two Reference Planes. (The exact size is not important yet)
- Edit the temporary dimensions to make the rectangle **5′-0″ [1500]** x **2′-6″ [750]** (see Figure 9–36).

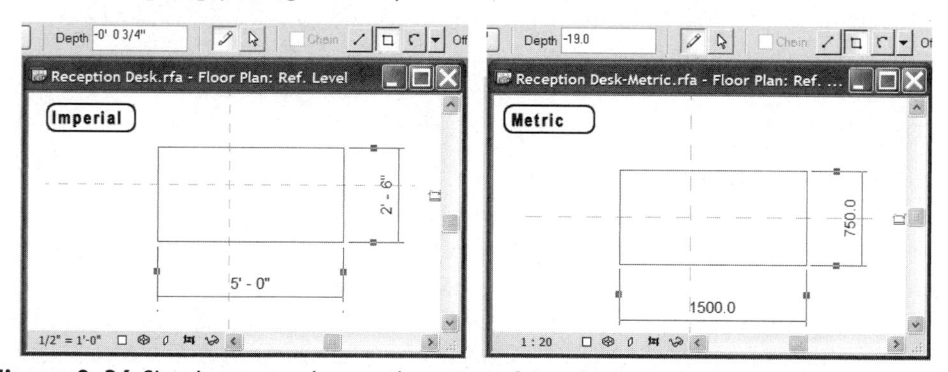

Figure 9–36 *Sketch a rectangle near the center of the plan and edit the temporary dimensions*

15. On the Design Bar, click the Modify tool.
    - Select all four of the sketch lines.

- On the toolbar, click the Move icon.
- For the Move Start Point, click the Midpoint of one of the edges.
- For the Move End Point, click the intersection with Reference Plane (see Figure 9–37).

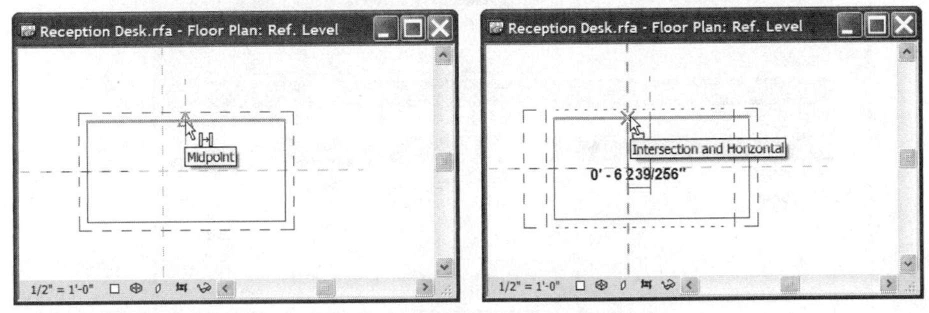

Figure 9–37 *Move the sketch to center it on the Reference Planes*

16. Repeat the Move operation in the other direction.

The rectangular sketch should now be centered on the Reference Planes. You can use the Tape Measure icon to verify if necessary.

17. On the Design Bar, click the Extrusion Properties button.
    - Beneath the "Identity Data" grouping, change Subcategory to **Worksurface** and then click OK (see Figure 9–38).

You could set a material here without having made the Subcategories, but using the Subcategory method give us more control in later changing materials, and once this Family is loaded into a project, we will have selective visibility control over the internal components of the Family.

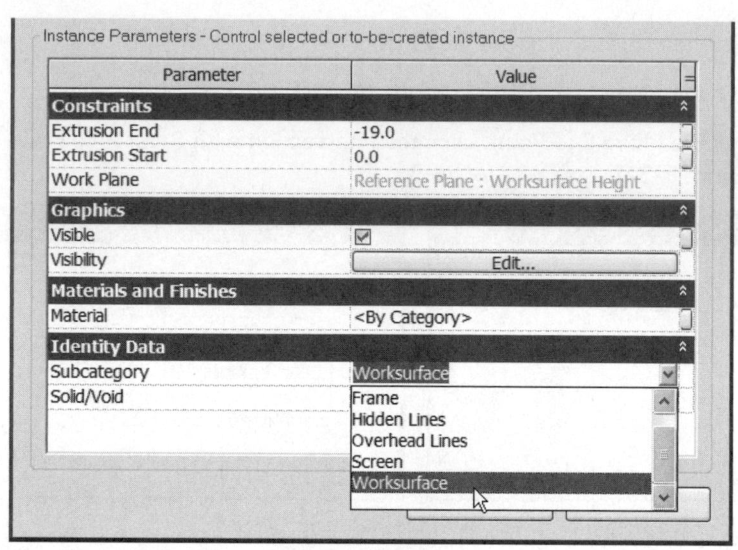

Figure 9–38 *Choose a Subcategory for the Solid Extrusion*

- On the Design Bar, click the "Finish Sketch" button.

ADJUST THE VIEW WINDOWS

Now that we have some geometry built, we can adjust our View windows to show it better.

18. From the View menu, choose **Zoom > Zoom All To Fit** (or type **ZA**).

Notice that the 3D View shows the model from the side.

- Click in the *3D View : View 1* window and then hold down the SHIFT key and drag the middle (wheel) button to spin the model.
- On the View Control Bar (for the 3D View), change the View Graphics to Shading with Edges (see Figure 9–39).

With the shading turned on, we can see the color from the assigned Subcategory's material designation.

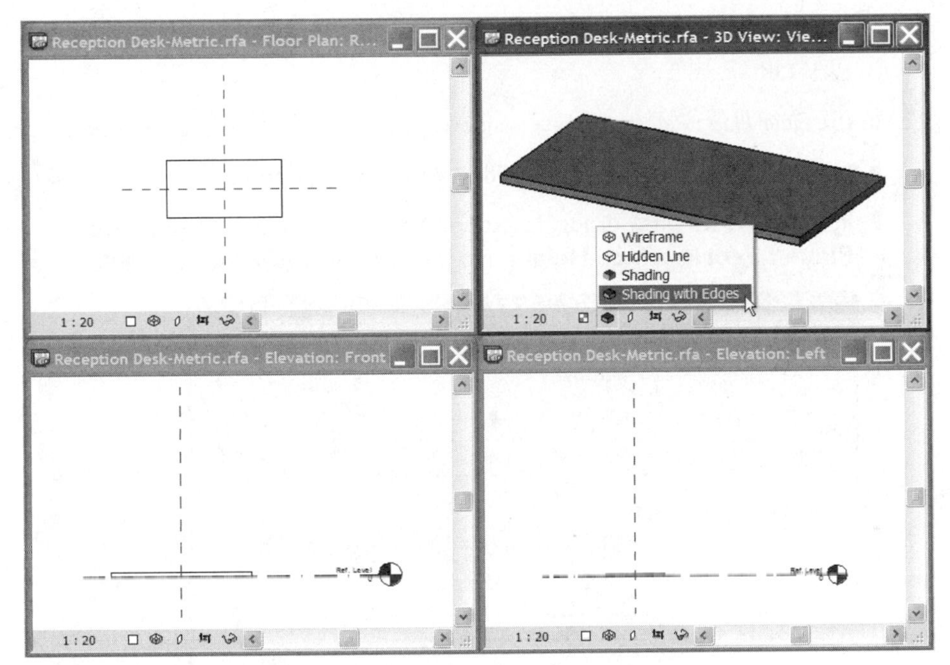

Figure 9–39 *Adjust all of the View windows to show the new geometry*

- Save the Family.

CREATE A REFERENCE PLANE

If you study the model, you will notice that the desktop surface is sitting on the floor. While it is possible to edit the parameters of the extrusion to set it at the correct height, we will get more control if we create a new Reference Plane and associate it as the Work Plane for the Solid Extrusion.

19. Click on the title bar of the *Elevation : Right* View window to make it active.
20. On the Design Bar, click the Ref Plane tool.
    - Click two points horizontally above the Ref. Level (floor) plane.
    - Edit the temporary dimension to **2′-6″ [750]**.
    - On the Design Bar, click the Modify tool or press the ESC key twice.

This will give you a Reference Plane that is 2′-6″ [750] off of the floor. This is the height of the top of our worksurface. This is the reason we used a negative Depth for the worksurface extrusion above. It will make it easier to edit the height of the worksurface later should it be required. We now need to name this Reference Plane and then make it the Work Plane for the extrusion.

21. Select the new Ref Plane and on the Options Bar, click the Properties icon.

- For the "Name" parameter, type **Worksurface Height** and then click OK.

22. In the *Floor Plan : Ref. Level* View window, select the worksurface extrusion.
    - On the Options Bar, click the "Edit Work Plane" button.
    - In the "Work Plane" dialog, from the "Name" list, choose **Reference Plane : Worksurface Height** and then click OK (see Figure 9–40).

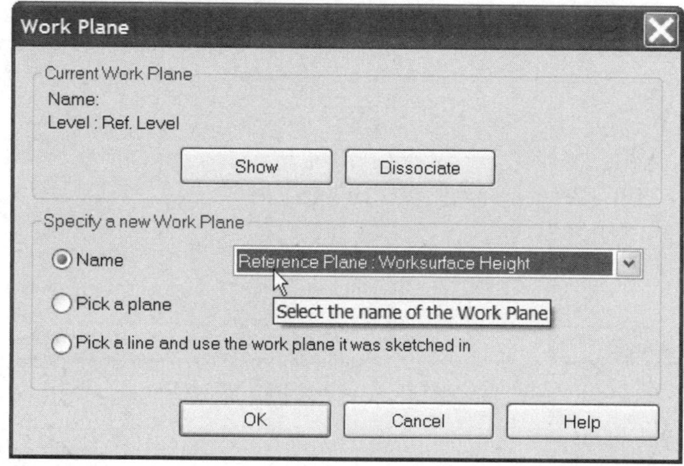

Figure 9–40 *Change the Work Plane of the Worksurface Extrusion to the new Reference Plane*

Notice the worksurface extrusion will shift up to the new height in the elevation and 3D Views.

23. Zoom all Viewports to Fit.
    - Save the Family.

CREATE A SOLID BLEND

Now that we have our desktop surface and it is sitting at the desired height, let's add some legs. For the main part of the leg, we will use a Solid Blend to give them a bit of a taper. We will square them off at the top using a simple extrusion.

24. Using Zoom In Region (View > Zoom menu or right-click) zoom in on the top left corner of the worksurface in the plan View window.
25. On the Design Bar, click the Solid Form tool and then from the flyout that appears, choose **Solid Blend**.
    - On the Options Bar, type **1′-10″ [550]** in the "Depth" field and then click the Rectangle icon.

- Sketch a simple square near the corner of the desk.
- Using the temporary dimensions and the left side of Figure 9–41 as a guide, make the shape 2″ [50] square set 2″ [50] away from the corner as shown.

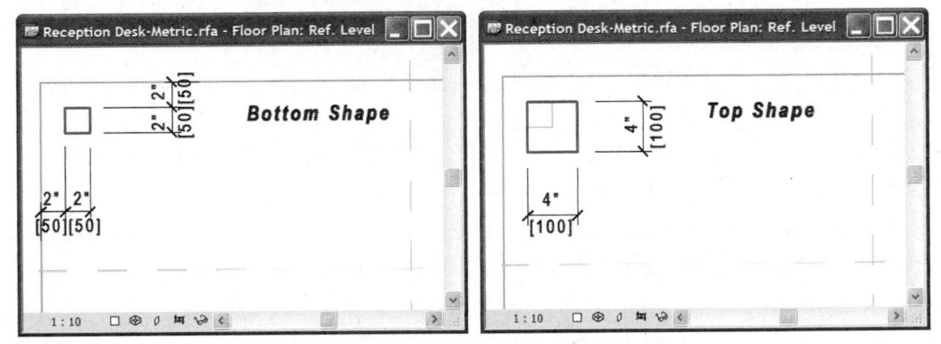

Figure 9–41 *Sketch the Base (Left) and Top (Right) of the Solid Blend*

A Solid Blend is basically an extrusion that transforms from one shape to another along the length of the extrusion. In this case, we will use two squares of varying sizes that share a common offset from the desktop corner. This will give a tapered shape to the final leg form. When you build a Blend, you first sketch the shape of the bottom, then the shape of the top. Use Blend Properties to configure other parameters like overall heights and Subcategory options.

26. On the Design Bar, click the Edit Top tool.
    - Using the Rectangle option again, sketch a **4″ [100]** square with the top left corner aligned to the square at the base (see the right side of Figure 9–41).
    - On the Design Bar, click the Blend Properties button.
    - Beneath the "Identity Data" grouping, change Subcategory to **Frame** and then click OK.
27. On the Design Bar, click the Finish Sketch button (see Figure 9–42).

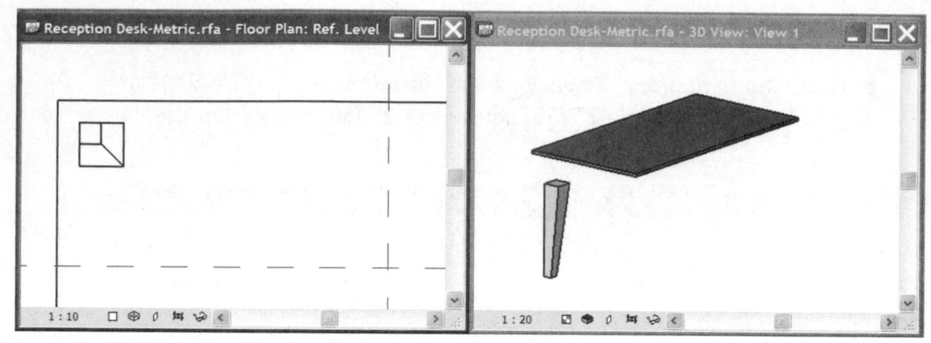

Figure 9–42 *Finish the Blend and view the results*

There is a noticeable gap between the top of the leg and the worksurface (spin the 3D View as necessary to see this). We will fill this in with an extrusion.

28. Click on the title bar of the *3D View : View 1* window to make it active.
29. On the Design Bar, click the Solid Form tool and then from the flyout that appears, choose **Solid Extrusion**.
    - On the Tools toolbar, click the Work Plane icon (see Figure 9–43).

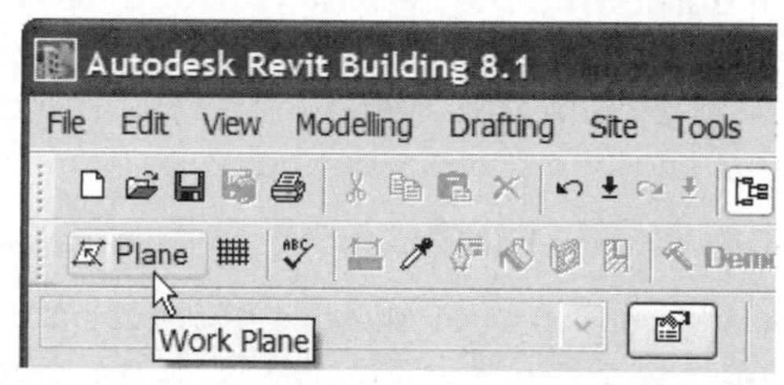

Figure 9–43 *The Work Plane icon allows you to change the Work Plane for the current Solid form*

Till now, we have used Reference Planes and Levels as Work Planes, but once you start building geometry, you can use an existing plane from the geometry in the model as a Work Plane. In this case, we will use the top of the Blend.

- In the "Work Plane" dialog, choose the "Pick a Plane" radio button and then click OK.
- In the 3D View, click on the top of the Solid Blend leg (see the left side of Figure 9–44).

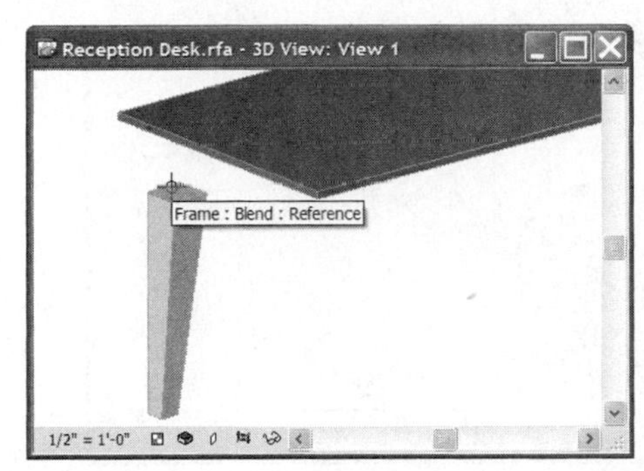

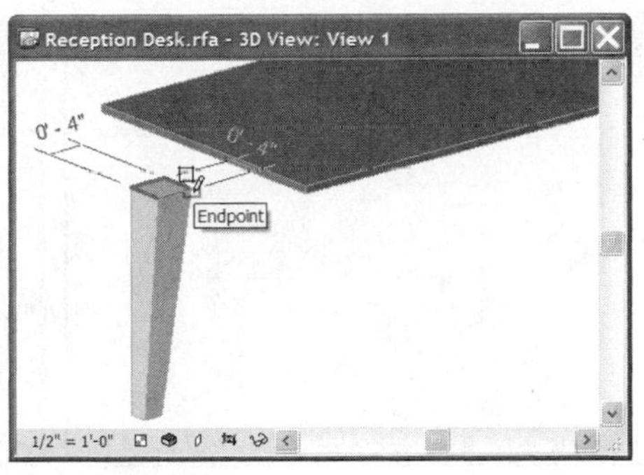

Figure 9–44 *Pick a Work Plane from the model in the 3D View*

- On the Design Bar, click the Lines tool.
- On the Options Bar, click the Rectangle icon and then sketch a rectangle on the top of the leg (see the right side of Figure 9–44).

After sketching this shape, a lock icon will appear on each edge. You can close these locks to constrain the shape of the Solid Extrusion to the top shape of the Blend. This can be handy if you anticipate making edits to the Blend. This would keep the extrusion at the top of the leg coordinated with these changes. However, in some cases, such a constraint can have adverse affects. This could occur for example if you changed the shape of the Blend top to something other than rectangular. In this case, Revit Building might have trouble maintaining the constraints. If this were to occur, a warning dialog would appear at the time of edit. For the purposes of this tutorial, we will leave the decision to the reader. Regardless of your choice for this exercise, do keep this option in mind for future reference in your own Families.

- On the Design Bar, click the Extrusion Properties button.
- In the "Element Properties" dialog, set the "Extrusion End" to **4″** **[100]**.
- Choose **Frame** from the Subcategory list and then click OK.
- On the Design Bar, click the Finish Sketch button.

The extrusion will appear at the top of the leg but will be too short to reach the worksurface. In the next step, we will stretch the height of this element and constrain it to the worksurface.

30. Click on the title bar of the *Elevation : Front* window to make it active.
    - Select the Extrusion and using the Shape Handle at the top, stretch it up to the bottom edge of the worksurface.
    - Click the lock icon that appears to apply the constraint (see Figure 9–45).

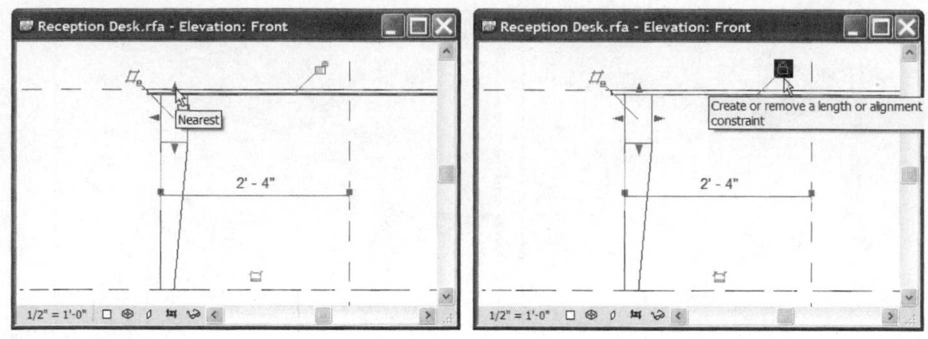

Figure 9–45 *Stretch the top edge of the leg and constrain it to the worksurface*

MIRROR THE REMAINING LEGS

Now that we have one completed leg, we can mirror it to create the others

31. Remaining in the *Elevation : Front* window select both pieces of the leg.
    - Click the Mirror icon on the toolbar.
    - Click the vertical Reference Plane (Center Front/Back) as the Axis of Reflection (see Figure 9–46).

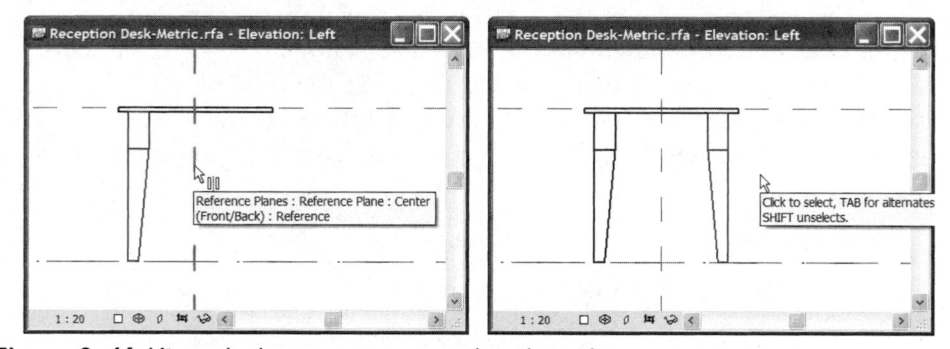

Figure 9–46 *Mirror the leg to create one on the other side*

32. Click on the title bar of the *Elevation : Left* window to make it active.
    - Select both legs and repeat the mirror process.
    - Save the Family file.

CREATE A PENCIL DRAWER

Using another extrusion, we can add some geometry to represent the structural support of the worksurface and a pencil drawer.

33. Click on the title bar of the *Floor Plan : Ref. Level* window to make it active.

Zoom as required to see the whole desktop.

34. On the Design Bar, choose Solid Form > Solid Extrusion.

    - Using the Work Plane icon, set the Work Plane to **Reference Plane : Worksurface Height**.
    - On the Options Bar, set the "Depth" to **-6 3/4″** [**-169**], click the Rectangle icon and type **-3″** [**-75**] in the "Offset" field.
    - Trace the desktop from top left corner to lower right corner (see Figure 9–47).

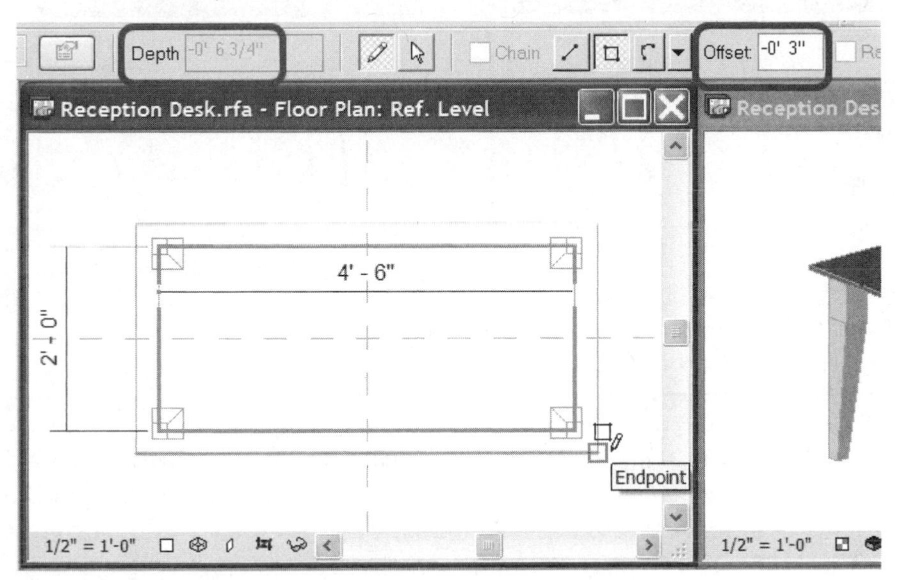

Figure 9–47 *Sketch a rectangle offset from the desktop*

    - On the Design Bar, click Element Properties, type **-3/4″** [**-19**] for the "Extrusion Start," set the Subcategory to **Frame** and then click OK.
    - On the Design Bar, click the Finish Sketch button.

35. Working in the *Elevation : Front* window create a 1/2″ [12] deep extrusion for a pencil drawer using the face of the extrusion just drawn as a Work Plane (see Figure 9–48). The exact size and position on the drawer are not important.

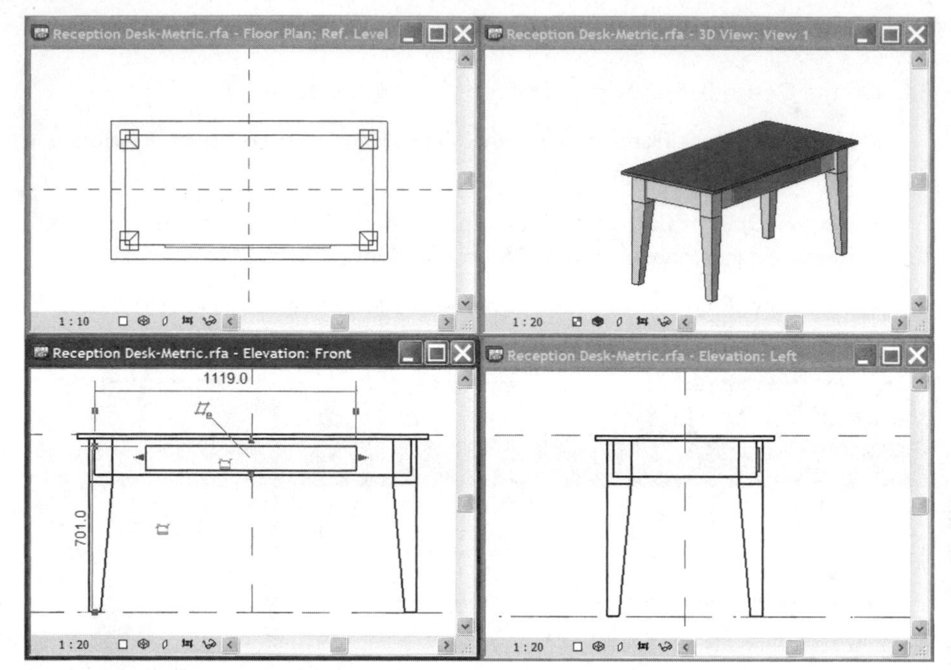

Figure 9–48 *Create a extrusion in the Front Elevation for a Pencil Drawer*

CREATE A SOLID REVOLVE

To add a drawer pull to the pencil drawer, we will use a Solid Revolve. A Solid Revolve is a form in which you sketch the profile of the form and then spin the profile around an axis. To create a simple drawer pull, you draw half of the cross section of the pull, and then place the revolution axis at the center of the pull.

36. Click on the title bar of the *Floor Plan : Ref. Level* window to make it active.
    - Zoom in on the Pencil Drawer in the plan View.
37. On the Design Bar, click the Solid Form tool and then from the flyout that appears, choose **Solid Revolve**.
    - On the Design Bar, click the Lines tool.
    - On the Options Bar, select "Chain."
    - Starting at the midpoint of the pencil drawer, sketch the form shown in Figure 9–49. The depth should be about 1″ [25] but the exact dimensions are not critical.

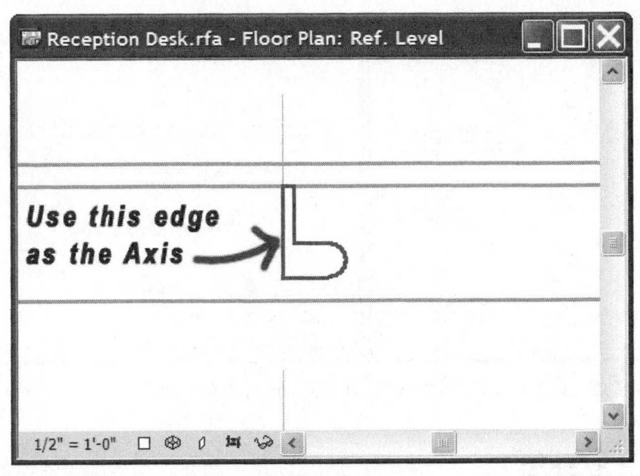

Figure 9–49 *Sketch the shape of the drawer pull*

38. On the Design Bar, click the Axis tool.
    - Trace the vertical edge of the shape drawn above.
    - On the Design Bar, click the Finish Sketch button.

If you study the result in all Views, you will notice that the drawer pull is at the height of the worksurface. This is because Revit Building used the "Worksurface Height" Reference Plane as the Work Plane for this element. As a result of this, you cannot move the drawer pull to the correct height—try it in the Front View to see for your self. If we had lots of hardware elements for this desk, we could create a new Reference Plane and associate the elements to it instead. In this case, we will simple disassociate the Work Plane from the Revolve which will allow us to move it freely.

39. Click on the title bar of the *Elevation : Front* window to make it active.
    - Select the drawer pull.

    A small Work Plane icon appears attached to the element.

    - Click the Work Plane icon to disassociate the Work Plane.
    - Move the drawer pull to center it on the height of the drawer (see Figure 9–50).

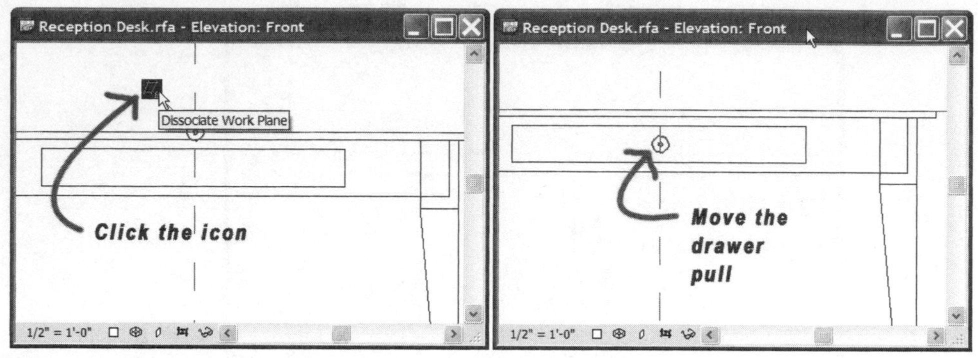

Figure 9–50 *Disassociate the Work Plane and move the drawer pull*

CREATE A SOLID SWEEP

Now let's create a privacy screen using the Solid Sweep command.

40. Click on the title bar of the *Floor Plan : Ref. Level* window to make it active.
    - Zoom out enough to see the whole desk.
41. On the Design Bar, click the Solid Form tool and then from the flyout that appears, choose **Solid Sweep**.
    - On the Design Bar, click the Sketch 2D Path tool.
    - On the Options Bar, click the "Pick Lines" icon.
    - In the "Offset" field, type **2″** **[50]**.

As with the "Offset" option in other commands, as you hover over a line in the model, a green dashed line will appear temporarily indicating the side to which the sketch line will offset. As you create sketch lines here, be sure to offset to the inside of the desk surface.

42. Offset the three outside edges of the desk (not the one with the pencil drawer) to create a "U' shaped sketch (see Figure 9–51).

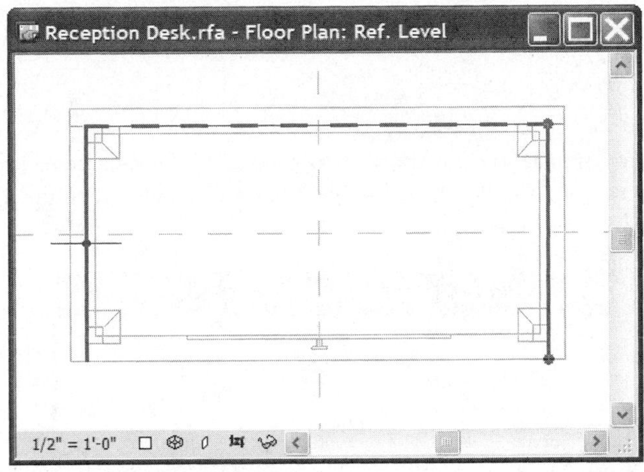

Figure 9–51 *Create Sketch Lines from the desktop*

- On the Design Bar, click the Finish Path button.

These three lines will be the "path" of the Sweep. We have seen Sweeps in other chapters. A Sweep "pushes" a profile along a path. It is similar to an extrusion except that the path does not have to be a simple straight line as it does in an extrusion.

43. On the Design Bar, click the Sketch Profile tool.

The profile travels perpendicular to and along the path. This means that since we sketched the path in plan, we must sketch the profile in elevation. When you choose the Sketch Profile tool, Revit Building will prompt us to select an appropriate View in which to sketch the profile. Please note that it is also possible to load an existing Profile when creating a Sweep instead of sketching it. This allows you to create "Profile" Families and store them in your library so that you can quickly create Sweeps from your most common profiles. In this example, we will sketch the profile as a custom shape just for this desk.

- In the "Go To View" dialog, choose *Elevation: Front* and then click Open View.
- Zoom the Front elevation View as necessary to center the red dot on screen.

The red dot indicates the start point of the profile. It appears automatically on the first segment of the Path. If you are not satisfied with the default location of the profile, you can move it, but be sure to move it before you start sketching the profile.

44. Using the Lines tool and the "Chain" option, create a profile similar to the one shown in Figure 9–52.

NOTE The exact shape of the profile is not critical. Just make sure to close the shape.

TIP Remember that you can zoom in to make sketch lines automatically snap to a smaller increment. The Profile in the figure has a material that snapped to 3/4″ [20] increment.

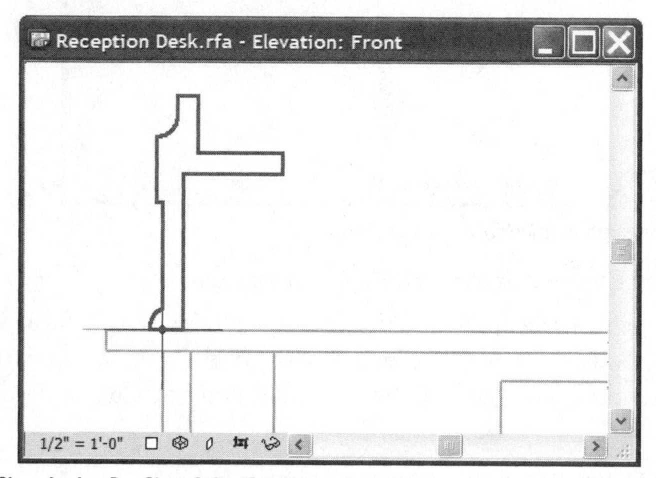

Figure 9–52 *Sketch the Profile of the Sweep*

45. On the Design Bar, click the Sweep Properties button.
    - Set the Subcategory to **Screen** and then click OK.
    - On the Design Bar, click the Finish Profile button.
    - On the Design Bar, click the Finish Sweep button.
46. Zoom All Viewports to Fit (see Figure 9–53).

TIP Try setting the Viewports to hidden line display to see the final product more clearly in plan and elevation.

Figure 9–53 *Completed Reception Desk with Privacy Screen*

47. Save the Family file.

LOAD THE CUSTOM FAMILY INTO THE PROJECT

At this point, we have completed the geometry of our Custom Family and are ready to load it into our commercial project. The process has been covered before. Let's review it now.

LOAD THE COMMERCIAL PROJECT

We closed the project above to make it easier to tile the windows in the Family Editor. We now need to reopen the project.

1. On the Standard toolbar, click the Open icon.

TIP The keyboard shortcut for Open is CTRL + O. **Open** is also located on the File menu.

- In the "Open" dialog box, click the *My Documents* icon on the left side.
- Double-click on the *MRB* folder, and then the *Chapter09* folder.

If you installed the dataset files to a different location than the one listed here, use the "Look in" drop down list to browse to that location instead.

2. Double-click *09 Commercial.rvt* if you wish to work in Imperial units. Double-click *09 Commercial Metric.rvt* if you wish to work in Metric units

   You can also select it and then click the Open button.

The project will open in Revit Building with the last opened View visible on screen.

3. On the Project Browser, double-click to open the *Level 3* floor plan View.
   - Zoom in on the reception space (in the middle of the plan).

LOAD A FAMILY

The Family file is still open. We can switch over to it and then load the Family back into the project. We could also use the "Load Family" command on the File menu, but since the Family file is still open, it is easier to load from there.

4. Hold down the CTRL key and then press TAB.

This will cycle through open windows.

- Repeat the CTRL + TAB until the Reception Desk Family file comes into view.

5. On the Design Bar, click the Load into Projects button.

The screen will switch back to the commercial project. If you look at the Furniture node beneath Families on the Project Browser, you will see that the Reception Desk Family is now loaded.

6. Add an Instance of the Family to the reception space (see Figure 9–54).

TIP Remember to use the space bar as you place it to rotate it correctly.

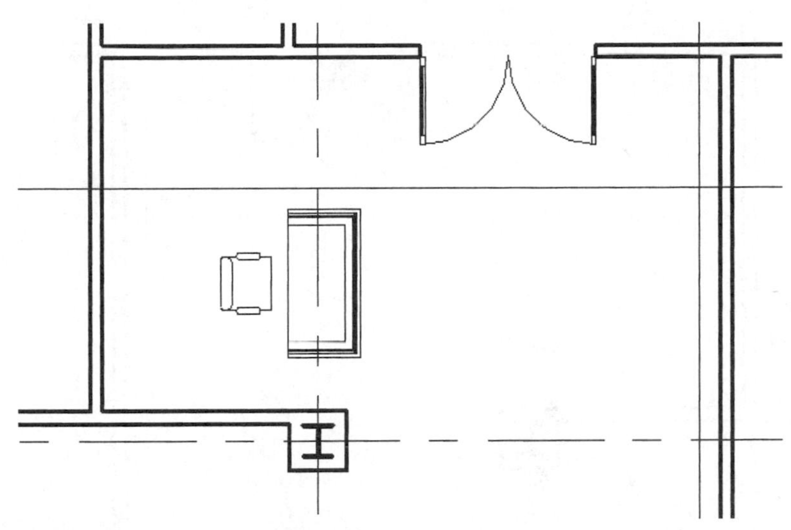

Figure 9–54 *Load the Family and then add an Instance to the commercial project*

Use the Component tool on the Design Bar to load and then place a task chair with the reception desk.

CONTROL DISPLAY OF SUBCATEGORIES

At the start of the custom Family creation above, we created some Subcategories and further assigned those Subcategories to Materials. However, Subcategories are also useful to selectively control the display of elements in the model. Using our custom Subcategories, we can turn on and off various parts of our reception desk.

7. Following the process used above, create a Camera View in the Reception area looking at the desk.

When you create your Camera View, there is a good chance that the far corner of the room will not appear in the View. This is controlled by a parameter of the Camera View called the Far Clip Offset. We can adjust this value to make the Camera "see" back further into the room.

- Right-click in the View window and choose View Properties.
- In the "Element Properties" dialog, beneath the "Identity Data" grouping, change the "View Name" to **Reception Camera**.
- Beneath the "Extents" grouping, change the "Far Clip Offset" to **30′-0″** **[9000]** and then click OK (see Figure 9–55).

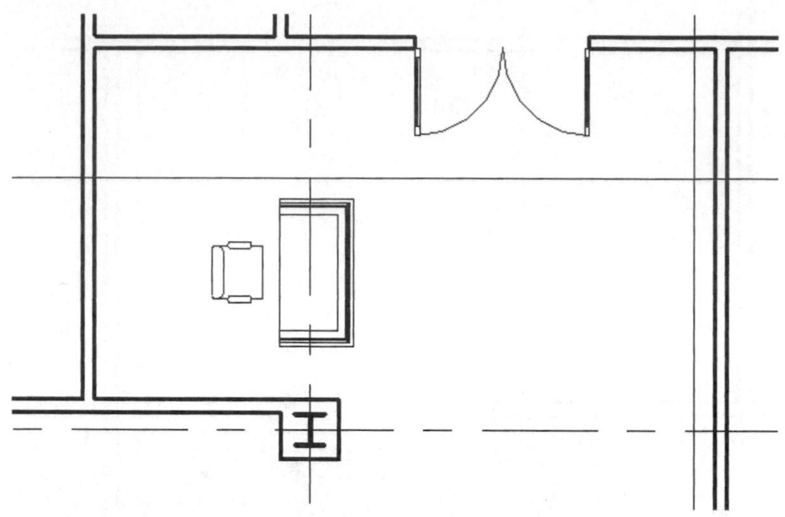

Figure 9–55 *Load the Family and then add an Instance to the commercial project*

8. From the View menu, choose **Visibility/Graphics**.

TIP The keyboard shortcut for Visibility/Graphics is **VG**. You can also right-click in the View, choose Properties, and then click the Edit button next to "Visibility."

Unlike above when we were in the Family Editor, here we see all of the possible Revit Building Model Categories because we are in a project.

- Locate the "Furniture" Category and expand it (see Figure 9–56).

Visibility/Graphic Overrides for 3D View: Reception Camera

Model Categories | Annotation Categories | Imported Categories | Linked RVT Categories | Design Options

Show model categories in this view — If a category is unchecked, it will not be visible.

Visibility	Line Style Projection	Line Style Cut	Halftone	Detail Level
Floors	By Category	By Category	☐	By View
Furniture	By Category		☐	By View
Frame	By Category			
Hidden Lines	By Category			
Overhead Lines	By Category			
Screen	Override...			
Worksurface	By Category			
Furniture Systems	By Category		☐	By View
Generic Models	By Category	By Category	☐	By View

Figure 9–56 *The Subcategories of Furniture have been added to the project*

Notice that each of the custom Subcategories created above have been added to the project.

9. Clear the checkmark from the "Screen" Subcategory checkbox and then click OK.

You will see that the screen portion of the reception desk has disappeared. In addition to the Subcategories themselves, the materials that we created above have also been imported with the Family. You can see this by going to the Settings menu and choosing **Materials**. This means that you can define Subcategories and Materials in a Family file in your library and then edit these items later in the specific projects in which they are used. This allows you to plan ahead while maintaining a good deal of flexibility. Try experimenting with editing the "MRB" Materials if you wish. Turn the "Screen" Subcategory back on when you are done experimenting.

For the final section of this chapter we will switch to the residential project. If you wish to add more furniture components or otherwise continue to practice in the commercial project, please feel free to do so before continuing.

10. Save the project.
11. Close the Reception Desk Family file.

BUILDING PARAMETRIC FAMILIES

Throughout the course of this chapter you have been exposed to a variety of Family editing and creation scenarios. As you have seen, Families can be very simple or very complex. When you are first learning their scope and potential, it is helpful to start with simple examples and work your way up to more complex ones. We have yet however to cover one of the most powerful and useful aspects of Families. That is their ability to be "parametric." Like much terminology that is associated with Building Information Modeling, the term *parametric* is often misused or misunderstood. Since it is nearly impossible to avoid hearing reference to the term in nearly any document, seminar or training session on Autodesk Revit Building, it will be helpful to take a moment to properly define the term.

Parametric is the adjective form of the noun "Parameter." Browsing "Merriam-Webster Online Dictionary" we find the following definition for the term:

Parameter—any of a set of physical properties whose values determine the characteristics or behavior of something.

This particular definition of "parameter" was selected from a few available variations because it is the most appropriate in the context of its use in describing the behavior of elements in Autodesk Revit Building. When we describe something in software such as Revit Building as being "parametric" we are therefore simply saying that the thing in question is characterized by its associated set of "physical properties" each of which hold "values determining characteristics of the element's behavior." In the specific case of Revit Building, each element has one to several available parameters. We input values into these parameters to determine the specific characteristics of the element (Wall, Door, Roof, Family, etc) which we are editing. Not specifically mentioned in the Webster's definition but implied by the use of the term *parametric* in software, is the ability to modify parameters at any time. Therefore, the ability for a particular parameter's value to determine the characteristics of behavior is not a limited to the point of creation, but rather is a "living" parameter with ongoing influence over the element. This behavior is what makes the notion of "parametric" so significant in software like Autodesk Revit Building. This dynamic interaction across the whole system, (not just with Model and Annotation Elements, but Views and Schedules as well), is often referred to in Revit Building as its "Parametric Change Engine."

Having outlined our definition of the term "parametric" a Parametric Family is simply a Family that has editable parameters. Typically such a Family will also have Types—though it is not required. Therefore, as we will see, Family Parameters can be Type- or Instance-based. We could certainly continue in our Reception Desk Family file and add parameters to the desk, but for simplicity, we will explore these topics, in a new Family file.

CREATE A NEW FAMILY

As we did above, we will create a new Family file based on one of the provided Revit Building Family templates. We are going to create a simple binder bin to hang on the wall behind the reception desk. This object will be geometrically simpler than the reception desk, so we can focus on adding and working with parameters.

1. On the Window menu, switch to the Reception Desk Family File and then from the File menu, choose **Close**.
   - In the Commercial project, on the Project Browser, double-click to open the *Level 3* floor plan View.
   - From the Window menu, choose **Close Hidden Windows**.

   The *Level 3* floor plan View should be the only open window now.
   - Minimize the *Level 3* floor plan View.
2. From the File menu, choose **New > Family**.
   - Browse to you Templates folder if necessary.
   - Select *Specialty Equipment wall based.rft* [*Metric Specialty Equipment wall based.rft*] and then click Open.

NOTE If your version of Autodesk Revit Building does not contain either of these templates, they have been provided in the *Out of the Box Templates* folder with the files installed from the Mastering Autodesk Revit Building CD ROM.

3. From the Window menu, choose **Tile**.

4. From the View menu, choose **Zoom > Zoom All To Fit**.

TIP The keyboard shortcut for Zoom All To Fit is **ZA**.

5. From the File menu, choose **Save**.
   - In the "Save as" dialog, type **Binder Bin** for the Name and then click Save.

Like the Furniture Family template used above, this Family opens with four View windows and some existing Reference Planes on screen. In addition, a temporary Wall element is included in this template. This Wall is used for reference while working in the Family Editor and will not be included with the binder bin when it is loaded into a project. We have chosen the "Specialty Equipment" template in this exercise for its simplicity. Because there are not many Reference Planes, Parameters or Settings in this template, we can learn to build them ourselves. When you build your own Families, study the list of provided templates carefully before making your choice.

ADD SUBCATEGORIES

There is an important point to note on the approach we have taken; we can change the category of the Family to Casework even though we did not start with that template. However, it is best to make this the first thing we do in the Family Editor.

6. As we did above, choose **Object Styles** from the Settings menu and note that "Specialty Equipment" is listed under "Model Objects" (see Figure 9–57).

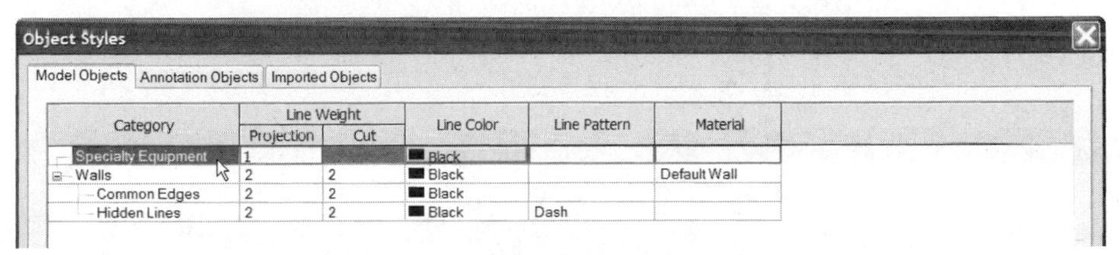

Figure 9–57 *View Object Styles to see the Family's current Category*

While we are in the Object Styles dialog, let's add a few Subcategories as we did above.

7. Repeat the steps above in the "Creating Sub-Categories" topic to create a Subcategory named: Bins.

- For the material assignment of this Subcategory, create a material named: **MRB - Natural Birch Wood** identical to the one of the same name created above.

NOTE Since this Family template included a Wall, the Subcategory list in Object Styles includes a Wall Category and its Subcategories. Be sure to select Casework when you add the new Subcategories, or they will be added to Walls instead.

- Create a second Subcategory named: **Hardware**.
- Assign it to a new Material following the procedure above. Name the new material **MRB – Chrome** (see Figure 9–58).

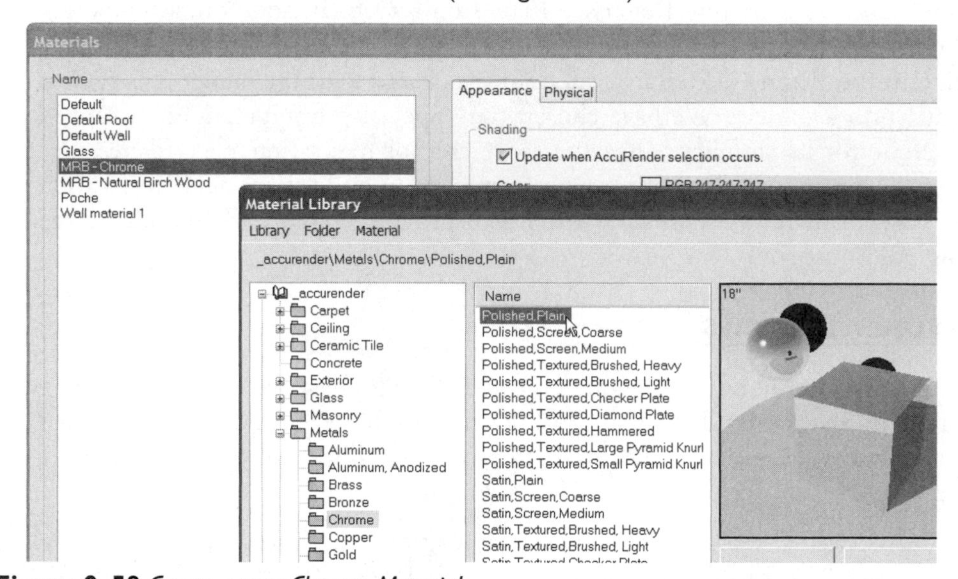

Figure 9–58 *Create a new Chrome Material*

- For its texture, browse to the *_accurender\Metals\Chrome* folder, and choose Polished:Plain (see Figure 9–59).

Object Styles

Model Objects | Annotation Objects | Imported Objects

Category	Line Weight		Line Color	Line Pattern	Material
	Projection	Cut			
Casework	1	1	Black		
Bins	1	1	Black	Solid	MRB - Natural Birch Wood
Hardware	1	1	Black	Solid	MRB - Chrome
Walls	2	2	Black		Default Wall
Common Edges	2	2	Black		
Hidden Lines	2	2	Black	Dash	

Figure 9–59 *Create two Subcategories for Casework and assign them custom Materials*

- When finished configuring the Subcategories, close the "Object Styles" dialog.

CREATE REFERENCE PLANES

We have used Reference Planes in several places throughout this book. Their inclusion in Families serves a vital purpose. If we wish to create a parametric Family, it is common practice to associate the parameters with various Reference Planes (or Reference Lines—which are similar but finite in length) and then lock the geometry to these Reference Planes. In this way we can vary the dimensions that control the locations of various Reference Planes and thereby manipulate the geometry constrained to them. Before we can demonstrate this here, we must add some Reference Planes to our Family.

8. Click on the title bar of the *Elevation : Placement Side* window to make it active.
9. On the Tools toolbar, click the Work Plane icon.
   - From the "Name" list, choose **Reference Plane : Back** and then click OK.
10. On the Design Bar, click the Ref Plane tool.
    - Create four Reference Planes as shown in Figure 9–60.

Do not be concerned with the precise locations. Simply sketch the Reference Planes in the approximate locations indicated in the figure. We will adjust the dimensions later.

- With the Modify tool select the top Reference Plane, on the Options Bar click the Properties icon, and then in the "Name" field type: **Top**.
- Click OK to complete the change, and then repeat for each of the other three Reference Planes, using the names indicated in the figure.

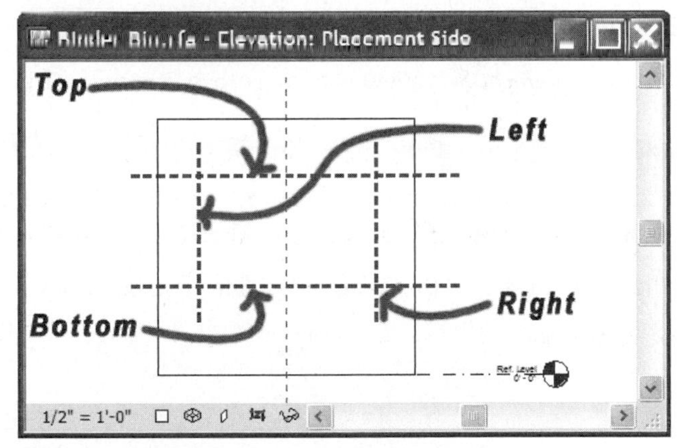

Figure 9–60 *Create four Reference Planes in elevation*

11. On the Design Bar, click the Solid Form > Solid Extrusion tool.

- On the Options Bar, set the "Depth" to **-1′-2″ [-350]** and then click the Rectangle icon.
- Snap to the intersection of the Top and Left Reference Planes for the first corner.
- Snap to the intersection of the Bottom and Right Reference Planes for the other corner (see Figure 9–61).

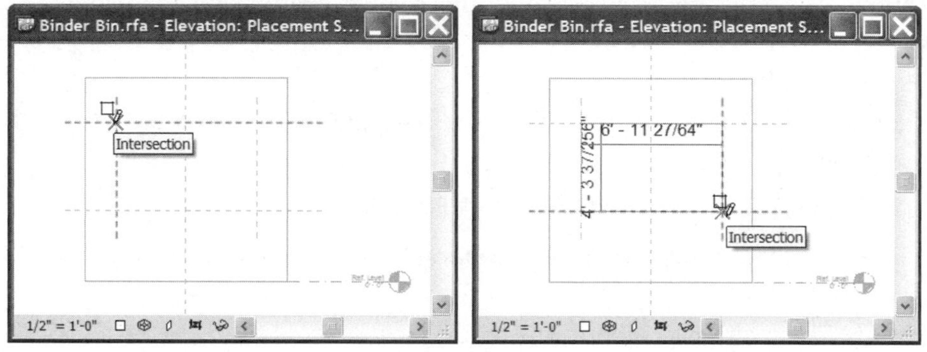

Figure 9–61 *Sketch a Rectangle snapping to the Reference Planes*

Four open padlock icons will appear—one on each edge of the rectangle.

- Close each of the lock icons to constrain the rectangle shape to the Reference Planes.
- On the Design Bar, click the Extrusion Properties button and change the Subcategory to **Bins**. Click OK to complete.
- On the Design Bar, click the Finish Sketch button.

Take note of the extrusion in each View window.

12. In the 3D View window, change the display to Shaded with Edges.
13. In the *Elevation: Placement Side* View window, move one of the Reference Planes.

The exact amount of the move is not important. What is important is that the shape of the Solid Extrusion will adjust when you move the Reference Planes. This is because we constrained the edges of the sketch to the Reference Planes.

- Undo the Move to return the Reference Plane to its previous position.
- Save the Family file.

CREATE DIMENSION PARAMETERS

The first step in creating our parametric Family is completed—we created some geometry that is constrained to our Reference Planes. The next task is to create Dimension elements and associate them to parameters.

14. Click on the title bar of the *Elevation : Placement Side* window to make it active.
    - Maximize the View window (you can do this with the icon in the top right corner of the window or just double-click the title bar).
    - Zoom the window to fit (type **ZF** on the keyboard or choose the command from the View > Zoom menu).
15. On the Design Bar, click the Dimension tool.
    - Highlight and then click the Ref. Level line.
    - Highlight and then click the Bottom Reference Plane (see Figure 9–62).

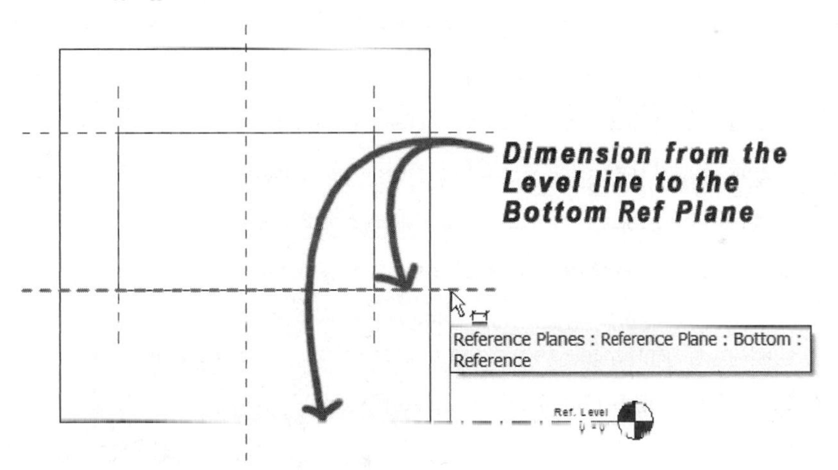

Figure 9–62 *Place the first Dimension element*

    - Click in a blank space between the two references to place the Dimension.
16. Repeat the process to add another Dimension between the Top and the Bottom Reference Planes (see Figure 9–63).

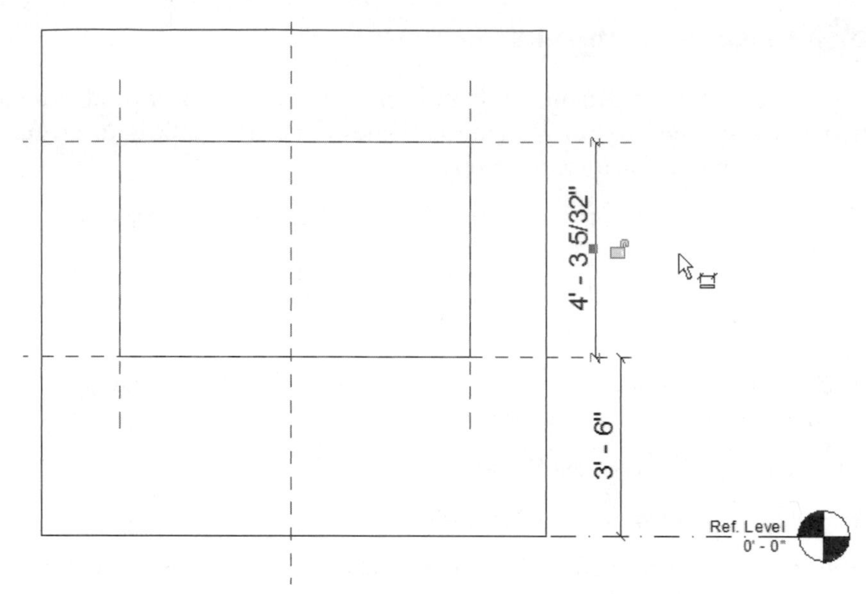

Figure 9–63 *Place another vertical Dimension between the Top and Bottom Reference Planes*

17. Begin a new horizontal Dimension.
    - Highlight and then click the Left Reference Plane.
    - Highlight and then click the Center Reference Plane.
    - Highlight and then click the Right Reference Plane.

    This will give you three witness lines and two dimensions in a continuous string.

    - Click in a blank space above the binder bin to place the Dimension.
    - Click the "Toggle Equality" icon (see Figure 9–64).

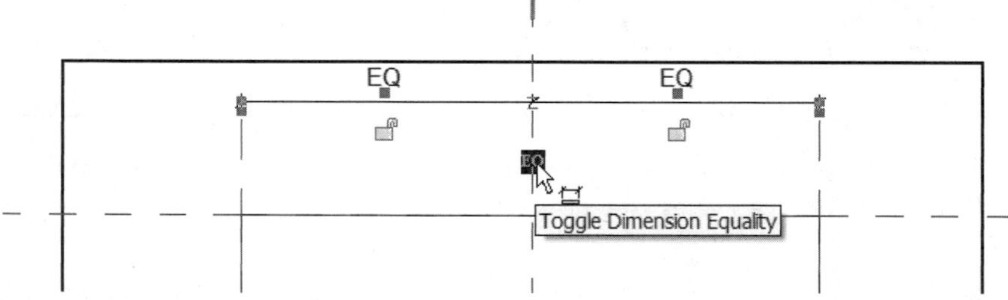

Figure 9–64 *Toggle on Dimension Equality for the horizontal string*

After you click this toggle, you should see the Center Reference Plane shift and the dimension values will be replaced with the letters "EQ." This applies a constraint that will keep the three Reference Planes equally spaced.

18. Add one final Dimension horizontally between the Left and Right Reference Planes only—do *not* include the Center one in this string.

   - On the Design Bar, click the Modify tool or press the ESC key twice.

Now that we have created several Dimensions, we will create parameters that will ultimately control them. Recall that you can manipulate model geometry in Revit Building by editing the values of permanent dimensions and temporary dimensions. When you apply parameters to Dimensions in a Family file, you are indicating which parts of the Family's geometry will be able to be edited dimensionally with the Family Type's parameter values. Let's take a look.

19. Select the overall horizontal Dimension element.

Take note of the "Label" item on the Options Bar. This drop-down list is used to assign parameters to the selected Dimension. You can also right-click and choose **Edit Label** if you prefer.

20. On the Options Bar, click the drop-down list next to "Label" (currently <none>) and choose **<Add parameter>**.

The "Parameter Properties" dialog will appear; verify that the "Family Parameter" radio button is selected at the top.

   - In the "Parameter Data" area, type **Width** in the "Name" field.
   - From the "Group parameter under" drop-down list, choose **Dimensions**.
   - Choose the "Type" radio button (see Figure 9–65).

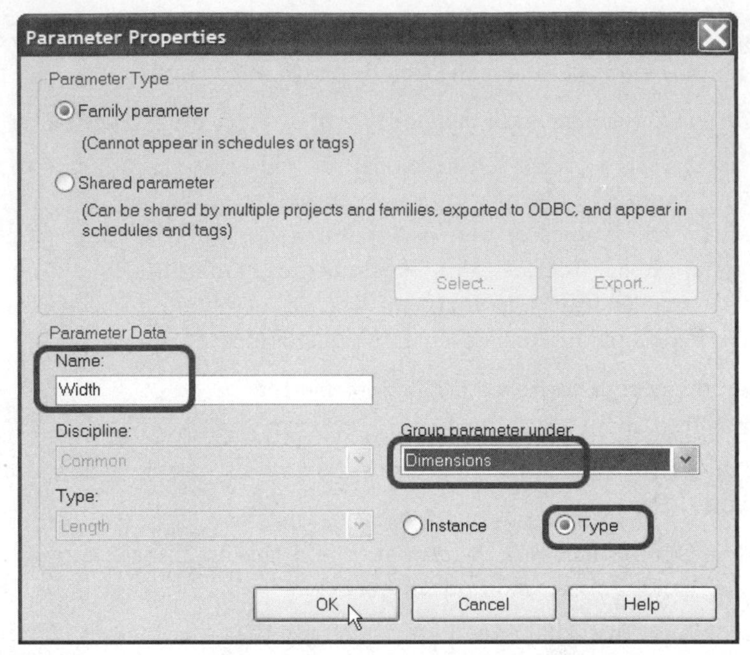

Figure 9–65 *Create a custom "Width" parameter for the Family's Types*

These settings create a parameter named "Width" that will appear beneath the "Dimensions" grouping of the "Type Properties" dialog box. If we chose "Instance" the parameter would apply instead to the instances of the Family and could be unique for each one—we will create an instance parameter below.

- Click OK to complete the parameter.

Notice that the Label "Width" now appears in front of the Dimension text. This tells us that this Dimension is controlled by this parameter.

21. Repeat the process on the vertical Dimension between the Top and Bottom Reference Planes.
    - Name the parameter **Height**, group it under Dimensions and make it a Type parameter as well.
22. Select the vertical Dimension between the Ref. Level and the Bottom Reference Plane and repeat the process again.
    - Name the parameter **Mounting Height**, group it under Dimensions and make it an Instance parameter this time (see Figure 9–66).

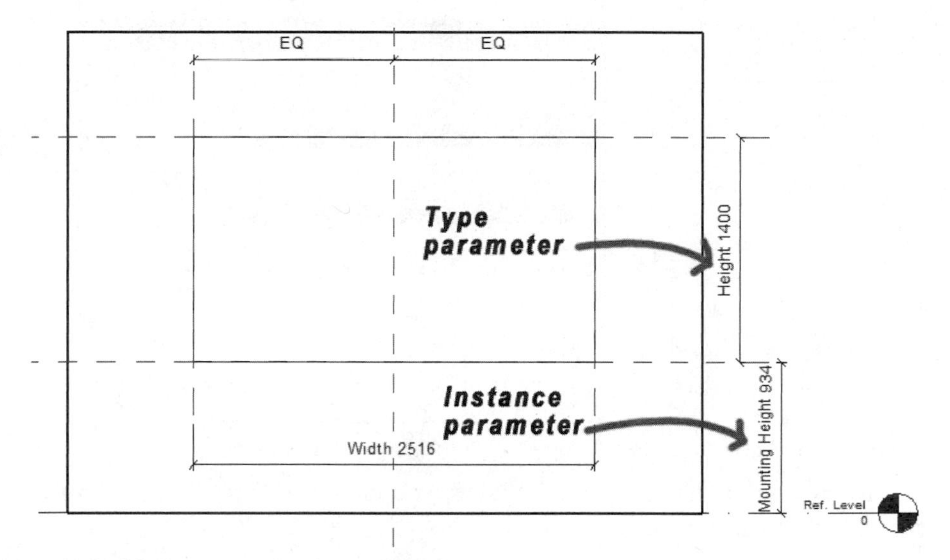

Figure 9–66 *Create another Type parameter and one Instance parameter*

23. Save the Family.

ADD FAMILY TYPES

All that remains now is for us to test our Family parameters. We do this with the Family Types command. This command allows us to not only test to see if our parameters are behaving properly, but to also create some Types that will load into projects when we load the Family. Remember, just like the out-of-the-box Families that we have used so far, users will always be able to add new Types to our custom Family later. But it is still a good idea to stock it with a few preferred Types ahead of time.

24. On the Design Bar, click the Family Types button.

Notice that the three parameters are listed at the top of the dialog beneath a "Dimensions" grouping just like we specified. To test the parameters, move the dialog to the side so you can see the View window in the background. Type numbers into the "Value" field next to each parameter and then click the Apply button. The Dimensions in the View window will adjust to the new sizes (see Figure 9–67).

NOTE In this exercise, we have waited until after adding several parameters to "test" them. It is very good practice, however, to perform this procedure immediately following the creation of each parameter. In this way, you can catch mistakes or issues before they become potentially compounded by other parameters.

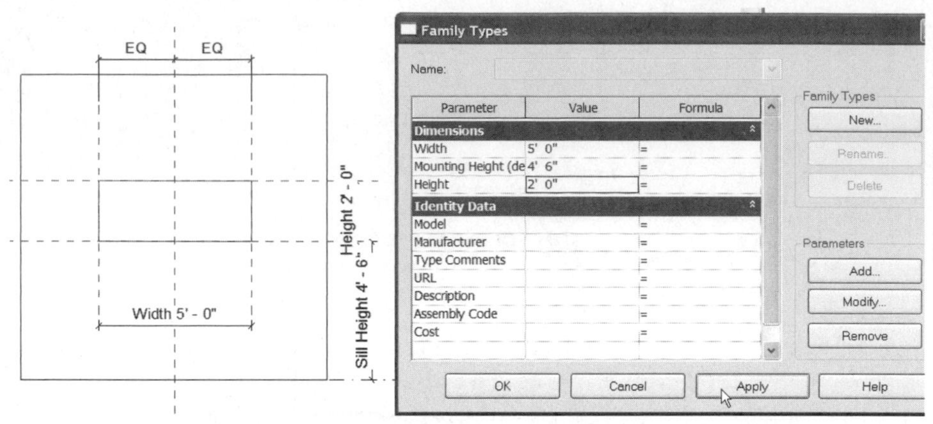

Figure 9–67 *View and test the custom parameters in the Family Types dialog*

On the right side of the "Family Types" dialog, you can add Types and add and edit Parameters. Let's add some Types.

25. On the right side of the dialog, click the "New" button beneath "Family Types."
    - In the "Name" dialog, type **60 wide 24 high [1500 wide 600 high]** and then click OK.

The new name will appear in a list at the top.

- Change the Width to **5′-0″ [1500]**.
- Change the Mounting Height to **4′-6″ [1350]**.
- Change the Height to **2′-0″ [600]**.

26. Click New again to create **60 wide 18 high [1500 wide 450 high]**.
    - Set the Height to **1′-6″ [450]**, leave the other dimensions the same.
    - Create additional Types if you wish. Click OK when finished.

ADD HARDWARE AND SYMBOLIC SWING LINES

We are almost ready to load our Family into our project. Above we created a Hardware Subcategory, but so far we have not made any hardware.

27. Tile the View windows again and create the hardware.

The specifics of creating the hardware will be left to you as a practice exercise. You can make a knob using a Solid Revolve like the one we created for the reception desk above, or you can make a wire pull from a Solid Sweep. The most important thing to remember however is that if you want to constrain the position of the hardware in any way, you must create and constrain it to a Reference Plane. Start by creating a Reference Plane 2″ [50] above the Bottom Reference Plane. Edit its properties and name it **Hardware Height**. Add a Dimension element between the Bottom Reference Plane and this new one and lock it to keep it constrained to that relative location in the Family. When you create the Sweep, start in the plan View, edit the Work Plane, and choose the Hardware Height Reference Plane from the list. Sketch the path centered on the bin in plan, then switch to the elevation and create the profile. A simple square profile is sufficient. Be sure to edit the Sweep Properties (on the Design Bar) and choose the **Hardware** Subcategory. When you have finished the Solid, open the Family Types dialog and choose each of your predefined Types in succession from the list to test it out. The wire pull should remain 2″ [50] from the bottom edge and centered on the bin (see Figure 9–68).

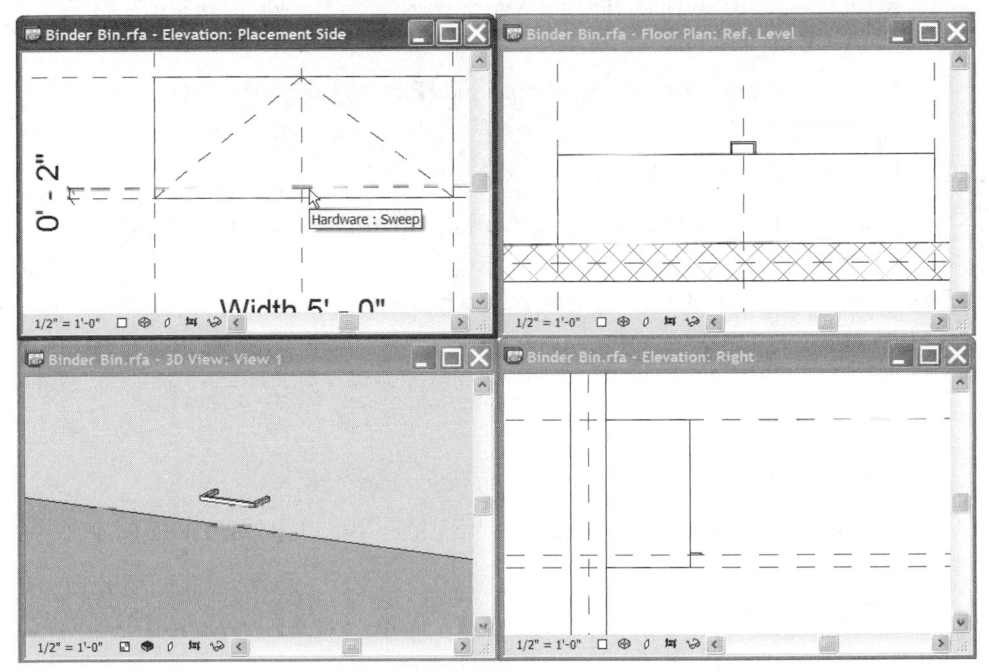

Figure 9–68 *Add a Solid Sweep wire pull for a handle*

28. Zoom all Views to Fit when finished.

Symbolic Lines are special Revit Building elements that can be added to Families to embellish the 3D Solids in certain Views. In projects, we have model elements and annotation elements. Model elements appear in all Views, and annotation (including detail) elements appear only in the View to which they were added. In Families, model elements behave exactly as they do in projects. However, the Views that we see in the Family Editor are not transferred over to the project when we load the Family. Therefore, drafting lines would not be transferred either. Instead, we use Symbolic Lines in the Family Editor. These elements share characteristics of both model and annotation/detail elements. Symbolic lines like Detail Lines appear only in certain Views, however, like model elements, they can appear in several Views based on circumstances. Symbolic Lines will appear in all Views parallel to the View in which they are created. So if you add a Symbolic Line to a front elevation, it will appear in all sections or elevations that face the same direction. The use of Symbolic Lines allows us to add detailing and embellishment that is View-specific like Detail Lines, but also somewhat parametric like model geometry. In this example, we will use Symbolic Lines to indicate the swing direction of the binder bin when it opens in the front facing elevation.

29. Click on the title bar of the *3D View : View 1* window to make it active.
    - On the Tools toolbar, click the Work Plane icon (shown in Figure 9–43 above).
    - In the "Work Plane" dialog, choose the "Pick a Plane" radio button and then click OK.
    - In the 3D View, click on the front face of the bin.
    - On the Tools toolbar, click the Work Plane Visibility icon (to the right of the Work Plane icon).

This allows you to see the plane that was selected.

30. From the Settings menu, choose Object Styles.
    - Create a new Subcategory of Specialty Equipment named **Elevation Swing**.
    - Change the Line Pattern to **Dash** and then click OK.
31. Click on the title bar of the *Elevation : Placement Side* window to make it active.
    - Zoom in on the bin.
32. On the Design Bar, click on the Symbolic Lines tool.
    - From the Type Selector, choose Elevation Swing.
    - On the Options Bar, select "Chain" and then sketch two lines as indicated in (see Figure 9–69).

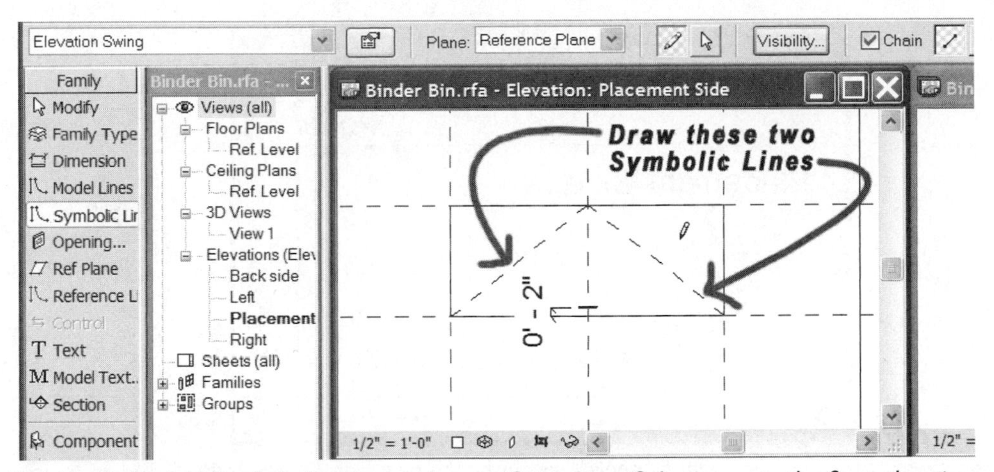

Figure 9–69 *Add Symbolic Lines using the new Swing Lines Subcategory to the front elevation*

- On the Design Bar, click the Modify tool or press the ESC key twice.

If you recall the default Mounting Heights that we assigned above were 4′-6″ [1350] for each Type. While we did make these Instance Parameters (which means that each bin we add to our model can have its own Mounting Height), initially when we place them, they will fail to show in most plan Views. This is because the default cut height in plan is 4′-0″ [1200] which is below the lowest point of our bin's geometry. To deal with this situation, we can add additional Symbolic Lines in the plan View that show an outline of our bin in a dashed line style to indicate it is mounted above the cut plane. We will also need to add an additional element to ensure that Revit Building recognizes that even though our geometry is not being cut by the cut plane that we still wish to have something drawn in plan Views.

33. Click on the title bar of the *Floor Plan : Ref. Level* window to make it active.
    - Check the Work Plane and verify that it is **Level : Ref. Level**. If not choose that and click OK.
34. Open Object Styles again and create a new Subcategory of Specialty Equipment named **Plan Representation**. Make its Line Pattern **Overhead**.
35. On the Design Bar, click on the Symbolic Lines tool.
    - From the Type Selector, choose Plan Representation.
    - On the Options Bar, click the Rectangle icon and then trace the bin geometry.
    - Close all four padlock icons (see Figure 9–70).

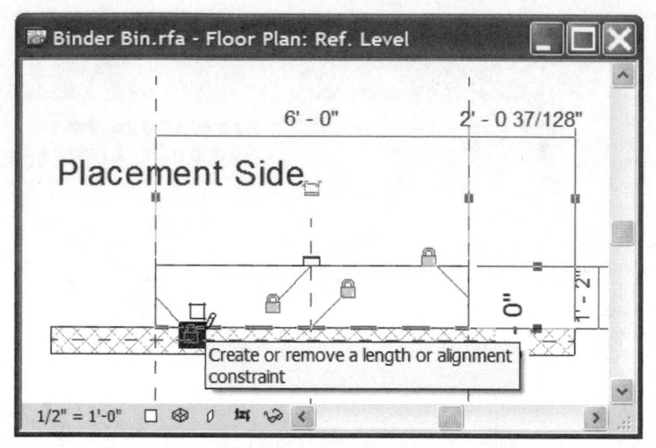

Figure 9–70 *Trace the plan with Symbolic Lines using the new Plan Representation Subcategory*

- On the Design Bar, click the Modify tool or press the ESC key twice.

We now have the Symbolic Lines that we need, but as noted above, we need to instruct Revit Building to show them. Since none of the geometry of the bins is being cut by the cut plane, Revit Building would simply not show any geometry in plan Views. To trigger it to display our Symbolic Lines, we can draw an invisible line element passing through the Cut Plane. In this way, Revit Building will recognize that we wish display something in plan.

36. Click on the title bar of the *Elevation : Placement Side* window to make it active.
    - On the Design Bar, click the Model Lines tool.
    - In the "Work Plane" dialog that appears, choose **Reference Plane : Back** and then click OK.
    - From the Type Selector, choose <Invisible lines>.
37. Draw a Model Line from the bottom of the bin to the Ref. Level.
    - On the Design Bar, click the Modify tool or press the ESC key twice.

With the placement of this Model Line, the cut plane cuts through it and Revit Building knows the bin is above. It then will display the bin's symbolic dashed lines (or anything else that it is told to display in plan/rcp).

The final step is to make sure that only the Symbolic Lines display in plan and not the bin geometry (which would appear like solid lines—instead of dashed overhead lines) is to instruct the bin and hardware geometry not to display in plan.

38. Select the Bins Extrusion.

- On the Options Bar, click the Visibility button.
- In the "Family Element Visibility Settings" dialog, clear the checkmark from the "Plan/RCP" box and then click OK (see Figure 9–71).

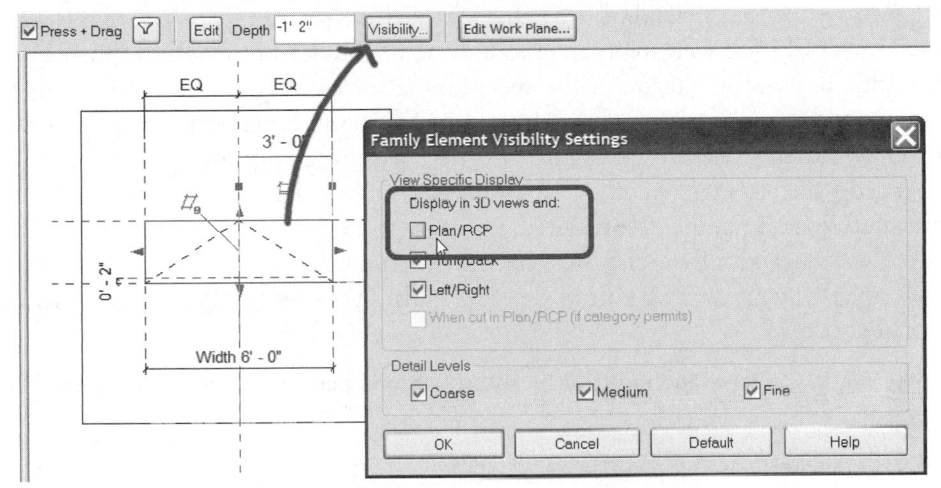

Figure 9–71 *Turn off visibility in elevation Views for the plan symbol*

39. Repeat the process for any hardware geometry that you added above.
    - Zoom all Views to fit.
40. Save the Family file.

Congratulations, you have just created a parametric Family from scratch! For the final test, let's load it into the project.

LOAD THE NEW FAMILY INTO THE PROJECT

Make sure you are satisfied with the geometry and have tested the various Types. Be sure to save the Family before you continue.

41. On the Design Bar, click the Load into Projects button.

This will restore the *Level 3* View of the Commercial Project. Check the Families branch of the Project Browser under *Specialty Equipment* for the newly loaded Family.

NOTE Ideally this Family might fit better in the Casework or Furniture categories. As was mentioned at the start of the lesson, the Specialty Equipment template was chosen for its educational value in being a very simple template that allowed us to build all of the elements we needed from scratch. As already noted, in your own Families, you can choose a more appropriate template from which to start.

42. Zoom in on the reception space.
43. From the Specialty Equipment node of the Families branch, drag one of the Family Types into the View.

At this point we can test several of the behaviors of this new Family. Open the Reception Camera 3D View created above to look at the binder bin in 3D. Return to the plan and cut a section through the reception space looking at the binder bin. Open the new Section View. You should see the Symbolic Elevation Swing Lines in that View. Open Object Styles and beneath Specialty Equipment, experiment with changing the settings—turn off components, change Line Styles and assign different materials. Remember you can also edit these settings per View by editing the View Properties and then clicking the Edit button next to "Visibility" (or just type VV). Finally, you can add, edit or delete Family Types on the Project Browser like we did above.

44. When you have finished experimenting with the binder bin, close the Family file.
45. Save and Close the commercial project.

ENHANCE FAMILIES WITH ADVANCED PARAMETERS

In the previous topic we made a basic parametric Family by adding labels to dimension strings which in turn controlled the size of the object. This is the most basic way to add parameters to a Family. Adding such parameters allows you to create multiple Types from a single Family.

Dimensional parameters are the most common parameters used in Families and are the most straightforward to define. But the potential of parameters goes well beyond the controlling of dimensions. In this topic, we will explore several "advanced" parameters. They require a little more planning and often a few more steps to create, but once you get the hang of them, they can add functionality and flexibility to the Families you create.

In this exercise, we will cover nested Families, visibility parameters, array parameters and the use of formulas in parameters.

CREATE A COAT HOOK FAMILY

For use in this exercise, a Family file has been provided with the files on the CD. The file contains a wall-based coat rack. It was built from scratch from the same template used above for the binder bin. Similar geometry, Reference Planes and Dimension Parameters have already been set up in that file. In this exercise, we will create a new Family for an individual coat hook. We will then open the provided wall coat rack Family file and load our coat hook Family into it. From there we will explore the various parameters mentioned above.

1. From the File menu, choose **New > Family**.
2. Browse to you Templates folder if necessary.
3. Select *Specialty Equipment.rft* [*Metric Specialty Equipment.rft*] and then click Open.

Be sure *not* to select the "wall based" template this time. The hook is not wall based, but by choosing "specialty equipment" again, we ensure that both Families share a common Category.

NOTE If your version of Autodesk Revit Building does not contain either of these templates, they have been provided in the *Out of the Box Templates* folder with the files installed from the Mastering Autodesk Revit Building CD ROM.

4. Click on the title bar of the *Floor Plan : Ref. Level* window to make it active.
   - Zoom in a bit toward the center.
5. Open Object Styles and create a Subcategory named **Hardware** following the procedure covered above.
6. From the Design Bar, choose **Solid Form > Solid Revolve**.
   - Draw a vertical line starting at the intersection of the two Reference Planes and moving up **4″** [**100**].
   - Draw a horizontal line to the left **1/2″** [**12**].
   - Draw a shallow arc vertically back down to the Reference Plane and then close the shape with a small **1/2″** [**12**] long horizontal line (see Figure 9–72).

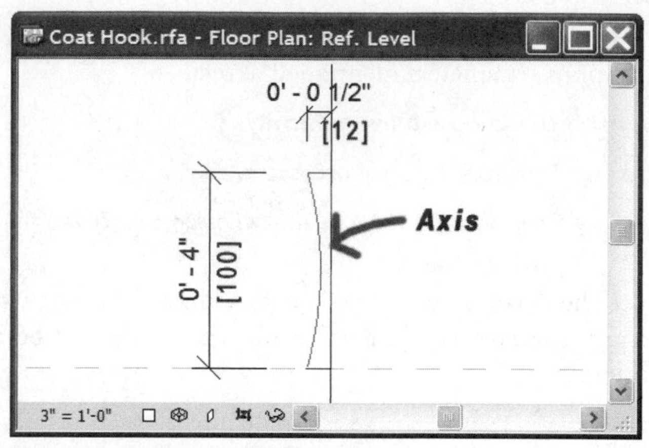

Figure 9–72 *Sketch the profile of the coat hook*

7. On the Design Bar, click the Axis tool.
   - On the Options Bar click the Pick Lines icon and then click the vertical Reference Plane.
   - On the Design Bar, click the Finish Sketch button (see Figure 9–73).

NOTE It will be necessary to zoom the various View windows to study the results.

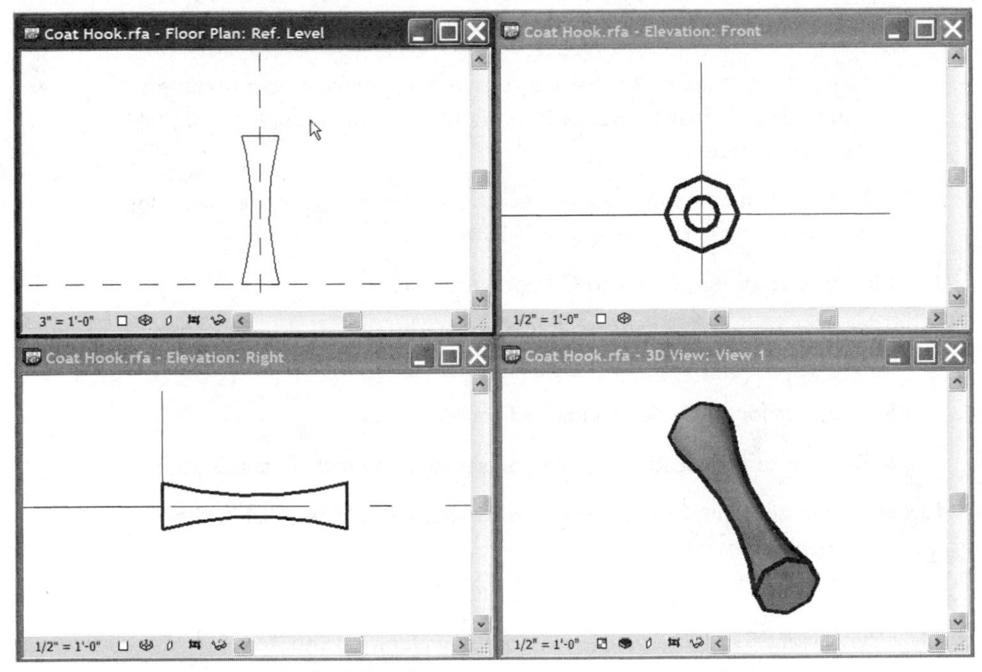

Figure 9–73 *Complete the Solid Revolve*

NOTE if you wish to use Zoom to Fit, you will have to resize the Reference Planes to make them shorter. Zoom to Fit includes these items in the extents of the model.

8. Select the element, edit its Properties and assign it to the Hardware Subcategory.
9. Save the Family as *Coat Hook.rfa* and then Close the file.

WORKING WITH NESTED FAMILIES

We now have the provided Wall Coat Rack Family and the Coat Hook Family that we just built. Let's add the hook to the rack Family. This will "nest" the hook Family into the rack Family.

10. In the *Chapter09* folder, open the *Wall Coat Rack.rfa* [*Wall Coat Rack-Metric.rfa*] Family file.
    - Click on the title bar of the *Elevation : Placement Side* window to make it active.

At the top of the shelf are two Dimensions, one on each side reading 5″ [125]. We need to create a new Dimension Parameter for these. We will create one parameter named "Hook Inset" and apply it to both dimensions.

11. Using the CTRL key, select both 5″ [125] dimensions at the top.
    - Using the procedure in the previous lesson, create a new parameter named **Hook Inset**. Include it in the Dimensions grouping and make it a Type parameter.
    - Use the Family Types tool on the Design Bar to check the Hook Inset parameter—set the value to **6″** [**150**].
12. On the Design Bar, click the Component tool.

An alert will appear indicating that no Families are loaded and requesting that you load one.

- In the dialog, click Yes to load a Family.
- Browse to the location where you saved *Coat Hook.rfa* and load it.

13. Place the hook anywhere on the Placement side of the Wall in the *Floorplan: Ref. Level* View window.

Place it randomly; do not snap it to anything at this time.

14. Using Figure 9–74 as a guide, click the Align icon on the Tools toolbar and align the hook to the back plate of the Coat Rack in plan. Lock the padlock to constrain it.

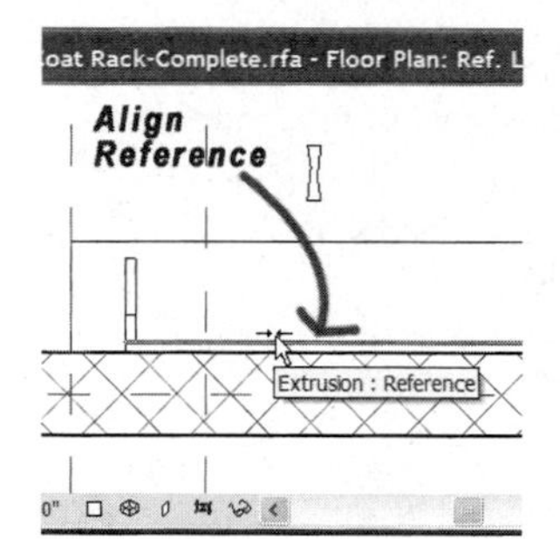

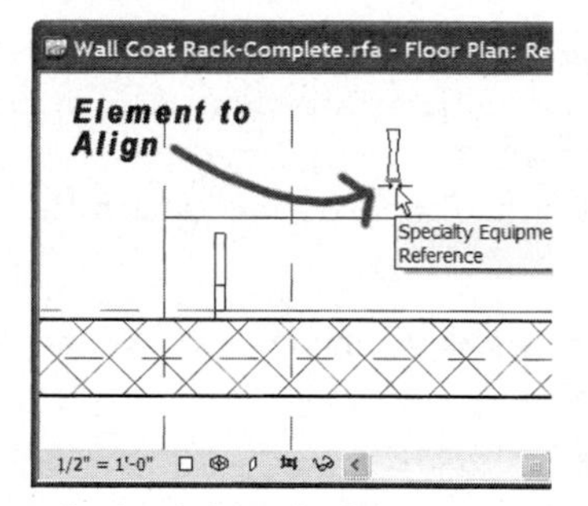

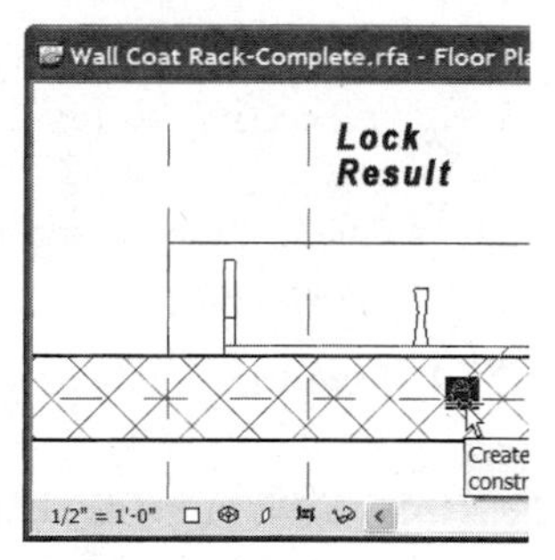

Figure 9–74 *Align the Hook to the back board of the Coat Rack*

15. Click on the title bar of the *Elevation : Placement Side* window to make it active.
    - Using Align again, align the hook to the "First Hook" Reference Plane on the left side of the rack. Identify it by highlighting it with the Modify tool (it will say "First Hook") and lock it.

You will have to look for the hook. It may be at the bottom of the Wall. When you align, be sure to click the Reference Plane first, then the middle of the hook. You may need to zoom in to the hook. When you zoom in on the hook, you will be able to highlight its centerline. Do *not* use align for the height of the hook. This we will achieve with a new formula parameter.

- On the Design Bar, click the Modify tool or press the ESC key twice.

CREATE A FORMULA PARAMETER

If you were to study the elevation and open the "Family Types" dialog, you would notice that the height of the back plate of the coat rack is controlled by a "Height" parameter. We can create a formula based upon this variable height that keeps our hooks centered vertically on the back plate.

16. On the Design bar, click on Family Types.
    - In the "Parameters" area, click the Add button.
    - Name it **Hook Height** and choose the leave it grouped under **Other**.
    - From the "Type" list, choose **Length** and then click OK.
17. Back in the "Family Types" dialog, click in the formula field to the right of the Hook Height value field (see Figure 9–75).
    - For the formula, type: **Mounting Height - (Height/ 2)**.

NOTE These label names *are* case-sensitive.

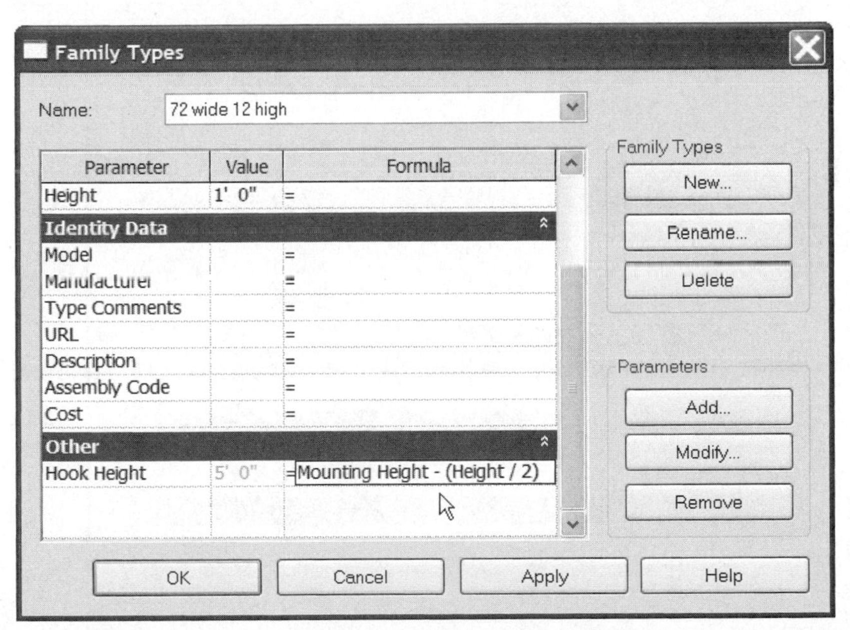

Figure 9–75 *Type a formula that subtracts half of the Height from the Mounting Height*

You will notice that once you click Apply, the result of the formula will be input read only in the Value field.

18. At the top of the "Family Types" dialog, choose 72 wide 8 high [1800 wide 200 high] from the Name list.

Note the change in the "Hook Height" automatically calculated Value.

- Click OK to dismiss the dialog.

19. Select the hook and then on the Options Bar, click the Properties icon.

Beneath the "Constraints" grouping, in the "Offset" row, in the far right column, is one more very narrow column. It has an equal sign (=) header there is a parameter button (see Figure 9–76).

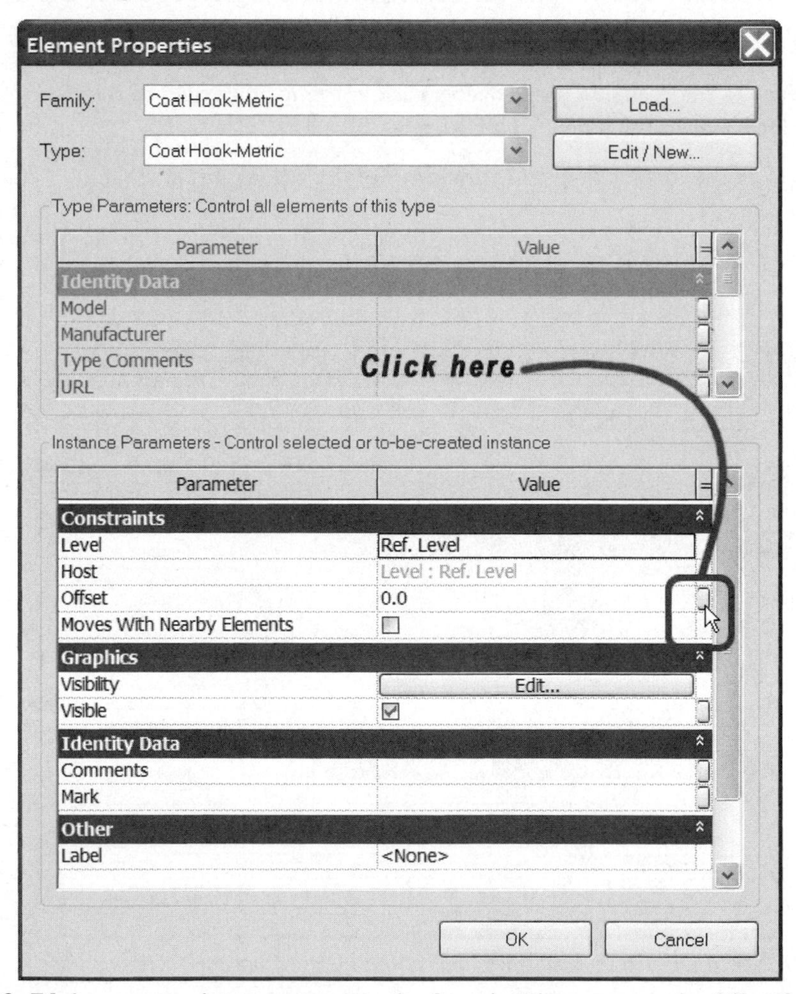

Figure 9–76 *A parameter button appears in the formula column next to the Offset field*

- Click this parameter button and in the "Associate Family Parameter" dialog, select the "Hook Height" parameter and then click OK.

Again notice the read only value of the resolved formula fills in automatically.

- Click OK again to dismiss the "Element Properties" dialog.

The hook will move to the center of the back plate height. You can re-open the "Family Types" dialog and test the formula by changing Types.

20. Save the Family file.

ADD A VISIBILITY PARAMETER

Sometimes you add details to a Family that you don't always want to display. We can create a Visibility parameter that will allow us to control the display of components in each Type within the Family. In this case, we will make it possible to turn the hooks on and off in each Family Type.

21. Select the Hook component and on the Options Bar, click the Properties icon again.
    - Beneath the "Graphics" grouping click the parameter button (in the same far right column) next to the "Visible" parameter (*not* "Visibility").
    - In the "Associate Family Parameter" dialog, click the "Add parameter" button.
    - Name the parameter: **Show Hook**, group it under **Graphics**, leave it a "Type" parameter and then click OK.
22. Return to Family Types, and create a new Type that does not show hooks. (uncheck Visible for this Type).

NOTE While you are working on the family, even if the visible checkbox is unchecked, it will still be visible. This parameter only becomes active when the Family is used in a project.

CREATE AN ARRAY OF HOOKS

By default, when you use the Array command in Revit Building it groups the arrayed elements and maintains some of the parameters used to create the array. We can utilize this feature to make a parametric array of hooks in our Family.

23. Click on the title bar of the *Elevation : Placement Side* window to make it active.
24. Select the hook and then on the Tools toolbar, click the Array icon.

- On the Options Bar, be certain that a checkmark appears in the "Group and Associate" checkbox.
- For the "Move To" option, choose the "Last" radio button.

Following the prompts at the Status Bar (bottom left corner of the Revit Building screen), we must indicate the start point and then the end point of the Array. We will use the Reference Planes for these.

- For the First Point, click the "First Hook" (left side) Reference Plane.
- For the End Point, click the "Last Hook" (right side) Reference Plane.

A temporary dimension will appear with the Array prompting you for the quantity of items.

- Type **3** for the "Array Count" and then press ENTER (see Figure 9–77).

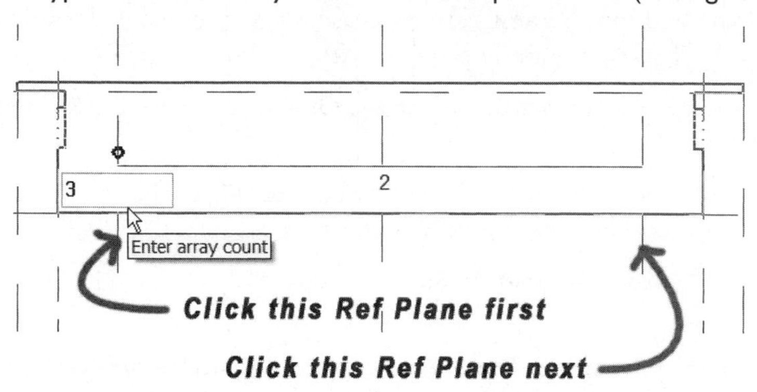

Figure 9–77 *Click each of the Reference Planes as the start and end of the Array and then type a quantity*

- On the Design Bar, click the Modify tool or press the ESC key twice.

There are now 3 hooks on the back plate of the coat rack. Click on any hook to edit the Array Count or to gain additional Group options on the Options Bar. To test this out, try changing the Array Count with the temporary dimension if you wish. Earlier, we constrained the hook on the left to the "First Hook" Reference Plane using the Align command. We need to also constrain the last hook (the one on the right) to the "Last Hook" Reference Plane. Otherwise, if you choose a different Family Type changes the width of the coat rack, the last hook will no longer line up properly.

25. Select the last hook (the one on the right) and then use the Align command (like above) to Align it to the "Last Hook" Reference Plane and lock it.

    Zoom in on the hook as required.

26. On the Design Bar, click the Family Types button.
    - Create a new Type named **84 wide 8 high [2100 wide 200 high]**.
    - Edit the Width and Height parameters accordingly, move the dialog out of the way and then click the Apply button to see the change (see Figure 9–78).

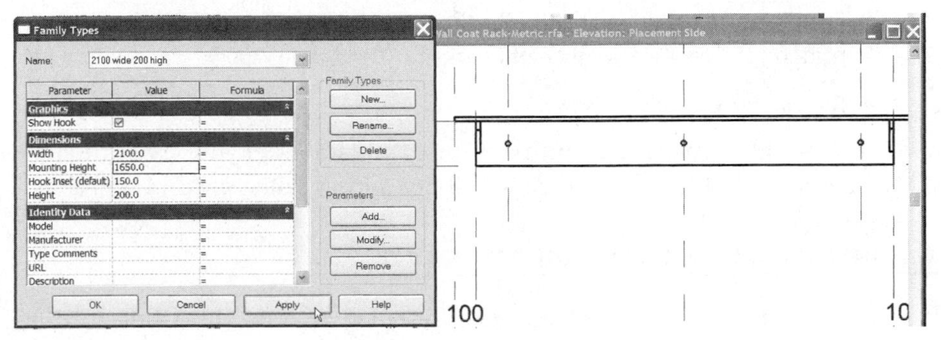

Figure 9–78 *Add a new Family Type to test the constraints*

27. Click OK to close the dialog and then Save the Family file.

MAKING THE ARRAY COUNT PARAMETRIC

At this point, we have a coat rack that keeps our three hooks equally spaced even as we change its dimension parameters. As a finishing touch, let's look at how we can make the Array Count determining the number of hooks be controlled by a parameter.

28. Select one of the hooks.

Like before, a temporary dimension indicating the current Array Count will appear. This dimension includes a horizontal line and a text field in which we can edit the count. If you select the horizontal line, rather than the text field, you expose some options on the Options Bar. This line is the Array itself.

- Click on the Array line to select it.
- On the Options Bar, next to "Label" choose **<Add parameter>** from the drop-down list.
- In the "Parameter Properties" dialog, name the parameter **Number of Hooks** and group under Graphics.
- Leave "Type" selected and then click OK.

29. Re-open the "Family Types," dialog and change the "Number of Hooks" value a few times.

- Click Apply after each change and watch the hooks appear and re-space automatically

This parameter can be different for each type.

30. For the 84 wide 8 high [2100 wide 200 high] Type, set the Number of Hooks to **5** and then Apply.
    - From the "Name" list at the top, choose 72 wide 12 high [1800 wide 300 high] Type, set the Number of Hooks to **4** and then Apply.
    - Repeat for the other Types.

Don't worry about changing the quantity on the "No Hooks" Type. Even though they show here in the Family Editor, they will not appear when used in a project.

USE A FORMULA TO SET THE ARRAY COUNT

Finally we can make the Number of Hooks a formula based on the Width of the coat rack.

31. If you closed "Family Types," re-open it now.
    - In the formula field next to the "Number of Hooks" parameter, type: **(Width - (Hook Inset * 2))/ 10″** [**(Width - (Hook Inset * 2))/ 250 mm**] (see Figure 9–79).

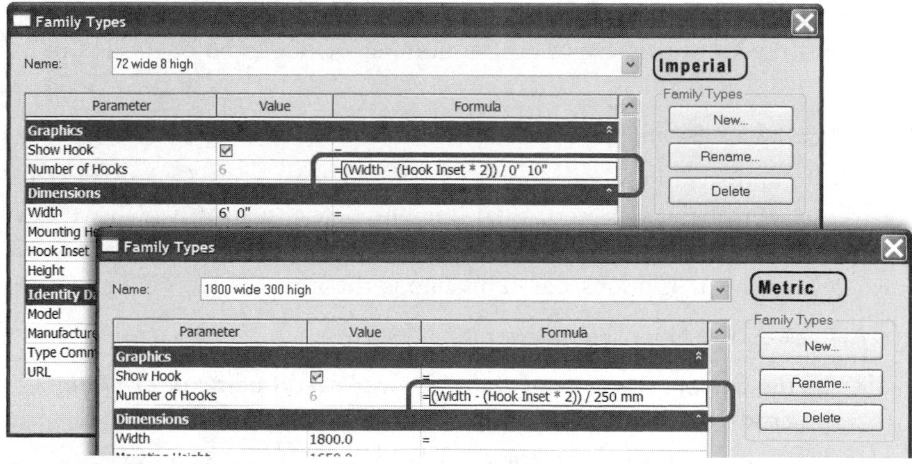

Figure 9–79 *Use a formula to calculate the Number of Hooks based on the Width*

You can test the formula in the "Family Types" dialog by changing the Width parameter and clicking Apply. Test out several values until you are satisfied. Be sure to return the Width to the value named in the Type name before you close the dialog.

32. Repeat the process above (used on the binder bins) to add an <Invisible Lines> Model Line to the Placement Side elevation View and then turn off the "Plan/RCP" visibility of the items above the cut plane.
33. Save the Family file.

LOAD THE RESIDENTIAL PROJECT

The Family file is complete and ready to load into a project. Let's load it into our residential project

1. On the Standard toolbar, click the Open icon.

TIP The keyboard shortcut for Open is CTRL + O. **Open** is also located on the File menu.

- In the "Open" dialog box, click the *My Documents* icon on the left side
- Double-click on the *MRB* folder, and then the *Chapter09* folder.

If you installed the dataset files to a different location than the one listed here, use the "Look in" drop down list to browse to that location instead.

2. Double-click *09 Residential.rvt* if you wish to work in Imperial units. Double-click *09 Residential Metric.rvt* if you wish to work in Metric units

You can also select it and then click the Open button.

The project will open in Revit Building with the last opened View visible on screen. By now we have learned a few ways to load Families into a project. You can either start from the residential project, click the Component tool on the Design Bar and then click the Load button on the Options Bar to browse to and load the *Wall Coat Rack* Family file, or you can return to one of its View windows from the Window menu (or CTRL + TAB) and then click the "Load into Projects" button on the Design Bar.

3. On the Project Browser, double-click to open the *First Floor* plan View.

- Load the *Wall Coat Rack* Family and place an Instance in the room on the left on the interior vertical Wall.
- On the Project Browser, double-click to open the *Entertainment Room Camera* 3D View (see Figure 9–80).

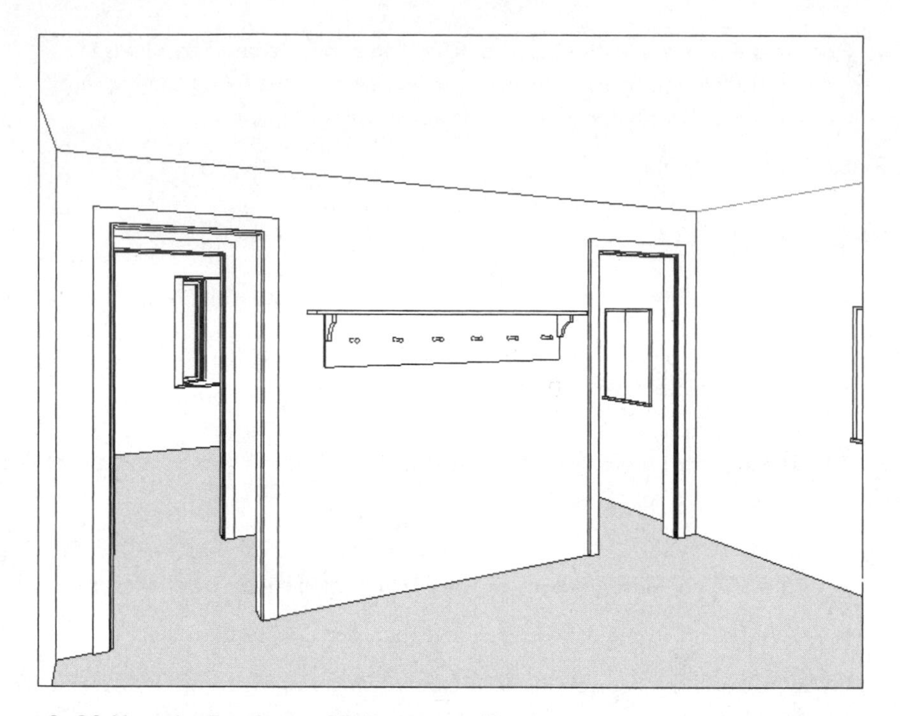

Figure 9–80 *View the Family in a 3D Perspective View*

From the Type Selector on the Options Bar, try the different Types to see them in the project. Note that the "No Hooks" Type now displays without hooks as we indicated. Remember as we noted above, to double-check your settings as you create each parameter and Type. With the hook Visibility parameter, you have to test it out in a project as we have done here to make sure it is working properly.

4. Save the project.

FAMILIES FROM MANUFACTURER'S CONTENT

As the final exercise for this chapter, let's remain in the residential project and create a custom Family for a new whirlpool tub on the second floor. Rather than build this Family from scratch, we will use drawing files (DWG) downloaded from a manufacturer's website. (In this exercise, the hypothetical manufacturer's files have been provided with the other CD files, but the procedure would be the same if you visit and download from actual manufacturer's web sites).

5. From the File menu, choose **New > Family**.
   - Choose the *Plumbing Fixture.rft* [*Metric Plumbing Fixture.rft*] Family template.
   - Tile the windows and make the plan active.
6. From the File menu, choose **Import/ Link > DWG, DXF, DGN, SAT**.

- Browse to the *Chapter09* folder and choose *Whirlpool_Tub_P.dwg* [*Whirlpool_Tub_P-Metric.dwg*].
- In the "Layer/Level Colors" area, choose "Black and White."
- In the "Positioning" area, choose "Origin to origin" and then click Open.

A two-dimensional drawing of the plan symbol will appear. You can import any AutoCAD or Microstation drawing this way. When you do, characteristics of the imported drawing, such as layers will be maintained. This particular file has only one layer, but if you link other drawing files, you can access their layers on the Options Bar and in the "Visibility/Graphics" dialog. We only want this symbol to appear in plan Views. To do this, we will edit the visibility of the imported element.

7. Select the imported drawing.
   - On the Options Bar, click the Visibility button.
   - In the "Family Element Visibility Settings" dialog, clear the checkmarks from the "Front/Back" and "Left/Right" boxes and then click OK (see Figure 9–81).

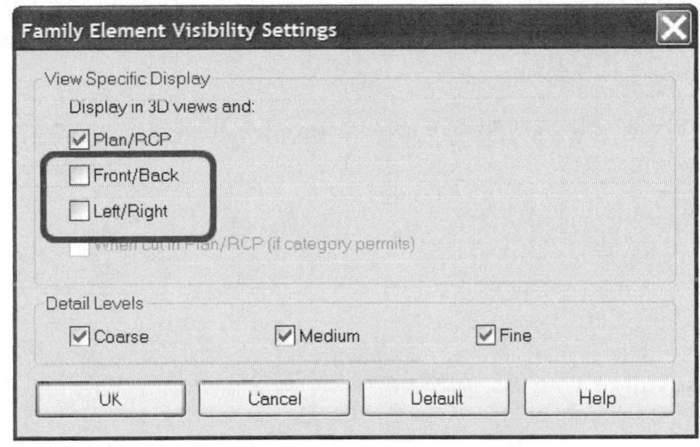

Figure 9–81 *Turn off visibility in elevation Views for the plan symbol*

8. Repeat the process to import another DWG file.
   - In the *Chapter09* folder, choose *Whirlpool_Tub_M.dwg* [*Whirlpool_Tub_M-Metric.dwg*] this time.

   Use the same color and positioning options.

This time the imported drawing contains a 3D model. We want this to show only in 3D and elevation Views.

9. Select the imported 3D drawing.
   - On the Options Bar, click the Visibility button.

- In the "Family Element Visibility Settings" dialog, clear the checkmark from the "Plan/RCP" box and then click OK (see Figure 9–82).

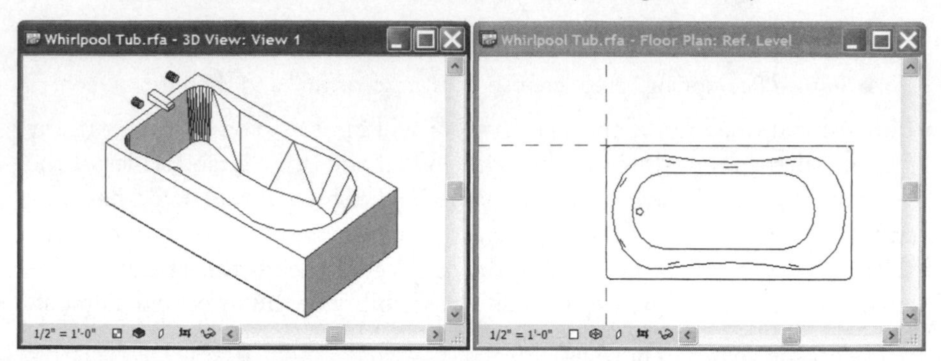

Figure 9–82 *Create a Family from imported geometry*

10. Save the Family with the name **Whirlpool Tub** and Close the Family file.
11. Back in the residential project, on the Project Browser, double-click to open the *Second Floor* plan View.
    - Click the Component tool, Load the Whirlpool Tub Family and place it in the bathroom at the top left of the plan. (Use the SPACEBAR if necessary to rotate it as you place it).
12. Save and Close the project and any open Family files.

SUMMARY

We have covered quite a bit of ground in this chapter. By now you have a good grasp of the power and flexibility of Families in Autodesk Revit Building. Despite the lengthiness of this chapter, we have only scratched the surface of the potential inherent in Families. Continue to explore and customize your own Families. If you have not already done so, read and work through the exercises in the tutorials provided with the software (access them from the Help menu). Each of the exercises provided explores different kinds of Families from the ones explored here. While the process is similar, the more different examples you work through, the more comfort and confidence you will gain with this critical and powerful part of the Revit Building software package.

An extensive Library of Family Content has been included with your Revit Building software.

Familiarize yourself with the provided Library before embarking upon any customization.

In addition to the included Libraries, extensive Libraries and resources are available on the World Wide Web.

Autodesk maintains a Web Library accessible directly from within Open dialogs in the software.

The simplest way to customize Families is to Add or Edit Types.

Before building a custom Family from scratch, determine if you can save a copy and modify an existing one first.

Editing an existing Family is accomplished by opening it in the Family Editor and making modifications and then saving a copy of the Family file.

You can build a completely custom Family from one of the many provided Family template files.

Family templates establish the basic framework and behaviors of the Families you create—choose your template carefully.

You can create a “singular” Family—a Family with only one Type, or a “extensible” Family—which has parameters allowing for multiple Types.

Add dimension parameters to your Families to allow for various “sizes” of the same basic Family.

Advanced parameters such as visibility controls, formulas and parametric Arrays enable you to make very complex and robust Families.

You can use manufacturer’s drawing files directly in Families to quickly create symbols and other items in your projects.

CONSTRUCTION DOCUMENTS

While the benefits of Building Information Modeling describe an architectural design and delivery process that may one day be "paperless" many firms and projects still rely heavily on traditional deliverables even as they look toward the future. Therefore, the creation of Construction Documents typically in the form of printed drawings will remain relevant and necessary for some time to come. In this section, we will explore tools that help us produce these deliverables in Autodesk Revit Building.

Section III is organized as follows:

Detailing and Annotation

INTRODUCTION

In this chapter, we will look at Detailing in Revit Building. We will explore the detailing process and tools in our residential project. As the Design Development phase gives way to Construction Documentation, details are created to clarify basic plan, section and elevation views of a project and assist in conveying overall design and construction intent. Typically such details have been drafted independently of the overall drawings with perhaps some tracing to help minimize redundant effort. In Revit Building, most detailing begins within a fully coordinated model View like the other Views (plans, sections, elevations) in a project.

The process is simple—create a callout or section View of the model at a large scale and then add additional drafted components, text, dimensions and other embellishments necessary to craft the detail and convey design intent. In most cases, such embellishments are at minimum drawn relative to an underlying building model View and in many cases remain automatically constrained or linked to model geometry in the View. Unlike the other Revit Building Views, all Drafting Components appear *only* in the View to which they are added.

OBJECTIVES

In this chapter, we will create detail drawings using several techniques. Working first from the Revit Building model, we add additional information to create a wall-floor-foundation Wall section detail. A variety of tools will be explored to assist in this process. We will also create a detail using a detail originally created in AutoCAD. This process allows you to utilize detail libraries commonly used in most firms today directly within Autodesk Revit Building. Our exploration will include coverage of Detailing Lines, Detail Components, Repeating Details, Filled Regions, and various annotations. After completing this chapter you will know how to:

- Modify Crop Regions and add View Breaks
- Add and modify Detail Lines.
- Add and modify Detail Components and Repeating Details
- Add and modify Text and Leaders

- Add and modify Filled Regions and Break Lines
- Work with Drafting Views
- Import legacy details into a Revit Building Drafting View

MODIFY WALL TYPES

The individual materials that make up the construction of a particular Wall Type are referred to in Revit Building as "Layers." In the previous chapter, we explored Component Families in great detail. Walls are also Families, but are a "System Family" rather than a Component Family. System Families are bound by predefined rules and parameters and can only be manipulated within the context of a project. Elements representing items that are typically "built on site" in a building (like Walls, Stairs, Roofs, etc) are considered System Families. You cannot open these and edit them in the Family Editor nor can you create new ones from scratch. Instead you must modify the existing Families and/or add and edit a System Family's Types.

To prepare us for the detailing tutorial that follows, we will modify the Wall Types currently in use in the residential project for the exterior walls. We will unlock the outer two material Layers of the brick Wall and add a brick ledge to the foundation Wall. Doing so will allow us to create a brick shelf on the concrete Wall and extend the brick down to sit on it. These steps will make the model more accurately represent the construction.

INSTALL THE CD FILES AND OPEN A PROJECT

The lessons that follow require the dataset included on the Mastering Revit Building CD ROM. If you have already installed all of the files from the CD, simply skip down to step 3 below to open the project. If you need to install the CD files, start at step 1.

1. If you have not already done so, install the dataset files located on the Mastering Revit Building CD ROM.

   Refer to "Files Included on the CD ROM" in the Preface for instructions on installing the dataset files included on the CD.

2. Launch Autodesk Revit Building from the icon on your desktop or from the **Autodesk** group in **All Programs** on the Windows Start menu.

   - From the File menu, choose **Close**.

This closes the empty new project that Revit Building creates automatically upon launch.

3. On the Standard toolbar, click the Open icon.

TIP The keyboard shortcut for Open is CTRL + O. **Open** is also located on the File menu.

- In the "Open" dialog box, click the *My Documents* icon on the left side.
- Double-click on the *MRB* folder, and then the *Chapter10* folder.

If you installed the dataset files to a different location than the one listed here, use the "Look in" drop down list to browse to that location instead.

4. Double-click *10 Residential.rvt* if you wish to work in Imperial units. Double-click *10 Residential Metric.rvt* if you wish to work in Metric units

You can also select it and then click the Open button.

The project will open in Revit Building with the last opened View visible on screen.

ADD A BRICK SHELF

Let's start with the foundation Wall. By modifying the Wall Type, we can create a brick shelf to receive the bricks from the exterior Wall above. This will be achieved by adding a Reveal to the Wall Type.

5. On the Project Browser, double-click to open the *Basement* floor plan View.

There are four foundation Walls, three bounding the outside Walls and another framing the right side of the passageway to the existing basement. We do not want to apply a brick shelf to this Wall in the passageway. The easiest way to prevent this is to create a new Type for the Walls with the brick shelf.

6. Using the CTRL key, select the three exterior foundation Walls (two vertical and one horizontal) (see Figure 10–1).

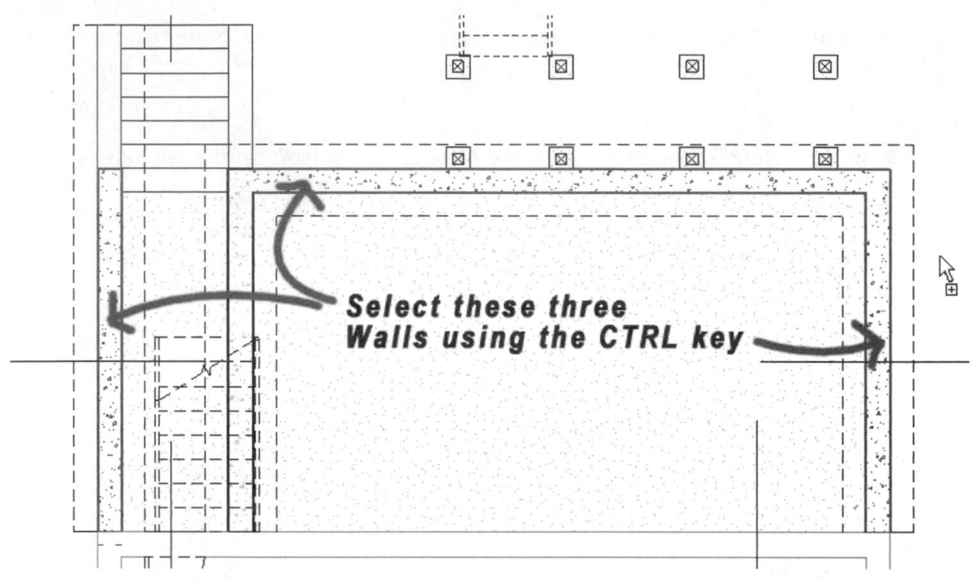

Figure 10–1 *Select the three exterior foundation Walls*

- On the Options Bar, click the Properties icon.
- Next to the Type list, click the Edit/New button.

TIP A shortcut to this is to press ALT + E.

The Type Properties dialog will appear.

- Next to the Type list, click the Duplicate button.

TIP A shortcut to this is to press ALT + D.

A new Name dialog will appear. By default "2" has been appended to the existing name.

- Change the name to: MRB - Foundation - 12″ Concrete (w Brick Shelf) [MRB - Foundation - 300mm Concrete (w Brick Shelf)] and then click OK.
- At the bottom of the dialog, click the Preview button>>.

A viewer window will appear to the left attached to the "Type Properties" dialog.

- From the "View" list (bottom left), choose **Section: Modify type attributes**.

- Zoom in to the top of the wall.
- On the right side of the dialog, at the top, click the Edit button next to Structure (see Figure 10–2).

This will open the "Edit Assembly" dialog and show the Wall Layers included in this Type. We can edit these Layers here as well as add other parameters such as Sweeps and Reveals (below we will use the same process to "unlock" some of the Wall Layers).

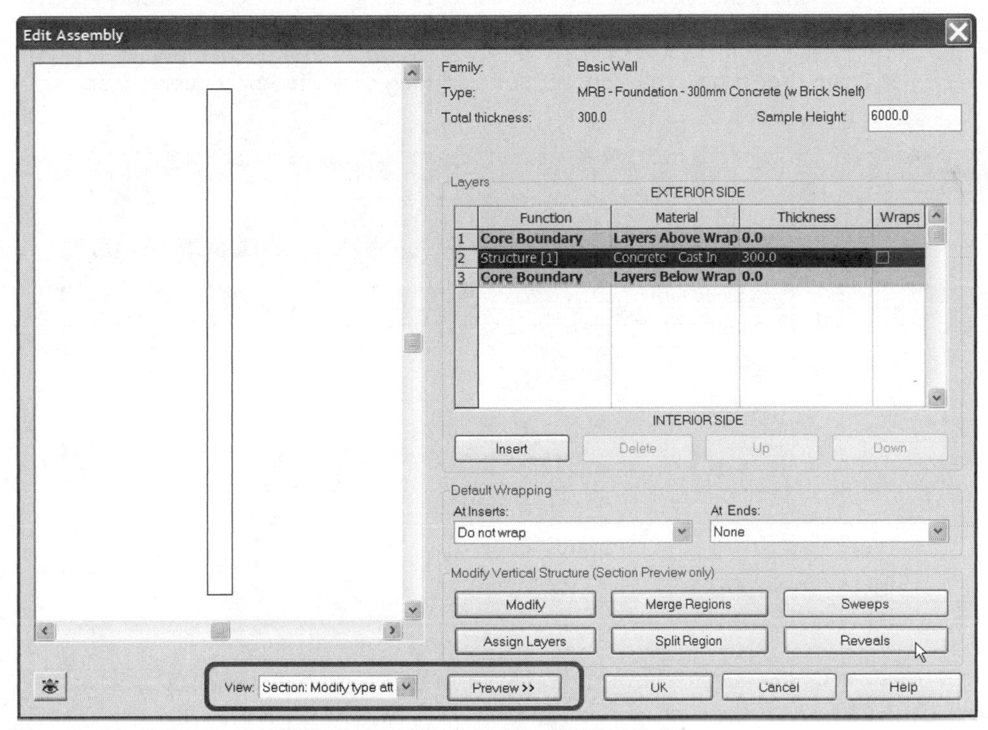

Figure 10–2 *Access the "Edit Assembly" dialog to edit the Wall Structure*

7. In the bottom right corner of the dialog, within the "Modify Vertical Structure" area, click the Reveals button.

NOTE The "Reveals" and other buttons in the "Modify Vertical Structure" area will not be available if you have not enabled the Section Preview as noted in the previous steps.

A Reveal is basically a profile-based sweep that cuts away from the mass of the Wall. In this case, we will use a Profile that has been provided with the files from the Mastering Autodesk Revit Building CD ROM.

- From the Reveals Dialog click the "Load Profile" button.

- Browse to the *Chapter10* folder where you installed the Mastering Autodesk Revit Building CD ROM files.
- Select the file named: *Brick Shelf Reveal.rfa [Brick Shelf Reveal-Metric.rfa]* and then click the Open button.

8. In the "Reveals" dialog, click the Add button to add a Reveal.
   - Click in the Profile cell and click again on the down arrow to display the Profile list, choose Brick Shelf Reveal : 12″ d x 6″ w [Brick Shelf Reveal-Metric : 300 d x 140 w].
   - From the "From" cell, choose **Top** and then click the Apply button (see Figure 10–3).

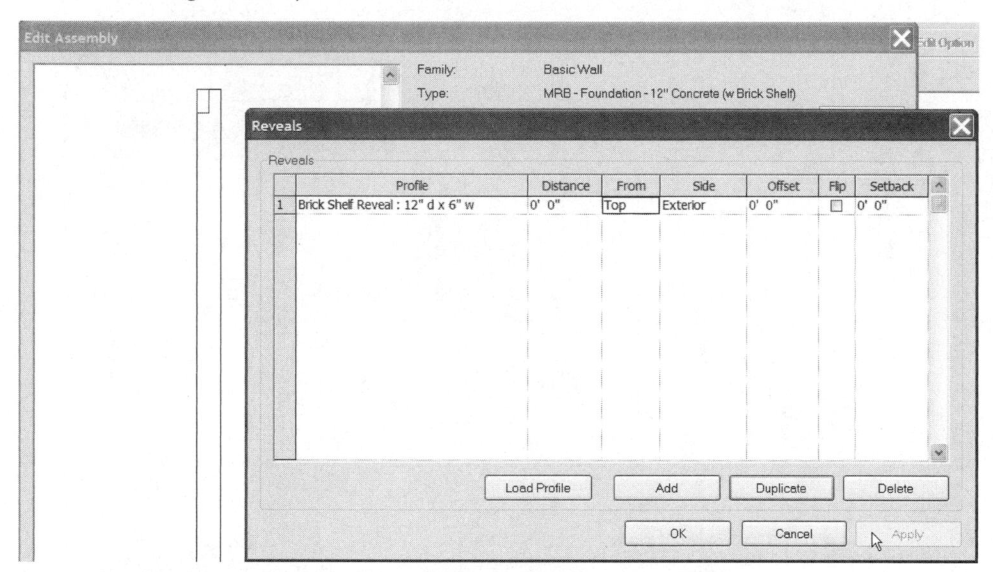

Figure 10–3 *Add a new Profile, set it to Top and then click Apply to see the result*

In the "Edit Assembly" dialog in the background, you should see the Reveal Profile appear at the top left edge of the Wall in the Viewer.

- Click OK to return to the "Edit Assembly" dialog.
- Click OK three more times to return to the model View window.

9. On the Project Browser, double-click to open the *Longitudinal* section View.
   - Zoom in on the left side to study the results (see Figure 10–4).

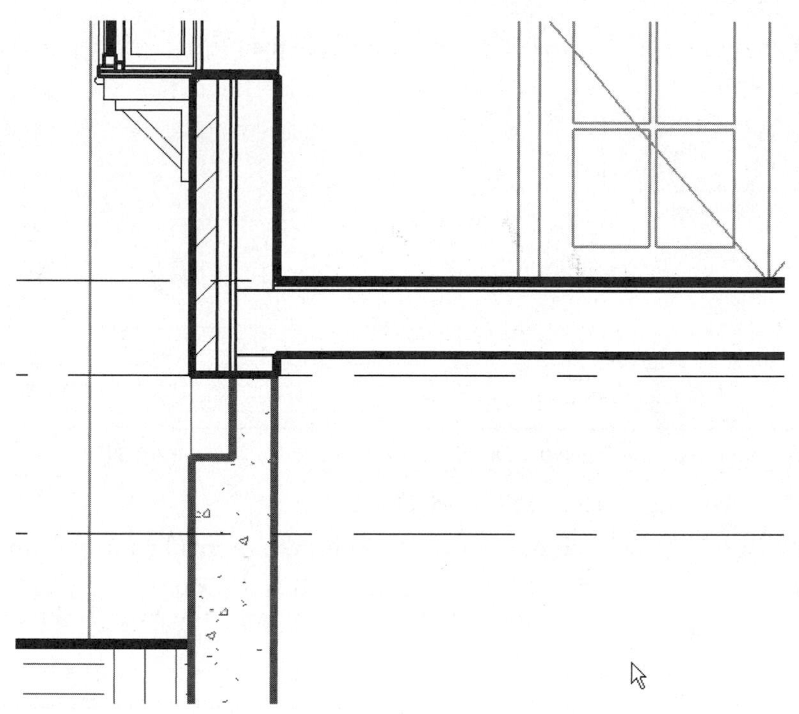

Figure 10–4 *Open the Longitudinal Section View to see the results*

10. On the Design Bar, click the Modify tool or press the ESC key twice.

UNLOCK WALL LAYERS

Now that we have a brick shelf in the foundation Wall Type, let's edit the brick Wall Type to unlock the outer material Layers (to allow their top and or bottom offsets to move freely from the Wall's top and base offsets) and then project them down to cover and sit on this foundation brick shelf.

11. Select the Left exterior brick Wall.
    - On the Options Bar, click the Properties icon.
    - Next to the Type list, click the Edit/New button.
    - Open the Preview window again (if it is not already open), and then click the Edit button next to Structure.

The preview window should be showing the section view. If it is not, choose **Section: Modify type attributes** from the "View" list. You can use standard navigation techniques such as the wheel of your mouse or the right-click menu to zoom and scroll the model in the viewer.

12. Zoom in Region on the lower portion of the Wall in the viewer window.

- In the "Modify Vertical Structure" area, click the Modify button (see Figure 10–5).

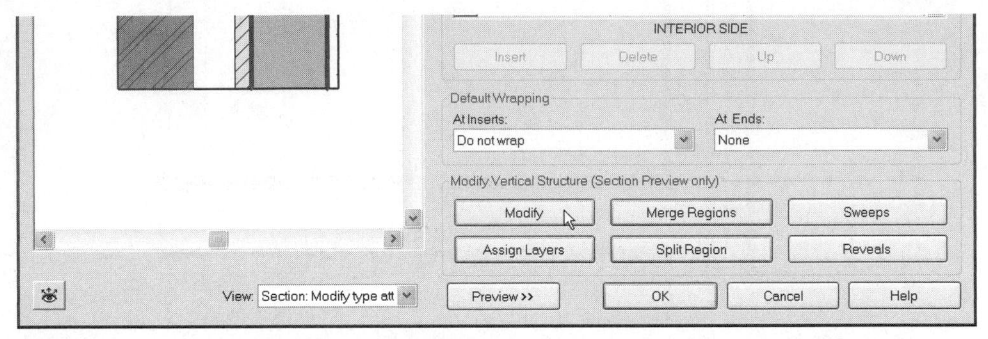

Figure 10–5 *Zoom in on the lower portion of the Wall and then click the Modify button*

13. Click the bottom edge of the Brick Layer.

The edge will highlight red to indicate that it is selected. A small padlock icon will appear on the edge. We can use this padlock icon to unlock the bottom edge of the Layer which will allow it to be moved independently from the Wall itself in the model.

- Click the padlock (to open it) and unlock the bottom edge of the Layer.
- Repeat the process to unlock the bottom edge of the Thermal/Air Layer (next to brick) (see Figure 10–6).

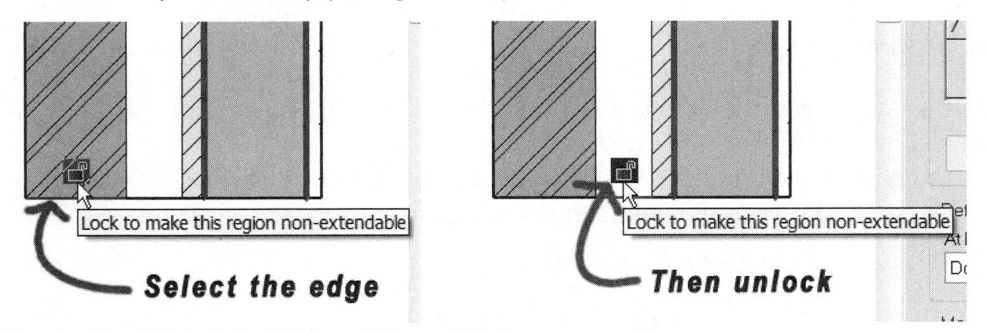

Figure 10–6 *Unlock the Brick and Thermal/Air Layers*

- Click OK three times to return to the model.

Upon returning to the model View window, you will note that the Wall now has two Shape Handles at the bottom edge (see Figure 10–7). You can use the second handle to modify the bottom edge of just the unlocked Layers. The other handle will continue to modify the Base Constraint of the entire Wall.

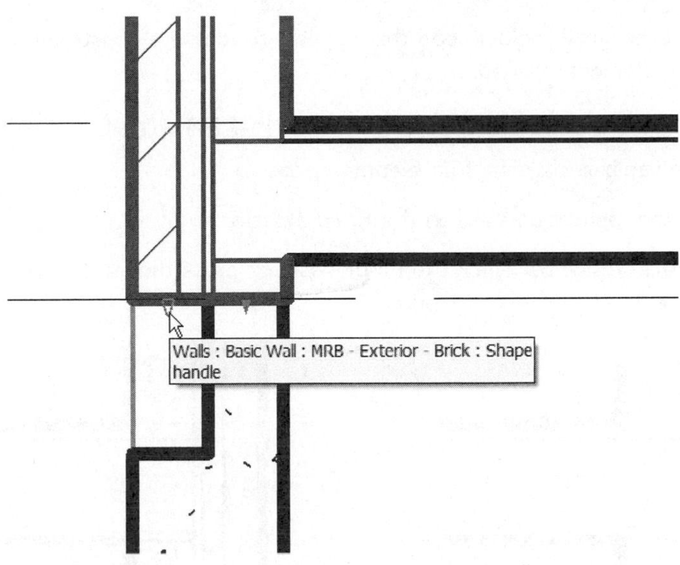

Figure 10–7 *Unlocking the Layers makes a second Shape Handle appear*

While the Shape Handle provides an easy way to edit the brick and air Layers, the Align tool gives a bit more control by allowing us to apply a constraint to the alignment after we make it.

14. On the toolbar, click the Align tool.
    - For the Reference line, click the bottom edge of the brick shelf on the foundation Wall (use the TAB key as necessary to make the proper selection).
    - For the Entity to Align, click the bottom edge of the brick Layer in the Wall above (see Figure 10–8).

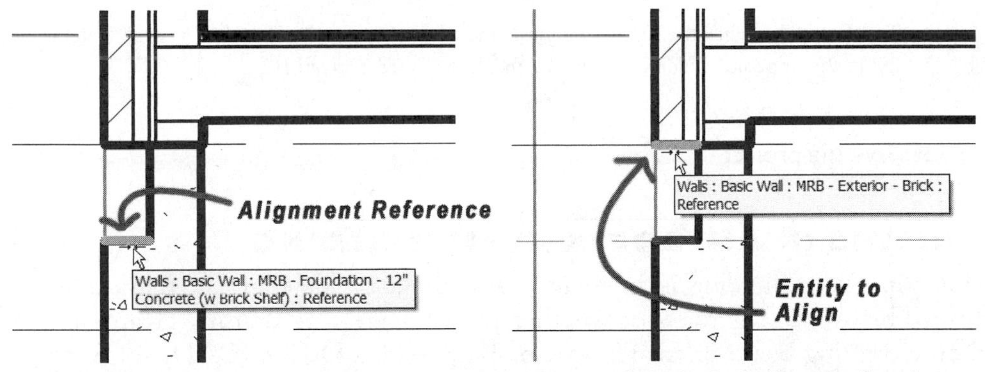

Figure 10–8 *Align the brick and air Layers to the bottom of the brick shelf on the foundation Wall*

- Click the small padlock icon that appears to lock the constraint and keep these elements aligned.
- On the Design Bar, click the Modify tool or press the ESC key twice.

15. On the toolbar, click the Join Geometry tool.
    - Join the foundation Wall to the Brick exterior Wall.
    - On the Design Bar, click the Modify tool or press the ESC key twice (see Figure 10–9).

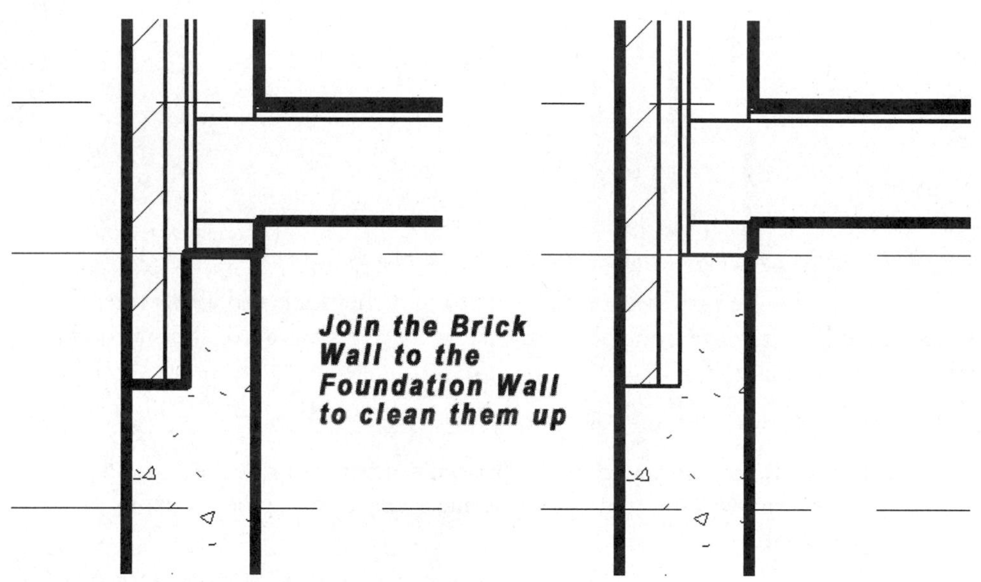
Join the Brick
Wall to the
Foundation Wall
to clean them up

Figure 10–9 *Join the two Walls to make them cleanup nicely*

TIP Be sure to click the Brick Wall first, and then join it to the Foundation Wall. Clicking in the opposite order will remove the customized bottom alignment.

16. Save the project.

DETAILING IN AUTODESK REVIT BUILDING

Detailing in Revit Building is in many ways similar to detailing in traditional drafting. This is true regardless of whether you compare it to drafting created by hand on a drafting board or created in Computer Aided Design (CAD) software on a computer. The major difference is that in Revit Building you rarely start from scratch and you can base your detail on Views that are automatically generated from your Revit Building model. The process involves leveraging data automatically captured from the three-dimensional model and additional two-dimensional View-specific embellishments that are added on top to more clearly communicate your design intent.

In other words, the model that we have been constructing so far includes enough data to generate a majority of the required architectural drawings that will be included in a document set with an appropriate level of detail and accuracy. This is true for most plans, sections and elevations. In the case of details however, while it is theoretically possible to embellish the level of detail in our model further to include all of the fasteners, hooks, and other items actually required for its successful construction, the amount of effort (in man-hours) and the sheer size of the resultant model (in computer memory and hard drive requirements) would typically not yield a sufficient return on investment. Instead, a process very similar to the traditional approach (noted above) is often the most appropriate course of action in Autodesk Revit Building.

Architectural communication has developed over many generations to include a variety of abstractions and drawing conventions. As Building Information Modeling evolves, new uses of the design and construction documentation artifacts are being formulated and considered which go beyond the building's design and construction phases. Despite this, in most firms, the primary goal remains the facilitation of the design and construction of a building. To efficiently realize this primary goal, it is imperative that the new paradigms enhance productivity, add value and most importantly do not lose sight of or compromise this primary need of architectural construction documentation. When all these factors are considered, the process of creating construction details—highly specialized, large scale diagrams explaining the intricate and specific connections, materials and construction specification requirements of some aspect of the building project—is best achieved using a hybrid between that which can be captured from the virtual three-dimensional building model and that which can be added as abstracted two-dimensional embellishment.

It is this process that will be discussed in detail in the following tutorial. In this exercise we will discuss the available tools and techniques using the wall and floor intersection from our residential project that we edited above.

ADDING A CALLOUT VIEW

Continue from the previous exercise in the *Longitudinal* section View. If you closed the project or this View, reopen them now. Details are typically presented at larger scales than the drawings from which they are referenced. The Callout tools in Revit Building will allow us to create a detailed View at a larger scale of any portion of the building model.

In the *Longitudinal* section View, be sure you can see the left exterior Wall from the first floor down to the footing. Zoom and Scroll as necessary.

1. On the Design Bar, click the View tab and then click the Callout tool.

   - Click a point outside the exterior Wall on the left above the first floor and then drag a callout around the Wall to beneath the foundation (see Figure 10–10).

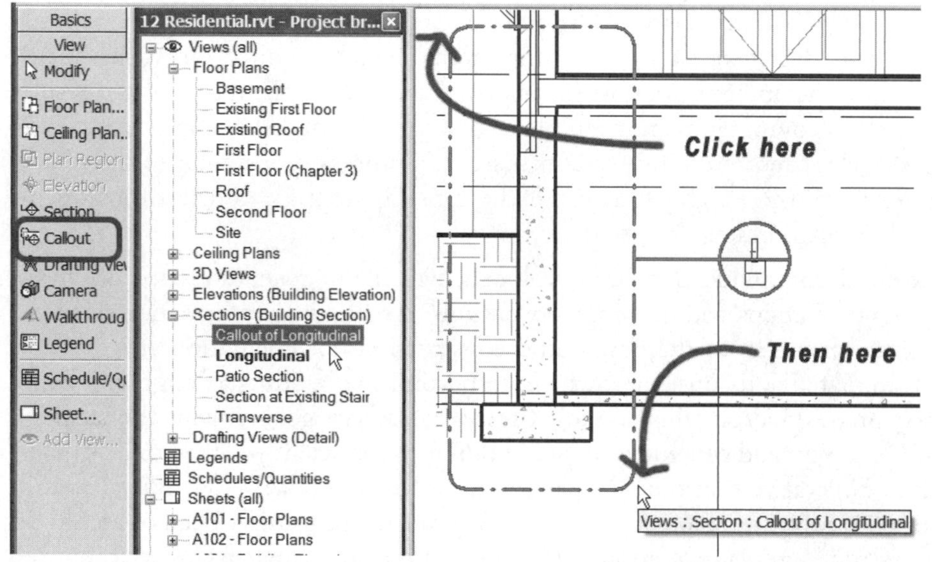

Figure 10–10 *Using the Callout tool, click two opposite corners to create the Callout View*

Revit Building will create a new View on the Project Browser called: *Callout of Longitudinal.*

Like section and elevation markers, a callout marker will appear. When you deselect all elements, this callout will remain blue. As with the others, this indicates that you can double-click it to jump to the referenced View. You can also open the View from the Project Browser.

2. On the Project Browser, beneath Sections right-click on *Callout of Longitudinal* and choose **Rename**.

   - Type: **Typical Wall Section** and then click OK.

3. Open up the callout View (you can double-click its name on Project Browser or its Callout symbol) (see Figure 10–11).

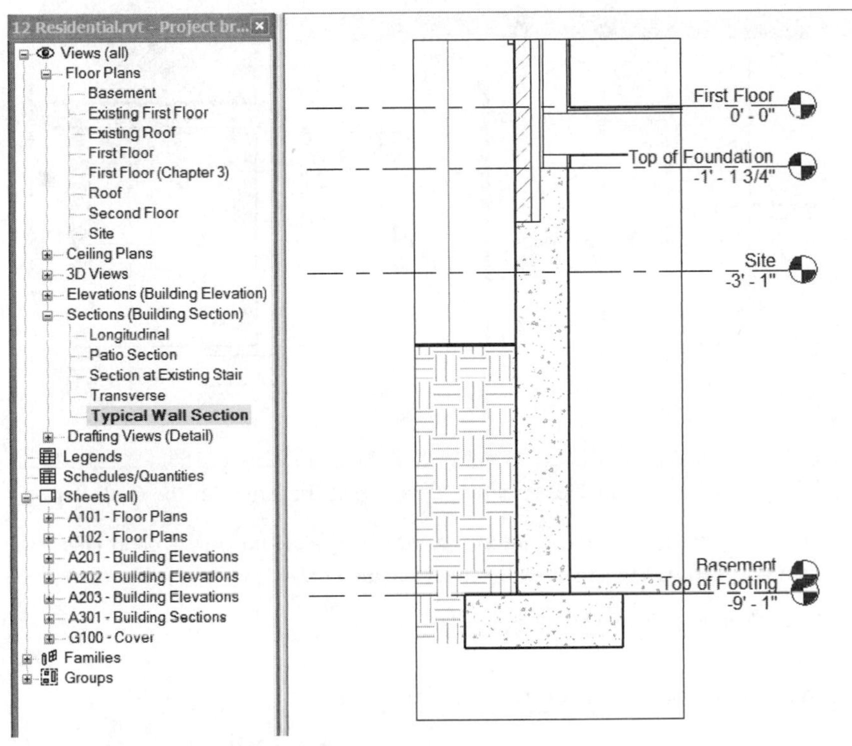

Figure 10–11 *Open the Callout View*

Notice that the boundaries of the crop region in the *Typical Wall Section* View match the extents of the callout boundary that we sketched in the *Longitudinal* section View. If this boundary is adjusted in either View, the boundaries in the other View automatically adjust. If you wish to see this, try tiling the *Longitudinal* section View and the *Typical Wall Section* View side by side and test it out. Remember to close hidden Views or minimize other Views first, so that when you tile, only the two sections will appear (see Figure 10–12).

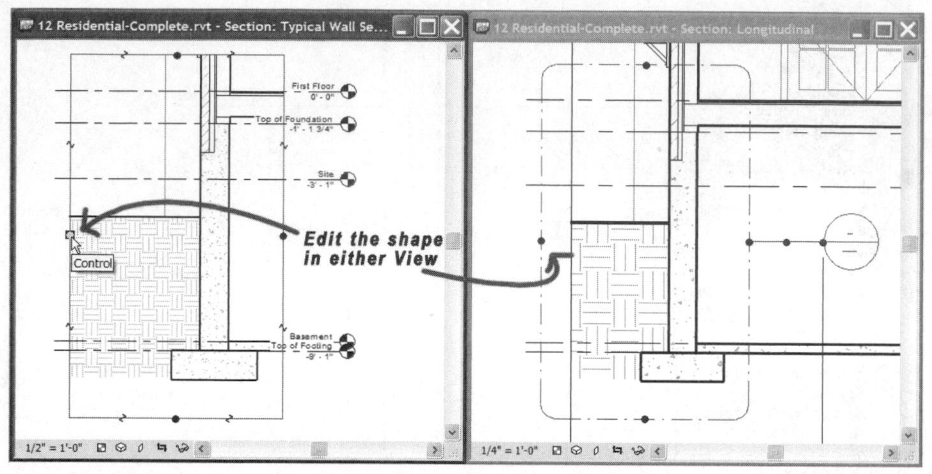

Figure 10–12 *Drag the Control Handles in either View to edit the extent of the Crop Region*

4. When you are finished experimenting, return the shape of the crop region to match approximately as shown above in Figure 10–11.

   ◗ Maximize the *Typical Wall Section* View.

ADJUST SCALE AND ANNOTATION VISIBILITY

In Revit Building, the annotation in a View is typically separate from the model geometry shown in a View. Thus, the Level lines and Callout indicators are not part of the model but rather automatically-created annotation that Revit Building maintains and displays for us in each appropriate View. Revit Building normally does a very good job at showing and positioning these elements in desirable locations. However, we can always make adjustments to the annotation in the active View, and unlike edits to the model, these changes will apply *only* to the active View. This very powerful feature affords us great flexibility in the composition of each View and what specific non-graphical data it conveys.

5. On the View Control Bar (bottom of the window) change the scale to **3/4″ = 1′-0″ [1:20]** (see Figure 10–13).

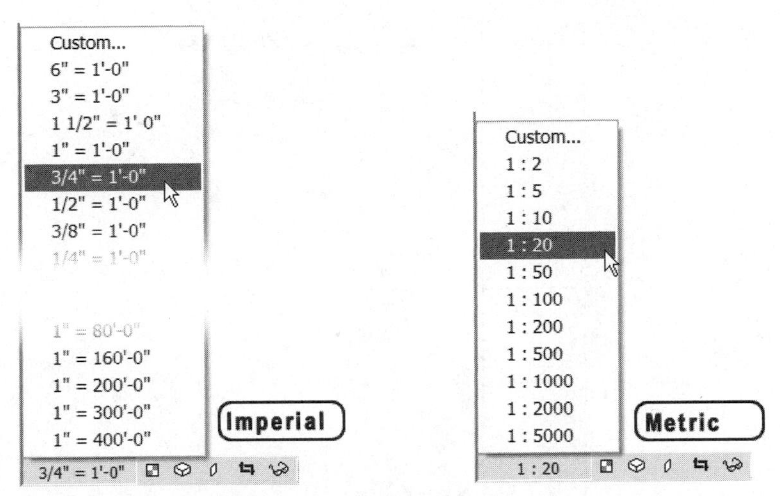

Figure 10–13 *Choose a larger scale for the detail View*

6. Select the Site Level Line.

   - Right-click and choose **Hide Annotation in View** from the menu that appears.

As noted above, this operation does not have any effect on any other View. We have hidden this Level line in only the current View. The Site Level is not really relevant in the current Callout View, so by removing the Level line, we eliminate potential clutter and confusion. In a similar fashion, we can adjust the location of the Level Heads and the length of the Level lines in this View and again, the edits will be confined to *only* this View.

7. Click on any Level line.

   - Using the Control Handle at the end, adjust the end points so the Level text is completely outside the Crop Region (see Figure 10–14).

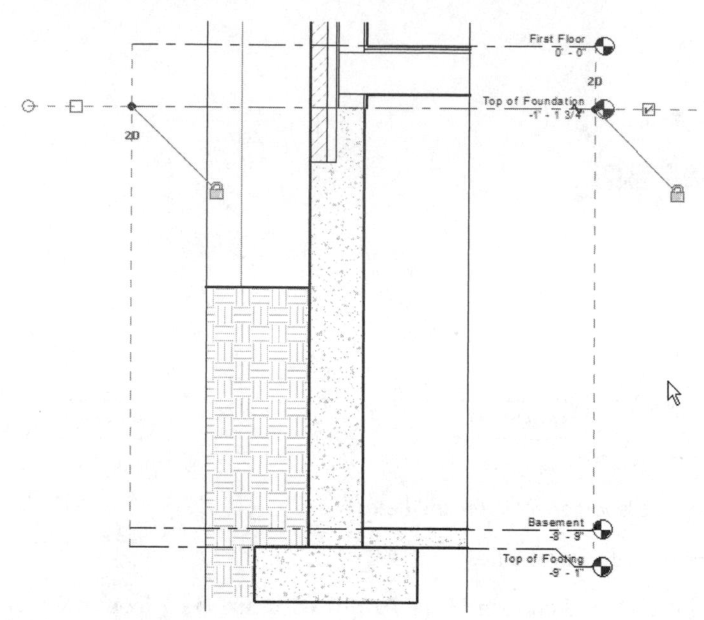

Figure 10–14 *Adjust the end points of the Level lines to move the text outside the Crop Region*

Notice that all Level lines move together when you drag one.

8. Save the project.

DETAIL LINES

Now that we have cut our detail callout and configured the Level lines and scale to our liking, we are ready to begin adding embellishments. We can draw a variety of View-specific elements directly on top of the section view of our model. We will start with Detail Lines. These are simple drafted elements much like the sketch lines with which you are already familiar. When you add a Detail Line, it appears only in the View to which you add it. If you wish to draft a line that appears in multiple Views, use a Model Line instead. We will use Detail Lines to sketch in some flashing at the bottom of the wall cavity.

9. On the Design Bar, click the Drafting tab and then click the Detail Lines tool.

Notice the choices on the Options Bar are very familiar and match those that we have seen in many sketch-based objects so far.

- From the Type Selector, choose Wide Lines.
- Be sure that the "Draw" icon is chosen.
- Place a checkmark in the "Chain" checkbox and choose the "Line" icon.

10. Zoom into the bottom of the wall cavity (at the brick shelf).
11. Sketch the line segments shown in Figure 10–15.

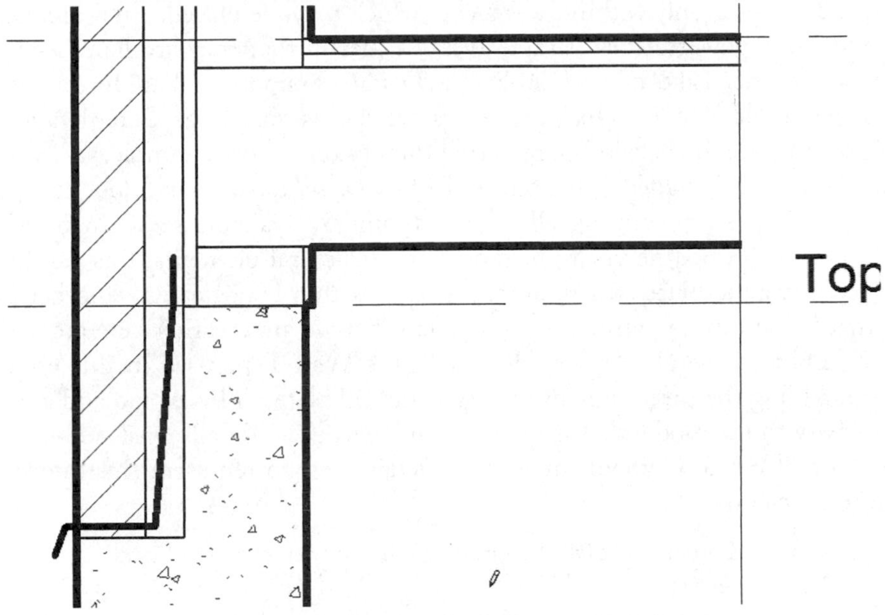

Figure 10–15 *Sketch Detail Lines to represent the flashing in the wall cavity*

Unlike the sketch lines that we drew in previous chapters, these lines are complete "as is." They do not describe the shape of a more complex element like a Floor or a Stair. These are simply drafted lines placed on top of a model View, much like drafting directly on top of a Mylar background in traditional hand drafting.

12. On the Project Browser, double-click to open the *Longitudinal* section View.
    - Zoom in to the same portion of the Wall and notice that this linework does not appear in this (or any other) View.
13. On the Project Browser, double-click to return to the *Typical Wall Section* View.

TIP You can hold down the CTRL key and then press TAB to cycle through the open Views.

All of the parts of the detail that we are going to create next could be created with Detail Lines following the same process. However, to speed up the process, several other detailing tools have been provided. Let's look at them now.

DETAIL COMPONENTS

Detail Components are simply two-dimensional View-specific elements that (like Detail Lines) appear only within the View in which they are placed. They are more useful and more powerful than simple Detail Lines in that they are Families and can be parametric. Like other Families, a Detail Component Family can have many Types built into it. The parameters can be as simple as Length and/or Depth, or include dozens of parametric dimensions. For Example, a "Wide Flange" Family file included in the out of the box *Detail Component* folder contains hundreds of Types representing all of the commonly-available sizes. In projects where it is not deemed necessary to model the structural elements, this can be a very efficient way to place structural steel shapes within your details. Another example that is a bit more pertinent to the detail that we are creating here are is dimension lumber. It would be possible to edit the Wall Type used in this project and begin adding the three-dimensional sole and sill plates. This would add a level of complexity to the model that is typically only needed in details. Instead, we can add predefined Detail Components to our Detail View to represent this information more efficiently.

14. On the Design Bar, click the Drafting tab and then click the Detail Component tool.

Note the options that appear on the Type Selector and the Options Bar. Currently, there are no "Dimension Lumber" Families loaded in the current project. Like other Components in Revit Building, we can simply load one from the library.

- On the Options Bar, click the Load button.
- In the "Open" dialog, browse to your default library folder (either the *Imperial Library* or the *Metric Library*) and then browse to the *Detail Components/Structural/Wood* folder.

NOTE If you do not have access to either of these libraries, the Family files mentioned in this tutorial have also been provided in the *Autodesk Web Library* folder with the files installed from the Mastering Autodesk Revit Building CD ROM.

- Select the file named: *Dimension Lumber-Section.rfa* [*M_Dimension Lumber-Section.rfa*].

NOTE Do not double-click this file, or simply open it. Unlike other Families, you must choose the specific Type from the matrix in the lower portion of the dialog to load the Type you need.

- From the matrix at the bottom of the dialog, select 2x6 [23x140] to highlight its row and then click Open (see Figure 10–16).

This allows you to load only this Type from this Family file into your project and not the other hundreds of Types associated with this Family file.

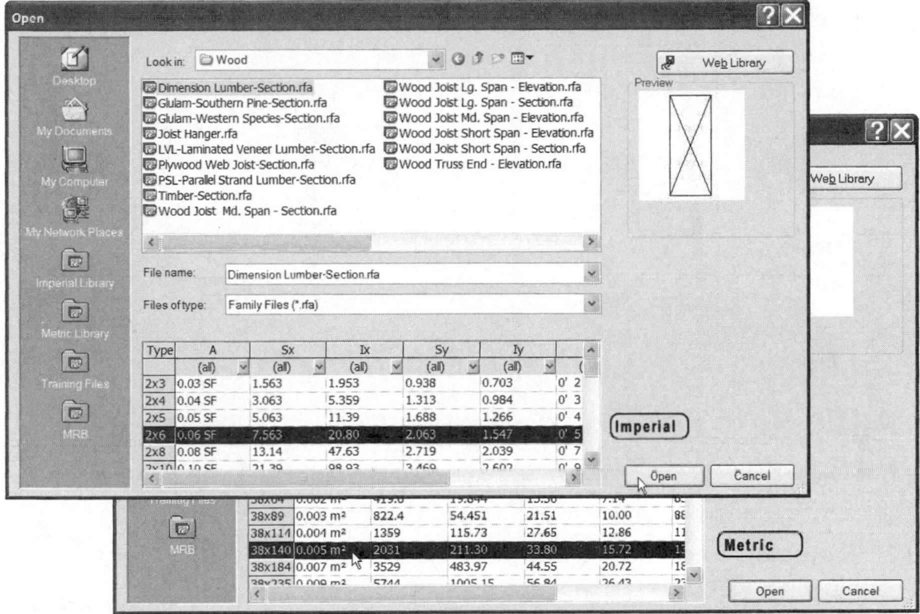

Figure 10–16 *Choose the Detail Component Family and Type that you wish to load*

15. From the Type Selector, choose Dimension Lumber-Section : 2 x 6 [M_ Dimension Lumber-Section : 38x140].

 ▶ Press the SPACEBAR three times.

This will rotate it so the placement point is at the top right of the 2x6 [38x140].

16. Place two plates in the space between the floor joist and the foundation Wall (see Figure 10–17).

TIP Use the Move or Align tools to assist in accurate placement.

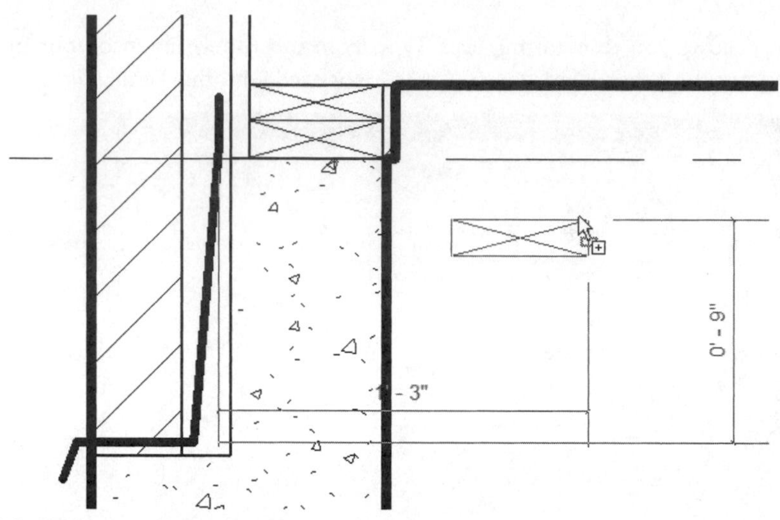

Figure 10–17 *Place a double top plate on the foundation Wall*

17. Repeat the process (or copy) to add a sill plate above the joist at the first floor.
18. Choose the Detail Component tool again.
    - Change the Type to Dimension Lumber-Section : 2 x 10 [M_Dimension Lumber-Section : 38x235] (Load it from the library if necessary).
    - Use the spacebar to rotate if necessary and place an end joist as shown in Figure 10–18.

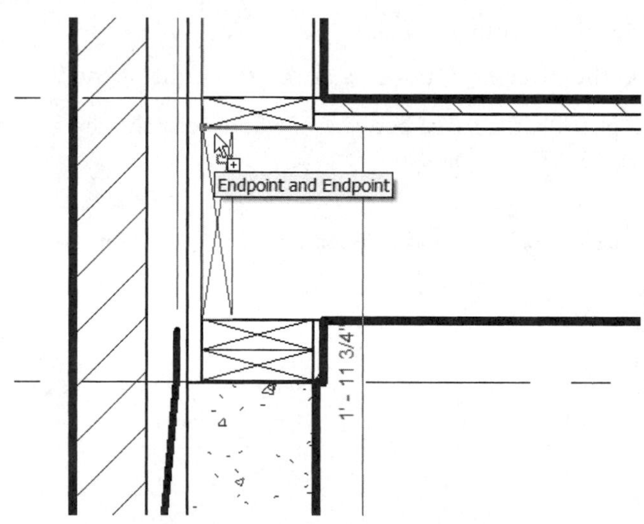

Figure 10–18 *Place an end joist using a 2x10 [38x235]*

Once you have loaded and placed one Detail Component, use the same process for any Detail Component. Revit Building ships with a very large collection of pre-made Detail Component Families. As we discussed in the previous chapter on Families, set aside some time to get acquainted with what is provided. You can use the components in the library, modify them or build your own. It is usually best to start with what has been provided before endeavoring to create your own. Let's continue to add to our detail by repeating the above process to load and add an anchor bolt.

19. On the Design Bar, choose the Detail Component tool again.
    - Click the Load button and browse to the *Detail Components/Structural* folder as before.
    - Open the *Hooked Anchor Bolt.rfa* [*M_Hooked Anchor Bolt.rfa*] file.
    - Place the anchor at the midpoint of the lower plate.
    - On the Design Bar, click the Modify tool or press the ESC key twice.
20. Select the bolt that you just placed, on the Options Bar, click the Properties icon and then click the Edit/New button.
    - In the "Type Parameters" dialog, change the "Embedment" parameter to **1′-6″** [**450**] and then click OK.
    - In the "Instance Parameters" dialog, change the "Bearing Proj" to **3″** [**76**] and then click OK (see Figure 10–19).

Remember, one of the things that make Detail Components powerful is that like other Component Families they can have Type and Instance parameters. We explored several examples of these in the previous chapter. Here in this example, the "Embedment" parameter is a Type parameter that determines how deeply the anchor is embedded in the concrete. The "Bearing Proj" parameter, on the other hand, is an instance parameter that can vary with each bolt we place. This can be easily adjusted to match the depth of one or two plates.

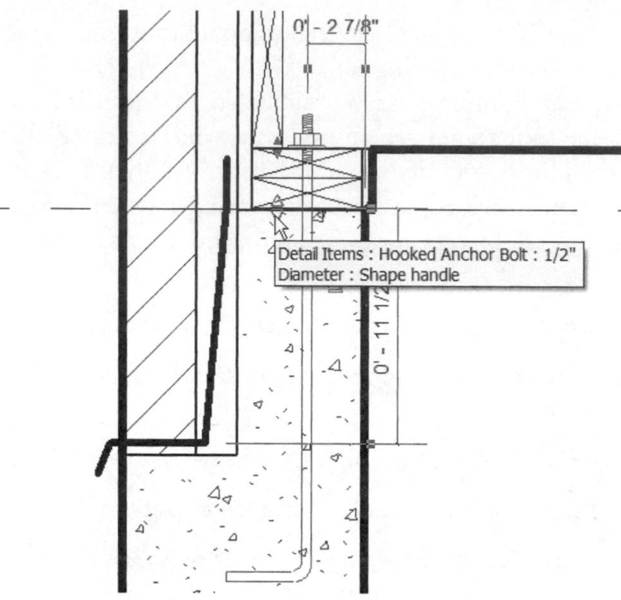

Figure 10–19 *Place an anchor bolt and adjust its parameters*

REPEATING DETAIL ELEMENTS

Repeating Detail Elements are Detail Components that get automatically repeated about an invisible sketch line. This allows more rapid placement of Detail Components like studs, CMU, Brick, etc. In the detail that we are constructing, Revit Building automatically generated the brick Layer of our Wall with the heavy and thin cut lines and the diagonal fill pattern. This rendition is fine for general scales and overall plans and sections. Adding mortar joints to our brick will better delineate the brick veneer and suggest the individual bricks. While we could place one mortar joint and then array or copy it, a Repeating Detail Component is more expedient.

21. On the Design Bar, choose the Detail Component tool again.
    - Click the Load button and browse to the *Detail Components/Architectural* folder.
    - Open the *Mortar Joint.rfa* [*M_Mortar Joint.rfa*].

NOTE An alternate version of the Mortar Joint Family is located in the *Chapter10* folder named: *Mortar Joint with concave joint.rfa*. You can use this one instead if you prefer.

You will see a single mortar joint attached to your pointer ready to place. Rather than place this one component, we will cancel and create a new Repeating Detail Type using this Detail Component.

- On the Design Bar, click the Modify tool or press the ESC key twice.

22. On the Design Bar, choose the Repeating Detail tool.

Only one Type is available on the Options Bar—Repeating Detail: Brick. We are going to use this as the basis for a new Type.

- On the Options Bar, click the Properties icon.
- In the "Element Properties" dialog, press ALT + E and then ALT + D to create a new Type.
- Name the new Type: **MRB Mortar** and then click OK.
- In the "Type Properties" dialog, choose Mortar Joint: Brick Joint from the "Detail" list.
- Leave all of the other settings unchanged and then click OK twice to return to the View window.

23. Click at the bottom left corner of the brick veneer and drag up past the top of the Crop Boundary and click again (see Figure 10–20).

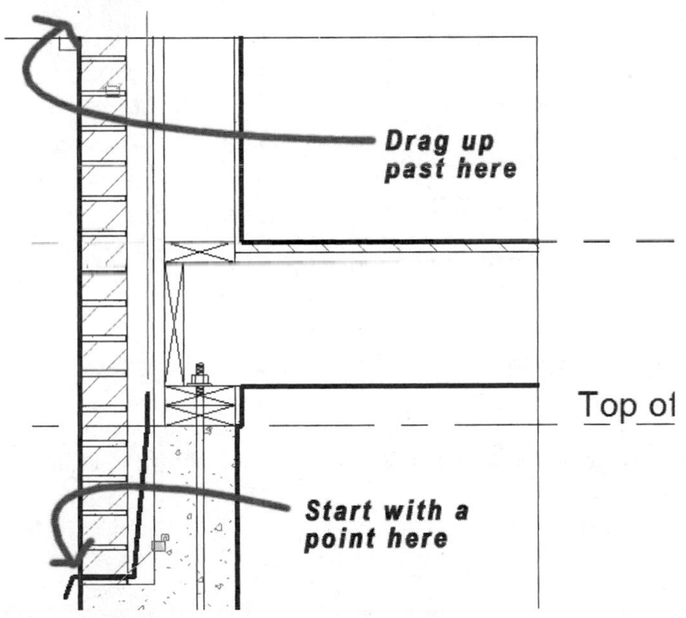

Figure 10–20 *Place a Repeating Detail for the Mortar Joints*

- On the Design Bar, click the Modify tool or press the ESC key twice.

- Save the project.

FILLED REGIONS

Filled Regions are two-dimensional shapes that contain boundary lines and fill patterns. You can draw them any shape you like and use them to shape, hatch or cover up parts of the detail drawing. We will use a Filled Region here to illustrate the filled trench on the exterior side of the foundation wall.

24. On the Design Bar, choose the Filled Region tool.

    The Design Bar changes to sketch mode and shows only the Filled Region tools.

    - From the Type Selector, choose Wide Lines.
    - Be sure that the "Draw" icon is chosen.
    - Place a checkmark in the "Chain" checkbox and choose the "Line" icon.

25. Zoom into the bottom of the foundation Wall near the footing.
26. Sketch the shape shown in Figure 10–21.

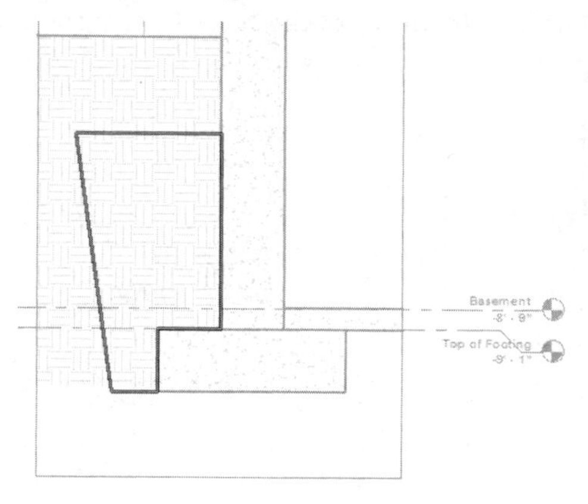

Figure 10–21 *Sketch a Filled Region*

27. On the Options Bar, change the shape to Circle.

    - Add a sketched circle with a **2″ [50]** radius as shown in Figure 10–22.

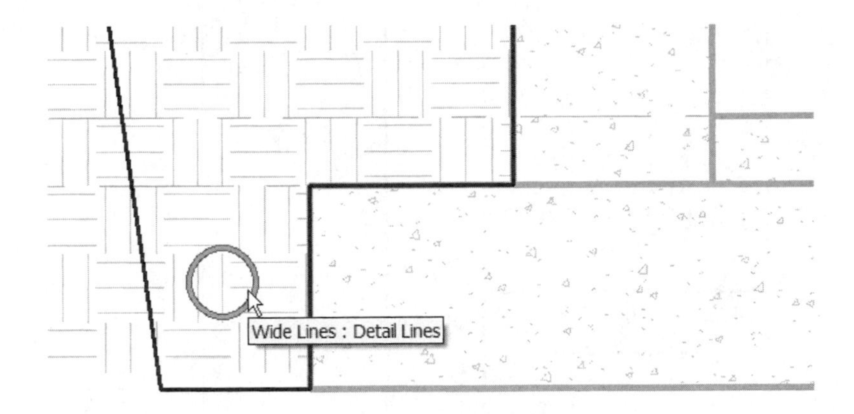

Figure 10–22 *Sketch a circle in the Filled Region shape*

28. On the Design Bar, click the Region Properties button.
    - From the Type list, choose River Rock and then click OK.
    - On the Design Bar, click the Finish Sketch button (See Figure 10–23).

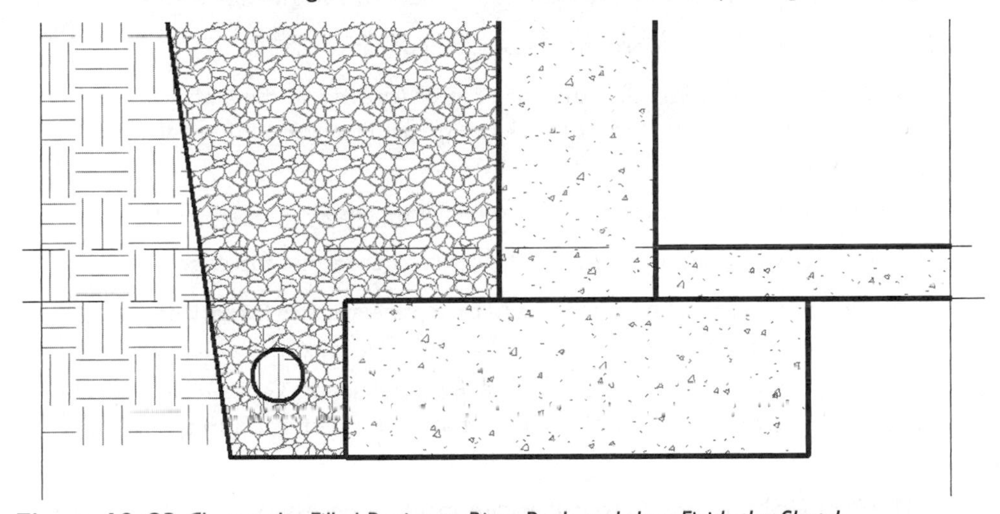

Figure 10–23 *Change the Filled Region to River Rock and then Finish the Sketch*

29. Repeat the same process for adding the finished grade with the Earth Disturbed Filled Region (See Figure 10–24).

Figure 10–24 *Add another Filled Region*

ADDING BREAK LINES

Next let's drop in some Break Line components to hide part of the model. Break Lines have Instance parameters so we can adjust their size to fit the Detail.

30. On the Design Bar, choose the Detail Component tool.

   - On the Options Bar, click the Load button, browse to the Detail Components folder, choose *Break Line.rfa* [*M_Break Line.rfa*] and then click Open.

31. On the Type Selector, verify that Break Line is chosen.

   - Place a Break Line at the top of the detail to cover the top edge.
   - Press the spacebar three times, and then place another Break Line covering part of the floor joist to the right (See Figure 10–25).

Breakline components contain invisible Filled Regions set to Opaque to hide or "mask" the model objects below. The concept of a mask is common in graphic design software and can be helpful in creating details.

   - Use the Shape Handles to make adjustments as necessary.

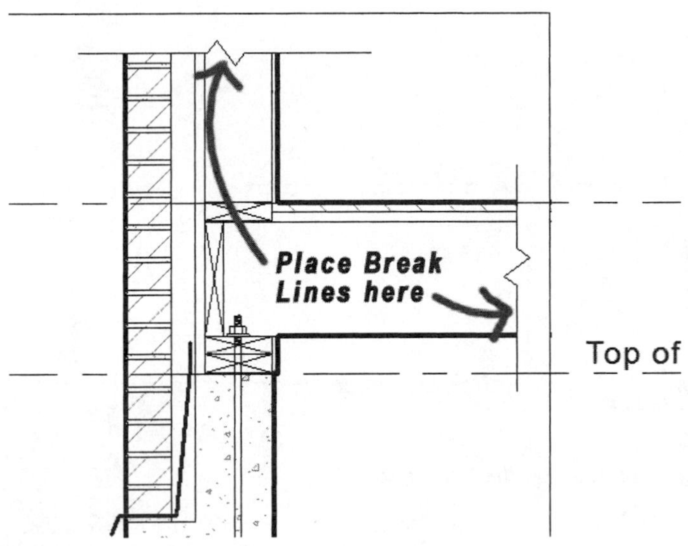

Figure 10–25 *Add Break lines with integral Filled Regions*

BATT INSULATION

Next we'll place some batt insulation in the wall and floor.

32. On the Design Bar, choose the Insulation tool.
    - On the Options Bar, set the Width to **5″ [130]**.
    - Click the first point at the bottom midpoint of the stud space, move up vertically and then pick the second point above the Crop Region (See Figure 10–26).

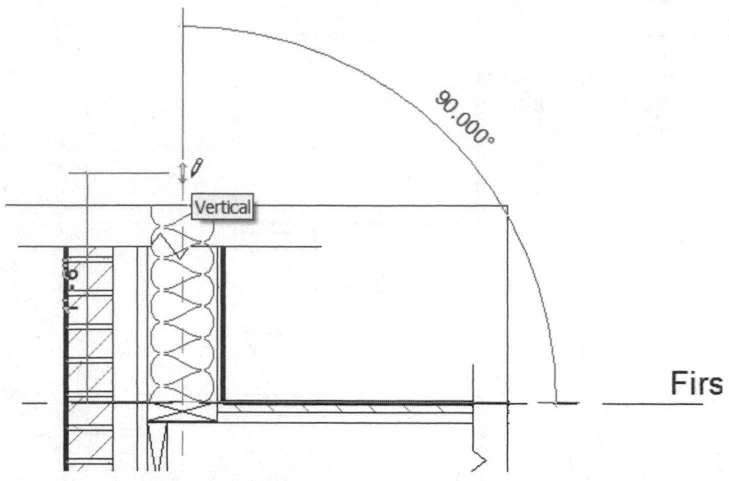

Figure 10–26 *Draw Insulation in the Stud cavity*

33. On the Options Bar, choose "to far side" from the drop down list.

   - Click a point on the inside of the end joist and drag to the right past the Crop Region (See Figure 10–27).

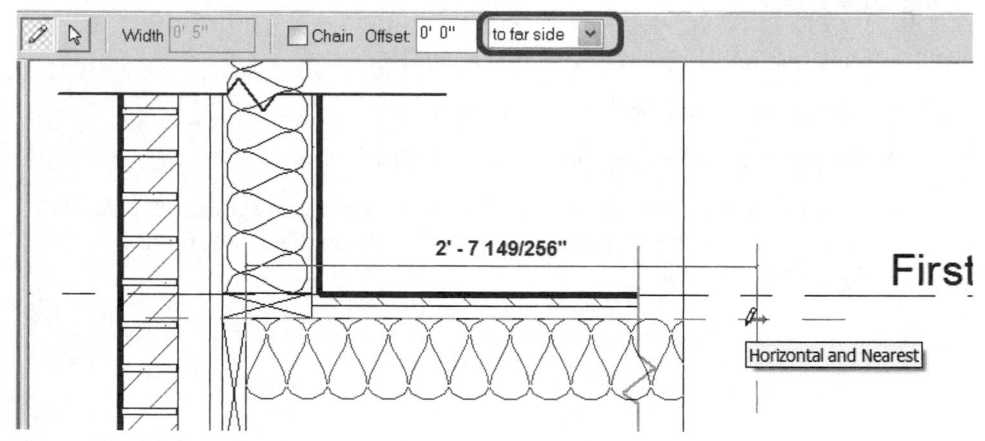

Figure 10–27 *Draw Insulation in the Floor cavity*

   - On the Design Bar, click the Modify tool or press the ESC key twice.

Notice that the insulation is not obscured by the Breaklines. This is because there is a display order for the View-specific Detail Components and the insulation is currently on top because it was added to the View after the Breaklines. We can shuffle the display order now.

34. Select the Breakline component on the right with the Modify tool.

   - On the Options Bar click the "Bring to Front" icon (See Figure 10–28).

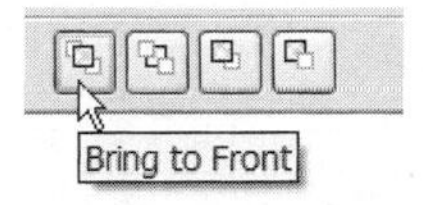

Figure 10–28 *Use the Display Order icons to shuffle the order of Components in the View*

- Repeat this process on the other Breakline and Insulation Components.
- Repeat once more on the flashing Detail Lines to bring them in front of the Mortar Joints.

EDIT CUT PROFILE

Sometimes you encounter a situation where the automatically-created graphics do not suit your specific needs. One such example is the keyway locking the foundation Wall to the Footing. Autodesk Revit Building provides us with the Edit Cut Profile tool. This tool gives us the ability to edit the path of the heavy cut line that Revit Building automatically generates. This type of edit is View-specific and two-dimensional. While it does not change the 3D shape of the model, it gives us a quick way to make the detail look the way we need without forcing us to model something that would have little or no benefit in other Views. For something like a key between the bottom of a foundation wall and the top of a footing, that would never be seen in any View other than a section or detail View, it would be difficult to justify the additional time or effort required to model it in 3D. Using the Edit Cut Profile tool we can make the section or detail appear as required without the extra modeling overhead.

35. From the Tools menu, choose **Edit Cut Profile**.

- On the Options Bar, select the "Boundary between Faces" option.

This option allows us to edit two boundaries—in this case the Footing's boundary and the foundation Wall's boundary—with one sketch. If we used the other option, Face, we would have to first edit the bottom face of the foundation and then go back and edit the top face of the footings.

- Select the boundary line between the foundation Wall and the footing (See Figure 10–29).

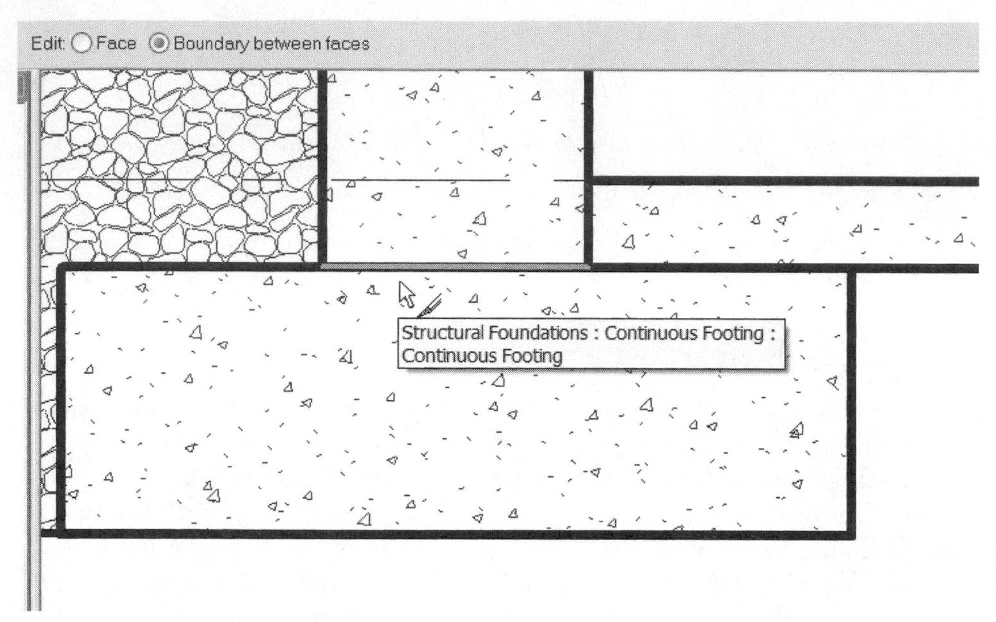

Figure 10–29 *Select the faces to edit*

The Design Bar changes to sketch mode.

36. Using the Lines tool, sketch the new path as indicated in Figure 10–30.

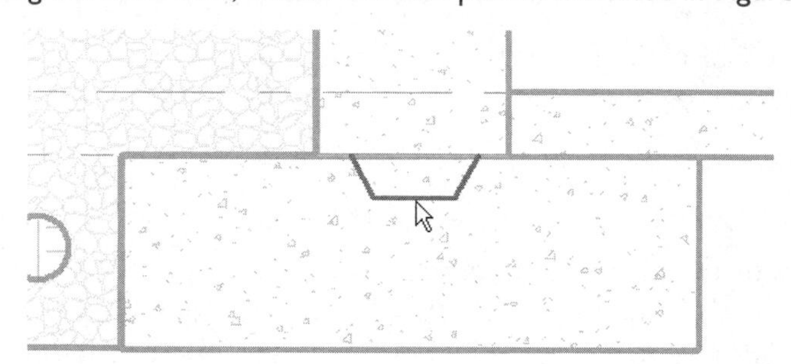

Figure 10–30 *Sketch the new shape using a Chain of Lines*

- On the Design Bar, click the Finish Sketch button.

In this case the fill pattern is the same on both sides of the Cut line, but if they were different you would notice that the fill pattern for the footing receded and the fill pattern for the foundation wall extended to fill the key.

VIEW BREAKS

It is common that a typical wall section will be too tall to fit on a Sheet. So it is typically broken into separate parts that crop out the repetitive areas. The Crop

Boundary includes "View Break" Controls and can be clipped to achieve this effect.

37. **Select the Crop Boundary surrounding the section callout (it appears as a rectangle surrounding the drawing)**

On each of the four edges of this Crop Region, a round Control Handle appears at the midpoint and a "zig zag" break Control appears on either side of it. These allow us to truncate the View into smaller parts facilitating placement on a Sheet (see the left side of Figure 10–31).

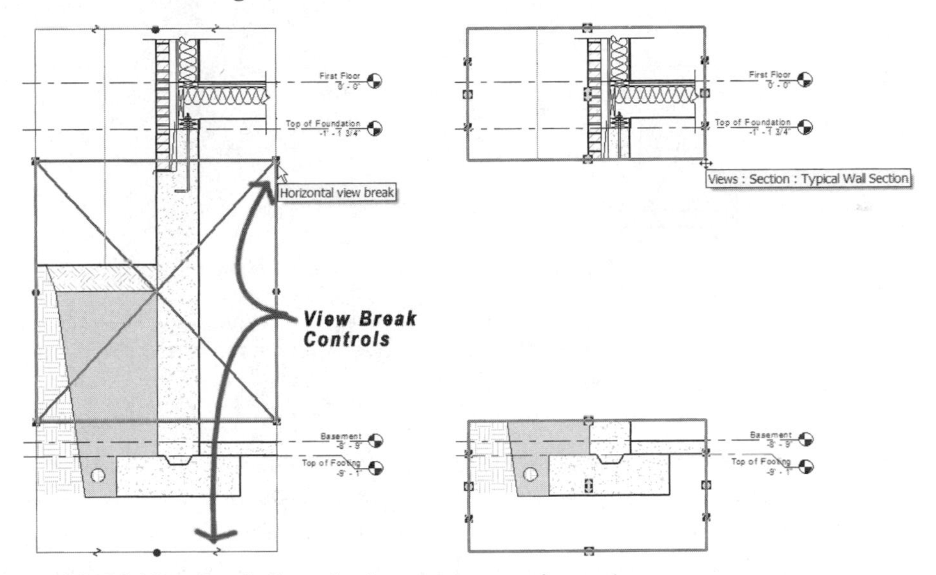

Figure 10–31 *View Break Controls allow you to crop the section*

- **Click one of the View Break Controls on a vertical edge (there are four total, you can pick any one) of the Crop Boundary (see the right side of Figurc 10 31).**

The View splits into two separate Crop Regions with a large gap in the middle. A blue arrow Control Handle appears in the middle of each View Break region. We can use these to move the two portions closer together. Notice that each of the two new Crop Boundaries includes the same types of Control Handles—but only on the vertical edges now. You can continue to break them into additional sub-Views as necessary. But all breaks must be along the same direction as the first one—vertical in this case. In this example two is enough so we will not break it any further. However, we need to adjust the top View Break so we can see the entire Anchor bolt.

- **Using the round Control Handle at the bottom of the upper View Break, drag the edge down a bit to show all of the anchor bolt.**

38. Click on the View Break Control arrow (in the middle) of the top View Break and drag it down so the Crop Region is a little above the upper crop boundary of the lower View Break (see Figure 10–32).

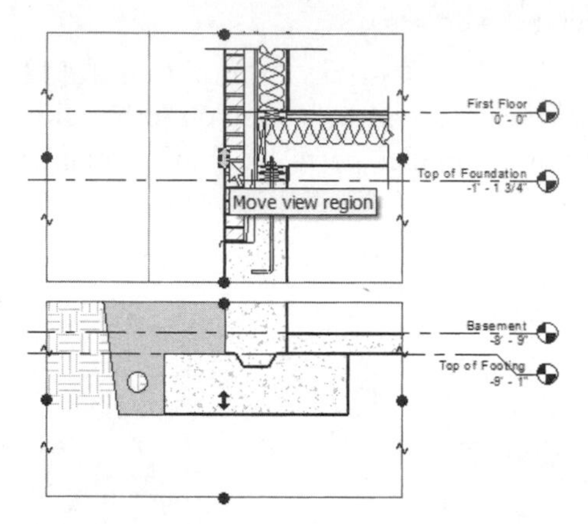

Figure 10–32 *Drag the upper View closer to the lower one using the Move View Region Control*

If you continue to drag so that you over lap the two View Breaks they will join back into one. This is how you "remove" the break.

CAUTION Be sure to move the portions of the View Breaks with the Control arrow in the middle. Do not drag the edge of the Crop Region. Doing so will actually move the area of the callout in both this View and the referring Longitudinal *section View.*

Although the sub-Views are truncated and closer together, distances are dimensionally correct. Look at the Level lines to the right of the Views. The First Floor is at elevation zero (0) and the Top of Footing is at elevation -9′-1″ [-2700]. Therefore if we were to add a dimension from the top of the finished floor on the first floor to the top of the footing, it should read a distance of 9′-1″ [2700]. Let's try it out.

39. On the Design Bar, click the Dimension tool.

- Move the Dimension tool over the top edge of the Floor at the First Floor.

Most likely the Level line will prehighlight. While we could dimension this point and still receive the correct value, we want to associate the dimension with the Floor element instead. We can use the TAB key here (like so many other places in Revit Building) to cycle to the element that we want.

- Press tab until the top edge of the Floor prehighlights and then click.

The selected edge will remain red while you complete the dimension operation.

- Move down with the Dimension tool and over the top cut line of the Footing.

If the top cut line of the Footing does not automatically prehighlight, use the TAB key again.

- With the top edge of the Footing prehighlighted, click to select it.
- Move to the left Crop Region Boundary and click next to it (in the white space) to place the Dimension string (see Figure 10–33).

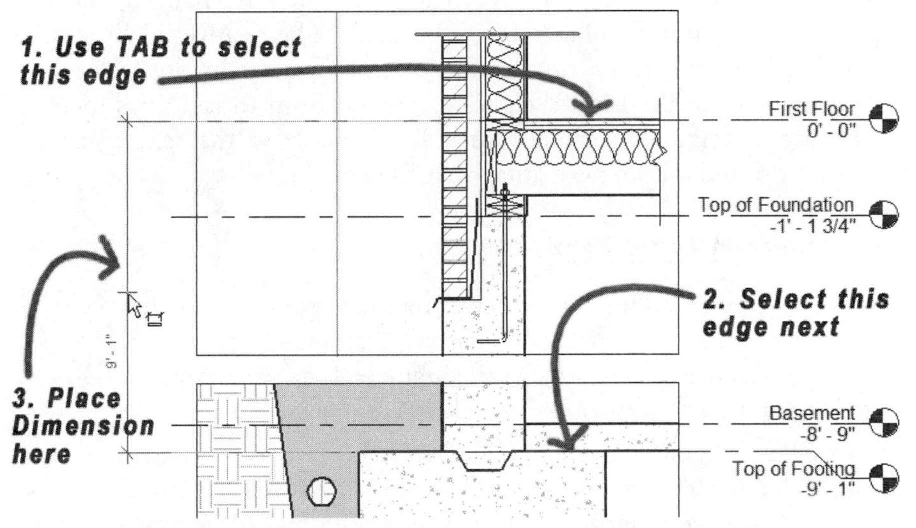

Figure 10–33 *Dimension the distance between the Floor and the Footing*

Notice that the Dimension displays the correct 9′-1″ [2700] value from the top of the Footing to top edge of the finish Floor. As you can see, applying a View Break is a graphical convention only and has no impact on the dimensional accuracy of the model being displayed in each portion of the Crop Boundary.

- On the Design Bar, click the Modify tool or press the ESC key twice.

40. Save the project.

ANNOTATION

Annotating a drawing with notes, dimensions, symbols, and tags is essential to communicating architectural design intent. Such annotations in Revit Building are View-specific elements. This means that these elements appear *only* in the View to which they are added. The exception to this is View indicators and Datums like section markers, elevation makers, Level Lines, Grids and Callouts. These items are purpose-built to appear in all appropriate Views and enhance the fully-coordinated nature of an Autodesk Revit Building project.

Each View in Revit Building has a "View Scale" parameter and all annotations added to a particular View will scale and adjust accordingly as necessary. View indicators and Datums are included in this behavior. This means that no matter what the scale of the drawing, the annotation, View indicators and Datums will be the correct size required for printed output. This behavior also applies to line weights and drafting patterns. Each graphical View you open can and often will have its own scale. Also, if the Scale parameter of a View is changed, the line weights and drafting patterns will automatically adjust. The relative thickness of a particular line, or line weight, is controlled in a matrix of common plot scales. You can edit this matrix with the **Line Weights** command from the Settings menu if desired. Drafting patterns will maintain their line spacing so the spacing always looks correct on printed output no matter what the scale is.

CREATE A CUSTOM TEXT TYPE

A text element in Revit Building like other elements is associated with a Type. A text "Type" in this case is simply a grouping of parameters that control the look and formatting of the text. There are several parameters, many of which are similar to text in other computer software and are likely familiar to you.

Text Types can be preconfigured and added to a Project Template. Additional Types can be added to the template or as a Project progresses. Like other Types, to create a new text Type, you simply duplicate an existing one and modify its parameters. Let's create a new text type for our project. In this example we will create a "general note" text Type that is 1/8″ [3] high and uses a different font. (Note: in Revit Building, the height is its final plotted height—you are not required to calculate text size relative to the model). The process for creating a new text Type is nearly identical to the one used for duplicating other element Types.

1. On the Design Bar, click the Basics tab and then click the Text tool.
   - On the Options Bar, click the Properties icon.
2. Next to the Type list, click the Edit/New button.

TIP A shortcut to this is to press ALT + E.

The Type Properties dialog will appear.

- Next to the Type list, click the Duplicate button.

TIP A shortcut to this is to press ALT + D.

A new Name dialog will appear. By default "(2)" has been appended to the existing name.

- Change the name to **MRB General Notes** and then click OK.

You can use any font that is installed on your system. Since the choice of fonts can vary widely from one computer to the next, your system may not have the same fonts as those indicated here. In an attempt to mitigate this, we will select a very commonly available Windows ™ standard font—Arial Narrow. Feel free to choose a different font if you prefer.

3. Beneath the "Text" grouping, from the "Text Font" list, choose **Arial Narrow**.

- In the "Text Size" field, type **1/8″** [**3**] (see Figure 10–34).

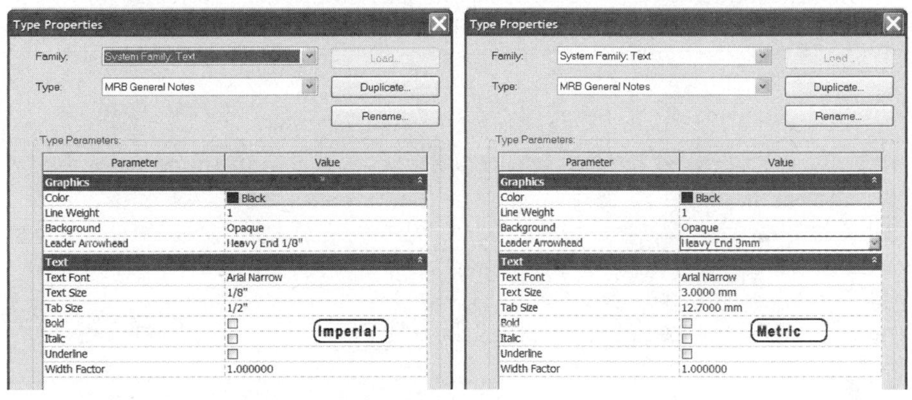

Figure 10–34 *Create a new Text Type*

This is the size the text will be when printed out. Beneath this, you can choose to make the text bold, italic or underline if desired. The "Width Factor" setting is used to compress or stretch the text. This is a multiplier. When set to 1, the text draws in the way it was designed in the font. A value less than 1 will compress the text and a value greater than 1 will stretch it out.

In the "Graphics" area, you can change the color of the text as well as assign an arrowhead to this Type. When you draw text, you can place just the text, or create it with a leader line attached. This is done on the Options Bar. The "Leader Arrowhead" parameter is used to assign an arrowhead Type to the Text Type.

- From the "Leader Arrowhead" list, choose Heavy End 1/8″ [Heavy End 3mm].

Arrowheads are System Families and you can add additional Types using the **Annotations > Arrowheads** command from the Settings menu.

- Click OK twice to complete the new Type.

The new Type will appear in the Type Selector list on the Options Bar.

PLACING AND MODIFYING TEXT

To place text in a View, simply click down or drag a rectangle at the location where you want the text to appear. If you click the point, the text will flow in one continuous line without wrapping. If you click and drag two points, it will wrap to the width between the points. Regardless of your choice, you can always edit the wrapping of a text element later using the control handles on the text element. Pressing the ENTER key within a text element will insert a "hard" Return. This will move the cursor to the next line regardless of the automatic wrapping. It is basically like using a word processor. Typically, you simply begin typing and your word processor will automatically wrap the text to the next line when you reach the edge of the paper. If you want to start a new paragraph, you press ENTER. It is the same in Revit Building.

4. On the left side of the detail, click and drag a text region near the top.

- Type: **Standard Face Brick Veneer- see specification for color** (see Figure 10–35).

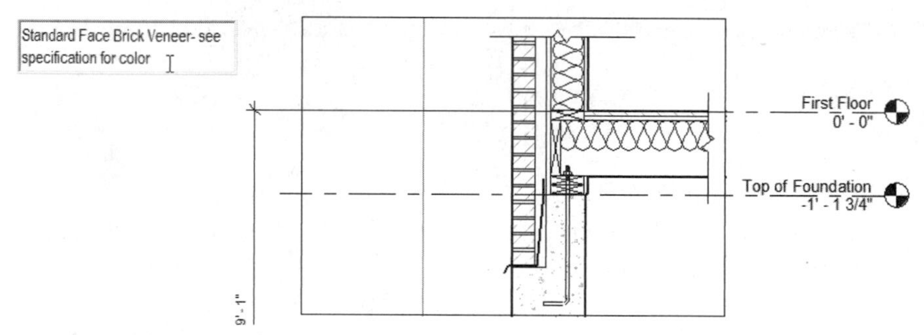

Figure 10–35 *Add a Text element and type in the desired note*

- Click next to the note (in the white space) to finish typing.

Some blue control handles will appear attached to the text element. You can use the one on the left to move the element (while leaving any arrow heads in place), the one on the right to rotate it and the two small round ones on either side to resize the and reshape the element and its word wrapping (see Figure 10–36).

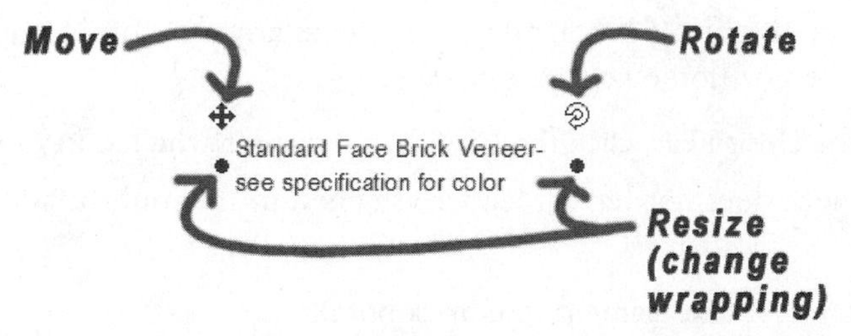

Figure 10–36 *Move, Rotate or Resize a Text Element with its Control Handles*

To place a leader and arrowhead with a note, you can choose the option on the Options Bar.

5. With the Text tool still active, on the Options Bar click the "One Segment" icon next to "Leader."

- Click near the middle of the double top plate on the foundation Wall.

This is the location of the arrowhead for the Leader.

- Drag to the left and click beneath the first note (a temporary guide line will appear to assist you).

This is the end of the Leader. A text object will appear. If you selected the "Two Segments" icon instead, you would place two segments of the Leader line before typing would begin.

- Type the next note: **Double Top Plate** and then click next to the note (in the white space) to finish typing (see Figure 10–37).

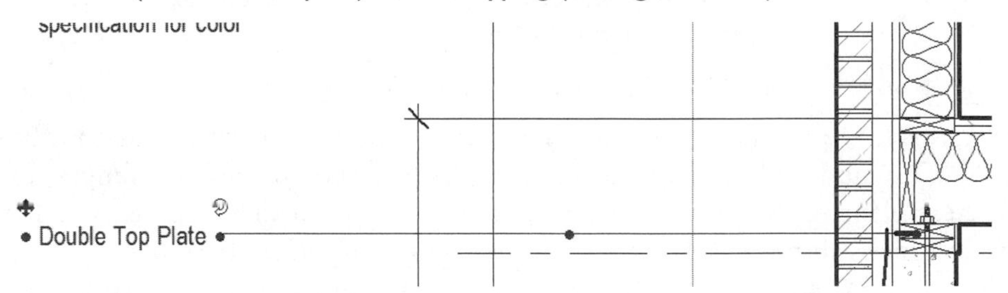

Figure 10–37 *Add a Text Element with a Leader*

- Using the Move handle, drag the text element to align it with the first one. A temporary guide line will appear to assist you.

When you drag a text element with a leader, be careful not to drag up or down as this will bend the leader line. This is because the leader's arrowhead stays attached to the element to which it points. If you want to move the entire thing (text and leader together), use the Move command (on the toolbar) or the arrow keys on the keyboard (to nudge) or the Modify tool and pick anywhere on the text Element's boundary except the double Move arrow icon. Make any fine-tuning adjustments that you wish to the position of either text element.

- On the Design Bar, click the Modify tool or press the ESC key twice.

The first text element does not have a leader attached to it. You can add leaders to text anytime.

6. Select the first text element (the brick note).

- On the Option bar, click the "Add Right Leader" icon (see Figure 10–38).

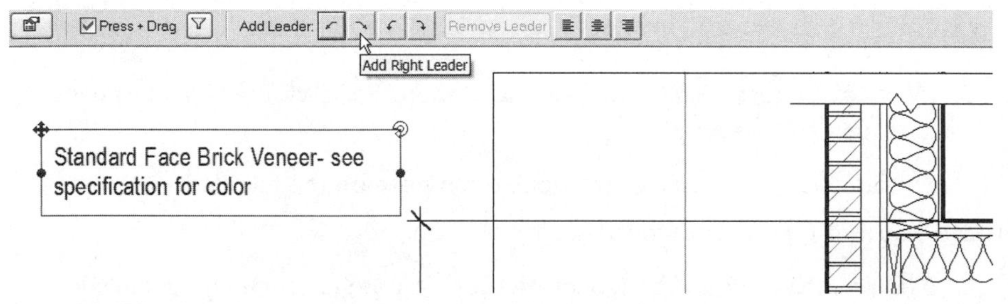

Figure 10–38 *Add a Leader to an existing Text Element*

A leader will appear attached to the text. You can then use the drag handles to modify its shape and adjust the location of the arrowhead.

- Make adjustments with the drag controls as necessary to move the arrowhead to point at the brick.

7. Using the Text tool with a leader option, add a note pointing to the batt insulation that reads: **Batt Insulation**.

- Adjust the position of the note and leader as required.

Sometimes you want to have the same note point to more than one location in the detail. You can add leaders to an existing text element. To do this, you simply select the text element and then click the appropriate icon on the Options Bar. To remove a leader you no longer need, click the "Remove Leader" button.

- On the Options Bar, click the "Add Right Leader" icon.
- Position the leader and its arrowheads as necessary to point at the insulation in the floor (see Figure 10–39).

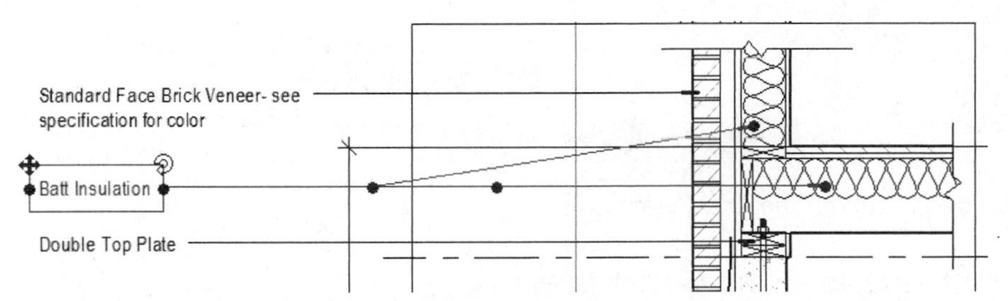

Figure 10–39 *Add a second Leader line to the Text Element*

8. Using the process covered above, add in two more Break Lines between the two portions of the split detail.
9. Using the process covered here; add additional notes to the detail (see Figure 10–40).

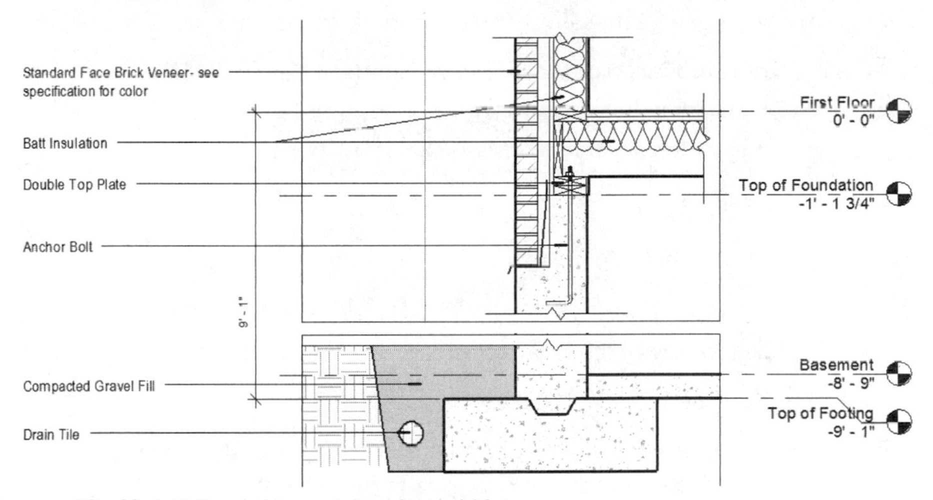

Figure 10–40 *Add Break Lines and additional Notes*

10. Using the Dimension tool add dimensions to the footing and foundation Walls.

TIP Remember to use your tab key as needed to select the required edges to dimension.

The Crop Regions around the detail are becoming a bit distracting. We can turn off their display.

11. From the View Control Bar (at the bottom of the View window) choose Hide Crop Region from the pop-up icon menu (see Figure 10–41).

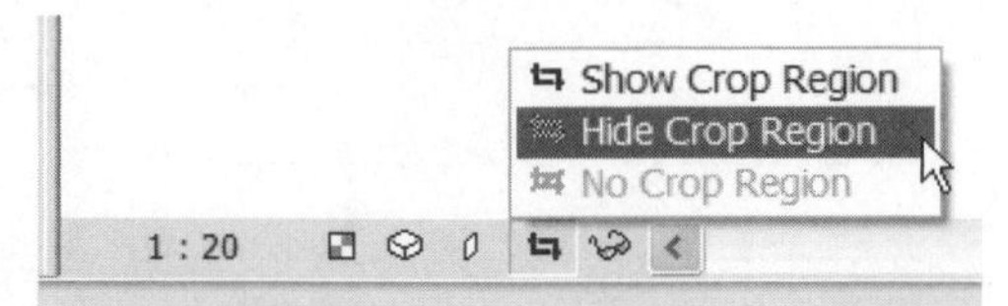

Figure 10–41 *Turn off the Crop Region*

MODIFYING LINEWORK IN THE VIEW

On the left side of this detail we see a gray vertical line. This is the edge of the chimney beyond. In cases like this, where some piece of the model displays that we would rather not see, or wish to display differently than its default behavior, we can use the Linework tool. The effect of the Linework tool is View specific. It is used to modify the display of any linework in the current View. You can use this tool to change heavy lines to thin, solid lines to dashed, or visible lines to invisible.

12. On the Tools toolbar, click the Linework icon (see Figure 10–42).

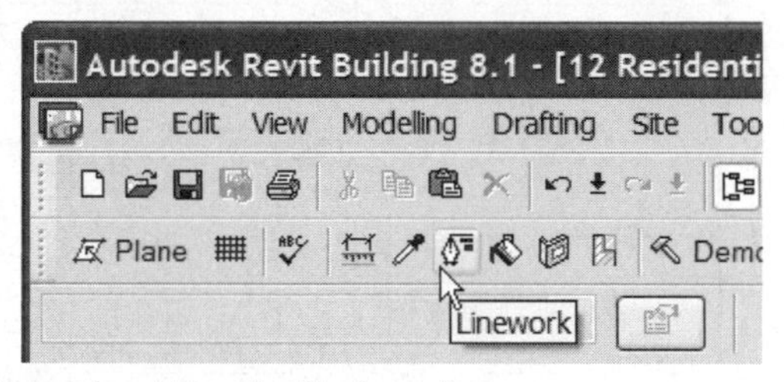

Figure 10–42 *The Linework tool on the Tools toolbar*

- From the Type Selector, choose <Invisible lines>.
- Move the pointer (now displaying a pencil cursor) over the gray vertical edge of the fireplace and then click (see Figure 10–43).

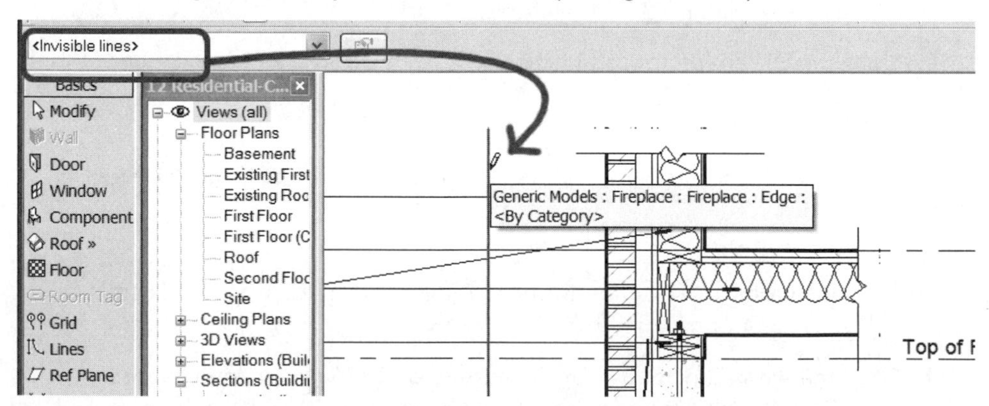

Figure 10–43 *Use the Linework tool to make the edge of the fireplace invisible*

When you click the edge of the fireplace, it will disappear. Later you can use the Linework tool and move it over the general area where the edge was displayed if you wish to change it to a different linetype or return it to its default setting. To restore the default, use the <By Category> option.

DETAIL THE REMAINDER OF THE WALL

To detail the rest of the Wall section, you can follow the same procedures as outlined here. Start by returning to the *Longitudinal* section View and create a new callout of the top portion. Use the View Breaks to crop the detail and remove the repetitive portions. Add break line components to each of the breaks. Hide the Crop Region when finished. Begin adding Detail Components on top of the section cut as we did above, add drafting lines and edit the linework as required. Complete the detail with dimensions and notes. Focus on the Wall connection at the second floor and the overall studs, rafters, joists and insulation. When you are finished, the detail should look something like Figure 10–44.

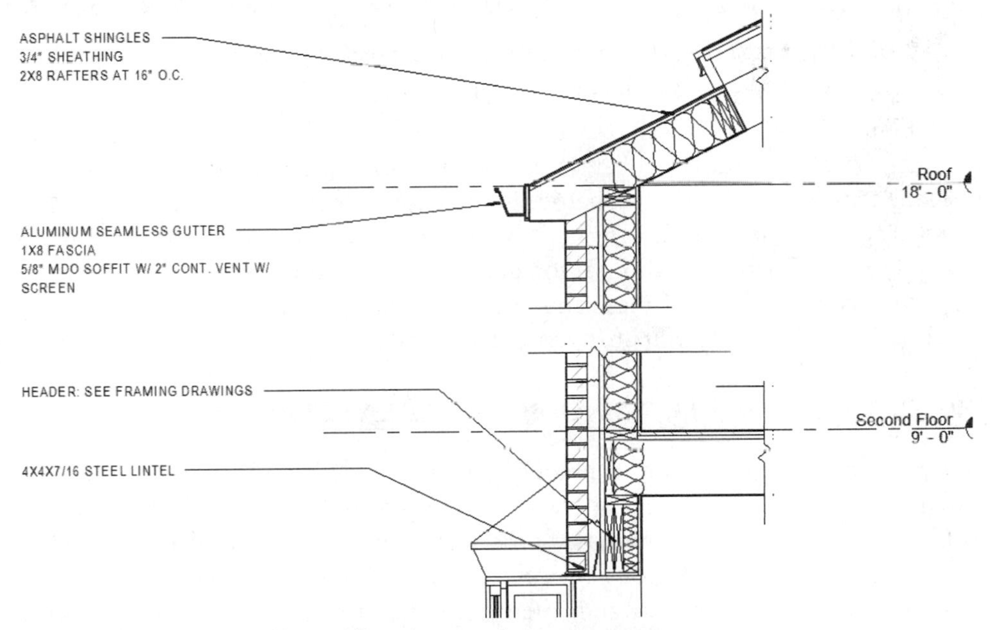

Figure 10–44 *Create additional Details using the same process*

Most of the components that you will need are already loaded into this project; however, for items like the steel angle at the Window lintel, you can simply load them in from the appropriate library. At the roof eave, you will need to rely more on Filled Regions, Drafting Lines and Edit Cut Profile. Using Figure 10–45 as a guide, add Filled Regions and Drafting Lines to create the blocking and roof vent.

TIP Remember to use the "Send to Back" and "Bring to Front" icons on the Options Bar as required to achieve the proper look.

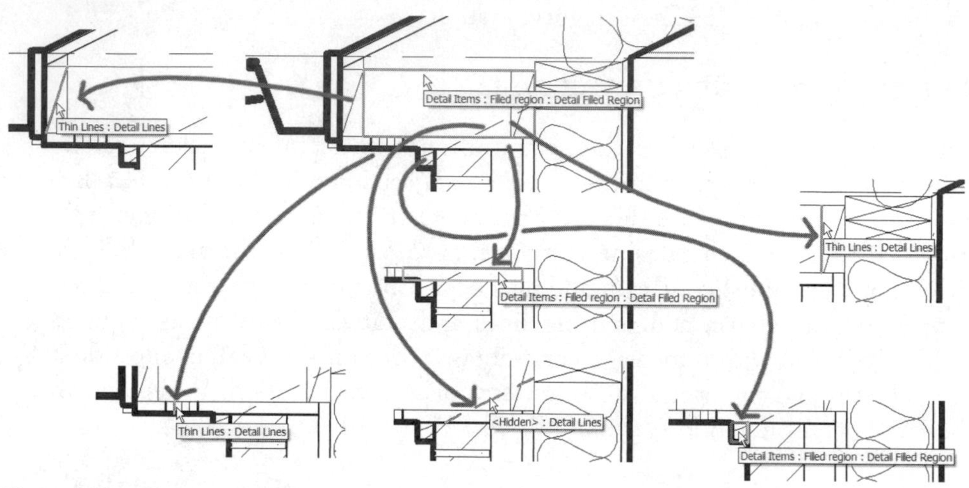

Figure 10–45 *Using Detail Lines and Filled Regions, create a roof vent condition*

Take notice of the <Hidden> : Detail Lines element shown in the bottom middle panel of Figure 10–45. This was drawn on top of the Filled Regions to represent the sloping rafter beyond. In most cases, you would use the Linework tool to change the actual edge drawn automatically to the <Hidden> Type. However, in this instance, since the Wall is attached to the Roof, the Roof does not actually draw any graphics in this location automatically. This is one of many examples where judicious use of Detail Lines can make the detail read correctly in a specific View without requiring any edits to the model geometry.

USING EDIT CUT PROFILE TO MODIFY WALL LAYERS

As you can see, you can use the technique of adding Filled Regions to cover unwanted geometry and then sketching Detail Lines on top almost exclusively. There is nothing inherently wrong with the procedure, but if the underlying model should change, the Filled Regions may no longer cover the intended geometry leading to errors in coordination and intent. Another approach is to modify the underlying geometry as it is displayed in this View. To do this, we use the Edit Cut Profile tool as we did above for the footing.

13. On the Tools toolbar, click the Edit Cut Profile icon (see Figure 10–46).

TIP Edit Cut Profile is also located on the Tools menu.

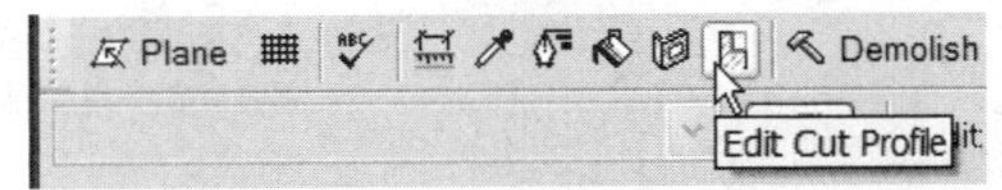

Figure 10–46 *Use the Edit Cut Profile command to edit the shape of the automatically created profile*

The cursor changes to a Cut Profile shape.

- Pass the cursor over the Wall and when the stud layer pre-highlights, click the mouse.

The Design Bar changes to sketch mode and shows some now familiar sketch tools. The existing boundary of the stud layer will show as an orange outline.

14. Using the Lines tool, and the Pick Lines option (on the Options Bar) click the bottom edge of the top plate to create a sketch line (see Figure 10–47).

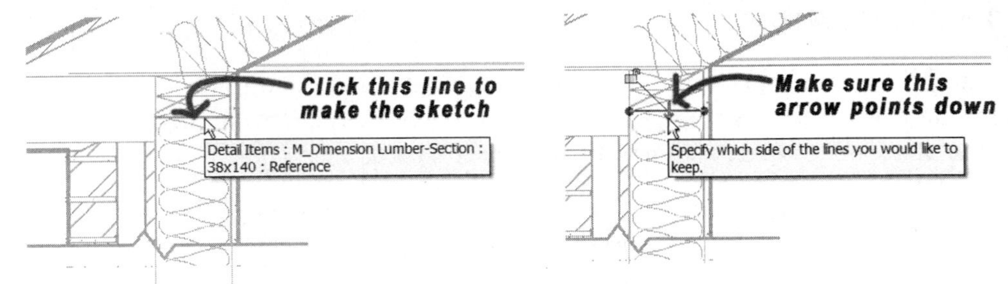

Figure 10–47 *Sketch the new edge of the Cut Boundary*

A sketch line will appear on top of the selected edge. A small arrow handle will also appear. It should be pointing down to indicate that you wish to keep everything below the sketch line. If it points up, click it to point it down.

- On the Design Bar, click the Finish Sketch button.

The result should like something like Figure 10–48.

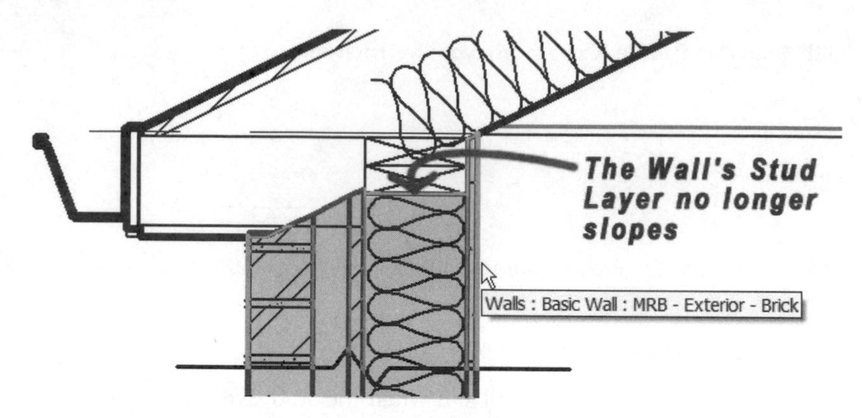
The Wall's Stud Layer no longer slopes
Walls : Basic Wall : MRB - Exterior - Brick

Figure 10–48 *After the Cut Profile Edit, the shape of the Wall reflects the change*

15. Repeat the procedure to modify the top of the brick, the air space and the sheathing as well.

16. Make any additional edits and then save the project.

ADD A DETAIL SHEET

Once we have created one or more Detail Views, we can add them to Sheets in the same fashion as other Views. We explored this process back in Chapter 4. Let's review the steps here to create a new Detail Sheet that contains our Typical Wall Section detail.

17. On the Project Browser, right-click the *Sheets (all)* node and choose **New Sheet**.

    - In the "Select a Titleblock" dialog, choose D 22 x 34 Horizontal [A1 metric] and then click OK.

This will create "G101 - Unnamed." This is because the last Sheet we created was Sheet G100.

18. On the Project Browser, right-click on *G101 – Unnamed* and choose **Rename**.

    - In the Name field, type **Details** in place of "unnamed."
    - In the Number field, replace the existing value with **A601** and then click OK.

TIP You can also click directly on the (blue text) values in the titleblock and edit them directly on screen without opening the Properties dialog.

19. From the Project Browser, drag the *Typical Wall Section* detail View and drop it on the Sheet.
    - Click a point to place the detail. Move it around as desired to fine-tune placement.
20. On the Project Browser, double-click to open the *Longitudinal* section View.

Notice that the callout annotation has automatically filled in to indicate that the detail is number one on Sheet A601. This will also remain coordinated automatically (see Figure 10–49).

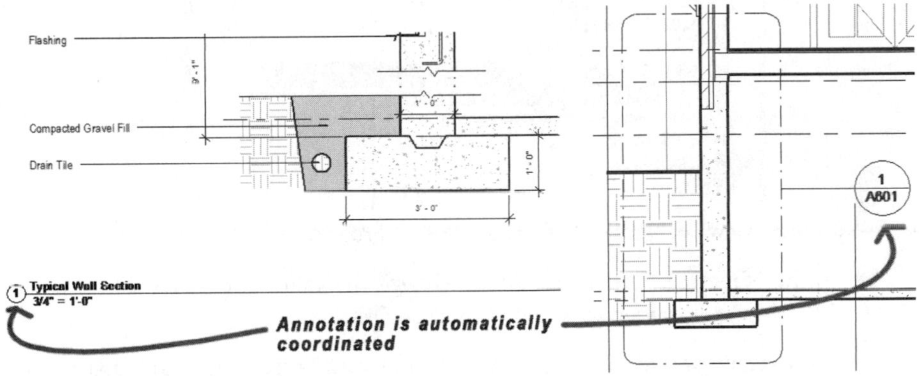

Figure 10–49 *Annotation will coordinate automatically after adding the detail View to a Sheet*

21. Repeat the process to add the other detail to this Sheet.

In some cases, you will add details to the Sheet and then later wish to reorganize or renumber them. To do this, you edit the View Properties of the View in question. Edit the value of the "Detail Number" parameter in the "Element Properties" dialog for the View. Be sure to type a number not yet in use—Revit Building will not allow you to duplicate an existing number. To swap the numbers of two details, first edit one to a unique value, edit the other to the value originally used by the first, and then edit the first to the number originally used by the second.

You can edit the View's Properties directly from the Sheet if you wish. Expand the Sheet entry on the Project Browser to see a listing of all Views already placed on a particular Sheet. Right-click the name listed and choose **Properties** to jump directly to the "Element Properties" for that View. You can also double-click the View from there to open it.

22. Expand the Sheet *A601 – Details* in the Project Browser.
    - Right-click the View *Section: Typical Roof Eave Condition* listed beneath it and choose **Properties** (see Figure 10–50).

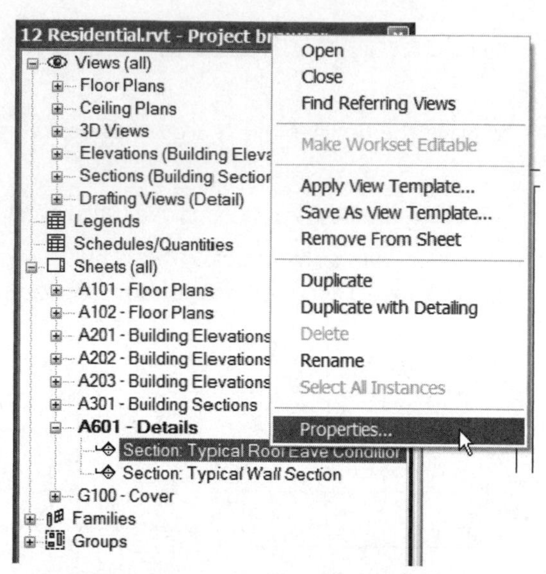

Figure 10–50 *Edit a View's Properties to from the Project Browser*

- In the "Element Properties" dialog, change the value of the "Detail Number" parameter to **3** and then click OK.
- Repeat this process for the *Section: Typical Wall Section* View and change its "Detail Number" parameter to **2** and then click OK.
- Return to the "Element Properties" dialog for the *Section: Typical Roof Eave Condition* View and change its "Detail Number" to **1** (see Figure 10–51).

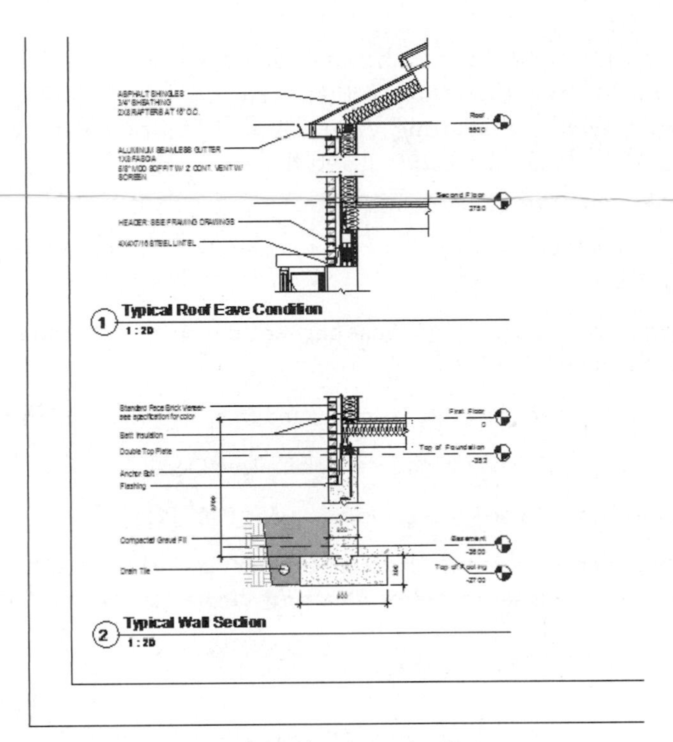

Figure 10–51 *Change the numbers of the details on the Sheet*

Open the *Longitudinal* section View and note that the new numbers are reflected there as well. A change in one location is a change everywhere in Autodesk Revit Building!

23. Save the project.

WORKING WITH LEGACY DETAILS

In most firms, details are reused from one project to the next. These "standard" details are often kept in libraries for easy reuse and retrieval. In the days before computer design and drafting software, such a library would be a three-ringed binder from which photo copies were made. With computers, these standard details are stored digitally. If your firm has been using such software for a while, you likely already have such a digital library of standard details. You can use these legacy files directly in your Revit Building projects. You simply import in the DWG or DGN files and place them on Sheets like other details.

CREATE A REFERENCED SECTION VIEW

In this tutorial we will assume that the handrail of the existing Stair will be replaced with a new one. To show this, we will create a "Referenced Section View" to create a Section marker callout of a handrail detail within a stair section View. However, instead of creating the actual section View in Revit Building, the Referenced Section will reference a Drafting View. On this Drafting View we will import an existing DWG detail of a handrail and then place the Drafting View on to a Sheet View.

1. On the Project Browser, double-click to open the *Section at Existing Stair* section View.
   - Zoom in on the area of the Stair between the First Floor Level and the Second Floor Level.
2. On the Design Bar, click the Basics tab and then click the Section tool.
   - From the Type Selector, choose Detail View: Detail.
   - On the Options Bar set the "Scale" to **6″=1′-0″** [**1:2**].
   - Place a checkmark in the "Reference other View" checkbox, and verify that the menu is set to **<New Drafting View>** (see Figure 10–52).

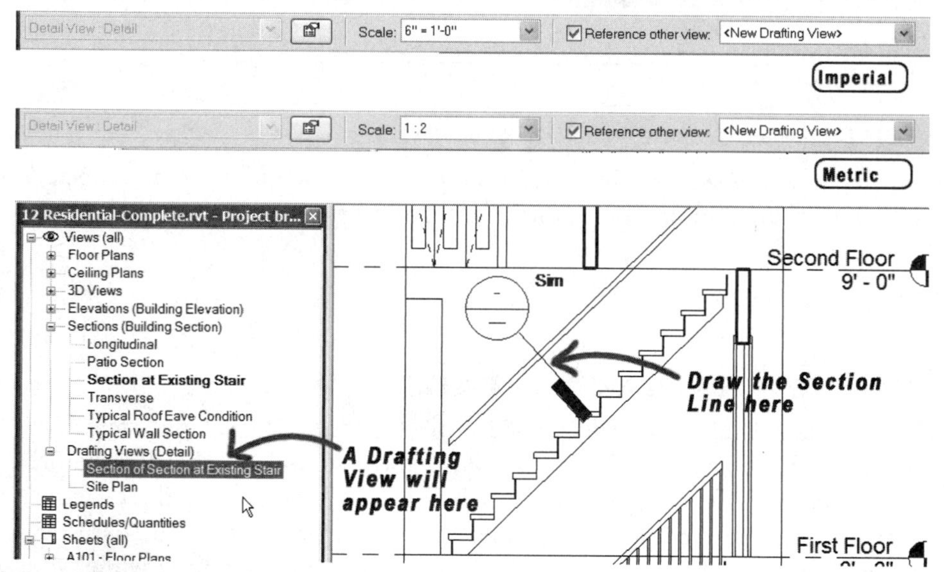

Figure 10–52 *Create a new Section View set to a Detail referencing a new Drafting View*

These settings instruct Revit Building to create a new Drafting View instead of the typical live section View of the model. The detail marker will point to this new drafting View.

- Drag the section line through the Railing as shown in the bottom of Figure 10–52.

Notice that a new Drafting View was created (on Project Browser beneath Drafting Views) as this marker was placed.

3. On the Project Browser, right-click the new Drafting View and choose **Rename**.
   - Name the View **New Railing Detail** and then click OK.
4. In the View window, double-click on the detail head to open this Drafting View (or double-click the name on the Project Browser instead).

When Revit Building opens the new Drafting View the most obvious characteristic is that the View is empty showing no model geometry. A Drafting View is like a blank sheet of paper. There are no automatically-generated graphics from the model. Drafting an image that makes sense relative to the detail cut location is up to you. The only reference back to the model is the callout. Any of the detailing and annotation tools (such as those used above) can be used to draft your detail or we can import a detail from our legacy detail library.

IMPORT A DETAIL DRAWING

5. From the File menu, choose **Import/Link > DWG, DXF, DGN, SAT**.
   - In the "Import/Link" dialog, browse to the *Chapter10* folder and choose *Typical Handrail Detail.dwg* [*Typical Handrail Detail-Metric.dwg*].
   - In the "Layer/Level Colors" area, choose "Black and white."
   - In the "Positioning" area, choose "Manually Place" and then the "Cursor at center" option.
   - In the "Import of Link" area, be sure that the "Link" checkbox is cleared (see Figure 10–53).

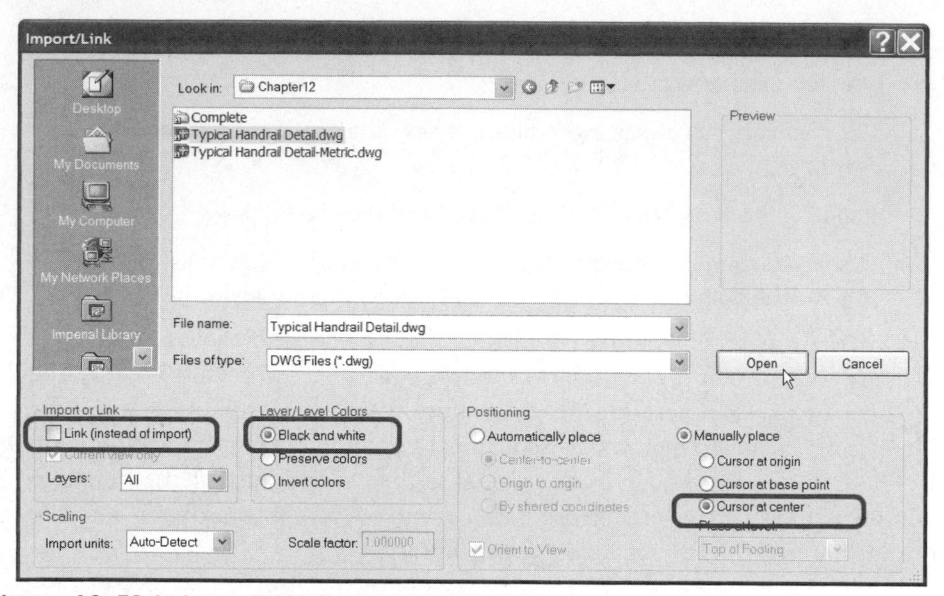

Figure 10–53 *Link in a DWG file for the handrail detail*

If you check the "Link" box, it will link the drawing file into the project as opposed to importing it. Linking maintains a connection to the original file. If it should change, you can refresh the link in Revit Building to reload and display the latest version of the file. Import embeds the file into the Revit Building project and does not maintain a link. If the file is changed outside of Revit Building, you would not be able to display those changes and would need to re-import the modified file. To reload a linked DWG file, use the same "Manage Links" dialog (File menu) that we used in previous chapters to reload linked RVT files.

6. Click Open to import the detail, and then click a point on screen to place the detail in the View.

- Verify that the scale of the current View is **6″=1′-0″** [**1:2**] as indicated above. If it is not, please change it.

Notice that if you change the scale, it has an impact on how the lineweights of the imported View display. If you wish to manipulate the way that the lineweights import, experiment with the **Import/Export Settings** command on the File menu.

This is a typical detail and there is no need for any changes. If we needed to make edits, we could select the detail, and then on the Options Bar, choose the Explode button to convert it to individual Revit Building Detail Lines and Text so we could edit it.

7. On the Project Browser, double-click to open the *A601 - Details* Sheet View.

- Drag the Drafting View and drop it on the Sheet (see Figure 10–54).

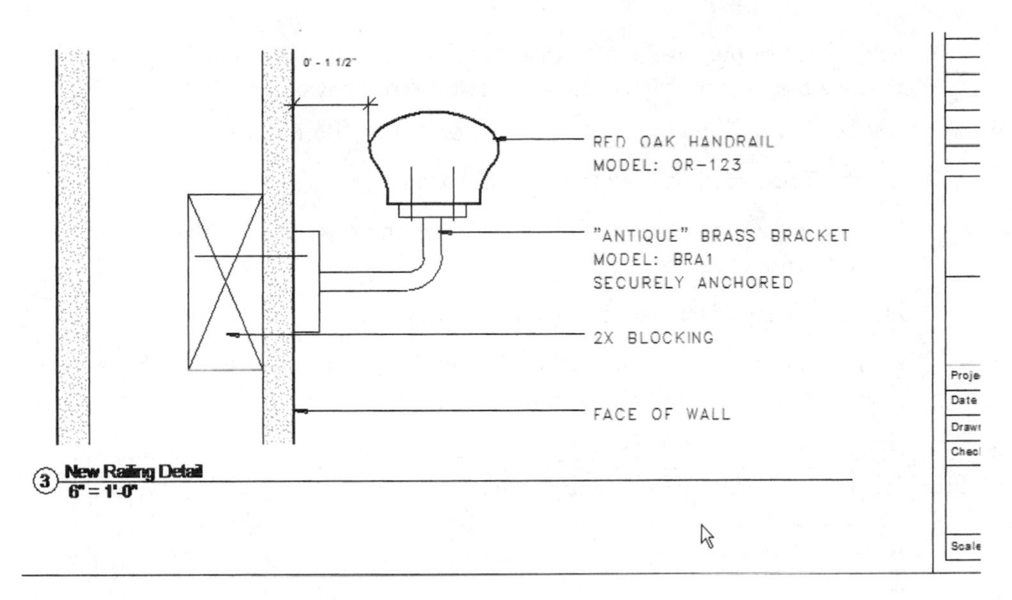

Figure 10–54 *Add the Detail View to the Details Sheet*

This will become detail 3 on the Sheet. If you return to the *Section at Existing Stair* section View you will see that this number and Sheet reference have appeared automatically in the callout.

8. Save the project.

EMBELISHING MODEL VIEWS

Except for Drafting Views, all Views in the Revit Building project are generated directly from the three-dimensional building model. While Revit Building does a very good job of interpreting this model geometry in two-dimensional representations such as plans and elevations, there are often items that we wish to create the Architectural drawings we are accustomed to producing. We have already seen all of the techniques that are used to perform such edits on the automatically-created Revit Building elevation Views. Remember, all "drafting" edits such as those made with the Linework tool, Filled Regions, Detail Components, Edit Cut Profile, and Drafting Lines can be done on any View and apply only to the View in which they are applied. Let's make a few enhancements to one of our elevation Views.

9. On the Project Browser, double-click to open the *East* elevation View.

One common architectural drafting convention is to show the new foundation in an elevation as dashed below grade. We can achieve this using a combination of the Linework tool and adding Drafting Lines. Let's start with the footing.

10. Select the Terrain element, and then on the View Control Bar, choose **Hide Object** from the Hide/Isolate menu.

11. Select each of the Level Heads that do not have associated Views (the ones that are black), right-click and choose **Hide Annotation in View**.

This will hide levels for "Top of Footing" or "Bottom of Stair" etc.

12. On the Tools toolbar, click the Linework icon.
    - Choose <Hidden> from the Type Selector and then click on the bottom edges of the footings.

    Do not do any of the vertical foundation lines yet.

    - Click on each of the edges of the footings to change them to <Hidden> lines (see Figure 10–55).

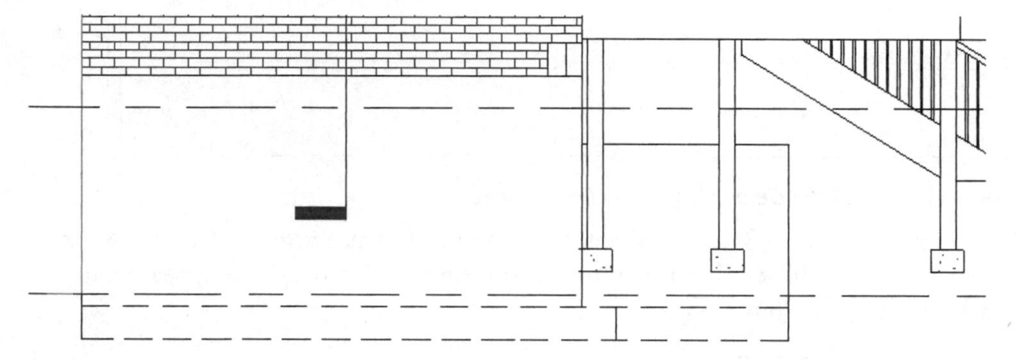

Figure 10–55 *Change the display of the footing lines to <Hidden> with the Linework tool*

You may need to pick more than once in the same general spot or a little to either side since there is more than one footing in the same spot in the elevation. If you are unhappy with the result, you can instead use the <Invisible lines> Type and then draw a continuous Drafting Line in on top. Be sure to lock the constraint padlock icon to keep the Drafting Line associated with the position of the footing

13. On the View Control Bar, choose **Reset Temporary Hide/Isolate**.

The terrain model will reappear.

14. On the Tools toolbar, click the Linework icon.
    - Choose <Hidden> from the Type Selector again.
    - Click on one of the vertical lines of the foundation Walls.

    With it still highlighted, a drag handle will appear at either end.

    - Drag the top handle down to the point where it intersects the terrain (see Figure 10–56).

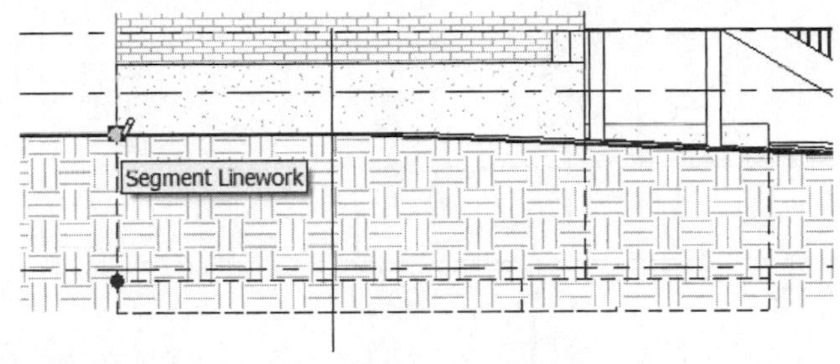

Figure 10–56 *Change the Linework of the foundation walls and edit the extent of the change with the drag handles*

- Repeat for other vertical foundation Wall edges.

If you wish to modify the way that the terrain displays, you can use a Filled Region to trace over it. Draft additional linework as desired to complete the elevation. You can add notes, dimensions and tags as required.

15. On the Design Bar, click the Drafting tab and then click the Tag tool.
    - On the Options Bar, clear the "Leader" checkbox.
    - Click on each of the Windows in the new addition. (Do not tag the Windows of the existing house).
16. Add some notes to the patio on the right or to indicate materials of the elevation such as brick veneer and roof shingles (see Figure 10–57).

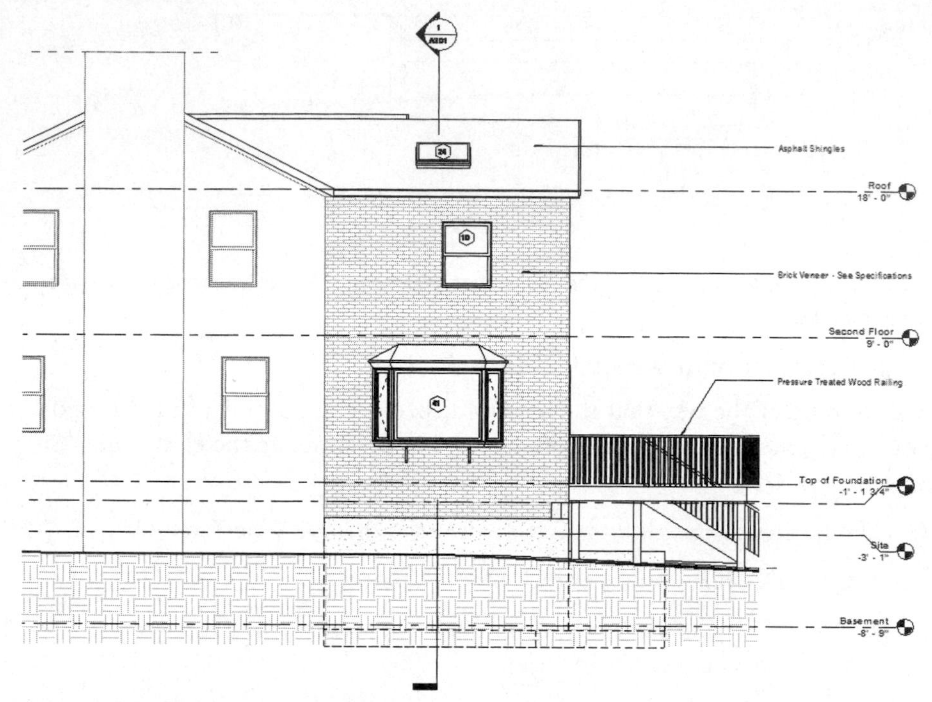

Figure 10–57 *Add tags and notes to complete the elevation*

17. Perform similar edits in other elevations if you wish.
18. Save the project.

SUMMARY

Understanding the relationship between modeled elements and drafted elements is an important concept in Autodesk Revit Building. Creating the basic model geometry can be accomplished in nearly any convenient View and as we have seen throughout this book, will remain coordinated as changes occur in all Views. Drafting and Annotation on the other hand occurs in only the currently active View. This means that we can apply additional embellishments on top of an automatically generated model to explain and clarify design intent. We can modify the display of underlying model geometry using the Linework tool or View-specific display settings. Finally, we can create Drafting Views which contain only drafting elements and no model geometry. Using a combination of these techniques, we can use any Revit Building View within our complete set of Architectural Construction Documents.

Detailing occurs at many levels in Autodesk Revit Building: as part of the model, as View-specific embellishment on top of the model and as completely independent Drafting Views.

You can use Reveals and other Wall Type edits to add details to the Walls throughout the model.

Create Callout Views of any overall View to begin detailing it.

Each View has its own scale and visibility settings.

Detail Components and Detail Lines are View-specific embellishments that are used to convey design intent.

Repeating Detail Components save time by adding an array of Detail Components at a set spacing.

Use Filled Regions in any View to mask unwanted portions of the model and apply patterns.

Add Break Lines and adjust Crop Regions to isolate "typical" portions of Detail Views.

Use Edit Cut Profile to modify the automatically created profile of building model elements in a particular View.

Adding Details to a Sheet automatically numbers them and keeps the annotation coordinated.

Import legacy CAD details and add them to Drafting Views.

Edit overall section and elevation Views using similar techniques used to create and modify Detail Views.

CHAPTER 11

Working with Schedules

INTRODUCTION

Schedules are an important part of any architectural document set. In traditional architectural practice (even when using computers to produce construction documents), generating schedules is a laborious process of manually tabulating the sometimes hundreds of items in a project that require presentation in a schedule. For example, in firms that do not use BIM software, a Door Schedule involves the painstaking process of manually listing each door specified in a project and then typing in detailed information about each entry—even those bits of information (such as size) that should be easily queried from the plans. In Autodesk Revit Building, Schedules are generated automatically from the building model data already in the project. A Schedule in Revit Building is just another View of the project that differs from plans and elevations only in its presentation as tabular information rather than graphics. You can create a Schedule from nearly any meaningful information in the model and like all Revit Building Views, you can edit in one View and see the change instantly in all Views, including the Schedule.

OBJECTIVES

In Chapter 4, we imported some typical Schedules and placed them in our commercial project on a Sheet. In this chapter, we will explore the workings of these Schedules as well as create additional Schedules not yet in our project. After completing this chapter you will know how to:

- Add and modify a Schedule View
- Edit model data from a Schedule View
- Place a Schedule on a Sheet
- Work with Tags
- Work with Rooms and Color Fills

CREATE AND MODIFY A NEW SCHEDULE VIEW

In this chapter, we will return to our commercial project and look deeper into the Scheduling tools in Autodesk Revit Building. As was mentioned above, we added

Schedules to this project back in Chapter 4. However, we simply imported those Schedules into the commercial project from an existing template project. Before we re-visit those Schedules, let's begin with a look at how to create a new Schedule View from scratch.

INSTALL THE CD FILES AND OPEN A PROJECT

The lessons that follow require the dataset included on the Mastering Revit Building CD ROM. If you have already installed all of the files from the CD, simply skip down to step 3 below to open the project. If you need to install the CD files, start at step 1.

1. If you have not already done so, install the dataset files located on the Mastering Revit Building CD ROM.

   Refer to "Files Included on the CD ROM" in the Preface for instructions on installing the dataset files included on the CD.

2. Launch Autodesk Revit Building from the icon on your desktop or from the **Autodesk** group in **All Programs** on the Windows Start menu.

   - From the File menu, choose **Close**.

This closes the empty new project that Revit Building creates automatically upon launch.

3. On the Standard toolbar, click the Open icon.

TIP The keyboard shortcut for Open is CTRL + O. **Open** is also located on the File menu.

   - In the "Open" dialog box, click the *My Documents* icon on the left side.
   - Double-click on the *MRB* folder, and then the *Chapter11* folder.

   If you installed the dataset files to a different location than the one listed here, use the "Look in" drop down list to browse to that location instead.

4. Double-click *11 Commercial.rvt* if you wish to work in Imperial units. Double-click *11 Commercial Metric.rvt* if you wish to work in Metric units

   You can also select it and then click the Open button.

The project will open in Revit Building with the last opened View visible on screen.

ADD A SCHEDULE VIEW

A Schedule is just another View in Revit Building. To add a new one, look for the Schedule/Quantitics tool on the View tab of the Design Bar.

5. On the Project Browser, double-click to open the *Level 3* floor plan View.

As you can see, on the third floor we have the beginnings of a furniture layout.

- Zoom In Region on the lower left portion of the plan (the part with furniture already in place).

In Chapter 9, we added some furniture to this plan. In that chapter the focus was on working with Component Families, we were not too concerned with actual furniture placement. As you can see now, some of the layout has been refined a bit, but it is not yet complete. The focus of this chapter will be on Schedules. But as we work on our first Schedule, we will make some adjustments to and complete the furniture layout of our tenant space on the third floor. Before we add the Schedule, let's take a quick look at what we have so far.

- Pre-highlight any item in the reception space (just below the elevators).

Each of these items, including the custom reception desk, is simply a Component Family placed in the model. You can see their names in the tool tip accompanying the pre-highlight.

- Pre-highlight any item in the upper office (with furniture in it) on the left.

Notice that this time, the entire collection of furniture in this office highlights together (see Figure 11–1). This is a Model Group named "Typical Office." You can also see it beneath the Groups node of the Project Browser.

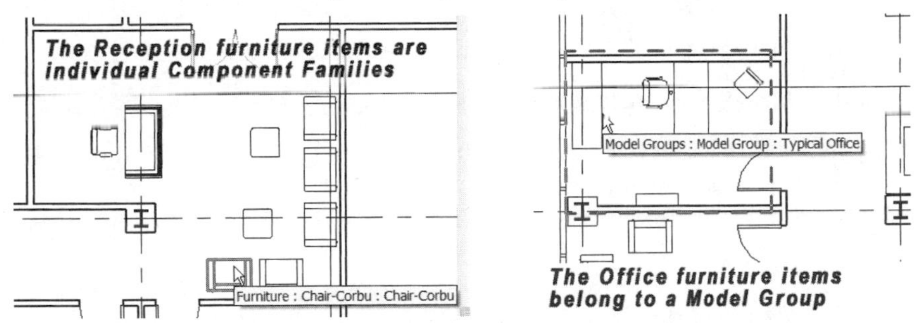

Figure 11–1 *Explore the composition of various furniture items in the plan*

We explored the concept of Model Groups briefly in Chapter 5. In that chapter we grouped the typical framing members and then copied them to each level. Now we will take a more practical look at the use of Model Groups in our project. The concept of Groups is simple, select one or more elements, click the Group icon on the toolbar and they become grouped. Later, if you wish to edit the composition of the Group, simply select it, click the Edit Group button on the Options Bar and then make changes. When you finish the changes and close the Group, it will update all instances of the Group in your project. A Group is not a Family, even though their behaviors are similar. Groups are simply a named collection of elements. They have no parameters of their own and cannot have "Types" as Families can. Groups are a good way to manage "Typical" conditions (such as this common furniture layout) in a project. By using Groups you can avoid having to make the same edit repeatedly when a change in layout occurs.

The Model Group is being re-introduced here because this is a good example of how useful they can be. It is also important to understand that Grouped elements will appear in Schedules. So let's begin making changes. As we make adjustments to the furniture layout, let's also begin tracking the various pieces of furniture in a Schedule.

The "Category" list is a fixed list of categories built into Revit Building. In the "Name" field you can type any suitable name to describe the contents of the Schedule—usually the default is acceptable. Schedules can list building elements directly from the model listing each one in a row of the Schedule. To do this, you choose the "Schedule building components" radio button. If you would rather group items by a common collection of parameters that a group of items in the model shares in common, you can instead choose the "Schedule keys" option. We will look at this option later. In a multi-phase project, you can limit what the Schedule includes by choosing a Phase from the list. It is most common to Schedule the New Construction Phase.

6. On the Design Bar, click the View tab and then click the Schedule/Quantities tool.

- In the "New Schedule" dialog, Select "Furniture" from the Category list.
- In the "Name" field, accept the default name of "Furniture Schedule."
- Accept the remaining defaults of "Schedule building components" and "New Construction" for the Phase (see Figure 11–2).

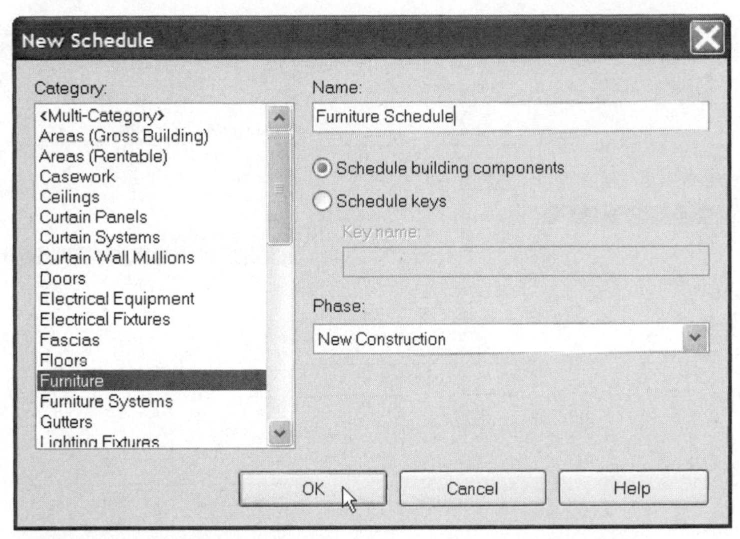

Figure 11–2 *Create a Furniture Schedule*

7. Click OK in the "New Schedule" dialog to proceed.

In the "Schedule Properties" dialog that appears, there is a list of "Available Fields" on the left side. This list includes the most common parameters available for the items you are scheduling (furniture in this case). To add a field to a Schedule, simple select it from the list and then click the Add button to move it to the "Scheduled fields" list on the right. Once you add fields, they can be organized in the desired display order on the right. Each field will become a column in the resultant Schedule View.

- In the "Available fields" list, choose Mark and then click the Add × button.
- Repeat for "Manufacturer," "Model," "Cost," "Family and Type" and "Comments."
- Be sure the order of fields is as listed in Figure 11–3. If it is not, select a field that you wish to move and then click the "Move Up" or "Move Down" buttons.

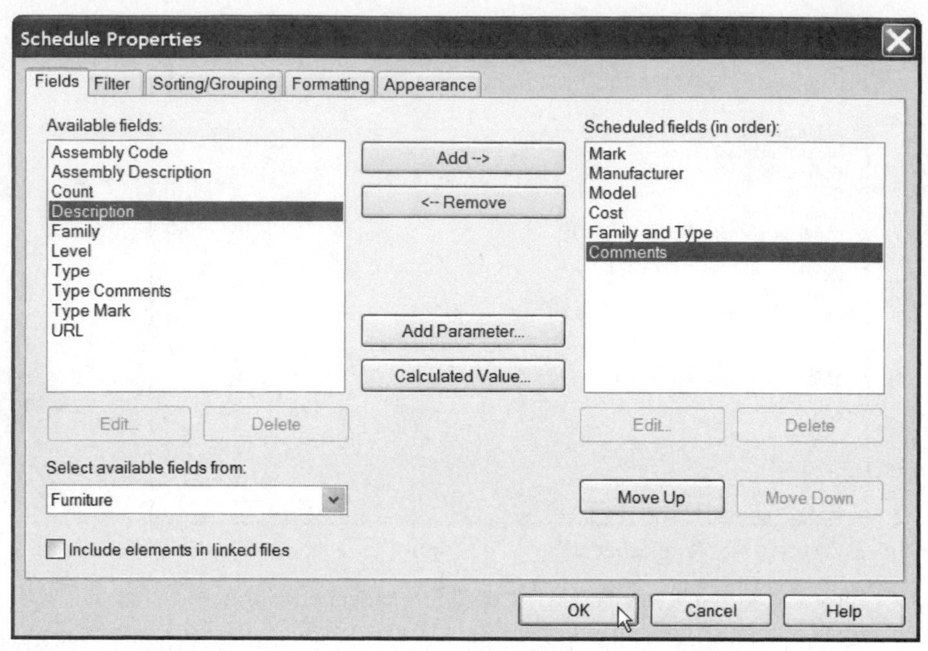

Figure 11–3 *Add Fields to the Furniture Schedule*

8. Click OK in the "Schedule Properties" dialog to create the Schedule.

A new Schedule View will appear beneath the *Schedules/Quantities* node of the Project Browser and this View will open on screen. The name of the Schedule will appear at the top and directly beneath it, each field will be listed as a column header. If you scroll the Schedule up or down, the title and header will remain at the top as hidden rows come in and out of view. The overall look of the Schedule View closely mimics a spreadsheet like Microsoft Excel. Like we did at the very beginning of the book in the Quick Start chapter, let's look at this Schedule View and the *Level 3* plan View tiled on screen together.

9. From the Window menu, choose **Close Hidden Windows.**
   - On the Project Browser, double-click to open the *Level 3* floor plan View.
   - From the Window menu, choose **Tile**.

Now that we can see the plan and the Schedule together on screen, we can explore some of the basic behaviors of Schedules in Revit Building.

EDIT MODEL ELEMENTS FROM THE SCHEDULE

This has been stated before, but it is worth repeating: elements in a Schedule are the same elements that we see graphically in the model. A Schedule is just another View of the model—a tabular View rather than graphical View.

10. Click on any filed in the Schedule View.

Notice how the corresponding furniture element highlights in the plan View.

- Repeat on as many elements as you wish.

11. Click in the "Mark" column for any element.

Notice how this is a simple text field into which you could begin typing a value.

- Click in the "Manufacturer" column next to "Mark."
- Click in each field in succession.

"Cost" is like "Mark" and is a simple text field. Notice that "Manufacturer" and "Model" however, appear with a drop-down list icon. However, you can still type in these fields. If you try to open the list, it will appear empty. The way these fields work is that you can type in any value, as in the plain text fields, but these values you type will begin populating a list that you can choose from in subsequent edits. Let's try it.

12. Click in the "Comments" field (at the right) for the third item in the Schedule (it should be a Secretary desk).

- Type: **Include Keyboard Tray and Footrest Option** and then press ENTER (see Figure 11–4).

Figure 11–4 *Input a value in a list field*

NOTE To better see the value in a particular field, you can place your cursor between two fields and then drag the column to widen it.

There is a second secretarial desk beneath the one we just edited.

- Click in the "Comments" field of the second secretarial desk and then click the drop-down list icon.

Notice that the note typed above now appears in the list. Each new item you type will be added automatically to this list making it easier to input existing values—simply choose them from the list. The parameter you are editing in the Schedule can be either an Instance parameter like the one edited here, or it can be a Type parameter. Instance parameters are applied independently to each instance of the item in the project. In this case, it would be possible to order a keyboard tray and footrest for one of the secretarial desks but not the other. With a Type parameter, the value applies to all instances of the Type in question. This is the case with "Manufacturer," "Model" and "Cost." The assumption here being that the Family and Type in question represents a particular item that can be ordered, purchased and installed in the project. Therefore, the Manufacturer, Model and Cost would, of course, be the same for all instances of that item. For our purposes, these behaviors are perfectly logical. However, in your own projects, if you wish to modify these behaviors, you must edit the Families of the elements in question. Let's look at Type parameter next.

13. Click in the "Manufacturer" field for the first item in the Schedule (it should be a desk).

   - Type: **Acme Furniture** and then press ENTER.

   A dialog will appear indicating that this change will apply to all instances of this Type (see Figure 11–5).

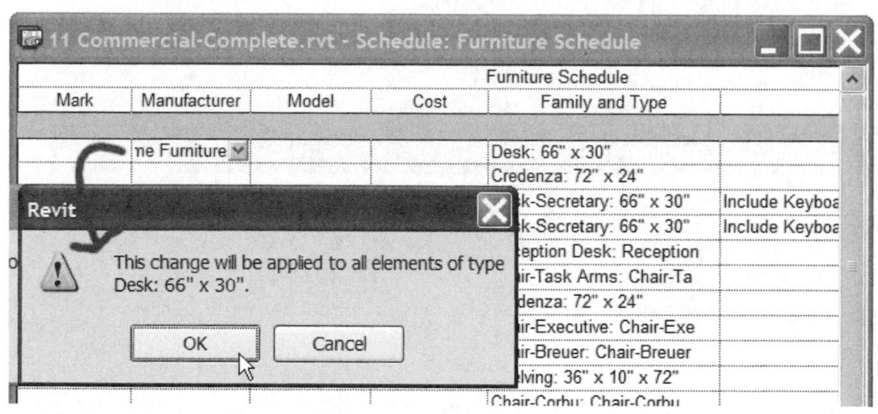

Figure 11–5 *Applying a change to a Type parameter*

   - Click OK to accept this change.
   - Scroll down in the Schedule and notice that the change has applied to several desks.
   - Repeat the process to add a Model designation and Cost to the same Desk.

14. From the "Manufacturer" column, the value "Acme Furniture" now appears. Apply this value to the Chair-Executive [M_Chair-Executive] Family.

   - Again, when you make this edit, it will apply at the Type level. Click OK to confirm the change.

- Add a Model and Cost to the Chair as well (see Figure 11–6).

11 Commercial-Complete.rvt - Schedule: Furniture Schedule

Furniture Schedule				
Mark	Manufacturer	Model	Cost	Family and Type
				Chair-Task Arms: Chair-Ta
				Chair-Task Arms: Chair-Ta
				Chair-Corbu: Chair-Corbu
				Chair-Corbu: Chair-Corbu
				Table-End: 24" x 24"
				Table-End: 24" x 24"
				Table-End: 24" x 24"
	Acme Furniture	ExecuDesk	1500.00	Desk: 66" x 30"
	Acme Furniture	ExecuChair	800.00	Chair-Executive: Chair-Exe
				Chair-Breuer: Chair-Breuer
				Chair-Breuer: Chair-Breuer
				Shelving: 36" x 10" x 72"
	Acme Furniture	ExecuDesk	1500.00	Desk: 66" x 30"
				Credenza: 72" x 24"
	Acme Furniture	ExecuChair	800.00	Chair-Executive: Chair-Exe
				Chair-Breuer: Chair-Breuer
				Shelving: 36" x 10" x 72"
	Acme Furniture	ExecuDesk	1500.00	Desk: 66" x 30"
				Credenza: 72" x 24"
	Acme Furniture	ExecuChair	800.00	Chair-Executive: Chair-Exe
				Chair-Breuer: Chair-Breuer
				Shelving: 36" x 10" x 72"

Figure 11–6 *Changes to Type parameters apply to all instances*

15. Using the same process, assign a Manufacturer, Model and Cost to the Corbu and Breuer chairs.

16. Save the project.

SORTING AND GROUPING SCHEDULE ITEMS

Since many of the values in Schedule are identical for all instances of a particular item, it might be nice to sort and group some of these values in the Schedule instead of listing each element separately and showing so much repetition.

17. Right-click anywhere in the Schedule View window and choose **View Properties**.

TIP This can also be done from the Project Browser.

Beneath the "Other" grouping in the "Element Properties" dialog, are five items each with an "Edit" button next to them. These five items correspond to the five tabs that were in the "Schedule Properties" dialog shown in Figure 11–3. We only looked at the "Fields" tab above. Each of these tabs can be accessed here anytime. Using the settings for Sorting/Grouping, we can modify the Schedule to group elements with common parameters into a single entry. We can also sort the Schedule in a variety of ways.

- Next to "Sorting/Grouping," click the Edit button.

NOTE You can see an example below in Figure 11–10.

This returns us to the "Schedule Properties" dialog (that we saw above) open to the Sorting/Grouping tab.

18. At the top of the dialog, from the "Sort by" list, choose **Manufacturer** and then click OK twice.

- Scroll to the bottom of the Schedule to see the results.

All of the blank fields (which are listed first in an alphabetic sort) are now at the top of the list, and the "Acme Furniture" items are next. The Desks and Chairs are still interspersed. If you wish, you can sort by more than one criterion.

19. Return to the Sorting/Grouping tab of "Schedule Properties" dialog (right-click and choose **View Properties**, then Edit Sorting/Grouping again).

- At the top of the dialog, beneath "Sort by" choose **Model** from the "Then by" list and then click OK twice (see Figure 11–7).

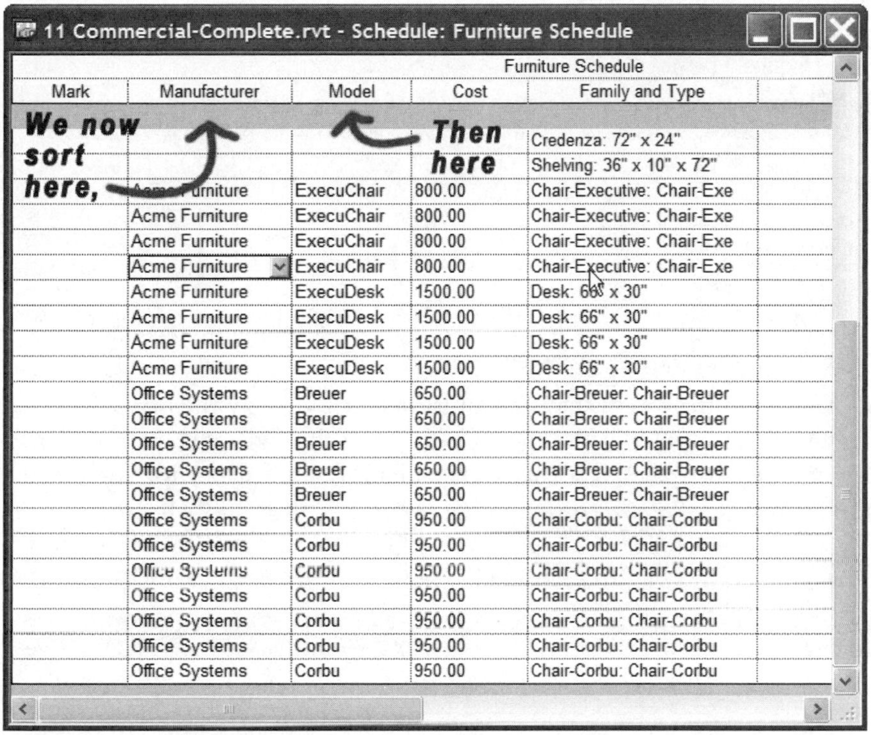

11 Commercial-Complete.rvt - Schedule: Furniture Schedule

We now sort here,

Then here

Furniture Schedule				
Mark	Manufacturer	Model	Cost	Family and Type
				Credenza: 72" x 24"
				Shelving: 36" x 10" x 72"
	urniture	ExecuChair	800.00	Chair-Executive: Chair-Exe
	Acme Furniture	ExecuChair	800.00	Chair-Executive: Chair-Exe
	Acme Furniture	ExecuChair	800.00	Chair-Executive: Chair-Exe
	Acme Furniture	ExecuChair	800.00	Chair-Executive: Chair-Exe
	Acme Furniture	ExecuDesk	1500.00	Desk: 66" x 30"
	Acme Furniture	ExecuDesk	1500.00	Desk: 66" x 30"
	Acme Furniture	ExecuDesk	1500.00	Desk: 66" x 30"
	Acme Furniture	ExecuDesk	1500.00	Desk: 66" x 30"
	Office Systems	Breuer	650.00	Chair-Breuer: Chair-Breuer
	Office Systems	Breuer	650.00	Chair-Breuer: Chair-Breuer
	Office Systems	Breuer	650.00	Chair-Breuer: Chair-Breuer
	Office Systems	Breuer	650.00	Chair-Breuer: Chair-Breuer
	Office Systems	Breuer	650.00	Chair-Breuer: Chair-Breuer
	Office Systems	Corbu	950.00	Chair-Corbu: Chair-Corbu
	Office Systems	Corbu	950.00	Chair-Corbu: Chair-Corbu
	Office Systems	Corbu	950.00	Chair-Corbu: Chair-Corbu
	Office Systems	Corbu	950.00	Chair-Corbu: Chair-Corbu
	Office Systems	Corbu	950.00	Chair-Corbu: Chair-Corbu
	Office Systems	Corbu	950.00	Chair-Corbu: Chair-Corbu
	Office Systems	Corbu	950.00	Chair-Corbu: Chair-Corbu

Figure 11–7 *Sorting the Schedule first by Manufacturer, then by Model*

Several items in the Schedule still do not have parameters assigned. Let's edit another value and see how the Schedule responds with our current sort settings.

20. Locate an instance of the "Chair-Task Arms" in the Schedule and edit the "Manufacturer" to **Furniture Concepts**.
    - Click OK in the dialog that appears.
    - Input a Model and Cost if you wish.

Notice that the Schedule immediately re-sorts to accommodate the new value. In addition to sorting the items in the Schedule, we can also group them when all of the values are the same like the variety of chairs we have here. It would be easier to work with the Schedule if we simply had one listing for each type of chair that reported how many of them there were instead of showing each as a separate item. We need to make two changes to display the Schedule this way.

21. Return to the Sorting/Grouping tab of "Schedule Properties" dialog.
    - At the bottom of the dialog, clear the "Itemize every instance" checkbox.
    - Click the Fields tab.

- Add the Count field to the bottom of the list and then click OK twice (see Figure 11–8).

11 Commercial-Complete.rvt - Schedule: Furniture Schedule

Furniture Schedule						
Mark	Manufacturer	Model	Cost	Family and Type	Comments	Count
						14
	Acme Furniture	ExecuChair	800.00	Chair-Executive: Chair-Exe		4
	Acme Furniture	ExecuDesk	1500.00	Desk: 66" x 30"		4
	Furniture Concept	ComfySeat	475.00	Chair-Task Arms: Chair-Ta		3
	Office Systems	Breuer	650.00	Chair-Breuer: Chair-Breuer		5
	Office Systems	Corbu	950.00	Chair-Corbu: Chair-Corbu		7

Figure 11–8 *Group items and show the count*

There are nearly limitless ways that we can format and display the same data. Notice the way the blank fields at the top (Count 14 in the figure) show none of the values even though we have data (particularly Family and Type) for some of the values. This is because we are sorting on the Manufacturer and Model fields and those fields are blank for the 14 items. If you edit the Schedule Properties and sort by Family and Type instead, the result will change dramatically (see Figure 11–9).

11 Commercial-Complete.rvt - Schedule: Furniture Schedule

Furniture Schedule						
Mark	Manufacturer	Model	Cost	Family and Type	Comments	Count
	Office Systems	Breuer	650.00	Chair-Breuer: Chair-Breuer		5
	Office Systems	Corbu	950.00	Chair-Corbu: Chair-Corbu		7
	Acme Furniture	ExecuChair	800.00	Chair-Executive: Chair-Exe		4
	Furniture Concept	ComfySeat	475.00	Chair-Task Arms: Chair-Ta		3
				Credenza: 72" x 24"		4
				Desk-Secretary: 66" x 30"	Include Keyboard Tray and Footr	2
	Acme Furniture	ExecuDesk	1500.00	Desk: 66" x 30"		4
				Reception Desk: Reception		1
				Shelving: 36" x 10" x 72"		4
				Table-End: 24" x 24"		3

Figure 11–9 *Sort by Family and Type instead of Manufacturer*

HEADERS, FOOTERS AND GRAND TOTALS

It is not necessary to add the Count Field as we have done here to see counts of the various items. You add the Count Field when you wish to have the count inline with the rows of the Schedule. If you instead wish to show totals above or beneath the items you are totaling, you can add Headers and Footers and Grand Totals to the Schedule instead.

22. Return to the Sorting/Grouping tab of "Schedule Properties" dialog.
    - If you changed the first "Sort by" criterion to Family and Type, change it back to Manufacturer.
    - Beneath Manufacturer, place a checkmark in the "Header" checkbox.

- Beneath Model, place a checkmark in the "Footer" checkbox and from the drop-down list that appears choose **Title, count and totals**.
- At the bottom of the dialog, place a checkmark in the "Grand totals" checkbox and accept the default of **Title, count and totals** (see Figure 11–10).

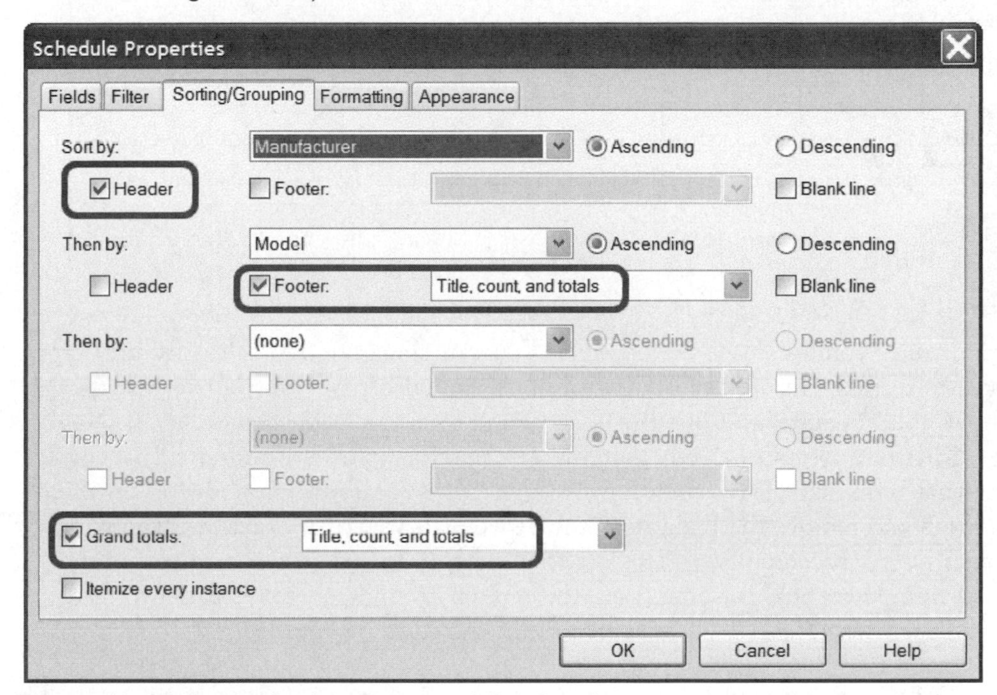

Figure 11–10 *Enable Headers, Footers and Grand totals*

23. Click OK twice to return to the Schedule View window and study the results (see Figure 11–11).

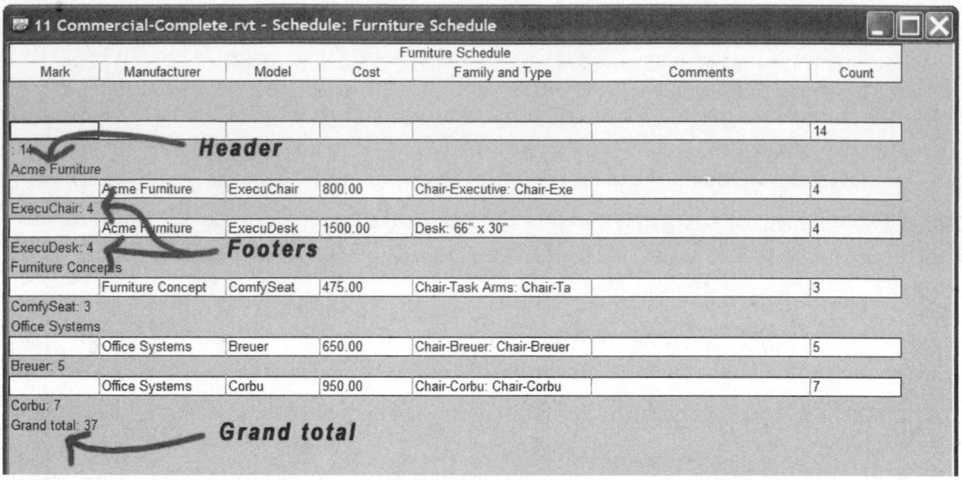

11 Commercial-Complete.rvt - Schedule: Furniture Schedule

Furniture Schedule

Mark	Manufacturer	Model	Cost	Family and Type	Comments	Count
						14
: 14						
Acme Furniture						
	Acme Furniture	ExecuChair	800.00	Chair-Executive: Chair-Exe		4
ExecuChair: 4						
	Acme Furniture	ExecuDesk	1500.00	Desk: 66" x 30"		4
ExecuDesk: 4						
Furniture Concepts						
	Furniture Concept	ComfySeat	475.00	Chair-Task Arms: Chair-Ta		3
ComfySeat: 3						
Office Systems						
	Office Systems	Breuer	650.00	Chair-Breuer: Chair-Breuer		5
Breuer: 5						
	Office Systems	Corbu	950.00	Chair-Corbu: Chair-Corbu		7
Corbu: 7						
Grand total: 37						

Figure 11–11 *Study the results of Headers, Footers and Grand totals in the View*

Several other combinations are possible. You can change either Manufacturer or Model to display Headers or Footers or both. You also have control over whether it shows totals, counts and/or titles. The Grand total can be turned on or off. Notice with these items enabled, that the information in the Count field is now somewhat redundant. You can return to the "Schedule Properties" dialog on the Fields tab and remove the Field from the Schedule if you wish. Finally, if you wish to put a bit of space between each sort criterion, you can place a checkmark in the "Blank line" checkbox. In some cases, this will make your sub-totals easier to read; particularly when there are many items in the Schedule.

DUPLICATING A SCHEDULE VIEW

The only problem with the current format of our Schedule View is that while it presents the data in a nice format for quickly grasping the salient information, it does not lend itself as easily to editing as the original format did. If we wished to complete the input of Manufacturers, Models and Costs begun above on the remaining 14 items, the current sorting does not allow use to edit them separately. We could edit the Schedule back to its previous format, but a better approach is to create a duplicate Schedule solely for the purposes of editing the model data.

24. On the Project Browser, right-click *Furniture Schedule* and choose **Duplicate**.
    - Right-click *Copy of Furniture Schedule* and choose **Rename**.
    - In the "Rename View" dialog, type **Furniture Schedule (Working)** and then click OK.
25. On the Project Browser, right-click *Furniture Schedule (Working)* and choose **Properties**.

- Click the Edit button next to Sorting/Grouping.
- Place a checkmark in the "Itemize every instance" checkbox and then click OK twice.

Notice that we did not need to remove the Headers, Footers or Totals to return to showing every instance separately. Now when we wish to modify an item, we can open the "Working" version of the Schedule to see each item listed separately which will make it easier to select and edit what we need. However, regardless of where we make the edit, it will show in all Views—both Schedules and the model.

26. Edit the remaining items to add Manufacturers, Models and Costs (see Figure 11–12).

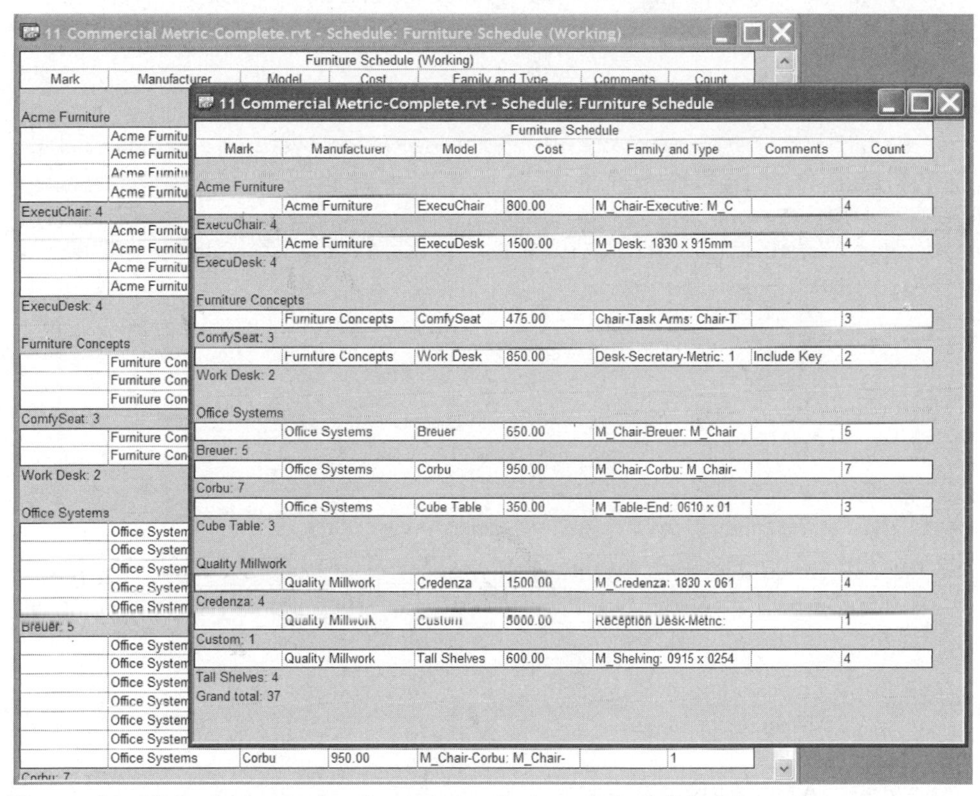

Furniture Schedule						
Mark	Manufacturer	Model	Cost	Family and Type	Comments	Count
Acme Furniture						
	Acme Furniture	ExecuChair	800.00	M_Chair-Executive: M_C		4
ExecuChair: 4						
	Acme Furniture	ExecuDesk	1500.00	M_Desk: 1830 x 915mm		4
ExecuDesk: 4						
Furniture Concepts						
	Furniture Concepts	ComfySeat	475.00	Chair-Task Arms: Chair-T		3
ComfySeat: 3						
	Furniture Concepts	Work Desk	850.00	Desk-Secretary-Metric: 1	Include Key	2
Work Desk: 2						
Office Systems						
	Office Systems	Breuer	650.00	M_Chair-Breuer: M_Chair		5
Breuer: 5						
	Office Systems	Corbu	950.00	M_Chair-Corbu: M_Chair-		7
Corbu: 7						
	Office Systems	Cube Table	350.00	M_Table-End: 0610 x 01		3
Cube Table: 3						
Quality Millwork						
	Quality Millwork	Credenza	1500.00	M_Credenza: 1830 x 061		4
Credenza: 4						
	Quality Millwork	Custom	5000.00	Reception Desk-Metric:		1
Custom: 1						
	Quality Millwork	Tall Shelves	600.00	M_Shelving: 0915 x 0254		4
Tall Shelves: 4						
Grand total: 37						

Figure 11–12 *Study the results of completing the input in both Schedules*

As you apply a Manufacturer the item will be removed from the group of non-defined items at the top and immediately sort in the groups below. If you wish, you can turn off the sorting to prevent this in the "Working" Schedule.

27. Save the project.

THE FILTER TAB

So far we have explored the high level of flexibility in our two Schedules from just the Fields and Sorting/Grouping tabs. There are other settings that we can apply in the remaining three tabs of the "Schedule Properties" dialog. For example, sometimes you only need part of the information available. Suppose you were on the phone with your sales representative from Office Systems. You could create a version of your furniture Schedule that listed only the items that you are specifying from this manufacturer.

28. Right-click on *Furniture Schedule (Working)* and choose **Properties**.
    - In the "Element Properties" dialog, click the Edit button next to "Filter."
    - From the "Filter by" list, choose **Manufacturer**.
    - Leave the next list set to "equals."
    - From the list below Manufacturer, choose **Office Systems** and then click OK twice (see Figure 11–13).

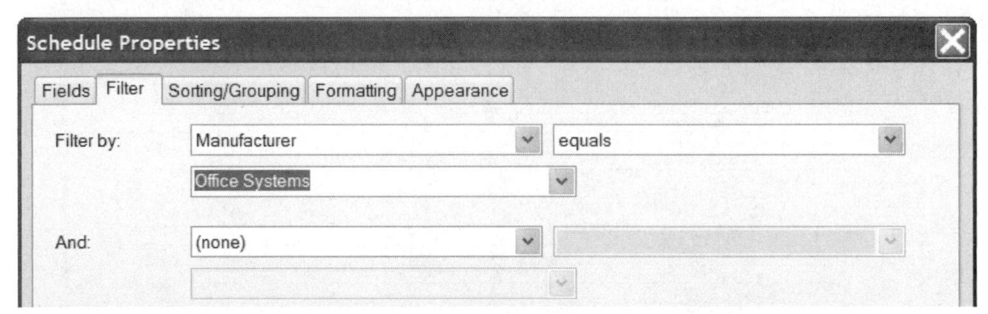

Figure 11–13 *Set the Schedule to Filter by a particular Manufacturer*

Notice that the Schedule now only shows the items from Office Systems. You can try other Filters if you wish. Note that the operator list includes many options like "not equal to," "contains" and "greater" or "less than.."

29. Return to the Filter tab and remove the Filters from the Schedule when you are done experimenting.

FORMATTING AND APPEARANCE

In the interest of completeness, let's take a brief look at the remaining two tabs. The settings on these tabs are straightforward. On the Formatting tab you can configure the orientation and alignment of each individual field. On the Appearance tab, you configure the look of the overall Schedule.

30. On Project Browser, right-click on *Furniture Schedule* and choose **Properties** (not the "Working" one this time).

- Click the Edit button next to Formatting.

31. Click on the Cost field in the list at the left.
    - Change the "Alignment" to **Right**.
    - Place a checkmark in the "Calculate totals" checkbox.
32. Click on the Family and Type field in the list at the left.
    - Place a checkmark in the "Hidden field" checkbox.
33. Click on the Count field in the list at the left.
    - Change the "Header orientation" to **Vertical** (see Figure 11–14).

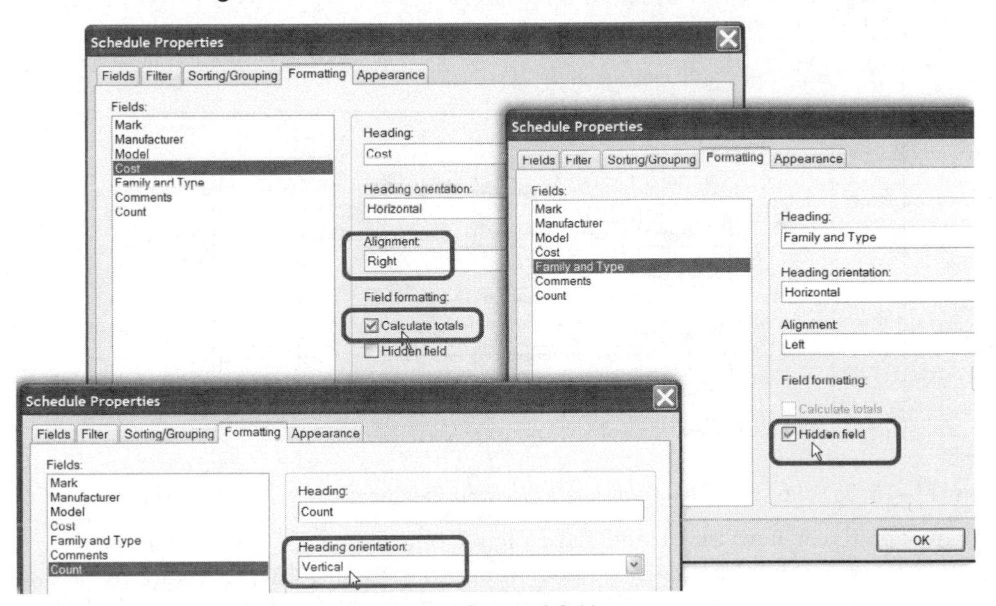

Figure 11–14 *Set the Formatting options of several fields*

34. Click OK twice to return to the Schedule View window.

Notice that most of the formatting shows immediately in the View window. However, the changing of the "Header orientation" to Vertical for Count will only appear when we add this Schedule to a Sheet.

ADD A SCHEDULE TO A SHEET

To see the vertical orientation of the Count field and to see the effects of most of the settings on the Appearance tab, we need to add the Schedule to a Sheet. You add a Schedule View to a Sheet the same way as any other View.

35. On the Project Browser, create a new Sheet named: **A602 – Schedules** and then open it.

We already have one Schedule Sheet (A601) but it has several Schedules on it already. Creating a new Sheet will give us plenty of room to work.

36. From Project Browser, drag *Furniture Schedule* and drop it on the Sheet.

- Click a point near the top left corner of the Sheet to place it.

Notice that with the Schedule still selected, there are control handles at the top of each column. You can use these to interactively change the size of each field for purposes of the Sheet (see Figure 11–15). Notice also that you can see that the Count field header is in fact oriented vertically.

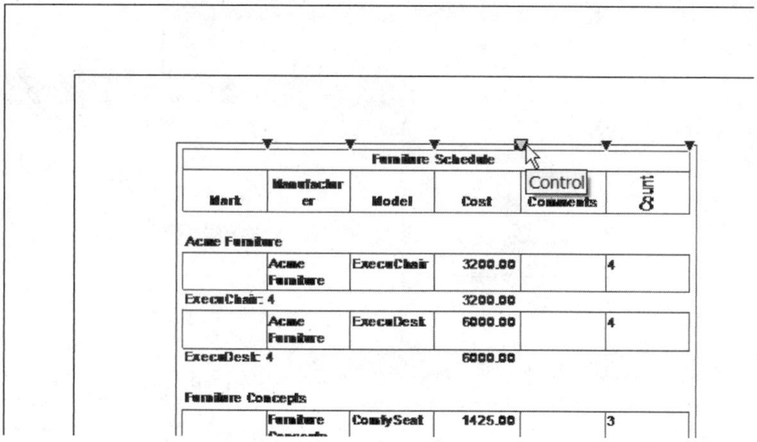

Figure 11–15 *Edit the size of the fields interactively on the Sheet*

Since we created a new Sheet and have plenty of room, let's resize each of the fields widths to prevent the data from wrapping to a second line as it does by default. To do this, you simply drag the control handles to the right. Work from left to right across the Schedule.

- Resize each column as required to prevent wrapping.

37. On the Project Browser, right-click the *Furniture Schedule* and choose **Properties**.

- Click the Edit button next to "Appearance."

If you wish to have the grid lines show continuously across the header and footer areas, check the "Grid in headers/footers/spacers" checkbox.

- Place a checkmark in the "Outline" checkbox and then next to it, choose **Wide Lines** from the list.
- Click OK twice to see the result (see Figure 11–16).

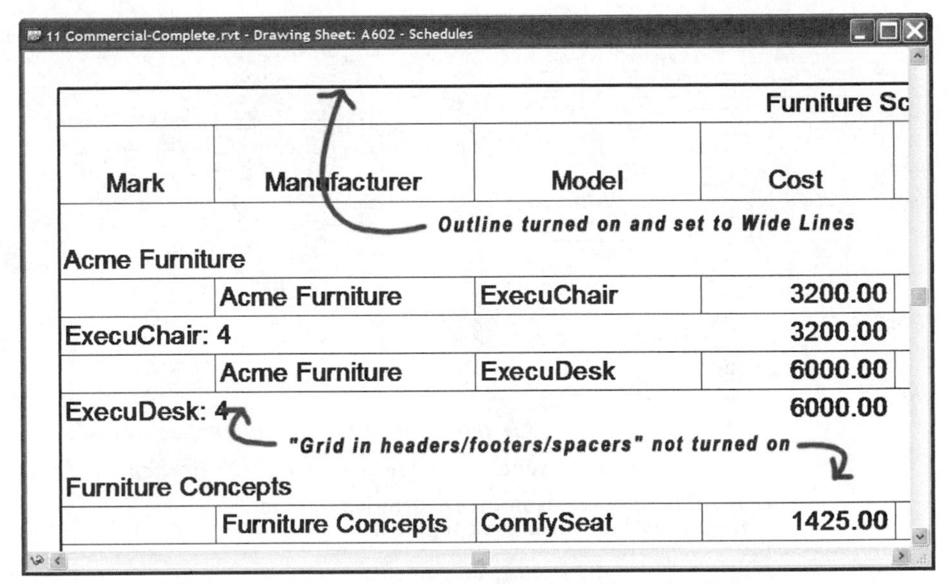

Figure 11–16 *Edits on the "Appearance" tab affect the look on the Sheet*

Return to the Appearance tab and make other edits if you wish to fine-tune the look of the Schedule on the Sheet. Notice that there are settings for the title, headers and body text. You can change the fonts, heights, and linework drawn for each component.

SPLIT A LONG SCHEDULE

In some cases, the data in the Schedule grows beyond the edge of the Sheet. You can break the Schedule into pieces to make it fit the Sheet better. (An example occurs on the existing *A601 – Schedules* Sheet).

38. On the Project Browser, double-click to open the *A603 - Schedules* Sheet View.

This Sheet contains two Schedules already, but the Wall Schedule overruns the Room Schedule beneath it. Let's split the Wall Schedule to make the Sheet legible.

39. Select the Wall Schedule on the Sheet.

   - On the right side, is a small control handle (looks like an "Z") click this control to split the Schedule.

Using the other control handles that appear, you can make additional adjustments as necessary. The Schedule can be split multiple times if required. To remove a split and put the two pieces back together, simply drag and drop one piece on top of another (see Figure 11–17).

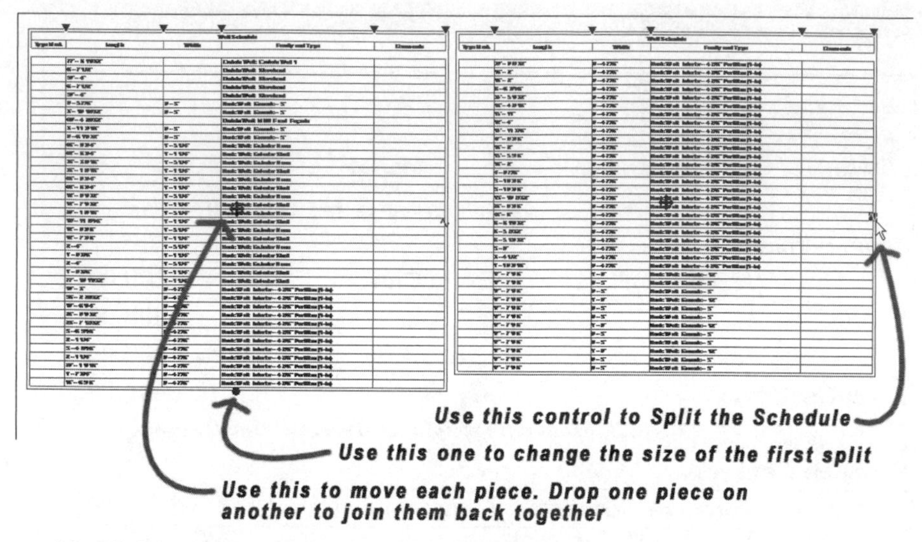
Use this control to Split the Schedule
Use this one to change the size of the first split
Use this to move each piece. Drop one piece on another to join them back together

Figure 11–17 *Split the Schedule and make other adjustments if needed*

If after splitting a Schedule it is still too long for the Sheet, you can use the Filter tab to edit the Schedule View. Filter by a number range or by floor to shorten the overall Schedule. Then duplicate this Schedule View and Filter the new copy to include the rest. Drag the two (or more Schedules) to different Sheets.

40. Save the project.

EDITING THE MODEL

Returning to our Furniture Schedule, let's see how it keeps up to date as we make modifications to the model. In this case, we will complete our third floor furniture layout and see how the Schedule updates automatically to keep pace.

ADD ELEMENTS TO THE MODEL

The space in the bottom middle of the plan is a conference room. It would be nice to place a table and chairs in this room and receive updated quantities in the Schedule.

1. Repeating the process from the start of the Chapter, open only the *Level 3* and *Furniture Schedule* Views (closing others) and then Tile them.
2. On the Design Bar, click the Basics tab and then click the Component tool.
   - On the Options Bar, click the Load button.
   - In the "Open" dialog, click the Web Library button.

This will load the Autodesk Web Library page over the Internet. If you do not have Internet access, the Family file in question has been provided with the files installed from the Mastering Autodesk Revit Building CD ROM. You can load the Family from there instead if you wish.

- Click the link for the "Revit 8.1 Library", and then click the "Furniture" link.
- Click the "Conference-Table2 w Chairs" link to download this Family file.
- In the "File Download" dialog, click the Open button.

The Family file will download and open in your Revit Building session.

3. On the Design Bar, click the Load into Projects button.

Revit Building will return to the commercial project and a "Reload shared family" dialog will appear. A Family file can contain other Families nested within it. In this case, the Task Chair Family that we already have in our project is also being used in the incoming Family file. When this occurs, you can use the Family as it is already defined in your current project, or you can replace the Family definition in the project with the one coming in. If you choose the "Family" option, items in your project could change if the two definitions do not match. On the other hand, if you choose the version in your project, it could have an impact on the items coming in and they may not look the way they did in the Family Editor. There is also a checkbox in this dialog where you can choose how any existing Family Types will behave. If you choose to override them (checking the box) then the Types in the project will be reset by the incoming Family. In this case, since we have applied Schedule parameters to the Types in our project already, we will choose to use the definitions already in our project.

- In the "Reload shared family" dialog, choose the "Project" version and then click OK to All (see Figure 11–18).

NOTE If you choose "OK" you may be prompted again if other nested Families existing in the incoming file. If you want your choice to apply to all nested elements, choose "OK to All."

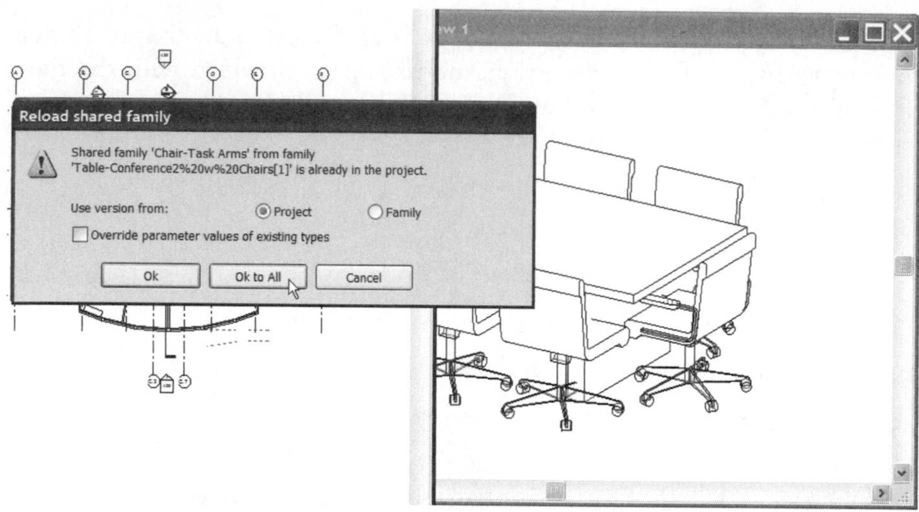

Figure 11–18 *Use the Project version of the shared Family*

The table and chairs will come in horizontally.

- Press the SPACEBAR to rotate it vertically.

4. Place the conference table in the conference room.

- On the Design Bar, click the Modify tool or press the ESC key twice (see Figure 11–19).

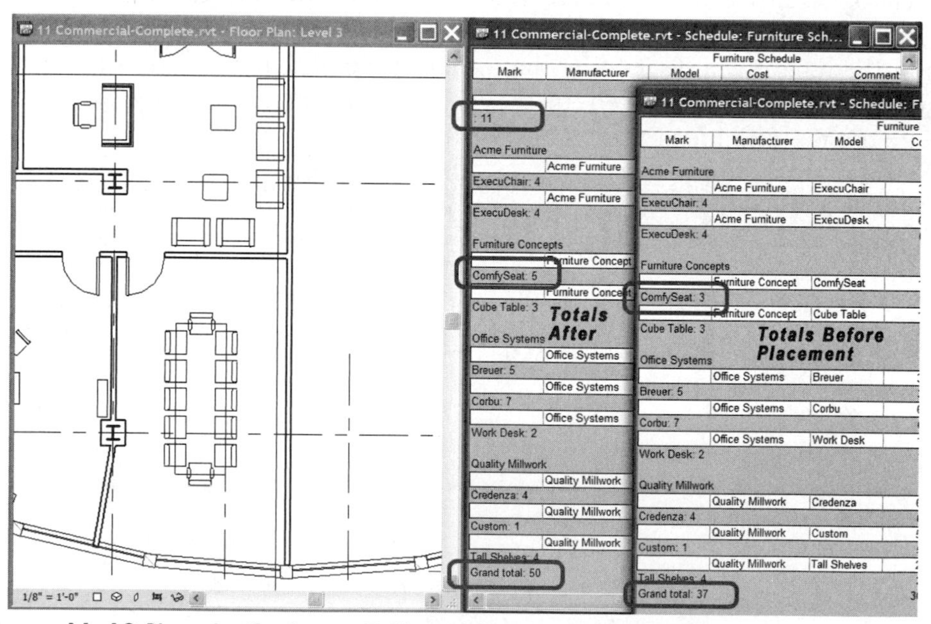

Figure 11–19 *Place the Conference Table and Chairs in the Conference Room*

You will notice that the new furniture items immediately appear in the Schedule. It may be necessary to undo and then redo to really see the change occur. Before placement, we had a total of 37 Furniture elements on our Schedule. After placement there are 13 more (1 table and 12 chairs) for a total of 50 elements. Two of the chairs, the ones at the ends, were added automatically to the "Furniture Concepts" group increasing the total for "ComfySeat" from 3 to 5. (Your model name may vary depending on what you specifically typed in the exercise above). These are the two chairs at the ends of the table and the ones we were prompted about in the dialog above. The remaining items—10 chairs and the table have no manufacturer, model or cost yet. So these list as a new blank line in the Schedule at the top with a total of 11 (see the right side near the top of Figure 11–19).

ADDING SCHEDULE PARAMETERS TO NESTED FAMILIES

If you select the new conference table, you will notice that it highlights the collection of chairs and table together. If you want to input manufacturer data for the chairs, you must select them individually. To do this is the plan View, you can use the TAB key; this can also be accomplished via the "Working" Schedule we created above, or on the Families branch of the Project Browser.

5. From the Project Browser, double-click the *Furniture Schedule (Working)* View to open it.

In this View, we can see each element listed separately and the Family and Type column is still visible. We can use the procedures listed above to add Manufacturer, Model and Cost data. We can also use the Project Browser. Since we have not tried this method yet, let's have a look now.

6. On the Project Browser, expand the *Families* branch and then the *Furniture* node (see Figure 11–20).

Beneath the Furniture node, all of the Furniture Families used in the project are listed. "Chair-Task" is the chair that repeats along the sides of the table. "Chair-Task Arms" is used at the ends and already has data assigned as we saw above. The table is "Table-Conference2 w Chairs[1]" (Note, when you download from the web, sometimes a space character is replaced with the html code %20 as seen in the figure).

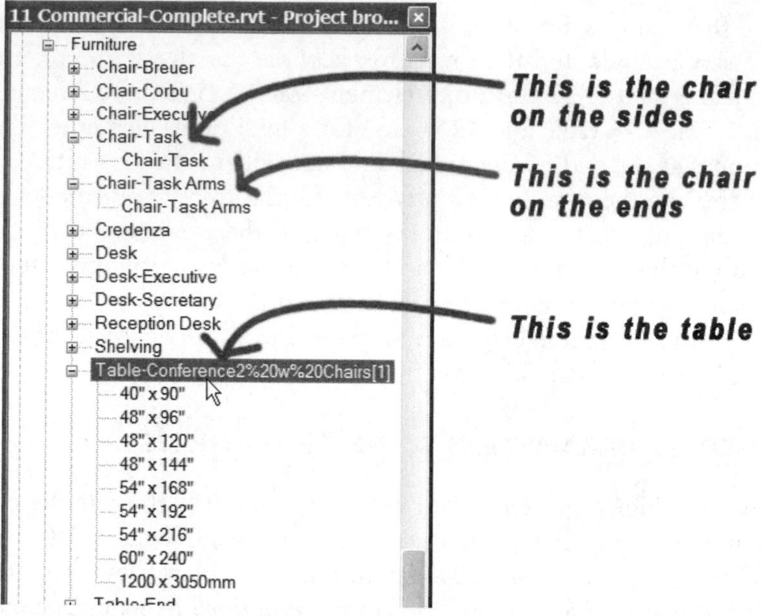

Figure 11–20 *Locate each of the items used in the Family*

7. Beneath the Chair-Task Family, right-click on the Chair-Task Type and choose **Properties**.

   - Input values for the Manufacturer, Model and Cost (see Figure 11–21).

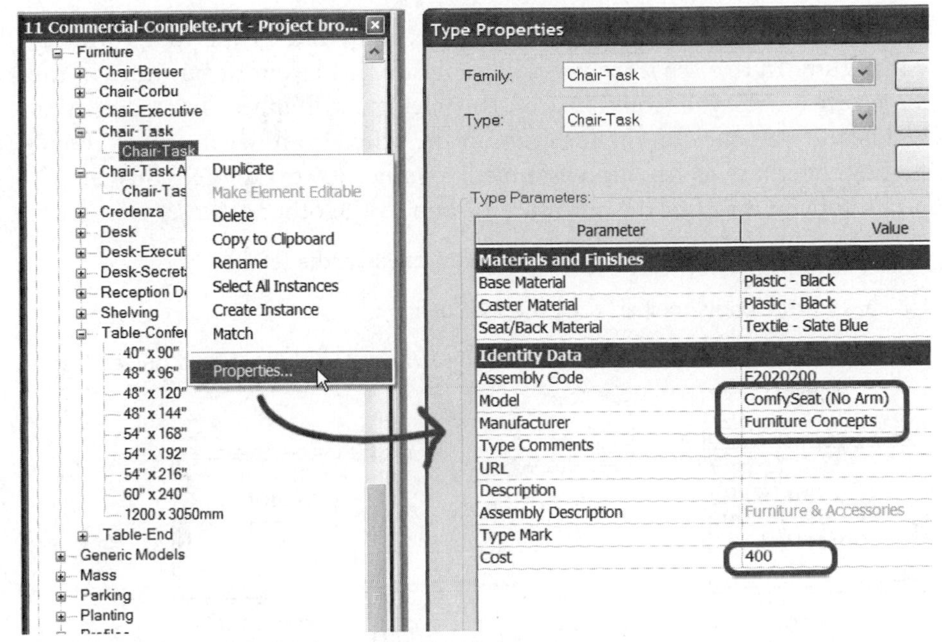

Figure 11–21 *Input data for the conference chairs*

8. Click OK to apply the change and update the model and Schedule.

SWAPPING TYPES AFFECTS THE SCHDULE

The Family that we just imported has several Types. Each Type includes a different quantity of chairs. If you choose a different Type, both the plan View and the Schedule will change together.

9. Select the conference table in the *Level 3* plan View.
   - From the Type Selector, choose a different Type of the Conference Table Family.

Notice the quantity of chairs will change in both the model and the Schedule. Experiment further if you wish.

- Settle on a Type for the project and then edit that Type to add a Manufacturer, Model and Cost.

10. Save the project.

COPY AND MIRROR GROUPS

On the left side of the plan View are five offices. The one at the bottom (the corner office) is unique, but the four above it are the same size and shape. Each of these four offices should have the same furniture. Only one of the four currently has any furniture. In this sequence, we will duplicate this layout of furniture to the other three offices. As you would expect, this change will appear immediately in the Schedules. If you click on the furniture in the office from which we will copy, it highlights together. It has been grouped together. We explored this briefly above. Let's copy and mirror this furniture Group to the other offices.

11. Select the Furniture Group in the typical office at the left.
    - On the Tools toolbar, click the Mirror icon.
    - For the "Axis of Reflection" click the Wall separating the two offices (see Figure 11–22).

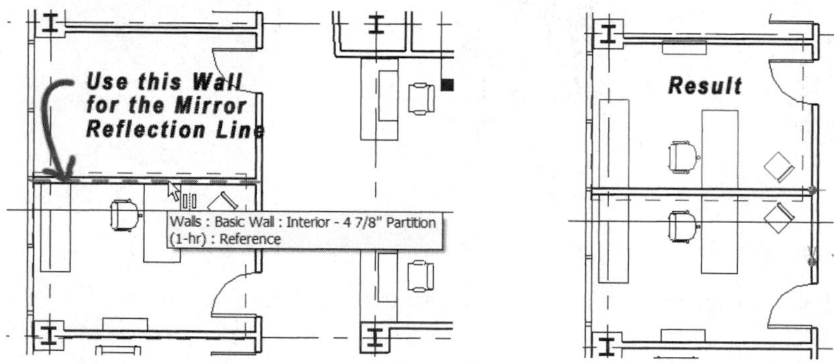

Figure 11–22 *Mirror the Group to the other office*

12. Select both Furniture Groups (the original and the one just copied) and mirror then up to the other two offices.
    - Check the Schedule to see the new quantities.

Since we copied items that already had Manufacturer data, all of the new items also have this data and merely update the existing quantities. We now have 65 items total.

NOTE Depending upon the specific Conference Table Type on which you settled above, your total may vary.

EDIT A GROUP

Suppose that after a presentation to the client, they decide to add an extra guest chair to each office. Since we used a Group, this change is easy.

13. Select one of the Typical Office furniture Groups in the plan View.
    - On the Options Bar, click the Edit Group button.

- Select the guest chair (the one at 45°) and mirror a copy relative to the center of the desk.
- Move it or otherwise fine tune the placement as necessary (see Figure 11–23).

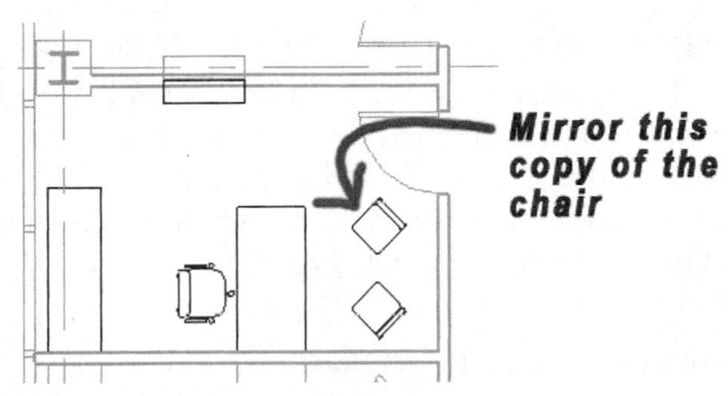
Mirror this
copy of the
chair

Figure 11–23 *Mirror a new copy of the guest chair*

14. On the Design Bar, click the Finish Group button.

Watch both the plan View and the Schedule as you do this. A new Breuer chair will appear in each office and the total on the Schedule will update to 14 to reflect this change.

15. Save the project.

WORKING WITH TAGS

Tags serve an important role in Architectural Documentation. They provide a means of easily and uniquely locating and identifying elements in a project. All sorts of tags are needed in Construction Document sets: Room Tags, Door Tags, Window Tags, etc. Autodesk Revit Building makes adding and managing Tags simple. Furthermore, the data displayed in the Tags comes directly from the objects in the same way as the data that feeds Schedules. Therefore, one need only add Tags to a View and let Revit Building handle the rest.

ADDING DOOR TAGS

We added a Door Schedule to this project back in Chapter 4. We have not yet added any Door Tags. Adding Tags to existing elements in the model is simple.

1. On the Design Bar, click the Drafting tab and then click the Tag tool.
   - Slowly move the pointer around the model pausing over different kinds of elements. Do not click yet.

Notice that the shape of the tag on your cursor changes with each element that you pre-highlight. Revit Building can have a different Tag loaded for nearly every type of element. Therefore, if you pre-highlight a Wall, it will automatically place a Wall Tag—pre-highlight a Door and you get a Door Tag and so on. You can place Tags with or without a leader attached, the default is with a leader.

- On the Options Bar, clear the "Leader" checkbox.
- Pre-highlight and then click on one of the Doors to the offices at the left.

Notice that when the Tag appears, it already has a number in it. The Doors were numbered automatically when they were added to the model. The tag simply queries the element (the door in this case) for the instance mark value and then displays the value in the tag. Tags can be created that query and report any of the parameters in the element. You can continue placing Tags manually using the same process. However, there is a quicker method.

2. On the Design Bar, click the Tag All Not Tagged tool.
   - In the "Tag All Not Tagged" dialog, click on Door Tag and then click OK (see Figure 11–24).

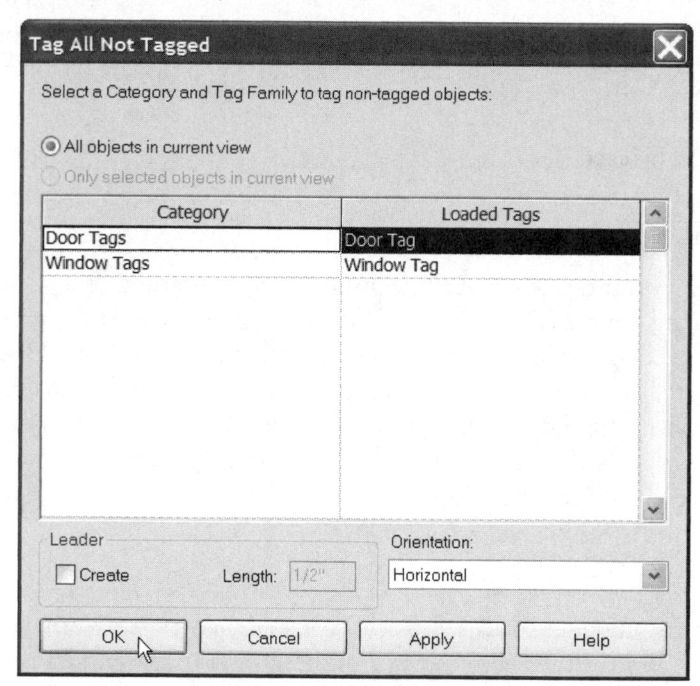

Figure 11–24 *Tag all Doors that are not currently Tagged*

Door Tags will appear on all Doors in the currently active View.

LOADING TAGS

Sometimes you want to Tag elements in your model, but an appropriate Tag is not loaded in your project. Like other Families, you can simply load a Tag from a library. Let's load a furniture Tag and then Tag some of our furniture.

3. From the Settings menu, choose **Annotations > Loaded Tags**.

Scroll through the list. All Tag types are listed. Tags that are already loaded are listed next to their respective type. For element types with no Tag loaded, it reads: "No Tag Loaded."

4. Click the Load button.
   - Browse to your library (Imperial or Metric) and open the *Annotations* folder.
   - Open the Architectural folder, select *Furniture Tag.rfa* [*M_Furniture Tag.rfa*] and then click Open.

NOTE As with other library content in this book, you can find these items with the files installed from the CD in the "*Autodesk Web Library*" folder.

5. Click OK to dismiss the "Tags" dialog.
6. On the Design Bar, click the Tag tool.
   - On the Options Bar, place a checkmark in the "Leader" checkbox.
   - Tag some of the chairs on the right in the reception area (see Figure 11–25).

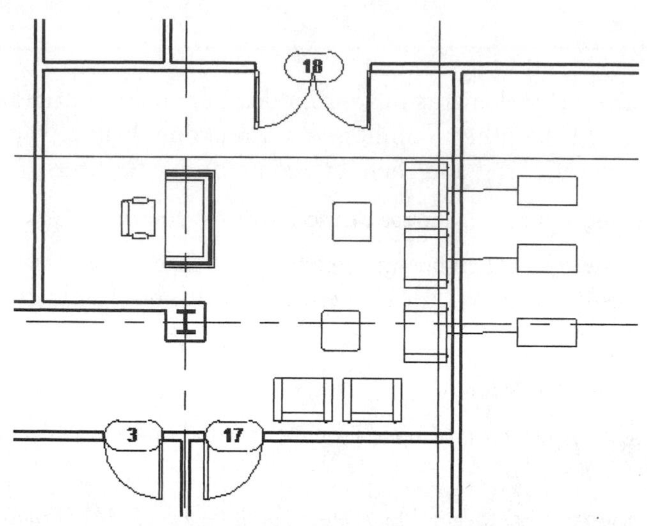

Figure 11–25 *Tag some chairs in the reception area*

Notice that the Tag appears blank. This is because we did not input a value for "Mark" in the sequence above. Furthermore, it turns out that there are two "Mark" parameters: "Mark" and "Type Mark." The names of these two parameters ought to make them easy to distinguish. The "Mark" parameter is an instance parameter and is the one we currently have in the Schedule. The "Type Mark" occurs as part of the Family Type and like other Type Parameters applies to all instances of that Type in the model. So for this sequence, let's edit the Type Mark of the chair we just tagged, and then edit the Schedule to show this field instead of the Mark field.

7. Using the procedures outlined above, edit the Type Mark parameter of the Corbu chair.

TIP Remember, you can edit from the Project Browser beneath the Families branch.

- Number this as item **1**.
- Tag the remaining Corbu chairs in the model.

Notice these newly Tagged chairs also report a Type Mark of 1 (see Figure 11–26).

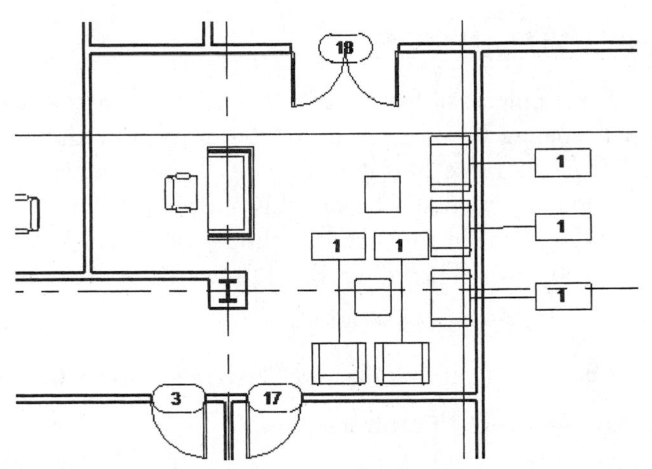

Figure 11–26 *Edit the Type Mark and tag the rest of the reception space chairs*

EDIT THE SCHEDULE

To remove the Mark field and add the Type Mark, we simply edit the Schedule View Properties. Making this edit on the *Furniture Schedule* will also appear on the Sheet. It will *not* appear on the "Working" version of the Schedule unless you also repeat the edit there. If you wish, you can simply add the Type Mark field to the "Working" Schedule instead of swapping it out.

8. On the Project Browser, right-click the *Furniture Schedule* and choose **Properties**.
   - Click the Edit button next to Fields.
   - On the right side, click on Mark and then click the ← Remove button.
   - On the left side, click on the Type Mark field and then click the Add → button.
   - On the right, with the Type Mark item selected, click the Move up button until it is at the top.
   - Click OK twice to return to finish.

Look for the Corbu chair entry in the Schedule, and notice that it appears with "1" in the newly added Type Mark column. Now that we have this column in the Schedule, it is even easier to add Type Marks to each furniture element.

9. Add Type Mark values to each item in the Schedule.
10. Tag a few other furniture items in the model.

CREATE A FURNITURE PLAN VIEW

You do not have to Tag all pieces of furniture in this View. In fact, this might be a good time to create a separate Furniture Plan for the third floor and perhaps even create an enlarged "Typical Office Furniture Layout" View. Doing so at a larger scale would eliminate lots of redundant tags in this View and allow us more room to place the Tags. At this point in the text, you should have the skills required to complete this task on your own. Therefore, we will review the overall steps only.

1. Delete the Furniture Tags added so far.
2. On Project Browser, right-click the Level 3 plan and choose **Duplicate**.
3. Rename this plan **Level 3 Furniture Plan**.
4. Tag all of the unique furniture in the View—reception, secretarial areas, conference room, etc.
5. Create a Callout plan at lager scale of one typical office—call it: **Typical Office Furniture Plan**.
6. Add Tags in the View to complete it (see Figure 11–27).

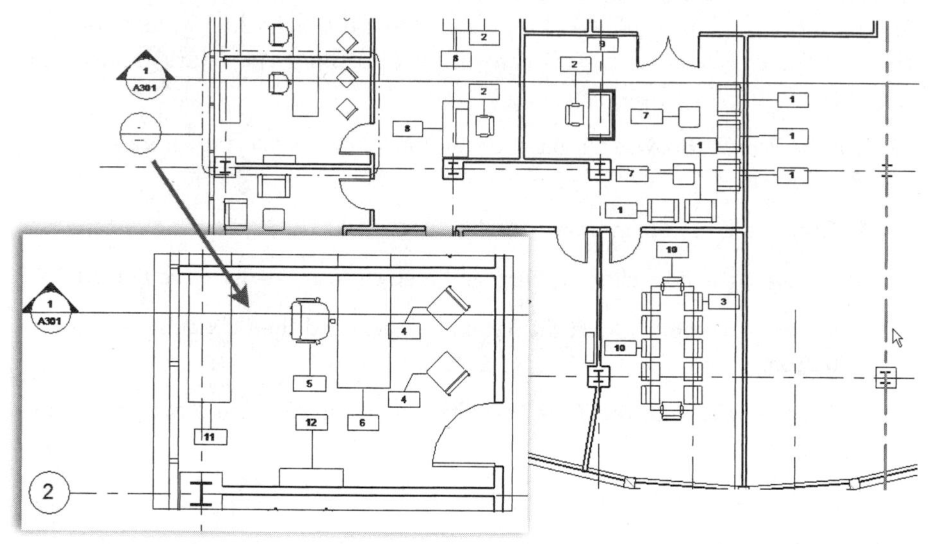

Figure 11–27 *Create a separate furniture plan and enlarged plan*

When adding Tags, you can move them around and reshape the leaders using the control handles. Now that we have a separate furniture plan, we might not want to see the furniture in the *Level 3* plan View any more.

7. On the Project Browser, double-click to open the *Level 3* floor plan View.

- On the keyboard, type **VG** (You can also choose **Visibility/Graphics** from the View menu).
- Clear the checkbox next to "Furniture."
- Expand the list next to "Specialty Equipment" and clear the "Bins," "Bins Above" and "Hardware" checkboxes.

This will hide the furniture in this View.

8. Save the project.

ROOMS AND ROOM TAGS

Adding Room Tags is basically the same as adding other Tags. However, there is a separate tool for this purpose. This is because to add Room Tags, you are actually also adding "Room" elements as well. A Room is a non-graphic element in your model. They can be created automatically in many instances from the bounding Walls and other elements. However, not all geometry can bound a Room, and in open plan situations Revit Building will be unable to correctly determine the proper shape and boundary of the Room. In these cases, you can manually sketch Room Separations. We'll explore both techniques here.

1. On the Design Bar, click the Drafting tab and then click the Room Tag tool.
2. Zoom In Region around the two spaces to the left of the Stair tower.
3. Move the cursor into the upper room directly to the left of the Stairs (see Figure 11–28).

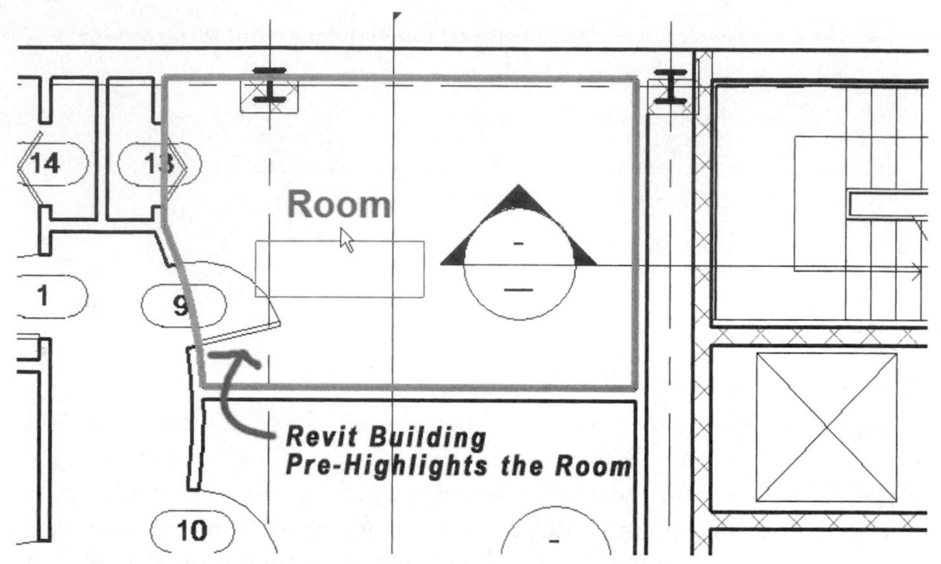

Figure 11–28 *Pre-Highlight the Room with the Room Tag Cursor*

NOTE In the illustration above, if you wish to include the closet as part of the room area, you can select the Wall between the room and the closet, edit its Properties and deselect the "Room Boundary" checkbox to have Revit Building ignore that Wall when finding the Room Boundary.

- Click in the space to Create the Room and add its Tag.
- Move into the space directly below and then click again.

Notice that the first Room was automatically numbered as "1" and the next one sequentially numbered to "2."

4. Do the small utility Room below the previous one next.
5. Zoom out a bit and pre-highlight some other Rooms. Don't click yet.

Notice how most of the pre-highlighting boundaries are not correct. The corridor flows into the reception and secretarial spaces in an open plan configuration. We really need three separate Rooms for these. Many of the offices also don't pre-highlight as expected either. This is because the Columns that define some of the corners cannot be used as boundaries automatically. If you are uncertain which elements can and cannot be used as boundaries, there is a tool on the Options Bar to help you see them.

6. With the Room Tag tool still active, click the "Show Bounding Elements" button on the Options Bar.

This will highlight all of the eligible elements in orange on screen. A message will also appear explaining what you see (see Figure 11–29).

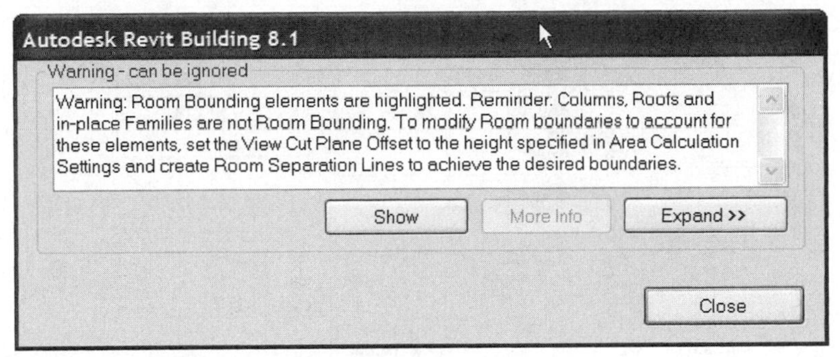

Figure 11–29 *Turn on the "Show Bounding Elements" Options*

As you can see, none of the Columns are highlighted orange. This is because these elements cannot be bounding elements. For these areas, we will have to sketch in Room Separations instead.

- Close the warning message.
- On the Design Bar, click the Modify tool or press the ESC key twice.

7. Zoom in on Column A2 (lower left corner).
8. On the Design Bar, click the Room Separation tool.
   - Sketch a Line segment across the Column (see Figure 11–30).

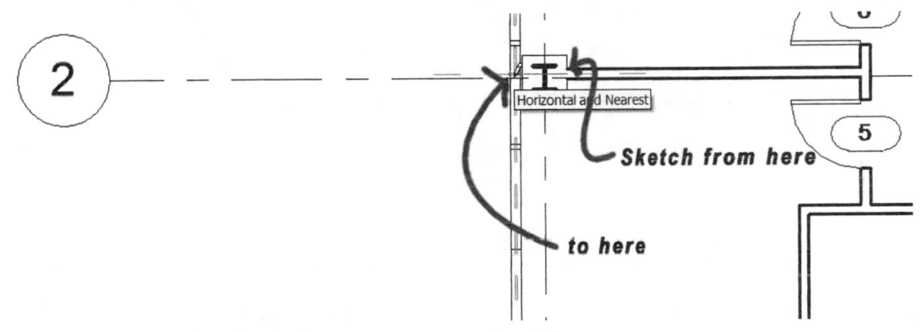

Figure 11–30 *Sketch a Room Separation across the Column*

9. Test your Room Separation by returning to the Room Tag tool and tagging the corner office.

It should now Tag only the one Room.

10. Repeat the process to add additional Room Separation lines at each of the columns between offices.
    - Add Room Tags to the remaining offices and the conference room (see Figure 11–31).

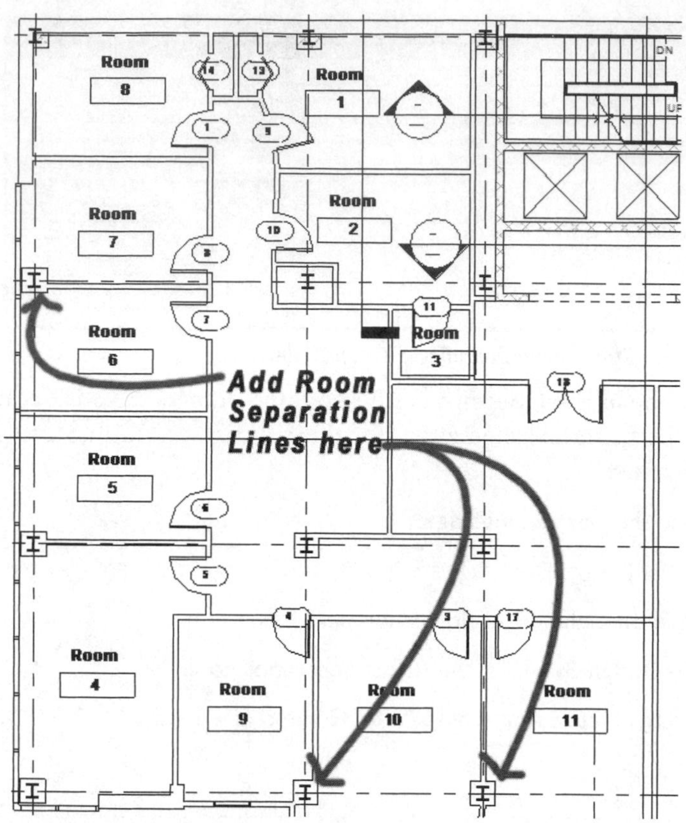

Figure 11–31 *Add Room Separations and then Tag the perimeter Rooms*

For the remaining Rooms, we need to add Room Separation lines to designate where the reception area ends and the corridor begins, and then where the secretarial area boundary is.

11. On the Design Bar, click the Room Separation tool.
    - Sketch Room Separations as indicated in Figure 11–32.

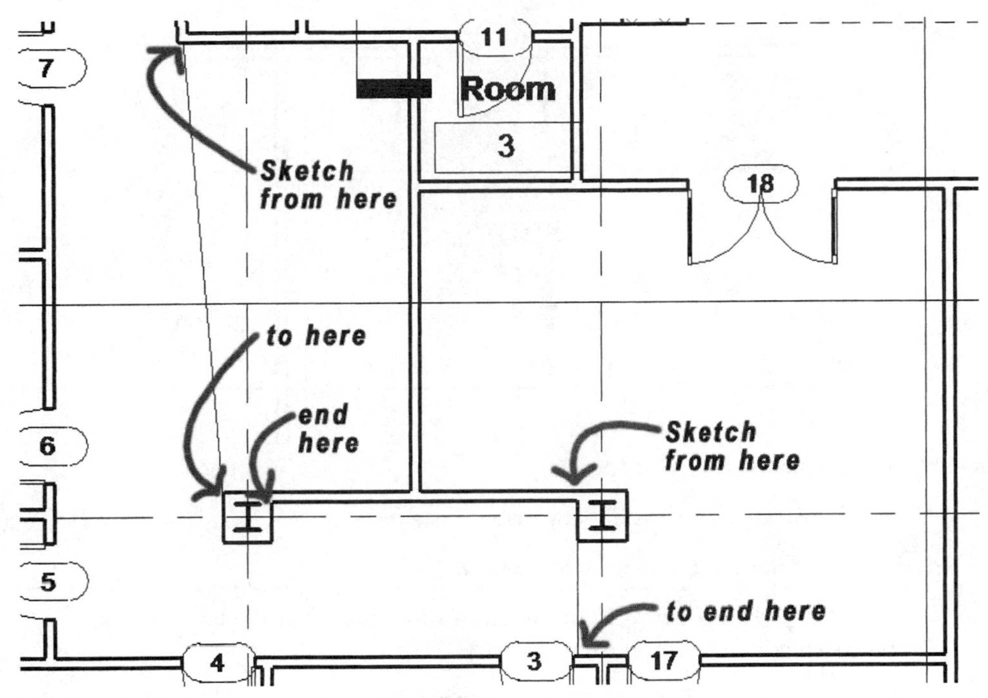

Figure 11–32 *Add Room Separations for the remaining spaces*

Remember that Columns do not define boundaries, so be sure your sketch closes any gaps left by the Columns.

12. Add Room Tags to the corridor, reception and secretarial spaces.

EDIT ROOM NAMES

To edit the Room names, simply click on the Tag and type in the new value.

13. Click on the room label of the Tag in the reception area.

An editable text field should appear on screen.

- Type **Reception** and then press ENTER (see Figure 11–33).
- Repeat the process to renumber the space to number **1**.

A warning box will appear when you do this indicating a duplicate number. It is safe to ignore it for now.

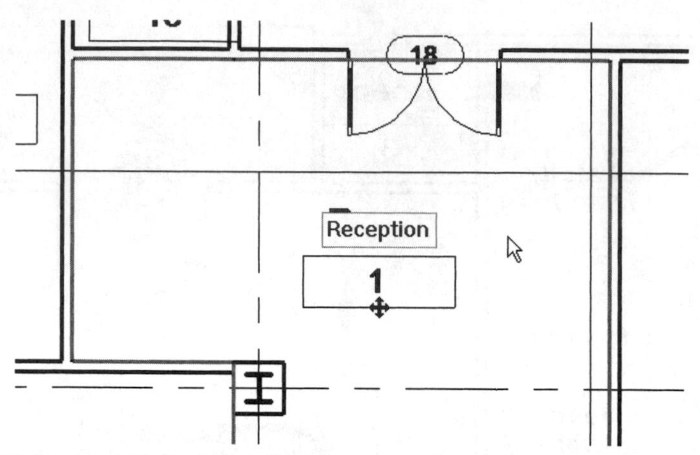

Figure 11–33 *Rename the Reception Space*

The Conference room is directly below Reception.

14. Rename it next and number it as room **2**.
15. Using the CTRL key, select all of the office room tags (2 at the bottom, 1 in the bottom left corner and 4 at the left).
    - On the Options Bar, click the Properties icon.
    - Beneath the "Identity Data" grouping type **Office** in the "Name" field and then click OK.
    - Using either technique, rename and renumber all Rooms as indicated in Figure 11–34.

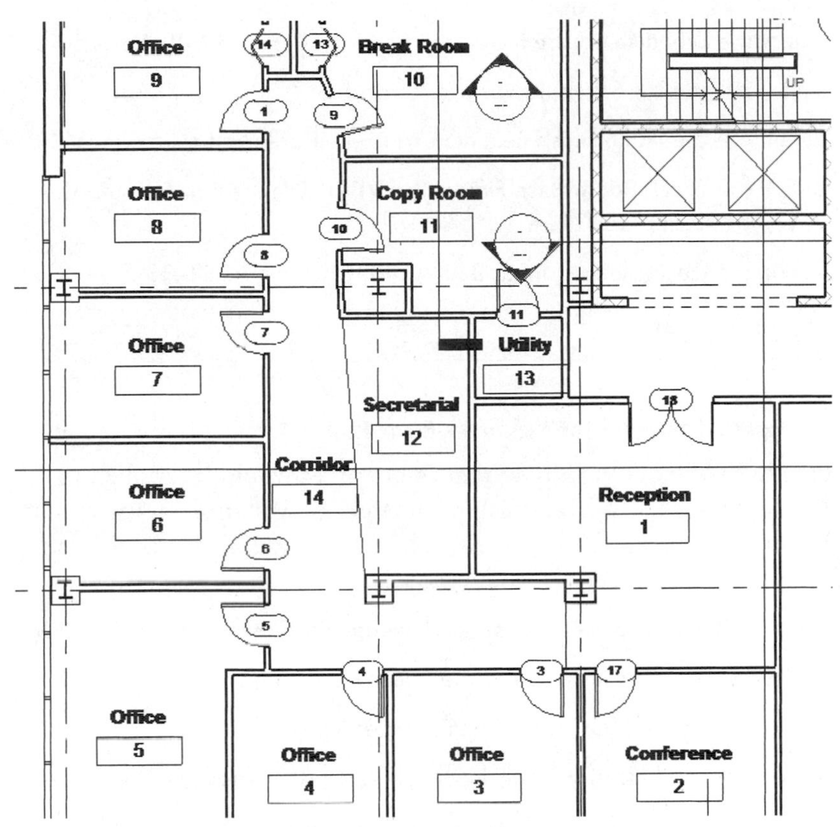

Figure 11–34 *Rename and renumber the remaining Rooms*

VIEWING A ROOM SCHEDULE

Back in Chapter 4 we added several Schedules to this project—among them a Room Schedule. Let's take a look at that Schedule now to see how it reflects the data that we just added to the model .

16. On the Project Browser, double-click to open the *A603 - Schedules* Sheet View.

17. Zoom In Region on the Room Schedule.

Earlier when we made adjustments to the Wall Schedule; also on this Sheet, the Room Schedule was empty. Now that we have added Rooms to our model, this Schedule has been filled in to list these Rooms. Notice that this is really a room finishes Schedule and that currently there is no finish information listed. We can add finish data directly in the Schedule as we did in the furniture Schedule above, or we can select a Room (by selecting a Room Tag) and edit the Properties in the model.

18. Select the Schedule on the Sheet, right-click and choose **Edit Schedule**.

    This will open the Schedule View on screen.

    - Click in the Floor Finish field next to Break Room and then type **VCT-1**.
    - For the Break Room Base Finish type **VB-1**, Wall Finish: **P-1** and the Ceiling Finish: **ACT-1**.
    - For the Ceiling Height input **8′-0″** **[2400]** (see Figure 11–35).

Room Schedule						
Room Number	Room Name	Floor Finish	Base Finish	Wall Finish	Ceiling Finish	Ceiling Height
10	Break Room	VCT-1	VB-1	P-1	ACT-1	2400.0
11	Copy Room					
13	Utility					

Figure 11–35 *Input values for the finishes in the Break Room*

As you can see, we were not prompted that these values would be applied to Types this time. This is because all of the finish parameters are instance parameters of the Rooms. This makes editing them via the Schedule View a little more time consuming than editing via Properties in the model.

19. On the Project Browser, double-click to open the *Level 3* floor plan View.
20. Using the CTRL key, select all of the Office Room Tags.
    - On the Options Bar, click the Properties icon.
    - Beneath the "Identity Data" input values for the various finishes (see Figure 11–36).

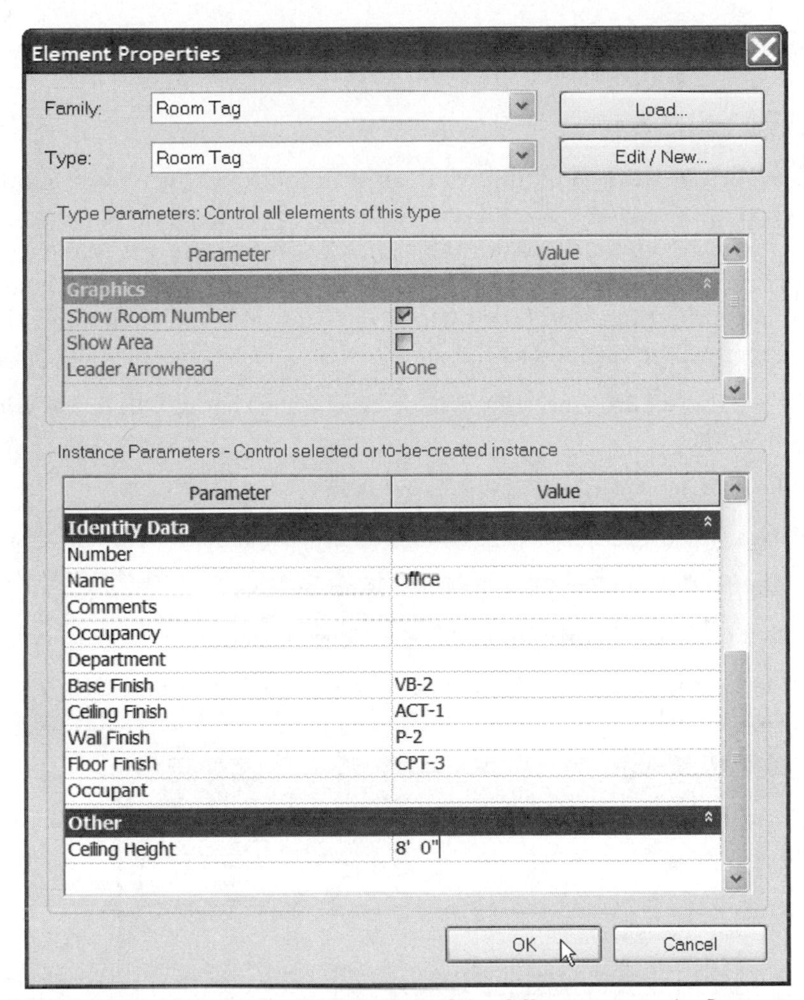

Figure 11–36 *Input values for the finishes of all of the Office spaces in the Properties dialog*

21. Click OK to dismiss the dialog. Return to the Schedule View to see the results.

Notice that all of the Offices have received the same values. When you select multiple Room Tags, you are actually editing several Rooms at once. If you wish, you can modify individual values of Rooms that vary from the "typical" condition directly in the Schedule.

22. Using either technique, finish editing the finish information of the remaining Rooms and make any desired edits.

23. Edit the Properties of the Schedule, and then click the Edit button for Sorting/Grouping.

- From the "Sort by" list, choose **Number** and then click OK twice.

24. Save the project.

MAKE AN AREA PLAN

Like most Families in Revit Building, you can swap out one Type with another. Above, we created a separate third floor furniture plan. In that sequence, we choose the "Duplicate" command from the right-click menu. In this sequence we will repeat the process but use the "Duplicate with Detailing" command instead. This will duplicate the third floor with its Room Tags. We can then swap these Tags for a different Type that shows the area as well.

25. On the Project Browser, right-click on Level 3 and choose **Duplicate with Detailing**.
    - Rename the new View **Level 3 Area Plan**.
26. Make a window selection large enough to highlight all Room Tags on screen.
    - On the Options Bar, click the Filter Selection icon.
    - In the "Filter" dialog, click the Check None button.
    - Place a checkmark in only the Room Tags box and then click OK (see Figure 11–37).

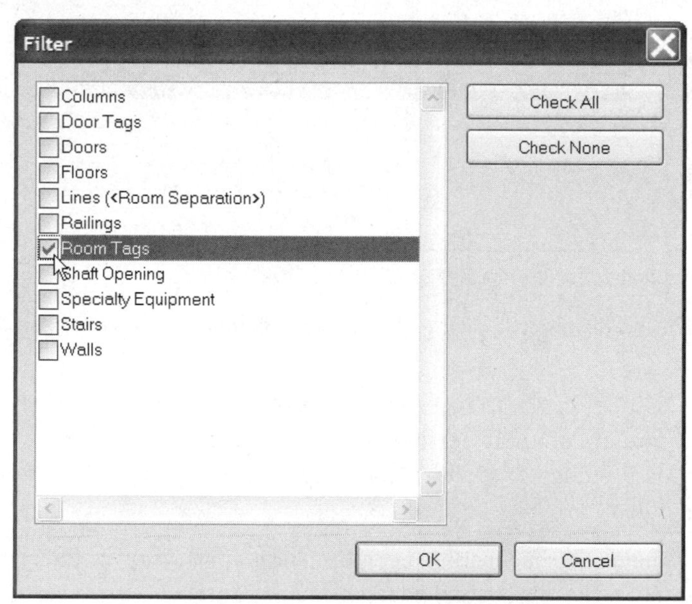

Figure 11–37 *Use the Filter Selection to select only the Room Tags*

27. From the Type Selector, choose Room Tag: Room Tag With Area [M_Room Tag: Room Tag With Area].

All of the Room Tags in this View will now display the area beneath the Tag (see Figure 11–38).

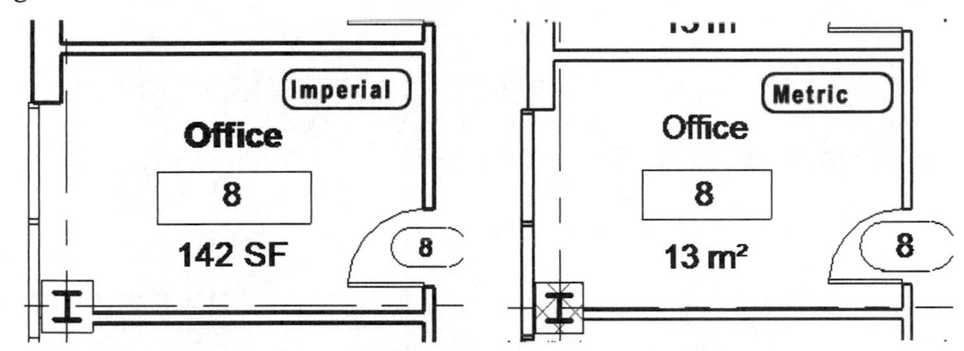

Figure 11–38 *Swap the Room Tags for a Type that displays the Area*

The Room Tag used here was part of the template from which the project was created. You can create a Room Tag (or any Tag) that displays any of the parameters you wish. While we will not create one here, building a Tag Family is not unlike creating any of the Families we build in Chapter 9. If you wish to experiment on your own, select one of the Tags in this project that is similar to the one you wish to create. On the Options Bar, click the Edit button to launch the Family Editor and load the Tag for editing. Save the file as a new name and manipulate it to suit your needs. The text values are special elements called "Labels." You can add or edit Labels in the Family editor and have them report any of the parameters available. Look up "Labels" in the online Help for more information. If you wish to add a parameter that is not included on the list available, you must create a custom Parameter. The best way to do this is to create a "Shared Parameter" in a Shared Parameter File (**Shared Parameters** on the File menu). A Shared Parameter File is simply a text file saved on a hard drive or network server. The advantage of this file is that you can store your custom parameters in this file, save it to a common network server location and then all users in the firm can access and use these parameters in their projects, Schedules and Tags. For more information on creating and working with Shared Parameters, see the online Help.

ADDING A COLOR FILL

Using the Room Tag with Area is a useful way to display the area of each Room directly in the plan, but in many cases it is useful to show this data more graphically using color coded shading. Adding a Color Fill element to a View will achieve this goal.

28. On the Design Bar, click the Drafting tab and then click the Color Fill tool.

- Zoom out a bit allowing room to the left side of the plan.

- Click a point to the left of the plan to place the Color Fill Legend (see Figure 11–39).

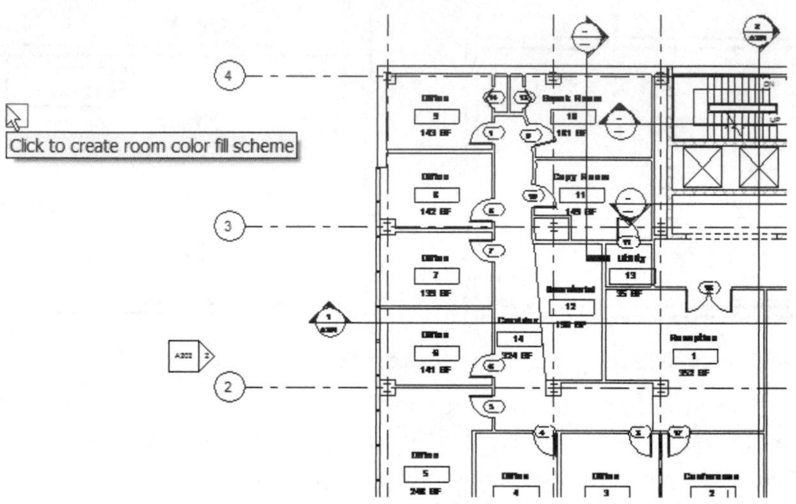

Figure 11–39 *Add a Color Fill to the* ***Level 3 Area Plan*** *View*

A message will appear indicating that some elements were turned off. This is OK for this View. This is one of the reasons that we created a new View before proceeding. If you need to see the elements that are being turned off in the normal floor plan View, they are still visible there, remember changes to visibility are View-specific.

- Click OK to accept and dismiss the message.

A single square symbol with a label reading "No colors defined" will appear. To display a color scheme, we need to specific what parameter we wish to have color-coded.

29. Select the Color Fill legend symbol.
    - On the Options Bar, click the Edit Color Scheme button.
    - In the "Edit Color Scheme" dialog, choose **Area** from the Color list (see Figure 11–40).

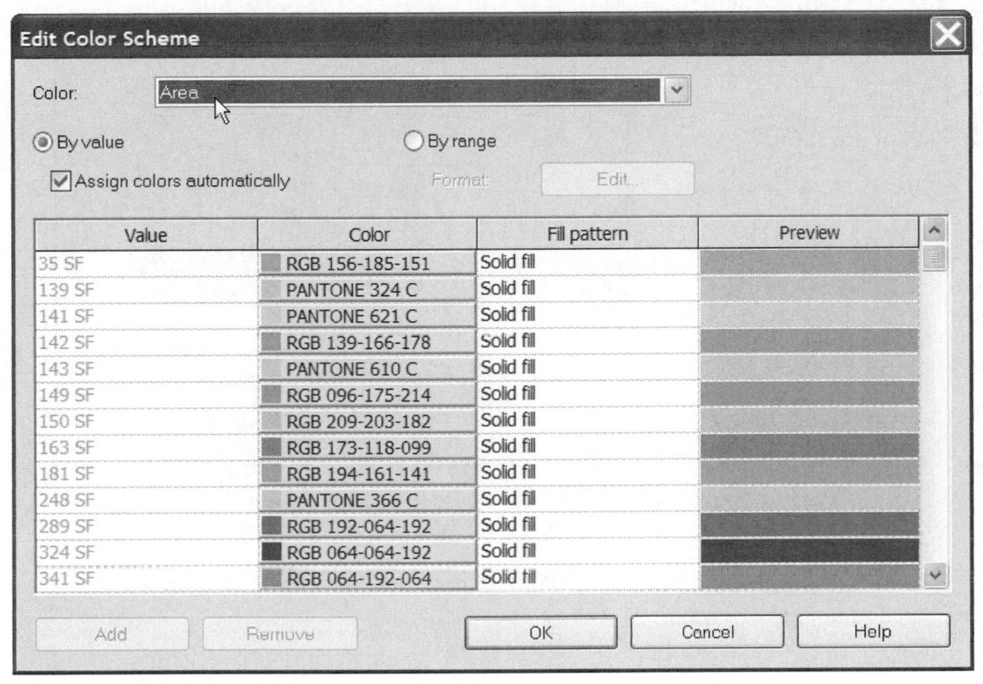

Figure 11–40 *Change the Color Scheme to color by Area*

By default, the "Assign colors automatically" checkbox is selected. This is the easiest way to see immediate results. If you prefer to designate your own colors, you can clear this checkbox and specify them in the table. We will not do that for this exercise.

30. Click OK to accept the default color scheme and display it in the View.

By default, the "By Value" radio button was selected and we therefore get a unique color for nearly every Room. If you would rather color-code by a range, you can edit the scheme and define the ranges.

31. Edit Color Scheme again.

- Choose the "By Range" radio button.

Two items appear in the table below.

- Select the value in the "At Least" column of the second value and change it to **300 [30]** and then press ENTER (see Figure 11–41).

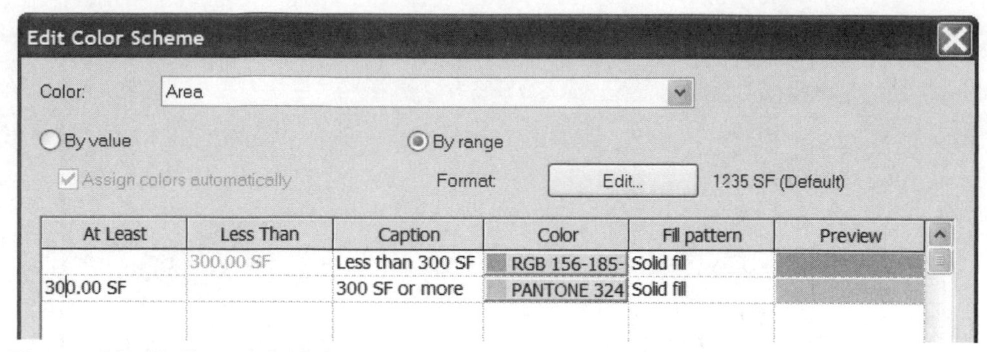

Figure 11–41 *Change the Scheme to "By Range" and edit the values*

32. Select the first item on the list and then click the Split button at the bottom left of the dialog.
    - In the "At Least" field of the new value, change the value to **200** [**20**] and then press ENTER.
    - Select the first entry again, and then click the Split button one more time (see Figure 11–42).

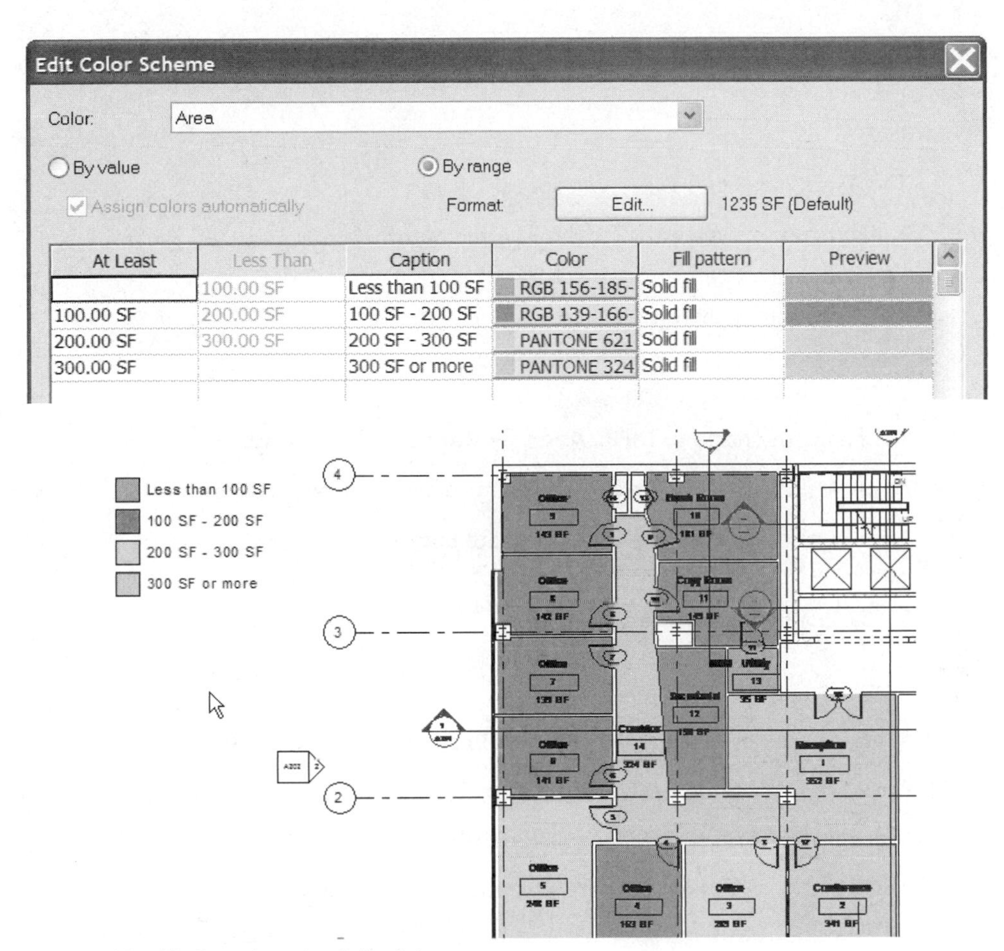

Figure 11–42 *Complete the Color Scheme*

33. Click OK to see the results.

SAVE THE NEW AREA SCHEME AS A TYPE

If you wish to experiment further with the Color Fill, you should first save the current scheme as a Type. This way, you can easily return to this scheme later if you wish without having to reconfigure it.

34. Select the Color Fill legend on screen.
    - On the Options Bar, click the Properties icon.
35. Next to the Type list, click the Edit/New button.

TIP A shortcut to this is to press ALT + E.

The Type Properties dialog will appear.

- Next to the Type list, click the Rename button.

TIP A shortcut to this is to press ALT + R.

- Change the name to **MRB Area by Range** and then click OK.

If you wish, change the Text parameters to suit your preferences.

36. Next to the Type list, click the Duplicate button.

TIP A shortcut to this is to press ALT + D.

- Name the new Type: **MRB Floor Finish** and then click OK.

37. Click the Edit button next to the Color Scheme.
    - From the "Color" list choose **Floor Finish**.
    - Click OK three times to return to the plan View and see the results.

In this case, the same four colors are used in both schemes. You may wish to clear the "Assign colors automatically" checkbox mentioned above and edit the colors to provide some variety.

38. Repeat the process again to create schemes that show the Wall Finish, the Base Finish and any other value you wish to explore.
39. Use the Type Selector to change from one Type to another back in the View (see Figure 11–43).

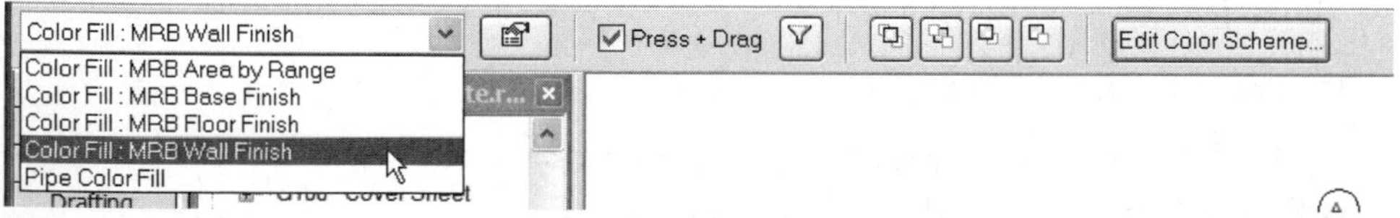

Figure 11–43 *New Types now appear in the Type Selector for easy retrieval*

CREATE A ROOM/AREA REPORT

In some jurisdictions detailed area analysis with triangulated area proofs are required. You can generate such a report from Revit Building in HTML format.

40. From the File menu, choose **Export > Room/Area Report**.

    - In the dialog that appears browse to a location where you wish to save the report, give it a name and then click Save.

41. Launch your web browser and open the HTM file created to view the report.

The potential of Color Fills is limited only by the parameters available on each Revit Building element. Feel free to experiment further with them and try other possibilities.

SUMMARY

A Schedule is simply a tabular View of your building model data.

You can edit elements referenced in a Schedule directly from the Schedule or in the model

Changes to model elements are reflected immediately in both the graphical Views and the Schedules.

Schedules can contain any combination of fields and be grouped, sorted and filtered.

Add headers, footers and totals to your Schedules to break them up and make the data more legible.

Adding Schedules to Sheets is a simple drag and drop process.

You can adjust the size and formatting of Schedule columns on the Sheet.

Add breaks to the Schedule on the Sheet to wrap a long Schedule into two or more columns.

New elements added to the model will immediately appear in Schedules as appropriate.

Groups allow you to efficiently manage repetitive elements and their components show in Schedules.

Tags report data in similar fashion to Schedules.

Add Tags manually or all at once using "Tag all not Tagged."

Create custom Views to show different Tags and graphics.

Create Room elements by adding Room Tags.

Use Room Separations to manually define the shape of Rooms.

Color Fills can be used to color-code nearly any data that appears in Room Tags or Schedules.

CHAPTER 12

Ceiling Plans and Interior Elevations

INTRODUCTION

The goal of this chapter is to round out our construction document set. Reflected ceiling plans will be the primary focus of the chapter with a brief exploration of interior elevations at the end. Reflected ceiling plan Views are included in the default project template file used to start both projects in this book. Therefore, we simply need to open these Views and add appropriate model data and annotation. In addition, we will need to indicate which rooms we wish to elevate and create the required interior elevation Views.

OBJECTIVES

Ceiling plans are very similar to other plans. The View Properties are the only major difference between the two. We will add Ceiling elements to model and ceiling specific annotation. After completing this chapter you will know how to:

- Add and modify Ceiling elements
- Understand Ceiling Types
- Add and modify Ceiling Component Families
- Manipulate the View Properties of the Ceiling Plan
- Create Interior Elevation Views

CREATING CEILINGS

Ceiling elements in Revit Building are used in Ceiling plans to convey the material of the ceiling plane—such as acoustical tile ceiling, gypsum board or other ceiling treatment. Ceiling elements are created basically the same way as Floors and Roofs. They are sketch-based elements, and they utilize the "Pick Walls" and other sketch methods on the Design Bar during creation and editing. In addition, Ceilings can also be created with a special tool (much like creating Rooms, demonstrated in the previous chapter) by picking a point within a closed space in the model. When using this method, Revit Building determines the boundary formed by the Walls, creates sketch lines and automatically finishes the sketch.

INSTALL THE CD FILES AND OPEN A PROJECT

The lessons that follow require the dataset included on the Mastering Revit Building CD ROM. If you have already installed all of the files from the CD, simply skip down to step 3 below to open the project. If you need to install the CD files, start at step 1.

1. If you have not already done so, install the dataset files located on the Mastering Revit Building CD ROM.

   Refer to "Files Included on the CD ROM" in the Preface for instructions on installing the dataset files included on the CD.

2. Launch Autodesk Revit Building from the icon on your desktop or from the **Autodesk** group in **All Programs** on the Windows Start menu.

   - From the File menu, choose **Close**.

This closes the empty new project that Revit Building creates automatically upon launch.

3. On the Standard toolbar, click the Open icon.

TIP The keyboard shortcut for Open is CTRL + O. **Open** is also located on the File menu.

- In the "Open" dialog box, click the *My Documents* icon on the left side.
- Double-click on the *MRB* folder, and then the *Chapter12* folder.

If you installed the dataset files to a different location than the one listed here, use the "Look in" drop down list to browse to that location instead.

4. Double-click *12 Commercial.rvt* if you wish to work in Imperial units. Double-click *12 Commercial Metric.rvt* if you wish to work in Metric units

   You can also select it and then click the Open button.

The project will open in Revit Building with the last opened View visible on screen.

CREATING AN ACOUSTICAL TILE CEILING

Suspended acoustical tile ceilings in commercial office buildings are constructed in one of two ways: either the Walls are built past the height of the ceiling tiles (to a fixed height or all the way to the deck) and each room contains its own ceiling; or the Walls are "underpinned" being built up to the height of the ceiling with the ceiling plane being continuous across the tops of the Walls.

When each room contains its own Ceiling and the Walls continue past the ceiling plane height, Ceiling elements can usually be created quickly with the "Auto Ceiling" function of the Ceiling tool. This is the default function of the Ceiling tool. When you have an underpinned Ceiling, you can instead sketch the shape of the overall Ceiling plane using any of the available sketching tools. We will explore both ceiling types as we continue to refine the third floor of our commercial project.

Revit Building usually creates both a floor plan and a ceiling plan View whenever you create a new Level in a project. The template from which we originally created the commercial project includes such Views. In this sequence, we will work in the ceiling plan Views for the first time. Be sure that you work in a Ceiling plan View when adding Ceiling elements. If you do not, you will receive a warning like the one shown in Figure 12–1.

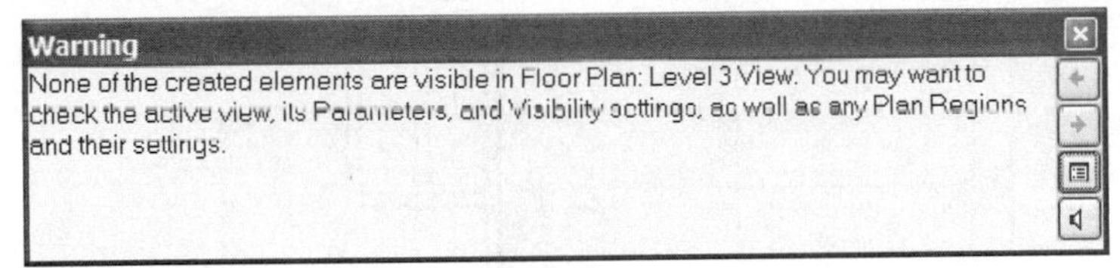

Figure 12–1 *Attempting to add Ceilings in floor plan Views yields a warning*

5. On the Project Browser, expand the *Ceiling Plans* node.

   - Double-click to open the *Level 3* ceiling plan View.

NOTE Again, please be sure that you are opening the *Level 3* ceiling plan View and *not* the *Level 3* floor plan View.

6. On the Design Bar, click the Modeling tab and then click the Ceiling tool.

   On the Status Bar, a message will appear "Click inside a room to create ceiling." On the Options Bar, a single button labeled "Sketch" will appear (see Figure 12–2).

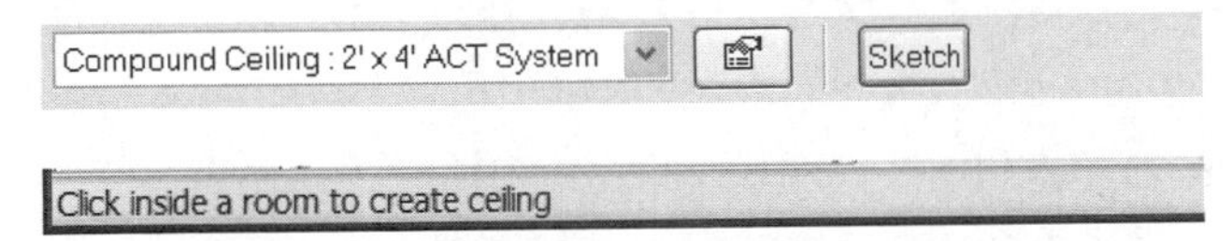

Figure 12–2 *Using the Ceiling Tool, "Sketch" is an option, not the default*

   - From the Type Selector, choose Compound Ceiling : 2′ x 4′ ACT System [Compound Ceiling : 600 x 1200mm grid].
   - Move the cursor around the screen pausing within various Rooms—do not click yet.

Notice that the behavior is similar to the Room Tag tool from the previous chapter. However, unlike the Room Tag tool, this tool does not have the "Room Separation" option, nor can it recognize those Room Separation lines already in the model. Therefore, if the auto-detecting routine does not automatically recognize the Room you need, you can use the "Sketch" option noted above. We will try this option a little later. For now, we will add a Ceiling in a Room that is recognized by the auto-detection routine.

- Move the cursor into the upper left corner office.

The Room boundary should highlight automatically.

- Click in the Room to add the Ceiling (see Figure 12–3).

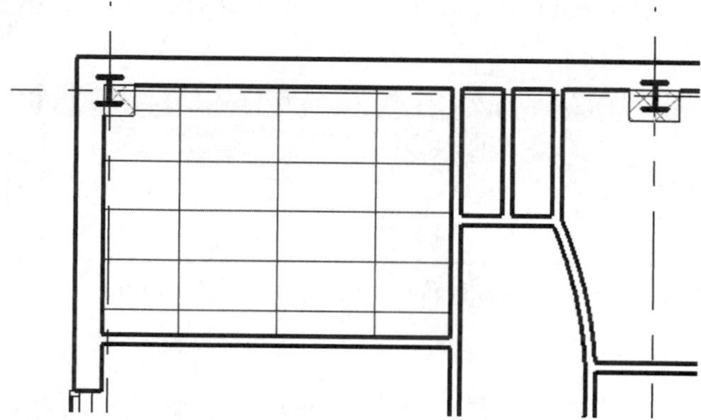

Figure 12–3 *Add a Ceiling to an office*

- On the Design Bar, click the Modify tool or press the ESC key twice.

EXPLORE CEILING TYPE PROPERTIES

7. Click on one of the Ceiling grid lines.

Notice that each line is individually selectable. However, they behave as a unit. Revit Building will attempt to center the grid in the room left to right and top to bottom. You can move any grid line if you like, and the entire grid pattern will move with it. Use this technique to apply custom centering (see below).

- On the Options Bar, click the Properties icon.

Notice that the "Height Offset From Level" is to 8′-0″ [2600] above the associated Level (see Figure 12–4).

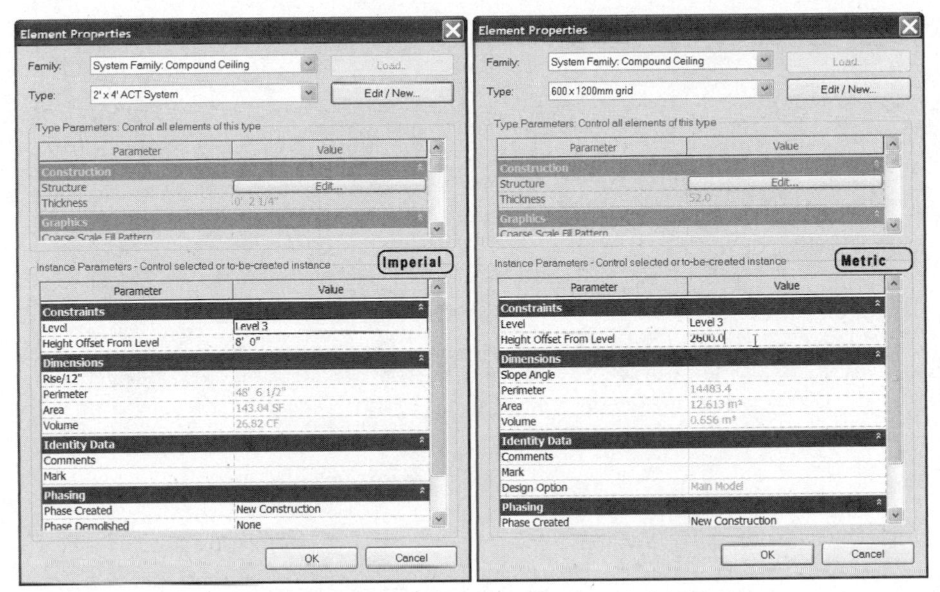

Figure 12–4 *Ceilings are set a default height above the current Level*

8. Next to the Type list, click the Edit/New button.

TIP A shortcut to this is to press ALT + E.

The Type Properties dialog will appear.

Notice that a Ceiling element's Type has an Edit "Structure" button like many other Types. If you were to edit this Structure, you would notice that it is comprised of a simple Core element with a Finish Layer on the bottom. The grid pattern comes from the Material assigned to this Finish Layer (see Figure 12–5).

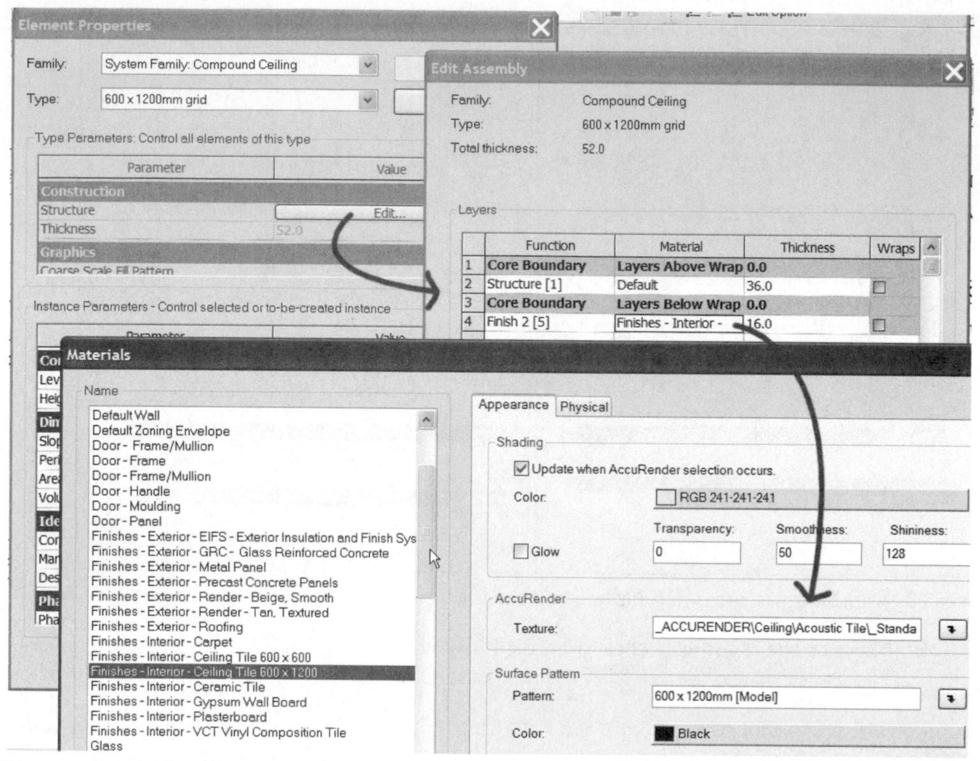

Figure 12–5 *The Type Properties of the ACT Ceiling uses a "Ceiling Tile" Material*

9. Cancel all dialogs when finished.
10. Add another Ceiling using the same technique in the Break Room (across the corridor on the right side).
    - On the Design Bar, click the Modify tool or press the ESC key twice.

MOVE, ROTATE AND ALIGN CEILING GRIDS

We could continue and add additional Ceilings in the other Rooms, but for now we will work with just these two.

11. Select one of the gird lines of the Ceiling again.
    - On the toolbar, click the Move icon and then move the grid line (see Figure 12–6).

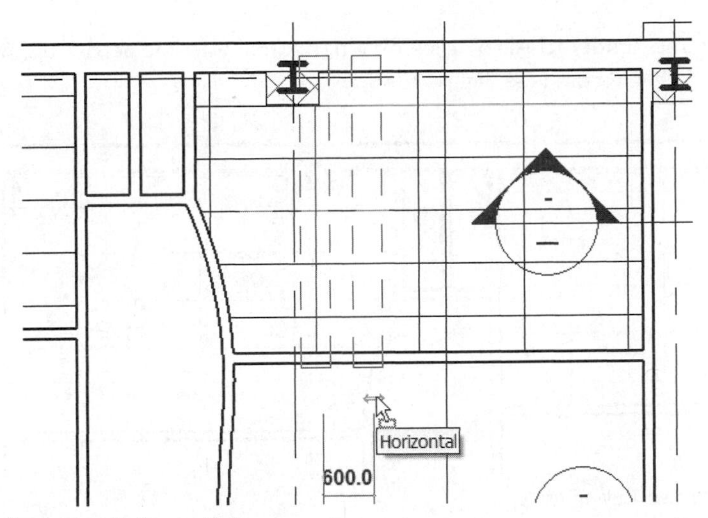

Figure 12–6 *Move a Ceiling grid line*

As noted above, notice that the entire grid will reposition with this move; not just the selected grid line and not the boundary of the ceiling element. You can use other typical modification techniques as well, such as Rotate. If you need your grid pattern at an angle other than 0° or 90°, simply click the Rotate icon and rotate the grid line. The rest of the grid will follow (see Figure 12–7).

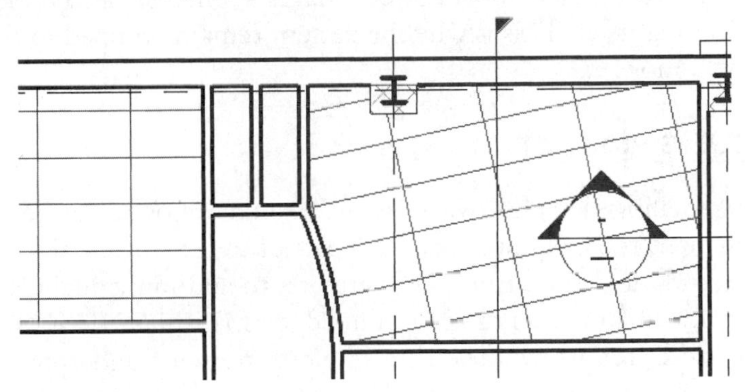

Figure 12–7 *Rotate a Grid*

If you rotated the grid, please undo the change now to return it to a horizontal orientation.

You can use the Align command to create alignment between the grids in two Rooms.

12. Select one of the horizontal grid lines in either Room.
    - On the toolbar, click the Align icon.
    - For the "Point of Reference" click on one of the horizontal grid lines.

- For the "Entity to Align" click on a horizontal grid line across the hall in the other Room (see Figure 12–8).

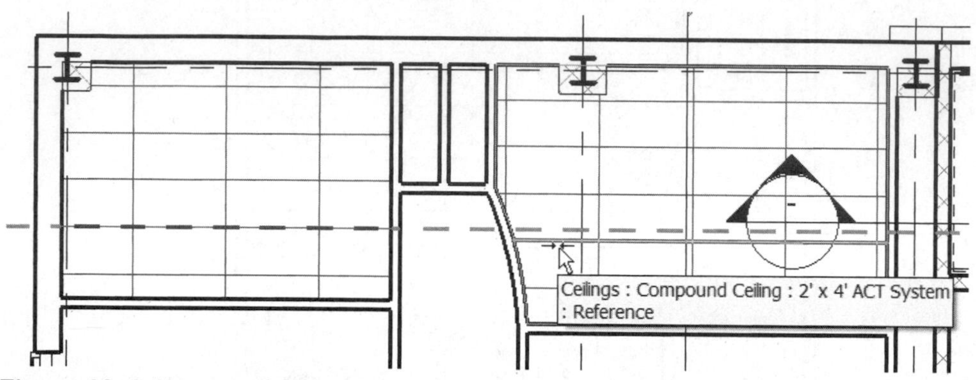

Figure 12–8 *Align two Ceiling grids*

13. Click the padlock icon that appears to constrain the alignment.
14. Select one of the horizontal grid lines.
    - Move it up half a tile (**1′-0″** [**300**]).

Notice that the Ceiling grids in both Rooms move together. When you move this way, you are simply moving the Model Pattern that is applied via a Material to the Ceiling, *not* the Ceiling itself. This is why the pattern remains clipped to the shape of the Room as you move it.

CREATE A CEILING VIA SKETCH

For the remaining Ceiling elements, we'll use the Sketch option. As we discovered in the previous chapter, the Columns that form part of the enclosure of the Rooms do not work as boundaries of the Rooms. Therefore, to add individual Ceilings to the remaining offices along the left and bottom edge of the plan, we must use the Sketch option. In addition, we will sketch a single continuous underpinned Ceiling for the remaining non-office spaces in the suite.

15. Zoom in on the office on left directly beneath the one to which we already added a Ceiling.
16. On the Design Bar, click the Modeling tab and then click the Ceiling tool.
    - From the Type Selector, choose Compound Ceiling : 2′ x 4′ ACT System [Compound Ceiling : 600 x 1200mm grid].
    - On the Options Bar, click the Sketch button.

    The Design Bar changes to sketch mode and shows only the Ceiling tools.

By now, the sketch tools in Revit Building should be very familiar to you. The basic steps are summarized here, but please feel free to use whatever techniques you prefer to sketch the inside shape of the Room.

- On the Design Bar, click the Pick Walls tool.
- Pick the Walls surrounding the Room.
- Use the Flip controls to change the side of the Sketch lines as required.
- Use Lines to sketch around the Column.
- Use Trim/Extend to cleanup the sketch (see Figure 12–9).

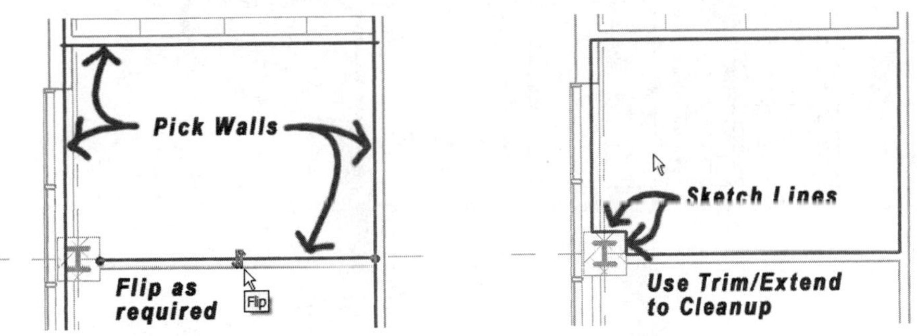

Figure 12–9 *Sketch the inside shape of the office space Ceiling*

17. On the Design Bar, click the Finish Sketch button.

A new Ceiling will appear in the office.

18. Repeat this process to add Ceilings to each of the remaining offices (see Figure 12–10).

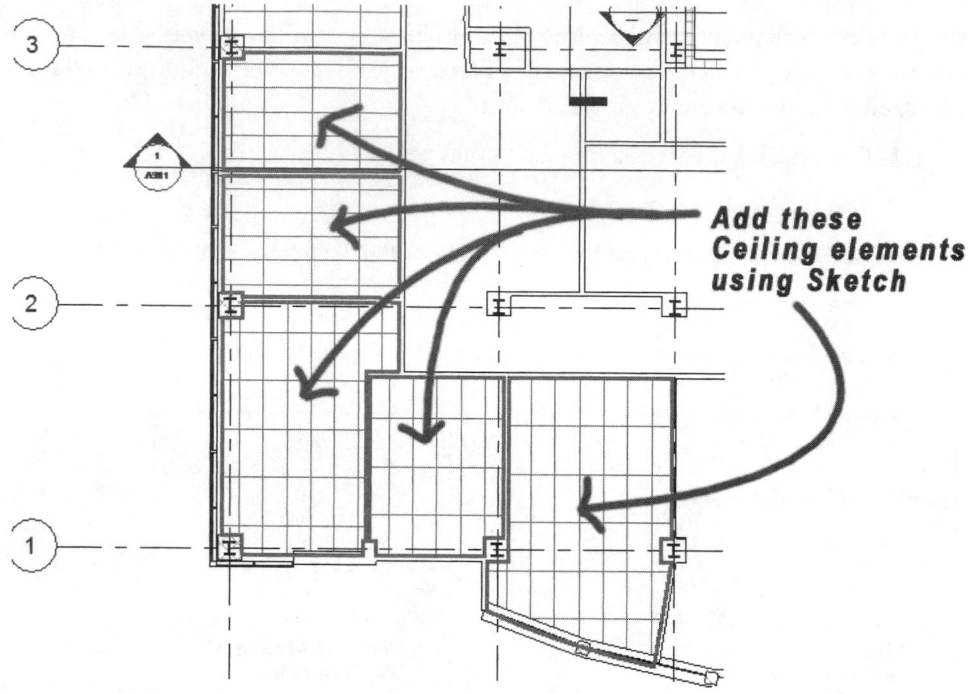

Figure 12–10 *Sketch the remaining office space Ceilings*

Notice that Revit Building will typically choose what it considers to be the best orientation for the ceiling grid. If you wish, you can rotate the grid 90° to orient the offices all the same. To do this, remember, simply select one of the gird lines, and then click the Rotate icon and rotate the line 90°. The remainder of the grid will match the new orientation. Also remember as noted above, that you are merely rotating the Model Pattern that is applied to the Ceiling element. If you should even need to select the actual Ceiling element, place the cursor near the edge of the Room and use the TAB key to highlight and then select the Ceiling element. Do this if you should ever need to delete a Ceiling element.

19. Using the process covered above, Align the Ceiling grids to one another in a logical pattern.
20. Save the project.

CREATE AN UNDERPINNED CEILING

The Ceilings that we just sketched were all contained wholly within single Rooms. You can use the same procedure, using sketch lines to encompass the overall perimeter of several spaces at once. When you do this, you will have a single continuous Ceiling grid that can represent an underpinned ceiling.

21. On the Design Bar, click the Ceiling Tool.
    - On the options Bar, click the Sketch button.
    - Sketch the outline shown in Figure 12–11.

    Don't include the Conference Room. We will do this one later.

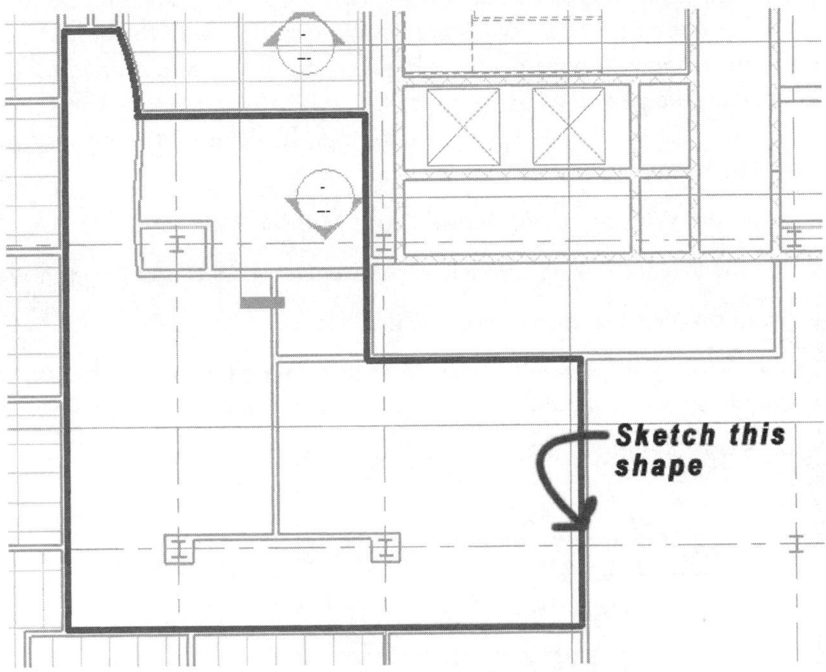

Figure 12–11 *Sketch an underpinned Ceiling across several Rooms*

22. On the Design Bar, click the Ceiling Properties button.
    - Change the "Height Offset From Level" to **9′-0″ [2800]** and then click OK.
    - On the Design Bar, click the Finish Sketch button.

STUDY A CEILING IN SECTION

At this point we have several Ceiling elements in our project. We have studied their Type parameters, moved, rotated and aligned them with one another and discussed the difference in selecting and manipulating the Model Pattern vs. the Ceiling itself.

23. On the Project Browser, double-click to open the *Longitudinal* section View.

TIP If you prefer, double-click the Section Head in the plan instead.

24. Zoom in on the third floor at the left side of the section.

Notice that the Ceiling in the offices at the left is a bit lower than the one in the public spaces. Also notice that the Walls currently go all the way to the deck and join with the Floor. Since we have made a single continuous underpinned Ceiling in the public spaces, we might want to adjust these Walls to stop at the Ceiling height. The section View helps us spot the problem, but the edit is best accomplished in the plan View.

- From the Window menu, choose **Close Hidden Windows**.
- On the Project Browser, double-click to open the *Level 3* ceiling plan View.
- From the Window menu, choose **Tile**.

25. Using Figure 12–12 as a guide, split Walls that bound both the full height and underpinned areas, and then select all of the Walls in the underpinned area.

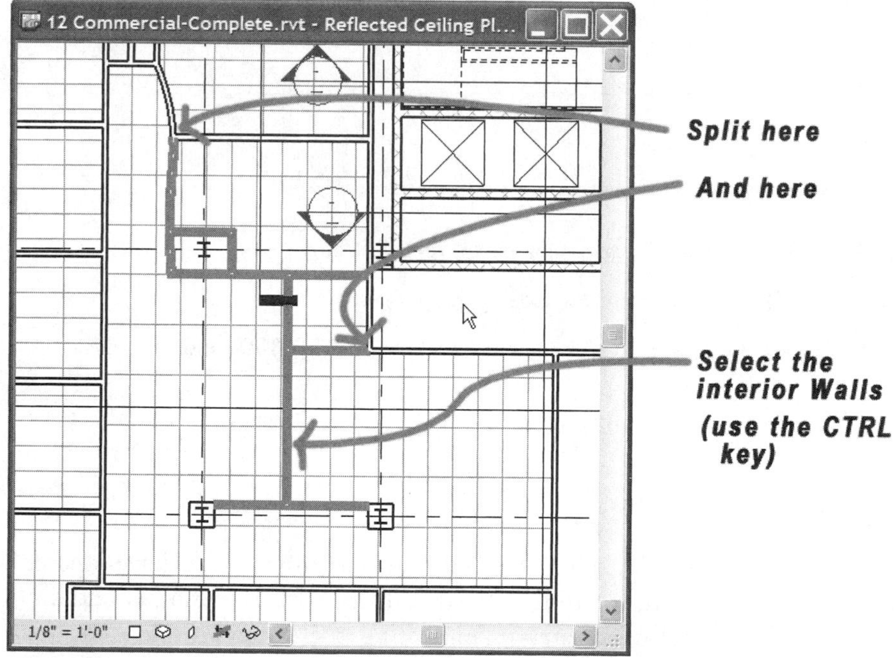

Figure 12–12 *Split Walls as shown and then select all interior Walls*

- On the Options Bar, click the Properties icon.

- Change the "Top Constraint" to **Unconnected** and set the "Unconnected Height" to **9′-0″** **[2800]**.
- Click OK to complete the edit.

NOTE If you anticipate that the Ceiling height will change, you can attach the Walls to the underside of the Ceiling instead (select the Walls and then click the "Attach" button on the Options Bar). In this way, the Wall's height will change with the Ceiling.

A dialog will appear alerting you that some of the elements edited no longer join. This is because the Walls which were previously joined to the Floor above, no longer touch the Floor. Therefore, we should unjoin the elements to prevent Revit Building from wasting computer resources trying to calculate how these elements should join.

26. In the warning dialog, click the Unjoin Elements button (see Figure 12–13).

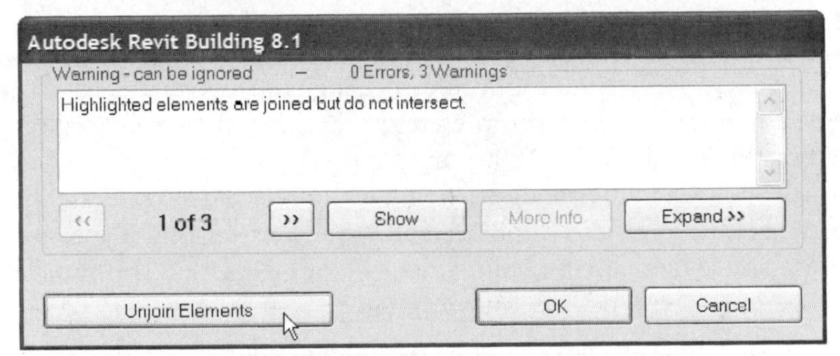

Figure 12–13 *Walls that no longer touch the Ceiling should be unjoined*

Now that we have adjusted the height of the Walls, the model more accurately reflects our design intent. You can see this best in the section View (see Figure 12–14).

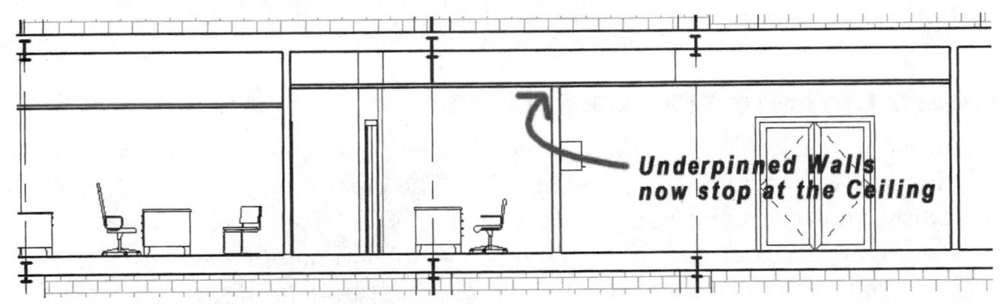

Figure 12–14 *Study the results it the Section View*

ADDING A DRYWALL CEILING

For the Conference Room, we will use the same basic process as the other Ceilings, but simply choose a different Type.

27. On the Design Bar, click the Modeling tab and then click the Ceiling tool.
    - From the Type Selector choose Compound Ceiling : GWB on Mtl. Stud [Compound Ceiling : Plain].
    - On the Options Bar, click the Sketch button.
    - Follow the process outlined above to create a Ceiling in the Conference Room.

As you can see, except for choosing a different Type, the process is exactly the same as with Ceiling grids.

SWITCH TO A DIFFERENT GRID SIZE

Like other elements in Revit Building, you can change the Type used on an existing element at any time. Suppose you preferred a 2′ x 2′ [600 x 600] ceiling grid layout instead of the 2′ x 4′ [600 x 1200] one we used; you can simply choose the Type from the Type Selector with an existing Ceiling selected. While we will not cover the steps here, you can also duplicate and modify an existing Type, assign a different pattern and achieve other Ceiling designs not included in this project file. Feel free to experiment with this on your own later if you wish.

28. Select one of the grid lines on the large underpinned Ceiling in the middle of the plan.
    - From the Type Selector, choose Compound Ceiling : 2′ x 2′ ACT System [Compound Ceiling : 600 x 600mm grid].
29. Save the project.

ADDING CEILING FIXTURES

Now that we have completed adding ceiling elements to our project, we can add some lights and other fixtures to the ceiling plan. The process for adding such elements is very similar to the detailing that we did back in Chapter 10. To achieve the optimal lighting layout, you may need to adjust the location of your grid lines. You can use your Move and Rotate commands as noted above to this. If you rotate a grid and an error appears indicating that constraints are no longer satisfied, simply click the “remove Constraints” button and then if desired, use the Align tool again to reapply constraints.

ADDING LIGHT FIXTURES

Let's start with a simple fluorescent lighting fixture in the offices.

1. Make adjustments to the position of gird lines in the offices at the left to accommodate a suitable lighting layout (see Figure 12–15).

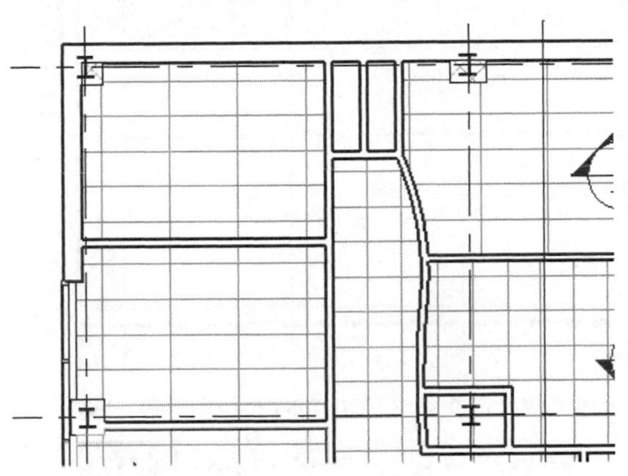

Figure 12–15 *Move grid half a tile to better accommodate the lighting*

2. On the Design Bar, click the Modeling tab and then click the Component tool.
   - On the Options Bar, click the Load button.
   - Browse to your library folder and open the *Lighting Fixtures* folder.
   - Select the *Troffer - 2x4 Parabolic.rfa* [*M_Troffer - Parabolic Rectangular.rfa*] Family file and then click Open.

NOTE As mentioned in previous chapters, if you do not have these Families in your installed libraries, you can find them in the *Autodesk Web Library* folder with the files installed from the Mastering Autodesk Revit Building CD ROM.

3. Place a light fixture in the office.
   - Use the Align command to align it to the grid lines.
   - Select and then Copy the light as appropriate for the space (see Figure 12–16).

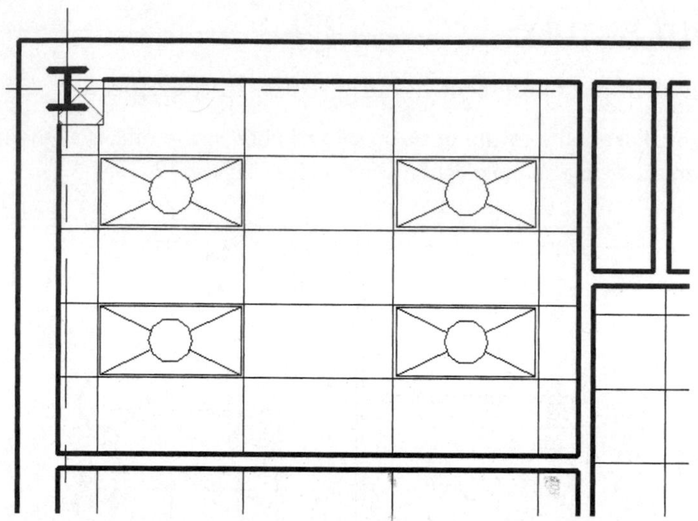

Figure 12–16 *Align the light to the grid and then copy it to create a lighting pattern*

If you are not satisfied with the position of your lights relative to the grid lines, you can move the grid line one-half a tile in either direction. The lights will move with them.

4. Select all of the lights in the office and copy them to the other offices (see Figure 12–17).

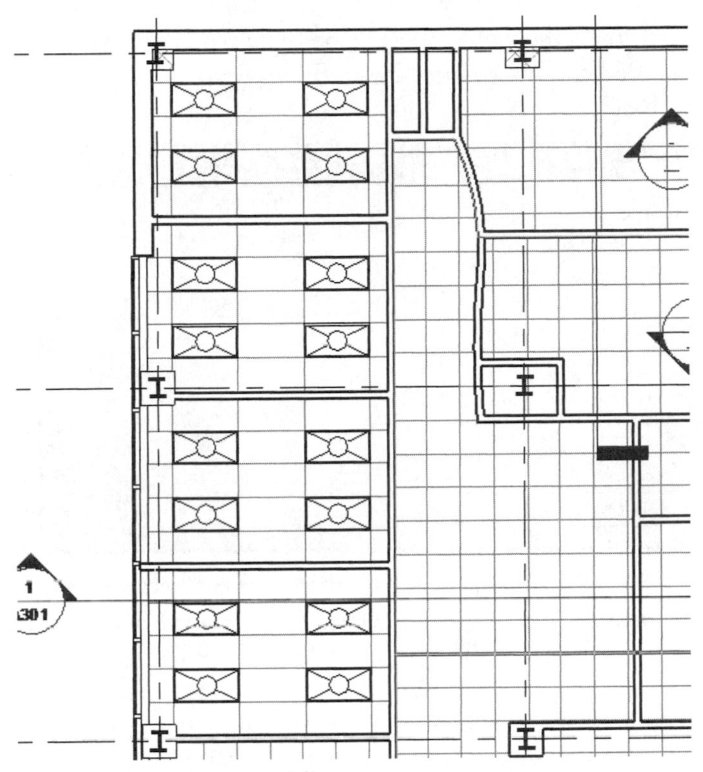

Figure 12–17 *Copy the lights to other Offices*

ADDING LIGHT SWITCHES

Some firms like to add light switches to ceiling plans instead of floor plans. You add switches the same as any Component Family.

5. On the Design Bar, click the Component tool.
   - Click the Load button again, browse to the *Chapter12* folder and then open the *Switch-Single (rcp).rfa* Family file.

This Family file is a modification of the standard Autodesk Revit Building library symbol with modifications to allow it to display in reflected ceiling plan. In Chapter 9 we explored the creation and editing of Families. We discussed Symbolic Lines in that chapter. However, in order for any elements to display in a plan View (ceiling plan in this case), they must be located "beyond" in the View or there must be some element that passes through the Cut Plane height that triggers Revit to display its 2D plan representation. The two ways to deal with the situation are to lower the Cut Plane of the ceiling plan View (currently set at 7′-6″ [2300]) or to add an element to the Family file that passes through this default Cut Plane. The Family file provided in the *Chapter12* folder has a simple Model Line drawn vertically down and set to the <Invisible Lines> Line Style passing through the reflected ceiling plan Cut Plane. This triggers Revit Building to display the graphics for this symbol in the RCP.

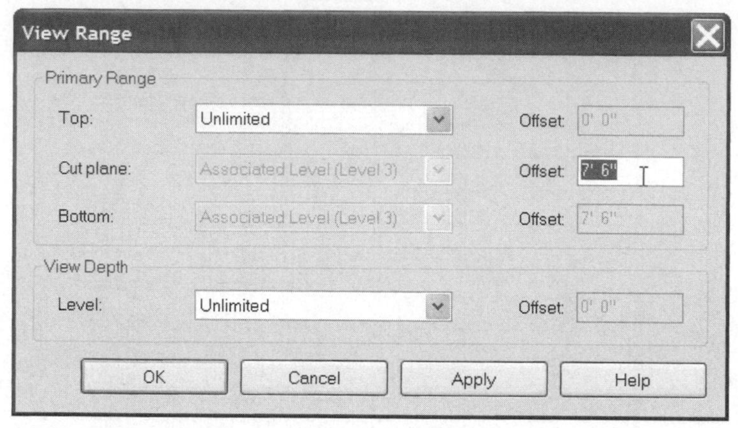

Figure 12–18 *Use the "View Range" dialog to edit the Cut Plane height*

If you prefer, you can edit the View Properties of the Level 3 ceiling plan View and then click the View Range Edit button. In the "View Range" dialog, edit the Offset to a lower height such a 4'-0" [1200] (see Figure 12–18). This would have the added benefit of showing the Doors in the reflected ceiling plan View. Some firms prefer for Doors to show in RCP to make it easier to locate light switches, exit signs and other RCP equipment relative to door swing. If you make this edit, you might also consider editing the Door graphic defaults in the Visibility/Graphics dialog.

6. Place a light switch adjacent to the latch side of the door.
   - Place or copy additional switches in the other offices.

ADDING WIRING

Revit Building does not include a wiring element. To show the wiring connection to the switches, simply draft Detail Lines on the View. Remember, Detail Lines are View-specific and will show only in this reflected ceiling plan.

7. On the Design Bar, click the Drafting tab and then click the Detail Lines tool.
   - From the Type Selector, choose <Hidden>.
   - On the Options Bar, click the "Arc passing through three points" icon.
8. For the first point, click near the switch.
   - For the end point, click near the middle of a light fixture, and then click a third point somewhere in between (see Figure 12–19).

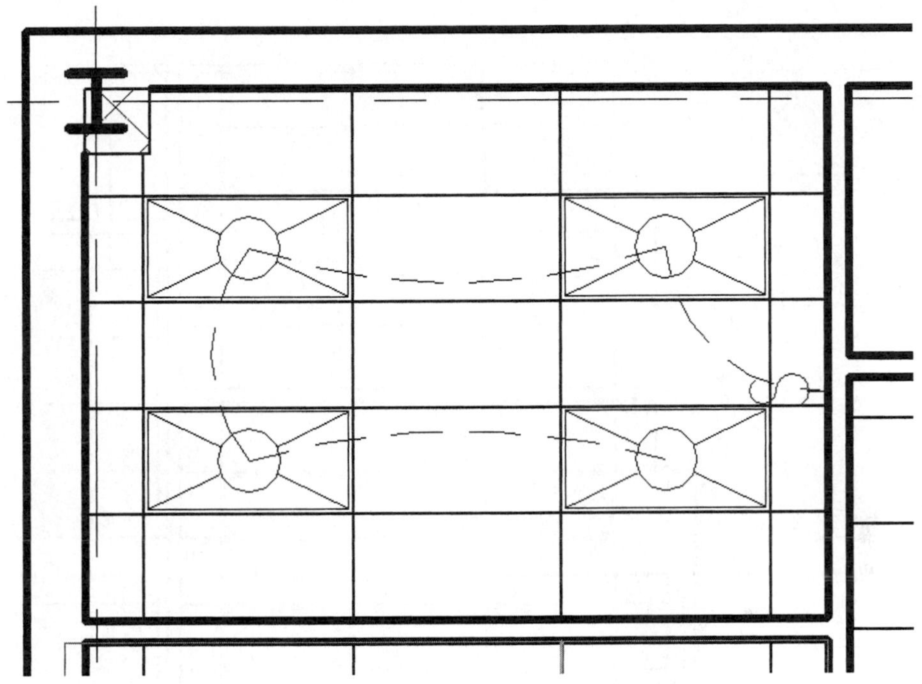

Figure 12–19 *Sketch a three-point arc Drafting Line for the wiring*

9. Continue drawing arcs to connect all the lights in the room to the switch.

   - On the Design Bar, click the Modify tool or press the ESC key twice.

COMPLETE THE RCP

At this point, you can continue adding elements to the reflected ceiling plan as needed. Following procedures covered in previous chapters, you can add text, dimensions, and tags as necessary. If you wish to add Room Tags, they will "see" the Rooms that we have already added to this model in the previous chapter including the Room Separation lines and report the room's parameter values, i.e. Room Name, Room Number, Area, etc. If the Room Tag does not fit comfortably in the Room, add it first within the space to properly associate it with the Room, then on the Options Bar, check the "Leader" checkbox to add a leader and enable the Room Tag to be moved outside the boundaries of the Room (see Figure 12–20).

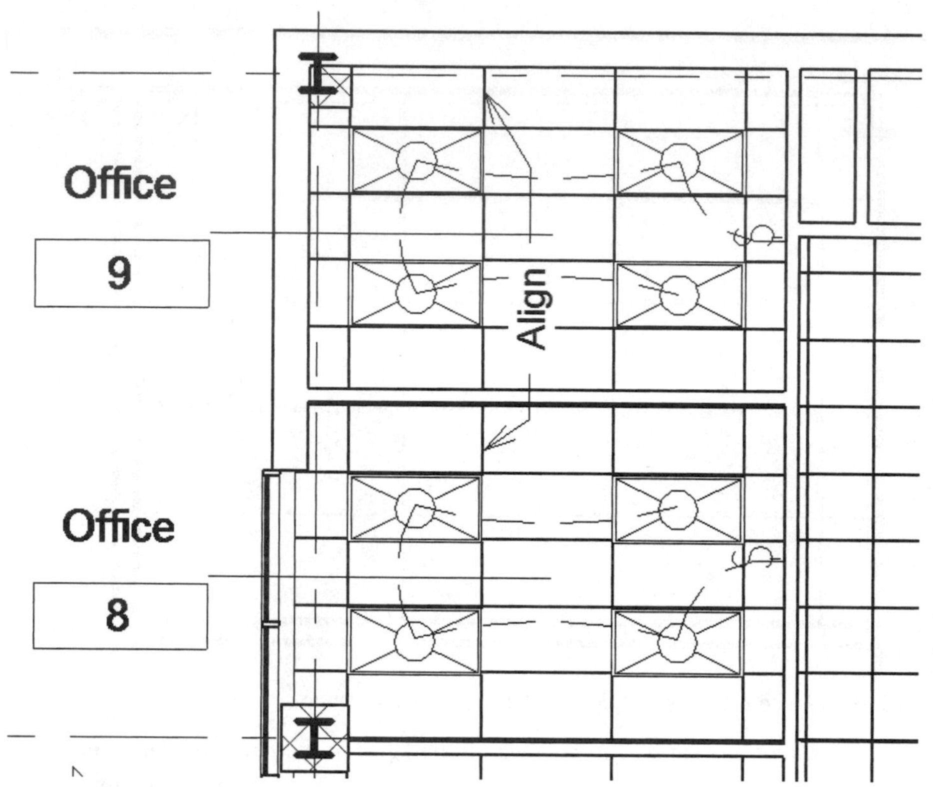

Figure 12–20 *Adding notes and tags*

10. When you are finished add elements to your reflected ceiling plan, save the project.

CREATING INTERIOR ELEVATIONS

No construction documents set would be complete without interior elevations. Adding such Views is easy to do. Our reception area could use some interior elevations, let's add some now.

ADD AN INTERIOR ELEVATION

1. On the Project Browser, double-click to open the *Level 3* floor plan View.
2. On the Design Bar, click the View tab and then click the Elevation tool.
   - From the Type Selector, choose Elevation : Interior Elevation.
   - On the Options Bar, from the "Scale" list, choose **1/4"=1'-0"** [**1:50**].
3. Move the cursor into the reception area—do not click yet.

- Move the cursor around the room and watch the orientation of the elevation head change dynamically (see Figure 12–21).

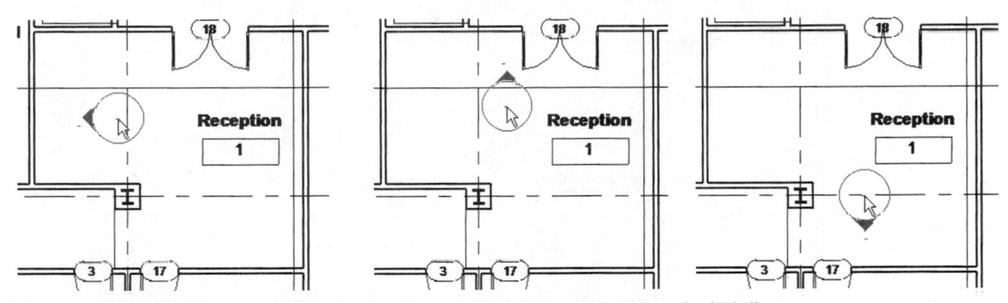

Figure 12–21 *Elevation Heads orient automatically to the nearby Walls*

The orientation of the elevation symbol automatically orients to a nearby Wall. This makes creating single elevations easy, especially to angled or curving walls. In this case, we want four elevations—one in each direction. So it does not matter which way it points for the first one, but pointing up is a good choice.

4. When the symbol is pointing up, click to place the elevation marking in the room.

- On the Design Bar, click the Modify tool or press the ESC key twice.

You will notice that a new node has appeared in the Project Browser for the Interior Elevations. In addition, if you expand it, a new View associated with the elevation marker has also appeared named *Elevation 1 - a*.

5. Double-click on triangle part of *Elevation 1 – a* to open it.

Notice that the View is cropped nicely to the size of the Room. You can adjust the cropping if you need to fine-tune it.

ADD ADDITIONAL INTERIOR ELEVATIONS

We can add additional elevations of the reception space very easily.

6. On the Project Browser, double-click to open the *Level 3* floor plan View again.

7. Click on the Elevation Marker (the circle part of the marker) in the middle of the reception space.

A series of checkboxes will appear surrounding the symbol and ghosted arrows pointing to up to four possible elevation directions. To add another elevation View, simply check one of the boxes.

- Place a checkmark in each of the three other checkboxes surrounding the elevation symbol (see Figure 12–22).

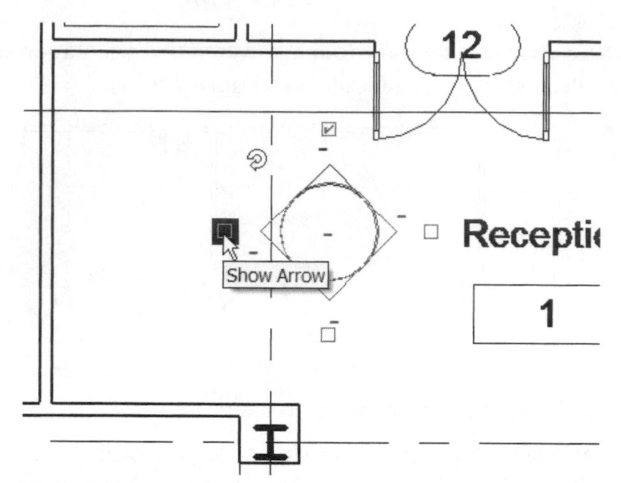

Figure 12–22 *Add new Elevation Views by checking the direction checkboxes*

The new Views will appear on the Project Browser.

8. On the Project Browser right-click on the *Elevation 1 – a* View and choose **Rename**.
    - Name it **Reception – North Elevation.**
    - Repeat for each of the other three elevations.

You can complete these Views in any way you wish. Add notes, dimensions or tags. Drop them onto a new Sheet to round out the set (see Figure 12–23).

To turn off the View name on the Elevation marker edit the Type properties of the marker. To do this go to the **Settings menu > View Tag > Elevation tags.**

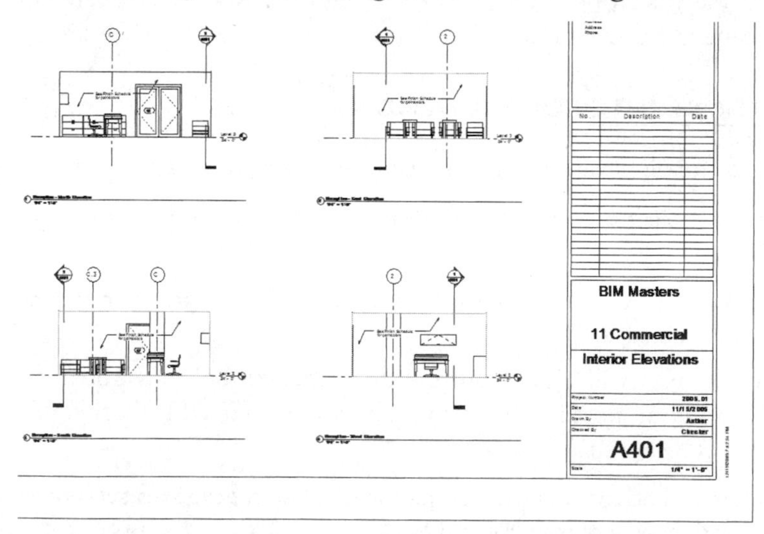

Figure 12–23 *Annotate Interior Elevations and add them to a new Sheet*

SUMMARY

Reflected ceiling plans and interior elevations are two important parts of a complete construction documentation set. In this chapter, we have taken a brief look at the steps to create both of these important document types.

Ceiling elements are easily created from existing Walls by picking points within a Room.

If bounding Walls are not available, or if you wish to create a custom Ceiling shape, you can Sketch the Ceiling.

Ceiling Types control the structure of a Ceiling element.

Acoustical Tile Ceilings are simply Ceiling Types that have a finish layer assigned to an appropriate Material and surface pattern.

The grid lines in the Material use a Model Pattern. The stippling uses a Drafter Pattern.

To create an underpinned Ceiling, use the Sketch option.

Adjust the height of Walls to coordinate with the underpinned Ceiling after creation.

Drywall Ceilings are created with the same process and simply use a different Type.

You can change grid size by swapping the Type.

Light fixtures and switches are component Families that you load and place in the Ceiling plan.

Add wiring using Detail Lines.

Add Interior Elevations using the Elevation View tool.

Check more than one box on the interior elevation symbol to add additional elevations.

Appendices

The Appendices include additional important resources. Worksharing is an important part of Revit Building workflow, enabling teams to work on projects efficiently and effectively. Key concepts and issues about Worksharing are introduced in Appendix A. Look to Appendix B for information on printing your Revit Building designs. Appendix C and D provide additional resources that you might find useful as you begin using Revit Building in your daily work. Also be sure to explore the contents of the files installed from the CD for other useful files and resources.

Section IV is organized as follows:

Workshariing

INTRODUCTION

The term "Worksharing" applies collectively to the various techniques used in Autodesk Revit Building to work in teams of multiple individuals and firms. Worksharing utilizes processes we have already seen in this book such as linked Revit (RVT) files and linked AutoCAD (DWG) files, as well as the internal segmentation of a project using Worksets. Worksets is the Revit Building toolset that allows a single project file to be organized into smaller pieces with which team members can work independently without impeding the work of others. Care must be taken when enabling Worksets to ensure that a strategy for their use appropriate to the team dynamics is established. In this appendix, we will introduce the concepts and key terminology used in Worksets as well as briefly discuss general Worksharing issues.

OBJECTIVES

Many resources are available to the reader for learning and understanding the concept of Worksets. This appendix will provide an overview of the salient concepts and suggestions for further reading in the Autodesk Revit Building Help system and online resources. After completing this appendix you will understand:

- Key Worksharing tools available
- Workset Terminology
- Where to find additional Worksharing Resources

TYPES OF WORKSHARING

Several forms of Worksharing are possible in Autodesk Revit Building. Architectural projects usually involve teams of professionals either within the same firm (under the same roof) or dispersed among several companies and physical locations. Whether you simply need to load a CAD file as a background for your own design work or you need to manage a fully-coordinated Revit Building model among several members of your firm, Revit Building has tools and capabilities suited to the task.

LINKING AND IMPORTING

In earlier chapters, we explored two forms of linking: linked RVT files and linked CAD files. If you need to simply keep track of work being done in another application such as AutoCAD or Microstation, then file linking is the appropriate solution. DWG (AutoCAD) and DGN (Microstation) files can be either linked or imported into your Revit Building model. If you wish to maintain the ability to update the file periodically as the original author of the file makes changes, choose to link the file. In this scenario, you simply reload the linked file when you receive an updated version from your consultant or teammate. If you instead need to use the geometry in the CAD file to assist you in creating your Revit Building model and have no need to reload changes in the future, you can Import the file instead. This places a static copy of the file within your Revit Building model. You can leave this imported file intact as a single element in the Revit Building model or even choose to explode it. If you explode it, it will convert the internal geometry to simple Revit Building Drafting and Model Lines. You should not explode these files unless necessary, as it will increase overhead and memory demands on your system depending on how large and extensive the imported files are. Examples of linking and importing DWG files can be found in Chapters 4, 9 and 10.

While linking to CAD files does help bridge the gap between your firm and those not using Autodesk Revit Building, it is always better if you can get the entire project team working in the same file format and software. Therefore, wherever possible, having all team members using Revit Building or Revit Structure is preferable. You can still work in separate models and utilize linking in this scenario as we did in Chapter 5. In the case of RVT files, you can only choose to Link. Importing an RVT file is not an option. Should you decide that it is desirable to "merge" two RVT files into a single file, you can use Copy and Paste or Group Save and Group Load instead. Open one of the files, copy all of the elements that you wish to merge, and then paste them into the other project. Group all of the model, annotation, and datum elements you wish to merge, save the group file out to a RVG file and then load that RVG file into the other project.

COORDINATION MONITOR

When you link two Revit Building files together, (such as a Revit Building and Revit Structure file) you can use the Coordination Monitor tool to keep track of duplicate elements or elements that rely upon one another in each file. For example, if the Structural Columns are in the Revit Structure file and the partitions and the Architectural Columns are in the Revit Building file, these elements can get out of synch as users in each model make changes. Using the **Copy/Monitor, Coordination Review** and **Interference Check** items on the Tools menu, you can copy elements between files, watch them for changes and make updates to keep both linked projects synchronized. Look for more information about these tools in the online help.

WORKSETS

Even if all work on a project is carried out within the physical walls of your own office, it is typical that more than one member of your firm will need to access and work in the project at the same time. In this situation, use "Worksets" to organize a project into smaller pieces, each enabling a different member of the project team to work simultaneously on the same project file. Each Workset remains linked back to a Central file that keeps all changes and team interactions coordinated. While Workset use is most common on a team project, it can be used in any Revit Building project and some Sole Practitioner firms find it useful as a tool for managing project data and visibility of elements. Whatever your specific situation, achieving success with Worksets requires a clear understanding of each tool, careful pre-planning and ongoing management.

NOTE The following is a brief introduction to the concept of Worksets. It is intended to introduce the concepts, define the terminology and suggest some common scenarios for their usage. A comprehensive tutorial-based exploration of the topic falls out of the scope of this text. Explore the online help and various online resources on the topic and consider attending a formal training session at a local training provider or Autodesk reseller.

GETTING STARTED WITH WORKSETS

Worksets must be enabled in a project and then set up. Typically this task should be performed by a single member of the project team knowledgeable in Worksets and their nuances. This person is often referred to as the "Project Coordinator" or some similar title. They might be a CAD Manager, or simply a Project Architect on the project who has good knowledge of Autodesk Revit Building.

To enable Worksets the first time, choose the command from the File menu, or display the appropriate toolbar (see Figure A–1).

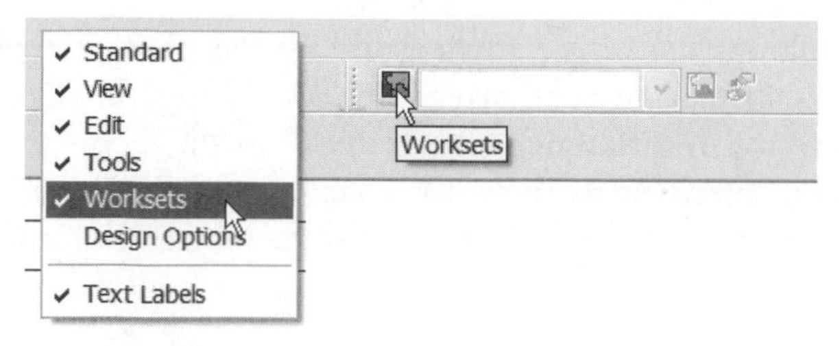

Figure A–1 *Display the Worksets toolbar*

Every member working on the team should have an understanding of the basic concepts of a Workset-enabled project. In addition, it is highly recommended that each member of the team completes the tutorial on the subject provided with the software and multiple team members practice together on a sample workset project. To access the tutorial, choose **Tutorials** from the Help menu. In the Help window, expand the "Using Advanced Features" topic, then select the "Sharing Projects" topic. Read and complete all exercises beneath this topic (see Figure A–2).

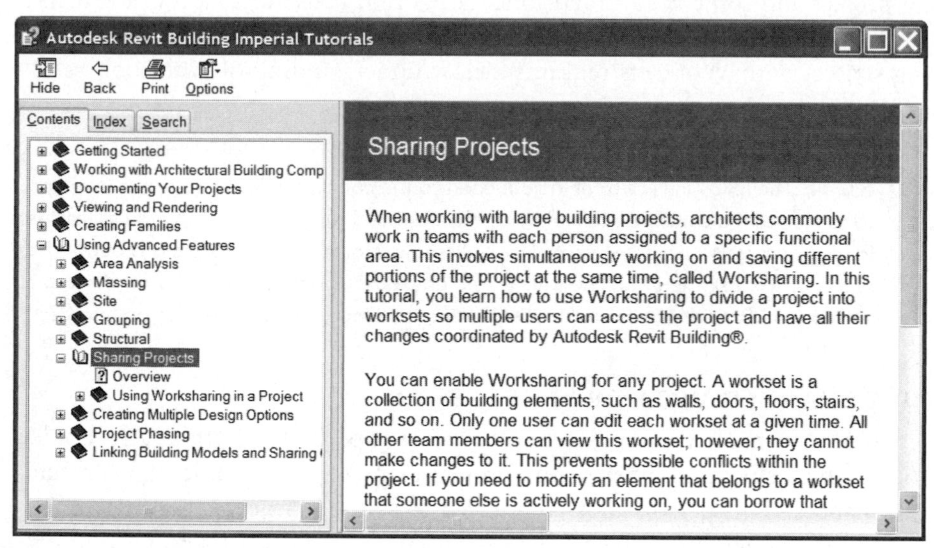

Figure A–2 *Worksharing Tutorials accessed from the Revit Building Help menu*

You may need to return to your installation CD to install the required tutorial files. Depending on the options that you chose when you installed your Revit Building software, these files may or may not already be on your system. Several other topics and tutorials are also included in the Help system. It is worthwhile to spend some time exploring this valuable resource.

NOTE An additional resource is included with the files from the Mastering Autodesk Revit Building CD ROM. A White Paper published by Autodesk is included in PDF form with the files from the CD. This document is packed with general information on the topic and is summarized in the topics that follow. Browse the CD files folder in the *Autodesk Documents* folder and open the *Multi_User_Collaboration_Revit_8-10.pdf* file.

UNDERSTANDING WORKSETS

The concept behind Worksets is simple. Normally, in a computer environment, only one user at a time can access a particular file. Since by default, Revit Building places all project elements within a single file, this would naturally impede teamwork. Using Worksets, there is still a single "Central" file that houses all project data, but from this Central file, each team member can save a local copy that maintains its association with the Central file. In this way, when any team member saves work from their local copy back to the Central, those changes are merged into the Central file and become available to all members of the team. Simply having the Central and Local files is not enough however, the vast collection of potentially editable elements within the project files must be managed. Worksets are used for this purpose. A Workset is basically a named group of elements that can be "checked out" by an individual user. Once checked out, a Workset becomes "read-only" to other members of the team. This allows those members to view any part of the project that they wish, but only edit those parts that they have explicitly checked out. When saves back to Central are made, users can choose to "relinquish" their Worksets and/or check out other ones. The concept is similar to a public library or video rental store. If you wish to check out a particular book or video, it must be in the store at the time of your visit. If someone else has already checked out the item, you must wait for them to return it before you can check it out.

The biggest challenge involved in a Revit Building Workset-enabled project is deciding how to organize the elements into Worksets. You want to have enough Worksets for the quantity of team members, but don't want to have so many that management of them becomes problematic. Certain Worksets are created and maintained automatically by Revit Building. These include "Views," "Families" and "Project Standards" Worksets.

- **View Worksets:** For each View, a dedicated View Workset is created. It automatically contains the View's parameters (scale, visibility, graphics style, etc.) and any View-specific elements such as text notes, dimensions, detail elements, etc. View-specific elements *cannot* be moved to another Workset.
- **Family Worksets:** one Workset is created for each loaded Family in the project.
- **Project Standards Worksets:** one Workset for each type of project setting; such as Materials, Line Styles, etc.

These Worksets are created and maintained automatically by Revit Building. In addition, Revit Building also allows "User Defined" Worksets. These contain all of the building model elements in a project. By default, Revit Building will create two such Worksets when Worksets is first enabled: "Shared Levels and Grids" and "Workset1." All of the Levels and Grids are moved automatically to the "Shared Levels and Grids" Workset and all of the rest of the geometry is moved to "Workset1."

An element can only belong to one Workset at a time. You can move an element from one Workset to another, but you cannot assign it to more than one at the same time. This is why careful planning is important. The Project Coordinator must therefore attempt to anticipate the needs of the project team and create Worksets to house model elements in a way that supports these needs. While there are no "rules" regarding this, there are common best practices.

The factors to consider include:

- Project size
- Team size
- Team member roles
- Default Workset visibility

Each of these factors may have an impact on the use and composition of each Workset. Larger projects and larger project teams will typically have more Worksets. This stands to reason. However, even small projects can benefit from Worksets, so this is not the only factor. When you open a project that has Worksets enabled, you can choose at that time which Worksets to load. Choosing to open only those Worksets needed for a particular task can help files load more quickly and preserve valuable computer resources. You can also take advantage of Workset visibility to hide entire Worksets in one or more Views as appropriate. Therefore, any or all of these factors can play an important role in determining the ultimate composition of Worksets in a particular project. Also remember that each project is unique and while you may follow many common strategies from one project to the next, you must always be flexible enough to allow for specific circumstances that may arise in a particular project.

ENABLING WORKSETS

As was mentioned above, it is typically advisable for a single experienced member of the project team to take responsibility for Workset management and setup. Worksets must be enabled before they can be used. Then a Central file must be saved in a location that all team members can readily access. This location is typically on a network server on the company LAN (Local Area Network).

NOTE While it is possible to host a shared project over a WAN (Wide Area Network) it is only advisable if the bandwidth available to all team members is up the task.

To Enable Worksets, choose the Worksets command from the File menu, or click the icon on the Worksets toolbar. A dialog like the one in Figure A–3 will appear.

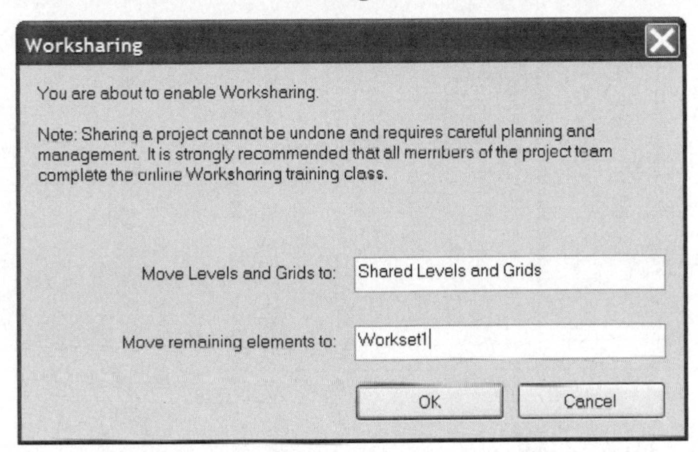

Figure A–3 *Worksharing must be enabled. Revit Building will suggest two User Defined Worksets to start*

By default, two User-Created Worksets will be suggested by Revit Building: One for Shared Levels and Grids containing those elements, and another called "Workset1" that will contain all of the other elements. You can accept these names or choose alternate names. You can also rename a Workset later (provided that no users are accessing the project). It will take a moment for Worksharing to be enabled; once complete, the "Worksets" dialog will appear and list each Workset in the project. Additional User-created Worksets can be added at this time, or later. You can also use this dialog to list all of the "Views," "Families" and "Project Standards" Worksets (see Figure A–4).

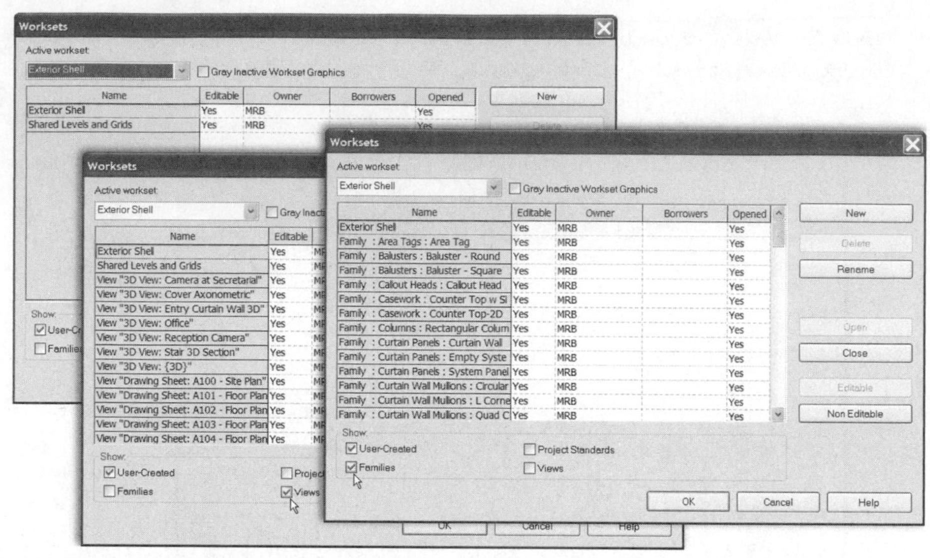

Figure A–4 *Use the "Worksets" dialog to View and edit Worksets*

CREATE A CENTRAL FILE

Before users can begin working in a Workset-enabled project, there must be a Central file. This file is the main "hub" of the project. It should be located on a network server accessible to all team members. After Worksharing has been enabled, you use the **Save as** command on the File menu to save a Central file. The first time that Save as is chosen after enabling Worksets, the file saved automatically becomes a Central file. After that, if you wish to create a new Central file for any reason, you must choose Save as from the File menu, click the Options button and then place a checkmark in the "Make this the Central location after save" checkbox (see Figure A–5).

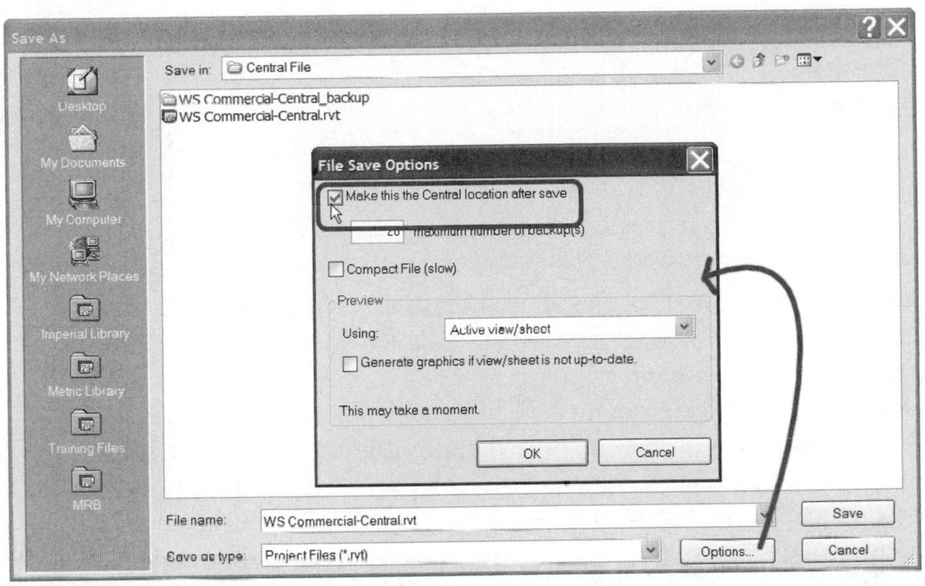

Figure A–5 *Making a file the Central file after save*

NOTE While it is possible to move or make a copy of the Central file (see the PDF While Paper on the CD ROM noted above for more information on this) it is highly recommended that you maintain only one Central File to avoid confusion to project team members and potential loss of work. You can and should regularly back up the Central file using whatever method is currently in place in your firm, but avoid simply making a separate copy to another location on the network as users may mistakenly open this file and create Local files from it.

CREATING ADDITIONAL WORKSETS

After the Central file has been saved, you can return to the "Worksets" dialog and create additional Worksets required by the project team. In the "Worksets" dialog, simply click the New button to add a Workset. You can name each descriptively. An important consideration when creating new Worksets is their default visibility. If you wish to have the Workset automatically visible in all project Views by default, then check the "Visible by default in all views" box when creating it. However, in many cases, it will be better to leave this setting disabled and allow users to control the visibility of each Workset manually. This can help increase performance on large projects (see Figure A–6).

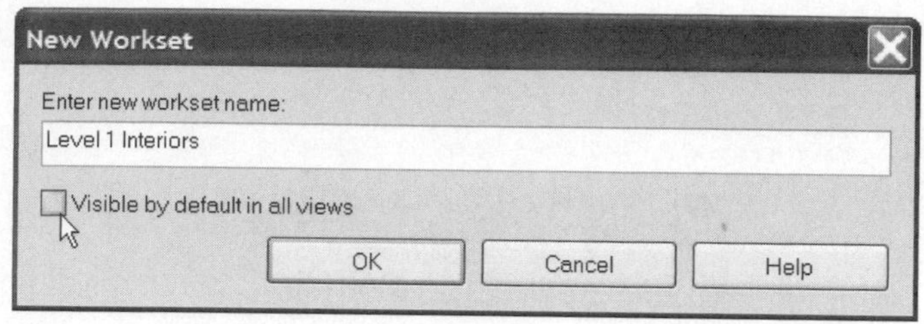

Figure A–6 *You can disable visibility by default of a new Workset if desired*

When setting up the initial Worksets, try to divide building elements into logical working groupings. These will often be "task-based" to support the work of the team member who will author its contents. For example, in a typical commercial office building like the project constructed in this book, you might create a Workset for the exterior shell of the building, the core elements, the lobby and one for each floor's interior elements. Furniture might also be separated as might other specialty equipment (see Figure A–7).

Worksets

Active workset: Exterior Shell

Gray Inactive Workset Graphics

Name	Editable	Owner	Borrowers	Opened
Core Elements	Yes	MRB		Yes
Exterior Shell	Yes	MRB		Yes
Level 1 Interiors	Yes	MRB		Yes
Level 2 Interiors	Yes	MRB		Yes
Level 3 Interiors	Yes	MRB		Yes
Level 4 Interiors	Yes	MRB		Yes
Lobby and Entrance	Yes	MRB		Yes
Shared Levels and Grids	Yes	MRB		Yes
Level 3 Furniture	Yes	MRB		Yes

New | Delete | Rename | Open | Close | Editable | Non Editable

Show: User-Created | Families | Project Standards | Views

OK | Cancel | Help

Figure A–7 *Some Worksets added to the commercial project*

You can add new Worksets at any time in a project. It is however a good idea to try to establish the basic Workset organization as early as possible. This will make it easier for project team members to become comfortable with the project and its organization. As new Worksets are added later in the project, be sure that these and their functions are clearly communicated to the project team.

MOVING ITEMS TO WORKSETS

Once Worksets have been created in the Central file, existing elements in the model must be moved to the appropriate Workset. For instance, if a Workset named "Level 1 Interiors" has been created, then any interior elements such as partitions, doors, etc. on Level 1 should be moved to this Workset. This way, when a team member opens their local file and checks out this Workset, they will gain access to all of these elements.

To move existing elements to a different Workset, select one or more elements, click the Properties icon on the Options Bar, and then choose the desired Workset from the "Workset" list (see Figure A–8). When Worksets are enabled in a Revit Building project a Workset parameter will be automatically added to every model element and appear on the Element Properties dialog.

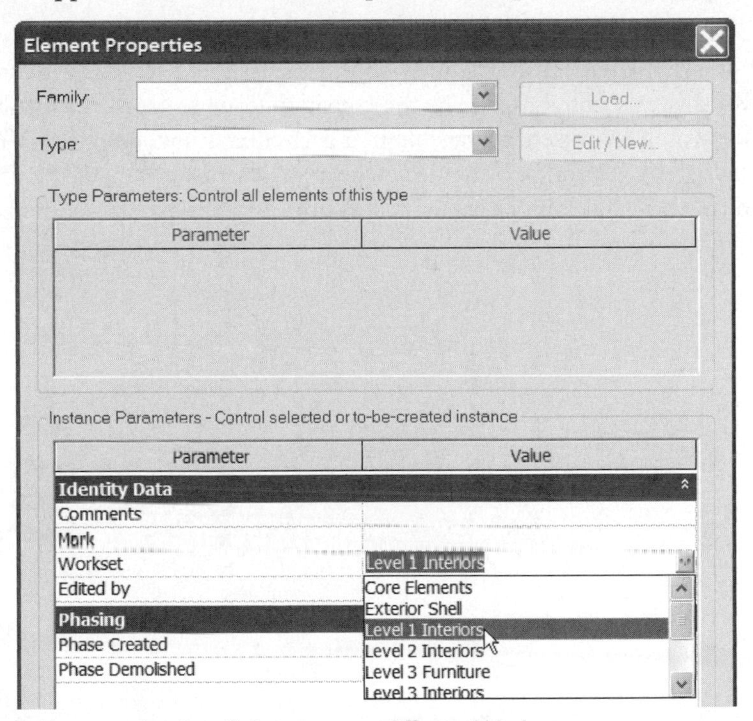

Figure A–8 *Move a selection of elements to a different Workset*

When you work in an Workset-enabled project, the tool tips that appear when elements are pre-highlighted will now report the Workset in front of the usual Element Class : Family : Type designation (see Figure A–9).

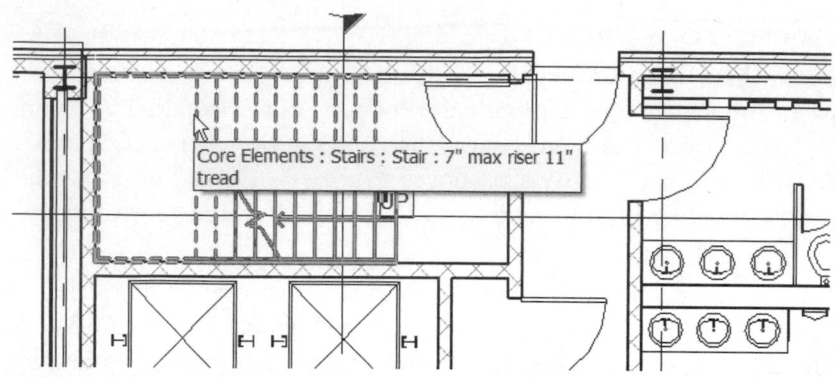

Figure A–9 *Tool tips in Workset projects will report Workset : Element Kind : Family : Type*

WORKSET VISIBILITY

Worksets can be set to display or be hidden in each View of the project. When Worksets are created, the default visibility setting is assigned to the Workset. In the topic above, this was noted as one of the considerations to factor into Workset creation. To see which Worksets are visible in a particular View, open the Visibility/Graphics dialog. In a Workset-enabled project, a Worksets tab will appear in the dialog. On this tab, you can choose which Worksets you wish to see in the current View (see Figure A–10).

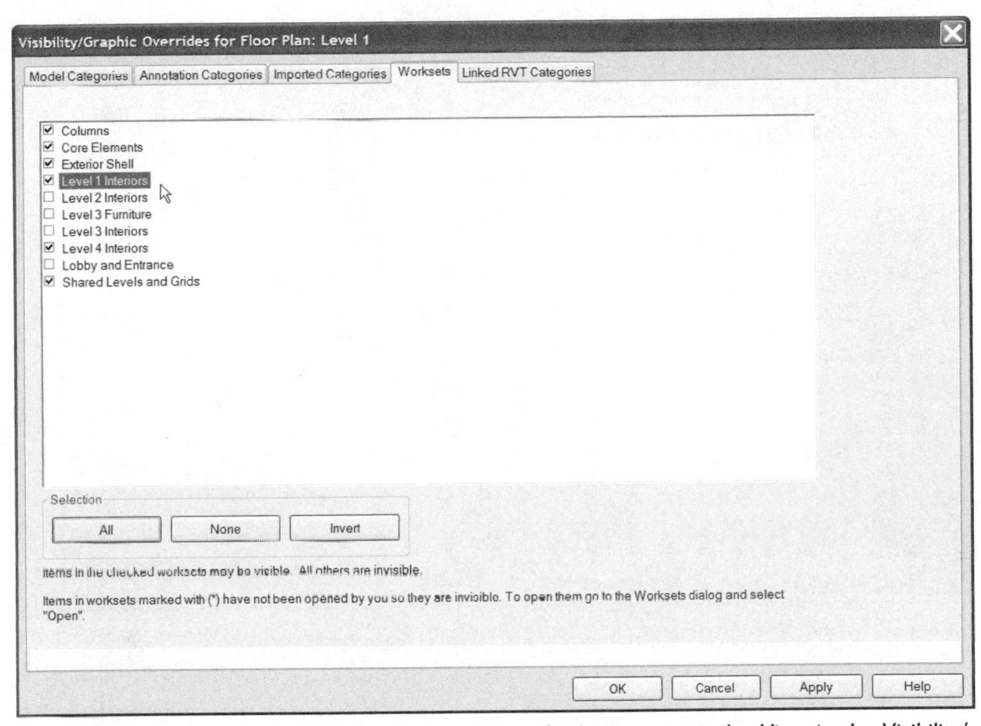

Figure A–10 *Choose which Worksets you wish to display in a particular View in the Visibility/ Graphics dialog*

This functionality offers a powerful way to manage the specific visibility of elements in a project. To fully realize the potential of this functionality, take care in the planning stages to determine the default visibility of each Workset in each View of the project. Users can change the settings later, but it is a good idea to establish effective defaults ahead of time.

CREATING A LOCAL FILE

Each member of the project team must create and work in a Local copy of the Central file. From this Local copy, users can check out the Worksets in which they need to work and can work disassociated from other team members. When a user checks out a Workset, it is locked in the Central file and becomes read-only to other team members until that user relinquishes it.

Creating a Local file is easy. Open the Central file and choose **Save as** from the file menu. Save the Local file in a convenient location with a unique name. The Local file can be saved to the hard disk of your local computer. This will increase performance when saving and loading the Local file. When you create your Local file, you can choose which Worksets you wish to open. In larger projects, this can make load times quicker (see Figure A–11).

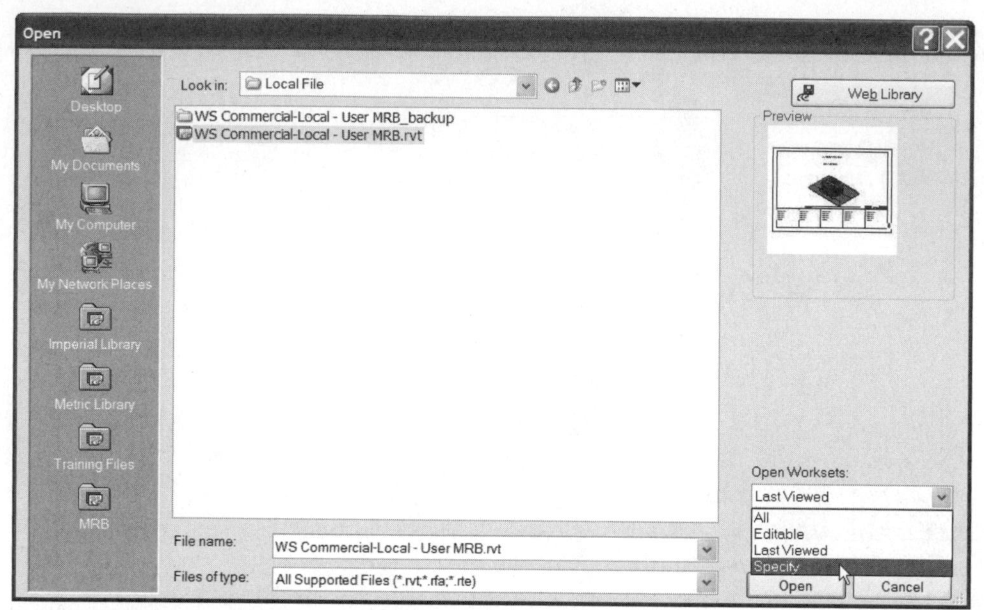

Figure A–11 *When opening a Local file, choose Specify to load a selection of Worksets*

After choosing "Specify" from the "Open Worksets" option in the "Open" dialog, an "Open Worksets" dialog will appear. Choose the Worksets that you wish to open by using the SHIFT and CTRL keys to select multiple items. With various Worksets highlighted, click the Open or Close buttons as desired (see Figure A–12).

Opening Worksets

Name	Editable	Owner	Borrowers	Opened
Columns	No			Yes
Core Elements	No			Yes
Exterior Shell	No			Yes
Level 1 Interiors	No			No
Level 2 Interiors	No			No
Level 3 Furniture	No			Yes
Level 3 Interiors	No			Yes
Level 4 Interiors	No			No
Lobby and Entrance	No			No
Shared Levels and Grids	No			Yes

New
Delete
Rename
Open
Close
Editable
Non Editable
Show:
User-Created
Project Standards
Families
Views
OK
Cancel
Help

Figure A–12 *Open and Close Worksets in the "Opening Worksets" dialog*

When you click OK on this dialog, the model will load with only those Worksets that you specified. You can see this by opening Views that require a Workset that you did not load. The required elements will not appear in that View. Should you realize that you need to open a Workset that you did not choose to open initially; you can simply launch the "Worksets" dialog, and then select the Workset you need and click the Open button. You can Close Worksets that you no longer need in the same fashion.

EDITING WORKSET ELEMENTS

When you first open your Local file and choose the Worksets you wish to Open, none will be editable. Opening a Workset does not automatically make it editable. This action is achieved in the "Worksets" dialog. There are a few approaches you can take to editing the model. In general, if an element that you wish to edit is not being edited by another user, Revit Building will allow you to edit it, even if you do not have the associated Workset checked out. If another user is actively editing this element, or has the Workset checked out, you will need to issue an "Editing Request" to that user. The other user can then choose to allow your edit or refuse it. In this case, you should contact the user via phone, instant messenger or face-to-face to coordinate the edits.

You can "check out" a Workset by making it "Editable" in the "Worksets" dialog. Launch the "Worksets" dialog, select one or more Worksets and click the Make Editable button (see Figure A–13).

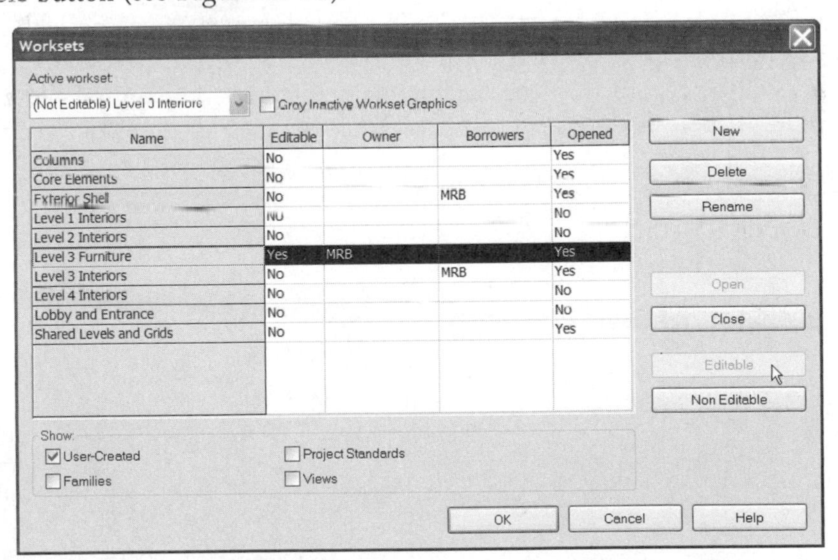

Figure A–13 *Make a Workset editable*

Notice in the figure that the user "MRB" is listed as a "Borrower" to some of the Worksets. This occurs when you edit an element in that Workset that no one else is editing at that time and was not locked for editing. Revit Building will "borrow" the element from the Workset and allow you to make changes. When you Save to Central, the element will be relinquished and the changes updated to the Central file.

Any new elements that you add to a model will be added to the active Workset. You can choose the active Workset from the drop-down list on the "Worksets" toolbar. You can make a Workset active even if it is not editable. This means that you will be adding elements to the Workset, but once you save to Central, you may not be able to edit them anymore. Therefore, if possible, make the active Workset editable when you launch your Local session.

SAVE TO CENTRAL

You should save your work at regular intervals regardless of whether you are working in a Workset-enabled project or not. When you work in a Local file, there are two types of save. Save and Save to Central. When you choose **File > Save**, you are simply saving your Local copy of the project file. As mentioned, it is a good idea to do this regularly, for example, every 15 minutes. Also at regular intervals, (but perhaps not quite as frequently) you should also **Save to Central**. Every 1 or 2 hours would be a practical choice. This command updates the Central file with all of the changes that you have made in your Local file. It also retrieves changes made to the Central file by other team members since you opened your Local copy (basically synchronization), or last Saved to Central, or performed a Reload Latest. When you Save to Central, you also have the option to relinquish your checked out Worksets and borrowed elements. You can add comments to your Save to Central as well. This will be recorded in the log file maintained by the Central file and might be useful if there is ever a reason to roll back changes to a previous version (see Figure A–14).

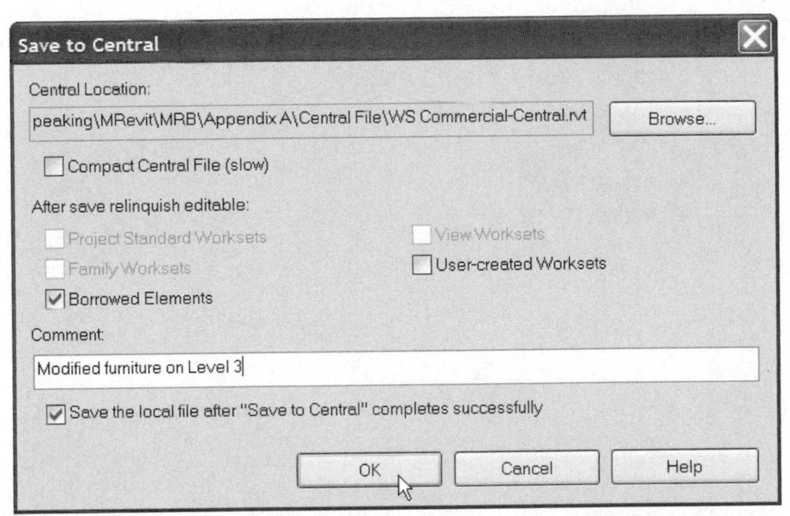

Figure A–14 *Save to Central with options to relinquish borrowed elements, Worksets and save your Local copy after save*

As a general rule of thumb, you should always check the 'Save the local file after "Save to Central" completes successfully' box in the "Save to Central" dialog to have Revit Building also save your Local copy when finished. This helps keep both files synchronized.

GOING FURTHER

This appendix explains the basic concepts of Worksharing and Worksets. The best way to learn and implement this multi-user mechanism is to practice with at least 1 other individual. Remember, you are encouraged to work through the tutorials provided with the software. If you can work with another individual on these tutorials, you will get a better sense of the nuances of working in a Workset-enabled project. A copy of the commercial project has been provided with the files from the Mastering Autodesk Revit Building CD ROM. In the *Appendix A* folder, are two sub-folders: one named *Central File* the other named *Local File.* The project has been saved as a Central file in the Central File folder. It has had Worksets added and all model elements have been moved to the appropriate Worksets. In the Local File folder, a Local copy has been created. If you choose to Open it with the "Last Viewed" Worksets option you can load the same selection of Worksets open before it was saved. You can also choose "Specify" if you wish. Open the "Worksets" dialog to see that Worksets and Borrowed elements were not relinquished on the previous save.

Feel free to experiment in these files. If you have a colleague who can help you, open the Central file, perform a Save as to create another Local copy and have that person work in that one while you work in the Local file that has been provided.

Each of you should make changes, Save to Central, then reload from the Central file. Try borrowing elements, and issuing requests.

SUMMARY

Working in teams in an important part of Architectural production. Using linked files and Worksets, Revit Building provides the means to accomplish sharing of data and managing coordinated team projects. When implemented with care, Worksets provide an invaluable toolset to the extended Revit Building project team.

AutoCAD and Microstation files can be linked to Revit Building models.

Import CAD files when you do not need them to update.

Worksets provide the means to sub-divide a project into parts so that multiple team members can work simultaneously.

Enable Worksets in the project and create User-created Worksets.

Move existing elements to appropriate Worksets.

Save the project file as a Central File on a network server accessible to all users.

Each user creates a Local file from the Central file in which to work.

Decide which Worksets to Open when opening the Local file.

Decide which Worksets to make Editable.

You can edit elements that are not in an editable Workset if they are not being edited by other users.

Editing an element in a non-editable Workset is called "borrowing."

Borrowed elements and Editable Worksets can be relinquished during Save to Central operations.

Save your Local file and Central files often.

APPENDIX B

Printing

INTRODUCTION

Printing (or plotting) from Autodesk Revit Building is nearly Identical to printing from any Windows software application. You choose **Print** from the File menu, configure your options and then print. In this appendix, we will look at the basic process of printing from Revit Building.

OBJECTIVES

Creating printed output from Revit Building is simple. After reading this appendix you will know how to:

- Configure Print Setup options
- Print to a hard copy printer
- Create a multi-sheet DWF file

DATASET

For the purposes of the topics covered in this appendix, you can open any Revit Building project. Both the final version of the commercial project and the residential project in Imperial units has been provided in a folder called Appendix B. You can practice and print from these projects if you wish, or open the "Complete" version of either project from any chapter folder instead.

PRINT SETUP

The "Print Setup" dialog box has many settings that can be configured to enhance the quality of printed output. Choose Print Setup from the File menu to access this dialog (see Figure B–1).

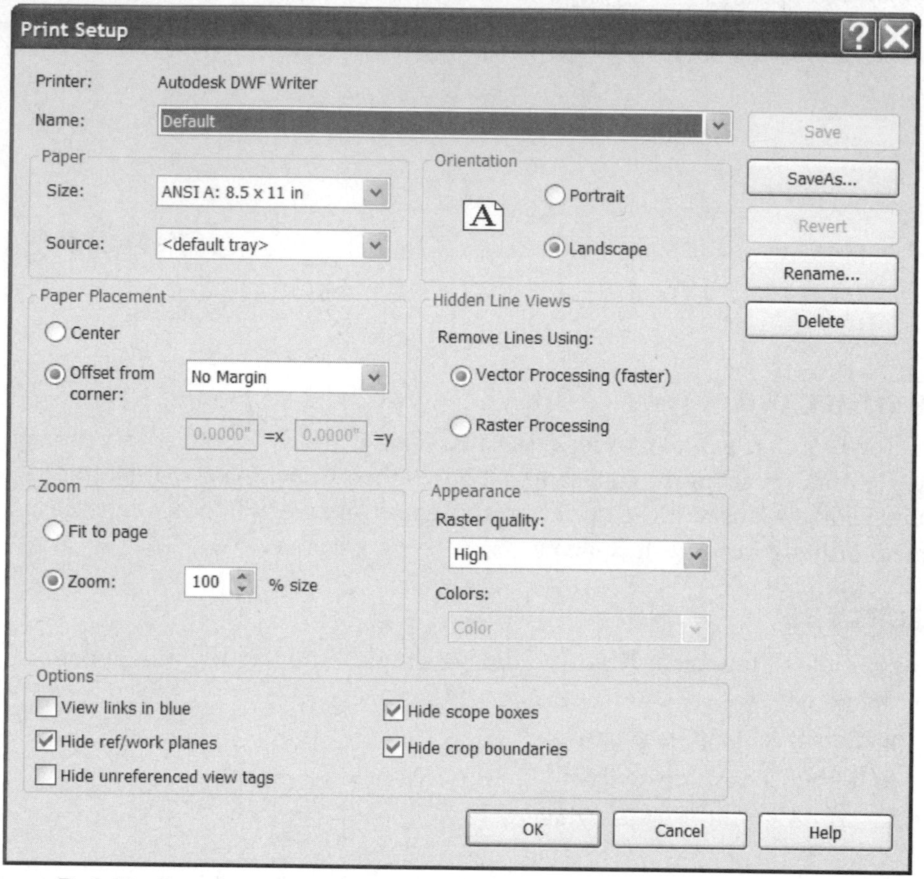

Figure B–1 *The Print Setup dialog*

If you make changes to any of the default settings in this dialog, you can click the SaveAs button on the right and give the configuration a new name. The Print Setups are stored in the project. This custom configuration will then be available to you when you print the project in the future. If you want to read a description on any element in the dialog, click the question mark (?) icon at the top right and then click it on the item in question. A help window will appear with a detailed description of the item. Be certain that the printer you wish to use is listed at the top of the dialog before you configure the options in the dialog. If it is not listed, click Cancel, choose Print from the File menu, choose the desired Printer from the list and then click the Close button. Reopen the Print Setup dialog and continue.

Paper: Set the correct paper size. Only sizes that the printer driver supports will be available.

Orientation: Set to Portrait or Landscape.

Paper Placement: Center will work for most printers. Otherwise use Offset from corner and its associated options. You may need to experiment with your printer and driver to find the right combination for correct placement.

Zoom: To print to the View's current scale it must be set to 100% of size. To print a half sized set change the value to 50%. Keep in mind that everything will be smaller including all annotations.

Fit to page will also affect the actual size of the printed output unless it happens to be 100%.

Hidden Line Views: For any view in Revit Building that is set to Hidden line, for example plan views, elevation views, sections views, and most construction document views will be set to Hidden Line, it is frequently faster to use Vector processing. The time it takes Revit Building to process the Vector print job depends on the size of the project, the type of model objects in the view, and the amount of hidden line removal that must be processed.

At some point Vector processing a View can be slower than Raster processing. If Revit Building's Vector processing time is taking more than a minute try setting it to Raster processing.

Raster processing breaks up the View into a grid of squares, takes a series of digital-type "pictures" of the view then sends the raster data to the printer. The number of squares in the grid depends of View size, paper size, and Raster Quality (set below in Appearance). Raster processing time will be basically the same whether the view is the world's largest building or whether it is blank.

Printer Processing time: Vector data will be processed faster by the printer. Raster data is normally slower. Both can produce very high quality.

If the view contains any Raster data, color fills, shaded surfaces, renderings, etc, Revit Building will automatically print it with Raster Processing.

Appearance: Set the appropriate Raster Quality and number of colors. If you are printing to a color printer and want black lines, remember to set the colors to black lines. This will result in all lines printing black—even ones that you intend to be gray. Setting the colors to grayscale will print any lines in the View that are black in solid black ink, and every other color will be a value of gray. If the gray printed output does not look exactly the way you want try setting the colors to Color and then set the printer driver to black and white. This often gives slightly different grayscale results that might be useful.

Options: This tells Revit Building to print or hide certain elements in the view.

Name: This is the name of the Print Setup settings. Click Save or Save As to create a new Print Setup. This is remembered in the project for future use.

PRINT

When you are ready to print, choose **Print** from the File menu. You can print from any Revit Building View. Most often you will want to print Sheets because they are designed for this purpose and include title blocks and borders, however, if you wish to print from another View you can do that as well.

Always choose your printer from the list first. Click the Properties button if necessary and make any edits to the printer's unique properties (these properties vary by printer device). For the "Print Range" you have the option of the current View, the "Visible Portion" of the current View, or a range of Views or Sheets. Click the Select button to choose a range of Sheets or Views to print (this is sometimes referred to as batch printing). Under Options choose "Reverse print order" if you want to reverse the print order. If a physical printer is used you may also set the number of copies to be printed and whether they are collated. Click the Setup button to open the "Print Setup" dialog and make additional edits. Click OK to print (see Figure B–2).

Current View: prints the entire extents of the currently active view or sheet view.

Visible portion of current window: prints only the part of the view that is displayed within the current window. To set what you want to print, close the Print dialog if it currently open and re-size the View's window accordingly. The proportion of the window is also important and should be similar to the paper size and layout (portrait vs. landscape). If the window proportions do not match the paper proportions the printed output may include or exclude part of the view.

Selected views/sheets: is a mechanism for batch printing. It allows you to pick multiple Views and or Sheet Views and print them at the same time. You can save the selection of Views and or Sheet Views in named sets that can be selected later.

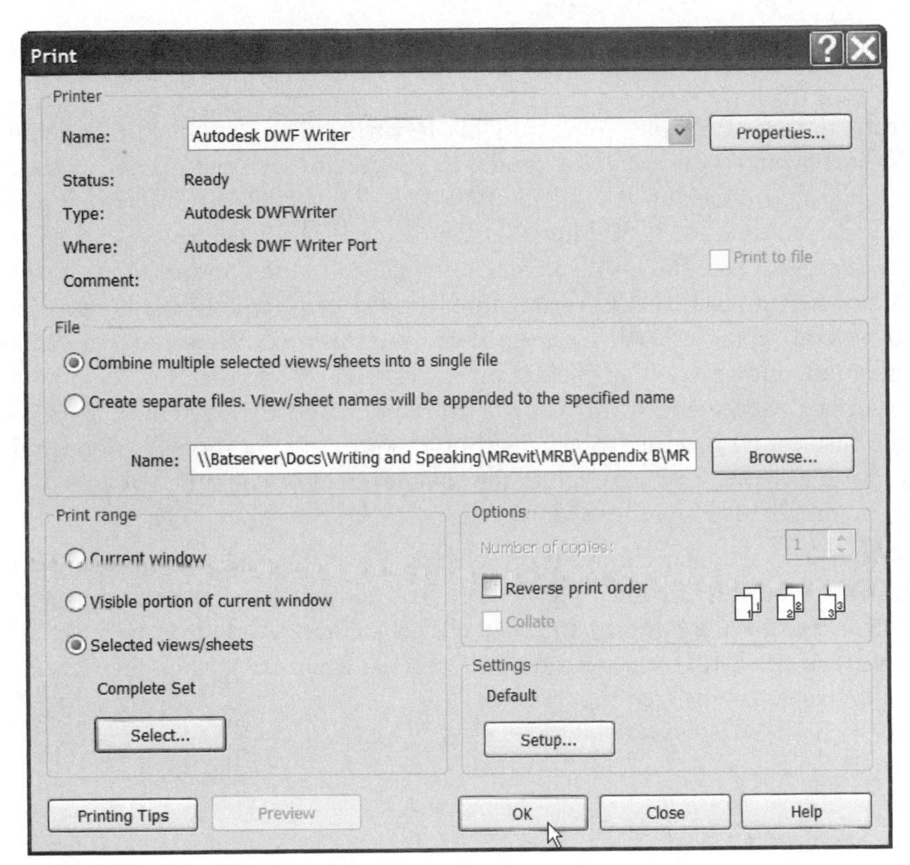

Figure B–2 *The Print dialog*

PRINT A MULTI-SHEET DWF FILE

While many projects require output to actual paper sheets, digital submissions such as DWF files are becoming more popular. A DWF (Design Web Format) file is a highly compressed, vector-based file format designed for viewing and distributing design files over the Internet and by email. What makes the DWF file so powerful is that it is a vector-based, high-quality graphics file that is read only. This means it can be distributed to consultants and clients without fear of unauthorized editing. DWF preserves access to critical design data and graphics from the original model. DWF files can also be embedded in Web pages for viewing in a browser with the plug-in provided free from Autodesk. Anyone with a copy of the Autodesk DWF Viewer software, available as a free download on the Autodesk Web site (*http://www.autodesk.com/*), can view, zoom, pan, query, and print the DWF file. If the recipient has a copy of Autodesk DWF Composer, they can add redline comments to the DWF file, which can then be loaded back into Revit

Building. Creating a DWF is simple, because it is the same as printing to a hard copy device.

In order to create a DWF file from Autodesk Revit Building, you must install the DWF Writer printer software. This printer driver creates a virtual printer in your *Printers* folder that is used to create the DWF files. To install this software, insert your original Autodesk Revit Building CD ROM install disk. You can find the installer on the disk. Once the DWF Writer software is installed, return to the Print dialog. Choose **Autodesk DWF Writer** from the Printer Name list. It is possible to create several separate DWF files—one for each printed View—in the same print operation. However, it is usually more desirable to choose the "Combine multiple selected views/sheets into a single file" option and thereby create a single DWF file that contains several Sheets within it. Verify the location for the saved file and give it a name. You can type in the "Name" field or click the Browse button (see the middle of Figure B–2 above).

In the "Print Range" area, choose the "Selected Views/Sheets" option and then click the Select button. In the "View/Sheet Set" dialog that appears, clear the "Views" checkbox (at the bottom) and then click the Check All button to select all of the Sheets in the list. If you wish to reuse this list again later, click the SaveAs button and give it a name (see Figure B–3).

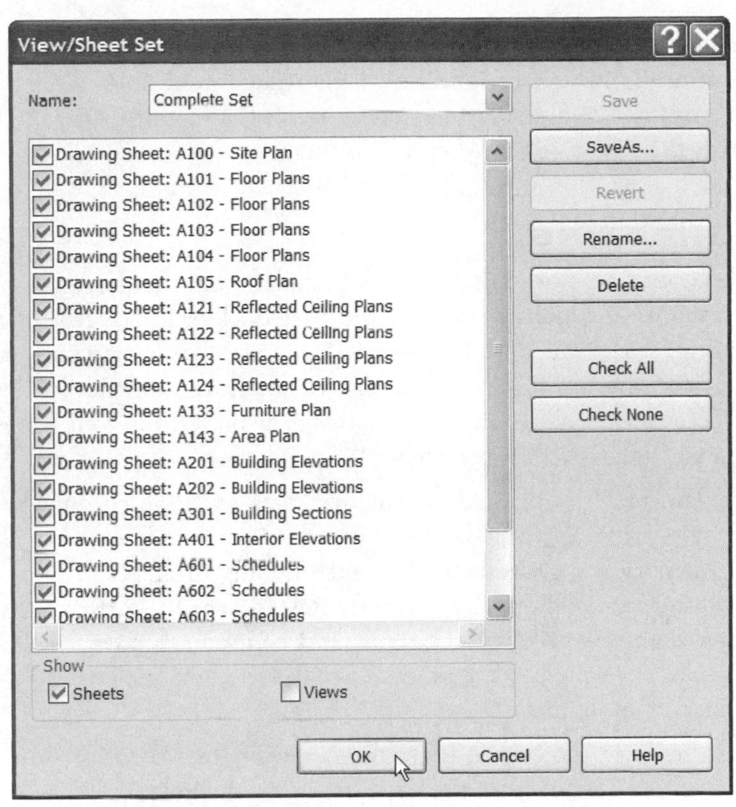

Figure B–3 *Create a Multi-sheet DWF file*

Click OK to return to the "Print" dialog and then click OK again to create the DWF.

When printing is complete, open the DWF file with Autodesk DWF Viewer or DWF Composer software and view the results. You can also create a DWF by choosing the **Export to DWF** option on the File menu.

PRINTER DRIVER CONFIGURATION

Since Revit Building prints using the Windows printing mechanism like most Windows software it is important that the correct printer driver be used. The correct driver is the driver specifically made for the physical printer and model used. To obtain the correct drivers for your printers contact the printer manufacturer and ask for the Windows printer driver for the model printer you have and follow the manufacturer's instructions for installing the driver. Most drivers are available for download from the printer manufacturer's web site.

Frequently Printing/Plotting Services or Bureaus will ask for a plot file in the HPGL format. This will not work with Revit Building. Inform the plotting service that you are printing from Windows-based Revit Building and must use the Windows printer driver for the printer model that they intend to use. Or you can ask if they can print a DWF file. Most service bureaus can print DWFs. In this case, simply print the DWF file and send it to them.

TROUBLE SHOOTING

If your printed output does not look correct or complete, the first thing to check is whether the Print Preview looks correct. If the Preview looks correct that tells us that Revit Building, Windows, and the Printer Driver are all working correctly and the problem is at the physical printer. Note that the resolution of the Preview may not be correct because the high resolution is being displayed on your screen at a lower resolution and that is to be expected.

A common problem includes printed output that does not look complete. Either all the text will be missing or whole portions will not be printed. This is indicative of the printer's memory being overloaded. Revit Building often sends a combination of vector, raster, and other data formats. Raster data in particular is more memory intensive than vector data. As a result, it is common to have a 100 MB or larger print job. Some printers/plotters do not have enough onboard RAM or a built in hard disk to handle this.

Some printer drivers have and option that says something like "Process print job in the computer's memory". To resolve the issue, try using this setting. If your printer driver does not have this setting you can either install more memory in the physical printer, use a printer that has a built-in hard disk and more memory or add a separate RIP (Raster Image Processor) device to aid the printer. Many currently available new large format printers have built in RIPs.

SUMMARY

Printing from Revit Building is simply and straight forward.

Like other Windows software, choose Print Setup to configure the Print options.

Choose Print from the File menu to print to paper or DWF file.

You can create a multi-sheet DWF file of your entire document set in one step.

Online Resources

In this appendix are listed several online Web sites and other resources that you can visit for information on Revit Building and related topics.

WEB SITES RELATED TO THE CONTENT OF THIS BOOK

http://www.paulaubin.com

Web site of the author. Includes information on ordering and updates related to this book and Aubin's other books like *Mastering Autodesk Architectural Desktop* and *Mastering VIZ Render – for Autodesk Architectural Desktop Users* (co-authored with James D. Smell). Check there for ordering information and addenda.

http://www.autodeskpress.com

Web site for Autodesk Press. Visit for information on other CAD titles, online resources, student software, and more.

http://www.rgmarchitecture.com

Web site of technical editor. Includes information on Revit consulting and implementation services.

AUTODESK SITES

http://www.autodesk.com

Autodesk main Web site. Visit often for the latest information on Autodesk products.

http://discussion.autodesk.com

Autodesk Discussion Groups main page. Online community of Autodesk users sharing comments, questions, and solutions about all Autodesk products.

http://revit.autodesk.com/Building/8.1/common/download.asp?en-us

Autodesk Revit download center. Download Autodesk Revit Building and other resources related to Revit Building from this site.

USER COMMUNITY

http://www.augi.com/revit/default.asp

Autodesk Users Group International Revit focused forums and information.

http://www.revitcity.com

Web site devoted to all things Revit.

http://cadence.advanstar.com//ubbthreads/ubbthreads.php?Cat=

Online user forum hosted by CADalyst Magazine and moderated by Paul F. Aubin.

http://www.cadalyst.com/cadalyst/

Main home page for CADalyst Magazine. View magazines online or subscribe to print edition.

http://modocrmadt.blogspot.com/2005/01/bim-what-is-it-why-do-i-care-and-how.html

Excerpts from Chapter 1 come from an essay on this Blog (Web Log) maintained online by Matt Dillon. Matt is a registered Architect and expert in Autodesk Architectural Desktop, Autodesk Revit Building and other BIM technologies. Matt was gracious enough to provide permission to quote his essay in this book.

ONLINE RESOURCES FOR PLUG-INS

www.GreenBuildingStudio.com

Green Building Studio is a free web-service that provides architects and engineers using Autodesk Revit, Autodesk Architectural Desktop, or Autodesk Building Systems with early design stage whole building energy analysis and product information appropriate for their building design

www.e-specs.com

Built around our e-SPECS technology which links the project drawings to the specification documents, InterSpec has a variety of products and services to help you manage your construction specifications in the most accurate, efficient and cost-effective manner possible.

Keyboard Shortcut Keys

USING THE KEYBOARD TO EXECUTE COMMANDS

Many Revit Building commands can be executed from the keyboard with a combination of two keystrokes. All commands can be accessed from the menus. Many are also available on the Design Bar and the toolbars. In this appendix, a keyboard shortcut map is provided. You can view the complete list of Autodesk Revit Building Keyboard Shortcuts in the file named: *KeyboardShortcuts.txt*. This file is located in the *C:\Program Files\Autodesk Revit Building 8.1\Program* folder. This file is the location where the keyboard shortcuts are stored and read by Revit Building. Therefore, be very careful not to change anything when reading it. You can edit this file if you wish. However please be sure to make a backup copy before proceeding so that you can restore it to the original state if necessary.

KEY BOARD SHORTCUT MAP

The following illustrations show where the default keyboard shortcuts are located. Six keys are used with a variety of second keys to branch to a separate set of related commands. For instance, all of the Zoom commands begin with the letter 'Z.'

NOTE If the keyboardshortcuts.txt is customized any two-key combinations may be used for many of the menu commands within Revit Building and therefore these illustrations may not apply.

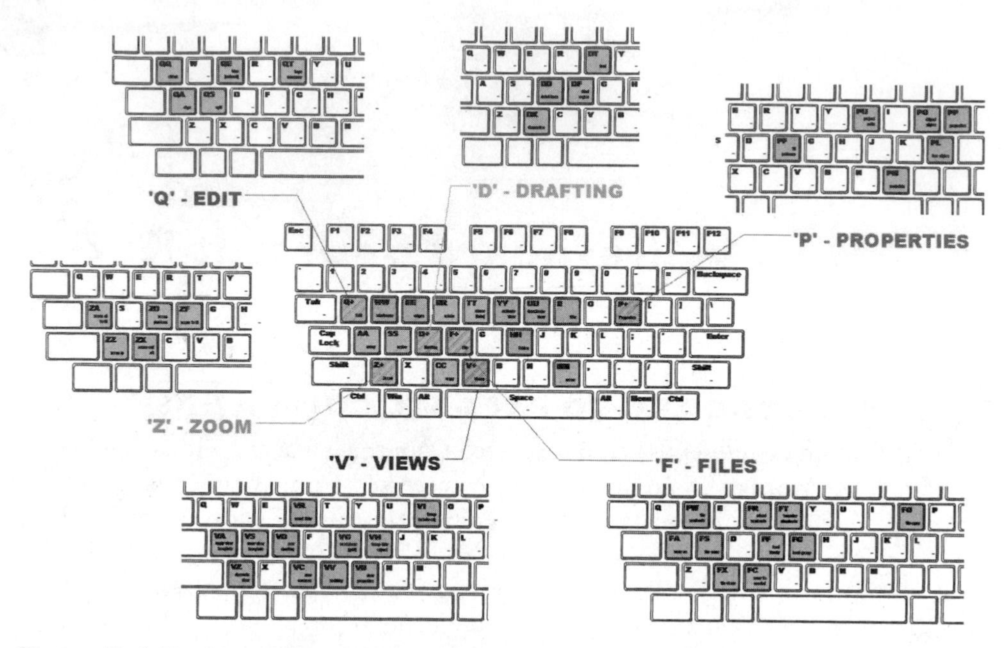

Figure D–1 Keyboard Shortcut map—Overall with sub-maps

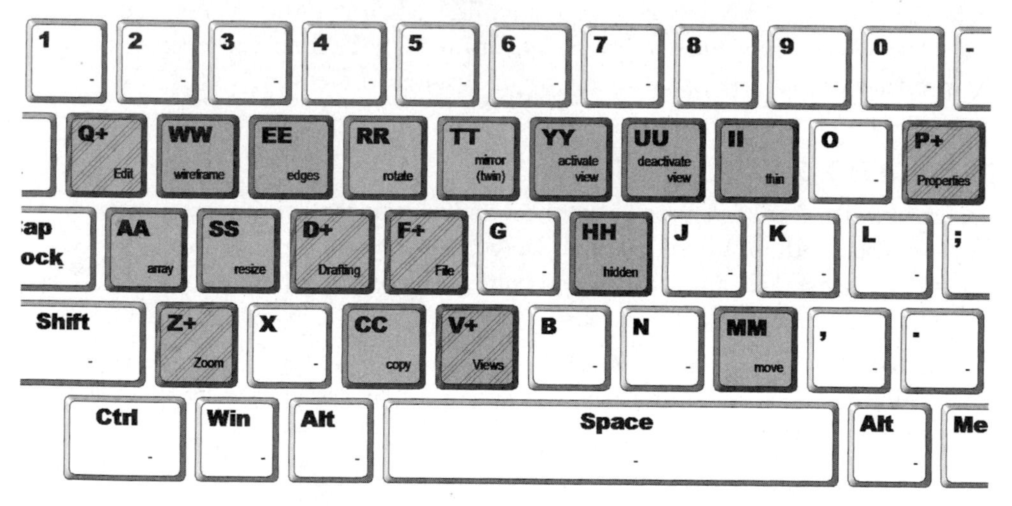

Figure D–2 Keyboard Shortcut map—Overall

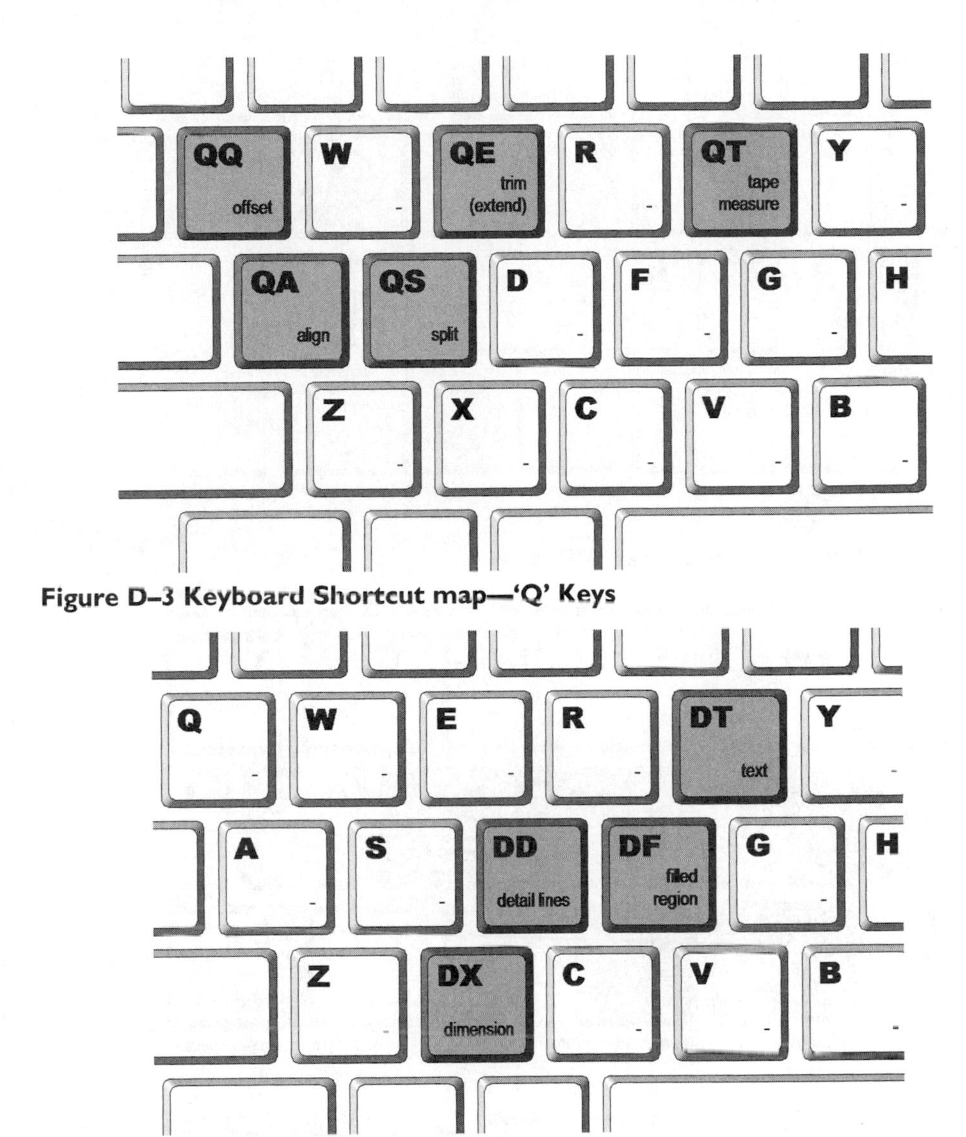

Figure D–3 Keyboard Shortcut map—'Q' Keys

Figure D–4 Keyboard Shortcut map—'D' Keys

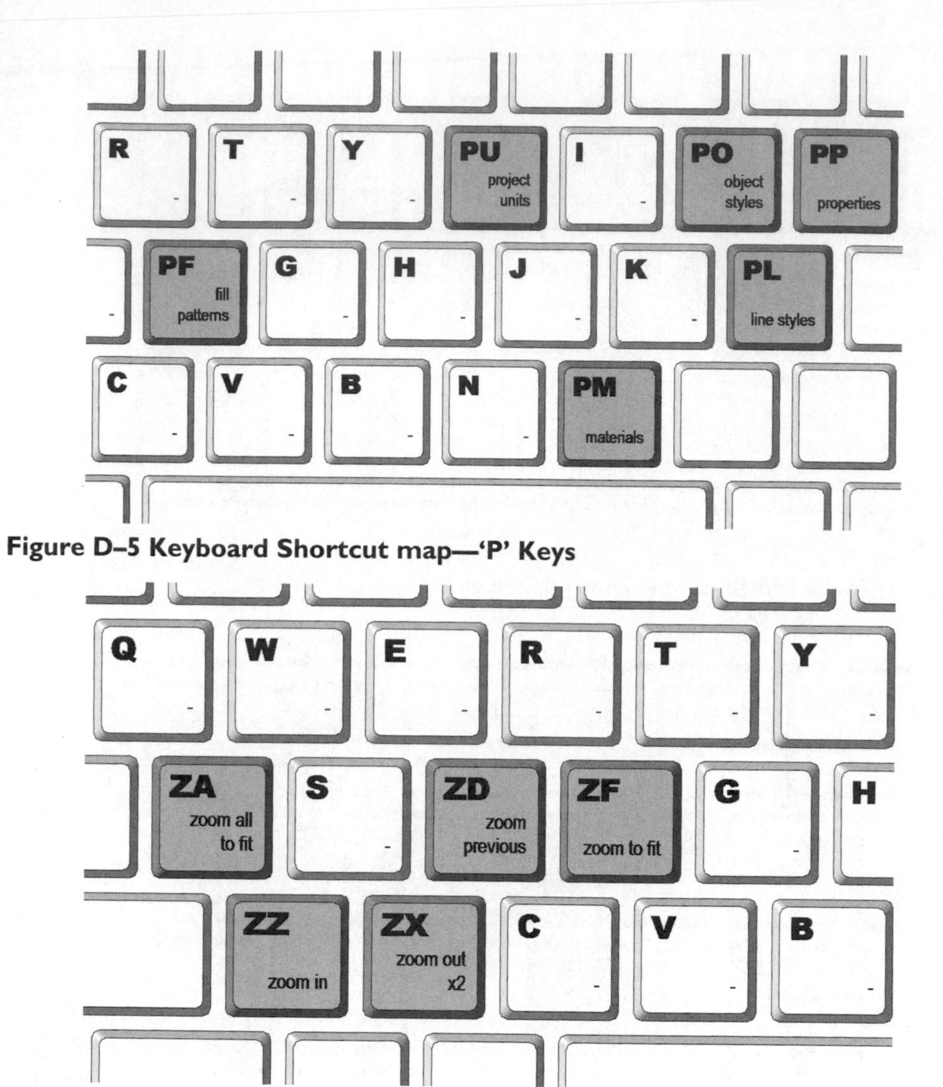

Figure D–5 Keyboard Shortcut map—'P' Keys

Figure D–6 Keyboard Shortcut map—'Z' Keys

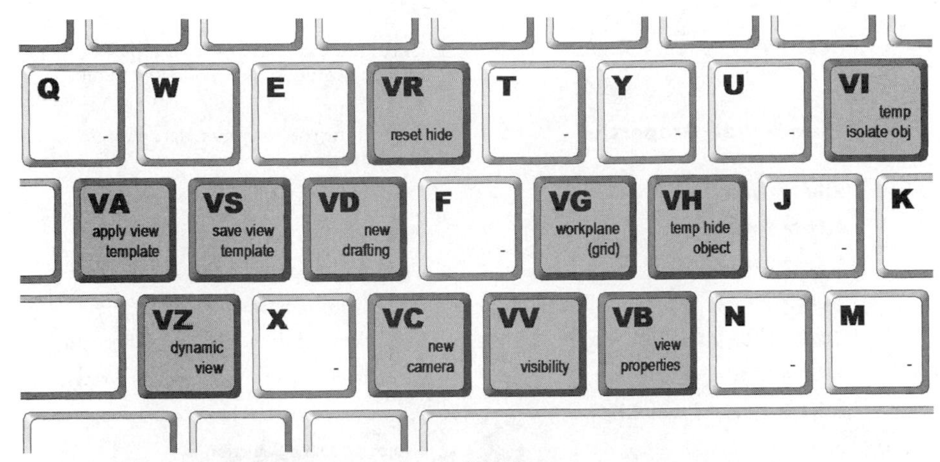

Figure D–7 Keyboard Shortcut map—'V' Keys

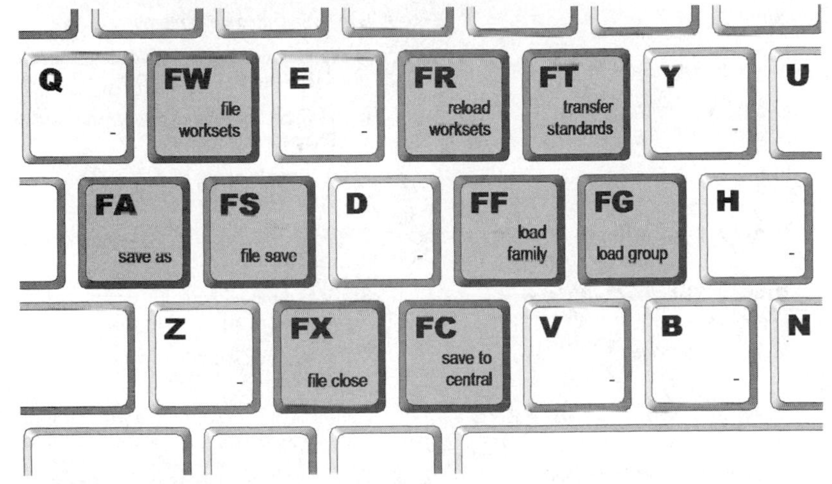

Figure D–8 Keyboard Shortcut map—'F' Keys

SHORTCUT KEY LIST

Shortcut Key	Menu Pick	Description
OF	**Tools > Offset**	Move or copy elements parallel to themselves
SL	**Tools > Split Walls and Lines**	Break a single Line or Wall into two or more segments
TR	**Tools > Trim/Extend**	Lengthen or shorten lines and Walls to a specified edge
MV	**Edit > Move**	Move elements from a base point to a new location
CO	**Edit > Copy**	Copy elements from a base point to a new location

AR	**Edit > Array**	Creates several equally spaced copies optionally grouped and associated to one another
VP	**View > View Properties**	Control the specific parameters of a particular view
CTRL + O	**File > Open**	Open a project
CTRL + S	**File > Save**	Save the project
CTRL + Z	**Edit > Undo**	Undo the last command
CTRL + Y	**Edit > Redo**	Redo the last command
CTRL + C	**Edit > Copy to Clipboard**	Copy elements to the Windows clipboard
CTRL + X	**Edit > Cut**	Cut elements to the Windows clipboard
CTRL + V	**Edit > Paste from Clipboard**	Paste elements from the Windows clipboard
CTRL	N/A	Use to add to a Selection
SHIFT	N/A	Use to remove from a Selection
TAB	N/A	Use to cycle to next element
ALT + E	N/A	In the Element Properties dialog, shortcut to the **Edit/New** button
ALT + D	N/A	In the Type Properties dialog, shortcut to the **Duplicate** button
ALT + R	N/A	In the Type Properties dialog, shortcut to the **Rename** button
F8	**View > Dynamically Modify View**	Choose between Scroll, Zoom and Spin to dynamically change the active view
SD	**View > Shading w/Edges**	Shade the model with all corners and edges highlighted in contrasting color.
WT	**Window > Tile**	Tiles the open windows within the workspace
ZA	**View > Zoom > Zoom All To Fit**	Zooms all active windows to fit within there window boundary
UN	**Settings > Project Units**	Set the unit of measure for the project in either Imperial or Metric.
GR	**Drafting > Grid**	Creates a Grid line.

Numeric

A

B

C

D

E

F

G

H

N

O

P

R

S

T

U

V

W

Z

LICENSE AGREEMENT FOR AUTODESK

A Thomson Learning Company

IMPORTANT-READ CAREFULLY: This End User License Agreement ("Agreement") sets forth the conditions by which Delmar Learning, a division of Thomson Learning Inc. ("Thomson") will make electronic access to the Thomson Delmar Learning-owned licensed content and associated media, software, documentation, printed materials and electronic documentation contained in this package and/or made available to you via this product (the "Licensed Content"), available to you (the "End User"). BY CLICKING THE "I ACCEPT" BUTTON AND/OR OPENING THIS PACKAGE, YOU ACKNOWLEDGE THAT YOU HAVE READ ALL OF THE TERMS AND CONDITIONS, AND THAT YOU AGREE TO BE BOUND BY ITS TERMS CONDITIONS AND ALL APPLICABLE LAWS AND REGULATIONS GOVERNING THE USE OF THE LICENSED CONTENT.

1.0 SCOPE OF LICENSE

1.1 Licensed Content. The Licensed Content may contain portions of modifiable content ("Modifiable Content") and content which may not be modified or otherwise altered by the End User ("Non-Modifiable Content"). For purposes of this Agreement, Modifiable Content and Non-Modifiable Content may be collectively referred to herein as the "Licensed Content." All Licensed Content shall be considered Non-Modifiable Content, unless such Licensed Content is presented to the End User in a modifiable format and it is clearly indicated that modification of the Licensed Content is permitted.

1.2 Subject to the End User's compliance with the terms and conditions of this Agreement, Thomson Delmar Learning hereby grants the End User, a non-transferable, non-exclusive, limited right to access and view a single copy of the Licensed Content on a single personal computer system for noncommercial, internal, personal use only. The End User shall not (i) reproduce, copy, modify (except in the case of Modifiable Content), distribute, display, transfer, sub-license, prepare derivative work(s) based on, sell, exchange, barter or transfer, rent, lease, loan, resell, or in any other manner exploit the Licensed Content; (ii) remove, obscure or alter any notice of Thomson Delmar Learning's intellectual property rights present on or in the License Content, including, but not limited to, copyright, trademark and/or patent notices; or (iii) disassemble, decompile, translate, reverse engineer or otherwise reduce the Licensed Content.

2.0 TERMINATION

2.1 Thomson Delmar Learning may at any time (without prejudice to its other rights or remedies) immediately terminate this Agreement and/or suspend access to some or all of the Licensed Content, in the event that the End User does not comply with any of the terms and conditions of this Agreement. In the event of such termination by Thomson Delmar Learning, the End User shall immediately return any and all copies of the Licensed Content to Thomson Delmar Learning.

3.0 PROPRIETARY RIGHTS

3.1 The End User acknowledges that Thomson Delmar Learning owns all right, title and interest, including, but not limited to all copyright rights therein, in and to the Licensed Content, and that the End User shall not take any action inconsistent with such ownership. The Licensed Content is protected by U.S., Canadian and other applicable copyright laws and by international treaties, including the Berne Convention and the Universal Copyright Convention. Nothing contained in this Agreement shall be construed as granting the End User any ownership rights in or to the Licensed Content.

3.2 Thomson Delmar Learning reserves the right at any time to withdraw from the Licensed Content any item or part of an item for which it no longer retains the right to publish, or which it has reasonable grounds to believe infringes copyright or is defamatory, unlawful or otherwise objectionable.

4.0 PROTECTION AND SECURITY

4.1 The End User shall use its best efforts and take all reasonable steps to safeguard its copy of the Licensed Content to ensure that no unauthorized reproduction, publication, disclosure, modification or distribution of the Licensed Content, in whole or in part, is made. To the extent that the End User becomes aware of any such unauthorized use of the Licensed Content, the End User shall immediately notify Delmar Learning. Notification of such violations may be made by sending an Email to delmarhelp@thomson.com.

5.0 MISUSE OF THE LICENSED PRODUCT

5.1 In the event that the End User uses the Licensed Content in violation of this Agreement, Thomson Delmar Learning shall have the option of electing liquidated damages, which shall include all profits generated by the End User's use of the Licensed Content plus interest computed at the maximum rate permitted by law and all legal fees and other expenses incurred by Thomson Delmar Learning in enforcing its rights, plus penalties.

6.0 FEDERAL GOVERNMENT CLIENTS

6.1 Except as expressly authorized by Delmar Learning, Federal Government clients obtain only the rights specified in this Agreement and no other rights. The Government acknowledges that (i) all software and related documentation incorporated in the Licensed Content is existing commercial computer software within the meaning of FAR 27.405(b)(2); and (2) all other data delivered in whatever form, is limited rights data within the meaning of FAR 27.401. The restrictions in this section are acceptable as consistent with the Government's need for software and other data under this Agreement.

7.0 DISCLAIMER OF WARRANTIES AND LIABILITIES

7.1 Although Thomson Delmar Learning believes the Licensed Content to be reliable, Thomson Delmar Learning does not guarantee or warrant (i) any information or materials contained in or produced by the Licensed Content, (ii) the accuracy, completeness or reliability of the Licensed Content, or (iii) that the Licensed Content is free from errors or other material defects. THE LICENSED PRODUCT IS PROVIDED "AS IS," WITHOUT ANY WARRANTY OF ANY KIND AND THOMSON DELMAR LEARNING DISCLAIMS ANY AND ALL WARRANTIES, EXPRESSED OR IMPLIED, INCLUDING, WITHOUT LIMITATION, WARRANTIES OF MERCHANTABILITY OR FITNESS OR A PARTICULAR PURPOSE. IN NO EVENT SHALL THOMSON DELMAR LEARNING BE LIABLE FOR: INDIRECT, SPECIAL, PUNITIVE OR CONSEQUENTIAL DAMAGES INCLUDING FOR LOST PROFITS, LOST DATA, OR OTHERWISE. IN NO EVENT SHALL DELMAR LEARNING'S AGGREGATE LIABILITY HEREUNDER, WHETHER ARISING IN CONTRACT, TORT, STRICT LIABILITY OR OTHERWISE, EXCEED THE AMOUNT OF FEES PAID BY THE END USER HEREUNDER FOR THE LICENSE OF THE LICENSED CONTENT.

8.0 GENERAL

8.1 Entire Agreement. This Agreement shall constitute the entire Agreement between the Parties and supersedes all prior Agreements and understandings oral or written relating to the subject matter hereof.

8.2 Enhancements/Modifications of Licensed Content. From time to time, and in Delmar Learning's sole discretion, Thomson Thomson Delmar Learning may advise the End User of updates, upgrades, enhancements and/or improvements to the Licensed Content, and may permit the End User to access and use, subject to the terms and conditions of this Agreement, such modifications, upon payment of prices as may be established by Delmar Learning.

8.3 No Export. The End User shall use the Licensed Content solely in the United States and shall not transfer or export, directly or indirectly, the Licensed Content outside the United States.

8.4 Severability.If any provision of this Agreement is invalid, illegal, or unenforceable under any applicable statute or rule of law, the provision shall be deemed omitted to the extent that it is invalid, illegal, or unenforceable. In such a case, the remainder of the Agreement shall be construed in a manner as to give greatest effect to the original intention of the parties hereto.

8.5 Waiver. The waiver of any right or failure of either party to exercise in any respect any right provided in this Agreement in any instance shall not be deemed to be a waiver of such right in the future or a waiver of any other right under this Agreement.

8.6 Choice of Law/Venue. This Agreement shall be interpreted, construed, and governed by and in accordance with the laws of the State of New York, applicable to contracts executed and to be wholly preformed therein, without regard to its principles governing conflicts of law. Each party agrees that any proceeding arising out of or relating to this Agreement or the breach or threatened breach of this Agreement may be commenced and prosecuted in a court in the State and County of New York. Each party consents and submits to the non-exclusive personal jurisdiction of any court in the State and County of New York in respect of any such proceeding.

8.7 Acknowledgment. By opening this package and/or by accessing the Licensed Content on this Website, THE END USER ACKNOWLEDGES THAT IT HAS READ THIS AGREEMENT, UNDERSTANDS IT, AND AGREES TO BE BOUND BY ITS TERMS AND CONDITIONS. IF YOU DO NOT ACCEPT THESE TERMS AND CONDITIONS, YOU MUST NOT ACCESS THE LICENSED CONTENT AND RETURN THE LICENSED PRODUCT TO THOMSON DELMAR LEARNING (WITHIN 30 CALENDAR DAYS OF THE END USER'S PURCHASE) WITH PROOF OF PAYMENT ACCEPTABLE TO DELMAR LEARNING, FOR A CREDIT OR A REFUND. Should the End User have any questions/comments regarding this Agreement, please contact Thomson Delmar Learning at delmarhelp@thomson.com.